T0388123

HETEROGENEOUS CATALYSIS

HETEROGENEOUS CATALYSIS

Materials and Applications

Edited by

MOISÉS ROMOLOS CESARIO

Department of Materials Engineering, Federal University of Paraíba, Cidade Universitária, João Pessoa, Paraíba, Brazil

DANIEL ARAÚJO DE MACEDO

Department of Materials Engineering, Federal University of Paraíba, Cidade Universitária, João Pessoa, Paraíba, Brazil

ELSEVIER

Elsevier
Radarweg 29, PO Box 211, 1000 AE Amsterdam, Netherlands
The Boulevard, Langford Lane, Kidlington, Oxford OX5 1GB, United Kingdom
50 Hampshire Street, 5th Floor, Cambridge, MA 02139, United States

ISBN: 978-0-323-85612-6

For information on all Elsevier publications
visit our website at https://www.elsevier.com/books-and-journals

Publisher: Susan Dennis
Acquisitions Editor: Anita Koch
Editorial Project Manager: Kathrine Esten
Production Project Manager: Sruthi Satheesh
Cover Designer: Matthew Limbert

Typeset by STRAIVE, India

Contents

6. The role of CO_2 sorbents materials in SESMR for hydrogen production

THAMYSCIRA H.S. DA SILVA, MURIEL CHAGHOURI, MOISÉS R. CESÁRIO, HAINGOMALALA LUCETTE TIDAHY, CÉDRIC GENNEQUIN, DANIEL A. MACEDO, AND EDMOND ABI-AAD

7. Catalysts for syngas production by dry reforming of methane

JORGE ÁLEF ESTEVAM LAU BOMFIM, JOSÉ FAUSTINO SOUZA CARVALHO FILHO, TÉRCIA DINIZ BEZERRA, FERNANDO CESÁRIO RANGEL, THIAGO ARAUJO SIMÕES, PEDRO NOTHAFT ROMANO, AND ROSENIRA SERPA DA CRUZ

8. Dry reforming of methane for catalytic valorization of biogas

MURIEL CHAGHOURI, SARA HANY, HAINGOMALALA LUCETTE TIDAHY, FABRICE CAZIER, CÉDRIC GENNEQUIN, AND EDMOND ABI-AAD

9. Catalysts for steam reforming of biomass tar and their effects on the products

MIRA ABOU RJEILY, CÉDRIC GENNEQUIN, HERVÉ PRON, EDMOND ABI-AAD, AND JAONA HARIFIDY RANDRIANALISOA

10. Heterogeneous catalysts for biomass-derived alcohols and acid conversion

GHEORGHITA MITRAN, OCTAVIAN DUMITRU PAVEL, AND DONG-KYUN SEO

11. Zinc oxide or molybdenum oxide deposited on bentonite by the microwave-assisted hydrothermal method: New catalysts for obtaining biodiesel

ANA FLÁVIA FELIX FARIAS, MARCOS ANTONIO GOMES PEQUENO, SUELEN ALVES SILVA LUCENA DE MEDEIROS, THIAGO MARINHO DUARTE, HERBET BEZERRA SALES, AND IEDA MARIA GARCIA DOS SANTOS

12. Assisted catalysis: An overview of alternative activation technologies for the conversion of biomass

C. COUTANCEAU, F. JÉRÔME, AND K. DE OLIVEIRA VIGIER

13. Regenerable adsorbents for SOx removal, material efficiency, and regeneration methods: A focus on CuO-based adsorbents

JULIE SCHOBING, MOISÉS R. CESÁRIO, SOPHIE DORGE, HABIBA NOUALI, DAVID HABERMACHER, JOËL PATARIN, BÉNÉDICTE LEBEAU, AND JEAN-FRANÇOIS BRILHAC

14. Solid oxide cells (SOCs) in heterogeneous catalysis

FRANCISCO J.A. LOUREIRO, ALLAN J.M. ARAÚJO, DANIEL A. MACEDO, MOISÉS R. CESÁRIO, AND DUNCAN P. FAGG

15. Electrocatalytic oxygen reduction and evolution reactions in solid oxide cells (SOCs): A brief review

ALLAN J.M. ARAÚJO, FRANCISCO J.A. LOUREIRO, LAURA I.V. HOLZ, VANESSA C.D. GRAÇA, DANIEL A. MACEDO, MOISÉS R. CESÁRIO, CARLOS A. PASKOCIMAS, AND DUNCAN P. FAGG

16. Catalysts for hydrogen and oxygen evolution reactions (HER/OER) in cells

VINICIUS DIAS SILVA, FABIO EMANUEL FRANÇA DA SILVA, ELITON SOUTO DE MEDEIROS, AND THIAGO ARAUJO SIMÕES

17. Zeolitic imidazolate framework 67 based metal oxides derivatives as electrocatalysts for oxygen evolution reaction

ANNAÍRES DE A. LOURENÇO AND FAUSTHON F. DA SILVA

18. Electrochemical ammonia synthesis: Mechanism, recent developments, and challenges in catalyst design

VANESSA C.D. GRAÇA, FRANCISCO J.A. LOUREIRO, LAURA I.V. HOLZ, SERGEY M. MIKHALEV, ALLAN J.M. ARAÚJO, AND DUNCAN P. FAGG

19. Non-faradaic electrochemical modification of catalytic activity: A current overview

LAURA I.V. HOLZ, FRANCISCO J.A. LOUREIRO, VANESSA C.D. GRAÇA, ALLAN J.M. ARAÚJO, DIOGO MENDES, ADÉLIO MENDES, AND DUNCAN P. FAGG

Contributors

Edmond Abi-Aad Environmental Chemistry and Life Interactions Unit (UCEIV), University of the Littoral Opal Coast, UR 4492, SFR Condorcet—FR CNRS 3417, Dunkerque, France

Samer Aouad Department of Chemistry, Faculty of Arts and Sciences, University of Balamand, Tripoli, Lebanon

Allan J.M. Araújo Centre for Mechanical Technology and Automation (TEMA), Department of Mechanical Engineering, University of Aveiro, Aveiro, Portugal; Materials Science and Engineering Postgraduate Program—PPGCEM, Federal University of Rio Grande do Norte—UFRN, Natal, Brazil

Anuradha M. Ashok PSG Institute of Advanced Studies, Coimbatore, India

Braulio Barros Department of Mechanical Engineering, Federal University of Pernambuco, Recife, Brazil

Tércia Diniz Bezerra Grupo Bioenergia e Meio Ambiente, Universidade Estadual de Santa Cruz—UESC, Ilhéus, BA, Brazil

Jorge Álef Estevam Lau Bomfim Materials Science and Engineering Postgraduate Program (PPCEM), Federal University of Paraíba (UFPB), João Pessoa; Grupo Bioenergia e Meio Ambiente, Universidade Estadual de Santa Cruz—UESC, Ilhéus, BA, Brazil

Jean-François Brilhac University of Upper Alsace (UHA), LGRE UR 2334, Mulhouse; University of Strasbourg, Strasbourg, France

Rita M. Brito Alves Department of Chemical Engineering, University of São Paulo, São Paulo, Brazil

Fabio M. Cavalcanti Department of Chemical Engineering, University of São Paulo, São Paulo, Brazil

Fabrice Cazier Environmental Chemistry and Life Interactions Unit (UCEIV), University of the Littoral Opal Coast, UR 4492, SFR Condorcet—FR CNRS 3417, Dunkerque, France

Moisés R. Cesário Materials Science and Engineering Postgraduate Program (PPCEM), Federal University of Paraíba (UFPB), João Pessoa, Brazil

Muriel Chaghouri Environmental Chemistry and Life Interactions Unit (UCEIV), University of the Littoral Opal Coast, UR 4492, SFR Condorcet—FR CNRS 3417, Dunkerque, France

C. Coutanceau IC2MP, University of Poitiers, UMR CNRS 7285, Poitiers, France

Patrick Da Costa Institut Jean Le Rondd'Alembert, Sorbonne Université, CNRS UMR 7190, Saint-Cyr-l'Ecole, France

Rosenira Serpa da Cruz Materials Science and Engineering Postgraduate Program (PPCEM), Federal University of Paraíba (UFPB), João Pessoa; Grupo Bioenergia e Meio Ambiente, Universidade Estadual de Santa Cruz—UESC, Ilhéus, BA, Brazil

Fabio Emanuel França da Silva Materials Science and Engineering Postgraduate Program (PPCEM), Federal University of Paraíba (UFPB), João Pessoa, Brazil

Fausthon F. da Silva Department of Chemistry, Federal University of Paraíba, UFPB, João Pessoa, PB, Brazil

Thamyscira H.S. da Silva Environmental Chemistry and Life Interactions Unit (UCEIV), University of the Littoral Opal Coast, UR 4492, SFR Condorcet—FR CNRS 3417, Dunkerque, France; Materials Science and Engineering Postgraduate Program (PPCEM), Federal University of Paraíba (UFPB), João Pessoa, Brazil

Sophie Dorge University of Upper Alsace (UHA), LGRE UR 2334, Mulhouse; University of Strasbourg, Strasbourg, France

Ieda Maria Garcia dos Santos NPE-LACOM, Federal University of Paraiba, João Pessoa, PB, Brazil

Thiago Marinho Duarte NPE-LACOM, Federal University of Paraiba, João Pessoa, PB, Brazil

Duncan P. Fagg Centre for Mechanical Technology and Automation (TEMA), Department of Mechanical Engineering, University of Aveiro, Aveiro, Portugal

Ana Flávia Felix Farias NPE-LACOM, Federal University of Paraiba, João Pessoa; UAEMa, Federal University of Campina Grande, Campina Grande, PB, Brazil

José Faustino Souza Carvalho Filho Escola de Química, Universidade Federal do Rio de Janeiro—UFRJ, Rio de Janeiro, RJ, Brazil

Cédric Gennequin Environmental Chemistry and Life Interactions Unit (UCEIV), University of the Littoral Opal Coast, UR 4492, SFR Condorcet—FR CNRS 3417, Dunkerque, France

Reinaldo Giudici Department of Chemical Engineering, University of São Paulo, São Paulo, Brazil

Alberth Renne Gonzalez Caranton Federal University of Rio de Janeiro—COPPE/PEQ/Nucat, Rio de Janeiro, Brazil; Universidad ECCI-Grupo de Investigación en Aprovechamiento Tecnológico de Materiales y Energía GIATME, Bogotá D.C., Colombia

Vanessa C.D. Graça Centre for Mechanical Technology and Automation (TEMA), Department of Mechanical Engineering, University of Aveiro, Aveiro, Portugal

Henrik Grénman Faculty of Science and Engineering, Johan Gadolin Process Chemistry Centre, Laboratory in Industrial Chemistry and Reaction Engineering, Åbo Akademi University, Turku, Finland

David Habermacher University of Upper Alsace (UHA), LGRE UR 2334, Mulhouse; University of Strasbourg, Strasbourg, France

Wim Haije Faculty of Applied Science, Department of Chemical Engineering, Materials for Energy Conversion and Storage Section, Delft University of Technology, Delft, The Netherlands

Sara Hany Environmental Chemistry and Life Interactions Unit (UCEIV), University of the Littoral Opal Coast, UR 4492, SFR Condorcet—FR CNRS 3417, Dunkerque, France

Laura I.V. Holz Centre for Mechanical Technology and Automation (TEMA), Department of Mechanical Engineering, University of Aveiro, Aveiro; LEPABE—Faculty of Engineering, University of Porto, Porto; Bondalti Chemicals, S.A., Quinta da Indústria, Estarreja, Portugal

F. Jérôme IC2MP, University of Poitiers, UMR CNRS 7285, Poitiers, France

Wiebren de Jong Faculty 3mE, Department of Process and Energy, Large-Scale Energy Storage Section, Delft University of Technology, Delft, The Netherlands

Camila E. Kozonoe Department of Chemical Engineering, University of São Paulo, São Paulo, Brazil

Bénédicte Lebeau University of Strasbourg, Strasbourg; University of Upper Alsace (UHA), CNRS, IS2M UMR 7361, Mulhouse, France

Francisco J.A. Loureiro Centre for Mechanical Technology and Automation (TEMA), Department of Mechanical Engineering, University of Aveiro, Aveiro, Portugal

Annaíres de A. Lourenço Department of Chemistry, Federal University of Paraíba, UFPB, João Pessoa, PB, Brazil

Daniel A. Macedo Materials Science and Engineering Postgraduate Program (PPCEM), Federal University of Paraíba (UFPB), João Pessoa, Brazil

Suelen Alves Silva Lucena de Medeiros NPE-LACOM, Federal University of Paraiba, João Pessoa, PB, Brazil

Adélio Mendes LEPABE—Faculty of Engineering, University of Porto, Porto, Portugal

Diogo Mendes Bondalti Chemicals, S.A., Quinta da Indústria, Estarreja, Portugal

Sergey M. Mikhalev Centre for Mechanical Technology and Automation (TEMA), Department of Mechanical Engineering, University of Aveiro, Aveiro, Portugal

Gheorghita Mitran Laboratory of Chemical Technology and Catalysis, Department of Organic Chemistry, Biochemistry & Catalysis, University of Bucharest, Bucharest, Romania

Habiba Nouali University of Strasbourg, Strasbourg; University of Upper Alsace (UHA), CNRS, IS2M UMR 7361, Mulhouse, France

K. De Oliveira Vigier IC2MP, University of Poitiers, UMR CNRS 7285, Poitiers, France

Carlos A. Paskocimas Materials Science and Engineering Postgraduate Program—PPGCEM, Federal University of Rio Grande do Norte—UFRN, Natal, Brazil

Joël Patarin University of Strasbourg, Strasbourg; University of Upper Alsace (UHA), CNRS, IS2M UMR 7361, Mulhouse, France

Octavian Dumitru Pavel Laboratory of Chemical Technology and Catalysis, Department of Organic Chemistry, Biochemistry & Catalysis, University of Bucharest, Bucharest, Romania

Marcos Antonio Gomes Pequeno NPE-LACOM, Federal University of Paraiba, João Pessoa, PB, Brazil

Hervé Pron Université de Reims Champagne-Ardenne, Institut de Thermique, Mécanique, Matériaux—ITheMM, EA 7548, SFR Condorcet - FR CNRS 3417, Reims Cedex 2, France

Jaona Harifidy Randrianalisoa Université de Reims Champagne-Ardenne, Institut de Thermique, Mécanique, Matériaux—ITheMM, EA 7548, SFR Condorcet - FR CNRS 3417, Reims Cedex 2, France

Fernando Cesário Rangel Materials Science and Engineering Postgraduate Program (PPCEM), Federal University of Paraíba (UFPB), João Pessoa; Grupo Bioenergia e Meio Ambiente, Universidade Estadual de Santa Cruz—UESC, Ilhéus, BA, Brazil

Mira Abou Rjeily Université de Reims Champagne-Ardenne, Institut de Thermique, Mécanique, Matériaux—ITheMM, EA 7548, SFR Condorcet - FR CNRS 3417, Reims Cedex 2, France

Pedro Nothaft Romano Universidade Federal do Rio de Janeiro—UFRJ, Campus Duque de Caxias, Rio de Janeiro, RJ, Brazil

Karina T. de C. Roseno Department of Chemical Engineering, University of São Paulo, São Paulo, Brazil

Herbet Bezerra Sales UAEMa, Federal University of Campina Grande, Campina Grande, PB, Brazil

Martin Schmal Department of Chemical Engineering, University of São Paulo, São Paulo; Federal University of Rio de Janeiro—COPPE/PEQ/Nucat, Rio de Janeiro, Brazil

Julie Schobing University of Upper Alsace (UHA), LGRE UR 2334, Mulhouse; University of Strasbourg, Strasbourg, France

Dong-Kyun Seo School of Molecular Sciences, Arizona State University, Tempe, AZ, United States

Vinicius Dias Silva Materials Science and Engineering Postgraduate Program (PPCEM), Federal University of Paraíba (UFPB), João Pessoa, Brazil

Thiago Araujo Simões Materials Science and Engineering Postgraduate Program (PPCEM), Federal University of Paraíba (UFPB), João Pessoa; Postgraduate Program in Science, Innovation and Modeling in Materials (PROCIMM), State University of Santa Cruz—UESC, Ilhéus, BA; Center for Science and Technology in Energy and Sustainability (CETENS), Federal University of the Recôncavo of Bahia (UFRB), Feira de Santana, Brazil

Eliton Souto de Medeiros Materials Science and Engineering Postgraduate Program (PPCEM), Federal University of Paraíba (UFPB), João Pessoa, Brazil

Chao Sun Institut Jean Le Rondd'Alembert, Sorbonne Université, CNRS UMR 7190, Saint-Cyr-l'Ecole, France

Haingomalala Lucette Tidahy Environmental Chemistry and Life Interactions Unit (UCEIV), University of the Littoral Opal Coast, UR 4492, SFR Condorcet—FR CNRS 3417, Dunkerque, France

T. Vijayaraghavan PSG Institute of Advanced Studies, Coimbatore, India

Liangyuan Wei Faculty 3mE, Department of Process and Energy, Large-Scale Energy Storage Section, Delft University of Technology, Delft, The Netherlands

CHAPTER

1

Understanding heterogeneous catalysis: A brief study on performance parameters

Moisés R. Cesário[a], Braulio Barros[b], Samer Aouad[c], Cédric Gennequin[d], Edmond Abi-Aad[d], and Daniel A. Macedo[a]

[a]Materials Science and Engineering Postgraduate Program (PPCEM), Federal University of Paraíba (UFPB), João Pessoa, Brazil [b]Department of Mechanical Engineering, Federal University of Pernambuco, Recife, Brazil [c]Department of Chemistry, Faculty of Arts and Sciences, University of Balamand, Tripoli, Lebanon [d]Environmental Chemistry and Life Interactions Unit (UCEIV), University of the Littoral Opal Coast, UR 4492, SFR Condorcet–FR CNRS 3417, Dunkerque, France

1.1 Introduction

The development of the catalysis phenomenon has gone through several phases. In 1835, Jöns Jakob Berzelius (1779–1848) suggested that the addition of small amounts of substances can affect transformations due to a catalytic force. Subsequently, in the early twentieth century, this concept has been modeled and associated with a kinetic nature. Wilhelm Oswald (1853–1932) introduced the concept of a catalyst as a substance capable of changing the rate of a chemical reaction without being consumed. Between the end of the nineteenth century and around 1920, Paul Sabatier (1854–1941) introduced a fundamental concept in developing catalysis in organic and inorganic synthesis: the formation of unstable intermediate compounds on the catalyst surface. The following period (until 1940) was marked by the development of the adsorption phenomenon. Hug Stott Taylor (1890–1974) suggested the existence of active sites on the surface of the catalyst. Irving Langmuir (1881–1957) presented his theory of adsorption, and in turn, Stephen Brunauer (1903–1986), Paul Hugh Emmett (1900–1985), and Edward Teller (1908–2003) formulated the BET equation to describe multilayer adsorption. Following this period, new interpretations about catalytic activity, new experimental techniques, and new catalytic processes have been developed (Ciola, 1981; Ross, 2018).

The use of catalysts for the most diverse purposes has intensified in the last decades since it is a powerful tool for controlling water and air pollutants and since it plays a vital role in many industrial processes. It is well known that catalysis plays an important role in the production of

Heterogeneous Catalysis
https://doi.org/10.1016/B978-0-323-85612-6.00001-2

the vast majority of chemicals and other goods used by our society. It is estimated that 85%–90% of all products manufactured in the chemical industry undergo catalytic processes (Niemantsverdriet & Schlögl, 2013). Heterogeneous catalysis is even more attractive when applied for the valorization of pollutant gases into high added-value chemicals. Moreover, industries increasingly require faster and more efficient processes making the use of catalysts fundamental. However, choosing an adequate catalytic material is essential to achieve the best performance in the considered reaction, even though, in many cases, it leads to the formation of more thermodynamically stable reaction intermediates. Along with temperature, pressure/composition, and contact time, the catalyst plays a vital role in controlling the rate and the direction of a chemical reaction (Ciola, 1981).

Many petroleum refineries use zeolite-based catalysts to break up long, complex, and diverse hydrocarbons present in crude oil into smaller fractions that can later be used as transportation fuels. Many chemicals, such as hydrogen, ammonia, urea, methanol, sulfuric acid, nitric acid, alcohols, ethylene oxide, ethylene glycol, ethylbenzene, and others, are mainly obtained through catalytic processes. Different catalysts have also been used for environmental protection, such as the abatement of CO, NO, hydrocarbons from automotive exhausts, and sulfur and nitrogen oxides from industrial exhaust streams (DeNOx/DeSOx technology) (Niemantsverdriet & Schlögl, 2013). Besides, catalysts have also been used to capture and recover CO_2 (main greenhouse gas) for its transformation into high value-added products.

This chapter is devoted to heterogeneous catalysis, which can be applied to several areas such as petrochemistry, environment, biomass, fine chemistry, etc. Clean energy and chemical production by the mean of heterogeneous catalysis are the main focus of this part as it is essential for the transition towards a more sustainable future. In this context, "Heterogeneous Catalysis" will be discussed in the scope of (1) the elimination of atmospheric pollutants, (2) the production of clean energy, and (3) the valorization of some molecules into value-added chemical products. To implement the aforementioned processes, it is essential to understand the properties of the catalytic materials and their role in the electrocatalytic/catalytic reactions, specifically the study of surface and interface chemistry.

This chapter is expected to become a "must-read" reference material to help in understanding the principles of heterogeneous catalysis (concepts and theories, physical adsorption, and chemical adsorption), the constitution of supported and/or promoted catalysts, and their properties (activity, selectivity, stability, and regenerability/deactivation). The study of the synthesis methods and the determination of several parameters (structural, microstructural, textural, surface acidity/basicity, reducibility, metallic dispersion, catalyst/support interactions, thermal, and mechanical) will also be highlighted whenever it is deemed crucial to the understanding of the catalytic performance in some specific reactions. Finally, this chapter will reinforce the ability to relate theoretical concepts to practical cases in the field of heterogeneous catalysis, with an emphasis on materials and their applications.

1.2 Catalysis fundamentals

Heterogeneous catalyst

Catalysis can be defined as a process of catalyst action. A catalyst increases the rate of a chemical reaction by providing a new and more comfortable pathway without modifying the thermodynamic factors.

The catalysis discipline is usually divided into heterogeneous, homogeneous, and enzymatic catalysis (or biocatalysis). In heterogeneous catalysis, also referred to as surface catalysis, the catalyst and the reactants are present in different phases. In general, the catalyst constitutes a solid phase while the reactants are in a fluid phase (solid-gas or solid-liquid). On the other hand, in homogeneous catalysis, the catalyst is generally an organometallic complex that is present with the reactant molecules in the same phase (liquid or gas). Finally, biocatalysis considers the use of enzymes to increase the rate of some types of chemical reactions.

This chapter will only consider heterogeneous catalysis. Fig. 1.1 shows the principle of operation of a heterogeneous catalyst.

This role of a heterogeneous catalyst can be described by the following cycle:

Diffusion of the reactants to the catalyst;
Adsorption of the reactants on the catalytic sites at the surface of the catalyst. The catalytic sites are specific "locals" on the surface of the solid where the reaction occurs (via heterogeneous catalysis);
Chemical reaction between the adsorbed reactants molecules;
Desorption of the products;
Regeneration of the catalytic site for the new catalytic cycle.

The catalytic operation consists of several steps, during which the catalyst is not consumed but assists in binding and breaking reactant molecules while allowing the intermediates to readily react until the formation and release of the products and the regeneration of the active site for it to start a new cycle (Niemantsverdriet & Schlögl, 2013). The critical question is: why is the reaction faster? In fact, the associated changes in the potential energy during the process are the main reason. To explain this, the literature highlights the practical and simple example of the oxidation of carbon monoxide, as shown in Fig. 1.2. Without the presence of a catalyst, this reaction would require high energy to break the strong bond of molecular oxygen to overcome the activation barrier (Niemantsverdriet & Schlögl, 2013).

The use of the catalyst in this process allows a more energetically favorable reaction path that requires lower activation energy and leads to an increase in the reaction rate.

Initially, the reactants CO and O_2 are adsorbed on the surface of the catalyst. This adsorption reduces the unbalance of the attractive forces on the surface and, consequently, the surface free energy ($\Delta G < 0$). This spontaneous process is accompanied by a loss in the rotation and translation degrees of freedom, and therefore the entropy change is negative. As

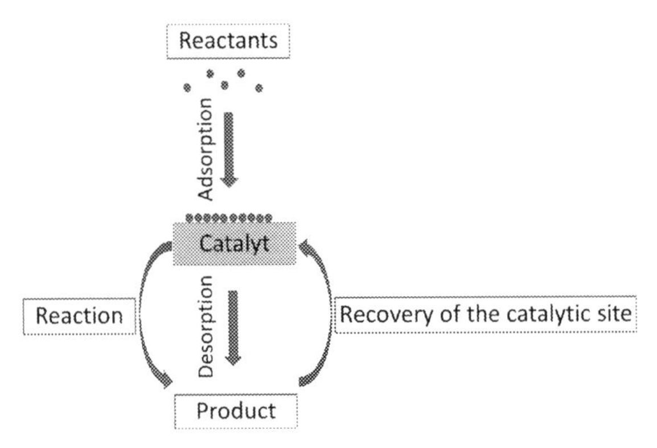

FIG. 1.1 Elementary steps of the heterogeneous catalytic reaction.

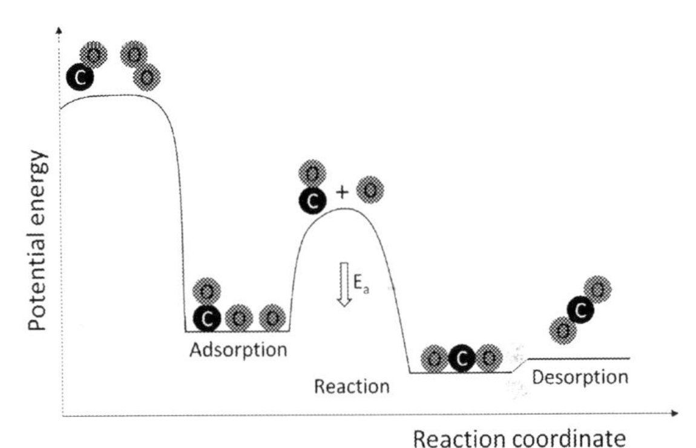

FIG. 1.2 Potential energy diagram of the CO oxidation catalytic reaction. *Modified from Niemantsverdriet, J. W., & Schlögl, R. (2013). Heterogeneous Catalysis: Introduction. In J. Reedijk & K. Poeppelmeier (Eds.),* Comprehensive Inorganic Chemistry II *(2nd ed., Vol. 7, pp. 1–6). Elsevier. https://doi.org/10.1016/B978-0-08-097774-4.00701-4.*

$\Delta G = \Delta H - T\Delta S$, ΔH must be negative (heat of adsorption), i.e., adsorption is an exothermic process (Ciola, 1981). Thus, the potential energy decreases with adsorption. According to the literature, the antibonding orbitals of the O_2 molecule are filled due to the interaction with the d-orbitals of metals such as Rh, Pd, and Pt, leading to its spontaneous dissociation. Thus, two O atoms bond more strongly to the metal surface of the catalyst. In general, the action of the heterogeneous catalyst is associated with breaking a strong bond and subsequently forming new bonds, resulting in the formation of intermediates and products (Niemantsverdriet & Schlögl, 2013). In short, the change in free energy is relatively similar for both catalyzed and non-catalyzed reactions. However, catalysts change the kinetics of the forward and reverse reactions under different conditions of pressure and temperature.

A suitable catalyst should follow Sabatier's Principle. According to this principle, the reactants and intermediates must moderately bind to the surface (Niemantsverdriet & Schlögl, 2013). If the adsorption of the reactants is very strong, the bond with the surface is difficult to break, and the substance can act as a poison to the catalyst. Besides, the potential energy of the reactants will be lowered, thus making it more challenging to overcome the energy barrier of the surface reaction. On the other hand, if the adsorption is very weak, the amount of the adsorbed molecules will be very low, and consequently, the reaction will be very slow (Ciola, 1981).

Adsorption phenomenon

As discussed above, adsorption is the first step of the catalytic reaction. The adsorption occurs whenever there is an interaction between the reactants (gas or liquid) and the surface of the catalyst. The residence time of the molecule on the surface is related to its retention energy or adsorption energy. The adsorption has large technological importance since many adsorbents are used for/as a desiccant, separating gases, catalysts or catalytic supports, liquid purification, pollution control, biological mechanisms, etc. Moreover, the adsorption techniques allow the determination of surface and texture properties, particularly surface area, volume, size, and pore distribution. The oldest absorbents used in the industry are activated carbon and silica gel, generally non-crystalline. Their porous surfaces and structures tend to be poorly defined and difficult to characterize. However, there is an increasing number of adsorbents with intracrystalline pore structures such as zeolites and aluminosilicates (Ciola, 1981; Ruthven, 1984).

The adsorption process may be classified into two different types based on the molecule-surface interactions: physisorption and chemisorption.

The physisorption involves weaker adsorbate-surface bond strength, while chemisorption provides a strong interaction with the solid surface. The interactions in physisorption and chemisorption are governed by Van der Waals forces and chemical bonding (covalent), respectively. The physisorption can lead to minor changes in the electronic structure of the adsorbate, while chemisorption induces significant changes in the molecular properties and can lead to the dissociation of the adsorbed molecule (Becker, 2018; Devred, Dulgheru, & Kruse, 2013). Besides, in physisorption, multiple layers can be formed, in which the adsorption force decreases as the number of layers increases. In chemisorption, a single layer is formed, and the adsorption force decreases as the extension of the occupied surface increases. The other properties of these types of adsorption are potentially referred to in the literature (Ruthven, 1984; Schmal, 2016; Thomas & Thomas, 2015). In a catalytic reaction, the molecules can initially undergo physisorption, i.e., the molecules are directed to the surface through a low-energy path. Thus, the initial physisorption is an essential step in the chemisorption. Then, the transition to chemisorption occurs after reaching the required activation energy (Ciola, 1981; Ruthven, 1984).

It is important to note that many industrial catalysts are porous and most of their surface is internal. The texture of a catalyst may be defined as the geometry of the empty spaces in the catalyst grains and determines its porosity (Ciola, 1981). From the adsorption isotherms, the BET method proposed by Brunauer, Emmett, and Teller allows the measurement of the specific surface area of a solid by monitoring the nitrogen adsorption at 77 K. Furthermore, the adsorption isotherms will allow determining the distribution of the size, volume, and shape of the pores of the materials. The adsorption isotherm establishes the relationship between the amount of substance absorbed (moles, grams, or cm^3) and the equilibrium pressure at a constant temperature. The adsorption isotherms can be determined experimentally via manometry of gas adsorption, gas flow techniques, or gravimetry of gas adsorption. In both cases, the sample is previously degassed. In the case of the gas adsorption manometry procedure, the method consists of measuring the gas pressure at constant volume and a known temperature. The temperature and pressure of the gas dose are measured, and the gas is inserted into the adsorption bulb. After the adsorption equilibrium is reached, the amount of gas adsorbed is calculated by the change of the pressure relative to the inlet pressure (Ciola, 1981).

Most adsorption isotherms result from physical adsorption, and they can be classified as (Ruthven, 1984; Schmal, 2016; Sing et al., 1985):

Type I: Characteristic of microporous solids (<2 nm), adsorption occurs in an adsorbed monolayer.
Type II and Type III: correspond to adsorption in multiple overlapping layers and occur in non-porous solids or with macropores (>50 nm).
Type IV and Type V: Characteristics of solids with mesopores (2–50 nm), where the capillary condensation occurs. The hysteresis phenomenon is associated with this capillary condensation, and it can be of the type H1 to H4.
Type VI: Occurs on a uniform non-porous surface and represents adsorption layer by layer.

The adsorption isotherm provides useful information about the adsorption mechanisms and nature of the adsorbent.

Catalyst properties

The performance of a catalyst is related to its activity, selectivity, stability, and regenerability (deactivation). The catalyst activity is a measure of its efficiency in converting *n* reactive molecules into *p* molecules of products. The activity can be expressed as a rate (specific activity) in $mol\,s^{-1}\,g^{-1}_{(cat)}$ or $mol\,h^{-1}\,cm^{-3}$; turnover frequency (TOF) in s^{-1}; conversion (%).

The same reactants can lead to several thermodynamically possible reactions ($\Delta G < 0$). In this case, a new catalyst property, selectivity, is fundamental. The selectivity of the catalyst is the ability to guide the reaction in one of the thermodynamically permitted paths, i.e., it expresses the preference in the formation of the desired product compared to all the obtained products. As an example, the reaction between carbon monoxide and hydrogen can be quoted. The use of different catalysts in this process gives rise to different products. If a nickel catalyst is used under the conditions of 100–200°C and 1–10 atm, CH_4 + H_2O will be the main product. However, when ZnO-Cr_2O_3 is used (400°C, 500 atm), a mixture of CH_3OH + H_2O products can be obtained (Ciola, 1981). Therefore, the catalyst is selective towards the formation of intended products.

The stability of a catalyst can be defined as the property to maintain its activity and selectivity.

The catalyst lifetime can be compromised throughout the reaction cycle. The catalyst deactivation is the loss of catalyst activity over the reaction time. The causes of this loss can be associated with aging, poisoning, or coke formation. Aging can be due to decreased porosity and specific surface, recrystallization, or mutual fusion (sintering). While poisoning consists of the adsorption of substances foreign to the reaction, due to impurities in the reactants. Finally, the formation of carbon blocks the catalyst's active sites.

The study of several parameters that influence the catalytic performance in certain reactions will be presented below.

1.3 Performance parameters

The performance of the catalyst depends on the interaction of a large number of factors, including synthesis method, shaping, the composition of the constituent phases (the metal, the support, the promoter), particles size, porosity, specific surface area, surface acid-basicity, metal-support interaction, active phase dispersion, redox potential, crystal structure, and oxygen transport properties. To better understand these factors, some brief examples of catalytic processes will be addressed, such as the dry and steam reforming of hydrocarbons or alcohols, SO_x adsorption, CO_2 methanation, and electrocatalytic reactions. These processes will be covered in detail in the following chapters.

The catalyst consists of an active phase and support and eventually a promoter. A bulk catalyst can be defined as an agglomerate formed from grains of the active phase, while a supported catalyst is made up of support over which an active phase is dispersed. The support provides several advantages: increasing the activity, providing a large surface area to support the active phase, maintaining the specific area of the active phase, increasing the mechanical resistance, among many other essential functions. The active phase is the component responsible for the chemical reaction. Finally, the promoter is a substance that has low or no activity, but if added in low contents to the catalyst, leads to better activity, stability, or selectivity in a given reaction.

Choosing the type of catalyst is an important initial step. In processes such as dry and steam reforming, the deactivation of the catalyst by carbon deposition is still a major problem and currently prevents its commercialization. The noble metals catalysts have been efficient for this purpose. However, the high cost and low availability limit their applicability in the industry. Nickel-based catalysts have been extensively investigated for these reactions due to their high catalytic activity, selectivity, and low cost (Zhai, Ding, Liu, Jin, & Cheng, 2011). However, Ni catalysts are prone to deactivation by carbon deposition and sintering (Aboonasr Shiraz, Rezaei, & Meshkani, 2016; Özdemir, Faruk Öksüzömer, & Ali Gürkaynak, 2010). Both factors led to the development of supported catalysts that are capable of offering high resistance to carbon deposition and also preventing catalyst sintering (Fan, Abdullah, & Bhatia, 2011; Meshkani & Rezaei, 2010; San-José-Alonso, Juan-Juan, Illán-Gómez, & Román-Martínez, 2009; Tanios et al., 2017; Zhang et al., 2015).

A crucial step before implementing fuel processing reactions is to understand the effect of the powder preparation method on the physicochemical properties of the catalysts.

The literature highlights the effect of the synthesis method on the properties of Ni-CaO-Mayenite catalysts for the CO_2 Sorption Enhanced Steam Methane Reforming (SE-SMR) (Cesário, Barros, Courson, Melo, & Kiennemann, 2015). The catalysts were obtained by wet impregnation of Ni on the support previously synthesized by two methods: microwave-assisted self-combustion (MM) and hydration and calcination process-based method. Then, the materials were subjected to reduction under H_2 flow before the catalytic tests.

The catalyst obtained by the MM method showed more promising results. The 5%Ni/ 75%CaO-25%$Ca_{12}Al_{14}O_{33}$ composition was the most active and stable for 50 h with H_2 yield of almost 90% in SE-SMR reaction at 650°C due to its high CO_2 absorption capacity, the reducibility at lower temperatures, and the particle size stability after reduction and reaction, according to Thermogravimetric analysis (TG), H_2-Temperature Programmed Reduction (TPR) and X-ray Diffraction (XRD).

The microwave-assisted self-combustion method is highlighted as the primary method for the preparation of the CaO-$Ca_{12}Al_{14}O_{33}$ supports. This method consists of the classical combustion reaction of reducing and oxidizing precursors, where the heat required for ignition is provided by heating the polar molecules when exposed to microwave radiation. The heat released by this combustion allows the decomposition of the precursors and the formation of new crystal structures. The microwave-assisted self-combustion method has several advantages compared to the traditional combustion method using a hot plate and muffle furnace: reducing the time required to reach the ignition temperature and uniformity of temperature distribution, which is generated inside the precursor suspension. Moreover, microwave heating allows better control of the synthesis conditions, since the intensity of the microwave emission can be quickly interrupted, reduced, or increased, allowing to obtain materials with very specific characteristics.

The wet impregnation method has been commonly used to deposit the active phase on the support. This method without interaction consists of impregnating the support with an excess of a solution of the active species (usually a salt) compared to the pore volume of the support (Ciola, 1981). In the case of deposition of more than one active species, each active species may be successively or simultaneously impregnated.

The literature also highlights the influence of the synthesis method of mixed oxides derived from Ru-Mg-Al hydrotalcite on the catalytic performance in Glycerol Steam Reforming (GSR) (Dahdah et al., 2018). Two methods have been proposed: the grafted catalyst prepared by co-precipitation at a constant pH and the impregnated catalyst prepared by wet impregnation of the calcined Mg-Al support. The glycerol conversion and product selectivity strongly depended on the Ru nanoparticle's size and the easier accessibility of the active phase. The reconstruction of the hydrotalcite structure during the impregnation step may have influenced the Ru particle size. The impregnated catalyst was the most promising, presenting a higher activity beyond 600°C attributed to the easier accessibility of the active phase.

The synthesis of mixed oxides-based catalysts from the controlled thermal decomposition of Layered Double Hydroxides (LDHs), called Hydrotalcites, can be carried out using the co-precipitation, sol-gel, ion exchange, or hydrothermal treatment methods and suitable calcination. Hydrotalcite-like compounds (LDHs) are natural or synthetic laminar materials, whose structure is similar to brucite ($Mg(OH)_2$) and natural hydrotalcite ($Mg_6Al_2(OH)\ 16CO_34H_2O$). The chemical composition of LDHs can be represented by $[M^{II+}{}_{1\text{-}x}M^{III+}{}_{x}(OH)_2]^{x+}\ (A^{n-})_{x/n}\cdot y\ H_2O$, where M^{II+} and M^{III+} are divalent (Ni^{2+}, Co^{2+}, Zn^{2+}, Cu^{2+}, etc.) and trivalent (Al^{3+}, Ce^{3+}, Fe^{3+}, etc.) metal cations, respectively; x the mole fraction of the trivalent cation, A^{n-} is the anion of compensation (CO_3^{2-} or NO_3); and y the degree of hydration. This method has resulted in catalysts with excellent properties such as small crystal size, basic character, high dispersion, and high specific surface area (Tanios et al., 2017).

Recent work (Herminio et al., 2020) reports the direct preparation of the Ni-CeO_2 (45 wt% Ni) fibrous catalyst and evaluation of its performance in CO_2 dry reforming of methane (DRM). For comparison, the Ni/CeO_2 catalyst was also prepared from the wet impregnation of the support with Ni. The fibrous catalyst showed

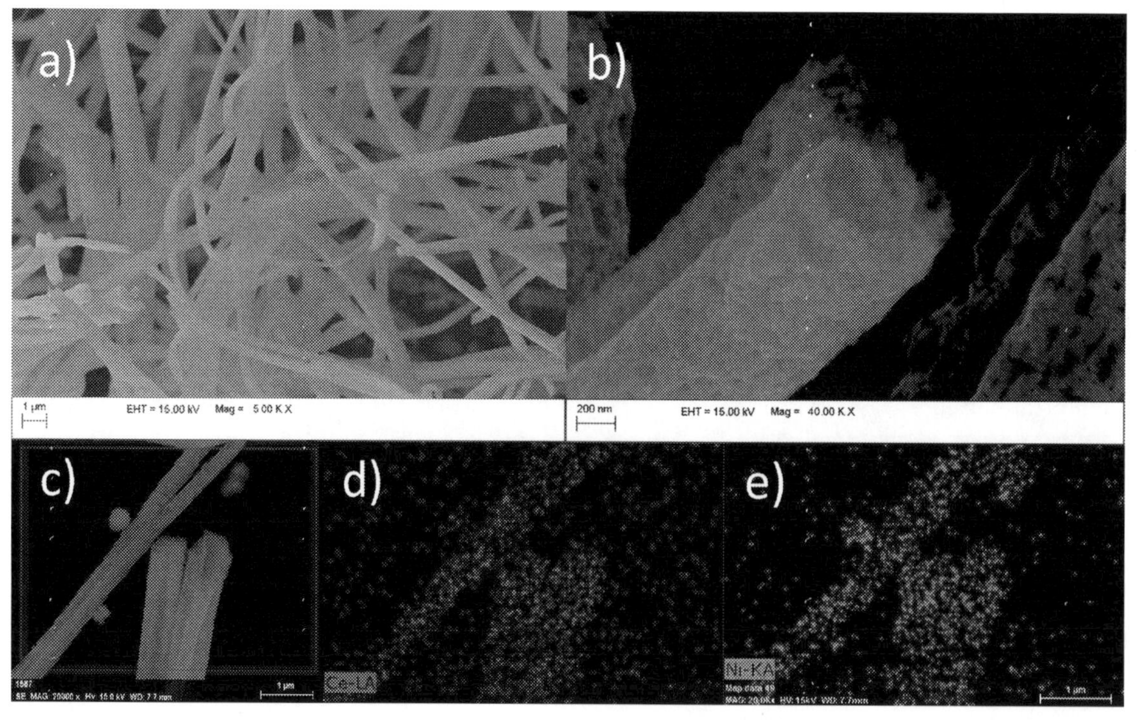

FIG. 1.3 Fibers morphology. Field emission electron microscopy images of calcined fibers show a uniform morphology (A); with a hallow circular cross-section (B); fibers surface mapping by EDS (C); EDS mapping of the uniform distribution of CeO_2 (D) and NiO (E).

surprising results. Fig. 1.3 shows the Field Emission Gun Electron Microscopy (FEG-SEM) micrograph of the nanofibers. This fiber hollow-type structure catalyst was active and stable for 30 h at 700°C. The Ni particle's size remains unchanged during the reaction and the good dispersion of the metallic phase on CeO_2 support allowed better CO_2 adsorption on support, reducing carbon deposits. The conventional Ni/CeO_2 catalyst suffers a severe deactivation of 84.2% after only 5 h.

The nanofibers were prepared by Solution Blow Spinning (SBS) technique. SBS is a simple technique, easy to set up, low cost, and without the need to use high voltages. The SBS configuration has been described in previous work (Medeiros, Glenn, Klamczynski, Orts, & Mattoso, 2009). SBS contains a system of concentric nozzles, where a precursor solution is injected in a controlled way into the inner nozzle; and in the outer nozzle, a flow of gas at high pressure promotes the evaporation of the solvent, shaping it into a fibrillar format, which is then deposited on the collector. The geometry of the nozzle creates a region of low pressure around the inner nozzle which helps to suck the precursor solution into a cone.

Lanthanides-based oxides have been commonly studied as supports for nickel catalysts in the DRM reaction (Barros, Melo, Libs, & Kiennemann, 2010).

Cerium oxide is well-known support due to its fluorite type structure, with large tolerance to high levels of atomic disorder, which can be introduced by doping, reduction, or oxidation (Inaba & Tagawa, 1996; Skinner & Kilner, 2003).

The redox potential (Ce^{3+}/Ce^{4+}) can facilitate the reduction of nickel species. Then, the presence of metallic sites (nickel for example) activates the hydrocarbon and leads to its decomposition into carbon and hydrogen (González, Asencios, Assaf, & Assaf, 2013; Perez-Lopez, Senger, Marcilio, & Lansarin, 2006; Rached et al., 2018).

The addition of dopants to CeO_2 results in the creation of charge-compensating oxygen vacancies, promoting the mobility of oxygen ions, according to the Kröger-Vink notation, as well as the creation of basic sites (Inaba & Tagawa, 1996; Skinner & Kilner, 2003). The presence of basic sites improves the CO_2 adsorption on the support and consequently allows faster activation of CO_2 in the catalyst (CO_2 + basic site = CO + O-basic site). The reaction between carbon + O-basic site also enables the production of CO and allows the removal of deposited carbon (González et al., 2013; Perez-Lopez et al., 2006; Rached et al., 2018).

Researchers evaluated the performance of Ni/$CeAlO_3$-Al_2O_3 catalysts in the DRM reforming (Chen, Zhao, Xue, Chen, & Lu, 2013). They concluded that adding Ce can significantly improve the resistance to carbon deposition on the catalyst surface. The support was also able to decompose CO_2 to form active oxygen on its surface.

A previous study (Hori et al., 1998) has also reported that the addition of zirconium can improve the properties inherent to cerium. The addition of Zr in the CeO_2 structure improved the oxygen storage and transport capacity.

The literature (Lovell, Horlyck, Scott, & Amal, 2017) also highlights the properties of the Si/CeZr-supported Ni catalyst synthesized by the Flame Spray method (FSP) in the DRM reforming. The addition of Ce-Zr improved the dispersion of Ni on support and allowed to minimize the decomposition of methane, promoting better activity and stability of the catalyst.

A previous study (García-Vargas, Valverde, Dorado, & Sánchez, 2014) has reported the influence of the type of support (alumina, ceria, β-silicon carbide, and yttria-stabilized zirconia) on the catalytic properties of Ni in the methane tri-reforming reaction. Methane tri-reforming is a synergistic combination of dry reforming, steam reforming, and partial oxidation which uses CO_2, H_2O, and O_2 as reforming agents, respectively. The CeO_2 and β-SiC catalysts produced a high reaction rate of methane without significant deactivation due to the high degree of reduction, which would increase the availability of the Ni^0 species. Furthermore, the catalyst Ni/CeO_2 showed the highest basicity, confirmed by CO_2-Temperature Programmed Desorption (TPD-CO_2).

The preparation of supports with excellent textural properties has been intensified. The support with high surface area and well-developed pore volume allows good dispersion of the active phase and easy diffusion of the gas toward the active phase. Also, the support with high oxygen mobility, high basicity, high redox potential can interact closely with the active phase, favoring its activity, and stability.

The study of the effect of ordered mesoporous SBA-15 and COK-12 supports on structural, textural, and SO_x adsorption and regeneration properties of the CuO/SBA-15 and CuO/COK-12 adsorbents has been reported in the literature (Schobing et al., 2021). SBA-15 has cylindrical and uniform pores, ordered in a hexagonal structure, with pore diameter from 5 to 30 nm (Rahmat, Abdullah, & Mohamed, 2010). SBA-15 has texture properties, such as high surface area and well-developed mesoporous volume, as well as being chemically inert and thermally stable in the SO_x adsorption process (Gaudin et al., 2016). However, the synthesis of SBA-15 involves expensive precursors, being carried out under strongly acidic conditions. The COK-12 synthesis with mesoporous architecture similar to SBA-15 at quasi-neutral pH, in the presence of citric acid/citrate buffer solution, can be an attractive alternative. XRD confirmed the hexagonal structure of the SBA-15 and COK-12 supports. The pore volume of both is slightly reduced after impregnation-calcination and modeling steps (Pelletized, Crushed, and Sieved). Besides, CuO/SBA-15 showed significantly reduced pore diameter after impregnation. The 15%CuO-COK adsorbent exhibited higher dynamic and total SO_x adsorption capacities than 15%CuO-SBA adsorbent during fifteen adsorption-regeneration cycles at 400°C, due to the better accessibility of SO_2 to the active sites. In fact, 15%CuO-COK had the highest porosity and shortest mesopores, and the most favorable particle morphology. Both adsorbents preserved their SO_2 adsorption capacity over the 15 cycles.

In another previous work (Cesario et al., 2020), the authors evaluated the impact of the binder content (9.1, 16.7, and 33.3 wt%) and shaping on the preparation of SBA-15 beads through shearing using an Eirich mixer. The porosity and mechanical residence are strongly dependent on the binder (bentonite) content and the beads' preparation process. The presence of the binder leads to an attack of the SBA-15 walls, due to the basic medium that increased the mesoporous size of the silica and, therefore, the mesoporous volume. In fact, SBA-15 beads containing 16.7 wt% of bentonite showed an increase in mesoporous volume by 25% and mesopores size (6.7 nm vs 6.2 nm) compared to the Sheared-SBA-15 (without binder). Beads containing 16.7 wt% of bentonite showed more promising results concerning the compromise between textural properties and mechanical resistance. Thus, these beads can be used as support for the desulfurization process, in which the porosity of the SBA-15 must be preserved to ensure good dispersion of the active phase and easy gas diffusion. The larger pore size of SBA-15 could lead to better accessibility of the active sites and, consequently, to better efficiency of the adsorbent.

The literature has also highlighted the effect of transition metal additions on the catalytic properties of Ni catalysts towards CO_2 reforming of methane when supported by various types of oxides such as Al_2O_3, Al_2O_3-ZrO_2, SBA-15, CeO_2, etc. (San-José-Alonso et al., 2009; Sharifi, Haghighi, Rahmani, & Karimipour, 2014; Turap et al., 2020; Wu, Liu, Liu, & He, 2019).

Cobalt-based catalysts can improve catalytic stability and resistance to coke when their content is increased, which probably results from their high affinity for oxygen species (Tanios et al., 2017). Ni-Co bimetallic catalysts are attractive since it provides high catalytic activity and stability as well as good resistance to carbon deposition.

The literature (Sharma & Dhir, 2020) addresses the preparation of TiO_2-supported nickel and cobalt bimetallic catalysts via wet impregnation and the catalytic activity in biogas reforming (model reaction). The catalytic performance was strongly dependent on the Ni/Co ratio, the synergy between Ni and Co particles, the formation of (Ni, Co) stable solid solutions, and their dispersion on the support. All of these factors are important to avoid sintering the active phase and carbon deposition while improving the catalytic properties and longevity of the catalyst.

$Co_xNi_yMg_zAl_2$ mixed oxide-based catalysts, synthesized by hydrotalcite route, were investigated towards dry reforming of methane for syngas production (Tanios et al., 2017). The Co/Ni ratio strongly influenced the catalytic activity and the resistance to carbon deposition. The $Co_2Ni_2Mg_2Al_2$ composition was more favorable to the formation of the Ni/Co alloy and stronger NiCo/support interactions (confirmed by TPR) which lead to its high performance in the reaction. On the other hand, catalysts with a molar ratio of Co/Ni higher than one ($Co_2Mg_4Al_2$ and $Co_3Ni_1Mg_2Al_2$) improved the resistance to carbon deposition, according to XRD and TG-Differential Scanning Calorimetry (TG-DSC).

Gadolinia-doped ceria (CGO) supported Ni catalysts, promoted with Cu and Co (20, 40, and 60 mol%), synthesized by a one-step synthesis route, were also investigated towards dry reforming of methane for hydrogen production (Cesario et al., 2018). The Co-containing catalysts exhibited higher surface area values (S_{BET}), higher reducibility, and stronger metal-support interactions in comparison to the Cu-containing catalysts. Thus, the NiCo (40 mol%) and NiCo (60 mol%) catalysts gave higher CH_4 and CO_2 conversions and increased resistance towards carbon deposition than Cu-based catalysts. NiCo (40 mol%) catalyst showed improved selectivity below 600°C. Overall, co-promoter is shown to be highly beneficial for the dry reforming of methane. The electrocatalytic performance of ($NiCo_{0.4}$/CGO/$NiCo_{0.4}$) screen-printed symmetrical cells from hydrogen and synthetic biogas (CH_4 + CO_2) electro-oxidation reaction at 650–850°C were studied by impedance spectroscopy. $NiCo_{0.4}$-CGO exhibited polarization resistances (Rp) of 0.96 and 36.10 $\Omega\,cm^2$ at 750°C for H_2 and biogas atmospheres, respectively. Moreover, the activation energy (Ea) was lower under H_2 (0.92 eV). Therefore, hydrogen electro-oxidation reaction occurs more easily (lower Rp and activation energy) than the biogas reforming.

The positive effect of cobalt on Ni catalysts has also been highlighted towards CO_2 methanation (Alrafei, Polaert, Ledoux, & Azzolina-Jury, 2020). The Co-containing catalysts improved the reducibility of Ni species and Ni dispersion on the support. The presence of Co enhanced the catalyst activity and selectivity to CH_4.

The impact of the Ce and La promoter on Ni-Al catalyst has been studied in the dry reforming of toluene (Rached et al., 2018). The cerium-promoted catalyst showed the most promising results. The cerium-promoted catalyst was reducible at lower temperatures and had high basicity (256 μmol $CO_2\,g^{-1}$ vs 38 μmol $CO_2\,g^{-1}$ for La or 27 μmol $CO_2\,g^{-1}$ for non-promoted) than La-promoted catalyst and non-promoted catalyst. These factors have played a key role in increasing catalytic activity and removing carbon deposits. The Ce-promoted catalyst was more selective and stable in the dry reforming reaction of toluene. In addition, the Ce-promoted catalyst showed less intense oxidation peaks that indicate lower carbon deposition, according to the Temperature Programmed Oxidation Analysis (TPO). This potential to inhibit carbon has been related to the ability of Ce_2O_3 and CeO_2 to coexist during the reforming process, resulting in electron transfer reactions. The following mechanism has been proposed: CO_2 is adsorbed on the basic or redox centers and then CO and CeO_2 in a redox cycle are produced. Subsequently, CeO_2 reacts with the carbon deposited by dehydrogenization of the C—H bond and oxidation to produce Ce_2O_3 and CO_2 again. This mechanism would also explain the increase of the total basicity of the Ce-promoted catalyst, due to the adsorption of CO_2 that can occur in the anionic vacancies of CeO_2. The La-promoted catalyst showed similar textural and physical–chemical properties compared to the non-promoted catalyst, however, it exhibited higher catalytic performance due to the presence of La_2O_3 that can influence the transport and adsorption of CO_2 on the active sites surface.

The effect of Ce promoter on the porous structure, reducibility, basicity, catalytic activity, and stability of Ru/KIT-6, Ni/KIT-6, and Ni-Ru/KIT-6 catalysts from DRM reaction has also been described in the literature (Mahfouz et al., 2021). Although the Ce promotion impacted the porous structure of ordered mesoporous silica (KIT-6), decreased the surface area and pore volume, Ce promotion favored the dispersion of RuO_2 species, enhanced the reducibility of the active phases and the strength of the basic sites. Non-promoted bimetallic catalyst (Ni-Ru/KIT-6) showed better performance than monometallic catalysts, taking into account its smaller particle size and higher basicity. After Ce addition, Ce-promoted Ni-Ru/KIT-6 exhibited much higher activity and stability for 12 h, despite the formation of carbon resulting from the decomposition of methane.

The influence of the lanthanum promoter on the physicochemical and catalytic properties of Ni/Mg-Al from glycerol steam reforming has been studied (Dahdah et al., 2020). The lanthanum content influenced the reducibility of Ni species, metal-support interactions, Ni dispersion, and basicity. $Ni/Mg_6Al_{1.6}La_{0.4}$ was the most active catalyst with higher hydrogen yields and glycerol conversion due to an appropriate balance of metal-support interactions, active metal dispersion, and basicity. The high content of lanthanum ($Ni/Mg_6Al_{1.2}La_{0.8}$) contributed to the coke gasification, which justifies its low coke formation (1.8 wt%). In fact, the presence of La_2O_3 assisted in coke gasification through the formation of surface carbonates ($La_2O_2CO_3$) that oxidized the deposited coke. Despite the slight initial increase in deposited carbon (2.5 wt%), the $Mg_6Al_{1.6}La_{0.4}$ catalyst is the most promising catalyst considering its improved stability (24 h) after a reduced flow rate (0.008 mL/min—water/glycerol). The operating conditions can be changed to further improve the activity and stability of this catalyst.

The development of catalytic materials via pyrolysis of metal-organic frameworks (MOFs) can be attractive for an application in heterogeneous catalysis.

According to the International Union of Pure and Applied Chemistry (IUPAC), MOF is defined as a coordination network with organic binders that have potentially empty cavities (Batten et al., 2013). MOFs are formed by Secondary Building Units (SBU), which are made up of metal ions and oxygen atoms. The organic binder is extremely important in the connection of the SBUs for the assembly of the MOF.

Different types of SBU binding with organic ligands can lead to the synthesis of MOFs with different structures.

The structured shape, metal centers, and organic binders are considered essential in the preparation of the MOFs. One of the characteristics inherent to MOFs is their high porosity, due to the expanded size of inorganic SBUs and dimensions of the network by large metal agglomerates. The possibility of using different ions or metallic centers becomes the preparation of MOFs flexible to change their physical and chemical characteristics (Hu et al., 2019). Different synthetic methods have been reported for the preparation of MOFs, including the diffusion self-assembly method, hydro and/or solvothermal methods, mechanochemical routes, ultrasonic and electrochemical methods (do Nascimento et al., 2017; Konnerth et al., 2020). Moreover, the proper choice of precursors can play a very important role in the crystallization and properties of MOFs (do Nascimento et al., 2017; Hu et al., 2019; Konnerth et al., 2020).

MOFs are an extensive class of crystalline materials that have high porosity, high thermal stability, a high surface area extending beyond $6000\,m^2\,g^{-1}$, easy to prepare which makes them competitive compared to the zeolites or activated carbon, for example. These properties of MOFs allow them to be applied in several areas (Gangu, Maddila, Mukkamala, & Jonnalagadda, 2016; Kaneti et al., 2017), particularly in Heterogeneous Catalysis (Konnerth et al., 2020). The catalytic activity of MOF originates from uncoordinated metal centers or functional groups attached to the structure's binders. The MOF provides stability to the active phase or may act as catalyst support (Kaneti et al., 2017).

Cerium-based MOFs are being increasingly studied, due to their high catalytic activity and high adsorption capacity (Zhou, Liu, Lin, Zhou, & Huang, 2020). Previous studies described the preparation of Ce-MOFs with 1,3,5-tricarboxylic benzene acid (H3BTC) by solvothermal methods, to evaluate the catalytic performance of MOF in the CO oxidation process. Ce-BTC MOF showed high catalytic activity and long-term stability (Zhang et al., 2018).

The thermal and chemical stability of cerium-based MOFs with different linker molecules has been studied (Lammert et al., 2015).

Cerium-based MOFs with terephthalic acid ligand (1,4-H2BDC), a structure similar to Fig. 1.4, showed higher chemical and thermal stability. Furthermore, Ce-BDC MOF can be used as a cocatalyst when combined with (2, 2,6,6-tetramethylpiperidin-1-yl)oxyl (TEMPO) for the aerobic oxidation of benzyl alcohol. Ce-BDC MOF/TEMPO showed 88% conversion of benzyl alcohol and complete selectivity to benzaldehyde at an activation temperature of the framework at 220°C.

Zirconium-based MOFs have also been widely studied using a variety of ligands, such as UiO-66 (with 1,4-H2BDC ligand) and MOF-808 (with 1,3,5-H3BTC ligand) (Ardila-Suárez, Perez-Beltran, Ramírez-Caballero, & Balbuena, 2018). Zr-MOFs also have large pore sizes, which allow their application in catalytic reactions. Fig. 1.5 shows the three-dimensional structure of Zr-1,4-H2BDC MOF.

Seeking to unite the properties of cerium and zirconium-based MOFs, researchers carried out a study to obtain bimetallic Ce/Zr compounds, with UiO-66 and MOF-808 structures. Ce/Zr-UiO-66 showed high thermal resistance, smaller particle size, and high acid resistance. While Ce/Zr-MOF-808 showed an increase in thermal stability and high surface area (Lammert, Glißmann, & Stock, 2017).

Despite presenting several advantages for their catalytic application, MOFs still have drawbacks such as lack of stability, especially

FIG. 1.4 Three-dimensional structure of Ce-BDC MOF. (*red spheres* (dark gray in print version): oxygen atoms, *gray spheres*: carbon atoms, and *yellow spheres* (light gray in print version): Ce atoms).

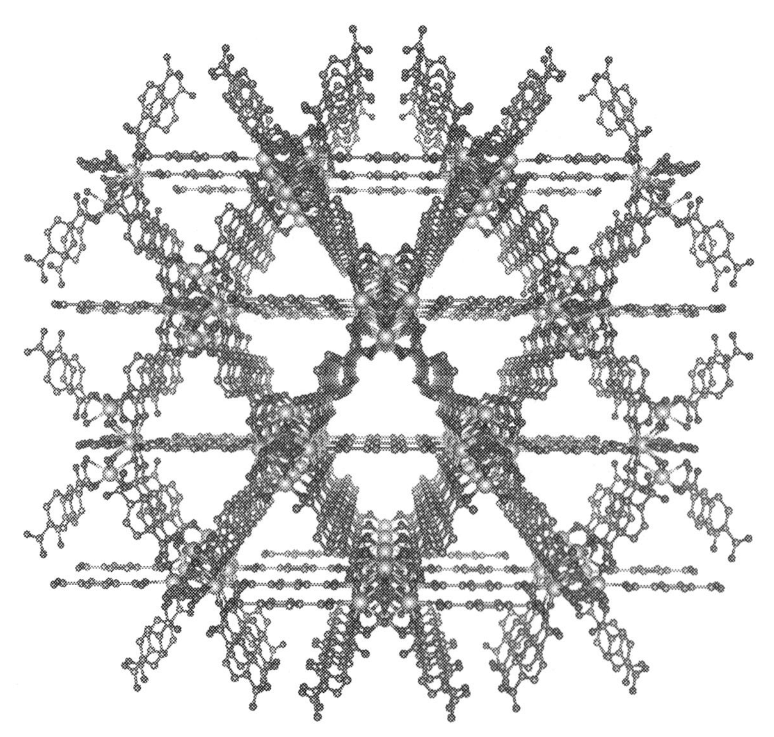

FIG. 1.5 Three-dimensional structure of Zr-BDC MOF. (*red spheres* (dark gray in print version): oxygen atoms, *gray spheres*: carbon atoms, and *blue spheres* (dark gray in print version): Zr atoms).

when it comes to organic transformations in adverse reaction conditions such as high temperature, presence of organic solvents, water, acidic or basic medium, which can impair catalytic activity and stability. Thus, it has been sought to carry out syntheses of materials derived from MOF, such as carbon and metallic oxides (Chaikittisilp, Ariga, & Yamauchi, 2013; Chen et al., 2018). One possibility of obtaining these derivatives is the use of the pyrolysis process.

The literature addresses the preparation of metal oxides as catalytic supports from the pyrolysis of the MOFs precursors. This recent study compared the catalytic activity of the CeO_2 support derived from MOFs, commercial CeO_2, and CeO_2 synthesized by precipitation method, in the combustion of toluene. The authors have obtained promising toluene conversion results when MOF-based support is used (Chen et al., 2018).

Another recent work (Jampaiah et al., 2020) reports the preparation of the $Ce_{0.8}Zr_{0.2}O_2$ support from the MOF pyrolysis. The performance of this catalyst in the CO_2 methanation reaction was compared with that of the $Ce_{0.8}Zr_{0.2}O_2$ support synthesized by the conventional co-precipitation (CP) method. Both were used as supports for the cobalt catalyst. The MOF-based catalyst showed improved redox properties compared to that obtained by CP. The MOF-based catalyst showed higher conversion of CO_2 (81.2%) and higher selectivity of CH_4 (99%) than the CP catalyst (48.7% and 97% respectively). Besides, the MOF-based catalyst was more resistant to sintering due to its high dispersion and metal-support interactions.

1.4 Conclusions

This chapter aimed to cluster knowledge on heterogeneous catalysis and make it available to the academic/industrial community. The chapter briefly highlighted the basic principles of heterogeneous catalyzes, such as the action of the heterogeneous catalyst, the phenomenon of adsorption, surface reactivity, and the properties of the heterogeneous catalyst. Besides, the chapter has emphasized previously published studies that showed how the interaction of various factors can influence the performance of the catalyst in certain processes such as dry and steam reforming of hydrocarbons or alcohols, SO_x adsorption, CO_2 methanation, and electrocatalytic reactions.

Several synthesis methods of powders or nanofibers were addressed with emphasis on microwave-assisted self-combustion and Solution Blow Spinning. The impact of the type of metal used as catalyst, of the support, of the promoter was thoroughly described in this study. The preparation method and the components of the catalyst played a significant role in particle size, porosity, specific surface area, acid-basicity surface, metal-support interaction, active phase dispersion, redox potential, crystalline structure, and oxygen transport properties of the catalyst.

Finally, metal-organic frameworks precursors have an essential and promising role in developing heterogeneous catalysts for specific reactions. This approach is left open to a range of future opportunities and for further explorations.

We believe that this chapter reinforces the ability to relate physicochemical properties and catalytic performance, as well as hope it will become a reference to academic/industrial scientists from chemistry, physics, and materials science interested in this field.

Conflict of interest

The authors declare no conflict of interest.

References

Aboonasr Shiraz, M. H., Rezaei, M., & Meshkani, F. (2016). Microemulsion synthesis method for preparation of mesoporous nanocrystalline γ-Al_2O_3 powders as catalyst carrier for nickel catalyst in dry reforming reaction. *International Journal of Hydrogen Energy*, *41*(15), 6353–6361. https://doi.org/10.1016/j.ijhydene.2016.03.017.

Alrafei, B., Polaert, I., Ledoux, A., & Azzolina-Jury, F. (2020). Remarkably stable and efficient Ni and Ni-Co catalysts for CO_2 methanation. *Catalysis Today*, *346*, 23–33. https://doi.org/10.1016/j.cattod.2019.03.026.

Ardila-Suárez, C., Perez-Beltran, S., Ramírez-Caballero, G. E., & Balbuena, P. B. (2018). Enhanced acidity of defective MOF-808: Effects of the activation process and missing linker defects. *Catalysis Science & Technology*, *8*(3), 847–857. https://doi.org/10.1039/C7CY02462B.

Barros, B. S., Melo, D. M. A., Libs, S., & Kiennemann, A. (2010). CO_2 reforming of methane over La_2NiO_4/α-Al_2O_3 prepared by microwave assisted self-combustion method. *Applied Catalysis A: General*, *378*(1), 69–75. https://doi.org/10.1016/j.apcata.2010.02.001.

Batten, S. R., Champness, N. R., Chen, X.-M., Garcia-Martinez, J., Kitagawa, S., Öhrström, L., et al. (2013). Terminology of metal–organic frameworks and coordination polymers (IUPAC recommendations 2013). *Pure and Applied Chemistry*, *85*(8), 1715–1724. https://doi.org/10.1351/PAC-REC-12-11-20.

Becker, C. (2018). From Langmuir to Ertl: The "Nobel" history of the surface science approach to heterogeneous catalysis. In K. Wandelt (Ed.), *Encyclopedia of interfacial chemistry* (pp. 99–106). Elsevier. https://doi.org/10.1016/B978-0-12-409547-2.13527-9.

Cesario, M., Schobing, J., Bruder, F., Dorge, S., Nouali, H., Habermacher, D., et al. (2020). Impact of bentonite content on the structural, textural and mechanical properties of SBA-15 mesoporous silica beads. *Journal of Porous Materials*, *27*, 905–910. https://doi.org/10.1007/s10934-020-00865-5.

Cesário, M. R., Barros, B. S., Courson, C., Melo, D. M. A., & Kiennemann, A. (2015). Catalytic performances of Ni–CaO–mayenite in CO_2 sorption enhanced steam methane reforming. *Fuel Processing Technology*, *131*, 247–253. https://doi.org/10.1016/j.fuproc.2014.11.028.

Cesario, M. R., Macedo, D. A., Souza, G. S., Loureiro, F. J. A., Tidahy, H. L., Gennequin, C., et al. (2018). NiCo and NiCu based-materials for hydrogen production and electro-oxidation reactions. In M. R. Cesario, D. A. Macedo, C. Gennequin, & E. Abi-Aad (Eds.), *Vol. 2. Catalytic materials for hydrogen production and electro-oxidation reactions* (pp. 129–161). Bentham Science https://doi.org/10.2174/9781681087580118020008.

Chaikittisilp, W., Ariga, K., & Yamauchi, Y. (2013). A new family of carbon materials: Synthesis of MOF-derived nanoporous carbons and their promising applications. *Journal of Materials Chemistry A*, *1*(1), 14–19. https://doi.org/10.1039/C2TA00278G.

Chen, W., Zhao, G., Xue, Q., Chen, L., & Lu, Y. (2013). High carbon-resistance Ni/$CeAlO_3$-Al_2O_3 catalyst for CH_4/CO_2 reforming. *Applied Catalysis B: Environmental*, *136–137*, 260–268. https://doi.org/10.1016/j.apcatb.2013.01.044.

Chen, X., Chen, X., Yu, E., Cai, S., Jia, H., Chen, J., et al. (2018). In situ pyrolysis of Ce-MOF to prepare CeO_2 catalyst with obviously improved catalytic performance for toluene combustion. *Chemical Engineering Journal*, *344*, 469–479. https://doi.org/10.1016/j.cej.2018.03.091.

Ciola, R. (1981). *Fundamentos da catálise*. Moderna.

Dahdah, E., Aouad, S., Gennequin, C., Estephane, J., Nsouli, B., Aboukaïs, A., et al. (2018). Glycerol steam reforming over Ru-Mg-Al hydrotalcite-derived mixed oxides: Role of the preparation method in catalytic activity. *International Journal of Hydrogen Energy*, *43*(43), 19864–19872. https://doi.org/10.1016/j.ijhydene.2018.09.042.

Dahdah, E., Estephane, J., Gennequin, C., Aboukaïs, A., Aouad, S., & Abi-Aad, E. (2020). Effect of La promotion on Ni/Mg-Al hydrotalcite derived catalysts for glycerol steam reforming. *Journal of Environmental Chemical Engineering*, *8*(5). https://doi.org/10.1016/j.jece.2020.104228, 104228.

Devred, F., Dulgheru, P., & Kruse, N. (2013). Elementary steps in heterogeneous catalysis. In J. Reedijk, & K. Poeppelmeier (Eds.), *Vol. 7. Comprehensive inorganic chemistry II* (2nd ed., pp. 7–38). Elsevier. https://doi.org/10.1016/B978-0-08-097774-4.00703-8.

do Nascimento, J. F. S., Barros, B. S., Kulesza, J., de Oliveira, J. B. L., Leite, A. K. P., & de Oliveira, R. S. (2017). Influence of synthesis time on the microstructure and photophysical properties of Gd-MOFs doped with Eu3+. *Materials Chemistry and Physics*, *190*, 166–174. https://doi.org/10.1016/j.matchemphys.2017.01.024.

Fan, M.-S., Abdullah, A. Z., & Bhatia, S. (2011). Hydrogen production from carbon dioxide reforming of methane over Ni–Co/MgO–ZrO_2 catalyst: Process optimization. *International Journal of Hydrogen Energy*, *36*(8), 4875–4886. https://doi.org/10.1016/j.ijhydene.2011.01.064.

Gangu, K. K., Maddila, S., Mukkamala, S. B., & Jonnalagadda, S. B. (2016). A review on contemporary metal–organic framework materials. *Inorganica Chimica Acta*, *446*, 61–74. https://doi.org/10.1016/j.ica.2016.02.062.

García-Vargas, J. M., Valverde, J. L., Dorado, F., & Sánchez, P. (2014). Influence of the support on the catalytic behaviour of Ni catalysts for the dry reforming reaction and the tri-reforming process. *Journal of Molecular Catalysis A: Chemical*, *395*, 108–116. https://doi.org/10.1016/j.molcata.2014.08.019.

Gaudin, P., Michelin, L., Josien, L., Nouali, H., Dorge, S., Brilhac, J.-F., et al. (2016). Highly dispersed copper species supported on SBA-15 mesoporous materials for SOx removal: Influence of the CuO loading and of the support. *Fuel Processing Technology*, *148*, 1–11. https://doi.org/10.1016/j.fuproc.2016.02.025.

González, A. R., Asencios, Y. J. O., Assaf, E. M., & Assaf, J. M. (2013). Dry reforming of methane on Ni–Mg–Al nano-spheroid oxide catalysts prepared by the sol–gel method from hydrotalcite-like precursors. *Applied Surface Science*, *280*, 876–887. https://doi.org/10.1016/j.apsusc.2013.05.082.

Herminio, T., Cesário, M. R., Silva, V. D., Simões, T. A., Medeiros, E. S., Macedo, D. A., et al. (2020). CO_2 reforming of methane to produce syngas using anti-sintering carbon-resistant Ni/CeO_2 fibers produced by solution blow spinning. *Environmental Chemistry Letters*, *18*, 895–903. https://doi.org/10.1007/s10311-020-00968-0.

Hori, C. E., Permana, H., Simon Ng, K. Y., Brenner, A., More, K., Rahmoeller, K. M., et al. (1998). Thermal stability of oxygen storage properties in a mixed CeO_2-ZrO_2 system. *Applied Catalysis B: Environmental*, *16*(2), 105–117. https://doi.org/10.1016/S0926-3373(97)00060-X.

Hu, C., Xiao, J.-D., Mao, X.-D., Song, L.-L., Yang, X.-Y., & Liu, S.-J. (2019). Toughening mechanisms of epoxy resin using aminated metal-organic framework as additive. *Materials Letters*, *240*, 113–116. https://doi.org/10.1016/j.matlet.2018.12.123.

Inaba, H., & Tagawa, H. (1996). Ceria-based solid electrolytes. *Solid State Ionics*, *83*(1–2), 1–16. https://doi.org/10.1016/0167-2738(95)00229-4.

Jampaiah, D., Damma, D., Chalkidis, A., Venkataswamy, P., Bhargava, S. K., & Reddy, B. M. (2020). MOF-derived ceria-zirconia supported Co_3O_4 catalysts with enhanced activity in CO_2 methanation. *Catalysis Today*, *356*, 519–526. https://doi.org/10.1016/j.cattod.2020.05.047.

Kaneti, Y. V., Dutta, S., Hossain, M. S. A., Shiddiky, M. J. A., Tung, K.-L., Shieh, F.-K., et al. (2017). Strategies for improving the functionality of zeolitic imidazolate frameworks: tailoring nanoarchitectures for functional applications. *Advanced Materials*, *29*(38), 1700213. https://doi.org/10.1002/adma.201700213.

Konnerth, H., Matsagar, B. M., Chen, S. S., Prechtl, M. H. G., Shieh, F.-K., & Wu, K. C.-W. (2020). Metal-organic framework (MOF)-derived catalysts for fine chemical production. *Coordination Chemistry Reviews*, *416*. https://doi.org/10.1016/j.ccr.2020.213319, 213319.

Lammert, M., Wharmby, M. T., Smolders, S., Bueken, B., Lieb, A., Lomachenko, K. A., et al. (2015). Cerium-based metal organic frameworks with UiO-66 architecture:

Synthesis, properties and redox catalytic activity. *Chemical Communications, 51*(63), 12578–12581. https://doi.org/10.1039/C5CC02606G.

Lammert, M., Glißmann, C., & Stock, N. (2017). Tuning the stability of bimetallic Ce(iv)/Zr(iv)-based MOFs with UiO-66 and MOF-808 structures. *Dalton Transactions, 46*(8), 2425–2429. https://doi.org/10.1039/C7DT00259A.

Lovell, E. C., Horlyck, J., Scott, J., & Amal, R. (2017). Flame spray pyrolysis-designed silica/ceria-zirconia supports for the carbon dioxide reforming of methane. *Applied Catalysis A: General, 546*, 47–57. https://doi.org/10.1016/j.apcata.2017.08.002.

Mahfouz, R., Estephane, J., Gennequin, C., Tidahy, L., Aouad, S., & Abi-Aad, E. (2021). CO_2 reforming of methane over Ni and/or Ru catalysts supported on mesoporous KIT-6: Effect of promotion with Ce. *Journal of Environmental Chemical Engineering, 9*(1). https://doi.org/10.1016/j.jece.2020.104662, 104662.

Medeiros, E. S., Glenn, G. M., Klamczynski, A. P., Orts, W. J., & Mattoso, L. H. C. (2009). Solution blow spinning: A new method to produce micro- and nanofibers from polymer solutions. *Journal of Applied Polymer Science, 113*(4), 2322–2330. https://doi.org/10.1002/app.30275.

Meshkani, F., & Rezaei, M. (2010). Nanocrystalline MgO supported nickel-based bimetallic catalysts for carbon dioxide reforming of methane. *International Journal of Hydrogen Energy, 35*(19), 10295–10301. https://doi.org/10.1016/j.ijhydene.2010.07.138.

Niemantsverdriet, J. W., & Schlögl, R. (2013). Heterogeneous catalysis: Introduction. In J. Reedijk, & K. Poeppelmeier (Eds.), *Vol. 7. Comprehensive inorganic chemistry II* (2nd ed., pp. 1–6). Elsevier. https://doi.org/10.1016/B978-0-08-097774-4.00701-4.

Özdemir, H., Faruk Öksüzömer, M. A., & Ali Gürkaynak, M. (2010). Preparation and characterization of Ni based catalysts for the catalytic partial oxidation of methane: Effect of support basicity on H_2/CO ratio and carbon deposition. *International Journal of Hydrogen Energy, 35*(22), 12147–12160. https://doi.org/10.1016/j.ijhydene.2010.08.091.

Perez-Lopez, O. W., Senger, A., Marcilio, N. R., & Lansarin, M. A. (2006). Effect of composition and thermal pretreatment on properties of Ni–Mg–Al catalysts for CO_2 reforming of methane. *Applied Catalysis A: General, 303*(2), 234–244. https://doi.org/10.1016/j.apcata.2006.02.024.

Rached, J. A., Cesario, M. R., Estephane, J., Tidahy, H. L., Gennequin, C., Aouad, S., et al. (2018). Effects of cerium and lanthanum on Ni-based catalysts for CO_2 reforming of toluene. *Journal of Environmental Chemical Engineering, 6*(4), 4743–4754. https://doi.org/10.1016/j.jece.2018.06.054.

Rahmat, N., Abdullah, A. Z., & Mohamed, A. R. (2010). A review: Mesoporous Santa Barbara amorphous-15, types, synthesis and its applications towards biorefinery production. *American Journal of Applied Sciences, 7*(12), 1579–1586. https://doi.org/10.3844/ajassp.2010.1579.1586.

Ross, J. (2018). *Contemporary catalysis: Fundamentals and current applications* (1st ed.). Elsevier. https://www.elsevier.com/books/contemporary-catalysis/ross/978-0-444-63474-0.

Ruthven, D. M. (1984). *Principles of adsorption and adsorption processes.* John Wiley & Sons.

San-José-Alonso, D., Juan-Juan, J., Illán-Gómez, M. J., & Román-Martínez, M. C. (2009). Ni, Co and bimetallic Ni–Co catalysts for the dry reforming of methane. *Applied Catalysis A: General, 371*(1–2), 54–59. https://doi.org/10.1016/j.apcata.2009.09.026.

Schmal, M. (2016). *Heterogeneous catalysis and its industrial applications.* Springer.

Schobing, J., Cesario, M., Dorge, S., Nouali, H., Patarin, J., Martens, J., et al. (2021). CuO supported on COK-12 and SBA-15 ordered mesoporous materials for temperature swing SOx adsorption. *Fuel Processing Technology, 211.* https://doi.org/10.1016/j.fuproc.2020.106586, 106586.

Sharifi, M., Haghighi, M., Rahmani, F., & Karimipour, S. (2014). Syngas production via dry reforming of CH_4 over Co- and Cu-promoted $Ni/Al_2O_3–ZrO_2$ nanocatalysts synthesized via sequential impregnation and sol–gel methods. *Journal of Natural Gas Science and Engineering, 21*, 993–1004. https://doi.org/10.1016/j.jngse.2014.10.030.

Sharma, H., & Dhir, A. (2020). Hydrogen augmentation of biogas through dry reforming over bimetallic nickel-cobalt catalysts supported on titania. *Fuel, 279.* https://doi.org/10.1016/j.fuel.2020.118389, 118389.

Sing, K. S. W., Everett, D. H., Haul, R. A. W., Moscou, L., Pieroti, R. A., Rouquerol, J., et al. (1985). Reporting physisorption data for gas/solid systems with special reference to the determination of surface area and porosity. *Pure and Applied Chemistry, 57*(4), 603–619.

Skinner, S. J., & Kilner, J. A. (2003). Oxygen ion conductors. *Materials Today, 6*(3), 30–37. https://doi.org/10.1016/S1369-7021(03)00332-8.

Tanios, C., Bsaibes, S., Gennequin, C., Labaki, M., Cazier, F., Billet, S., et al. (2017). Syngas production by the CO_2 reforming of CH_4 over Ni–Co–Mg–Al catalysts obtained from hydrotalcite precursors. *International Journal of Hydrogen Energy, 42*(17), 12818–12828. https://doi.org/10.1016/j.ijhydene.2017.01.120.

Thomas, J. M., & Thomas, W. J. (2015). *Principles and practice of heterogeneous catalysis* (2nd ed.). Wiley-VCH.

Turap, Y., Wang, I., Fu, T., Wu, Y., Wang, Y., & Wang, W. (2020). Co–Ni alloy supported on CeO_2 as a bimetallic catalyst for dry reforming of methane. *International Journal of Hydrogen Energy, 45*(11), 6538–6548. https://doi.org/10.1016/j.ijhydene.2019.12.223.

Wu, H., Liu, J., Liu, H., & He, D. (2019). CO_2 reforming of methane to syngas at high pressure over bi-component

Ni-Co catalyst: The anti-carbon deposition and stability of catalyst. *Fuel, 235*, 868–877. https://doi.org/10.1016/j.fuel.2018.08.105.

Zhai, X., Ding, S., Liu, Z., Jin, Y., & Cheng, Y. (2011). Catalytic performance of Ni catalysts for steam reforming of methane at high space velocity. *International Journal of Hydrogen Energy, 36*(1), 482–489. https://doi.org/10.1016/j.ijhydene.2010.10.053.

Zhang, X., Yang, C., Zhang, Y., Xu, Y., Shang, S., & Yin, Y. (2015). Ni–Co catalyst derived from layered double hydroxides for dry reforming of methane. *International Journal of Hydrogen Energy, 40*(46), 16115–16126. https://doi.org/10.1016/j.ijhydene.2015.09.150.

Zhang, X., Hou, F., Li, H., Yang, Y., Wang, Y., Liu, N., et al. (2018). A strawsheave-like metal organic framework Ce-BTC derivative containing high specific surface area for improving the catalytic activity of CO oxidation reaction. *Microporous and Mesoporous Materials, 259*, 211–219. https://doi.org/10.1016/j.micromeso.2017.10.019.

Zhou, J., Liu, H., Lin, Y., Zhou, C., & Huang, A. (2020). Synthesis of well-shaped and high-crystalline Ce-based metal organic framework for CO_2/CH_4 separation. *Microporous and Mesoporous Materials, 302*. https://doi.org/10.1016/j.micromeso.2020.110224, 110224.

CHAPTER

2

Use of CO_2 as a source for obtaining value-added products

Martin Schmal[a,b], Alberth Renne Gonzalez Caranton[b,c], Camila E. Kozonoe[a], Karina T. de C. Roseno[a], Fabio M. Cavalcanti[a], Rita M. Brito Alves[a], and Reinaldo Giudici[a]

[a]Department of Chemical Engineering, University of São Paulo, São Paulo, Brazil [b]Federal University of Rio de Janeiro—COPPE/PEQ/Nucat, Rio de Janeiro, Brazil [c]Universidad ECCI-Grupo de Investigación en Aprovechamiento Tecnológico de Materiales y Energía GIATME, Bogotá D.C., Colombia

2.1 Introduction

Global warming is related to the emission of greenhouse gases, with carbon dioxide (CO_2) being the gas with the greatest contribution, reaching 76%. Within this amount, only 14% of the emission comes from the use of land and forests, with the remainder attributed to power generation, industrial processes, and the use of fossil fuels (Intergovernmental Panel on Climate Change, 2015).

The increase in the emission of this gas into the atmosphere contributes to global warming. This has been occurring since the industrial revolution, when there was an increase in demand and consumption of fossil fuels, abruptly increasing the emission of gases and unbalancing the carbon cycle. Global warming is more than climate change; it also involves the development of extreme weather conditions, rising sea levels, and, with this, changes in the marine ecosystem (Intergovernmental Panel on Climate Change, 2015).

Fig. 2.1 illustrates the average temperature rise with industrial development between 1850 and 2017, making the influence of emissions on the climate clear with the increase in energy demand on the global stage (Ritchie & Roser, 2020).

According to Świrk et al., the energy sector produced two-thirds of the total anthropogenic greenhouse gas (GHG) emissions in 2014, of which more than 80% were CO_2 (Świrk, Grzybek, & Motak, 2016). Nevertheless, CO_2 emissions have increased with the increased consumption of fossil fuels, which has alarmed climate change experts. Approximately 1 ton of carbon present in nonrenewable fuels produces 3.5 tons of CO_2, or the average atmospheric CO_2 concentration had increased from

Heterogeneous Catalysis
https://doi.org/10.1016/B978-0-323-85612-6.00002-4

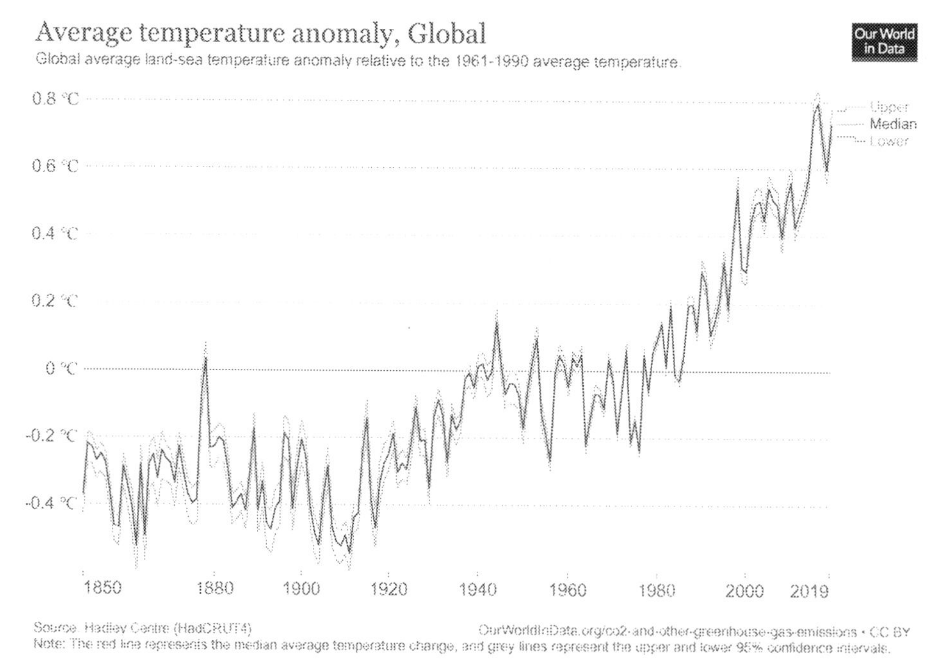

FIG. 2.1 Increase in average temperature after the industrial revolution over the years. *Reprinted with permission from Ritchie, H., & Roser, M. (2020).* CO_2 and greenhouse gas emissions. *OurWorldInData.Org. https://ourworldindata.org/co2-and-other-greenhouse-gas-emissions.*

280 ppm in 1900 to 400 ppm in 2015. This increase includes CO_2 emissions from fossil fuels, industrial processes, and CO_2 emissions from land use for agriculture (Raupach et al., 2007). The predominant emissions are industrial emissions from the combustion of solid, liquid, and gaseous fuels, petroleum gas flaring, cement industries, and hydrocarbon production and oxidation (Raupach et al., 2007).

The use of fossil fuels for energy supply from chemical conversions inherently modifies carbon bonds producing other molecules that accumulate in the atmosphere, such as methane and CO_2. The reuse of these molecules requires knowledge of the market potential of CO_2 chemicals and their derivatives (González, Asencios, Assaf, & Assaf, 2013).

The pollution caused by the growth in energy consumption of society and industry has promoted the search for alternative, less-polluting energy sources, such as hydrogen and energy obtained from fuel cells (Centi, Perathoner, Park, Chang, & Lee, 2004; Chagas, de Souza, de Carvalho, Martins, & Schmal, 2016).

According to Tundo et al., it is an alternative for developing and applying new chemical products and processes, which reduce or eliminate the use and generation of dangerous materials harmful to human health and the environment (Tundo et al., 2000).

Bearing in mind that nanotechnology can boost environmental protection through the control of emissions and the development of new technologies that minimize the production of byproducts, the development of new processes and the manufacture of new materials on a nanoscale should follow the following guidelines: reduce/eliminate material waste, reduce the number of resources used as much as possible, minimize the use of energy and guarantee environmental, safety and ethical aspects (Centi, Quadrelli, & Perathoner, 2013).

One of the challenges present in the field of research in Chemical Engineering concerns the treatment of tailings and products resulting from numerous existing industrial processes. In addition, the environmental aspect reinforces the urgent need to develop processes called "green," in which the products are totally or often reused in the process or directed to another application (Silva, 2017).

The main strategy is to capture CO_2 from the emission gases. A means of capturing CO_2 of long duration was by submerged tanks, yet it is remote. The other strategy is the adsorption on reactive substrates, such as zeolites, metal oxides, and new materials such as carbon nanotubes and graphenes. Alternatively, it is necessary to know the form and the amount of CO_2 released into the atmosphere. Galhotra presents different materials used for sequestering CO_2 (Galhotra, 2010). In this study, CO_2 adsorption in the presence and absence of co-adsorbed H_2O was investigated in different nanomaterials.

Several studies have been published with the aim of studying such reactions and developing catalysts that enable their occurrence. Whang et al. published a review of catalysts for converting CO_2 into various high value-added products (Whang, Lim, Choi, Lee, & Lee, 2019). The key point for the conversion of CO_2 is the activation of CO_2 or co-reagent under different conditions. For instance, activating CO_2 with the presence of substances rich in electrons or hydrogen and, in some cases, converting CO_2 with a coupling reaction. The conversion of CO_2 was carried out by different methodologies, including the reforming of methane to produce synthesis gases with bifunctional catalysis, synthesis of methanol hydrogenation with nanostructured catalyst, and synthesis of carbonates from CO_2. This review presents the recent progress of the catalytic conversion of CO_2 and clarifies the chemical fixation of CO_2.

The development of processes and use of CO_2, through the use and capture of CO_2 and natural gas by hydrocarboxylation of methane with CO_2, hydrogenation of CO_2 to obtain chemical products, such as vinyl acetate, and dimethyl ether, acetic acid, and obtaining light hydrocarbons and the search for new materials, nanostructured catalysts and synthesize metallic nanoparticles are themes of this chapter.

There are several motivations for producing chemicals from CO_2 (Centi et al., 2013), which are given as follows:

1. CO_2 is cheap, even when considering possible taxes related to the production of emissions.
2. CO_2 is also regarded as a nontoxic raw material that could replace highly harmful products such as phosgene and isocyanate.
3. Chemical processes using CO_2 can lead to the production of important products for industry, such as polycarbonates, adipic acid, vinyl acetate, and dimethyl ether (DME).
4. CO_2 reuse may be unrepresentative for the industry in general, but it has positive impacts on the global carbon balance.
5. The use of greenhouse gases as feedstock, mainly CO_2, is a challenge that stimulates new approaches to industrial chemistry.

2.2 Background

CO_2 is a thermodynamically stable compound; whose conversion requires an external energy source and frequent use of catalysts to promote activation. Many products are obtained by heterogeneous gas-solid catalysis involving CO_2, such as methylamines, carboxylic acids, synthesis gas, and vinyl compounds, in addition to the hydrocarbons traditionally synthetized by Fischer-Tropsch synthesis (Fig. 2.2). Other CO_2 transformation processes have been studied, such as hydrogenation, oxidative dehydrogenation of ethane to produce ethylene, conversion of

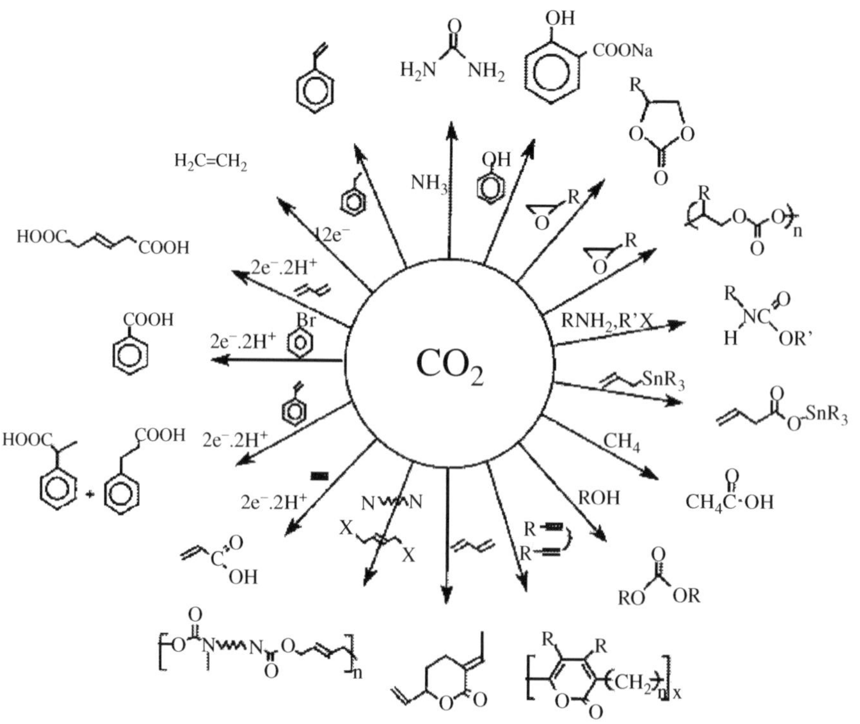

FIG. 2.2 Pathway of CO_2 utilization to produce chemical compounds and derivatives. *Reprinted with permission from Joshi, P. (2014). Carbon dioxide utilization: A comprehensive review.* International Journal of Chemical Sciences, *12(4), 1208–1220. https://www.tsijournals.com/abstract/carbon-dioxide-utilization-a-comprehensive-review-11696.html.*

alkanes, and electrocatalysis processes (Fiorani, Guo, & Kleij, 2015).

Recent studies show that there are several routes for CO_2 conversion into other useful products due to the electronic interactions with the catalytic surfaces, showing different possible reaction paths (Fiorani et al., 2015). CO_2 conversion is of great industrial and academic interest, with imminent importance to create and promote sustainability (Joshi, 2014).

It is also interesting to study some compounds, aiming to produce new products and the development of renewable chemical processes, minimizing environmental impact.

CO_2 is currently used in the synthesis of fuels and other compounds through the formation of various chemical bonds, including C–C, C–H, C–O, and C–N (nitrogenated graphene) (Gonzalez Caranton et al., 2018; Gonzalez Caranton, da Silva Pinto, Stavale, Barreto, & Schmal, 2020; Song et al., 2021). Fig. 2.3 shows the compounds obtained from the different chemical routes used in the industry, highlighting the formation of carbonates, carboxylic acids, urea, and liquid hydrocarbon derivatives.

The other renewable carbon source that contributes to global warming is methane gas, which can be generated from renewable biomass and can also be obtained from nonrenewable sources, such as methane from petroleum. From an industrial perspective, the CO_2 conversion routes can be differentiated into two groups: (i) conversion with H_2 to form hydrocarbons and alcohols, such as methanol and olefins; (ii) reaction with other molecules, to form chemicals. The first approach integrates renewable

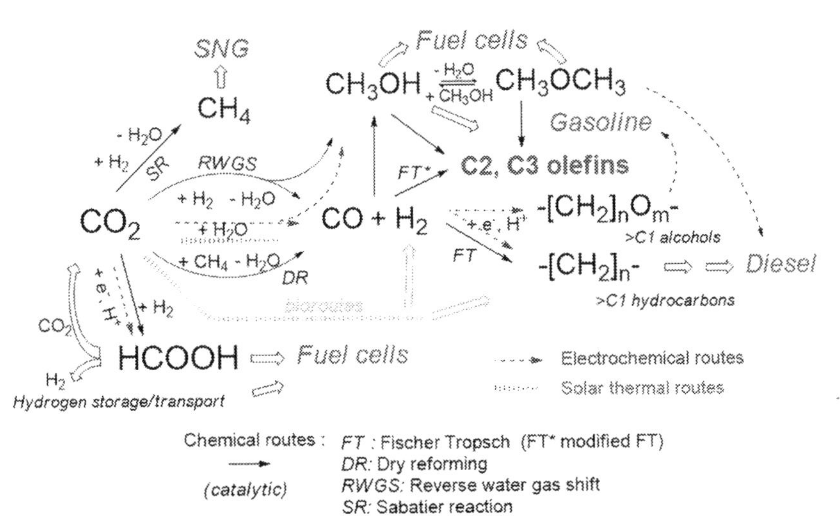

FIG. 2.3 Alternative routes for CO_2 usage and production of renewable compounds derived from synthesis gas. *Reprinted with permission from Centi, G., Quadrelli, E. A., & Perathoner, S. (2013). Catalysis for CO_2 conversion: A key technology for rapid introduction of renewable energy in the value chain of chemical industries.* Energy and Environmental Science, *6(6), 1711–1731. https://doi.org/10.1039/C3EE00056G.*

energy into chemical industry processes, such as the use of synthesis gas obtained in reforming processes and its subsequent use in the production of methanol and DME. The second approach is more attractive from an industrial perspective since a greater variety of products can be obtained on an industrial scale with simpler routes (Joshi, 2014). Direct routes involve selective oxidations such as high-temperature coupling to produce formaldehyde, methanol, long-chain hydrocarbons, carboxylic acids, vinyl acetate, and aromatics (Xu, Xiu, Liu, Liang, & Wang, 2020).

2.3 Catalytic transformations of CO_2

The processes for converting natural gas and CO_2 are, in general, heterogeneous catalysts and depend, fundamentally, on highly efficient catalysts. All the different uses and definitions in relation to nanotechnology reflect its applicability in an extensive range of research fields and the need for multidisciplinary to understand it (Centi et al., 2013; Fiorani et al., 2015; Gonzalez Caranton et al., 2018, 2020; Joshi, 2014; Song et al., 2021).

Several strategies have been studied for methane conversion, based on direct synthesis in the production of chemical compounds, such as obtaining syngas from reforming with CO_2, water vapor, and oxygen. The direct synthesis of acetic acid, vinyl acetate via methane carboxylation, and hydrogenation with CO_2 are complex and possible, but for all cases through the simultaneous activation of the two molecules in the same catalyst, which is a great challenge compared to the usual methods and due to the thermodynamic barrier.

Rabie et al. claim that catalytic conversion of methane and CO_2 and the surface activation sites are the key problems to overcome the thermodynamic barrier and the kinetic ability for these reactions to occur and thermodynamically very unfavorable (Rabie, Betiha, & Park, 2017; Wilcox et al., 2001). Moreover, even under drastic test conditions (at 725°C, 100 atm), the thermodynamic equilibrium conversion is low 1.6×10^{-6}. Moreover, the simultaneous activation of both molecules looks attractive but even tougher (Liu et al., 2020; Xu et al., 2020). In fact, it is important to understand the C–H bonding activation of methane and that the CO_2 activation needs a reductant to lose oxygen.

The methane molecule is very stable, with strong C–H bonds (425 kJ/mol), and the fission of such bonds requires high temperatures and the use of oxidizing agents. The symmetry of the molecule and the absence of functional groups, magnetic or polar moment hinder the chemical attack. As the strength of the C–H bond is greater in methane than in its possible products, it is known that the products will be more reactive than the methane itself. This means that the direct conversion of methane is not necessarily limited to a reactivity problem, but also covers the question of the selectivity of the product formed, considering that the instability of the product leads to other possibly undesirable byproducts. In this way, catalysts have an important role in the objective of overcoming these limitations through the stabilization of an intermediate compound on its surface that will direct the reaction in a determined route (Holmen, 2009).

Huang et al. and Wu et al. demonstrated the importance of activating methane in products obtained by direct conversion (Huang et al., 2001; Wu et al., 2013). First, there is a methane activation in CH_x, with x ranging from 0 (carbon formation) to 3 (methyl radicals), obtaining a wide variety of products. Subsequently, achieved CH_3-Zn formation, obtaining a higher selectivity in the production of acetic acid. Then, not only if the dissociation of methane occurs, but how it happens and what by-products are formed are aspects extremely important in this chapter.

The first studies of methane interaction with transition metals were done by Morikawa, Benedict, and Taylor (Morikawa, Benedict, & Taylor, 1935; Morikawa, Trenner, & Taylor, 1937), through the exchange of hydrogen-deuterium (HD) atoms. They concluded that the first step of the reaction is the activated dissociative adsorption of methane. However, the quantitative activated methane adsorption was reported by Kubokawa on metallic Ni and observed the formation of CH_x species (Kubokawa, 1938). Later, other authors investigated more deeply about these dissociations, examining films and single crystals, with IR spectroscopic, which allowed to interpret quantitatively the surface reaction and intermediates.

The literature has reported the methane-metal interactions on different metals and oxides (Amariglio, Paréja, & Amariglio, 1995; Wright, Ashmore, & Kemball, 1958) and found activation energies of the order of 42 kJ/mol, and verified that adsorbed radicals and nonionic species are the probable intermediate species in the metal (Kemball, 1959). Catalysts with metals well dispersed on supports proved to be more reactive than larger particles and methane dissociated at low temperatures, such as 30°C (Kuijpers, Breedijk, van der Wal, & Geus, 1981; Kuijpers, Breedijk, van der Wal, & Geus, 1983; Kuijpers, Jansen, van Dillen, & Geus, 1981).

Bradford and Vannice claimed that the CH_x fragments at the surface tend to occupy adsorption sites that complete the tetravalence of the carbon atom (Bradford & Vannice, 1999). Thus, CH_3 tends to bond to one metal atom, CH_2 binds to two metal atoms, and so on. So, each CH_x species occupies sites that have different surface geometries, and therefore, the process of dissociation of methane is sensitive to the structure.

Wei and Iglesia studied the mechanism and the sites required for the activation and chemical conversion of methane on metals supported and observed that the activation of the C–H bond is the only kinetically relevant step (Wei & Iglesia, 2004). This means that the reaction of CH_x with other species is very quick and kinetically irrelevant. Viñes et al. observed the critical role of different sites of Pt nanoparticles in the activation of methane and found both theoretically and experimentally that the presence of low coordination sites (edges and corners, for example) in Pt nanoparticles reduces the energy barriers of individual reaction steps and stabilizes the reaction intermediates (Viñes et al., 2010). Schmal et al. studied the nonoxidative methane activation and evidenced the significant formation of methane coupling products over the Ni-O_x/SiO_2 catalyst during methane chemisorption (Moya, Martins, & Schmal, 2011). The Ni-O_x/SiO_2 catalyst produced C_2 and CH_4 for different temperatures. Opposite to the NiB catalyst, a higher C_2 yield was obtained than C_{2+} in the first step. The hydrogenation of carbon species in the 2nd step produced the majority of methane.

The carbon dioxide molecule is linear with strong bonds (532 kJ/mol) and is thermodynamically stable. The high positive value of Gibbs free energy of formation ($\Delta G_f^0 = +394.6\,kJ/mol$) shows a high inert capacity of CO_2 and makes its reactions energetically unfavorable. Because of this, reactions with CO_2 require an effectively designated catalytic system that overcomes the respective existing thermodynamic barriers (Havran, Duduković, & Lo, 2011).

Several studies for methane conversion, based on direct synthesis in the production of chemical compounds, such as obtaining syngas from reforming with CO_2, water vapor, and oxygen. Much effort has been invested in research related to reforming methane with CO_2, according to the scheme from Fig. 2.4 (Centi et al., 2013; Chen, Zaffran, & Yang, 2020; Chen, Zhang, et al., 2020; Gonzalez Caranton et al., 2018; Xu et al., 2020).

This process can be carried out with natural gas, containing high concentrations of CO_2; however, supported metal catalysts deactivate rapidly by Boudouard and disproportionalization reactions (Erdőhelyi, 2021).

The main advantage of dry reforming is the use of greenhouse gases; in contrast, the main disadvantages are the high reaction temperatures and the deactivation of catalysts by coke formation (Eqs. 2.1–2.2) (Al Abdulghani et al., 2020). A wide variety of noble metal-based catalysts have been proposed in the literature. In order of reactivity, the most used metals are Ru, Rh ≈ Ni ≫ Ir > Pt > Pd (Liu, Jiang, Liu, Li, & Suib, 2013). There is evidence related to the effect of zirconia on γ-alumina nanofibers, showing that Pt catalysts supported on ZrO_2-Al_2O_3 exhibiting an increase in catalyst stability, also increasing the metal dispersion on the support and oxygen vacancies in the lattice structure, which improve

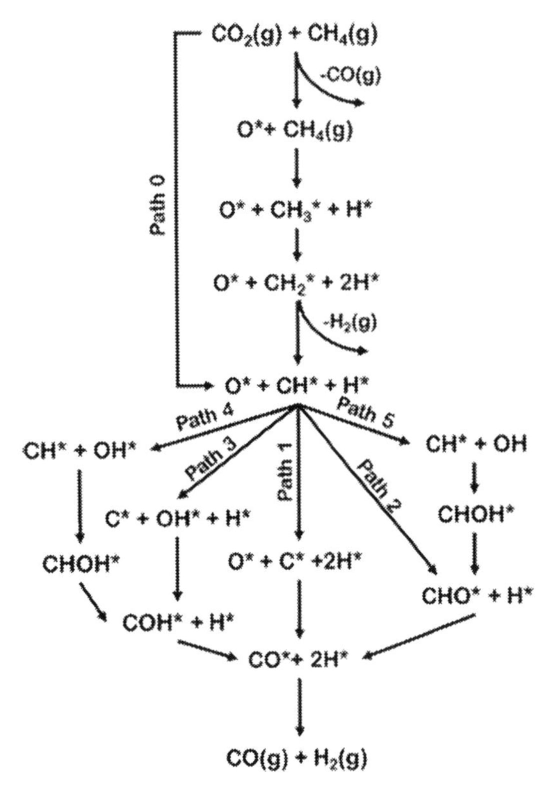

FIG. 2.4 Reaction pathway for methane dry reforming over nickel catalysts. *Reprinted with permission from Chen, S., Zaffran, J., & Yang, B. (2020). Descriptor design in the computational screening of Ni-based catalysts with balanced activity and stability for dry reforming of methane reaction.* ACS Catalysis, *10(5), 3074–3083. https://doi.org/10.1021/acscatal.9b04429.*

the reducibility and thermal stability of the materials (El-Salamony et al., 2021).

$$CH_4 \rightarrow C + 2H_2 \quad \Delta H^0 = 74\,\text{kJ/mol} \tag{2.1}$$

$$2CO \rightarrow CO_2 + C \quad \Delta H^0 = -172\,\text{kJ/mol} \tag{2.2}$$

From a perspective of energy efficiency, partial oxidation represents a more promising process since it is thermodynamically more favorable, and higher methane conversions are obtained, when being used for synthesis gas suitable for fuel cells.

Other processes, such as tri-reforming of methane, take advantage of the exothermic nature of one of the reactions to provide energy in the endothermic reaction, as shown in Eqs. (2.3)–(2.6) (Zhang, Zhang, Gossage, Lou, & Benson, 2014).

$$CH_4 + H_2O \Leftrightarrow 3H_2 + CO \quad \Delta H^0 = \frac{206.3\,\text{kJ}}{\text{mol}} \tag{2.3}$$

$$CH_4 + CO_2 \Leftrightarrow 2H_2 + 2CO \quad \Delta H^0 = \frac{247\,\text{kJ}}{\text{mol}} \tag{2.4}$$

$$CH_4 + \frac{1}{2}O_2 \Leftrightarrow 2H_2 + CO \quad \Delta H^0 = -\frac{35.6\,\text{kJ}}{\text{mol}} \tag{2.5}$$

$$CH_4 + 2O_2 \Leftrightarrow 2H_2O + CO_2 \quad \Delta H^0 = -880\,\text{kJ/mol} \tag{2.6}$$

Oxidative dehydrogenation (ODH) processes of alkanes and alkenes have now become an important way to increase the production of olefins. As an alternative, oxidation with CO_2 as a mild oxidant has emerged (Eqs. 2.7–2.9), which inhibits the oxidation of the background products and increases the selectivity of olefins (Bawah, Lucky, & Hossain, 2021).

$$C_2H_6 + CO_2 \rightarrow C_2H_4 + CO + H_2O \tag{2.7}$$

$$C_3H_8 + CO_2 \rightarrow C_3H_6 + CO + H_2O \tag{2.8}$$

$$iC_4H_{10} + CO_2 \rightarrow iC_4H_8 + CO + H_2O \tag{2.9}$$

Nonconventional processes, such as dehydrogenation of ethane to ethylene with CO_2, have been reported (Qiu, Zhang, Liu, & Zhang, 2021), focused on the study of the active metal and catalytic promoters, mainly precious metals supported on cerium, silica, and zeolites. The most efficient processes have achieved ethane conversions and ethylene selectivities between 68.6% and 92.3% over SiO_2 (Koirala, Safonova, Pratsinis, & Baiker, 2017).

Methane tri-reforming is an important way for CO_2 usage aiming at hydrogen and syngas production without separation of carbon dioxide. Świrk et al., claims that tri-reforming has several advantages, due to the H_2O and O_2 addition, which would inhibit the coke formation, and therefore increase the lifetime and product purity (Świrk et al., 2016). Another critical characteristic of the tri-reforming process is that it is not necessary to handle pure oxygen, producing synthesis gas directly with a desirable H_2/CO ratio (e.g., $H_2/CO = 1.5$–2). Literature about tri-reforming has been first reviewed by Song and Pan (2004). Complete reviews for different processes for methane conversion have been published recently by Świrk et al., Ghoneim et al., and many others (Cañete, Gigola, & Brignole, 2017; de Sousa, Toniolo, Landi, & Schmal, 2017; Ghoneim, 2016; Gómez-Cuaspud, Perez, & Schmal, 2016; Świrk et al., 2016) including thermodynamics of the process, the possible conversions of methane and carbon dioxide for the tri-reforming and the single processes (SMR, DMR, and POM). Eqs. (2.3)–(2.8) show the basic reforming reactions for synthesis gas production. The main challenges in bringing this reforming technology to practice rest in catalysts development and the understanding of the reaction mechanisms for kinetic and process optimization (Schmal, Toniolo, & Kozonoe, 2018).

Noteworthy is the work of García-Vargas et al., who studied a novel β-SiC support and the influence of the order of addition of the metal Ni and Mg as promoters on the catalytic activity and stability for tri-reforming (García-Vargas, Valverde, Díez, Dorado, & Sánchez, 2015). Higher reduction temperatures were needed when Mg was firstly loaded or when both metals, Ni and Mg, were simultaneously loaded, which was attributed to the interactions between Ni and Mg. The Ni-Mg/SiC (1Ni:1Mg) is apparently a good catalyst for tri-reforming due to its high catalytic activity and stability (low coke formation). The highest conversion reached was 97.9% at 800°C and very stable after running 25h with time on stream (García-Vargas et al., 2015). Pino et al. studied the La-Ce-O-NiO mixed oxide, with different compositions ($Ce_{1-3x}La_{2x}Ni_xO_{2-\delta}$, $x=0.1$; 0.2 and 0.25) (Pino, Vita, Cipitì, Laganà, & Recupero, 2011; Pino, Vita, Laganà, & Recupero, 2013).

Kozonoe and Schmal studied the tri-reforming reaction with new material, synthesizing the Ni@MWCNT/Ce catalyst, and evaluated the influence of feed rate and testing variables on the tri-reforming of methane, varying temperatures, space velocities, and feed rate concentrations of water and oxygen, using experimental design (Kozonoe, Brito Alves, & Schmal, 2020). The highest CH_4 conversions were obtained for similar feed conditions (CH_4:CO_2:H_2O:O_2:N_2) (1,0.34:0.23:0.5:2.1) but for different space velocities and temperatures. The results are presented in Fig. 2.5.

The highest conversions (96.8% CH_4, 38.7% of CO_2, and the ratio $H_2/CO=1.88$) were obtained for a higher temperature (750°C) and lower space velocity 1250mL/gmin, which indicates the influence of residence time and kinetics.

Finally, Sedighi et al. studied the direct conversion of carbon dioxide into hydrocarbons, which is a very desirable but difficult approach

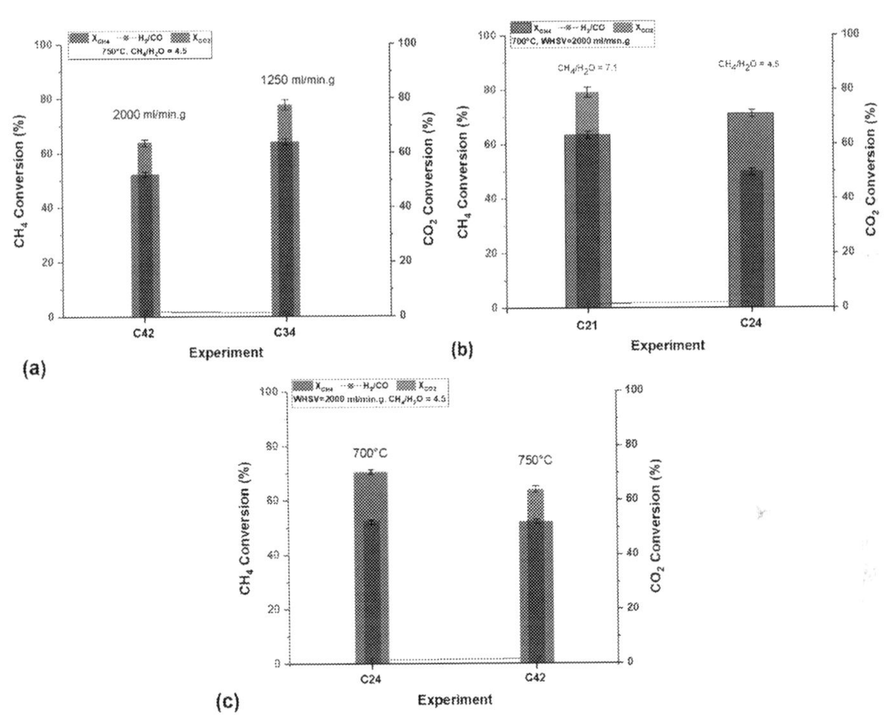

FIG. 2.5 Conversions of CH_4, CO_2, and ratio H_2/CO for the Ni@MWCNT/Ce catalysts in the evaluation of the effects of the ratio H_2O/O_2 (A), space velocity (B), and temperature (C). *From Kozonoe, C. E., Brito Alves, R. M., & Schmal, M. (2020). Influence of feed rate and testing variables for low-temperature tri-reforming of methane on the Ni@MWCNT/Ce catalyst.* Fuel, 281, 118749. *https://doi.org/10.1016/j.fuel.2020.118749.*

for achieving lower value-added olefins with minimal CO selectivity (Sedighi & Mohammadi, 2019). In this effort, they report the direct conversion of CO_2 into light olefins on a Cu/CeO_2 hybrid catalyst mixed with SAPO-34 zeolite. A high olefin selectivity of 70.4% has been obtained on CuCe/SAPO-34 at $H_2/CO_2=3$, 10h, 382.46°C, 17.33L/gh, and 20bar, which seems to be a promising catalyst.

The catalytic reduction of CO_2 to methanol

Methanol can be used as a fuel in internal combustion engines, in direct methanol fuel cells (DMFCs), and as a feedstock for light olefin production and subsequent hydrocarbon production. The catalytic reduction of CO_2 to methanol has been considered the most economical way to mitigate the greenhouse effect caused by CO_2 (Centi et al., 2013; Chen, Zaffran, et al., 2020; Gonzalez Caranton et al., 2018; Xu et al., 2020). In addition to conventional catalytic reduction with hydrogen at high temperatures, photocatalysis has also become an important research focus, using water and other reductants under illumination and room temperature (Aresta, 2010).

Copper-based catalysts, typically $Cu\text{-}Zn\text{-}Al_2O_3$ and $Cu\text{-}Zn/ZrO_2$, are the most active materials for the catalytic reduction of CO_2 with hydrogen. The formation of methanol is an exothermic reaction that is favored by decreasing temperature and increasing pressure operating conditions. At temperatures above 513K, the activation of CO_2 is favored, and methanol is formed. The maximum conversion occurs at 240°C, with increasing pressure (from 1000 to 1800psi) and at high GHSV ($8300 h^{-1}$). The optimal molar ratio of H_2/CO_2 is 3 (Dibenedetto, Angelini, & Stufano, 2014), as shown in Eq. (2.10)

$$CO_2 + H_2 \rightarrow C_2H_4 + CH_3OH + H_2O \quad (2.10)$$

Olah, Goeppert, and Prakash (2009) developed a catalytic technology for methanol synthesis based on the direct conversion of syngas mixtures obtained by steam methane reforming and CO_2 (Eq. 2.11). This process, called bi-reforming, occurs at temperatures of 800–1000°C and pressures of 0.5–4MPa over nickel-based catalysts (Abd El-Hafiz, Sakr, & Ebiad, 2021).

$$3CH_4 + CO_2 + 2H_2O \rightarrow 4CO + 8H_2 \rightarrow 4CH_3OH \quad (2.11)$$

The effective conversion of methane and CO_2 can be optimized by reducing process temperatures and improving catalytic structure, resulting in minimized process costs and reduced catalyst deactivation. It is now also an environmental problem within production plants and refineries. Another limiting factor in the conversion of methane and CO_2 is that methane can be activated easily, although CO_2 requires a greater amount of energy (Zoeller, 2008).

The activation of CO_2 depends on the $Cu\text{-}Zn\text{-}Al_2O_3$ and $Cu\text{-}Zn/ZrO_2$ catalyst. In fact, the ZnO in this catalyst guarantees the spacing between the copper nanoparticles, promoting a better dispersion of the active copper metal and, consequently, a greater number of active sites exposed to gaseous reagents (Chen, Lin, Lee, & Huang, 2012). ZnO can also store excess hydrogen for the successive hydrogenation steps of the intermediates adsorbed from the reactions (Spencer, 1998) and induces greater copper activity due to the strong metal-support interaction between the Cu and ZnO nanoparticles, while alumina increases the stability by inhibiting the thermal sintering of copper particles and assisting in the dispersion of copper through the trivalent ions Al^{3+} (Behrens et al., 2012).

Yang et al. proposed two possible routes depending on the reagent to be hydrogenated. In the case of carbon dioxide, hydrogenation can follow the formate or carboxyl route as intermediates (Yang, Mims, Mei, Peden, & Campbell, 2013). Hydrogenation of carbon monoxide can also produce methanol via carboxyl, but its other

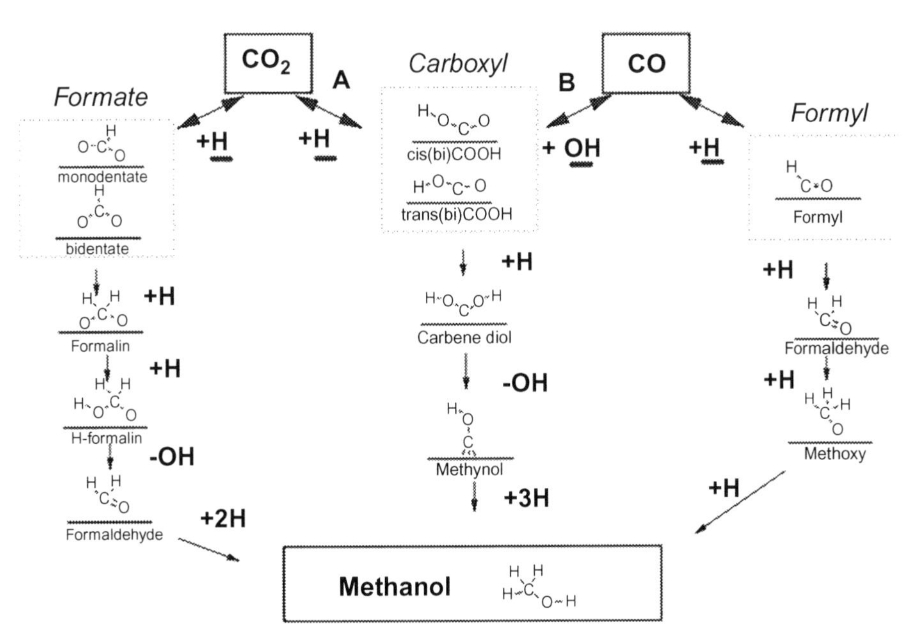

FIG. 2.6 Mechanisms of methanol synthesis from CO_2/CO hydrogenation. *From Yang, Y., Mims, C. A., Mei, D. H., Peden, C. H. F., & Campbell, C. T. (2013). Mechanistic studies of methanol synthesis over Cu from CO/CO_2/H_2/H_2O mixtures: The source of C in methanol and the role of water.* Journal of Catalysis, 298, 10–17. *https://doi.org/10.1016/j.jcat.2012.10.028.*

route corresponds to the formation of formates as an intermediate (Sato, 2019). Fig. 2.6 shows a schematic representation.

Arena et al. proposed that the carbon dioxide is adsorbed on the surface of the ZnO nanoparticles and the ZrO_2 support (Arena et al., 2008). Subsequent reactions lead to the formation of hydrogenated bicarbonate species, generating the formate groups, which in turn, through the spillover of hydrogen adsorbed on copper, producing methanol, as shown in Fig. 2.7.

Gao et al. showed through TPD of CO_2 on the pre-reduced catalysts the presence of weak, moderate, and strong basic sites (Gao et al., 2013). The hydroxyl groups present on the catalyst surface are attributed to the weak sites; the metal-oxygen pairs, such as Zn-O, Al-O, and X-O, are responsible for the moderate basic sites, while the strong sites are associated with coordinated unsaturated oxygen ions. CO_2 molecules are adsorbed on such species, forming weak, moderate, and strong bicarbonate, bidentate, and monodentate species on basic sites, respectively. The reaction mechanisms are shown in Fig. 2.8.

The adsorption and desorption of hydrogen occur on the copper sites and of CO_2 adsorption on basic sites, which depend on the strength of the basic sites. The bicarbonate species adsorbed on weak basic sites α are easily desorbed to form carbon dioxide. The adsorbed CO_2 on moderate β and strong γ basic sites undergoes several hydrogenation steps with the hydrogen "spillover" provided by copper. The carbonyl bond of aldehyde adsorbed on the strong basic sites C-γ can react with hydrogen to form methanol. The C-β bond is weak, making the C=O bond relatively stable, which causes the preferential route of dehydrogenation of formaldehyde to form CO (Gao et al., 2013).

The authors studied the influence of the basic sites on the methanol selectivity for different

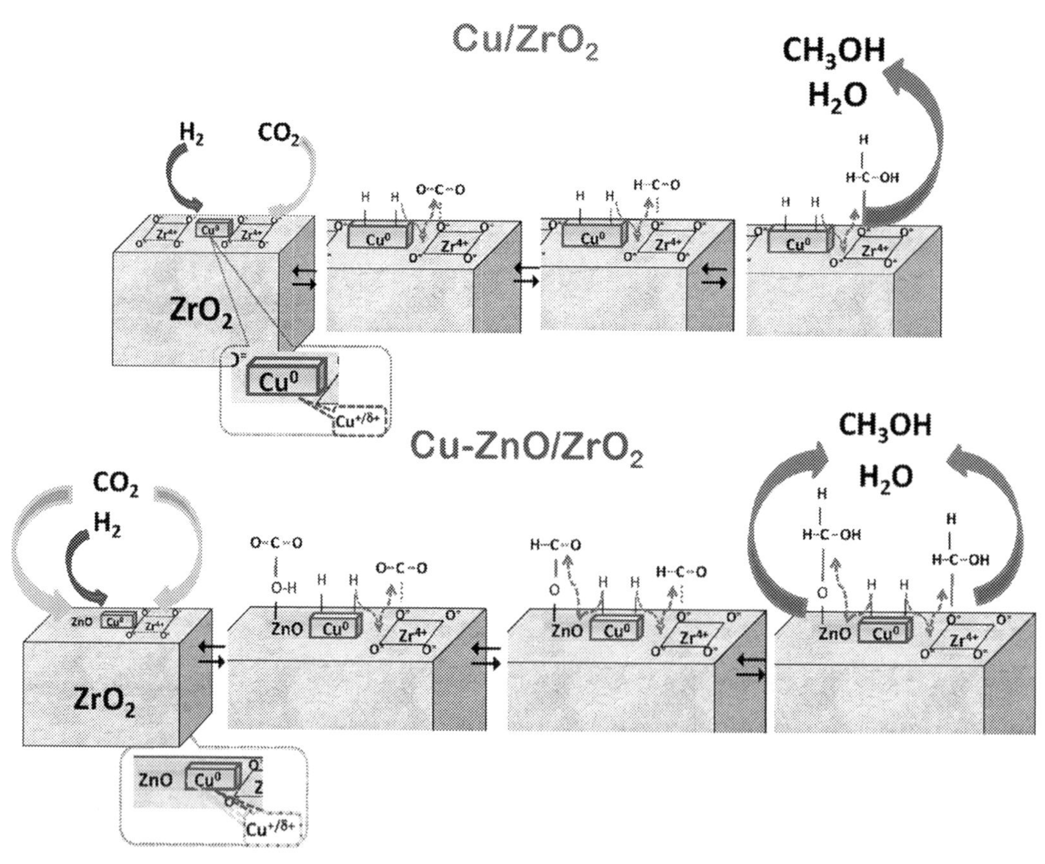

FIG. 2.7 Schematic representation of methanol synthesis with Cu-ZnO/ZrO_2 catalyst. *From Arena, F., Italiano, G., Barbera, K., Bordiga, S., Bonura, G., Spadaro, L., & Frusteri, F. (2008). Solid-state interactions, adsorption sites and functionality of Cu-ZnO/ZrO₂ catalysts in the CO₂ hydrogenation to CH₃OH.* Applied Catalysis A: General, *350(1), 16–23. https://doi.org/10.1016/j.apcata.2008.07.028.*

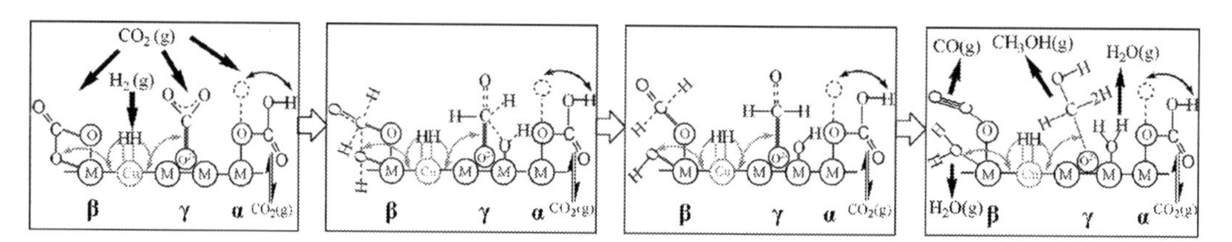

FIG. 2.8 Schematic representation of the functionality of the various surface sites for the hydrogenation of CO_2. *M*—metallic cations; α, β, γ, weak, moderate, and strong basic sites, respectively. *From Gao, P., Li, F., Zhao, N., Xiao, F., Wei, W., Zhong, L., & Sun, Y. (2013). Influence of modifier (Mn, La, Ce, Zr and Y) on the performance of Cu/Zn/Al catalysts via hydrotalcite-like precursors for CO₂ hydrogenation to methanol.* Applied Catalysis A: General, *468, 442–452. https://doi.org/10.1016/j.apcata.2013.09.026, with permission.*

promoters. The CZA-Zr has high number of basic sites and the highest selectivity of methanol (Table 2.1). The addition of quadrivalent cations (Zr^{4+}) generated a higher net electric charge compared to bivalent (Mn^{2+}) and trivalent (Al^{3+}, La^{3+}, Ce^{3+}, Y^{3+}), which resulted in a higher amount of oxygen atoms with low coordination and, therefore, higher conversion of CO_2 into methanol instead of CO as presented in Table 2.1.

In conclusion, besides the activation of methane and CO_2 the metallic and basic sites played an important role in the direct conversion of CO_2 and yield to methanol (Sato, 2019).

Chemical production of monomers and derivatives from CO_2

Recent studies have shown that the direct conversion of methane to oxygenates, such as acetic acid, acetates, and formaldehyde, is also limited to the high CH_4 binding energy (ΔH_{C-H}= 439.57 kJ mol^{-1}), and the difficulty of finding catalysts and kinetic parameters involving the reaction mechanism. Data on the kinetics and mechanism for CH_4 carboxylation are crucial. Mono and bifunctional catalysts were prepared, showing that double active sites preferentially adsorb CH_4 and CO_2 to form acetic acid directly (Joshi, 2014).

In view of the entire spectrum of research on the direct reaction of methane with carbon dioxide, it can be highlighted that the role of dissociation of methane on the catalyst surface is vital for the reaction so that the radicals formed will influence the products obtained. It was also concluded that an understanding of the reaction path on the surface is essential for developing the experimental procedure to be applied, for instance, the addition of reagents at reactor, the reaction temperature range analyzed, the screening of products expected, and intermediates, among others. It can be said that the works cited in this section describe experiments at low pressures, unlike the works developed so far with homogeneous catalysts. There are theoretical and practical confirmations of the formation of acetic acid from the direct reaction of methane with carbon dioxide. Despite the performance reported by previous works and knowing that, certainly, the research on this topic is recent, each catalyst proposed must be analyzed for various experimental conditions in order not to restrict the research, despite conclusions established based on experiments carried out with other catalysts.

TABLE 2.1 Catalytic performance for the synthesis of methanol from the hydrogenation of CO_2 with CZA and CZA-X catalysts.

Catalyst	CO_2 conversion (%)	Selectivity (mol%)		CH_3OH yield ($g\,gcat^{-1}\,h^{-1}$)
		CH_3OH	CO	
CZA	19.7	39.7	59.7	0.34
CZA-Mn	22.3	43.0	56.5	0.42
CZA-La	23.3	43.8	55.7	0.44
CZA-Ce	23.6	45.9	53.6	0.45
CZA-Zr	24.7	48.0	51.5	0.49
CZA-Y	26.9	47.1	52.4	0.52

Reaction conditions: T = 250°C, P = 5.0 MPa, GHSV = 12,000 mL gcat^{-1} h^{-1}, H_2/CO_2 = 3: 1 (molar ratio) (Gao et al., 2013).

Acetic acid, acetic anhydride, acetaldehyde, and vinyl acetate (VAM) make up the family of acetyl compounds. The annual production of acetic acid is 6×10^9 kg VAM/year only in the United States. Vinyl acetate consumes 33% of the acetic acid produced worldwide, making it the most important substance of the acetyl family (Gonzalez Caranton et al., 2018). VAM is a colorless, flammable, and volatile liquid with a boiling point between 72°C and 73°C. It is used as a monomer in the production of polyvinyl acetate copolymers (PVACP) and in the production of derivatives such as polyvinyl alcohol and ethylene vinyl alcohol (Zhang et al., 2021), as shown in Fig. 2.9 (Zoeller, 2008).

In the last 6 years, the demand for VAM has increased by 5% in China, and projections show that this growth should be confirmed, with the appearance of new applications in the next 5 years (Hayashi, Kawamura, & Takayama, 1970). Currently, global production involves more than 6 million tons, mainly for the production of polymers and vinyl acetate copolymers. More than 80% of the annual VAM production is obtained by gas-phase acetoxylation of ethylene using Pd-Au catalysts. According to the OECD database, in addition to the applications already mentioned, VAM is also used as a graft for lignin, which is one of the renewable resources currently used in paper derived and biofuel industry (Zhang et al., 2014).

Carboxylation of methane with CO_2

From the standpoint of homogeneous catalysis, major advances have been made in several distinct areas of chemical feedstock production. Historically, alkanes have been activated with CO and CO_2 molecules in a homogeneous phase for the catalytic production of acetic acid, under high pressures and moderate temperatures (20 atm, 80°C in an autoclave), using Pd-Cu catalysts, obtaining acetic acid as a single product (Eq. 2.12) with a selectivity of 99% (Ganesh, 2012). Industrially, acetic acid is produced in the gas phase from the hydrocarboxylation of methanol (Cui, Qian, Zhang, Chen, & Han, 2017).

Recent advances have shown that acetic acid can also be obtained from the hydrocarboxylation reaction of methanol with hydrogen and

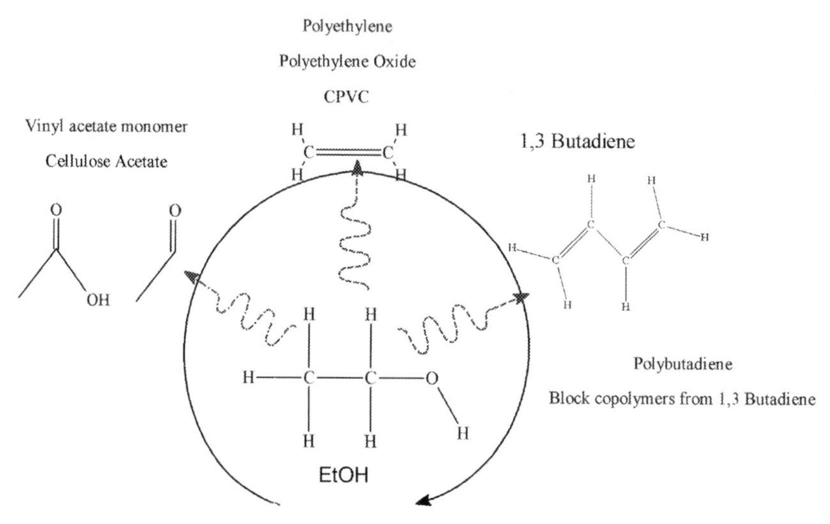

FIG. 2.9 Monomers derived from ethanol. *Adapted from Zoeller, J. R. (2008). Evolving production of the acetyls (acetic acid, acetaldehyde, acetic anhydride, and vinyl acetate): A mirror for the evolution of the chemical industry. In* Innovations in industrial and engineering chemistry *(Vol. 1000, pp. 365–387). American Chemical Society. https://doi.org/10.1021/bk-2009-1000.ch010.*

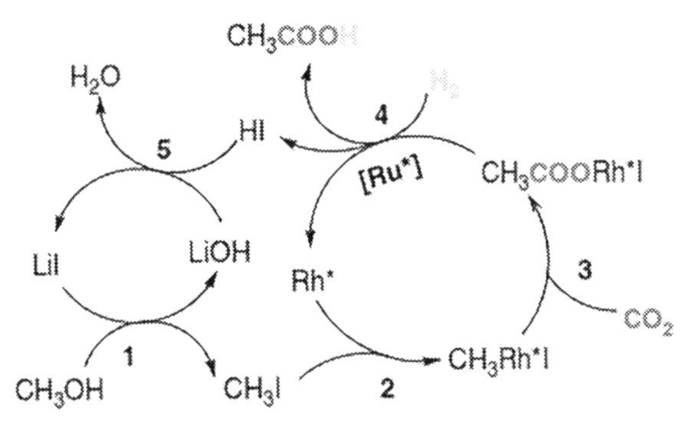

FIG. 2.10 Synthesis of acetic acid over Ru-Rh catalysts under mild conditions. *Reprinted with permission from Qian, Q., Zhang, J., Cui, M., & Han, B. (2016). Synthesis of acetic acid via methanol hydrocarboxylation with CO_2 and H_2.* Nature Communications, *7(1), 11481. https://doi.org/10.1038/ncomms11481.*

CO_2 on bimetallic Ru-Rh structures, Qian et al. reported that the turnover frequency for this particular reaction increases with temperature and those other products such as ethyl acetate and methyl acetate are observed at the reactor outlet as shown Fig. 2.10 (Qian, Zhang, Cui, & Han, 2016).

Another process that has been patented in the United States is the direct synthesis of acetic acid from methane and CO_2 (Eq. 2.12) over group VIA and VIIA metals, supported on alumina, under temperatures of 100–600°C, pressures up to 20 MPa and with reported selectivities of up to 95% (Hong, Li, Zhang, Sun, & Mo, 2019).

$$CO_2 + CH_4 \Leftrightarrow CH_3COOH_{(g)} \quad \Delta G^0 = 71\,\text{kJ/mol} \tag{2.12}$$

The direct formation of acetic acid eliminates the need to use other molecules, e.g., methanol, and synthesis gas, as intermediate compounds. However, this reaction is thermodynamically unfavorable at equilibrium ($\Delta G \gg 0$), obtaining low concentrations of acetic acid (Rabie et al., 2017; Wilcox et al., 2001). Wilcox et al. examined the synthesis of acetic acid by diffuse reflectance spectroscopy (DRIFT) over Pd and Pt catalysts supported on alumina (Wilcox et al., 2001). For this reason, they used equimolar mixtures of methane and CO_2, finding yields for acetic acid of 1.5% and acetic acid formation above 623 K for the palladium as active phase. For platinum, acetic acid formation occurs above 473 K. Several activation methods were used on the oxide: methane activation with hydrogenation, activation with hydrogen, and activation with CO_2 on the catalyst. The formation of acetic acid only occurred upon activation of the catalyst with CO_2, achieving equilibrium conversions of $3E^{-7}$ at pressures of 200 atm. Wilcox also analyzed the effect of acetic acid formation from CO, CH_4, by adding O_2 to the reactor, obtaining equilibrium conversions around 99% for any pressure of reactor (Eq. 2.13).

$$CO_{(g)} + \frac{1}{2}O_{2(g)} + CH_{4(g)} \xrightarrow[T \geq 600^\circ C]{Pt/Al_2O_3} CH_3COOH_{(g)}$$

$$\Delta G^{298\,K} = -187\,\text{kJ/mol} \tag{2.13}$$

This reaction is thermodynamically more favorable but presents less selectivity because of the reversible nature of the process, exhibiting immediate decomposition of acetic acid to CH_4, CO, and O_2 (Neurock et al., 1996). In contrast, the direct reaction between methane and CO_2 (Eq. 2.15), although thermodynamically unfavorable, is favorable from an economic perspective because it is a process operated under mild temperature and pressure conditions.

One of the limitations related to experimental methods for the conversion of methane and CO_2 to acetic acid is related to the unknown of the possible reaction path involved over solid catalysts. Wilcox et al. suggested the formation of CH_3COO- species as preliminary products, analyzed techniques of infrared spectroscopy (FTIR) and diffuse reflectance spectroscopy (DRIFT) (Spivey, Wilcox, & Roberts, 2007). However, the identification of these critical intermediates is rapid and ambiguous since they present low selectivities and yields.

Shavi et al. studied the direct conversion of CH_4 and CO_2 into acetic acid using CeO_2-ZnO, MnO_2-ZnO, and CeO_2-MnO_2 catalysts supported by MMT (montmorillonite) (Shavi, Ko, Cho, Han, & Seo, 2018). MMT was used as a support to aid in the dissociation of CH_4, as according to the literature (Weng, Ren, Chen, & Wan, 2014) the activation of CH_4 occurs on surfaces rich in oxygen. The CeO_2-ZnO/MMT catalyst showed the best catalytic activity in terms of acetic acid yield (0.35 mmol). The atomic size of the active site and the presence of the Lewis acid site (ZnO) were significant in the CH_4 carbonylation reaction for the formation of acetic acid. DFT (density functional theory) calculations showed that ZnO promotes the formation of acetate species on the surface through the migration of CO_2 adsorbed on CeO_2 to the ZnO site.

The influence of the Bronsted acid sites

In fact, the Brønsted acid site plays a critical role in the final formation of acidic acid by proton transference to the surface and formation of acetate species. It forms a methyl radical intermediate that reacts with CO_2, likely initiated by the interaction with the Brønsted acid sites.

Rabie et al. reported an experimental study for the direct synthesis of acetic acid from CH_4 and CO_2 in one step, evidencing the influence of Brønsted sites with copper exchanged M-ZSM-5 ($M = Li^+$, Na^+, K^+, and Ca^{2+}) and the role of metallic Cu species and basic cations as a bifunctional catalyst for the simultaneous activation of CH_4 and CO_2 for the direct formation of acetic acid, as shown in Fig. 2.11 (Rabie et al., 2017).

These authors showed that the activation of CO_2 is related to the alkali cations, which allow inserting onto CH-activated species from methane. The CO_2 adsorption capacity increases with the cation size $Li < Na < K$. In fact, the CO_2-TPD

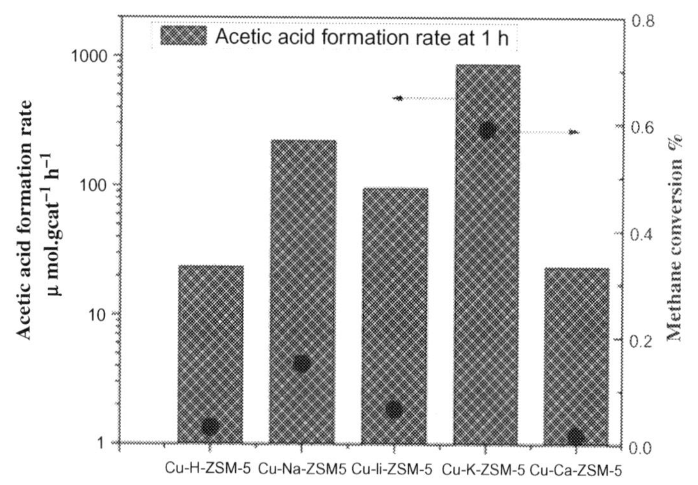

FIG. 2.11 Acetic acid formation rate over different cation exchanged Cu-M-ZSM-5 catalysts at 500°C. *From Rabie, A. M., Betiha, M. A., & Park, S.-E. (2017). Direct synthesis of acetic acid by simultaneous co-activation of methane and CO_2 over Cu-exchanged ZSM-5 catalysts.* Applied Catalysis B: Environmental, 215, *50–59. https://doi.org/10.1016/j.apcatb.2017.05.053, with permission.*

exhibited the highest concentration of adsorbed CO_2 of the Cu-K-ZSM-5 catalysts among all catalysts. There is an effective charge density of the cations inside the zeolite pores, which increased the CO_2 adsorption.

Limtrakul et al. developed a theoretical study using DFT techniques regarding the structures of ZMS-5 zeolites with gold cations, revealing the effect of the Au-ZMS-5 structure and *Brønsted* sites on methane and CO_2 conversion (Panjan, Sirijaraensre, Warakulwit, Pantu, & Limtrakul, 2012). The authors reported the first evidence related to the reaction mechanism, which involved the activation of methane by adsorption on the catalytic sites of Au-ZMS-5, forming an *Au-Methyl* intermediate compound, which by inserting CO_2 into the catalytic site produced a surface intermediate (CH_3COO-δ $Au^{+\delta}$). Finally, the presence of *Brønsted* sites of the zeolite allowed the formation of acetic acid at the active site. Theoretical reaction rates were calculated based on the mechanisms computationally studied by DFT, as shown in Table 2.2 (Panjan et al., 2012).

Jian-Feng Wu et al. proposed the reaction path on Zn-ZMS-5 zeolites under temperature conditions of 523–773 K, proposing as a strategy the activation of methane and the insertion of CO_2 for the formation of intermediate species

TABLE 2.2 Values of the reaction kinetic constants, considering single and bifunctional mechanisms on Au-ZSM structures (Panjan et al., 2012).

Reaction rate constants (k) (S^{-1})			
T (K)	$k_{bifunctional}$	k_{Au}	$k_{bifunctional}/k_{Au}$
398	1.17×10^{-15}	6.42×10^{-24}	1.82×10^{8}
498	2.71×10^{-10}	9.30×10^{-17}	2.92×10^{6}
598	9.84×10^{-7}	5.46×10^{-12}	1.80×10^{5}
698	3.36×10^{-4}	1.39×10^{-8}	2.42×10^{4}
798	2.63×10^{-2}	4.98×10^{-6}	5.28×10^{3}

that are soon protonated, by the effect of the Brønsted sites of the zeolite, releasing acetic acid formed in the active site, as shown in Fig. 2.12 (Wu et al., 2013).

The reaction steps were confirmed using gas-phase nuclear magnetic resonance (NMR) techniques coupled with mass spectrometry, claiming the formation of acetic acid follows the following mechanism (Fig. 2.12):

Simultaneous formation of methyl-Zn species and Brønsted protonation via methane activation.

Insertion of CO_2 atoms and reaction with surface intermediate to produce zinc acetate species (–Zn–$OOCCH_3$) as a key intermediate.

Brønsted protonation of the surface species (–Zn–$OOCCH_3$); release of acetic acid and restoration of the Zn/H-ZMS-5 catalytic site.

Moreover, Qian et al. claim turnover frequency was optimized due to the main effect of the catalytic precursor used, and the ligand showed an increase in the stability of ruthenium catalyst (Fig. 2.13) (Qian et al., 2016). The performance of the ruthenium complex indicates that intermediates like ethanol, ethyl acetate, methyl acetate are crucial for the formation of acetic acid with higher productivity. Were computed acetate complexes and transition states, well defined for the best condition of cluster metal, improving high catalytic activity as shown in Fig. 2.13.

It is possible to predict the reaction pathway of acetic acid production from CO_2 and CH_4, based on geometrical factors in metal clusters (Cui et al., 2017), the main hysteresis effects in single metals (Liu et al., 2013), the ion-exchange properties of metal clusters (Yang et al., 2019), and the use of RMN spectroscopic (Wu et al., 2019).

The DFT analysis for CH_4 and CO_2 molecules over Zn, Ga, and ZrO_2 oxides supported in In_2O_3 surfaces reported by Zhao et al., showed the DFT model coupled to Langmuir-Hinshelwood kinetics, providing surface stability (Zhao et al., 2019). First, they studied the binding energy of In_2O_3 surfaces, revealing that lattice (110) was

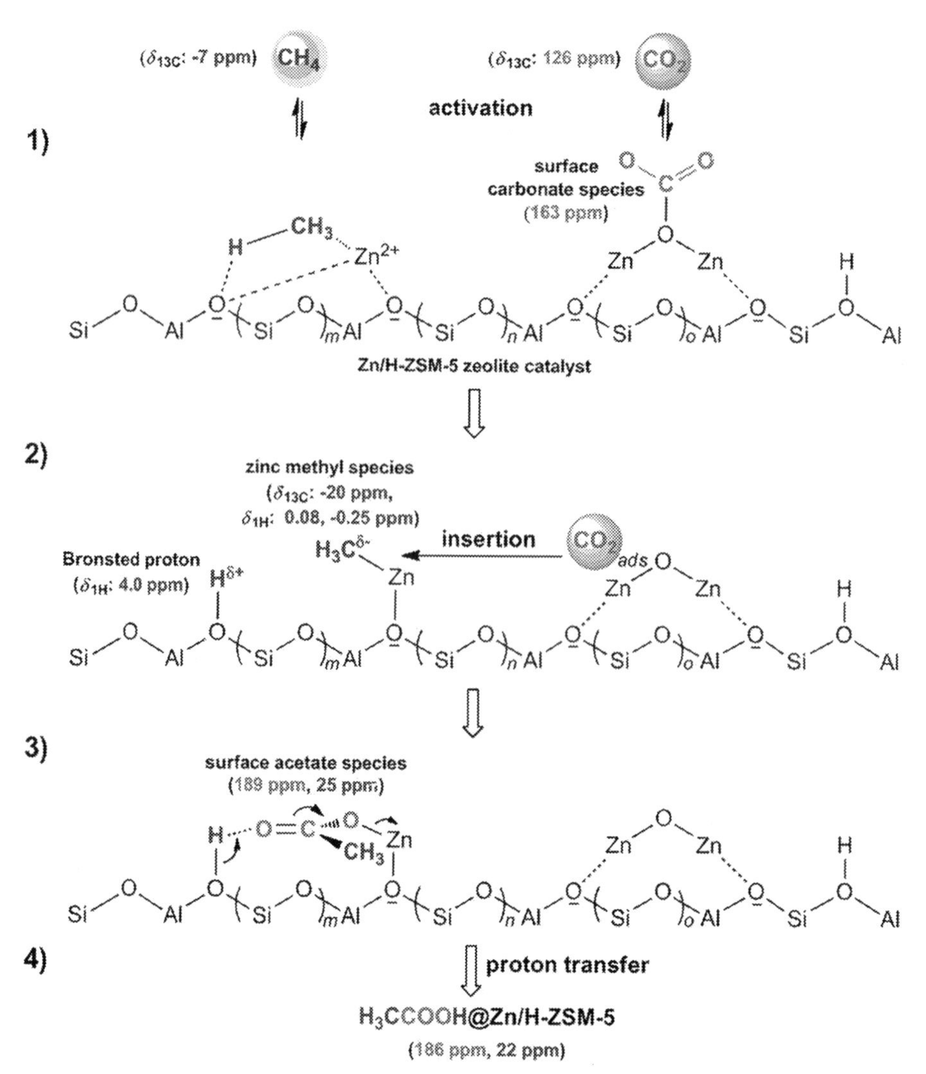

FIG. 2.12 Methane and CO_2 activation over Zn/H-ZSM-5 catalyst to produce acetic acid. *From Wu, J.-F., Yu, S.-M., Wang, W. D., Fan, Y.-X., Bai, S., Zhang, C.-W., Gao, Q., Huang, J., & Wang, W. (2013). Mechanistic insight into the formation of acetic acid from the direct conversion of methane and carbon dioxide on zinc-modified H–ZSM-5 zeolite.* Journal of the American Chemical Society 135*(36), 13567–13573. https://doi.org/10.1021/ja406978q.*

more stable when compared to (111) and (100) lattice, improving O_2 mobility and increasing the electronic vacancies.

Finally, the author showed remarkable differences in surface barrier energy for Zn, Ga, and ZrO_2, suggesting a reaction path as shown in Fig. 2.14 for $(ZnO)_3/In_2O_3$ surfaces, concluding that for such surfaces, Zn was more active in the catalytic coupling of CH_4 and CO_2. In contrast, the energy barrier for intermediates suggests a metal support interaction more stable for Zn, exhibited C=C cleavage, Zn-C-C-O species. Fig. 2.14 shows the reaction path and reaction coordinate for $(ZnO)_3/In_2O_3$ surfaces.

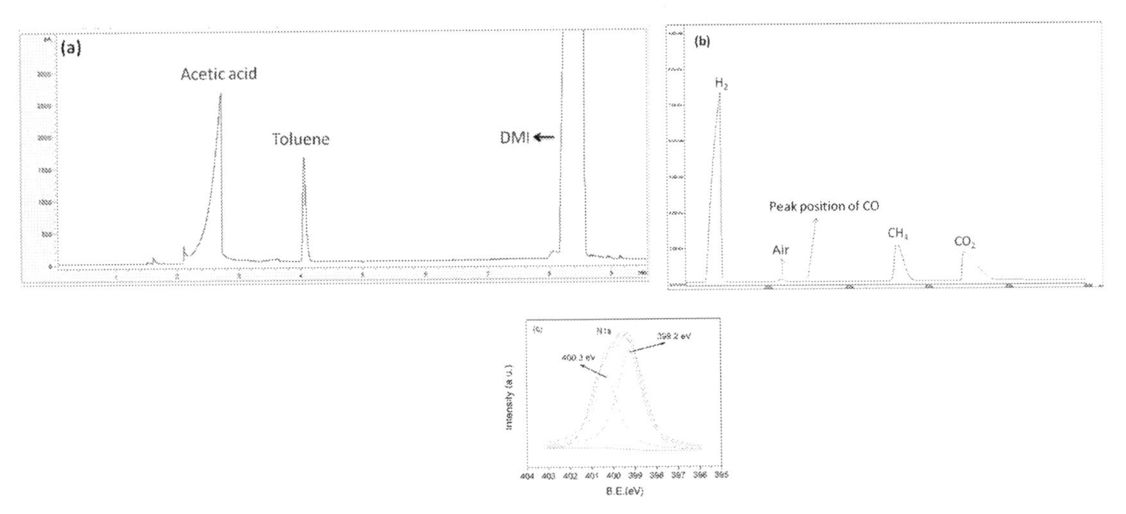

FIG. 2.13 Acetate complex formed and analyzed with XPS(a) and GC profile of reactants and products obtained in the catalytic reaction of CH_3CCOH(g) formation. (a) liquid sample (toluene as internal standard), (b) gaseous sample. Condition: 40 µmol Ru_3 (CO) 12 and 40 µmol Rh_2 $(OAc)_4$ (based on metals), 0.75 mmol imidazole, 3 mmol LiI, 2 mL DMI, 12 mmol MeOH, 4 MPa CO_2 and 4 MPa H_2 (at room temperature), 200°C, and 12 h. (c) XPS spectra of midazole-Rh_2 $(OAc)_4$ compound. *Reproduced with permission from Qian, Q., Zhang, J., Cui, M., & Han, B. (2016). Synthesis of acetic acid via methanol hydrocarboxylation with CO_2 and H_2.* Nature Communications, *7(1), 11481. https://doi.org/10.1038/ncomms11481.*

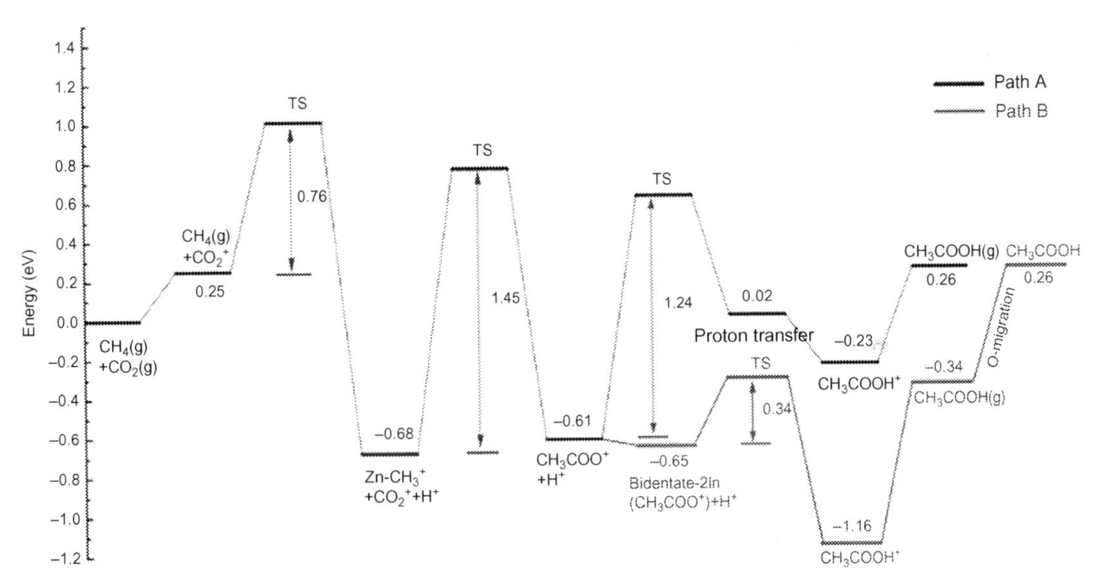

FIG. 2.14 Potential energy profile for CH_4 and CO_2 activation to CH_3COOH over $(ZnO)_3/In_2O_3$ surfaces. *Reprinted with permission from Zhao, Y., Wang, H., Han, J., Zhu, X., Mei, D., & Ge, Q. (2019). Simultaneous activation of CH_4 and CO_2 for concerted C–C coupling at oxide–oxide interfaces.* ACS Catalysis, *9(4), 3187–3197. https://doi.org/10.1021/acscatal.9b00291.*

Catalytic synthesis of vinyl acetate monomer from CH_4 and CO_2

Many experimental studies have been carried out, attempting to convert methane and CO_2 into products of interest, using homogeneous and heterogeneous catalysts, and some results with homogeneous catalysts have been reported. Wilcox et al. obtained successfully and for the first-time acetic acid, using heterogeneous catalysts (Spivey et al., 2007). However, the direct conversion of methane and CO_2 into highly valued products is difficult due to the thermodynamic limitations of this reaction, despite the use of different types of metallic catalysts and supports, according to the reaction mechanisms suggested in the literature (Spivey et al., 2007). In fact, this subject is still in progress, with the search for new catalysts and reaction conditions to achieve industrial application.

Spivey et al. proposed a different method to overcome the low yield of the direct reaction of methane with carbon dioxide, based on coupling this reaction with others (Wilcox, Roberts, & Spivey, 2003). The idea consisted of adding a reaction that consumes acetic acid, thus shifting the reaction equilibrium and increasing methane and carbon dioxide consumption. The authors added acetylene as a reactant, which favored the production of vinyl acetate. The catalyst was Pt/Al_2O_3 to produce acetic acid and zinc acetate supported on carbon (Zn/C) to produce vinyl acetate.

The zinc-based catalysts were investigated and revealed some insights on the reaction mechanism and the potential use of this metal in newly developed catalysts. Spectroscopic studies of zinc-modified zeolite (Zn/H-ZSM-5) with methane and CO_2 showed highly selective activation of methane and the insertion of CO_2 as carbonate species with the formation of acetate and acetic acid (Wu et al., 2013). Similar results were found with zinc-modified heteropoly acid catalyst ($Zn\text{-}HPW/SiO_2$). In both catalysts, zinc metal was present in the ionic form.

Catalytic formation of acetic acid is a thermodynamically unfavorable process, presenting low conversions and selectivities at chemical equilibrium (Eq. 2.14)

$$CO_2 + CH_4 \rightarrow CH_3COOH_{(g)} \quad \Delta G^0 = 71\,\text{kJ/mol} \tag{2.14}$$

Studies done on Pd monocrystals were critical to determine simple reaction models as a function of the metal content, allowing us to determine values for the activation energies and reaction frequency factors and important mechanistic aspects of the transformation. The combustion of ethylene is preponderant when the charge of Pd is high, while the combustion of AcOH is the dominant effect when the charge of Pd is low, and the reaction occurs under high pressures (Han, Kumar, Sivadinarayana, & Goodman, 2004).

Two fundamental reaction mechanisms have been found on these surfaces. One mechanism is based on the concept of a mobile active site, formed by palladium acetate intermediates with different conformational states and responsible for the formation of VAM. The second mechanism is based on the dehydrogenation of AcOH and C_2H_4 and the subsequent formation of hydrogenated vinyl acetate. The reactivity depends on the structure of the network, and for active Pd sites (100) the acetoxidation reactions are very sensitive to the interatomic distance of the Pd-Pd pair, while for Pd sites (111) the reactivity is given by VAM formation from VAM hydrogenated intermediates on the surface.

In these two scenarios, the geometric factor, the degree of coverage (which depends on the concentration of the reagents), and the direction of the crystallographic planes affect the reaction mechanism and enable several catalytic routes.

Table 2.3 shows the main catalytic systems used in the ethylene acetoxidation reaction.

However, Goodman et al. found that the use of a single metal is not efficient for the reaction

TABLE 2.3 Catalysts used for ethylene acetoxidation.

Catalyst	Reaction conditions	S_{VAM}	References
Pd-Au (3% Pd) nanoparticles	150–170°C, 8 bar	54.1	Kumar, Chen, and Goodman (2007)
Pd-Au/K-SiO_2 (2% Pd)	165°C, 8 bar	93.1	Hanrieder, Jentys, and Lercher (2015a)
Pd-Au (1.5% Pd) nanodendrimer	150°C, 9 bar	99	Kuhn et al. (2014)
Pd/SiO_2 wet impregnation	150–180°C, 1 bar	5	Han et al. (2004)
Pd-Au allow	0.25 ML/1E-4 Torr/150°C	80	Calaza, Mahapatra, Neurock, and Tysoe (2014)

due to the formation of palladium carbides on all sides of the Pd, decreasing the reaction rate, as shown in Fig. 2.15 (Kumar et al., 2007).

The formation of carbides is relevant for high levels of Pd. The specific reactivity of the sites is superior for low levels, but increases when bimetallic alloys are used, as in the case of Pd-Au-based catalysts.

Alloy-type Pd-Au catalysts have been studied recently, due to the good synergy and low segregation they present (Gao & Goodman, 2012) due to the characteristic distances between the active centers. However, Pd-Au sites still have some segregation effects related to the spatial distribution of active Pd centers on gold (Bachiller-Baeza, Iglesias-Juez, Agostini, & Castillejos-López, 2020).

Chen et al. developed experiments on bimetallic surfaces, showing that pairs of palladium atoms on gold surfaces form the most active catalytic sites for the synthesis of VAM (Chen & Goodman, 2008). In particular, the

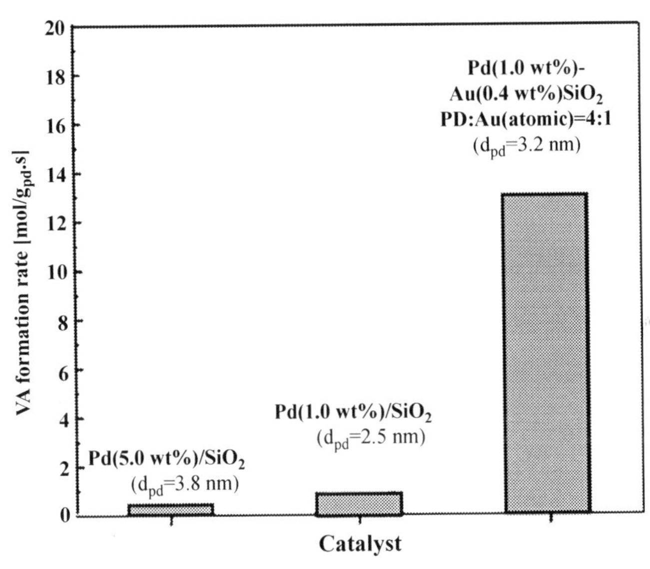

FIG. 2.15 Vinyl acetate formation over Pd and PdAu based catalysts. *Reprinted with permission from Kumar, D., Chen, M. S., & Goodman, D. W. (2007). Synthesis of vinyl acetate on Pd-based catalysts.* M. Albert Vannice Festschrift, 123(1), 77–85. *https://doi.org/10.1016/j.cattod.2007.01.050.*

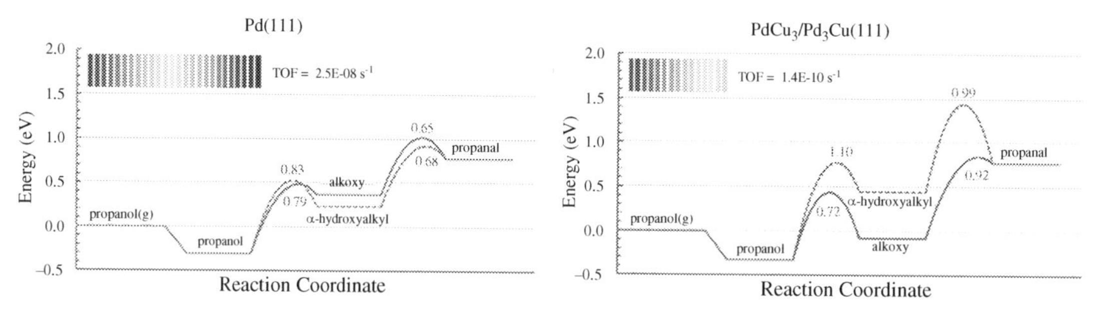

FIG. 2.16 Potential energy diagrams for sequential propane to propanal dehydrogenation on Pd (111) and Pd_3Cu (111) surfaces. *From Goulas, K. A., Sreekumar, S., Song, Y., Kharidehal, P., Gunbas, G., Dietrich, P. J., Johnson, G. R., Wang, Y. C., Grippo, A. M., Grabow, L. C., Gokhale, A. A., & Toste, F. D. (2016). Synergistic effects in bimetallic palladium–copper catalysts improve selectivity in oxygenate coupling reactions.* Journal of the American Chemical Society, 138*(21), 6805–6812. https://doi.org/10.1021/jacs.6b02247, with permission.*

acetoxidation reaction is sensitive to the structure of the catalyst.

In Pd-Cu alloys, the transfer of charge in the d band from Pd to O adsorbed can increase the O-metal bond in the active sites of Pd-Cu, as shown in Fig. 2.16. According to Goulas et al., the displacement of the d band from the active center of Pd allows the formation of different arrangements of this same element along with the network (111) in the centered cubic unit cell (fcc), with which the catalytic activity is influenced (Goulas et al., 2016).

The methane and CO_2 conversions reported experimentally for this process are so low that have used spectroscopy techniques to quantify the products (Erdőhelyi, 2021; Moya et al., 2011; Neurock et al., 1996; Wu et al., 2013). A more favorable thermodynamical reaction is the industrial production of vinyl acetate from acetylene and acetic acid using Zn catalysts supported on active carbon (Huang et al., 2001; Wu et al., 2013). The chemical removal of acetic acid in Eq. (2,15) would theoretically allow for increased acid conversions at equilibrium. Alternatively, the acetic acid produced can be consumed with acetylene to produce VAM, as shown in Eq. (2.15):

$$CH_3COOH + C_2H_2 \Leftrightarrow CH_3CO_2CH = CH_2 \ \Delta G^0 = -65\,\text{kJ/mol} \tag{2.15}$$

At equilibrium, the resulting transformation is (Eq. 2.16):

$$CO_2 + CH_4 + C_2H_2 \Leftrightarrow CH_3CO_2CH = CH_2 \ \Delta G^0 = +7\,\text{kJ/mol} \tag{2.16}$$

Although the standard free energy change for Eq. (2.16) is positive, the equilibrium is more favorable with respect to Eq. (2.14). Thermodynamic calculations have been reported in the literature using Aspen Plus software and Peng Robinson's thermodynamic model (Wilcox, 2005). For an equimolar feed of CH_4, CO_2, and C_2H_2. The study showed that at equilibrium, the conversion increases with increasing pressure and that it decreases with increasing temperature. For lower temperatures, the formation of vinyl acetate is more favorable than the formation of acetic acid. As the temperature increases, the formation of acetic acid becomes more favorable (Fig. 2.17).

It was not previously reported the thermodynamic equilibrium for CO_2. The authors proposed an adsorption-desorption cycle to study the activation effects of the intermediate step of the process (see Fig. 2.18). This optimization strategy is valid, considering the adsorbed and desorbed species as intermediates that partially drive the surface reaction equilibrium. However, studies on catalytic surfaces showed that

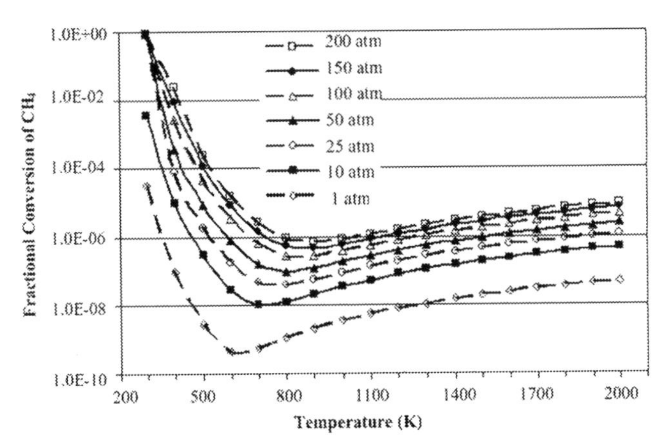

FIG. 2.17 Methane conversion at chemical equilibrium in the range $T = 300$–2000 K and $P = 1$–200 atm, using the Peng Robinson thermodynamic model and equimolar fractions of CH_4, CO_2. *From Wilcox, E. M. (2005).* Direct synthesis of acetic acid from carbon dioxide and methane. *PhD Dissertation. North Carolina State University. https://repository.lib.ncsu.edu/handle/1840.16/5624.*

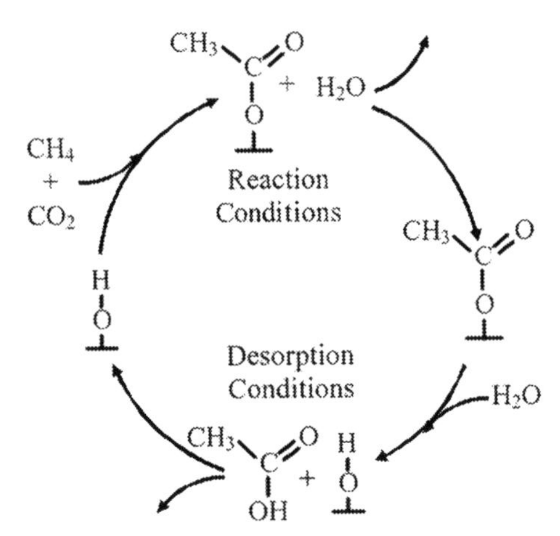

FIG. 2.18 Catalytic cyclic adsorption and elementary steps over Pt surfaces. *Reprinted with permission from Wilcox, E. M. (2005).* Direct synthesis of acetic acid from carbon dioxide and methane. *PhD Dissertation, North Carolina State University. https://repository.lib.ncsu.edu/handle/1840.16/5624.*

acetic acid did not adsorb molecularly on the active site, and monodentate configuration adsorption was more common. The adsorption-desorption cycle is divided into the following steps:

Spivey et al. showed that it is possible to develop a catalytic evaluation methodology for VAM production by activating the catalyst ($Pt/Al_2O_3+Zn/C$) (Spivey et al., 2007). The catalytic activity was compared concerning traditional VAM processes, finding a higher selectivity in the process involving greenhouse gases (CH_4, CO_2) and acetylene. At 400°C, the conversion was maximum for VAM and the acetic acid intermediate, being observed a higher selectivity for the formation of the first one. The decrease in activity for VAM occurs due to the decrease in the formation rate of acetic

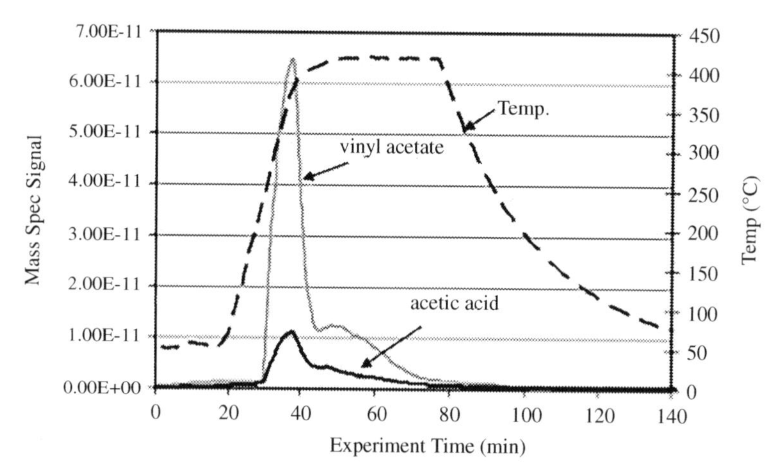

FIG. 2.19 Acetic acid and vinyl acetate mass spectrometer signals and sample temperature versus time for an admixture of 5% Pt/Al_2O_3 and Zn acetate/C exposed to an equimolar mixture of CO_2, CH_4, and C_2H_2; 10 cm^3/min-gcat, 1 at. *From Wilcox, E. M. (2005).* Direct synthesis of acetic acid from carbon dioxide and methane. *PhD Dissertation. North Carolina State University. https://repository.lib.ncsu.edu/handle/1840.16/5624.*

acid from 400°C. Spivey et al. reported that when monometallic Pt/Al_2O_3 catalyst was used, the signals of the products in the mass spectrometer were lower than those observed with the physical mixture of Pt/Al_2O_3 and Zn/C. As shown in Fig. 2.19, vinyl acetate first appeared at a temperature of about 225°C, the same temperature at which the acetic acid and acetylene signals decrease. This is also the same temperature at which vinyl acetate was formed from $CO_2+CH_4+C_2H_2$ over Pt/Al_2O_3 and admixtures of Pt/Al_2O_3 with Zn acetate/C. The mass spectrometer signal for vinyl acetate rapidly increased until it reached a maximum value.

The maximum signal found for VAM was 6.5×10^{-11}, corresponding to 10% molar. The signal of acetic acid was 1.1×10^{-11}, corresponding to 0.2% molar. Preliminary experiments were performed using Zn/C and Pt/Al_2O_3 catalysts for the reaction between CO_2, CH_4, and C_2H_2. The results showed that the active phase of Zn supported on carbon did not catalyze the formation of VAM; however, the formation of ethylene, CO, and H_2 was observed. In the case of VAM formation from acetic acid and acetylene, typically acetic acid is bubbled into the process, mixing with acetylene to produce VAM. When the catalyst of the physical mixture of Zn/C-Pt/Al_2O_3 was exposed to the reactants (acetic acid and acetylene), the formation of vinyl acetate was observed as the majority product, and two regions of the reaction temperature were found, in which the formation of VAM. They found, concentrations for acetic acid and maximum values for VAM at a retention time of $t=39$ min and $t=108$ min, related to adsorption/desorption phenomena that occur simultaneously with VAM formation. The almost instantaneous formation of acetic acid and resulted in an abrupt drop in concentration, shown by the minimum signal values for the acid. According to recent studies on catalytic processes involving the production of vinyl acetate (VAM) on Pt and Pd bimetallic surfaces with other metals, e.g., Zn, Cu, and Au, mechanisms of methane carboxylation and acetoxylation of acetylene-ethylene occur for vinyl acetate formation (Bonarowska, Machynskyy, Łomot, Kemnitz, & Karpiński, 2014; Hanrieder et al., 2015a). These mechanisms depend strongly on the nature of the metal-support interaction influenced by the conditions or variables of catalyst preparation of the crystalline phases and stability of these phases, besides the redox nature of the catalysts used, especially during the formation of key intermediates, such as acetic acid.

The role of supports

The commercial Pd-Au catalyst exhibits high catalytic activity caused by the synergy and the spatial distribution of Pd-Au clusters (Baber, Tierney, & Sykes, 2010). Usually, AcOH and C_2H_4 suffer partial dehydrogenation due to C=C cleavage for later through β-H elimination with the surface oxygen, producing vinyl acetate monomer (Huang, Dong, & Yu, 2016; Wu et al., 2013). Pd-Au-based catalysts showed that the ensemble promoted with K^{2+} contribute to dynamic self-organization of Pd active site, exchanging electrons during AcOH and ethylene dehydrogenation step (Bonarowska et al., 2014; Hanrieder, Jentys, & Lercher, 2015b). Godman and Kumar et al. claim that Au in the Pd-Au bimetallic system inhibits the Pd interactions with interstitial carbon for low Pd content (1%) (Wei & Iglesia, 2004). ZrO_2 surfaces own hydroxyls groups, which are determinants for oxygen in the vacancies, caused by dehydration mechanisms and the presence of unsaturated Zr^{4+} (Crisci, Dou, Prasomsri, & Román-Leshkov, 2014; Kouva, Honkala, Lefferts, & Kanervo, 2015; Paulidou & Nix, 2005). The ZrO_2 modified by metals like Ti^{4+} and Al^{3+} lead to the formation of nanostructured materials (Anzures et al., 2015; Gonzalez Caranton et al., 2018) of tetragonal/cubic nature, increasing its catalytic reactivity in terms of surface oxygen mobility.

Caranton et al. studied the production of vinyl acetate (VAM) on Pd-Cu cubic nanostructures, supported on modified ZrO_2-Al^{3+} and ZrO_2-Ti^{4+}, for the vinyl acetate monomer, using acetic acid (AcOH), ethylene, and oxygen (Gonzalez Caranton et al., 2018). It is shown that the combustion of ethylene and acetic acid can be inhibited below 180°C, maximizing the rates

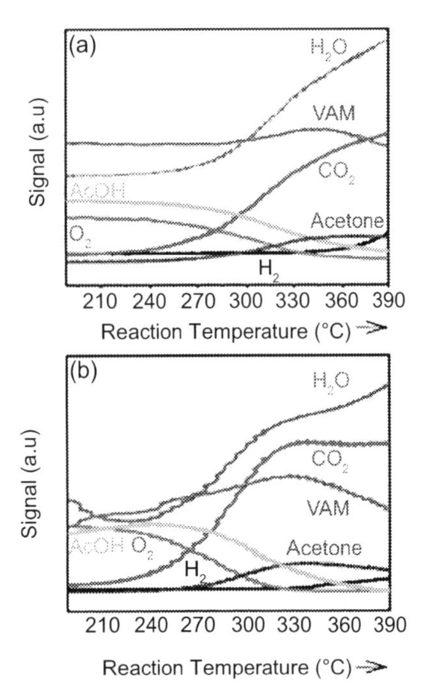

FIG. 2.20 TPSR results at atmospheric pressure for PCZT (A) and PCZA (B) samples: 0.55g/9.61mL/min C_2H_4/1.25mL/min O_2/Tsaturation = 24.5°C (2% AcOH). *Reprinted with permission from Gonzalez Caranton, A. R., Dille, J., Barreto, J., Stavale, F., Pinto, J. C., & Schmal, M. (2018). Nanostructured Pd – Cu catalysts supported on Zr–Al and Zr–Ti for synthesis of vinyl acetate.* ChemCatChem, *10(22), 5256–5269. https://doi.org/10.1002/cctc.201801083.*

of VAM formation when the catalysts are modified with Ti^{4+}. The effects of AcOH concentration on rates of VAM formation show that higher AcOH concentrations favor the formation of undesired byproducts, while lower AcOH concentrations favor effects related to O_2 mobility, which can lead to surface decomposition. VAM formation is favored, with selectivities ranging from 0.8 to 1.0. Fig. 2.20 displays the TPSR profiles of the PCZA and PCZT catalysts after reduction at 258°C and 400°C, respectively.

The PCZA catalyst under similar reduction conditions (Fig. 2.20B) showed high consumption of O_2 and consumption of AcOH at 200°C, with simultaneous formation of CO_2 and

H_2O. However, there is a continuous and significant formation of VAM.

Catalytic tests showed that for higher gas space velocity, the conversion increases for the PCZT samples. Probably, acetic acid reacts with C_2H_4 and forms acetoxy species, which correspond to the cleavage of C–C and C–O bindings at the interface and the formation of intermediate species on Pd-acetate.

From DRIFTS results, we observed that the nature of $PdCu_x$ active sites, the oxygen storage capacity (through OSC), and the hydroxyl species presented on catalysts surface suggest the β-elimination of vinyl acetate from the surface. It shows the surface hydroxyl groups, which allows the VAM desorption and H_2O formation. In Fig. 2.21, we propose the possible reaction path for VAM over PdCu catalysis.

Caranton et al. synthesized the Pd-Cu catalysts over ZrO_2 mixed oxides incorporating Al^{3+} and Ti^{4+} like promoters (Gonzalez Caranton et al., 2020). The catalytic evaluation was performed by experimental planning, varying the O_2 flow in the feed and the nature of the catalyst, aiming the production of vinyl acetate synthesis from C_2H_4, AcOH, and O_2 and the effects on the reactivity, proposing possible scenarios for the path reaction. Catalytic tests were performed through experimental planning. Statistical analyses allowed to find the main correlations of the products. The PCA coefficients revealed that vinyl acetate monomer formation was influenced by the AcOH and O_2 consumption at low ethylene coverage. Acetaldehyde emerged as an important intermediate for vinyl acetate monomer synthesis.

The AcOH and O_2 conversions predominate, followed by C_2H_4 conversion at low coverage of the active sites. Acetaldehyde and vinyl acetate monomer (VAM) are formed simultaneously by the AcOH partial oxidation and by hydrogenation on copper active sites to form VAM and water. The copper-like second active phase enables excessive AcOH hydrogenation on the Pd-Cu particles.

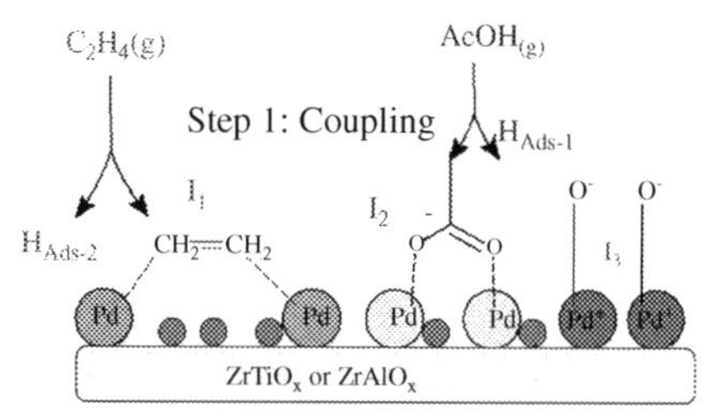

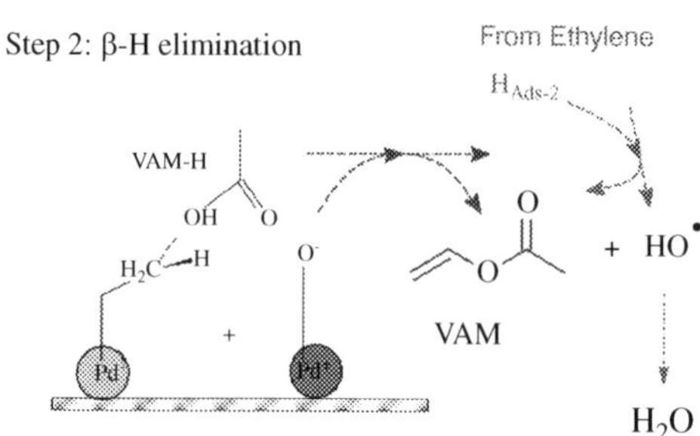

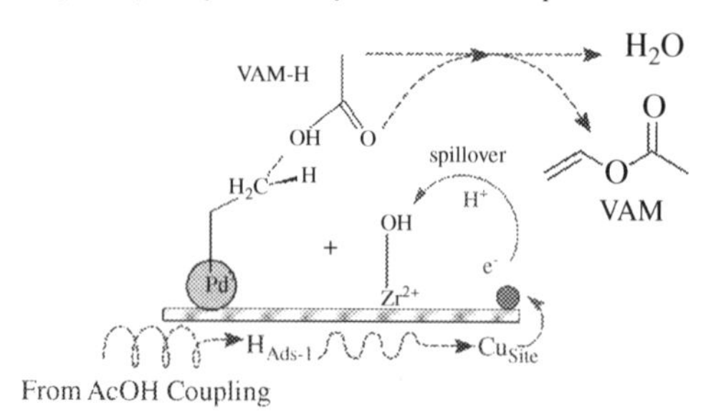

FIG. 2.21 Reaction pathway of ethylene acetoxidation over Pd-Cu/ZrO_2-based catalysis. *From Gonzalez Caranton, A. R., Dille, J., Barreto, J., Stavale, F., Pinto, J. C., & Schmal, M. (2018). Nanostructured Pd – Cu catalysts supported on Zr–Al and Zr–Ti for synthesis of vinyl acetate.* ChemCatChem, 10(22), 5256–5269. *https://doi.org/10.1002/cctc.201801083, with permission.*

We can propose and predict a possible aleatory reaction path, as shown in Fig. 2.22. The low ethylene coverage over the active sites showed an inverse correlation relative to acetaldehyde formation, considering that hydrogen released in the dehydrogenation step allowed

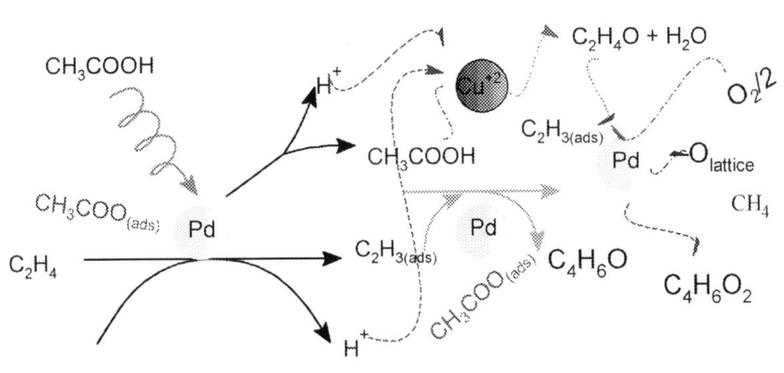

FIG. 2.22 Aleatory reaction path for VAM synthesis over Pd-Cu catalysts from experimental fluctuations. *From Gonzalez Caranton, A. R., da Silva Pinto, J. C. C., Stavale, F., Barreto, J., & Schmal, M. (2020). Statistical analysis of the catalytic synthesis of vinyl acetate over Pd-Cu/ZrO$_2$ nanostructured based catalysts.* SI: Catalysis: Academy and Industry, 344, *108–117. https://doi.org/10.1016/j.cattod.2018.10.034.*

$$CH_3COOH_{(g)} \xrightarrow{Pd\text{-}Cu} CH_3COO_{ads} + H^+ \quad \text{(AcOH coupling)}$$
$$C_2H_{4(g)} \xrightarrow{Pd\text{-}Cu} C_2H_{3ads} + H^+ \quad (C_2H_4 \text{ coupling})$$
$$C_2H_{3ads} + 0.5O_2 + H^+ \xrightarrow{Pd\text{-}Cu} C_2H_4O \quad \text{(Ethylene oxidation and } H_2 \text{ Spillover)}$$
$$C_2H_4O + CH_3COO_{ads} \xrightarrow{Pd\text{-}Cu} C_4H_6O_2 + OH \quad (\beta-\text{elimination})$$
$$OH + H^+ \xrightarrow{Pd\text{-}Cu} H_2O \quad \text{(Hydroxil recovery)}$$

FIG. 2.23 Elementary steps for VAM synthesis over Pd-Cu catalysts from correlations matrix for replicates. *From Gonzalez Caranton, A. R., da Silva Pinto, J. C. C., Stavale, F., Barreto, J., & Schmal, M. (2020). Statistical analysis of the catalytic synthesis of vinyl acetate over Pd-Cu/ZrO$_2$ nanostructured based catalysts.* SI: Catalysis: Academy & Industry, 344, *108–117. https://doi.org/10.1016/j.cattod.2018.10.034.*

the acetaldehyde formation and water, as shown in Fig. 2.22. In addition, Fig. 2.23 synthesizes a possible elementary reaction mechanism.

Graphene support with ZnO

Nanomaterials supported on carbon have gained high interest because they fulfill conditions such as high surface area, resistance, thermal stability, and easier recovery or recycling (Julkapli & Bagheri, 2014). These materials are also resistant to acid and basic solutions, which can be used as the functionalization of heteroatoms at the surface and induce the interaction of the metallic phase with the carbon structure at the surface (Julkapli & Bagheri, 2014). Graphene is a carbon material constituted of a monolayer formed by carbon atoms with hybridization sp^2, arranged in a hexagonal form. The thickness is the physical limit of a bi-dimensional surface (2D). Therefore, it presents a high specific surface area (Navalon, Dhakshinamoorthy, Alvaro, & Garcia, 2016). Ideally, it is a one-atom layer material, but in reality, there are two or more atom layers. Moreover, graphene presents thermal and electronic properties that are important for catalysis and other applications (Machado & Serp, 2012). The main obstacle is the synthesis of these materials. In fact, Novoselov and Geim were the first to synthesize one layer of graphene (Novoselov et al., 2004). However, several other new methods have been proposed in the literature, like chemical vapor deposition (CVD), but in particular, the chemical methods, starting from graphite to graphite oxide to obtain sheets of graphene oxide and the subsequent reduction of the sheets to graphene are of great interest. The use of graphene as catalyst support has been widely studied in recent years (Machado & Serp, 2012).

Dutra et al. proposed a different catalyst with zinc oxide and graphene support (Dutra, Schmal, & Guardani, 2018). Results showed that graphene presented thermal stability and maintained its structure under heat treatment at temperatures of 500°C. The catalyst performance was evaluated for the reaction of CH_4+CO_2 and O_2 by surface reaction at programmed temperature. TPSR and DRIFTS coupled to a mass spectrometer evidenced methane activation on ZnO/rGO-T, due to the evolution of H_2 and CO_2 traces of water and hydrocarbons, such as ethane (C_2H_6). Less sensitive but present was the signal 60, which can be assigned to the formation of acetic acid (CH_3COOH) at 300°C. Fig. 2.24 shows the images of zinc oxide supported on thermally treated graphene at 500°C (ZnO/rGO-T) and the mean particle size distribution.

Fig. 2.25 shows the in situ DRIFT spectra for ZnO/rGO-T catalysts in CH_4 and CO_2 activation.

Fig. 2.25 displays the spectra of the simultaneous adsorption of CH_4 and CO_2. Bands around 3200–2800 cm^{-1} show methane adsorption with temperature and a slight band shift of methane with increasing temperature from 3016 cm^{-1} at 100°C to 3008 cm^{-1} at 350°C, which is the same pattern reported in the literature (Scarano, Bertarione, Spoto, Zecchina, & Otero Areán, 2001). No changes of the CO_2 bands around 2500–2250 cm^{-1} were observed. However, bands around 2173 and 2107 cm^{-1} were observed, which are attributed to CO formation at the surface (Hussain, 2009). The influence of the graphene rGO-T on the reaction was compared with the catalyst ZnO/rGO-T and showed catalytic activity. Infrared spectra indicate that methane possibly dissociates at the surface. TPSR and DRIFTS coupled to a mass spectrometer evidenced the methane activation on ZnO/rGO-T, as observed by the evolution of H_2 and CO_2, besides traces of water and hydrocarbons, such as ethane (C_2H_6). A weak signal was detected at m/e = 60, which can be assigned to the formation of acetic acid (CH_3COOH) at 300°C.

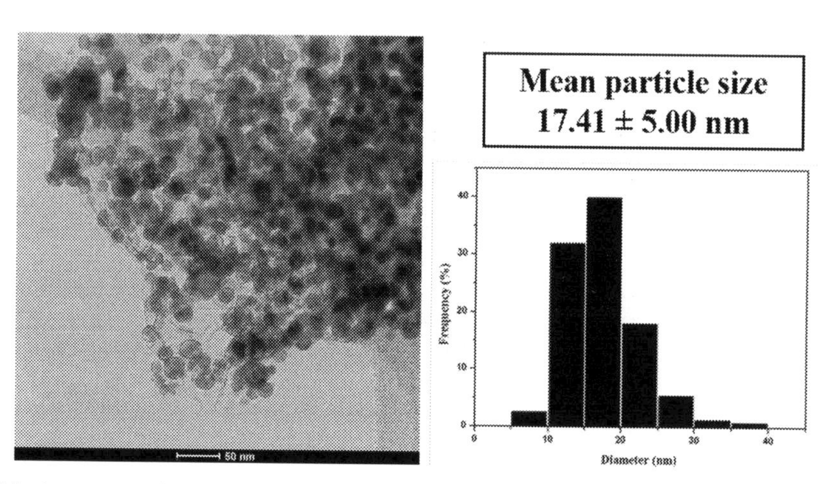

FIG. 2.24 TEM microscopy and particle distribution for ZnO/rGO-T samples. *Reproduced from permission from Dutra, M., Schmal, M., & Guardani, R. (2018). Syntheses and Characterization of zinc oxide nanoparticles on graphene sheets: Adsorption-reaction in situ DRIFTS of methane and* CO_2. Catalysis Letters, 148*(11), 3413–3430.https://doi.org/10.1007/s10562-018-2538-6.*

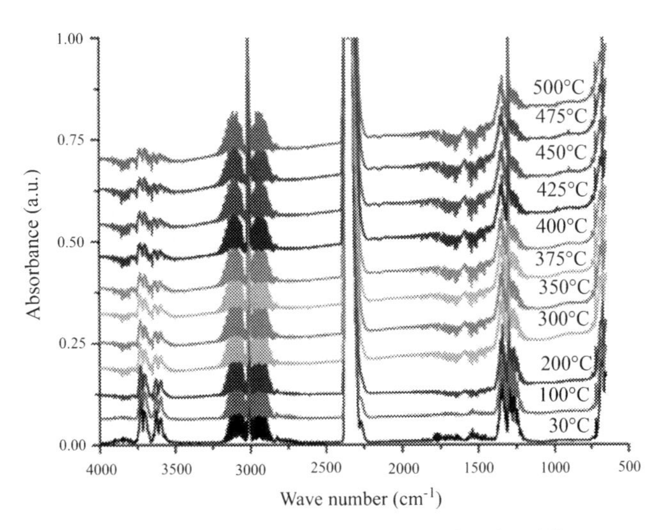

FIG. 2.25 IR spectra of $CH_4/CO_2/He$ interaction on ZnO/rGO-T catalyst for different temperatures. *From Dutra, M., Schmal, M., & Guardani, R. (2018). Syntheses and characterization of zinc oxide nanoparticles on graphene sheets: Adsorption-reaction in situ DRIFTS of methane and CO_2.* Catalysis Letters, 148*(11), 3413–3430. https://doi.org/10.1007/s10562-018-2538-6.*

Finally, different from the literature (Patil et al., 2014; Wu et al., 2013), the graphene support exhibited catalytic activity for methane and CO_2 reaction, but to a smaller degree than ZnO/rGO-T catalyst. Then, it should be stated that both supports (rGO-T) and metal (ZnO), present reactivity for the analyzed reaction. In summary, based on the available information, the catalytic activity of ZnO on graphene as support for this reaction system showed for the first time the interaction of CO_2 with the surface and the formation of different products and traces of acetic acids, as far as is known. However, the ZnO and rGO-T interactions and their effects on the studied reaction still need to be clarified in a way to understand the relationships of each part of the catalyst with the obtained results.

2.4 Neural kinetic models

Neural kinetic models have been developed for some chemical reactions in the last few years, followed by the rapid progress of machine learning algorithms to analyze large volumes of data in the Industry 4.0 context (Cavalcanti, Kozonoe, Pacheco, & Alves, 2021). Although most kinetic modeling works are focused on phenomenological models derived from Power-Law and Langmuir-Hinshelwood approaches (Schmal et al., 2018), these data-driven models are based on artificial neural networks (ANNs)—a powerful nonlinear regression approximator that "learns" the studied phenomenon from available experimental data (Hornik, Stinchcombe, & White, 1989). These networks can easily describe the kinetics of catalytic chemical reactions (computationally talking), which are frequently complex and present nonlinearities in their first-principles modeling, as most chemical engineering processes (Alves & Nascimento, 2002). However, instead of estimating preexponential and activation energies, the parameters obtained through the so-called training procedure are the ANN weights, which characterize each interconnection between the neurons in the network (Nascimento, Giudici, & Guardani, 2000).

In this section, we derived a neural kinetic model for the CO_2 hydrogenation reaction. This chemical process has several well-known phenomenological models developed in the literature that describes its kinetics (Nestler et al., 2020), highlighting the ones created by Graaf, Stamhuis, and Beenackers (1988) and Bussche and Froment (1996). The former considers the stepwise hydration of CO^* and CO_2^* species on two different active sites, while the latter was based on the carbonate species on one active site. Nevertheless, few are those works that reported an approach based on ANNs, taking advantage of its lower computational effort and rapid response without knowing the laws that govern the kinetics (Sun, Yang, Wen, Zhang, & Sun, 2018). Such models are crucial to reduce the time for process control and real-time optimization (Nascimento & Giudici, 1998). In addition, reactor design tasks can be easily carried out without dealing with high nonlinearity, stiffness, and singularity found in phenomenological kinetic models.

A neural kinetic model application to CO_2 hydrogenation is presented in this chapter, using the data of the study from Park et al. of methanol synthesis through the following chemical reactions (Eqs. 2.17–2.19) over the commercial catalyst of $Cu/ZnO/Al_2O_3$ (Park, Park, Lee, Ha, & Jun, 2014):

CO_2 hydrogenation:

$$CO_2 + 3H_2 \rightleftharpoons CH_3OH + H_2O \tag{2.17}$$

Reverse water-gas shift reaction:

$$CO_2 + H_2 \rightleftharpoons CO + H_2O \tag{2.18}$$

CO hydrogenation:

$$CO + 2H_2 \rightleftharpoons CH_3OH \tag{2.19}$$

The dataset is composed of 74 experimental points (NE = 74). The selection of the input variables of the ANN was the same as determining the classical kinetics through a differential reactor when macroscopic variables are varied, i.e., temperature (T), pressure (P), gas hourly space velocity (GHSV), and feed composition (y_{COin}, y_{CO_2in}, y_{H_2in}) (Schmal, 2016). The output variables are those related to the conversion of the main reactants, X_{CO} and X_{CO_2in}. Moreover, a bias neuron (equal to 1) was introduced in each layer, except in the last one. Fig. 2.26 exhibits the ANN topology used for the neural kinetic model with its weights presented in each neuron interconnection.

The neural kinetic model was developed using the neural net package (version 1.44.2) available in the R statistical and computational environment. For verifying the predictive capacity of the model, the dataset was randomly split into two subsets: the training set (80%) for estimating the ANN weights through the Resilient Backpropagation algorithm, and the testing set (20%) for validating the neural kinetic model developed.

The hyperparameters of the ANN were the sigmoid activation function and the number of neurons in the hidden layer (NH), which was determined by the k-fold cross-validation technique. In this case, we chose $k = 10$, i.e., the ANN calculations were performed 10 times so that all points were used in both training and testing settings, reducing sampling noise. Fig. 2.27 shows the mean squared error (MSE) for each subset. For the training set, the MSE decreases with the additional neurons, as expected for an overparameterized model. Alternatively, for the testing set, the MSE oscillates, indicating possible overfitting. The NH was selected by the minimum MSE value for the testing set that leads to a number of ANN weights lower than the number of experimental points (NE = 74), i.e., NH = 5, which corresponds to 47 ANN weights.

Figs. 2.28 and 2.29 exhibit the parity plots and residue distributions for the training and testing sets, respectively. As expected, the neural kinetic model was satisfactorily adjusted to the samples of the former group with an R^2 value of 0.928 and an almost symmetrical residue distribution zero-centered. The R^2 value for the latter group was also acceptable (0.869), mainly because it predicts the data not selected to

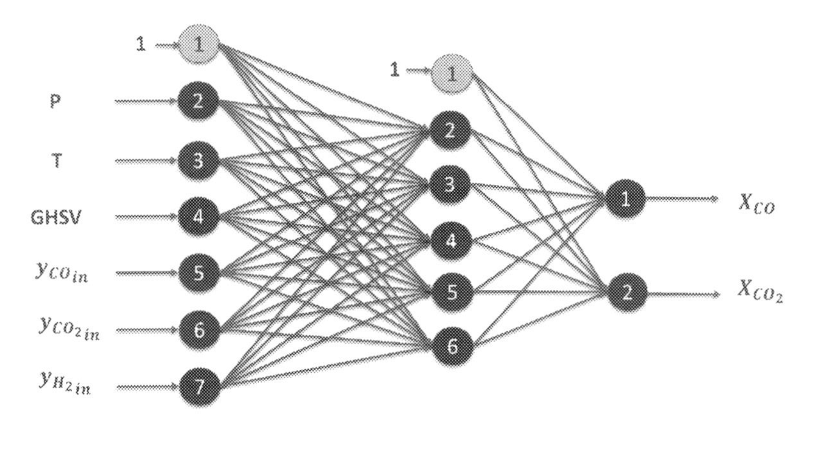

ANN weights:

$$W_{1^{st}-2^{nd}\ layer} = \begin{bmatrix} 0.77 & 1.22 & -1.15 & 1.67 & 2.18 \\ 1.60 & -1.53 & 3.15 & -0.73 & -6.86 \\ -2.63 & 0.85 & -5.65 & -0.22 & 13.48 \\ 0.47 & -0.62 & -0.43 & -0.52 & 1.24 \\ -1.37 & 4.06 & 3.66 & 1.24 & 15.43 \\ 1.52 & 0.60 & 3.96 & 0.78 & -16.86 \\ 1.14 & -0.51 & -1.05 & 0.08 & -0.20 \end{bmatrix} \quad W_{2^{nd}-3^{rd}\ layer} = \begin{bmatrix} 0.09 & -0.13 \\ 0.76 & 0.50 \\ 0.63 & -1.33 \\ 0.02 & -0.30 \\ -1.39 & 1.70 \\ 0.82 & 0.11 \end{bmatrix}$$

FIG. 2.26 ANN topology with its weight values.

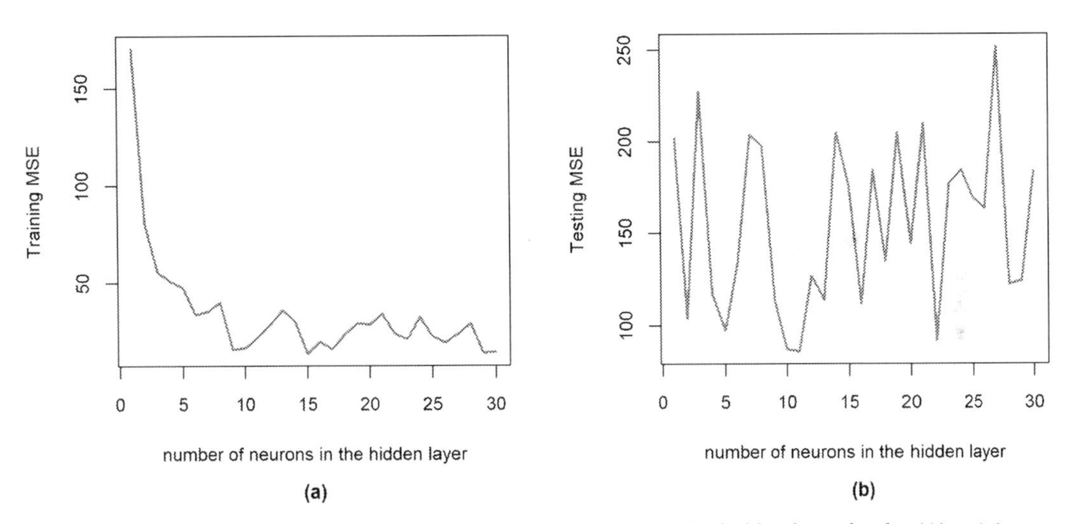

FIG. 2.27 Mean squared error (MSE) versus the number of neurons in the hidden layer for the (A) training set and (B) test set.

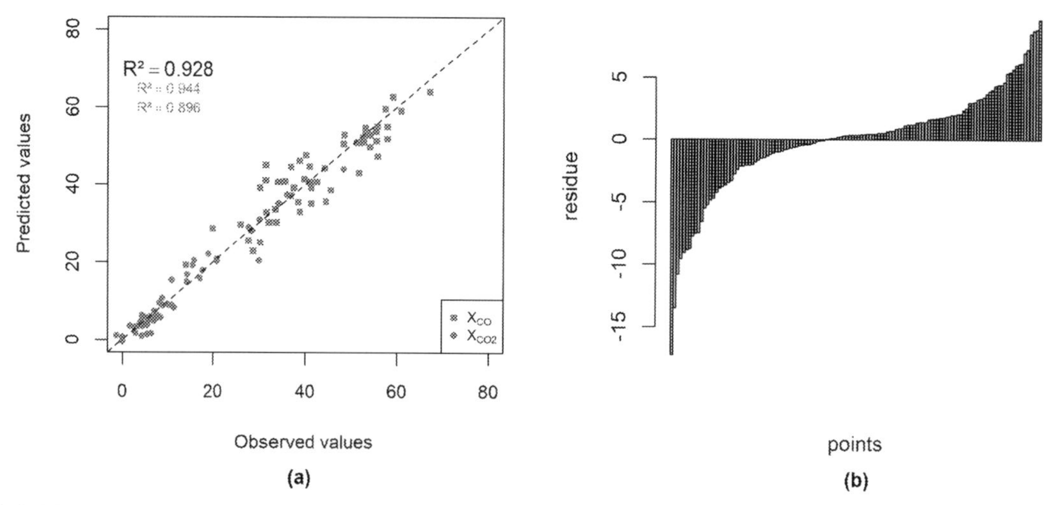

FIG. 2.28 (A) Parity plot and (B) residue distribution for the training set.

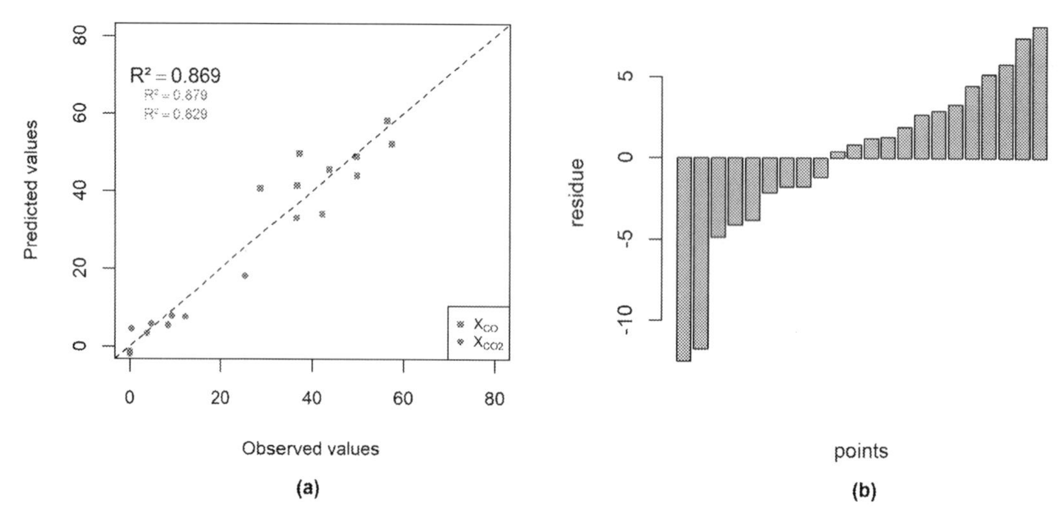

FIG. 2.29 (A) Parity plot and (B) residue distribution for the testing set.

estimate the ANN weights, and also it presents the upper and lower residue areas with approximately the same module. Also, it is noteworthy that the ANN model is more accurate for CO conversion than CO_2 conversion ($R^2_{CO} > R_{CO_2}^{\,2}$), which may be due to a more consistent acquisition/calibration of the CO output concentration in the chromatograph measurements and the dissolution of some amounts of CO_2 in water and methanol at the reactor outlet.

As the neural kinetic model developed can adequately describe the CO_2 hydrogenation reaction, we can use it to observe useful trends in its responses with the variation of its input parameters by performing a sensitivity analysis (Cavalcanti, Schmal, Giudici, & Brito Alves, 2019). Figs. 2.30 and 2.31 illustrate the surface plots of CO and CO_2 conversions, respectively, as a function of temperature and inlet concentrations of CO, CO_2, and H_2, which were varied

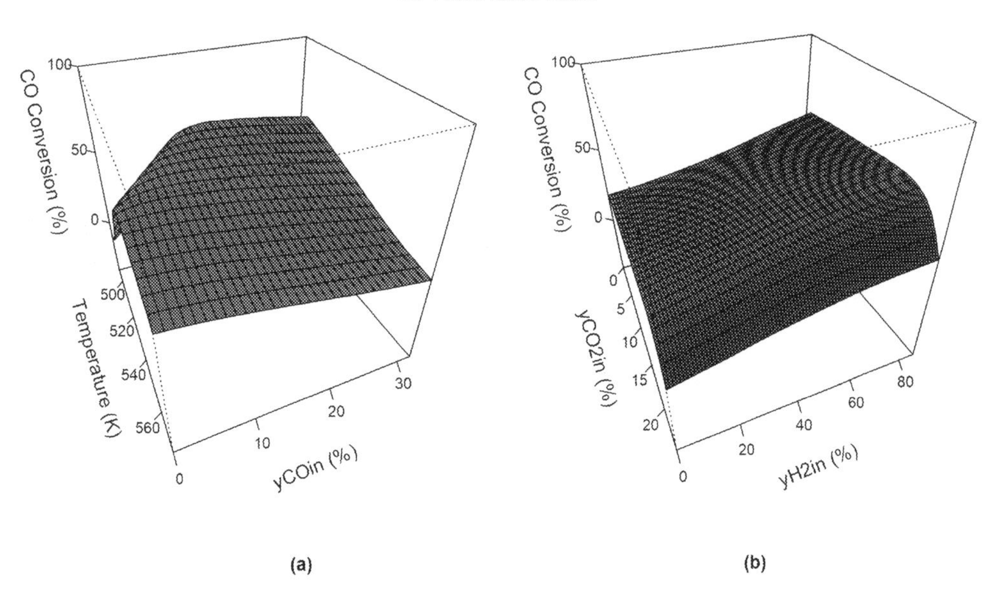

FIG. 2.30 CO conversion as function of (A) temperature and inlet CO concentration, and of (B) inlet CO_2 and H_2 concentrations.

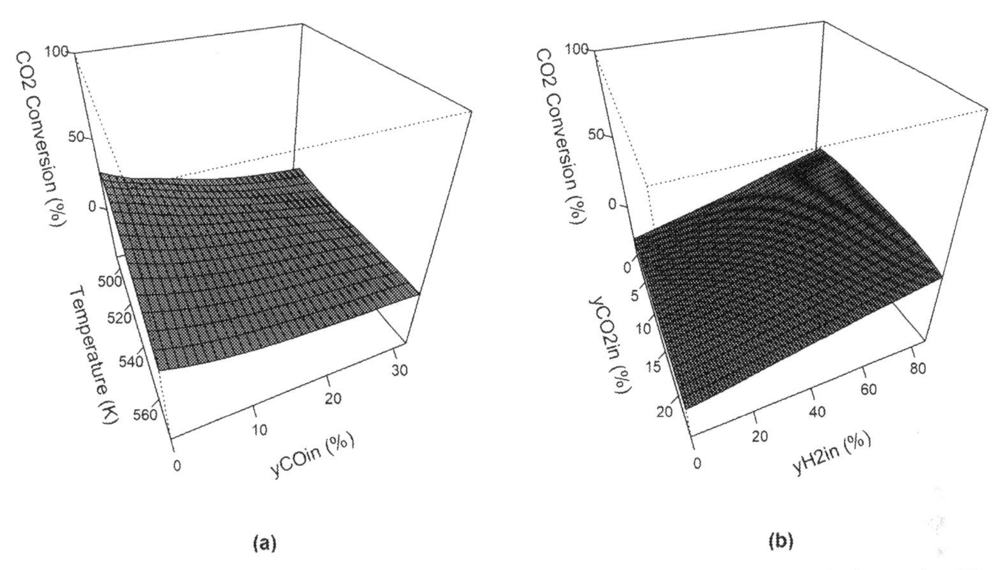

FIG. 2.31 CO_2 conversion as function of (A) temperature and inlet CO concentration, and of (B) inlet CO_2 and H_2 concentrations.

inside the domain of the dataset (extrapolation can lead to inappropriate conclusions).

Some unexpected effects of temperature can be observed: CO conversion decreasing a little with it and CO_2 conversion not changing with it. These facts can be explained by the proximity of the exothermic reaction system to equilibrium. Moreover, CO_2 conversion diminishes slightly with the CO inlet concentration since more CO hydrogenation can occur by consuming hydrogen that could react with CO_2. Another fact is that CO_2 conversion increases somewhat with the CO_2 inlet concentration because it can be consumed in both CO_2 hydrogenations and reverse water-gas shift reactions. However, CO conversion decays with a high content of CO_2 since H_2 will not be available for CO hydrogenation, but for CO_2 hydrogenation. It is also important to note that the higher the H_2 inlet concentration, the higher CO and CO_2 conversion, as more H_2 will be available for both hydrogenation reactions.

All these conclusions can be easily extracted with the neural kinetic model developed. Its lower computational cost can rapidly provide helpful information about this important chemical process of producing value-added products from CO_2, thus avoiding the time-consuming calculation of complex kinetic models based on detailed mechanisms.

2.5 Outlook

From the perspective of renewability and environmental impact, it is necessary to develop catalytic processes for the integral usage of CO_2 in industrial sources, that continuously emit this gas. For instance, flue gas stacks and oil & gas industry. Important limitations such as the barrier energy required to couple these molecules, without generating excessive carbon deposits over catalytic surfaces need to be studied, as well as more efficient capture systems that go hand in hand with eco-friendly conversion technologies. The development of new nanomaterials and alloys is fundamental in the design of compatible catalytic structures for CO_2 activation and coupling processes; for the production of polymer precursors, fuels, and mainly synthesis gas. Computational mechanics are important in the design of superstructures such as zeolites and MOFs promoted with metal clusters, these studies reveal that it is possible to obtain different products and intermediates, undertheorized conditions they would transform the CO_2.

Acknowledgments

The authors gratefully acknowledge support from FAPESP and Shell through the "Research Centre for Gas Innovation—RCGI" (Fapesp Proc. 2014/50279-4), hosted by the University of Sao Paulo, and the strategic importance of the support given by ANP (Brazil's National Oil, Natural Gas, and Biofuels Agency) through the R&D levy regulation. Camila Emilia thanks FUSP for the scholarship received during this work, Fabio Cavalcanti to FAPESP for his doctoral scholarship (2017/11940-5), and Martin Schmal to FAPESP for the researcher support (2014/00833-5). Thanks to Paul A. Sato for the supplementary review report.

References

Abd El-Hafiz, D. R., Sakr, A. A.-E., & Ebiad, M. A. (2021). Methane Bi-reforming for direct ethanol production over smart Cu/Mn-ferrite catalysts. *Renewable Energy, 167*, 236–247. https://doi.org/10.1016/j.renene.2020.11.078.

Al Abdulghani, A. J., Park, J.-H., Kozlov, S. M., Kang, D.-C., AlSabban, B., Pedireddy, S., et al. (2020). Methane dry reforming on supported cobalt nanoparticles promoted by boron. *Journal of Catalysis, 392*, 126–134. https://doi.org/10.1016/j.jcat.2020.09.015.

Alves, R. M. B., & Nascimento, C. A. O. (2002). Gross errors detection of industrial data by neural network and cluster techniques. *Brazilian Journal of Chemical Engineering, 19*, 483–489. https://doi.org/10.1590/S0104-66322002000400018.

Amariglio, H., Paréja, P., & Amariglio, A. (1995). Periodic operation of a catalyst as a means of overcoming a thermodynamic constraint. The case of methane homologation on metals. *Catalysis Today, 25*(2), 113–125. https://doi.org/10.1016/0920-5861(95)00102-L.

Anzures, F. M., Rivas, F. C., Ventura, J. H., Hernández, P. S., Berlier, G., & Zacahua-Tlacuatl, G. (2015). Spectroscopic characterization of CuO_x/TiO_2–ZrO_2 catalysts prepared by a-step sol–gel method. *Applied Catalysis A: General, 489*, 218–225. https://doi.org/10.1016/j.apcata.2014.10.003.

Arena, F., Italiano, G., Barbera, K., Bordiga, S., Bonura, G., Spadaro, L., et al. (2008). Solid-state interactions, adsorption sites and functionality of Cu-ZnO/ZrO_2 catalysts in the CO_2 hydrogenation to CH_3OH. *Applied Catalysis A: General, 350*(1), 16–23. https://doi.org/10.1016/j.apcata.2008.07.028.

Aresta, M. (2010). Carbon dioxide: Utilization options to reduce its accumulation in the atmosphere. In *Carbon dioxide as chemical feedstock* Wiley. https://doi.org/10.1002/9783527629916.ch1.

Baber, A. E., Tierney, H. L., & Sykes, E. C. H. (2010). Atomic-scale geometry and electronic structure of catalytically important Pd/Au alloys. *ACS Nano*, *4*(3), 1637–1645. https://doi.org/10.1021/nn901390y.

Bachiller-Baeza, B., Iglesias-Juez, A., Agostini, G., & Castillejos-López, E. (2020). Pd–Au bimetallic catalysts supported on ZnO for selective 1,3-butadiene hydrogenation. *Catalysis Science and Technology*, *10*(8), 2503–2512. https://doi.org/10.1039/C9CY02395J.

Bawah, A.-R., Lucky, R. A., & Hossain, M. M. (2021). Oxidative dehydrogenation of propane with CO_2—A green process for propylene and hydrogen (syngas). *International Journal of Hydrogen Energy*, *46*(5), 3401–3413. https://doi.org/10.1016/j.ijhydene.2020.10.168.

Behrens, M., Studt, F., Kasatkin, I., Kühl, S., Hävecker, M., Abild-Pedersen, F., et al. (2012). The active site of methanol synthesis over $Cu/ZnO/Al_2O_3$ industrial catalysts. *Science*, *336*(6083), 893. https://doi.org/10.1126/science.1219831.

Bonarowska, M., Machynskyy, O., Łomot, D., Kemnitz, E., & Karpiński, Z. (2014). Supported palladium–copper catalysts: Preparation and catalytic behavior in hydrogen-related reactions. *Recent Developments in Catalyst Design and Activation*, *235*, 144–151. https://doi.org/10.1016/j.cattod.2014.01.029.

Bradford, M. C. J., & Vannice, M. A. (1999). CO_2 reforming of CH_4. *Null*, *41*(1), 1–42. https://doi.org/10.1081/CR-100101948.

Bussche, K. M. V., & Froment, G. F. (1996). A steady-state kinetic model for methanol synthesis and the water gas shift reaction on a commercial $Cu/ZnO/Al_2O_3$ catalyst. *Journal of Catalysis*, *161*(1), 1–10. https://doi.org/10.1006/jcat.1996.0156.

Calaza, F., Mahapatra, M., Neurock, M., & Tysoe, W. T. (2014). Disentangling ensemble, electronic and coverage effects on alloy catalysts: Vinyl acetate synthesis on Au/Pd(111). *Journal of Catalysis*, *312*, 37–45. https://doi.org/10.1016/j.jcat.2014.01.003.

Cañete, B., Gigola, C. E., & Brignole, N. B. (2017). Enhancing the potential of methane combined reforming for methanol production via partial CO_2 hydrogenation. *Industrial and Engineering Chemistry Research*, *56*(22), 6480–6492. https://doi.org/10.1021/acs.iecr.6b04961.

Cavalcanti, F. M., Kozonoe, C. E., Pacheco, K. A., & Alves, R. M. B. (2021). Application of artificial neural networks to chemical and process engineering. In *Artificial neural networks and deep learning—Applications and perspective* IntechOpen. https://doi.org/10.5772/intechopen.96641.

Cavalcanti, F. M., Schmal, M., Giudici, R., & Brito Alves, R. M. (2019). A catalyst selection method for hydrogen production through Water-Gas Shift reaction using artificial neural networks. *Journal of Environmental Management*, *237*, 585–594. https://doi.org/10.1016/j.jenvman.2019.02.092.

Centi, G., Perathoner, S., Park, S.-E., Chang, J.-S., & Lee, K.-W. (2004). Heterogeneous catalytic reactions with CO_2: Status and perspectives. In *Vol. 153. Carbon dioxide utilization for global sustainability* (pp. 1–8). Elsevier. https://doi.org/10.1016/S0167-2991(04)80212-X.

Centi, G., Quadrelli, E. A., & Perathoner, S. (2013). Catalysis for CO_2 conversion: A key technology for rapid introduction of renewable energy in the value chain of chemical industries. *Energy and Environmental Science*, *6*(6), 1711–1731. https://doi.org/10.1039/C3EE00056G.

Chagas, C. A., de Souza, E. F., de Carvalho, M. C. N. A., Martins, R. L., & Schmal, M. (2016). Cobalt ferrite nanoparticles for the preferential oxidation of CO. *Applied Catalysis A: General*, *519*, 139–145. https://doi.org/10.1016/j.apcata.2016.03.024.

Chen, M., & Goodman, D. W. (2008). Promotional effects of Au in Pd-Au catalysts for vinyl acetate synthesis. *Chinese Journal of Catalysis*, *29*(11), 1178–1186. https://doi.org/10.1016/S1872-2067(09)60021-8.

Chen, W. H., Lin, B.-J., Lee, H.-M., & Huang, M.-H. (2012). One-step synthesis of dimethyl ether from the gas mixture containing CO_2 with high space velocity. *Applied Energy*, *98*, 92–101. https://doi.org/10.1016/j.apenergy.2012.02.082.

Chen, S., Zaffran, J., & Yang, B. (2020). Descriptor design in the computational screening of Ni-based catalysts with balanced activity and stability for dry reforming of methane reaction. *ACS Catalysis*, *10*(5), 3074–3083. https://doi.org/10.1021/acscatal.9b04429.

Chen, F., Zhang, P., Zeng, Y., Kosol, R., Xiao, L., Feng, X., et al. (2020). Vapor-phase low-temperature methanol synthesis from CO_2-containing syngas via self-catalysis of methanol and Cu/ZnO catalysts prepared by solid-state method. *Applied Catalysis B: Environmental*, *279*. https://doi.org/10.1016/j.apcatb.2020.119382, 119382.

Crisci, A. J., Dou, H., Prasomsri, T., & Román-Leshkov, Y. (2014). Cascade reactions for the continuous and selective production of isobutene from bioderived acetic acid over zinc-zirconia catalysts. *ACS Catalysis*, *4*(11), 4196–4200. https://doi.org/10.1021/cs501018k.

Cui, M., Qian, Q., Zhang, J., Chen, C., & Han, B. (2017). Efficient synthesis of acetic acid via Rh catalyzed methanol hydrocarboxylation with CO_2 and H_2 under milder conditions. *Green Chemistry*, *19*(15), 3558–3565. https://doi.org/10.1039/C7GC01391D.

de Sousa, L. F., Toniolo, F. S., Landi, S. M., & Schmal, M. (2017). Investigation of structures and metallic environment of the Ni/Nb_2O_5 by different in situ treatments—Effect on the partial oxidation of methane. *Applied Catalysis A: General*, *537*, 100–110. https://doi.org/10.1016/j.apcata.2017.03.015.

Dibenedetto, A., Angelini, A., & Stufano, P. (2014). Use of carbon dioxide as feedstock for chemicals and fuels: Homogeneous and heterogeneous catalysis. *Journal of Chemical Technology and Biotechnology*, *89*(3), 334–353. https://doi.org/10.1002/jctb.4229.

Dutra, M., Schmal, M., & Guardani, R. (2018). Syntheses and characterization of zinc oxide nanoparticles on graphene sheets: Adsorption-reaction in situ DRIFTS of methane and CO_2. *Catalysis Letters*, *148*(11), 3413–3430. https://doi.org/10.1007/s10562-018-2538-6.

El-Salamony, R. A., El-Temtamy, S. A., El Naggar, A. M. A., Ghoneim, S. A., Abd El-Hafiz, D. R., Ebiad, M. A., et al. (2021). Valuation of catalytic activity of nickel–zirconia-based catalysts using lanthanum co-support for dry reforming of methane. *International Journal of Energy Research*, *45*(3), 3899–3912. https://doi.org/10.1002/er.6043.

Erdőhelyi, A. (2021). Catalytic reaction of carbon dioxide with methane on supported noble metal catalysts. *Catalysts*, *11*(2). https://doi.org/10.3390/catal11020159.

Fiorani, G., Guo, W., & Kleij, A. W. (2015). Sustainable conversion of carbon dioxide: The advent of organocatalysis. *Green Chemistry*, *17*(3), 1375–1389. https://doi.org/10.1039/C4GC01959H.

Galhotra, P. (2010). *Carbon dioxide adsorption on nanomaterials*. PhD (Doctor of Philosophy) thesis University of Iowa.

Ganesh, I. (2012). Conversion of carbon dioxide into several potential chemical commodities following different pathways—A review. *Materials Science Forum*, *764*, 1–82. https://doi.org/10.4028/www.scientific.net/MSF.764.1.

Gao, F., & Goodman, D. W. (2012). Pd–Au bimetallic catalysts: Understanding alloy effects from planar models and (supported) nanoparticles. *Chemical Society Reviews*, *41*(24), 8009–8020. https://doi.org/10.1039/C2CS35160A.

Gao, P., Li, F., Zhao, N., Xiao, F., Wei, W., Zhong, L., et al. (2013). Influence of modifier (Mn, La, Ce, Zr and Y) on the performance of Cu/Zn/Al catalysts via hydrotalcite-like precursors for CO_2 hydrogenation to methanol. *Applied Catalysis A: General*, *468*, 442–452. https://doi.org/10.1016/j.apcata.2013.09.026.

García-Vargas, J. M., Valverde, J. L., Díez, J., Dorado, F., & Sánchez, P. (2015). Catalytic and kinetic analysis of the methane tri-reforming over a Ni–Mg/β-SiC catalyst. *International Journal of Hydrogen Energy*, *40*(28), 8677–8687. https://doi.org/10.1016/j.ijhydene.2015.05.032.

Ghoneim, S. A. (2016). Review on innovative catalytic reforming of natural gas to syngas. *World Journal of Engineering and Technology*, *4*(1), 116–139. https://doi.org/10.4236/wjet.2016.41011.

Gómez-Cuaspud, J. A., Perez, C. A., & Schmal, M. (2016). Nanostructured $La_{0.8}Sr_{0.2}Fe_{0.8}Cr_{0.2}O_3$ perovskite for the steam methane reforming. *Catalysis Letters*, *146*(12), 2504–2515. https://doi.org/10.1007/s10562-016-1885-4.

González, A. R., Asencios, Y. J. O., Assaf, E. M., & Assaf, J. M. (2013). Dry reforming of methane on Ni–Mg–Al nano-spheroid oxide catalysts prepared by the sol–gel method from hydrotalcite-like precursors. *Applied Surface Science*, *280*, 876–887. https://doi.org/10.1016/j.apsusc.2013.05.082.

Gonzalez Caranton, A. R., da Silva Pinto, J. C. C., Stavale, F., Barreto, J., & Schmal, M. (2020). Statistical analysis of the catalytic synthesis of vinyl acetate over Pd-Cu/ZrO_2 nanostructured based catalysts. *Catalysis Today*, *344*, 108–117. https://doi.org/10.1016/j.cattod.2018.10.034.

Gonzalez Caranton, A. R., Dille, J., Barreto, J., Stavale, F., Pinto, J. C., & Schmal, M. (2018). Nanostructured Pd–Cu catalysts supported on Zr–Al and Zr–Ti for synthesis of vinyl acetate. *ChemCatChem*, *10*(22), 5256–5269. https://doi.org/10.1002/cctc.201801083.

Goulas, K. A., Sreekumar, S., Song, Y., Kharidehal, P., Gunbas, G., Dietrich, P. J., et al. (2016). Synergistic effects in bimetallic palladium–copper catalysts improve selectivity in oxygenate coupling reactions. *Journal of the American Chemical Society*, *138*(21), 6805–6812. https://doi.org/10.1021/jacs.6b02247.

Graaf, G. H., Stamhuis, E. J., & Beenackers, A. A. C. M. (1988). Kinetics of low-pressure methanol synthesis. *Chemical Engineering Science*, *43*(12), 3185–3195. https://doi.org/10.1016/0009-2509(88)85127-3.

Han, Y.-F., Kumar, D., Sivadinarayana, C., & Goodman, D. W. (2004). Kinetics of ethylene combustion in the synthesis of vinyl acetate over a Pd/SiO_2 catalyst. *Journal of Catalysis*, *224*(1), 60–68. https://doi.org/10.1016/j.jcat.2004.02.028.

Hanrieder, E. K., Jentys, A., & Lercher, J. A. (2015a). Atomistic engineering of catalyst precursors: Dynamic reordering of PdAu nanoparticles during vinyl acetate synthesis enhanced by potassium acetate. *ACS Catalysis*, *5*(10), 5776–5786. https://doi.org/10.1021/acscatal.5b01140.

Hanrieder, E. K., Jentys, A., & Lercher, J. A. (2015b). Impact of alkali acetate promoters on the dynamic ordering of PdAu catalysts during vinyl acetate synthesis. *Journal of Catalysis*, *333*, 71–77. https://doi.org/10.1016/j.jcat.2015.10.019.

Havran, V., Duduković, M. P., & Lo, C. S. (2011). Conversion of methane and carbon dioxide to higher value products. *Industrial and Engineering Chemistry Research*, *50*(12), 7089–7100. https://doi.org/10.1021/ie2000192.

Hayashi, S., Kawamura, C., & Takayama, M. (1970). Color reactions of poly vinyl acetate and its derivatives with iodine. *Bulletin of the Chemical Society of Japan*, *43*(2), 537–542. https://doi.org/10.1246/bcsj.43.537.

Holmen, A. (2009). Direct conversion of methane to fuels and chemicals. *Catalysis Today*, *142*(1), 2–8. https://doi.org/10.1016/j.cattod.2009.01.004.

Hong, J., Li, M., Zhang, J., Sun, B., & Mo, F. (2019). C–H bond carboxylation with carbon dioxide. *ChemSusChem*, *12*(1), 6–39. https://doi.org/10.1002/cssc.201802012.

Hornik, K., Stinchcombe, M., & White, H. (1989). Multilayer feedforward networks are universal approximators. *Neural Networks*, *2*(5), 359–366. https://doi.org/10.1016/0893-6080(89)90020-8.

Huang, W., Xie, K.-C., Wang, J.-P., Gao, Z.-H., Yin, L.-H., & Zhu, Q.-M. (2001). Possibility of direct conversion of CH_4 and CO_2 to high-value products. *Journal of Catalysis*, *201*(1), 100–104. https://doi.org/10.1006/jcat.2001.3223.

Huang, Y., Dong, X., & Yu, Y. (2016). Surface carbon species formation from ethylene decomposition on Pd(100): A first-principles-based kinetic Monte Carlo study. *RSC Advances*, *6*(70), 65349–65354. https://doi.org/10.1039/C6RA13977A.

Hussain, G. (2009). Oxidation of CO by O_2 over ZnO studied by FTIR spectroscopy. *Journal of the Chemical Society of Pakistan*, *31*(5), 718–725.

Intergovernmental Panel on Climate Change. (2015). *Climate change 2014: Mitigation of climate change—Working group III contribution to the fifth assessment report of the intergovernmental panel on climate change*. Cambridge University Press. https://doi.org/10.1017/CBO9781107415416.

Joshi, P. (2014). Carbon dioxide utilization: A comprehensive review. *International Journal of Chemical Sciences*, *12*(4), 1208–1220. https://www.tsijournals.com/abstract/carbon-dioxide-utilization-a-comprehensive-review-11696.html.

Julkapli, N. M., & Bagheri, S. (2014). Graphene supported heterogeneous catalysts: An overview. *International Journal of Hydrogen Energy*, *40*(2), 948–979. https://doi.org/10.1016/j.ijhydene.2014.10.129.

Kemball, C. (1959). The catalytic exchange of hydrocarbons with deuterium. In *Vol. 11. Advances in catalysis* (pp. 223–262). Academic Press. https://doi.org/10.1016/S0360-0564(08)60419-8.

Koirala, R., Safonova, O. V., Pratsinis, S. E., & Baiker, A. (2017). Effect of cobalt loading on structure and catalytic behavior of CoO_x/SiO_2 in CO_2-assisted dehydrogenation of ethane. *Applied Catalysis A: General*, *552*, 77–85. https://doi.org/10.1016/j.apcata.2017.12.025.

Kouva, S., Honkala, K., Lefferts, L., & Kanervo, J. (2015). Review: Monoclinic zirconia, its surface sites and their interaction with carbon monoxide. *Catalysis Science and Technology*, *5*(7), 3473–3490. https://doi.org/10.1039/C5CY00330J.

Kozonoe, C. E., Brito Alves, R. M., & Schmal, M. (2020). Influence of feed rate and testing variables for low-temperature tri-reforming of methane on the Ni@MWCNT/Ce catalyst. *Fuel*, *281*. https://doi.org/10.1016/j.fuel.2020.118749, 118749.

Kubokawa, M. (1938). The activated adsorption of methane on reduced nickel. *Proceedings of the Imperial Academy*, *14*(2), 61–66. https://doi.org/10.2183/pjab1912.14.61.

Kuhn, M., Jeschke, J., Schulze, S., Hietschold, M., Lang, H., & Schwarz, T. (2014). Dendrimer-stabilized bimetallic Pd/Au nanoparticles: Preparation, characterization and application to vinyl acetate synthesis. *Catalysis Communications*, *57*, 78–82. https://doi.org/10.1016/j.catcom.2014.08.009.

Kuijpers, E. G. M., Breedijk, A. K., van der Wal, W. J. J., & Geus, J. W. (1981). Chemisorption of methane on silica-supported nickel catalysts: A magnetic and infrared study. *Journal of Catalysis*, *72*(2), 210–217. https://doi.org/10.1016/0021-9517(81)90003-8.

Kuijpers, E. G. M., Breedijk, A. K., van der Wal, W. J. J., & Geus, J. W. (1983). Chemisorption of methane on Ni SiO_2 catalysts and reactivity of the chemisorption products toward hydrogen. *Journal of Catalysis*, *81*(2), 429–439. https://doi.org/10.1016/0021-9517(83)90181-1.

Kuijpers, E. G. M., Jansen, J. W., van Dillen, A. J., & Geus, J. W. (1981). The reversible decomposition of methane on a $NiSiO_2$ catalyst. *Journal of Catalysis*, *72*(1), 75–82. https://doi.org/10.1016/0021-9517(81)90079-8.

Kumar, D., Chen, M. S., & Goodman, D. W. (2007). Synthesis of vinyl acetate on Pd-based catalysts. *Catalysis Today*, *123*(1), 77–85. https://doi.org/10.1016/j.cattod.2007.01.050.

Liu, G., Ariyarathna, I. R., Ciborowski, S. M., Zhu, Z., Miliordos, E., & Bowen, K. H. (2020). Simultaneous functionalization of methane and carbon dioxide mediated by single platinum atomic anions. *Journal of the American Chemical Society*, *142*(51), 21556–21561. https://doi.org/10.1021/jacs.0c11112.

Liu, L., Jiang, H., Liu, H., Li, H., & Suib, S. L. (2013). Chapter 7—Recent advances on the catalysts for activation of CO_2 in several typical processes. In *New and future developments in catalysis* (pp. 189–222). Elsevier. https://doi.org/10.1016/B978-0-444-53882-6.00008-5.

Machado, B. F., & Serp, P. (2012). Graphene-based materials for catalysis. *Catalysis Science and Technology*, *2*(1), 54–75. https://doi.org/10.1039/C1CY00361E.

Morikawa, K., Benedict, W. S., & Taylor, H. S. (1935). Catalytic exchange of deuterium and methane. *Journal of the American Chemical Society*, *57*(3), 592–593. https://doi.org/10.1021/ja01306a509.

Morikawa, K., Trenner, N. R., & Taylor, H. S. (1937). The activation of specific bonds in complex molecules at catalytic surfaces. III. The carbon—hydrogen and carbon—carbon bonds in propane and ethylene. *Journal of the American Chemical Society*, *59*(6), 1103–1111. https://doi.org/10.1021/ja01285a042.

Moya, S. F., Martins, R. L., & Schmal, M. (2011). Monodispersed and nanostructrured Ni/SiO_2 catalyst and its activity for non oxidative methane activation. *Applied*

Catalysis A: General, *396*(1), 159–169. https://doi.org/10.1016/j.apcata.2011.02.007.

Nascimento, C. A. O., & Giudici, R. (1998). Neural network based approach for optimisation applied to an industrial nylon-6,6 polymerisation process. *Computers & Chemical Engineering*, *22*, S595–S600. https://doi.org/10.1016/S0098-1354(98)00105-7.

Nascimento, C. A. O., Giudici, R., & Guardani, R. (2000). Neural network based approach for optimization of industrial chemical processes. *Computers and Chemical Engineering*, *24*(9), 2303–2314. https://doi.org/10.1016/S0098-1354(00)00587-1.

Navalon, S., Dhakshinamoorthy, A., Alvaro, M., & Garcia, H. (2016). Metal nanoparticles supported on two-dimensional graphenes as heterogeneous catalysts. *Coordination Chemistry Reviews*, *312*, 99–148. https://doi.org/10.1016/j.ccr.2015.12.005.

Nestler, F., Schütze, A. R., Ouda, M., Hadrich, M. J., Schaadt, A., Bajohr, S., et al. (2020). Kinetic modelling of methanol synthesis over commercial catalysts: A critical assessment. *Chemical Engineering Journal*, *394*. https://doi.org/10.1016/j.cej.2020.124881, 124881.

Neurock, M., Provine, W. D., Dixon, D. A., Coulston, G. W., Lerou, J. J., & van Santen, R. A. (1996). First principle analysis of the catalytic reaction pathways in the synthesis of vinyl acetate. *Chemical Reaction Engineering: From Fundamentals to Commercial Plants and Products*, *51*(10), 1691–1699. https://doi.org/10.1016/0009-2509(96)00028-0.

Novoselov, K. S., Geim, A. K., Morozov, S. V., Jiang, D., Zhang, Y., Dubonos, S. V., et al. (2004). Electric field effect in atomically thin carbon films. *Science*, *306*(5696), 666. https://doi.org/10.1126/science.1102896.

Olah, G. A., Goeppert, A., & Prakash, G. K. S. (2009). Chemical recycling of carbon dioxide to methanol and dimethyl ether: From greenhouse gas to renewable, environmentally carbon neutral fuels and synthetic hydrocarbons. *Journal of Organic Chemistry*, *74*(2), 487–498. https://doi.org/10.1021/jo801260f.

Panjan, W., Sirijaraensre, J., Warakulwit, C., Pantu, P., & Limtrakul, J. (2012). The conversion of CO_2 and CH_4 to acetic acid over the Au-exchanged ZSM-5 catalyst: A density functional theory study. *Physical Chemistry Chemical Physics*, *14*(48), 16588–16594. https://doi.org/10.1039/C2CP42066J.

Park, N., Park, M.-J., Lee, Y.-J., Ha, K.-S., & Jun, K.-W. (2014). Kinetic modeling of methanol synthesis over commercial catalysts based on three-site adsorption. *Fuel Processing Technology*, *125*, 139–147. https://doi.org/10.1016/j.fuproc.2014.03.041.

Patil, U., Saih, Y., Abou-Hamad, E., Hamieh, A., Pelletier, J. D. A., & Basset, J. M. (2014). Low temperature activation of methane over a zinc-exchanged heteropolyacid as an entry to its selective oxidation to methanol and acetic acid. *Chemical Communications*, *50*(82), 12348–12351. https://doi.org/10.1039/C4CC04950K.

Paulidou, A., & Nix, R. M. (2005). Growth and characterisation of zirconia surfaces on Cu(111). *Physical Chemistry Chemical Physics*, *7*(7), 1482–1489. https://doi.org/10.1039/B418693A.

Pino, L., Vita, A., Cipitì, F., Laganà, M., & Recupero, V. (2011). Hydrogen production by methane tri-reforming process over Ni–ceria catalysts: Effect of La-doping. *Applied Catalysis B: Environmental*, *104*(1), 64–73. https://doi.org/10.1016/j.apcatb.2011.02.027.

Pino, L., Vita, A., Laganà, M., & Recupero, V. (2013). Hydrogen from biogas: Catalytic tri-reforming process with Ni/LaCeO mixed oxides. *Applied Catalysis B: Environmental*, *148–149*, 91–105. https://doi.org/10.1016/j.apcatb.2013.10.043.

Qian, Q., Zhang, J., Cui, M., & Han, B. (2016). Synthesis of acetic acid via methanol hydrocarboxylation with CO_2 and H_2. *Nature Communications*, *7*(1), 11481. https://doi.org/10.1038/ncomms11481.

Qiu, B., Zhang, Y., Liu, Y., & Zhang, Y. (2021). The state of Pt active phase and its surrounding environment during dehydrogenation of ethane to ethylene. *Applied Surface Science*, *554*. https://doi.org/10.1016/j.apsusc.2021.149611, 149611.

Rabie, A. M., Betiha, M. A., & Park, S.-E. (2017). Direct synthesis of acetic acid by simultaneous co-activation of methane and CO_2 over Cu-exchanged ZSM-5 catalysts. *Applied Catalysis B: Environmental*, *215*, 50–59. https://doi.org/10.1016/j.apcatb.2017.05.053.

Raupach, M. R., Marland, G., Ciais, P., Le Quéré, C., Canadell, J. G., Klepper, G., et al. (2007). Global and regional drivers of accelerating CO_2 emissions. *Proceedings of the National Academy of Sciences of the United States of America*, *104*(24), 10288–10293. https://doi.org/10.1073/pnas.0700609104.

Ritchie, H., & Roser, M. (2020). *CO_2 and greenhouse gas emissions*. OurWorldInData.Org https://ourworldindata.org/co2-and-other-greenhouse-gas-emissions.

Sato, P. A. (2019). *Síntese de dme por meio da hidrogenação catalítica de CO_2*. Masters dissertation Universidade de Sao Paulo.

Scarano, D., Bertarione, S., Spoto, G., Zecchina, A., & Otero Areán, C. (2001). FTIR spectroscopy of hydrogen, carbon monoxide, and methane adsorbed and co-adsorbed on zinc oxide. *Thin Solid Films*, *400*(1), 50–55. https://doi.org/10.1016/S0040-6090(01)01472-9.

Schmal, M. (2016). *Heterogeneous catalysis and its industrial applications* (1st ed.). Cham: Springer. https://doi.org/10.1007/978-3-319-09250-8.

Schmal, M., Toniolo, F. S., & Kozonoe, C. E. (2018). Perspective of catalysts for (Tri) reforming of natural gas and flue gas rich in CO_2. *Applied Catalysis A: General*, *568*, 23–42. https://doi.org/10.1016/j.apcata.2018.09.017.

Sedighi, M., & Mohammadi, M. (2019). CO_2 hydrogenation to light olefins over Cu-CeO_2/SAPO-34 catalysts: Product distribution and optimization. *Journal of CO_2 Utilization, 35*, 236–244. https://doi.org/10.1016/j.jcou.2019.10.002.

Shavi, R., Ko, J., Cho, A., Han, J. W., & Seo, J. G. (2018). Mechanistic insight into the quantitative synthesis of acetic acid by direct conversion of CH_4 and CO_2: An experimental and theoretical approach. *Applied Catalysis B: Environmental, 229*, 237–248. https://doi.org/10.1016/j.apcatb.2018.01.058.

Silva, M. D. (2017). *Estudo de catalisadores nanoestruturados para carboxilação de metano com CO_2 para a produção de produtos de alto valor agregado*. Dissertação (Mestrado em Engenharia Química)-Escola Politécnica Universidade de São Paulo. https://doi.org/10.11606/D.3.2017.tde-26092017-145937.

Song, B., Lin, R., Lam, C. H., Wu, H., Tsui, T.-H., & Yu, Y. (2021). Recent advances and challenges of inter-disciplinary biomass valorization by integrating hydrothermal and biological techniques. *Renewable and Sustainable Energy Reviews, 135*. https://doi.org/10.1016/j.rser.2020.110370, 110370.

Song, C., & Pan, W. (2004). Tri-reforming of methane: A novel concept for catalytic production of industrially useful synthesis gas with desired H_2/CO ratios. *Catalysis Today, 98*(4), 463–484. https://doi.org/10.1016/j.cattod.2004.09.054.

Spencer, M. S. (1998). Role of ZnO in methanol synthesis on copper catalysts. *Catalysis Letters, 50*(1), 37–40. https://doi.org/10.1023/A:1019098414820.

Spivey, J. J., Wilcox, E. M., & Roberts, G. W. (2007). Direct utilization of carbon dioxide in chemical synthesis: Vinyl acetate via methane carboxylation. *Catalysis Communications, 9*(5), 685–689. https://doi.org/10.1016/j.catcom.2007.08.023.

Sun, Y., Yang, G., Wen, C., Zhang, L., & Sun, Z. (2018). Artificial neural networks with response surface methodology for optimization of selective CO_2 hydrogenation using K-promoted iron catalyst in a microchannel reactor. *Journal of CO_2 Utilization, 24*, 10–21. https://doi.org/10.1016/j.jcou.2017.11.013.

Świrk, K., Grzybek, T., & Motak, M. (2016). Tri-reforming as a process of CO_2 utilization and a novel concept of energy storage in chemical products. *Energy and Fuels, E3S Web of Conferences, 14*(02038), 1–10. https://doi.org/10.1051/e3sconf/20171402038.

Tundo, P., Anastas, P., Black, D. S. C., Breen, J., Collins, T. J., Memoli, S., et al. (2000). Synthetic pathways and processes in green chemistry. Introductory overview. *Pure and Applied Chemistry, 72*(7), 1207–1228. https://doi.org/10.1351/pac200072071207.

Viñes, F., Lykhach, Y., Staudt, T., Lorenz, M. P. A., Papp, C., Steinrück, H. P., et al. (2010). Methane activation by platinum: Critical role of edge and corner sites of metal nanoparticles. *Chemistry—A European Journal, 16*(22), 6530–6539. https://doi.org/10.1002/chem.201000296.

Wei, J., & Iglesia, E. (2004). Mechanism and site requirements for activation and chemical conversion of methane on supported Pt clusters and turnover rate comparisons among noble metals. *Journal of Physical Chemistry B, 108*(13), 4094–4103. https://doi.org/10.1021/jp036985z.

Weng, X., Ren, H., Chen, M., & Wan, H. (2014). Effect of surface oxygen on the activation of methane on palladium and platinum surfaces. *ACS Catalysis, 4*(8), 2598–2604. https://doi.org/10.1021/cs500510x.

Whang, H. S., Lim, J., Choi, M. S., Lee, J., & Lee, H. (2019). Heterogeneous catalysts for catalytic CO_2 conversion into value-added chemicals. *BMC Chemical Engineering, 1*(1), 9. https://doi.org/10.1186/s42480-019-0007-7.

Wilcox, E. M. (2005). *Direct synthesis of acetic acid from carbon dioxide and methane*. PhD dissertetion North Carolina State University. https://repository.lib.ncsu.edu/handle/1840.16/5624.

Wilcox, E. M., Gogate, M. R., Spivey, J. J., Roberts, G. W., Iglesia, E., Spivey, J. J., et al. (2001). Direct synthesis of acetic acid from methane and carbon dioxide. In *Vol. 136. Natural gas conversion VI* (pp. 259–264). Elsevier. https://doi.org/10.1016/S0167-2991(01)80313-X.

Wilcox, E. M., Roberts, G. W., & Spivey, J. J. (2003). Direct catalytic formation of acetic acid from CO_2 and methane. *Environmental Catalysis and Reaction Engineering, 88*(1), 83–90. https://doi.org/10.1016/j.cattod.2003.08.007.

Wright, P. G., Ashmore, P. G., & Kemball, C. (1958). Dissociative adsorption of methane and ethane on evaporated metal films. *Transactions of the Faraday Society, 54*, 1692–1702. https://doi.org/10.1039/TF9585401692.

Wu, J.-F., Gao, X.-D., Wu, L.-M., Wang, W. D., Yu, S.-M., & Bai, S. (2019). Mechanistic insights on the direct conversion of methane into methanol over Cu/Na–ZSM-5 zeolite: Evidence from EPR and solid-state NMR. *ACS Catalysis, 9*(9), 8677–8681. https://doi.org/10.1021/acscatal.9b02898.

Wu, J.-F., Yu, S.-M., Wang, W. D., Fan, Y.-X., Bai, S., Zhang, C.-W., et al. (2013). Mechanistic insight into the formation of acetic acid from the direct conversion of methane and carbon dioxide on zinc-modified H–ZSM-5 zeolite. *Journal of the American Chemical Society, 135*(36), 13567–13573. https://doi.org/10.1021/ja406978q.

Xu, L., Xiu, Y., Liu, F., Liang, Y., & Wang, S. (2020). Research progress in conversion of CO_2 to valuable fuels. *Molecules, 25*(16), 3653. https://doi.org/10.3390/molecules25163653.

Yang, Y., Mims, C. A., Mei, D. H., Peden, C. H. F., & Campbell, C. T. (2013). Mechanistic studies of methanol synthesis over Cu from CO/CO_2/H_2/H_2O mixtures: The source of C in methanol and the role of water. *Journal of Catalysis, 298*, 10–17. https://doi.org/10.1016/j.jcat.2012.10.028.

Yang, Y., Yang, B., Zhao, Y.-X., Jiang, L.-X., Li, Z.-Y., Ren, Y., et al. (2019). Direct conversion of methane with carbon dioxide mediated by $RhVO_3$–cluster anions. *Angewandte Chemie International Edition*, *58*(48), 17287–17292. https://doi.org/10.1002/anie.201911195.

Zhang, N., Liu, P., Yi, Y., Gibril, M. E., Wang, S., & Kong, F. (2021). Application of polyvinyl acetate/lignin copolymer as bio-based coating material and its effects on paper properties. *Coatings*, *11*(2). https://doi.org/10.3390/coatings11020192.

Zhang, Y., Zhang, S., Gossage, J. L., Lou, H. H., & Benson, T. J. (2014). Thermodynamic analyses of tri-reforming reactions to produce syngas. *Energy and Fuels*, *28*(4), 2717–2726. https://doi.org/10.1021/ef500084m.

Zhao, Y., Wang, H., Han, J., Zhu, X., Mei, D., & Ge, Q. (2019). Simultaneous activation of CH_4 and CO_2 for concerted C–C coupling at oxide–oxide interfaces. *ACS Catalysis*, *9*(4), 3187–3197. https://doi.org/10.1021/acscatal.9b00291.

Zoeller, J. R. (2008). Evolving production of the acetyls (acetic acid, acetaldehyde, acetic anhydride, and vinyl acetate): A mirror for the evolution of the chemical industry. In *Vol. 1000. Innovations in industrial and engineering chemistry* (pp. 365–387). American Chemical Society. https://doi.org/10.1021/bk-2009-1000.ch010.

CHAPTER

3

Transition metal-based catalysts for CO_2 methanation and hydrogenation

Chao Sun and Patrick Da Costa

Institut Jean Le Rondd'Alembert, Sorbonne Université, CNRS UMR 7190, Saint-Cyr-l'Ecole, France

3.1 Introduction

The anthropogenic CO_2 emissions derived mostly from the combustion of fossil fuels (mainly coal, petroleum, and natural gas) had caused a sharp increase in CO_2 concentration in the atmosphere (Aresta, Dibenedetto, & Angelini, 2014; Wang et al., 2018). The excess CO_2 concentration resulted in global warming and climate change, e.g., extreme climate, rising sea level, species extinction, etc. (Sreedhar, Varun, Singh, Venugopal, & Reddy, 2019). Such environmental issues and the depletion of conventional fuels are severe risks for the sustainable and environmentally friendly development of the future society (Younas et al., 2016). Therefore, human society must limit CO_2 emissions into the atmosphere. The CO_2 reduction routes consist of two types of technology: (1) CO_2 capture and storage (CCS) technologies which capture the CO_2 in the atmosphere and store a large quantity of CO_2 in geological subsurface or the ocean (Ahn et al., 2019; Younas et al., 2016); (2) CO_2 capture and utilization (CCU) technologies that use the CO_2 from the atmosphere or factory to produce fuel or chemicals (Ahn et al., 2019). The CCS technology still had difficulties like cost, potential transportation leakage, and appropriate storage sites. The CCU technologies had caused more interest as they can not only reduce CO_2 emission but also produce high-valued fuel or chemicals (Aresta et al., 2014; Wang et al., 2018). A variety of CO_2 utilization routes have already been developed, such as CO_2 methane reforming, CO_2 hydrogenation to chemicals, e.g., methane (Sun, Beaunier, & Da Costa, 2020; Sun et al., 2020; Wang & Gong, 2011), methanol (Kattel, Ramírez, Chen, Rodriguez, & Liu, 2017; Wang et al., 2017), ethanol (Bai et al., 2017), formate, etc. (Álvarez et al., 2017), with the methods of thermal catalysis (Aresta et al., 2014), photocatalysis (Wang, Himeda, Muckerman, Manbeck, & Fujita, 2015), electrochemical catalysis (Wang, Himeda, et al., 2015), and plasma-assisted catalysis (Mikhail et al., 2020) (Fig. 3.1).

These CO_2 utilization technologies establish the CO_2 recycling process, which is also called the carbon recycling economy (Aresta et al., 2014). Among them, a mature solution, the CO_2 hydrogenation to methane catalyzed by heterogeneous catalysts had been widely

Heterogeneous Catalysis
https://doi.org/10.1016/B978-0-323-85612-6.00003-6

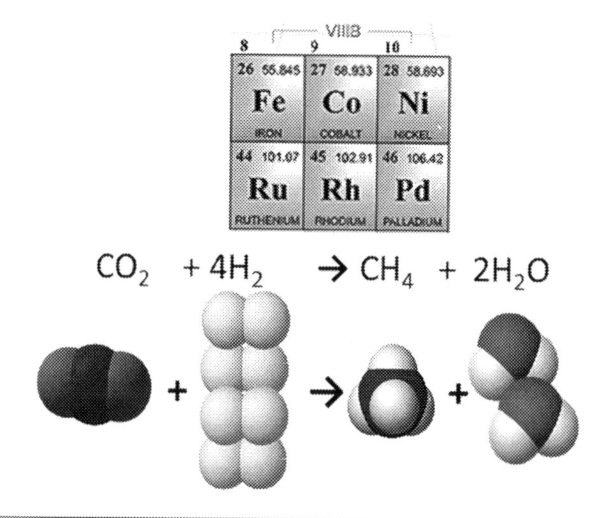

FIG. 3.1 A diagram for the overview of the chapter.

studied because it has many benefits compared to other CO_2 hydrogenation routes. The CO_2 hydrogenation to methane, i.e., CO_2 methanation (Eq. 3.1), first reported by Stangeland, Kalai, Li, and Yu (2017), is an exothermal reaction and is thermodynamically favorable at low and medium temperatures (Gao et al., 2012).

$$CO_2 + 4H_2 \leftrightarrow CH_4 + 2H_2O; \quad \Delta H = -165.15\,\text{kJ/mol} \tag{3.1}$$

The required conditions for the reaction are also not too critical. From the thermodynamic equilibrium, the reaction can reach high activity and selectivity at low temperature and ambient pressure, which also can prevent the catalyst from sintering and carbon formation (Bian, Chan, Yu, & Kawi, 2020; Sun, Świrk, Wierzbicki, et al., 2020). Meanwhile, the synthesized methane can also be stored easily and transported by a well-established natural gas system. Thus, CO_2 methanation reaction can be applied in Power-to-Gas (PtG) process for energy storage, in which excess produced energy from intermittent renewable sources, e.g., wind, solar, and water was used for the electrolysis of water to produce renewable H_2. Power-to-Gas (PtG) allows connecting the power plant from renewable energy to synthetic gas. It consists of a water electrolyzer, CO_2 production equipment, and a methanation reactor (Ghaib & Ben-Fares, 2018; Hoekman, Broch, Robbins, & Purcell, 2010). In this way, the produced H_2 reacts with CO_2 to produce methane, which can be used as fuel or chemical feedstock (Thema, Bauer, & Sterner, 2019). The PtG technology is nowadays considered the most cost-efficient long-term storage method for power with already many plants established in the world (Ghaib & Ben-Fares, 2018; Thema et al., 2019).

The CO_2 hydrogenation to methane has been widely studied and it was found that the reaction was feasible with appropriate catalysts under supplementary means like thermal, plasma-assisted, photo- or electrochemical technologies. Among them, thermal catalysis of CO_2 hydrogenation was mostly studied because of the low cost and easy set-up. The main barrier for this technology is to find an appropriate catalytic system. The transition metal-based catalytic systems had been widely developed like Ru (Guo et al., 2018; Sharma, Hu, Zhang, McFarland, & Metiu, 2011), Rh (Karelovic & Ruiz, 2013a), Pd (Kim, Lee, & Park, 2010; Park & McFarland, 2009), Pt (Maria, Gabriela, Monica, Lucian, & Lazar, 2019), Ni (Sun, Beaunier, & Da Costa, 2020; Sun, Świrk, Wierzbicki, et al., 2020; Wang & Gong, 2011), Co (Beaumont et al., 2014; Li et al., 2019), Fe (Baysal & Kureti, 2020), etc. The noble metal-based catalysts such as Ru, Rh, and Pd were found to have superior catalytic activity and selectivity in the CO_2 hydrogenation reaction. Especially, Ru-based catalysts performed high activity and selectivity at low temperature and moderate conditions (Wang, Li, et al., 2015; Wang et al., 2016). However, the high cost and low availability of such metals are the main barriers to the large-scale application of noble metals in CO_2 hydrogenation. Considering these later points, non-noble metal-based catalysts (e.g., Ni, Co, and Fe) were also intensively developed. Cobalt is a low-cost and high available metal and has been applied in catalysis in many different reactions including CO_2 methanation (Li et al., 2018; Storsæter, Tøtdal, Walmsley, Tanem, & Holmen, 2005; Yin & Ge, 2012; Zhang, Jacobs, Sparks, Dry, & Davis, 2002; Zhou, Wu, Zhang, Xie, & Feng, 2014). For example, Li et al. (2018) studied Co/ZrO_2 and Co/Al_2O_3 catalysts with different metal loading in CO_2 methanation under 3 MP and 673 K. It was then found that the Co/ZrO_2 of 10 wt% Co showed the highest CO_2 conversion of 92.5%, CH_4 selectivity of 99.9%, and high stability during the 300 h time on stream test. Meanwhile, Co/Al_2O_3 showed lower activity, selectivity, and stability compared to Co/ZrO_2 catalyst. Zhou et al. (2014) synthesized Co catalysts supported on mesoporous KIT-6 material and applied them in CO_2 methanation. It was found that the Co/KIT-6 catalyst led to a superior activity and selectivity during the giving experimental conditions when compared to Co/SiO_2. Based on these findings, the Co-based catalysts showed high activity and selectivity at high pressure, high temperature, and low gas hourly space velocity (GHSV). Under atmospheric pressure and high GHSV, the CO_2 conversion and CH_4 selectivity decreased sharply.

Concerning Fe-based catalysts in CO_2 methanation, the reported Fe-based catalysts cannot compete with Ni-based catalysts in terms of both CO_2 conversion and CH_4 selectivity (Baysal & Kureti, 2020). However, Fe could be used as a promoter in Ni catalysts. Burger, Koschany, Thomys, Köhler, and Hinrichsen (2018) studied the effect of Fe and Mn promoters in Ni-Al catalysts in CO_2 methanation. It was found that the doping of Fe and Mn could improve the activity and stability of $NiAlO_x$ catalysts. In the case of Fe promotion, the positive effects were attributed to the formation of Ni-Fe alloy, leading to a slight increase of CO_2 adsorption, and the higher interaction between Ni and Fe. Meanwhile, the Mn promoter increased CO_2 adsorption capacity and stable specific Ni metal surface area.

Furthermore, Ni-based catalysts are regarded as the most promising catalytic systems toward CO_2 methanation. Although the benefits of Ni-based catalysts such as low cost, high availability, and high activity, there are still some barriers to the industrial application in catalysis. Thus, the existence of side reactions and catalyst deactivation after a long-time run and possible carbon deposition are limiting their industrialization (Abelló, Berrueco, Gispert-Guirado, & Montané, 2016; Ewald, Kolbeck, Kratky, Wolf, & Hinrichsen, 2019). Ewald et al. (2019) reported the deactivation of $Ni/\gamma\text{-}Al_2O_3$ catalysts in long-term CO_2 methanation tests and could be attributed to metal sintering, a decrease of surface area, the loss of basicity, and structural changes of the mixed oxides. Meanwhile, the grown Ni particle size and decreased catalyst surface area resulted in the deactivation of the catalyst made by impregnation. Abelló, Berrueco, and Montané (2013) prepared the high loading Ni-Al catalysts and partially reduced the catalysts at relatively low temperatures, thus attaining the small Ni nanoparticles. The catalyst preserved high activity and selectivity even after a 500h long-term test. Over $Ni/CaO\text{-}Al_2O_3$ catalysts, Mutz, Carvalho, Mangold, Kleist, and Grunwaldt (2015) demonstrated that there was fast bulk oxidation of Ni metallic nanoparticles during the reaction when the H_2 was removed from reaction gas, resulting in the fast decrease of catalytic activity in the subsequent reaction due to the presence of oxidized Ni. The operando Raman spectroscopy and high-resolution transmission electron microscopy (HRTEM) revealed that the graphitic carbon species could be observed only in pure CH_4 atmosphere under the pressure of 1bar and the formed carbonaceous species would be removed during the oxidizing atmosphere (Mutz et al., 2018).

Considering these limitations so far, it is then essential to develop highly efficient and stable Ni catalysts toward CO_2 methanation. Meanwhile, recently, other technologies such as assisted catalysis by plasma or photocatalysis were also developed to achieve the high activity, selectivity of CH_4, and stability toward CO_2 methanation. Mostly the same active catalytic phases were reported, only the way of activated of the molecules was changed. To activate the molecules, instead of thermal treatment, the use of solar energy, electrons, or a plasma discharge was investigated (Suib, 2013). Until now, most of the researches still focuses on promoting the heterogeneous catalysts to obtain the desired activity, CH_4 selectivity, and stability of the catalysts in CO_2 methanation.

3.2 Thermodynamic aspects of CO_2 methanation and CO_2 hydrogenation

The effect of temperature in CO_2 methanation simulation under 1 bar

The effects of temperature toward CO_2 conversion and CH_4 selectivity in CO_2 methanation were simulated by the HSC Chemistry 5.0 software using the Gibbs free energy minimization method. The CO_2 conversion and CH_4 selectivity are calculated following Eqs. (3.2), (3.3):

$$XCO_2\,(\%) = \frac{FCO_2,in - FCO_2,out}{FCO_2,in} \times 100\% \quad (3.2)$$

$$SCH_4\,(\%) = \frac{FCH_4,out}{FCO_2,in - FCO_2,out} \times 100\% \quad (3.3)$$

TABLE 3.1 The reactions that are probably involved in the CO_2 methanation reaction.

Type	Formula	ΔH_{298K} ($kJ\,mol^{-1}$)	Definition
1	$CO_2+4H_2=CH_4+2H_2O$	−165.0	CO_2 methanation
2	$CO_2+H_2=CO+H_2O$	41.2	Reverse water-gas shift
3	$CO_2+2H_2=C+2H_2O$	−90.1	CO_2 reduction
4	$CO+3H_2=CH_4+H_2O$	−206.2	CO methanation
5	$2CO+2H_2=CH_4+CO_2$	−247.3	Reverse CH_4 dry reforming
6	$2CO=C+CO_2$	−172.4	Boudouard reaction
7	$CO+H_2=C+H_2O$	−131.3	CO reduction
8	$CH_4=2H_2+C$	74.8	CH_4 cracking

The X_{CO2} and S_{CH4} represent the CO_2 conversion and CH_4 selectivity. The $F_{CO2,in}$, $F_{CO2,out}$, and $F_{CH4,out}$ represent the molar flow of CO_2 or CH_4 in which the "in" and "out" mean inlet and outlet, respectively.

The possible reactions in CO_2 methanation reactions were displayed in Table 3.1. We can List eight reactions that probably happened during the CO_2 methanation process. Two of them are endothermic reactions and six reactions are exothermic.

The results of the effect of temperature at 1 bar in methanation were displayed in Fig. 3.2. The molar ratio of H_2/CO_2 was controlled at 4 with a gas hourly space velocity (GHSV) of 12,000. As shown in Fig. 3.2, the CO_2 conversion decreases with the increase of temperature until 839 K and then increases gradually at atmospheric pressure. As displayed in Eq. (3.4), the methanation reaction is an exothermic reaction, which is favored under low temperatures. In Fig. 3.3, the CH_4 selectivity decreases gradually with the increase of temperature up to 773 K and then decreases sharply until nearly zero. Meanwhile,

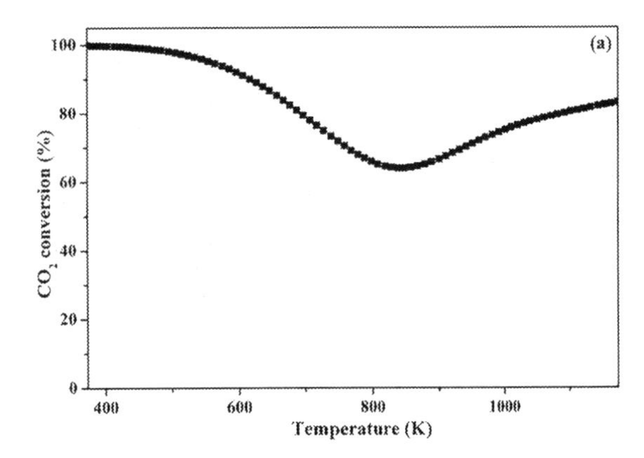

FIG. 3.2 The effect of temperature on CO_2 conversion (A), Condition: 1 bar, $H_2/CO_2/Ar$ (molar ratio) = 60/15/25, GHSV = 12,000 h^{-1}.

the selectivity of CO increases to nearly 100% as the temperature increases up to 1173 K. The presence of CO, which is the main by-product in methanation reaction, was caused by the reversed water-gas-shift reaction (Table 3.1, $CO_2+H_2=H_2O+CO$) (Gao et al., 2012).

The molar ratio of the species were shown in Fig. 3.4. It can be seen from Fig. 3.4 that the molar fraction of CH_4 decreases to nearly zero as the temperature increases up to 973 K and then keep stable. A similar evolution trend is also found over the molar fraction of H_2O, with a molar

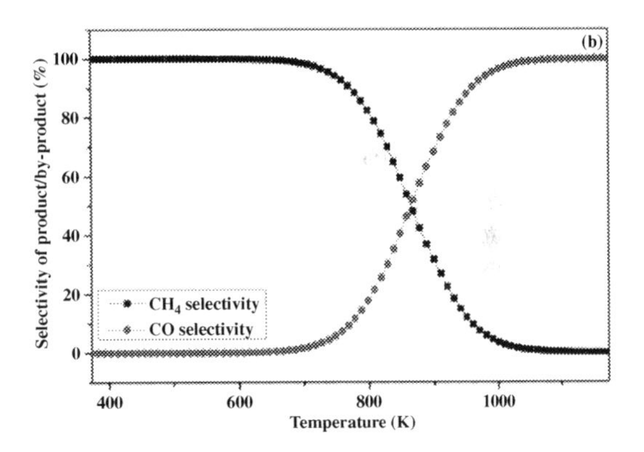

FIG. 3.3 The effect of temperature on CH_4/CO selectivity (B), Condition: 1 bar, $H_2/CO_2/Ar$ (molar ratio) = 60/15/25, GHSV = 12,000 h^{-1}.

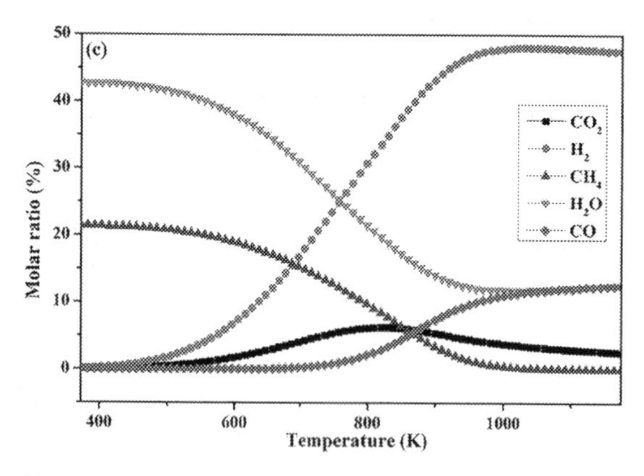

FIG. 3.4 The effect of temperature on the molar ratio of the species (C) in methanation; Condition: 1 bar, $H_2/CO_2/Ar$ (molar ratio) = 60/15/25, GHSV = 12,000 h^{-1}.

ratio of 12% found after 973 K. The change in the concentration of CH_4 and H_2O with the increase of temperature also proves that methanation reaction is favored in low temperature. The molar fraction of CO_2 increases gradually up to 823 K and then decreases with the increase of temperature. The concentration trends of H_2 and CO experience a slow, sharp, and stable increasing trend individually during the temperature range. In the calculation, nearly no carbon was formed in the chosen conditions. As reported by Park and McFarland (2009), the methanation reaction may experience the CO route, with CO as the main intermediate, in which the CO_2 reacts with H_2 to produce CO by reverse water-gas-shift, and the CH_4 is formed by CO methanation. The thermodynamic results reveal that the reaction temperature should not be very high to reach a high CO_2 conversion at ambient pressure. However, the kinetic barriers determine that the CO_2 activation is too difficult at low temperatures, which means the presence of an appropriate catalyst is indispensable for the reaction to take place (Gao et al., 2012; Thampi, Kiwi, & Grätzel, 1987).

The effect of the pressure in CO_2 methanation

The thermodynamic effect of pressure on CO_2 conversion and CH_4 selectivity in CO_2 methanation were calculated and displayed in Figs. 3.5 and 3.6. As shown in Fig. 3.5, the CO_2 conversion at each pressure presents a gradually decreasing trend below 873 K. The increase of pressure causes the increase of CO_2 conversion because the CO_2 methanation reaction is a volume-reducing reaction, with the gas volume changing from 5 to 3 units. Thus, the increase of pressure can promote the reaction thermodynamically.

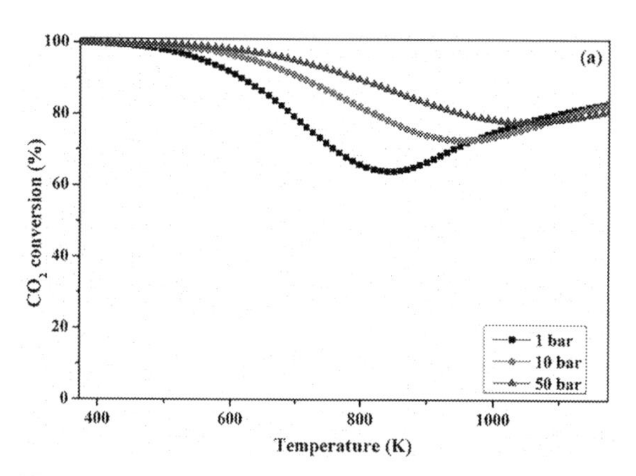

FIG. 3.5 The effect of pressure on CO_2 conversion in methanation; Condition: $H_2/CO_2/Ar$ (molar ratio) = 60/15/25, GHSV = 12,000 h^{-1}.

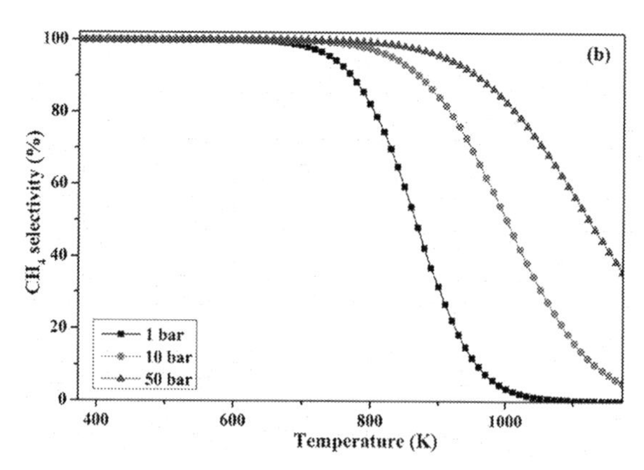

FIG. 3.6 The effect of pressure on CH_4 selectivity in methanation; Condition: $H_2/CO_2/Ar$ (molar ratio) = 60/15/25, GHSV = 12,000 h^{-1}.

However, when the reaction temperature exceeds 873K at 1bar, the CO_2 starts to be converted through the reverse water-gas-shift reaction (Table 3.1, $CO_2+H_2=CO+H_2O$) dominates the reaction (Gao et al., 2012). The effect on CH_4 selectivity is shown in Fig. 3.6. It is found that the CH_4 selectivity decreases slowly when the temperature increases until 773K because of the exothermicity of methanation. Beyond 773K, the sharp decrease of CH_4 selectivity is registered at each pressure, which is attributed to the effect of reverse water-gas-shift reaction. However, the increase of pressure can significantly promote the CH_4 selectivity, which is attributed to the volume-reducing property of methanation. Meanwhile, no carbon was registered in the studied pressure range. This can be explained by the effect of H_2O formed in the reaction, which can suppress the coke formation (Gao et al., 2012).

The calculation results in this study are well consistent with the calculation results carried out elsewhere (Beuls et al., 2012; Gao et al., 2012; Ocampo, Louis, & Roger, 2009).

The effect of H_2/CO_2 ratio in CO_2 methanation simulation

The effect of the H_2/CO_2 ratio on methanation reaction was calculated and shown in Fig. 3.7 and 3.8. It can be seen from Fig. 3.7 that the

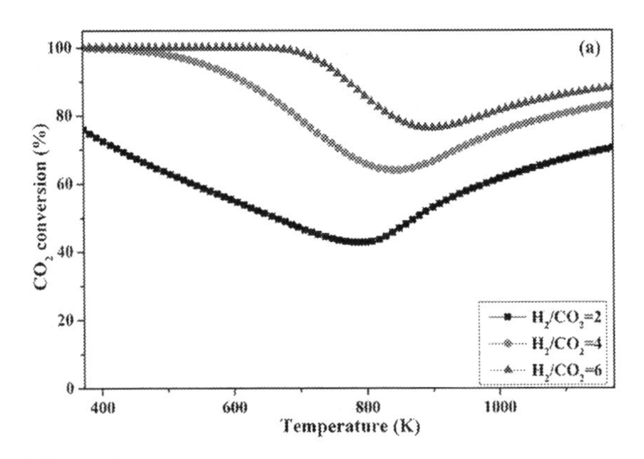

FIG. 3.7 The effect of H_2/CO_2 ratio on CO_2 conversion in methanation; Condition: 1bar, GHSV=12,000h^{-1}.

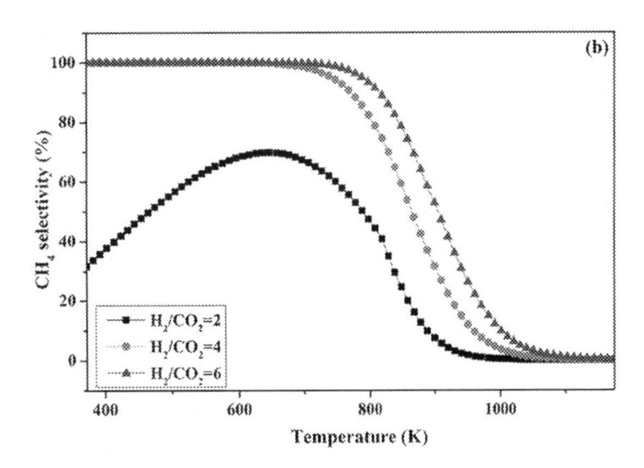

FIG. 3.8 The effect of H_2/CO_2 ratio on CH4 selectivity (B) in methanation; Condition: 1bar, GHSV=12,000h^{-1}.

H_2/CO_2 ratio notably affects the CO_2 conversion. The increase in H_2/CO_2 ratio leads to a significant increase in CO_2 conversion. The same phenomenon is also found for CH_4 selectivity (Fig. 3.8). In the case of the H_2/CO_2 ratio being 2, the CO_2 conversion and CH_4 selectivities are just 75% and 30%, separately. Meanwhile, the CO_2 conversion and CH_4 selectivity are nearly 100% when H_2/CO_2 ratio reaches 4. A lot of carbon was formed below 773K when the H_2/CO_2 ratio is 2. Almost no carbon deposition is registered when the H_2/CO_2 ratio exceeds 2. The CO_2 reduction reaction (Table 3.1, $CO_2+2H_2=C+2H_2O$) has probably happened when the H_2/CO_2 ratio is 2, which results in the formation of abundant carbon.

Thermodynamic aspect of CO_2 hydrogenation

Apart from the CO_2 methanation reaction, the other CO_2 hydrogenation reactions, e.g., CO_2 hydrogenation to methanol, dimethyl ether, and formic acid were also widely studied. In this chapter, the CO_2 hydrogenation to methanol will be discussed thoroughly in comparison to the CO_2 methanation reaction.

The reactions involved in CO_2 hydrogenation to methanol were listed in Table 3.2. The

TABLE 3.2 The reactions that are probably involved in CO_2 hydrogenation to methanol.

Type	Formula	ΔH_{298K} ($kJ\,mol^{-1}$)	Definition
1	$CO_2 + 3H_2 = CH_3OH + H_2O$	−49.4	CO_2 methanolization
2	$CO + 2H_2 = CH_3OH$	−90.4	CO methanolization
3	$CO_2 + H_2 = CO + H_2O$	41.0	Reverse water-gas shift

thermodynamic CO_2 and CO conversion as a function of temperature and yields of CH_3OH/CO as a function of temperature or H_2/CO_2 ratio were cited and displayed in Fig. 3.9 (Jiang, Nie, Guo, Song, & Chen, 2020). It can be seen from Table 3.2 that both the CO_2 methanolization and CO methanolization are exothermic reactions, which are favored at low temperatures. The reverse water-gas shift reaction, which is endothermic, also exists in the CO_2 hydrogenation to methanol process. As shown in Fig. 3.9A, the CO_2 conversion equilibrium in 2 MP decreases with the increase of temperature until 513K, which is caused by the exothermic property of the reaction. But a slight increase trend after 513K can be found because of the occurrence of reverse water-gas shift reaction (Table 3.2) at high temperatures. As the increase of pressure, the CO_2 conversion also increases because of the volume reduction of CO_2 methanolization. The CO methanolization shows much higher CO_2 conversion as compared with that of CO_2 methanolization at a low-temperature range. The product and by-product yields are shown in Fig. 3.9B, illustrating that the methanol yield decreases with the increase of temperature. Furthermore, the methanol yield increases with the increase of pressure. The CO yield increases as the temperature increases and exceeds the value of methanol yield after 535K, in which the reverse water-gas shift reaction dominates the reaction. In Fig. 3.9C, the yield of methanol increases significantly with the increase of the H_2/CO_2 ratio at each pressure. However, only a slight increase of CO yield as the increase of H_2/CO_2 ratio is registered.

From the thermodynamic calculation results of CO_2 methanolization, one can conclude that the reaction is favored at low temperature, high pressure, and high H_2/CO_2 ratio. However, the actual CO_2 methanolization reaction also needs a temperature of more than 323K generally and the presence of a good catalyst (Liu et al., 2020). Even in harsh conditions, the CO_2 conversion and CH_3OH selectivity were still very low, which

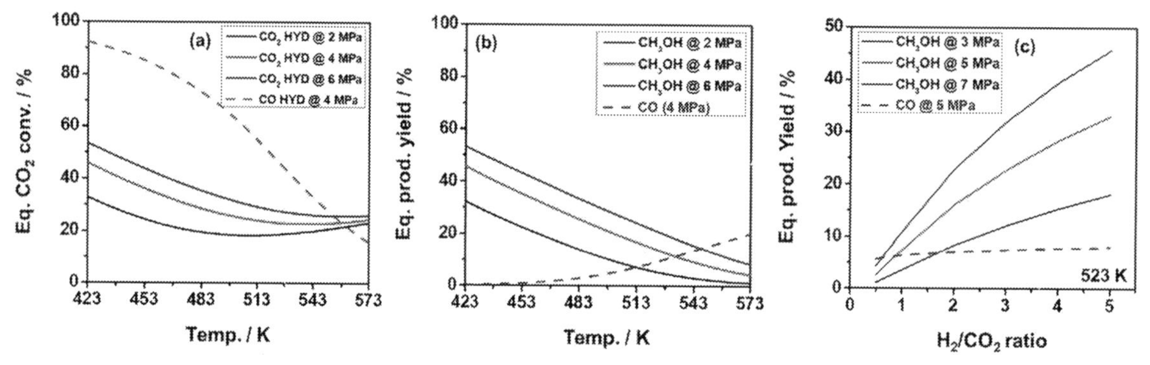

FIG. 3.9 The equilibrium of CO_2/CO conversion in CO_2/CO methanolization under different pressure (A), yields of CH_3OH and CO as functions of temperature (B) and H_2/CO_2 ratio (C). *From Jiang, X., Nie, X., Guo, X., Song, C., & Chen, J. G. (2020). Recent advances in carbon dioxide hydrogenation to methanol via heterogeneous catalysis.* Chemical Reviews, *120(15), 7984–8034. https://doi.org/10.1021/acs.chemrev.9b00723.*

makes the commercialization of such a reaction difficult to achieve (Liu, Xu, et al., 2020). Compared to CO_2 methanolization, CO_2 methanation has a better potential for large-scale applications.

3.3 Transition metal-based catalysts in CO_2 methanation and CO_2 hydrogenation

Ni-based catalysts for CO_2 methanation

Introduction: High activity, barrier, e.g., deactivation, and stability of Ni catalysts

The Ni-based catalysts were widely investigated in CO_2 methanation regarding their superiorities like good activity and CH_4 selectivity, low cost, high availability, and high accessibility making them appropriate in potential commercial applications (Aziz, Jalil, Triwahyono, & Saad, 2015; Sun, Beaunier, & Da Costa, 2020). A variety of Ni catalysts with good activity and selectivity, e.g., Ni/CeO_2 (Sun, Beaunier, La Parola, Liotta, & Da Costa, 2020), Ni/ZrO_2 (Zhao, Li, Zhu, Wang, & Wang, 2016), Ni/CeO_2-ZrO_2 (Ocampo et al., 2009), Ni/SBA-15 (Wang, Zhu, Zhuo, Zhu, & Wang, 2019), Ni/Al_2O_3 (Li, Hu, & Hill, 2006), and Ni/SBA-16 (Sun, Beaunier, & Da Costa, 2020) have been developed for CO_2 methanation. Tada et al. (Tada, Shimizu, Kameyama, Haneda, & Kikuchi, 2012) synthesized the Ni/CeO_2 catalyst by wet impregnation method and compared its activity and selectivity with Ni/α-Al_2O_3, Ni/TiO_2, and Ni/MgO. It was found that the Ni/CeO_2 catalyst showed the highest CO_2 conversion with CH_4 selectivity reaching nearly 1. Zhao et al. (Zhao, Wang, & Li, 2016) prepared the Ni/ZrO_2 catalysts via combustion method using different combustion mediums and tested their activity in CO_2 methanation. It was shown that the combustion medium could significantly affect the physiochemical properties regarding Ni metal size, Ni dispersion, the crystal structure of the support, and Ni-support interaction of the catalyst, which significantly affected the activity and stability of the Ni/ZrO_2 catalyst. The CO_2 conversion of Ni/ZrO_2 synthesized by the urea medium reached 60% at 573 K under ambient pressure.

Although the superior performance of Ni-based catalysts, the problems of deactivation and poor stability still exist (Liu, Hong, & Liu, 2020). Thus, it is still a challenge to develop a highly stable Ni catalyst for CO_2 methanation (Ye et al., 2019). At a low-temperature range lower than 573 K, the deactivation of Ni catalysts in CO_2 methanation was mostly caused by the sintering of Ni metal during the reaction process, which was caused by the formation of mobile nickel carbonyl originated from the interaction between Ni and CO (Agnelli, Kolb, Nicot, & Mirodatos, 1991; Agnelli, Kolb, & Mirodatos, 1994; Falbo et al., 2018; Wang, Wang, Ma, & Gong, 2011). At a high-temperature range between 623 and 723 K, the coke may form owing to the Boudouard reaction and methane cracking. Apart from sintering of Ni metal and coke, the formation of Ni-hydroxide at low temperature was also found, which caused the decrease of the active phase of Ni (Mebrahtu et al., 2019). Possible solutions toward tackling the problem were: (i) in adding promoter to improve the dispersion of Ni over the support or promote the formation of solid solution, (ii) in doping a metal to form an alloy, (iii) in modifying the support to obtain high surface area and pore volume, (iv) in making the catalyst with special structure, (v) in enhancing the metal-support interaction, and (vi) in changing the preparation method (Ye et al., 2019).

In the literature, it was also reported the addition of a second metal (Fe, Co, Cu) on Ni/ZrO_2 catalysts for CO_2 methanation (Ren et al., 2015). Thus, the Fe-doped Ni/ZrO_2 catalyst significantly enhanced the activity of Ni/ZrO_2 catalyst at a low-temperature range when compared to the Co and Cu metals. The promotion of Fe could improve the dispersion and reducibility of Ni species and even promote the reduction of ZrO_2, which enhanced the adsorption and dissociation of CO_2 and H_2. Thus, the activity of Ni/ZrO_2 was improved. Moreover, the cerium had been reported to have positive effects on the mesoporous materials supported by Ni catalysts in CO_2 methanation (Bian, Zhang, Zhu, & Li, 2018).

Ewald et al. (2019) studied the deactivation behavior of Ni-Al catalysts at 8bar in CO_2 methanation. In the latter study, it was found that the Ni/γ-Al_2O_3 catalyst synthesized by the impregnation method exhibited lower stability compared with the Ni-Al catalysts prepared by the co-precipitation method. The stability of the Ni-Al mixed oxides increased with the decrease of Ni content, but the activity of the mixed oxides showed a reversed trend. The deactivation mechanisms found over the Ni-Al catalysts were caused by the Ni metal sintering, a loss of specific surface area, a decrease of CO_2 adsorption capacity, decreasing medium basic sites, and structure change of the mixed oxide phase (Ewald et al., 2019). Moreover, the main deactivation reasons of impregnated catalysts were attributed to increased Ni particle size and decreased surface area. Also, the structure and metal-support interaction could play predominant effects over some structure-sensitive catalysts in methanation. The special structure like hydrotalcite, phyllosilicate, and perovskite was reported to have good activity and stability toward CO_2 methanation (Hongmanorom et al., 2021; Lim et al., 2021; Sun, Świrk, Wierzbicki, et al., 2020; Wierzbicki et al., 2017; Wierzbicki et al., 2018; Wierzbicki, Motak, Grzybek, Gálvez, & Da Costa, 2018). By tuning the metal-support interaction, a stronger metal-support interaction can promote the Ni dispersion and thus promote the activity and stability of Ni catalysts in CO_2 methanation (Bukhari et al., 2019).

Ye et al. (2019) synthesized a series of Ni/SiO_2 catalysts by different methods, i.e., wet impregnation and ammonia evaporation method. The Ni/SiO_2 (AEM) catalyst prepared by the ammonia evaporation method exhibited Ni phyllosilicate with the lamellar structure, which enhanced the Ni-support interaction of the catalyst. It also had a high BET surface area, good dispersion of Ni, and small Ni particle size. The stability test for CO_2 methanation showed a higher yield of methane and better long-term stability than the impregnated Ni/SiO_2 catalyst. Liu and Tian (2017) prepared the Ni/SBA-15-op catalyst by one-pot hydrothermal method and compared its activity with the Ni/SBA-15-im synthesized by conventional wet impregnation method. It was revealed that the Ni/SBA-15-op catalyst showed higher surface area, larger pore volume, and higher Ni dispersion compared to Ni/SBA-15-im. Thus, both better activity and stability were obtained over Ni/SBA-15-op catalyst.

Effect of the Ni loading and Ni dispersion

The performance of the Ni-based catalysts in CO_2 methanation depends both on Ni metal size and Ni dispersion (Aziz, Jalil, Triwahyono, & Saad, 2015; Chen et al., 2017). The nickel particle size can be affected by the Ni loading, the preparation method, and the support properties. Thus, the Ni particle size of the Ni/SiO_2 catalyst prepared by the deposition-precipitation method was affected by the parameters of the preparation method (Burattin, Che, & Louis, 1999). The average Ni particle size of the catalyst derived from the Ni phyllosilicate phase was smaller than that of the Ni-hydroxide-derived catalyst. The Ni loading also has a significant effect on many Ni-based catalysts. Generally, the Ni particle size of the supported catalysts is smaller at low Ni loading and tends to increase for high Ni loadings (Aziz, Jalil, Triwahyono, & Saad, 2015). For example, Ni nanoparticles supported over mesoporous SBA-16 reported in literature showed a decreasing trend toward Ni particle size as the increase of Ni loading (Chen, Britun, et al., 2017). In the same study, the catalyst with Ni loading of 5.9wt% showed the smallest Ni particle size and exhibited the highest turnover frequency (TOF) in the CO_2 methanation reaction.

Herein, we have listed the effect of Ni loading on the Ni particle sizes of some supported catalysts and illustrated the influence of Ni particle size on the activity of the catalyst in methanation. As shown in Figs. 3.10 and 3.11, the Ni particle size of Ni/MSN catalyst (mesostructured silica nanoparticle) increases with the increase of Ni loading, with the biggest particle obtained on the sample of 10 wt% Ni loading (Aziz, Jalil, Triwahyono, & Ahmad, 2015; Sun, Beaunier, & Da Costa, 2020). The TOF and CO_2 conversion showed sharply increased trends with the increase of Ni loading until 5 wt% and then only increased slightly.

A similar trend was also found on Ni/γ-Al_2O_3 catalyst (Rahmani, Rezaei, & Meshkani, 2014). The increase of Ni loading resulted in the increase of Ni crystalline size from 10 to 25 wt%, with the highest CO_2 conversion obtained over 20Ni/γ-Al_2O_3 at 573 K. Generally, the increase of Ni loading can promote the activity of the catalyst due to more active metals for the reactant, but a high Ni loading may result in the aggregation of Ni species and decrease of Ni dispersion, which leads to the decrease of activity. Nevertheless, the effect of Ni content on the Ni crystalline size depends on the type of support. The increase of Ni content in hydrotalcite-derived catalysts can lead to the decrease of Ni crystalline size for the reduced and spent catalysts (Wierzbicki et al., 2017). In the CO_2 methanation reaction, the activity of the catalyst increased as an increase of Ni content until 42.5 wt% at the temperature range of 523–923 K.

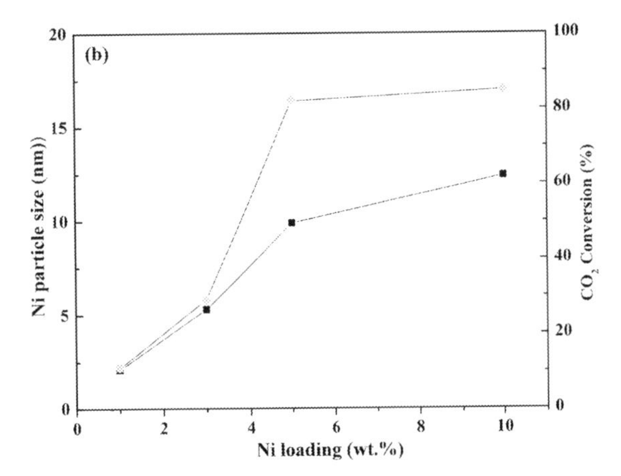

FIG. 3.11 The effect of Ni loading on the particle size and activity of Ni/MSN catalysts (B). Condition: GHSV = 50,000 mL $gcat^{-1}$ h^{-1}; H_2/CO_2 = 4/1; reaction temperature: 623 K.

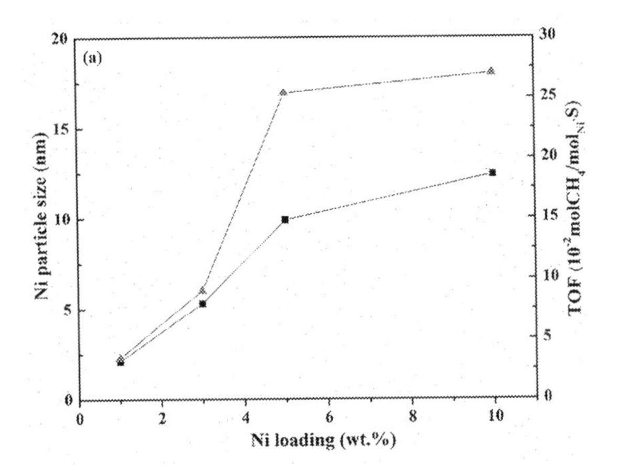

FIG. 3.10 The effect of Ni loading on the particle size and activity of Ni/MSN catalysts (A). Condition: GHSV = 50,000 mL $gcat^{-1}$ h^{-1}; H_2/CO_2 = 4/1; reaction temperature: 623 K.

The preparation method also plays an important role in the dispersion of Ni catalysts. The Ni/Al_2O_3 catalysts prepared by the co-precipitation method showed a higher Ni dispersion compared to the one prepared by the impregnation method due to the formation of surface $NiAl_2O_4$ spinel phase (Beaumont et al., 2014; Li et al., 2019).

Effect of the support on the performance of Ni-based catalysts

The support itself has a significant effect on the performance and stability of catalysts. The dispersion of Ni, the basicity, the metal-support interaction, and the morphology of the catalyst can be affected by the used support (Aziz, Jalil, Triwahyono, & Ahmad, 2015; Pan, Peng,

Sun, Wang, & Wang, 2014; Yan, Dai, Yang, & Lapkin, 2018). The support with high surface area and pore volume can confine the Ni particles inside the porous structure, and thus improve the dispersion of Ni (Sun, Beaunier, & Da Costa, 2020). The mesoporous support like SBA-15, SBA-16, or KIT-6 with high surface area and pore volume had been reported to promote the dispersion of Ni and lead to small Ni particles (Kim et al., 2010; Park & McFarland, 2009).

The basicity of supported catalyst plays an important role because the support can also participate in the reaction due to its CO_2 adsorption/desorption capacity. It has been reported that the $Ni/Ce_{0.5}Zr_{0.5}O_2$ catalyst exhibited more medium basic sites compared to that of Ni/γ-Al_2O_3 catalyst. This medium basicity promoted the formation of monodentate carbonate, thus resulting in the formation of monodentate formate, which was hydrogenated to methane at a higher rate (Pan et al., 2014). Also, Ni/γ-Al_2O_3 catalysts showed lower performance than the $Ni/Ce_{0.5}Zr_{0.5}O_2$ catalysts due to the lack of medium basic sites. CeO_2 supported nickel catalysts were intensively studied in CO_2 methanation because of their superior performance and selectivity toward CH_4, which could be attributed to the unique redox properties of cerium oxide (Tada et al., 2012). The cerium oxide can be reduced partially to form defective sites on the support, which resulted in the formation of oxygen vacancies, which will promote CO_2 adsorption and dissociation (Zhou et al., 2017).

The metal-support interaction is also controlling Ni particle size and oxygen availability of Ni-based catalysts. The small particle size could be favored under strong metal-support interaction (Singha et al., 2016; Singha, Shukla, Yadav, Sivakumar Konathala, & Bal, 2017). Lin et al. (2021) investigated the effect of activation treatment on the metal-support interactions (MSI) of Ni/CeO_2 catalysts. It was found that the Ni/CeO_2 catalyst pre-treated under H_2 atmosphere showed better activity and high selectivity compared to catalysts pre-treated under N_2 or air atmosphere, which was attributed to the appropriate MSI generated in the pre-treatment, leading to the high dispersion of Ni, high amount of oxygen vacancies, and abundant weak basic sites on the catalyst (Lin et al., 2021). Regarding regulating the MSI of the catalyst, the preparation method also plays a key role. Yan et al. reported the Ni/Y_2O_3 catalyst prepared by impregnating Ni onto different Y precursors like Y_2O_3, $Y_4O(OH)_9(NO_3)$, and $YO(NO_3)$ for CO_2 methanation. The MSI of the catalyst was changed because of different Y precursors, in which the $YO(NO_3)$ material supported Ni catalyst showed the moderate metal-support interaction between Ni and Y_2O_3, leading to superior activity for CO_2 methanation and good stability in CO-containing reactants (Yan et al., 2018). In conclusion, a moderate metal-support interaction can thus improve the dispersion of Ni, and increase basicity and oxygen vacancies of Ni catalysts, leading then to a better activity and stability in CO_2 methanation.

The morphology of Ni-based catalysts also contributes to the activity and stability of catalysts in CO_2 methanation. Bian et al. (2020) prepared the Ni/CeO_2 catalyst with different CeO_2 morphologies and carried it out in low-temperature CO_2 methanation. The Ni/CeO_2 (NR) catalyst with nanorod structure exhibited superior performance than Ni/CeO_2 (NC) catalyst with nanocube structure. This high activity was attributed to the higher Ce^{3+} ratio on the surface of the catalyst. Zhou et al. (2017) synthesized the CeO_2 materials by hard-template, soft-template, and precipitation methods and used them to prepare Ni/CeO_2 catalysts. It was shown that the hard-template method prepared sample (NCT) showed a well-developed mesoporous structure with the highest SSA. Meanwhile, the soft-template prepared sample (NCS) displayed a disordered structure with a smaller pore diameter. The precipitation method prepared sample (NCP) only showed the nanoparticles with disordered structures. The NCT catalyst performed the highest CO_2 conversion among the studied catalysts in methanation reactions. NCP catalyst showed then lower activity, which could be attributed to the low surface area and low porosity.

The most intensively used support such as CeO_2 (Bian et al., 2020; Moon et al., 2019; Tada et al., 2012), ZrO_2 (Le, Van, Nguyen, & Nguyen, 2017), SiO_2 (Le, Van, et al., 2017), Al_2O_3 (Muroyama et al., 2016), MgO (Loder, Siebenhofer, & Lux, 2020; Tada et al., 2012), Y_2O_3 (Muroyama et al., 2016) (Yan et al., 2018), $Ce_{0.5}Zr_{0.5}O_2$ (Wang, Pan, Peng, & Wang, 2014), TiO_2 (Beaumont et al., 2014; Li, Liu, et al., 2019), carbon nanotube (Romero-Sáez et al., 2018), zeolite (Bacariza, Amjad, Teixeira, Lopes, & Henriques, 2020), clays (Hongmanorom et al., 2021; Lu, Ju, Abe, & Kawamoto, 2015; Lv, Xin, Meng, Tao, & Bian, 2018; Park et al., 2004; Zhang, Muratsugu, Ishiguro, & Tada, 2013), SBA-15 (Liu & Tian, 2017), SBA-16 (Budi, Wu, Chen, Saikia, & Kao, 2016; Sun, Beaunier, & Da Costa, 2020), and hydrotalcite (Hongmanorom et al., 2021; Lim et al., 2021; Marocco et al., 2018; Sun, Świrk, Wierzbicki, et al., 2020; Wierzbicki et al., 2017; Wierzbicki, Baran, et al., 2018) were widely reported in the literature. Other transition metal oxide supports such as Sm_2O_3 (Muroyama et al., 2016), ZrO_2 (Zhao, Li, et al., 2016), CeO_2 (Zhou et al., 2017), $Ce_xZr_{1-x}O_2$ (Luciano et al., 2019), Y_2O_3 (Italiano et al., 2020), have shown superior activity and stability when compared to Al_2O_3-based catalyst. However, these supports contain VIII group metals which are not abundant and have a high cost.

Muroyama et al. (2016) synthesized the Ni catalysts supported on various supports by conventional incipient wet impregnation method and investigated their performance in CO_2 methanation. The CO_2 conversion comparison at 548 K under atmospheric pressure was displayed in Fig. 3.12. The Y_2O_3 supported Ni catalyst showed the highest CO_2 conversion compared to other metal oxide-supported Ni catalysts, which could be partly attributed to the effects of basicity and surface intermediates during the reaction that presented on the materials (Muroyama et al., 2016). By FTIR technology, the Ni/Al_2O_3 catalyst proceeded with a CO methanation process. Only carbonates were found on Ni/La_2O_3 catalyst. Meanwhile, the formate species from the carbonates were detected on Ni/Y_2O_3 catalyst.

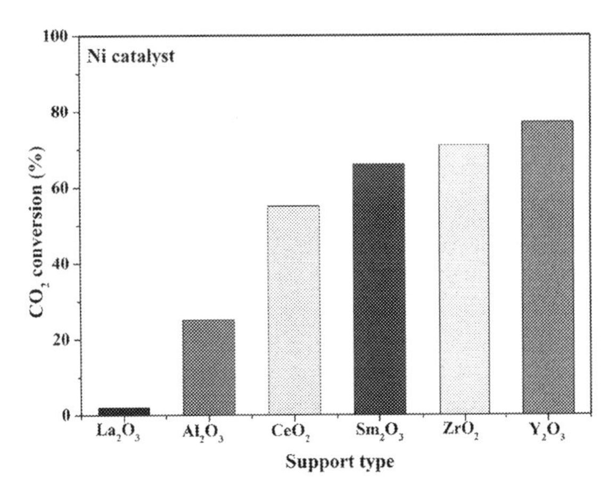

FIG. 3.12 The comparison of CO_2 conversion for Ni catalysts supported on different metal oxides; Condition: 548 K, atmospheric pressure, WHSV = 20,000 kg^{-1} h^{-1}.

The CeO_2 supported Ni catalysts reported in other works of literature exhibit superior activity and stability due to their effects of morphology and oxygen mobility property (Bian et al., 2020). Furthermore, the ZrO_2 and $Ce_{0.5}Zr_{0.5}O_2$ also showed good activity because they also had oxygen storage capability. Jia, Zhang, Rui, Hu, and Liu (2019) prepared the Ni/ZrO_2 catalyst by plasma decomposition of Ni precursor and compared its activity to that of Ni Ni/ZrO_2 catalyst obtained by a conventional thermal decomposition. It was revealed that the plasma decomposition derived catalyst higher Ni dispersion, high concentration of Ni(111) plane, and stronger coordination between Ni and interfacial sites, leading to a fast H_2 dissociation and more oxygen vacancies, thus a higher activity was obtained compared to the thermally processed catalyst. Also, the mechanism in methanation was different regarding different catalysts; the plasma processed catalyst experienced a carbon oxide hydrogenation route while the thermally processed catalyst had a formate hydrogenation pathway. Regarding $Ni/Ce_xZr_yO_2$ catalyst, it also proceeds through the formate species originating from carbonate hydrogenation with the CO species as a by-product (Wang et al., 2014).

Ordered mesoporous silica such as SBA-15/16 can also be potential thanks to their high surface area and confinement of porous structure, making it possible to obtain small nanoparticles of Ni. The Ni nanoparticles over Ni/SBA-16 catalyst can reach 4.3nm with a high surface area (Sun, Beaunier, & Da Costa, 2020). With the incorporation of Ce, good activity and stability can be obtained over the Ni/Ce/SBA-16 catalyst.

The hydrotalcite-derived mixed oxides catalysts also attracted attention because they cannot only generate small particles of active metals but also maintain high stability in the high-temperature range. For instance, the low loading of yttrium doped hydrotalcite-derived Ni catalyst reached a CO_2 conversion of 81% at 523K under atmosphere pressure (Sun, Świrk, Wierzbicki, et al., 2020). Its higher performance was attributed to the strong metal-support interaction, the high ratio of medium basic sites, and the small Ni particle size.

Other novel materials were also applied as Ni-based catalysts for methanation. Thus, metal–organic framework (MOF) materials were used as support to prepare catalysts due to their high surface area of more than $1000\,m^2\,g^{-1}$. Lin et al. (2019) synthesized the Ni@C (designated name by author) catalyst from MOF-derived hierarchical hollow spheres and used it in CO_2 methanation. The as-prepared catalyst possessed a high surface area and rich isolated active sites. The methanation test showed that the catalyst achieved a CO_2 conversion of 100% and CH_4 selectivity of 99.9% at 598K under atmospheric pressure and it also exhibited good stability at 523K.

In conclusion, the activity, selectivity, and stability of the Ni catalysts deeply depended on the properties of the support. Thus, the basicity, porosity, and redox property of the support can substantially influence the methanation reaction. For example, over hydrotalcite-derived Ni catalysts, a linear correlation was found between the basicity of the catalysts and CO_2 activity in methanation (Wierzbicki et al., 2017).

On the effect of the promoter on the performance of Ni-based catalysts

It is well reported that Ni catalysts without additives suffered from low CH_4 selectivity, poor stability, and deactivation problems due to the metal sintering at high temperatures (Beuls et al., 2012; Gao et al., 2012; Ocampo et al., 2009; Ocampo, Louis, Kiwi-Minsker, & Roger, 2011). Thus, promoter elements such as Mg (Hongmanorom et al., 2021), La (Wang, Zhu, et al., 2019; Wierzbicki et al., 2016; Zhang & Liu, 2020), Ce (Aresta et al., 2014; Kim et al., 2020; Wang, Han, et al., 2018; Westermann et al., 2017), Zr (Iglesias, Quindimil, Mariño, De-La-Torre, & González-Velasco, 2019), Y (Sun, Beaunier, La Parola, et al., 2020; Takano et al., 2016), Co (Ewald et al., 2019; Liu & Tian, 2017), Fe (H. Lu et al., 2016; Ren et al., 2015) and Cu (Ren, Qin, et al., 2015; Varun, Sreedhar, & Singh, 2020), etc. had been applied to improve the performance of Ni catalysts. The Ni/Al_2O_3, as a commonly studied catalyst in CO_2 methanation due to its economic advantage, was doped by CeO_2, La_2O_3, Sm_2O_3, Y_2O_3, and ZrO_2. The catalytic performance of the synthesized catalysts was investigated by a series of technologies (Guilera, Del Valle, Alarcón, Díaz, & Andreu, 2019). It was found that the addition of these promoters had a positive effect on the performance of the Ni/Al_2O_3 catalyst, and the La-promoted catalyst showed the best activity and stability. At 573 K, the CO_2 conversion decreased as the following sequence: NiLa > NiCe > NiSm > NiY > NiZr > Ni/Al_2O_3. the authors attributed these results to the improvement of nickel reducibility, nickel dispersion, and the presence of moderate basic sites. The La additive is a very promising promoter for $Ni/\gamma\text{-}Al_2O_3$ catalysts in CO_2 methanation. Furthermore, the effect of La content on the performance of $Ni/La\text{-}\gamma\text{-}Al_2O_3$ catalysts for CO_2 methanation was also studied, in which the 14 wt% La-doped $Ni/La\text{-}\gamma\text{-}Al_2O_3$ catalyst showed the best performance with nearly 100% of CH_4 selectivity at low temperature (Garbarino et al., 2019). This latter result was assigned to the enhanced basicity of $La\text{-}\gamma\text{-}Al_2O_3$ support and the reducibility of NiO.

The presence of La also showed a good improvement in CO_2 methanation for the Ni/SBA-15 catalyst, which was also a widely investigated catalytic system (Wang, Zhu, et al., 2019). The perovskite structure of $LaNiO_3$ was formed on $Ni\text{-}La_2O_3$/SBA-15(C) catalyst prepared by citrate complex method, leading to the enhanced interaction between La_2O_3 and Ni. Thus, high dispersion of Ni and small Ni particles (less than 5 nm) were obtained. In CO_2 methanation, the $Ni\text{-}La_2O_3$/SBA-15(C) catalyst showed a high CO_2 conversion of 90.7% and CH_4 selectivity of 99.5% at 593 K (Wang, Zhu, et al., 2019). The La-modified mesostructured cellular foam (MCF) supported Ni catalyst with phyllosilicate structure also showed a good performance in CO_2 methanation as reported in a new study (Abelló et al., 2013; Zhang & Liu, 2020). This is attributed to the small Ni particle size, high capacity of H_2 and CO_2 adsorption, and low activation energy.

Besides the aforementioned catalytic systems, La was also used to improve the activity of hydrotalcite-derived Ni catalysts (HTN) in CO_2 methanation (Wierzbicki et al., 2016; Wierzbicki, Motak, et al., 2018). It was found that the 2 wt% of La-doped HTN catalyst showed the higher CO_2 conversion at low temperature, which could be attributed to the increase of basicity caused by La incorporation. The different methods of promoter incorporation also have a significant effect on the performance of the La-doped HTN catalyst. The catalyst prepared by the ion-exchange method showed both the higher CO_2 conversion and CH_4 selectivity at low temperature (573 K) as well as superior stability. This remarkable activity was attributed to the increase of medium basic sites and the presence of smaller Ni crystal size (Wierzbicki, Motak, et al., 2018).

Ceria (cerium oxide) is also known to be an important additive for Ni-based catalysts in various CO_2 utilization reactions (Kim et al., 2010; Park & McFarland, 2009). As for Ni/Al_2O_3 catalyst, it has been reported that the adding of Ce on Ni/Al_2O_3 catalyst can significantly promote the activity and CH_4 selectivity in methanation. The optimum promotion was found with 15 wt% of CeO_2. This catalyst showed the best CO_2 conversion and high CH_4 selectivity at the temperature range from 523 to 673 K. In such catalysts, the presence of ceria promoted the higher dispersion of Ni, the smaller Ni particle size, the higher amount of medium basic sites, the higher oxygen deficiencies, and finally the higher surface ratio of Ni^0 and Ce^{3+}, all which contribute to a better activity and stability of the catalyst. Also, in such catalyst, the Ce^{4+} cation can be reduced to Ce^{3+} state under the pre-treatment of hydrogen, leading to the formation of oxygen vacancies, which can then lead to a promotion of the adsorption and dissociation of CO_2 species (Sun, Beaunier, La Parola, et al., 2020). Due to the unique redox and oxygen deficiency properties, the ceria is widely used as a promoter in the preparation of Ni catalysts.

In the literature, the Ce-doped Ni/SBA-16 catalysts were prepared and their catalytic performance was evaluated in CO_2 methanation (Sun, Beaunier, & Da Costa, 2020). The results indicated that the presence of Ce led to the formation of oxygen vacancies, which could significantly promote the CH_4 selectivity. The mesoporous SBA-16 can disperse the Ni and Ce species well and confine the Ni and Ce particles inside the mesoporous channels. The highest ratio of Ce^{3+}, highest amount of medium basic sites, and good dispersion of Ce have been obtained over 10 wt% Ce-doped Ni/SBA-16 catalyst, leading to the best activity and high CH_4 selectivity.

Zirconia or zirconium oxide also can act as the oxygen storage material and has been intensively studied in many CO_2 utilization reactions (Iglesias et al., 2019; Meng, Zhu, Zhu, Reubroycharoen, & Wang, 2020). Meng et al. (2020) studied the effect of Zr on MCM-41 supported Ni catalyst for CO_2 methanation. It was found that the adding of Zr strengthened interaction between Ni and Zr, leading to the spillover of H species adsorbed on the surface of Ni to the support, which is favorable for the hydrogenation of CO_2-derived intermediates (Meng et al., 2020). Also, the addition of Zr promoted the increase of oxygen species at the surface, and the decrease of strong basicity, which contribute to hindering the carbon deposition and the removal of deposited coke under reaction. However, an excess Zr on the support would be harmful to the activity of the catalyst due to the destruction of the ordered structure of the support.

The alkaline earth-like Mg can also be a good additive alternative for Ni catalysts (Liu, Hong, & Liu, 2020). Hongmanorom et al. (2021) Ni-Mg phyllosilicate mesoporous SBA-15 catalysts with different Mg loading by ammonia evaporation method and applied them in CO_2 methanation reaction. It was found that the presence of Mg of 5 wt% onto Ni/SBA-15 catalyst could improve the increase of medium basic sites, which enhanced the methanation activity at low temperatures. Also, the phyllosilicate structure could enhance the metal-support interaction, which suppressed the metal sintering, leading to good stability of the catalysts. The other alkaline earth metal oxides like Ca, Sr, and Ba were also studied for Ni catalysts (Liu, Xu, et al., 2020). The incorporation of alkaline earth on Ni/CeO_2 catalysts can promote the dispersion of Ni, an increase of moderate basic sites, and more oxygen deficiencies of the catalyst, thus improving the activity of the catalysts. The activity test showed that the Ca doped Ni/CeO_2 catalyst performed the best activity due to the highest Ni dispersion and largest moderate basicity. Besides the promotion of individual additives, the addition of two additives like Ce-Zr is also widely studied for Ni catalyst, which shows a good modification (Le, Kim, Lee, Kim, & Park, 2017; Sun, Beaunier, & Da Costa, 2020; Sun, Beaunier, La Parola, et al., 2020; Sun, Świrk, Wierzbicki, et al., 2020; Ye et al., 2020). Thus, the addition of Ce-Zr solid solution on activated carbon (AC) supported Ni catalyst can result in stronger interaction between Ni and Ce-Zr, higher dispersion of Ni, and high CO_2 adsorption ability, leading to high CO_2 conversion and CH_4 selectivity (Le, Van, et al., 2017). Li et al. (2020) studied the effect of ceria promotion on the performance of Ni-La/ZrO_2 catalyst in CO_2 methanation. The results revealed that the Ni-Ce-La/ZrO_2 catalyst showed better activity and high stability than the Ce or La-modified Ni/ZrO_2 and the non-modified Ni/ZrO_2 catalyst. The adding of Ce on Ni-La/ZrO_2 catalyst could promote the activation of CO_2 and the Ni nanoparticles of Ni-Ce-La/ZrO_2 catalyst had a high resistance to metal sintering.

The additives with more than two metal oxides also had attracted intensive attention. The Ni catalysts supported on γ-Al_2O_3 based on composite oxide were employed in CO_2 methanation (Abate et al., 2016). The γ-Al_2O_3-ZrO_2-TiO_2-CeO_2 (equivalent loading for the additive, 15%) composite oxide-supported Ni catalyst possessed higher Ni dispersion and better reducibility, leading to higher activity compared to that of Ni/γ-Al_2O_3 catalyst.

The other transition metal-based catalysts in CO_2 methanation

Noble metal-based catalysts

The noble metal-based catalysts such as Pt (Panagiotopoulou, 2017), Rh (Karelovic & Ruiz, 2012), Ru (Guo et al., 2018; Sharma et al., 2011; Zağli & Falconer, 1981), and Pd (Kim et al., 2010; Park & McFarland, 2009) were also studied in CO_2 methanation. Panagiotopoulou (2017) studied the catalytic performance of supported noble metal-based catalysts for CO_2 methanation. It was then showed that the CO_2 conversion of TiO_2 supported noble metal catalysts decreased following the order $Rh > Ru > Pt > Pd$, with the Rh/TiO_2 catalyst having the higher activity and selectivity of CH_4. The Pt and Pd catalysts were mainly known to promote the RWGS reaction to produce CO. The Ru crystallite size had a significant effect on the performance of Ru/TiO_2 and Ru/Al_2O_3 catalysts for CO_2 methanation. Also, the CO acted as an important intermediate for Ru catalysts during the formation of methane (Panagiotopoulou, 2017).

For noble metal-based catalysts, the CO_2 methanation is structural sensitive, which means the metal crystallite size has significant effects on the performance of the catalysts. The investigation of $Rh/\gamma\text{-}Al_2O_3$ for CO_2 methanation showed that the larger particle size could promote the activity of the catalyst at a temperature lower than 458 K (Karelovic & Ruiz, 2012). However, the activity of the catalyst could not be related to the particle size of the metal. The DRIFTS experiments showed that the CO was an active intermediate while the formate did not involve in the formation of methane. The doping of the additive can change the pathway of CO_2 methanation for noble metal-based catalysts. It has been reported that the doping of Ce on Ru/Al_2O_3 catalyst can significantly improve the activity and selectivity of the catalysts, 30% Ce modified catalyst possessing the best performance, which could be attributed to the formation of intermediates like formate and carbonate (Tada, Ochieng, Kikuchi, Haneda, & Kameyama, 2014). These species can then react with H_2 faster than CO.

The addition of Mg onto Pd/SiO_2 catalyst could promote the formation of methane in the CO_2 methanation reaction due to the stabilization of CO_2 on MgO (Park & McFarland, 2009). Also, CO acts as an important intermediate in the reaction.

In conclusion, the Ru and Rh-based catalysts are most active for the noble metal catalysts in CO_2 methanation. The Ru and Rh catalysts are even active at a very low temperature near to ambient temperature with a very high CH_4 selectivity (Karelovic & Ruiz, 2013b). The CO acts as a main intermediate in the methanation process (Jacquemin, Beuls, & Ruiz, 2010). Although the excellent properties of Ru/Rh catalysts, their potential commercialization is limited due to their low availability and high cost.

The other non-noble metal-based catalysts

Besides Ni catalyst, the Fe (Baysal & Kureti, 2020) and Co (Jampaiah et al., 2020; Razzaq et al., 2013; Wang et al., 2017; Zhou et al., 2014) catalysts had also been investigated in CO_2 methanation. Baysal and Kureti (2020) studied Fe catalyst promoted by different additives in CO_2 methanation. It was found that the bare Fe catalysts showed very low activity and CH_4 selectivity and the Fe catalysts doped by promoters did not perform a comparable activity and CH_4 selectivity as a comparison to Ni-based catalyst. Among the promoters, the Mg of 2 wt% modified Fe catalyst exhibited the best positive effect. At a harsh condition (8 bar), the highest CH_4 yield reached 32% at 673 K for $2Mg/Fe_2O_3$ sample.

Co-based catalysts can exhibit comparable activity in comparison to Ni ones in CO_2 methanation. Liang et al. (2020) investigated the Ni/Al_2O_3 and Co/Al_2O_3 catalysts in CO_2 methanation. It was found that the Co/Al_2O_3 catalyst showed higher activity than that of Ni/Al_2O_3. This is attributed to the better coordination effect between Co and alumina when compared to Ni on Al_2O_3. This coordination effect can promote the hydrogenation of intermediates like bicarbonate, formate, and carbonate species. The CO_2 can be activated on the surface of Co metal and subsequently mitigate and react with the hydroxyl in the alumina to form the intermediates. Such intermediates can be hydrogenated to methane over cobalt sites. Also, ZrO_2 supported cobalt catalysts were also employed to catalyze the CO_2 methanation reaction. It was reported that the Co/ZrO_2 catalysts prepared by the organic acid-assisted impregnation method could exhibit superior activity compared to the catalyst prepared without organic acid (Li, Liu, et al., 2019). Within these organic acids, carboxylic acids with a longer chain containing more carboxyl and hydroxyl groups showed good improvement for the catalytic performance of Co/ZrO_2 catalysts as compared to the convention method prepared catalyst. Meanwhile, the amino acid could change the charge properties of the surface of ZrO_2 by changing the pH and increasing the Co dispersion, thus exhibiting better performance than the conventional one. In all the organic acids, the Co/ZrO_2 catalysts derived from the citric acid-assisted impregnation exhibited the best activity and the highest TOF value with a CO_2 conversion of 85% and a CH_4 selectivity of 99% as well as TOF of $1116 h^{-1}$ at 673K and 30 bars. This high activity was attributed to the improved Co dispersion and enhanced Co-ZrO_2 interaction, resulting in more oxygen vacancies and high CO_2 adsorption capacity.

The Co-based catalysts prepared by the liquid reduction method can even reach high activity that is comparable to the noble metal-based catalysts at low temperature (<473K) in recent literature (Jinghui et al., 2021).

Apart from being a catalyst itself, cobalt can also play an important role as a promoter in improving the performance of Ni-based catalysts. Liu, Bian, Fan, and Yang (2018) reported that the addition of Co could improve the activity of ordered mesoporous Ni-Co/Al_2O_3 catalyst, which was attributed to the enhanced H_2 sorption capacity. Also, the ordered mesoporous structure contributed to achieving high stability. In conclusion, the Co-based catalysts can exhibit good performance in CO_2 methanation. However, Ni-based ones remained the most promising.

Catalysts for CO_2 hydrogenation to methanol

As reported in the literature, the CO_2 hydrogenation reactions consist of many reactions, e.g., CO_2 methanation (Stangeland et al., 2017), reverse water-gas-shift (Lu & Kawamoto, 2013), CO_2 hydrogenation to formic acid (Álvarez et al., 2017), methanol (Jiang et al., 2020), or dimethyl ether (DME) (Stangeland, Li, & Yu, 2018), and CO_2 hydrogenation to higher hydrocarbons, etc., (Yang et al., 2017). In this subchapter, only the CO_2 hydrogenation to methanol is presented. In Section 2.3, the thermodynamic research was already discussed thoroughly.

In Industry, methanol can be used as fuel directly for transportation or act as a chemical feedstock. CO_2 hydrogenation to methanol technology can not only alleviate the CO_2 emissions but also produce value-added methanol. The heterogeneous catalysis on this reaction received the most intensive research. The catalytic systems mainly consist of transition metal-based catalysts and main group metal-based catalysts. The transition metal-based catalysts mainly focus on Cu-based catalysts and noble metal-based catalysts, e.g., Pt and Pd (Kattel et al., 2017; Matsumura, Shen, Ichihashi, & Okumura, 2001; Men et al., 2019). Meanwhile, the main group of metal-based catalysts is made up of In_2O_3 catalysts and Ga-based catalysts (Tsoukalou et al., 2019; Wang, Tang, et al., 2019; Wang, Zhu, et al., 2019).

The studies for Cu-based catalysts mostly focused on the active sites of catalyst, the effect of catalyst structure, and activation and deactivation mechanisms (Jiang et al., 2020). It was found that the synergistic effect between Cu and ZnO contributed to the formation of methanol over $Cu/ZnO/Al_2O_3$ catalyst in CO_2 hydrogenation to methanol, in which the formate was considered as the reaction intermediate (Kattel et al., 2017). Natesakhawat et al. (2012) studied the active sites of Cu catalysts in methanol synthesis. The XPS characterization revealed that only metallic Cu species existed on the surface of the catalysts. The Cu^0 particle size and ZnO crystallinity were directly linked with the activity of such catalysts. Smaller Cu particles resulted in a higher TOF value in methanol production. The Cu-ZnO synergy was also a key parameter for the methanol formation. Also, the Cu structure was demonstrated to be an important factor in methanol synthesis (Jiang et al., 2020). Thus Cu atomic plane can affect the catalytic activity of Cu catalysts in methanol synthesis in which the Cu(110) orientation plane was found more active in methanol formation rate compared to that on Cu(100) and polycrystalline copper, indicating the structural sensitivity of Cu catalysts (Yoshihara & Campbell, 1996). Over the time on stream (TOS) process, as for methanation, a deactivation can occur for methanolation reaction. Thus, on $Cu/ZnO/Al_2O_3$ catalyst after a long TOS run, the agglomeration of ZnO particles and partial oxidation of metallic Cu were observed and were presented as key parameters for the deactivation of the studied catalysts in methanol synthesis (Liang et al., 2019).

Furthermore, to improve the catalytic performance of catalysts in CO_2 hydrogenation to methanol reaction, intensive research had been done regarding promoters, support, and preparation methods. Alkali and alkaline earth oxides had been reported to have a positive effect on Cu catalysts in methanol production (Bansode, Tidona, Von Rohr, & Urakawa, 2013). It was found that the Ba promoter can improve the activity of Cu/Al_2O_3 catalyst toward methanol synthesis while the potassium (K) promoter prefers to RWGS reaction (Bansode et al., 2013). Support oxides such as Al_2O_3, ZrO_2, Ga_2O_3, and Cr_2O_3 were also used as modifiers or promoters for Cu/ZnO catalysts (Saito, Fujitani, Takeuchi, & Watanabe, 1996). The other studies on additives mainly concentrate on the study of rare earth oxides (La or Ce) (Ban, Li, Asami, & Fujimoto, 2014), amphoteric oxides (e.g., TiO_2 and ZrO_2) (Nomura, Tagawa, & Goto, 1998), noble metal like Ag or Au (Pasupulety et al., 2015; Tada et al., 2017), main group oxides (e.g., SiO_2) (Samei, Taghizadeh, & Bahmani, 2012), and materials with semiconductor properties, e.g., g-C_3N_4, etc. (Deng, Hu, Lu, & Hong, 2017).

Regarding support, metal oxides like Al_2O_3, ZrO_2, CeO_2, and SiO_2 are then the most studied materials. The properties of the support, which consist of texture, structure, acidity/basicity, and electronic property, can have a significant effect on the performance of the catalyst (Jiang et al., 2020). Moreover, it has been reported that the porosity of the support with ordered or unordered porosities can significantly affect the stability of for example $CuZnO/SiO_2$ methanol synthesis catalysts (Prieto, Shakeri, De Jong, & De Jongh, 2014). The narrowest pore constrictions of the support can mitigate the growth of metal particle size and improve catalyst stability. Other materials such as layered double hydroxides (LDHs) and carbon nanotubes (CNTs) have also received more attention recently.

Thus, as reported for methanation, for methanolation also the preparation method can change the catalyst structure and distribution of active sites over methanol synthesis catalysts. The most common method is the precipitation method, which allows high metal loadings and well-defined metal particles. Other methods such as the sol-gel method, liquid reduction, ammonia evaporation, microwave, and combustion method, etc. were also studied in these years (Dong et al., 2017; Jiang et al., 2020; Ramli, Syed-Hassan, & Hadi, 2018).

Although many efforts have been devoted to the development of methanol synthesis catalysts, the activity and yield of methanol are still not very high due to the thermodynamic and dynamic barriers. Compared to methanol synthesis, the CO_2 methanation reaction can be more potential in the future because it can be carried under moderate conditions.

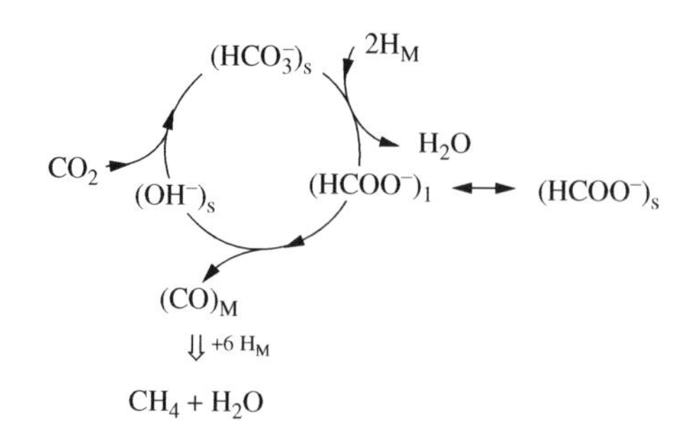

FIG. 3.13 The mechanism of CO_2 methanation with CO as an intermediate (M: metal, S: support). *From Marwood, M., Doepper, R., & Renken, A. (1997). In-situ surface and gas-phase analysis for kinetic studies under transient conditions: The catalytic hydrogenation of CO_2.* Applied Catalysis A: General, *151(1), 223–246. https://doi.org/10.1016/S0926-860X(96)00267-0.*

3.4 Proposed reaction mechanisms of CO_2 methanation

The CO hydrogenation mechanism

Nowadays the mechanism of CO_2 methanation remains unclear. Regarding the intermediate that account for the formation of methane, there are mainly two mechanisms proposed (Wang et al., 2011). One of them is the mechanism that carbon oxide plays as an intermediate in which CO_2 is reduced to CO and subsequently the methane is produced by CO methanation (Lapidus et al., 2007). The density functional theory on Ni(111) surfaces proposed three mechanisms in which the mechanism with a rate-determining step of $CO \rightarrow C+O$ has the lowest energy barrier (Liu et al., 2018; Ren, Qin, et al., 2015). The diagram of the scheme is shown in Fig. 3.13 (Scheme 1).

The experimental studies on the Ni catalysts supported on activated carbon for CO_2 methanation also confirmed that the CO was an indispensable intermediate in the formation process of methane (Lapidus et al., 2007).

Zhou et al. (Li et al., 2018; Storsæter et al., 2005; Yin & Ge, 2012; Zhang et al., 2002; Zhou et al., 2014) studied the effect of TiO_2 structure on Ni/TiO_2 for CO_2 methanation. The methanation over Ni/TiO_2 catalyst with principle Ni (111) facet experienced the CO intermediate pathway. However, the catalyst with multi-planes of Ni followed the mechanism of formate species as an intermediate pathway in which the Ni only accounted for H_2 dissociation.

The mechanism study on the noble metal-based catalyst ($Rh/\gamma\text{-}Al_2O_3$) at low temperature and ambient pressure revealed that the methanation process followed the pathway without the formation of formate species (Beuls et al., 2012; Gao et al., 2012; Fabien Ocampo et al., 2009). Three steps can be listed as follows:

$$CO_2\,(g) \rightarrow CO_2\,(ads) \tag{3.4}$$

$$CO_2\,(ads) \rightarrow CO\,(ads) + O\,(ads) \tag{3.5}$$

$$CO\,(ads) + H\,(ads) \rightarrow \ldots \rightarrow CH_4\,(ads) + H_2O\,(ads) \tag{3.6}$$

The H(ads) species were formed in the dissociation process of H_2. However, it was also noted that CO_2 adsorption could be a complicated process (Beuls et al., 2012). The oxidation part of Rh was also registered, which was attributed to the oxidation by O(ads).

The formate intermediate mechanism

The other proposed mechanism involves the formation of formate species during the reaction; this latter species is subsequently hydrogenated to methane (Cárdenas-Arenas et al., 2020) (Moon et al., 2019). This mechanism needs both the participation of support and active metal. It was reported that many intermediates, e.g., hydrogen carbonate, bidentate carbonate, monodentate carbonate, bridged carbonate, formate, and nickel carbonyl hydride can be formed (Moon et al., 2019). The properties of support can affect the variety of intermediates. As presented in Fig. 3.14 (Scheme 2) for the mechanism over Ni/$Ce_{0.5}Zr_{0.5}O_2$ and Ni/γ-Al_2O_3 catalysts, the H_2 gas is adsorbed and dissociated on the surface of the Ni atom to produce H while the CO_2 molecules are adsorbed and dissociated on the support surface to form carbonate species. The carbonate (bidentate or monodentate) and hydrocarbonate species subsequently react with the H to form formate (bidentate or monodentate) species, which will be hydrogenated by H to produce methane as reported elsewhere (Moon et al., 2019). For the Ni/γ-Al_2O_3 catalyst, it was demonstrated by FTIR that more hydrogen carbonates formed

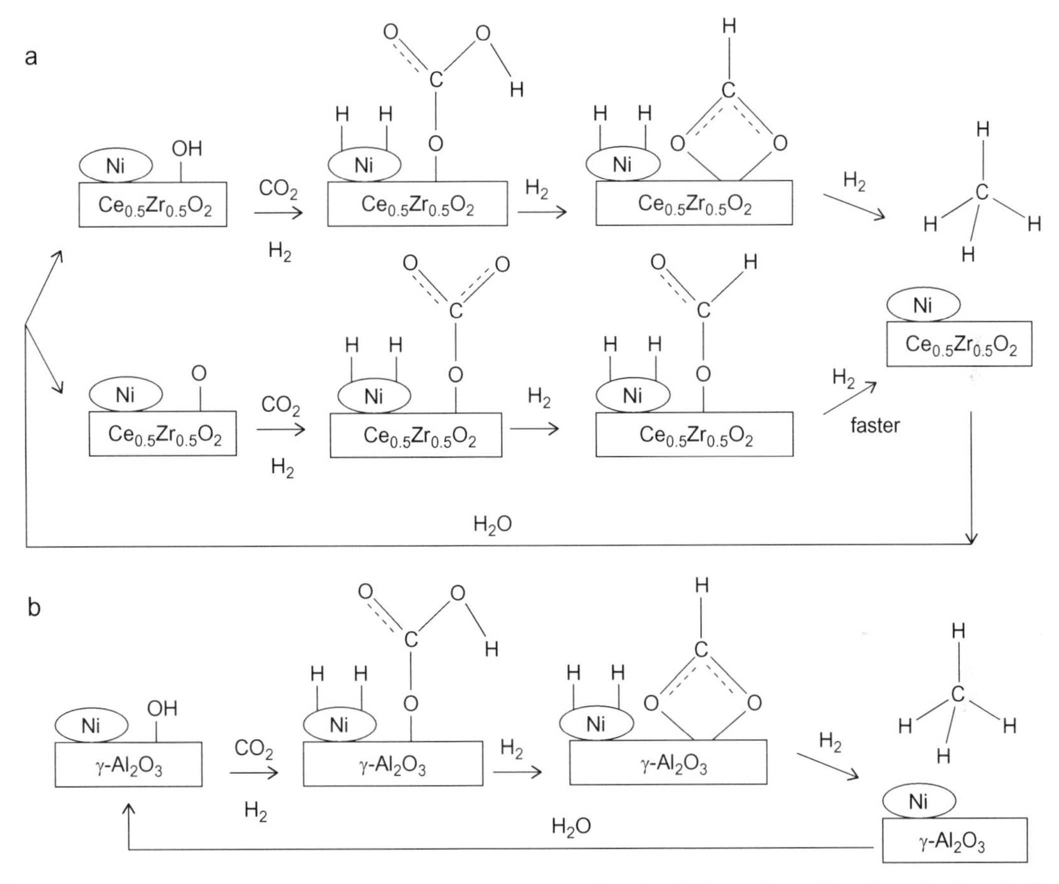

FIG. 3.14 The proposed mechanisms over Ni/Ce0.5Zr0.5O_2 and Ni/γ-Al_2O_3 catalysts. *From Pan, Q., Peng, J., Sun, T., Wang, S., & Wang, S. (2014). Insight into the reaction route of CO_2 methanation: Promotion effect of medium basic sites.* Catalysis Communications, 45, *74–78. https://doi.org/10.1016/j.catcom.2013.10.034.*

over the catalyst (Pan et al., 2014). The hydrogen carbonate species were hydrogenated to bidentate formate species, which were hydrogenated to methane. For $Ni/Ce_{0.5}Zr_{0.5}O_2$ catalyst, more monodentate carbonate rather than hydrogen carbonate species formed, thus leading to more monodentate formate species, which were hydrogenated to methane. The pathway of methane formation over $Ni/Ce_{0.5}Zr_{0.5}O_2$ is faster than that over $Ni/\gamma\text{-}Al_2O_3$ catalyst (Pan et al., 2014). This is attributed to the difference of basicity, in which more medium basic sites over $Ni/Ce_{0.5}Zr_{0.5}O_2$ are favorable for the formation of monodentate carbonate.

Another important parameter is the Ni particle size. In the literature, the study on the effect of Ni particle size over Ni/SiO_2 catalysts showed that the particle size of Ni also had significant effects on the pathway of methane formation (Wu et al., 2015).

The presence of side reactions in CO_2 methanation reaction

In the thermodynamic analyses of CO_2 methanation reported in Part 2, it has been demonstrated that the species formed during the methanation process depend significantly on the reaction conditions. Under atmospheric pressure, the main by-product in thermodynamic analyses is carbon monoxide (CO), which originates from the RWGS reaction ($H_2 + CO_2 \rightarrow CO + H_2O$) or reforming reactions (Wang, Zhao, Wang, Hu, & Da Costa, 2020; Wu et al., 2015). Besides, the CO species can also be formed by the hydrogenation of formate species as intermediates (Wu et al., 2015). The presence of CO can also cause the sintering of active metals (Ni, etc.), which is caused by the formation of $Ni(CO)_x$ species, leading to the agglomeration of Ni particles (Aziz, Jalil, Triwahyono, & Ahmad, 2015; Chen, Budi, Wu, Saikia, & Kao, 2017).

The H_2O formed in methanation can induce the formation of Ni-hydroxide at low temperatures, which results in the consequent decrease of metallic nickel. The Ni-hydroxide species cannot be reduced to Ni metal fast and contribute to the sintering of Ni and the formation of hardly reducible Ni-aluminate, leading to partial deactivation of catalysts (Mebrahtu et al., 2019).

When the ratio H_2/CO_2 decreases, the carbon can also be formed due to parallel reactions such as the Boudouard reaction or/and CO_2 reduction. It is worth noting that carbon deposition also occurs at high pressure over some catalytic systems (Baysal & Kureti, 2020).

3.5 Future prospects: Assisted nickel catalysts for CO_2 reduction: From photocatalysis to assisted plasma catalysis

Ni-based catalysts for photocatalytic methanation

These Last years, used as a possible CO_2 reduction to CH_4 at room temperature, the CO_2 methanation through a photocatalytic process has gained interest as a new sustainable pathway for CO_2 utilization (Bueno-Alejo, Arca-Ramos, Hueso, & Santamaria, 2020; Ulmer et al., 2019). This photocatalytic reduction, using solar energy, offers a promising way to produce CH_4 at a low cost and an environment-friendly production (Yuan, Yang, Du, & Yang, 2014; Zhao, Wang, & Li, 2016). Among all the semiconductors, photocatalytic CO_2 hydrogenation to methane using titanium oxide (TiO_2) has been the most used material due to its high activity and stability, relatively low-cost, and negligible toxicity (Liu et al., 2015). However, the methane selectivity over raw support as over the pure TiO_2 was very low. This is mainly due to the rapid recombination of photo-generated charges and the inability to utilize visible light as solar light. Thus, TiO_2 was used as support, and metals were impregnated on TiO_2 leading to the creation of defects in the lattice of the support and then can enhance the physicochemical properties of the support itself and overcome the problem of its low efficiency (Tahir, Tahir, Amin, & Muhammad, 2016). Among them, Nickel was found to be a good candidate to improve the activity in photocatalysis.

Moreover, Ni-promoted TiO_2 catalysts are the most reported due to low cost, high stability, high selectivity, and good photocatalytic activity (Chen et al., 2015; He, Tang, Shen, Chen, & Song, 2016; Kho, Scott, & Amal, 2016). In recent years, Ni-promoted TiO_2 has been tested for photocatalytic CO_2 reduction by water to methane with a high photoactivity and selectivity (Kwak et al., 2015; Ola & Mercedes Maroto-Valer, 2014). Many other groups also reported that NiO could be an efficient promoter for the improvement of the photoactivity of TiO_2 or other supports such as other mixed oxides for CO_2 reduction (Hongmanorom et al., 2021; Lu et al., 2015; Lv et al., 2018; Lv et al., 2020; Park et al., 2004; Wang et al., 2010; Yu et al., 2015; Zhang et al., 2013).

Finally, in addition to photocatalysts themselves, the choice of a reducing agent is also very important since the overall efficiency of the photo-CO_2 reduction process depends on the type of reductant and photocatalyst (Tahir et al., 2016). Water has been the most commonly used reducing agent in CO_2 photoreduction processes. Only recently, hydrogen was reported as a potentially viable reducing agent for CO_2 reduction (Tahir, Tahir, & Amin, 2015). According to thermodynamics, the reduction potential to generate H_2 from water is lower (0 V) than CO_2 standard reduction potential (−1.9 V) to reduce it (Tahir et al., 2016). However, a real competition between H_2O and CO_2 can be drawn and photoreactions can probably be more favorable for H_2O reduction instead of CO_2 (Tahir & Amin, 2013). However, the reaction mechanisms remained unclear and more investigations are needed to fully understand the role of water and hydrogen used as reducing agents on the catalytic activity and product selectivity (Tahir et al., 2016). Therefore, it is appropriate to study the combined effects of H_2O and H_2 reductants for a photocatalytic CO_2 methanation reaction over nickel-based catalysts (Tahir et al., 2016).

Ni-based catalysts for assisted plasma-catalytic methanation

The plasma-catalytic-assisted methanation started for CO_2 methanation in early 2010th. For such a reaction, the most common types of plasma reported in the literature are Non-thermal plasmas (NTP). An NTP is a plasma that is not in thermal equilibrium. The NTP is also called cold plasma. Among these plasmas, dielectric barrier discharges (DBD), microwave (MW), and glow discharge (GD) are the most reported in the literature (Chen, Qiu, Liu, & Zhang, 2020; Dębek, Azzolina-Jury, Travert, Maugé, & Thibault-Starzyk, 2019). The plasma-assisted methanation is generally carried out at low temperature and pressure and performs good activity and selectivity of the aimed reaction. Compared to conventional thermal catalysis, plasma-assisted methanation indeed does not need a high reaction temperature and a high running pressure, but it can reach high activity and selectivity in methanation reaction (Wang et al., 2021).

Among the presented NTP, in the literature, only a few studies dealt with MW and GD coupled with Ni catalysts. Indeed, using MW plasma, only Chen et al. presented the beneficial effect of the combination of NiO/TiO_2 catalyst and a plasma microwave (915 MHz) in CO_2 dissociation (Aziz, Jalil, Triwahyono, & Ahmad, 2015; Chen, Budi, et al., 2017). More recently, Debek et al. reported (Bogaerts et al., 2020; Chen et al., 2020; Dębek, Azzolina-Jury, Travert, & Maugé, 2019) the effect of metal oxide support effect on the activity of Ni-based catalysts for CO_2 methanation. Three supports were tested: SiO_2, CeO_2-ZrO_2, and Al_2O_3. Ni/Al_2O_3 catalysts showed the best performance in terms of CH_4 production.

Furthermore, the first study on CO_2 methanation with a DBD assisted with nickel catalysts was performed by Jwa, Mok, and Lee (2011) using Ni/Al_2O_3 and Ni/TiO_2-Al_2O_3 catalyst. They were the first group showing the synergetic effect between a Ni catalyst and the DBD plasma for methanation reaction. In 2016, Zeng and Tu (2016) studied the plasma-catalytic CO_2 hydrogenation over a Ni/Al_2O_3 catalyst for the cogeneration of CO and CH_4 in a dielectric barrier discharge (DBD) reactor at 150°C. The presence of the Ni catalyst in the DBD reactor has demonstrated a plasma-catalytic synergistic effect at low temperatures (Nizio et al., 2016; Zeng & Tu, 2016). Nizio et al. studied the effect of the composition of the CeO_2-ZrO_2 support plasma-catalytic methanation of CO_2. For the first time, a relatively high CO_2 conversion (80%) and a selectivity in CH_4 of 100% was reported in adiabatic conditions at low temperatures (Nizio et al., 2016). To obtain a similar conversion in the thermal condition a reaction temperature of 573 K is needed. Moreover, Benrabbah et al. (2017) reported on a Ni/CeO_2-ZrO_2 plasma coupled system that if the catalyst is pre-treated in plasma condition, a better activity in CO_2 methanation is obtained. This latter result was attributed to the strengthened Ni-support interaction. The plasma-assisted CO_2 methanation, which reached the same activity as conventional CO_2 methanation, normally required lower temperature (Benrabbah et al., 2017; Mikhail et al., 2020; Nizio et al., 2016). It was reported that the excited CO*, O*, and H* species could be formed in the presence of plasma, which led to the production of CH_4 at low temperatures (Mikhail et al., 2020). Jwa also reported that Ni/Zeolite was a promising catalyst for plasma-catalytic hydrogenation of CO_2, with ca. 80% CO_2 converted at 493 K (Jwa, Lee, Lee, & Mok, 2013). This was confirmed recently by Bacariza et al. (2018). The activity of this kind of zeolite-based catalysts is improved when Ce is added as reported by Biset-Peiró, Guilera, Zhang, Arbiol, and Andreu (2019). Furthermore, the positive impact of Ce was also reported on Ni/Al_2O_3 catalyst (Chen et al., 2020).

More recently, other Ni-based catalysts were reported as potential catalysts for plasma-catalytic methanation. Thus, hydrotalcite-derived catalysts impregnated by Ni and Fe showed interesting results at low temperatures (Wierzbicki et al., 2020). The effect of metal dispersion and support structure of Ni/silicalite-1 was recently presented (Wang, Zhao, et al., 2020; Wu et al., 2015).

Although the plasma technology has a good potential for future application in CO_2 methanation, the complexity of the plasma set-up may limit its large-scale application (Manthiram, Beberwyck, & Alivisatos, 2014; Wang et al., 2020), but it could be a good alternative for small-scale industries.

Finally, the electrochemical methods for CO_2 methanation can be another CO_2 reduction route (Banholzer & Jones, 2013). Compared to heterogeneous catalysis, the electrochemical methanation can directly convert CO_2 to methane in a single step at ambient temperature and pressure. But this process has the same problem as photomethanation of CO_2, the low activity and selectivity, which makes it not feasible to date (Sun, Beaunier, & Da Costa, 2020; Sun, Świrk, Wierzbicki, et al., 2020; Wang & Gong, 2011).

Apart from the plasma technology, the photomethanation of CO_2 also was reported by many researchers (Aziz, Jalil, Triwahyono, & Ahmad, 2015; O'Brien et al., 2018; Pan et al., 2014; Yan et al., 2018). The Ni and Ru-based catalysts had been used in the photomethanation of Gaseous CO_2 (Yan et al., 2019) (O'Brien et al., 2014). However, the methanation rate of the catalyst in photomethanation of CO_2 is too low compared to thermal methanation and plasma-assisted methanation, which indicates that this technology is still in the stage of scientific research.

3.6 Conclusion

Global warming originated from anthropogenic activities had caused serious environmental problems. Carbon dioxide (CO_2) is the main component of greenhouse gases, which account for global warming. Thus, the reduction of CO_2 has become the consensus in human society. The CO_2 capture and utilization technology (CCU) is more potent as compared to the CO_2 capture and storage technology (CCS). Because it can not only reduce CO_2 emissions but also produce valuable chemicals. Thus, the CO_2 hydrogenation route can play a significant role in the CCU route.

In this chapter, the heterogeneously catalytic methanation of CO_2, CO_2 hydrogenation to methanol, and the assisted catalysis of CO_2 reduction including plasma-assisted catalysis, photocatalysis, and electrocatalysis were discussed with the catalytic methanation of CO_2 on transition metal-based catalysts as an emphasis.

Different CO_2 reduction strategies were compared and the catalytic methanation of CO_2 was regarded as the most potential route toward commercialization. The thermodynamic analyses were discussed in detail. Also, the catalytic systems were thoroughly investigated including active metal, additive, support, and structure properties, etc. It is concluded that the noble metal-based catalysts show superior performance in the low-temperature CO_2 methanation reaction. However, the high cost and low availability limit the large-scale application of noble metal catalysts. For the non-noble-based catalysts, many kinds of research focus on the Ni and Co-based catalysts, which showed superior activity and selectivity in CO_2 methanation. Due to the low price, high availability, and good catalytic performance, Ni-based catalysts were regarded as the most potential catalysts toward commercialization. The influence factors of the performance of Ni catalysts were seriously analyzed with the promoter and support being intensively discussed. Nevertheless, the Ni-based catalysts still encounter the problems like poor activity at low temperatures and deactivation at high temperatures. In such a case, the deactivation mechanisms were also discussed, in which the main reason was attributed to Ni sintering. Multi-metallic catalysts showed better performance compared to monometallic catalysts, with a real promotion in adding a second metal. Furthermore, for nickel-based catalysts, the mixed oxide supports seem to have better promotion effects for Ni catalysts due to their physical and chemical properties. Also, Co-based catalysts as potential catalysts for CO_2 methanation were discussed in detail.

The mechanisms of CO_2 methanation were also presented based on results obtained both by theoretical computation and spectroscopic studies such as DRIFT. Mainly, two mechanisms were introduced based on different catalytic systems. In one of them, the CO molecule is considered as intermediate in the formation of methane. This type of mechanism is mostly reported over the Rh or Ru catalysts. The second mechanism is based on the formate intermediate way. Thus, in the latter one, the formate formation and its subsequent hydrogenation to methane are the crucial steps. This mechanism is mostly presented for Ni or Co-based catalysts. The limit of the catalysis is the activation of the molecules that need medium temperature. Thus assisted catalyst was developed in the last decade. And these other technologies such as plasma-assisted catalysis, photocatalysis, and electrocatalysis regarding CO_2 reduction presented here started to be potential solutions for the near future industrial applications.

References

Abate, S., Mebrahtu, C., Giglio, E., Deorsola, F., Bensaid, S., Perathoner, S., ... Centi, G. (2016). Catalytic performance of γ-Al_2O_3-ZrO_2-TiO_2-CeO_2 composite oxide supported Ni-based catalysts for CO_2 methanation. *Industrial and Engineering Chemistry Research*, *55*(16), 4451–4460. https://doi.org/10.1021/acs.iecr.6b00134.

Abelló, S., Berrueco, C., & Montané, D. (2013). High-loaded nickel-alumina catalyst for direct CO_2 hydrogenation into synthetic natural gas (SNG). *Fuel*, *113*, 598–609. https://doi.org/10.1016/j.fuel.2013.06.012.

Abelló, S., Berrueco, C., Gispert-Guirado, F., & Montané, D. (2016). Synthetic natural gas by direct CO_2 hydrogenation on activated takovites: Effect of Ni/Al molar ratio. *Catalysis Science and Technology*, *6*(7), 2305–2317. https://doi.org/10.1039/c5cy01200g.

Agnelli, M., Kolb, M., Nicot, C., & Mirodatos, C. (1991). Sintering of a Ni-based catalyst during CO hydrogenation: Kinetics and modeling. *Studies in Surface Science and Catalysis*, *68*, 62690–62697. https://doi.org/10.1016/S0167-2991(08.

Agnelli, M., Kolb, M., & Mirodatos, C. (1994). Co hydrogenation on a nickel catalyst: 1. Kinetics and Modeling of a low-temperature sintering process. *Journal of Catalysis*, *148*(1), 9–21. https://doi.org/10.1006/jcat.1994.1180.

Ahn, J. Y., Chang, S. W., Lee, S. M., Kim, S. S., Chung, W. J., Lee, J. C., ... Nguyen, D. D. (2019). Developing Ni-based honeycomb-type catalysts using different binary oxide-supported species for synergistically enhanced CO_2 methanation activity. *Fuel*, *250*, 277–284. https://doi.org/10.1016/j.fuel.2019.03.123.

Álvarez, A., Bansode, A., Urakawa, A., Bavykina, A. V., Wezendonk, T. A., Makkee, M., ... Kapteijn, F. (2017). Challenges in the greener production of formates/formic acid, methanol, and DME by heterogeneously catalyzed CO_2 hydrogenation processes. *Chemical Reviews*, *117*(14), 9804–9838. https://doi.org/10.1021/acs.chemrev.6b00816.

Aresta, M., Dibenedetto, A., & Angelini, A. (2014). Catalysis for the valorization of exhaust carbon: From CO_2 to chemicals, materials, and fuels. Technological use of CO_2. *Chemical Reviews*, *114*(3), 1709–1742. https://doi.org/10.1021/cr4002758.

Aziz, M. A. A., Jalil, A. A., Triwahyono, S., & Saad, M. W. A. (2015). CO_2 methanation over Ni-promoted mesostructured silica nanoparticles: Influence of Ni loading and water vapor on activity and response surface methodology studies. *Chemical Engineering Journal*, *260*, 757–764. https://doi.org/10.1016/j.cej.2014.09.031.

Aziz, M. A. A., Jalil, A. A., Triwahyono, S., & Ahmad, A. (2015). CO_2 methanation over heterogeneous catalysts: Recent progress and future prospects. *Green Chemistry*, *17*(5), 2647–2663. https://doi.org/10.1039/c5gc00119f.

Bacariza, M. C., Biset-Peiró, M., Graça, I., Guilera, J., Morante, J., Lopes, J. M., ... Henriques, C. (2018). DBD plasma-assisted CO_2 methanation using zeolite-based catalysts: Structure composition-reactivity approach and effect of Ce as promoter. *Journal of CO_2 Utilization*, *26*, 202–211. https://doi.org/10.1016/j.jcou.2018.05.013.

Bacariza, M. C., Amjad, S., Teixeira, P., Lopes, J. M., & Henriques, C. (2020). Boosting Ni dispersion on zeolite-supported catalysts for CO_2 methanation: The influence of the impregnation solvent. *Energy and Fuels*, *34*(11), 14656–14666. https://doi.org/10.1021/acs.energyfuels.0c02561.

Bai, S., Shao, Q., Wang, P., Dai, Q., Wang, X., & Huang, X. (2017). Highly active and selective hydrogenation of CO_2 to ethanol by ordered Pd-Cu nanoparticles. *Journal of the American Chemical Society*, *139*(20), 6827–6830. https://doi.org/10.1021/jacs.7b03101.

Ban, H., Li, C., Asami, K., & Fujimoto, K. (2014). Influence of rare-earth elements (La, Ce, Nd and Pr) on the performance of Cu/Zn/Zr catalyst for CH3OH synthesis from CO_2. *Catalysis Communications*, *54*, 50–54. https://doi.org/10.1016/j.catcom.2014.05.014.

Banholzer, W. F., & Jones, M. E. (2013). Chemical engineers must focus on practical solutions. *AICHE Journal*, *59*(8), 2708–2720. https://doi.org/10.1002/aic.14172.

Bansode, A., Tidona, B., Von Rohr, P. R., & Urakawa, A. (2013). Impact of K and Ba promoters on CO_2 hydrogenation over Cu/Al_2O_3 catalysts at high pressure. *Catalysis Science and Technology*, *3*(3), 767–778. https://doi.org/10.1039/c2cy20604h.

Baysal, Z., & Kureti, S. (2020). CO_2 methanation on Mg-promoted Fe catalysts. *Applied Catalysis B: Environmental*, *262*(2). https://doi.org/10.1016/j.apcatb.2019.118300, 118300.

Beaumont, S. K., Alayoglu, S., Specht, C., Michalak, W. D., Pushkarev, V. V., Guo, J., ... Somorjai, G. A. (2014). Combining in situ NEXAFS spectroscopy and CO_2 methanation kinetics to study Pt and CO nanoparticle catalysts reveals key insights into the role of platinum in promoted cobalt catalysis. *Journal of the American Chemical Society*, *136*(28), 9898–9901. https://doi.org/10.1021/ja505286j.

Benrabbah, R., Cavaniol, C., Liu, H., Ognier, S., Cavadias, S., Gálvez, M. E., & Da Costa, P. (2017). Plasma DBD activated ceria-zirconia-promoted Ni-catalysts for plasma catalytic CO_2 hydrogenation at low temperature. *Catalysis Communications*, *89*, 73–76. https://doi.org/10.1016/j.catcom.2016.10.028.

Beuls, A., Swalus, C., Jacquemin, M., Heyen, G., Karelovic, A., & Ruiz, P. (2012). Methanation of CO_2: Further insight into the mechanism over Rh/γ-Al_2O_3

catalyst. *Applied Catalysis B: Environmental, 113–114*, 2–10. https://doi.org/10.1016/j.apcatb.2011.02.033.

Bian, L., Zhang, L., Zhu, Z., & Li, Z. (2018). Methanation of carbon oxides on Ni/Ce/SBA-15 pretreated with dielectric barrier discharge plasma. *Molecular Catalysis, 446*, 131–139. https://doi.org/10.1016/j.mcat.2017.12.027.

Bian, Z., Chan, Y. M., Yu, Y., & Kawi, S. (2020). Morphology dependence of catalytic properties of Ni/CeO_2 for CO_2 methanation: A kinetic and mechanism study. *Catalysis Today, 347*, 31–38. https://doi.org/10.1016/j.cattod.2018.04.067.

Biset-Peiró, M., Guilera, J., Zhang, T., Arbiol, J., & Andreu, T. (2019). On the role of ceria in Ni-Al_2O_3 catalyst for CO_2 plasma methanation. *Applied Catalysis A: General, 575*, 223–229. https://doi.org/10.1016/j.apcata.2019.02.028.

Bogaerts, A., Tu, X., Whitehead, J. C., Centi, G., Lefferts, L., Guaitella, O., ... Carreon, M. (2020). The 2020 plasma catalysis roadmap. *Journal of Physics D: Applied Physics, 53*(44), 443001. https://doi.org/10.1088/1361-6463/ab9048.

Budi, C. S., Wu, H. C., Chen, C. S., Saikia, D., & Kao, H. M. (2016). Ni nanoparticles supported on cage-type mesoporous silica for CO_2 hydrogenation with high CH_4 selectivity. *ChemSusChem, 9*(17), 2326–2331. https://doi.org/10.1002/cssc.201600710.

Bueno-Alejo, C. J., Arca-Ramos, A., Hueso, J. L., & Santamaria, J. (2020). LED-driven continuous flow carbon dioxide hydrogenation on a nickel-based catalyst. *Catalysis Today, 355*, 678–684. https://doi.org/10.1016/j.cattod.2019.06.022.

Bukhari, S. N., Chong, C. C., Setiabudi, H. D., Ainirazali, N., Aziz, M. A. A., Jalil, A. A., & Chin, S. Y. (2019). Optimal Ni loading towards efficient CH_4 production from H_2 and CO_2 over Ni supported onto fibrous SBA-15. *International Journal of Hydrogen Energy, 44*(14), 7228–7240. https://doi.org/10.1016/j.ijhydene.2019.01.259.

Burattin, P., Che, M., & Louis, C. (1999). Metal particle size in Ni/SiO_2 materials prepared by deposition-precipitation: Influence of the nature of the Ni(II) phase and of its interaction with the support. *The Journal of Physical Chemistry. B, 103*(30), 6171–6178. https://doi.org/10.1021/jp990115t.

Burger, T., Koschany, F., Thomys, O., Köhler, K., & Hinrichsen, O. (2018). CO_2 methanation over Fe- and Mn-promoted CO-precipitated Ni-Al catalysts: Synthesis, characterization and catalysis study. *Applied Catalysis A: General, 558*, 44–54. https://doi.org/10.1016/j.apcata.2018.03.021.

Cárdenas-Arenas, A., Quindimil, A., Davó-Quiñonero, A., Bailón-García, E., Lozano-Castelló, D., De-La-Torre, U., ... Bueno-López, A. (2020). Isotopic and in situ DRIFTS study of the CO_2 methanation mechanism using Ni/CeO_2 and Ni/Al_2O_3 catalysts. *Applied Catalysis B: Environmental, 265*. https://doi.org/10.1016/j.apcatb.2019.118538, 118538.

Chen, W. T., Chan, A., Sun-Waterhouse, D., Moriga, T., Idriss, H., & Waterhouse, G. I. N. (2015). Ni/TiO_2: A promising low-cost photocatalytic system for solar H_2 production from ethanol-water mixtures. *Journal of Catalysis, 326*, 43–53. https://doi.org/10.1016/j.jcat.2015.03.008.

Chen, C.-S., Budi, C. S., Wu, H.-C., Saikia, D., & Kao, H.-M. (2017). Size-tunable Ni nanoparticles supported on surface-modified, cage-type mesoporous silica as highly active catalysts for CO_2 hydrogenation. *ACS Catalysis, 7*(12), 8367–8381. https://doi.org/10.1021/acscatal.7b02310.

Chen, G., Britun, N., Godfroid, T., Georgieva, V., Snyders, R., & Delplancke-Ogletree, M. P. (2017). An overview of CO_2 conversion in a microwave discharge: The role of plasma-catalysis. *Journal of Physics D: Applied Physics, 50*(8). https://doi.org/10.1088/1361-6463/aa5616, 084001.

Chen, Y., Qiu, B., Liu, Y., & Zhang, Y. (2020). An active and stable nickel-based catalyst with embedment structure for CO_2 methanation. *Applied Catalysis B: Environmental, 269*. https://doi.org/10.1016/j.apcatb.2020.118801, 118801.

Chen, H., Mu, Y., Xu, S., Xu, S., Hardacre, C., & Fan, X. (2020). Recent advances in non-thermal plasma (NTP) catalysis towards C1 chemistry. *Chinese Journal of Chemical Engineering, 28*(8), 2010–2021. https://doi.org/10.1016/j.cjche.2020.05.027.

Chen, H., Goodarzi, F., Mu, Y., Chansai, S., Mielby, J. J., Mao, B., ... Fan, X. (2020). Effect of metal dispersion and support structure of Ni/silicalite-1 catalysts on non-thermal plasma (NTP) activated CO_2 hydrogenation. *Applied Catalysis B: Environmental, 272*. https://doi.org/10.1016/j.apcatb.2020.119013, 119013.

Dębek, R., Azzolina-Jury, F., Travert, A., & Maugé, F. (2019). A review on plasma-catalytic methanation of carbon dioxide–Looking for an efficient catalyst. *Renewable and Sustainable Energy Reviews, 116*. https://doi.org/10.1016/j.rser.2019.109427.

Dębek, R., Azzolina-Jury, F., Travert, A., Maugé, F., & Thibault-Starzyk, F. (2019). Low-pressure glow discharge plasma-assisted catalytic CO_2 hydrogenation—The effect of metal oxide support on the performance of the Ni-based catalyst. *Catalysis Today, 337*, 182–194. https://doi.org/10.1016/j.cattod.2019.03.039.

Deng, K., Hu, B., Lu, Q., & Hong, X. (2017). Cu/g-C_3N_4 modified ZnO/Al_2O_3 catalyst: Methanol yield improvement of CO_2 hydrogenation. *Catalysis Communications, 100*, 81–84. https://doi.org/10.1016/j.catcom.2017.06.041.

Dong, X., Li, F., Zhao, N., Tan, Y., Wang, J., & Xiao, F. (2017). CO_2 hydrogenation to methanol over Cu/Zn/Al/Zr catalysts prepared by liquid reduction. *Chinese Journal of Catalysis*, *38*(4), 717–725. https://doi.org/10.1016/S1872-2067(17)62793-1.

Ewald, S., Kolbeck, M., Kratky, T., Wolf, M., & Hinrichsen, O. (2019). On the deactivation of Ni-Al catalysts in CO_2 methanation. *Applied Catalysis A: General*, *570*, 376–386. https://doi.org/10.1016/j.apcata.2018.10.033.

Falbo, L., Martinelli, M., Visconti, C. G., Lietti, L., Bassano, C., & Deiana, P. (2018). Kinetics of CO_2 methanation on a Ru-based catalyst at process conditions relevant for power-to-gas applications. *Applied Catalysis B: Environmental*, *225*, 354–363. https://doi.org/10.1016/j.apcatb.2017.11.066.

Gao, J., Wang, Y., Ping, Y., Hu, D., Xu, G., Gu, F., & Su, F. (2012). A thermodynamic analysis of methanation reactions of carbon oxides for the production of synthetic natural gas. *RSC Advances*, *2*(6), 2358–2368. https://doi.org/10.1039/c2ra00632d.

Garbarino, G., Wang, C., Cavattoni, T., Finocchio, E., Riani, P., Flytzani-Stephanopoulos, M., & Busca, G. (2019). A study of Ni/La-Al_2O_3 catalysts: A competitive system for CO_2 methanation. *Applied Catalysis B: Environmental*, *248*, 286–297. https://doi.org/10.1016/j.apcatb.2018.12.063.

Ghaib, K., & Ben-Fares, F. Z. (2018). Power-to-methane: A state-of-the-art review. *Renewable and Sustainable Energy Reviews*, *81*, 433–446. https://doi.org/10.1016/j.rser.2017.08.004.

Guilera, J., Del Valle, J., Alarcón, A., Díaz, J. A., & Andreu, T. (2019). Metal-oxide promoted Ni/Al_2O_3 as CO_2 methanation micro-size catalysts. *Journal of CO_2 Utilization*, *30*, 11–17. https://doi.org/10.1016/j.jcou.2019.01.003.

Guo, Y., Mei, S., Yuan, K., Wang, D. J., Liu, H. C., Yan, C. H., & Zhang, Y. W. (2018). Low-temperature CO_2 Methanation over CeO_2-supported Ru single atoms, nanoclusters, and nanoparticles competitively tuned by strong metal-support interactions and H-spillover effect. *ACS Catalysis*, *8*(7), 6203–6215. https://doi.org/10.1021/acscatal.7b04469.

He, Z., Tang, J., Shen, J., Chen, J., & Song, S. (2016). Enhancement of photocatalytic reduction of CO_2 to CH_4 over TiO_2 nanosheets by modifying with sulfuric acid. *Applied Surface Science*, *364*, 416–427. https://doi.org/10.1016/j.apsusc.2015.12.163.

Hoekman, S. K., Broch, A., Robbins, C., & Purcell, R. (2010). CO_2 recycling by reaction with renewably-generated hydrogen. *International Journal of Greenhouse Gas Control*, *4*(1), 44–50. https://doi.org/10.1016/j.ijggc.2009.09.012.

Hongmanorom, P., Ashok, J., Zhang, G., Bian, Z., Wai, M. H., Zeng, Y., … Kawi, S. (2021). Enhanced performance and selectivity of CO_2 methanation over phyllosilicate structure derived Ni-Mg/SBA-15 catalysts. *Applied Catalysis B: Environmental*, *282*. https://doi.org/10.1016/j.apcatb.2020.119564, 119564.

Iglesias, I., Quindimil, A., Mariño, F., De-La-Torre, U., & González-Velasco, J. R. (2019). Zr promotion effect in CO_2 methanation over ceria supported nickel catalysts. *International Journal of Hydrogen Energy*, *44*(3), 1710–1719. https://doi.org/10.1016/j.ijhydene.2018.11.059.

Italiano, C., Llorca, J., Pino, L., Ferraro, M., Antonucci, V., & Vita, A. (2020). CO and CO_2 methanation over Ni catalysts supported on CeO_2, Al_2O_3 and Y_2O_3 oxides. *Applied Catalysis B: Environmental*, *264*. https://doi.org/10.1016/j.apcatb.2019.118494, 118494.

Jacquemin, M., Beuls, A., & Ruiz, P. (2010). Catalytic production of methane from CO_2 and H_2 at low temperature: Insight on the reaction mechanism. *Catalysis Today*, *157*(1–4), 462–466. https://doi.org/10.1016/j.cattod.2010.06.016.

Jampaiah, D., Damma, D., Chalkidis, A., Venkataswamy, P., Bhargava, S. K., & Reddy, B. M. (2020). MOF-derived ceria-zirconia supported Co_3O_4 catalysts with enhanced activity in CO_2 methanation. *Catalysis Today*, *356*, 519–526. https://doi.org/10.1016/j.cattod.2020.05.047.

Jia, X., Zhang, X., Rui, N., Hu, X., & Liu, C. J. (2019). Structural effect of Ni/ZrO_2 catalyst on CO_2 methanation with enhanced activity. *Applied Catalysis B: Environmental*, *244*, 159–169. https://doi.org/10.1016/j.apcatb.2018.11.024.

Jiang, X., Nie, X., Guo, X., Song, C., & Chen, J. G. (2020). Recent advances in carbon dioxide hydrogenation to methanol via heterogeneous catalysis. *Chemical Reviews*, *120*(15), 7984–8034. https://doi.org/10.1021/acs.chemrev.9b00723.

Jinghui, T., Haihong, W., Qingli, Q., Shitao, H., Mengen, C., Shuaiqiang, J., … Buxing, H. (2021). Low temperature methanation of CO_2 over an amorphous cobalt-based catalyst. *Chemical Science*, *12*, 3937–3943. https://doi.org/10.1039/d0sc06414a.

Jwa, E., Mok, Y. S., & Lee, S. B. (2011). Nonthermal plasma-assisted catalytic methanation of CO and CO_2 over nickel-loaded alumina. *WIT Transactions on Ecology and the Environment*, *143*, 361–368. https://doi.org/10.2495/ESUS110311.

Jwa, E., Lee, S. B., Lee, H. W., & Mok, Y. S. (2013). Plasma-assisted catalytic methanation of CO and CO_2 over Ni-zeolite catalysts. *Fuel Processing Technology*, *108*, 89–93. https://doi.org/10.1016/j.fuproc.2012.03.008.

Karelovic, A., & Ruiz, P. (2012). CO_2 hydrogenation at low temperature over Rh/γ-Al_2O_3 catalysts: Effect of the metal particle size on catalytic performances and reaction mechanism. *Applied Catalysis B: Environmental*, *113–114*, 237–249. https://doi.org/10.1016/j.apcatb.2011.11.043.

Karelovic, A., & Ruiz, P. (2013a). Improving the hydrogenation function of Pd/γ-Al_2O_3 catalyst by Rh/γ-Al_2O_3 addition in CO_2 methanation at low temperature. *ACS Catalysis, 3*(12), 2799–2812. https://doi.org/10.1021/cs400576w.

Karelovic, A., & Ruiz, P. (2013b). Mechanistic study of low temperature CO_2 methanation over Rh/TiO_2 catalysts. *Journal of Catalysis, 301,* 141–153. https://doi.org/10.1016/j.jcat.2013.02.009.

Kattel, S., Ramírez, P. J., Chen, J. G., Rodriguez, J. A., & Liu, P. (2017). Active sites for CO_2 hydrogenation to methanol on Cu/ZnO catalysts. *Science, 355*(6331), 1296–1299. https://doi.org/10.1126/science.aal3573.

Kho, E. T., Scott, J., & Amal, R. (2016). Ni/TiO_2 for low temperature steam reforming of methane. *Chemical Engineering Science, 140,* 161–170. https://doi.org/10.1016/j.ces.2015.10.021.

Kim, H. Y., Lee, H. M., & Park, J. N. (2010). Bifunctional mechanism of CO_2 methanation on Pd-MgO/SiO_2 catalyst: Independent roles of MgO and Pd on CO_2 methanation. *Journal of Physical Chemistry C, 114*(15), 7128–7131. https://doi.org/10.1021/jp100938v.

Kim, M. J., Youn, J. R., Kim, H. J., Seo, M. W., Lee, D., Go, K. S., … Jeon, S. G. (2020). Effect of surface properties controlled by Ce addition on CO_2 methanation over Ni/Ce/Al_2O_3 catalyst. *International Journal of Hydrogen Energy, 45*(46), 24595–24603. https://doi.org/10.1016/j.ijhydene.2020.06.144.

Kwak, B. S., Vignesh, K., Park, N. K., Ryu, H. J., Baek, J. I., & Kang, M. (2015). Methane formation from photoreduction of CO_2 with water using TiO_2 including Ni ingredient. *Fuel, 143,* 570–576. https://doi.org/10.1016/j.fuel.2014.11.066.

Lapidus, A. L., Gaidai, N. A., Nekrasov, N. V., Tishkova, L. A., Agafonov, Y. A., & Myshenkova, T. N. (2007). The mechanism of carbon dioxide hydrogenation on copper and nickel catalysts. *Petroleum Chemistry, 47*(2), 75–82. https://doi.org/10.1134/S0965544107020028.

Le, M. C., Van, K. L., Nguyen, T. H. T., & Nguyen, N. H. (2017). The impact of Ce-Zr addition on nickel dispersion and catalytic behavior for CO_2 methanation of Ni/AC catalyst at low temperature. *Journal of Chemistry, 2017.* https://doi.org/10.1155/2017/4361056, 4361056.

Le, T. A., Kim, M. S., Lee, S. H., Kim, T. W., & Park, E. D. (2017). CO and CO_2 methanation over supported Ni catalysts. *Catalysis Today, 293–294,* 89–96. https://doi.org/10.1016/j.cattod.2016.12.036.

Li, G., Hu, L., & Hill, J. M. (2006). Comparison of reducibility and stability of alumina-supported Ni catalysts prepared by impregnation and co-precipitation. *Applied Catalysis A: General, 301*(1), 16–24. https://doi.org/10.1016/j.apcata.2005.11.013.

Li, W., Nie, X., Jiang, X., Zhang, A., Ding, F., Liu, M., … Song, C. (2018). ZrO_2 support imparts superior activity and stability of co catalysts for CO_2 methanation. *Applied Catalysis B: Environmental, 220,* 397–408. https://doi.org/10.1016/j.apcatb.2017.08.048.

Li, W., Liu, Y., Mu, M., Ding, F., Liu, Z., Guo, X., & Song, C. (2019). Organic acid-assisted preparation of highly dispersed Co/ZrO_2 catalysts with superior activity for CO_2 methanation. *Applied Catalysis B: Environmental, 254,* 531–540. https://doi.org/10.1016/j.apcatb.2019.05.028.

Li, J., Lin, Y., Pan, X., Miao, D., Ding, D., Cui, Y., … Bao, X. (2019). Enhanced CO_2 methanation activity of Ni/anatase catalyst by tuning strong metal-support interactions. *ACS Catalysis, 9*(7), 6342–6348. https://doi.org/10.1021/acscatal.9b00401.

Li, S., Liu, G., Zhang, S., An, K., Ma, Z., Wang, L., & Liu, Y. (2020). Cerium-modified Ni-La_2O_3/ZrO2 for CO_2 methanation. *Journal of Energy Chemistry, 43,* 155–164. https://doi.org/10.1016/j.jechem.2019.08.024.

Liang, B., Ma, J., Su, X., Yang, C., Duan, H., Zhou, H., … Huang, Y. (2019). Investigation on deactivation of Cu/ZnO/Al_2O_3 catalyst for CO_2 hydrogenation to methanol. *Industrial and Engineering Chemistry Research, 58*(21), 9030–9037. https://doi.org/10.1021/acs.iecr.9b01546.

Liang, C., Tian, H., Gao, G., Zhang, S., Liu, Q., Dong, D., & Hu, X. (2020). Methanation of CO_2 over alumina supported nickel or cobalt catalysts: Effects of the coordination between metal and support on formation of the reaction intermediates. *International Journal of Hydrogen Energy, 45*(1), 531–543. https://doi.org/10.1016/j.ijhydene.2019.10.195.

Lim, H. S., Kim, G., Kim, Y., Lee, M., Kang, D., Lee, H., & Lee, J. W. (2021). Ni-exsolved $La_{1-x}Ca_xNiO_3$ perovskites for improving CO_2 methanation. *Chemical Engineering Journal, 412.* https://doi.org/10.1016/j.cej.2020.127557, 127557.

Lin, X., Wang, S., Tu, W., Hu, Z., Ding, Z., Hou, Y., … Dai, W. (2019). MOF-derived hierarchical hollow spheres composed of carbon-confined Ni nanoparticles for efficient CO_2 methanation. *Catalysis Science and Technology, 9*(3), 731–738. https://doi.org/10.1039/c8cy02329h.

Lin, S., Hao, Z., Shen, J., Chang, X., Huang, S., Li, M., & Ma, X. (2021). Enhancing the CO_2 methanation activity of Ni/CeO_2 via activation treatment-determined metal-support interaction. *Journal of Energy Chemistry, 59,* 334–342. https://doi.org/10.1016/j.jechem.2020.11.011.

Liu, Q., & Tian, Y. (2017). One-pot synthesis of NiO/SBA-15 monolith catalyst with a three-dimensional framework for CO_2 methanation. *International Journal of Hydrogen Energy, 42*(17), 12295–12300. https://doi.org/10.1016/j.ijhydene.2017.02.070.

Liu, E., Qi, L., Bian, J., Chen, Y., Hu, X., Fan, J., ... Wang, Q. (2015). A facile strategy to fabricate plasmonic Cu modified TiO_2 nano-flower films for photocatalytic reduction of CO_2 to methanol. *Materials Research Bulletin*, *68*, 203–209. https://doi.org/10.1016/j.materresbull.2015.03.064.

Liu, Q., Bian, B., Fan, J., & Yang, J. (2018). Cobalt doped Ni based ordered mesoporous catalysts for CO_2 methanation with enhanced catalytic performance. *International Journal of Hydrogen Energy*, *43*(10), 4893–4901. https://doi.org/10.1016/j.ijhydene.2018.01.132.

Liu, K., Xu, X., Xu, J., Fang, X., Liu, L., & Wang, X. (2020). The distributions of alkaline earth metal oxides and their promotional effects on Ni/CeO_2 for CO_2 methanation. *Journal of CO_2 Utilization*, *38*, 113–124. https://doi.org/10.1016/j.jcou.2020.01.016.

Liu, T., Hong, X., & Liu, G. (2020). In situ generation of the Cu@3D-ZrOx framework catalyst for selective methanol synthesis from CO_2/H_2. *ACS Catalysis*, *10*(1), 93–102. https://doi.org/10.1021/acscatal.9b03738.

Loder, A., Siebenhofer, M., & Lux, S. (2020). The reaction kinetics of CO_2 methanation on a bifunctional Ni/MgO catalyst. *Journal of Industrial and Engineering Chemistry*, *85*, 196–207. https://doi.org/10.1016/j.jiec.2020.02.001.

Lu, B., & Kawamoto, K. (2013). Preparation of monodispersed NiO particles in SBA-15, and its enhanced selectivity for reverse water gas shift reaction. *Journal of Environmental Chemical Engineering*, *1*(3), 300–309. https://doi.org/10.1016/j.jece.2013.05.008.

Lu, B., Ju, Y., Abe, T., & Kawamoto, K. (2015). Grafting Ni particles onto SBA-15, and their enhanced performance for CO_2 methanation. *RSC Advances*, *5*(70), 56444–56454. https://doi.org/10.1039/c5ra07461d.

Lu, H., Yang, X., Gao, G., Wang, J., Han, C., Liang, X., ... Chen, X. (2016). Metal (Fe, Co, Ce or La) doped nickel catalyst supported on ZrO_2 modified mesoporous clays for CO and CO_2 methanation. *Fuel*, *183*, 335–344. https://doi.org/10.1016/j.fuel.2016.06.084.

Luciano, A., Elisabetta, R., Daniela, M., Roberto, M., Franca, S. M., & Giorgia, C. M. (2019). Nanostructured $Ni/CeO_2–ZrO_2$ catalysts for CO_2 conversion into synthetic natural gas. *Journal of Nanoscience and Nanotechnology*, 3269–3276. https://doi.org/10.1166/jnn.2019.16612.

Lv, Y., Xin, Z., Meng, X., Tao, M., & Bian, Z. (2018). Ni based catalyst supported on KIT-6 silica for CO methanation: Confinement effect of three dimensional channel on NiO and Ni particles. *Microporous and Mesoporous Materials*, *262*, 89–97. https://doi.org/10.1016/j.micromeso.2017.06.022.

Lv, C., Xu, L., Chen, M., Cui, Y., Wen, X., Li, Y., ... Shou, Q. (2020). Recent progresses in constructing the highly efficient Ni based catalysts with advanced low-temperature activity toward CO_2 methanation. *Frontiers in Chemistry*, *8*. https://doi.org/10.3389/fchem.2020.00269.

Manthiram, K., Beberwyck, B. J., & Alivisatos, A. P. (2014). Enhanced electrochemical methanation of carbon dioxide with a dispersible nanoscale copper catalyst. *Journal of the American Chemical Society*, *136*(38), 13319–13325. https://doi.org/10.1021/ja5065284.

Maria, M., Gabriela, B., Monica, D., Lucian, B.-T., & Lazar, M. D. (2019). Pt/UiO-66 nanocomposites as catalysts for CO_2 Methanation process. *Journal of Nanoscience and Nanotechnology*, 3187–3196. https://doi.org/10.1166/jnn.2019.16607.

Marocco, P., Morosanu, E. A., Giglio, E., Ferrero, D., Mebrahtu, C., Lanzini, A., ... Centi, G. (2018). CO_2 methanation over Ni/Al hydrotalcite-derived catalyst: Experimental characterization and kinetic study. *Fuel*, *225*, 230–242. https://doi.org/10.1016/j.fuel.2018.03.137.

Matsumura, Y., Shen, W. J., Ichihashi, Y., & Okumura, M. (2001). Low-temperature methanol synthesis catalyzed over ultrafine palladium particles supported on cerium oxide. *Journal of Catalysis*, *197*(2), 267–272. https://doi.org/10.1006/jcat.2000.3094.

Mebrahtu, C., Perathoner, S., Giorgianni, G., Chen, S., Centi, G., Krebs, F., ... Abate, S. (2019). Deactivation mechanism of hydrotalcite-derived Ni-AlO: X catalysts during low-temperature CO_2 methanation via Ni-hydroxide formation and the role of Fe in limiting this effect. *Catalysis Science and Technology*, *9*(15), 4023–4035. https://doi.org/10.1039/c9cy00744j.

Men, Y. L., Liu, Y., Wang, Q., Luo, Z. H., Shao, S., Li, Y. B., & Pan, Y. X. (2019). Highly dispersed Pt-based catalysts for selective CO_2 hydrogenation to methanol at atmospheric pressure. *Chemical Engineering Science*, *200*, 167–175. https://doi.org/10.1016/j.ces.2019.02.004.

Meng, Y., Zhu, L., Zhu, X., Reubroycharoen, P., & Wang, S. (2020). CO_2 methanation over nickel-based catalysts supported on MCM-41 with in situ doping of zirconium. *Journal of CO_2 Utilization*. https://doi.org/10.1016/j.jcou.2020.101304, 101304.

Mikhail, M., Da Costa, P., Amouroux, J., Cavadias, S., Tatoulian, M., Ognier, S., & Gálvez, M. E. (2020). Electrocatalytic behaviour of CeZrO: X-supported Ni catalysts in plasma assisted CO_2 methanation. *Catalysis Science and Technology*, *10*(14), 4532–4543. https://doi.org/10.1039/d0cy00312c.

Moon, L. S., Hwan, L. Y., Hyun, M. D., Yoon, A. J., Duc, N. D., Woong, C. S., & Su, K. S. (2019). Reaction mechanism and catalytic impact of Ni/CeO_{2-x} catalyst for low-temperature CO_2 methanation. *Industrial & Engineering Chemistry Research*. https://doi.org/10.1021/acs.iecr.9b00983.

Muroyama, H., Tsuda, Y., Asakoshi, T., Masitah, H., Okanishi, T., Matsui, T., & Eguchi, K. (2016). Carbon dioxide methanation over Ni catalysts supported on various metal oxides. *Journal of Catalysis*, *343*, 178–184. https://doi.org/10.1016/j.jcat.2016.07.018.

Mutz, B., Carvalho, H. W. P., Mangold, S., Kleist, W., & Grunwaldt, J. D. (2015). Methanation of CO_2: Structural response of a Ni-based catalyst under fluctuating reaction conditions unraveled by operando spectroscopy. *Journal of Catalysis, 327*, 48–53. https://doi.org/10.1016/j.jcat.2015.04.006.

Mutz, B., Sprenger, P., Wang, W., Wang, D., Kleist, W., & Grunwaldt, J. D. (2018). Operando Raman spectroscopy on CO_2 methanation over alumina-supported Ni, Ni3Fe and NiRh0.1 catalysts: Role of carbon formation as possible deactivation pathway. *Applied Catalysis A: General, 556*, 160–171. https://doi.org/10.1016/j.apcata.2018.01.026.

Natesakhawat, S., Lekse, J. W., Baltrus, J. P., Ohodnicki, P. R., Howard, B. H., Deng, X., & Matranga, C. (2012). Active sites and structure-activity relationships of copper-based catalysts for carbon dioxide hydrogenation to methanol. *ACS Catalysis, 2*(8), 1667–1676. https://doi.org/10.1021/cs300008g.

Nizio, M., Albarazi, A., Cavadias, S., Amouroux, J., Galvez, M. E., & Da Costa, P. (2016). Hybrid plasma-catalytic methanation of CO_2 at low temperature over ceria zirconia supported Ni catalysts. *International Journal of Hydrogen Energy, 41*(27), 11584–11592. https://doi.org/10.1016/j.ijhydene.2016.02.020.

Nomura, N., Tagawa, T., & Goto, S. (1998). Effect of acid-base properties on copper catalysts for hydrogenation of carbon dioxide. *Reaction Kinetics and Catalysis Letters, 63*(1), 21–25. https://doi.org/10.1007/BF02475425.

Ocampo, F., Louis, B., & Roger, A.-C. (2009). Methanation of carbon dioxide over nickel-based $Ce_{0.72}Zr_{0.28}O_2$ mixed oxide catalysts prepared by sol–gel method. *Applied Catalysis A: General, 369*(1), 90–96. https://doi.org/10.1016/j.apcata.2009.09.005.

Ocampo, F., Louis, B., Kiwi-Minsker, L., & Roger, A. C. (2011). Effect of Ce/Zr composition and noble metal promotion on nickel based $Ce_xZr_{1-x}O_2$ catalysts for carbon dioxide methanation. *Applied Catalysis A: General, 392*(1–2), 36–44. https://doi.org/10.1016/j.apcata.2010.10.025.

Ola, O., & Mercedes Maroto-Valer, M. (2014). Role of catalyst carriers in CO_2 photoreduction over nanocrystalline nickel loaded TiO_2-based photocatalysts. *Journal of Catalysis, 309*, 300–308. https://doi.org/10.1016/j.jcat.2013.10.016.

O'Brien, P. G., Sandhel, A., Wood, T. E., Jelle, A. A., Hoch, L. B., Perovic, D. D., ... Ozin, G. A. (2014). Photomethanation of gaseous CO_2 over Ru/silicon nanowire catalysts with visible and near-infrared photons. *Advanced Science, 1*(1). https://doi.org/10.1002/advs.201400001.

O'Brien, P. G., Ghuman, K. K., Jelle, A. A., Sandhel, A., Wood, T. E., Loh, J. Y. Y., ... Ozin, G. A. (2018). Enhanced photothermal reduction of gaseous CO_2 over silicon photonic crystal supported ruthenium at ambient temperature. *Energy and Environmental Science, 11*(12), 3443–3451. https://doi.org/10.1039/c8ee02347f.

Pan, Q., Peng, J., Sun, T., Wang, S., & Wang, S. (2014). Insight into the reaction route of CO_2 methanation: Promotion effect of medium basic sites. *Catalysis Communications, 45*, 74–78. https://doi.org/10.1016/j.catcom.2013.10.034.

Panagiotopoulou, P. (2017). Hydrogenation of CO_2 over supported noble metal catalysts. *Applied Catalysis A: General, 542*, 63–70. https://doi.org/10.1016/j.apcata.2017.05.026.

Park, J. N., & McFarland, E. W. (2009). A highly dispersed Pd-Mg/SiO_2 catalyst active for methanation of CO_2. *Journal of Catalysis, 266*(1), 92–97. https://doi.org/10.1016/j.jcat.2009.05.018.

Park, Y., Kang, T., Lee, J., Kim, P., Kim, H., & Yi, J. (2004). Single-step preparation of Ni catalysts supported on mesoporous silicas (SBA-15 and SBA-16) and the effect of pore structure on the selective hydrodechlorination of 1,1,2-trichloroethane to VCM. *Catalysis Today, 97*(2–3), 195–203. https://doi.org/10.1016/j.cattod.2004.03.070.

Pasupulety, N., Driss, H., Alhamed, Y. A., Alzahrani, A. A., Daous, M. A., & Petrov, L. (2015). Studies on Au/Cu-Zn-Al catalyst for methanol synthesis from CO_2. *Applied Catalysis A: General, 504*, 308–318. https://doi.org/10.1016/j.apcata.2015.01.036.

Prieto, G., Shakeri, M., De Jong, K. P., & De Jongh, P. E. (2014). Quantitative relationship between support porosity and the stability of pore-confined metal nanoparticles studied on CuZnO/SiO_2 methanol synthesis catalysts. *ACS Nano, 8*(3), 2522–2531. https://doi.org/10.1021/nn406119j.

Rahmani, S., Rezaei, M., & Meshkani, F. (2014). Preparation of highly active nickel catalysts supported on mesoporous nanocrystalline γ-Al_2O_3 for CO_2 methanation. *Journal of Industrial and Engineering Chemistry, 20*(4), 1346–1352. https://doi.org/10.1016/j.jiec.2013.07.017.

Ramli, M. Z., Syed-Hassan, S. S. A., & Hadi, A. (2018). Performance of Cu-Zn-Al-Zr catalyst prepared by ultrasonic spray precipitation technique in the synthesis of methanol via CO_2 hydrogenation. *Fuel Processing Technology, 169*, 191–198. https://doi.org/10.1016/j.fuproc.2017.10.004.

Razzaq, R., Zhu, H., Jiang, L., Muhammad, U., Li, C., & Zhang, S. (2013). Catalytic methanation of CO and CO_2 in coke oven gas over Ni-CO/ZrO_2-CeO_2. *Industrial and Engineering Chemistry Research, 52*(6), 2247–2256. https://doi.org/10.1021/ie301399z.

Ren, J., Guo, H., Yang, J., Qin, Z., Lin, J., & Li, Z. (2015). Insights into the mechanisms of CO2 methanation on Ni(111) surfaces by density functional theory. *Applied Surface Science, 351*, 504–516. https://doi.org/10.1016/j.apsusc.2015.05.173.

Ren, J., Qin, X., Yang, J. Z., Qin, Z. F., Guo, H. L., Lin, J. Y., & Li, Z. (2015). Methanation of carbon dioxide over Ni-M/ZrO_2 (M = Fe, Co, Cu) catalysts: Effect of addition of a second metal. *Fuel Processing Technology*, *137*, 204–211. https://doi.org/10.1016/j.fuproc.2015.04.022.

Romero-Sáez, M., Dongil, A. B., Benito, N., Espinoza-González, R., Escalona, N., & Gracia, F. (2018). CO_2 methanation over nickel-ZrO_2 catalyst supported on carbon nanotubes: A comparison between two impregnation strategies. *Applied Catalysis B: Environmental*, *237*, 817–825. https://doi.org/10.1016/j.apcatb.2018.06.045.

Saito, M., Fujitani, T., Takeuchi, M., & Watanabe, T. (1996). Development of copper/zinc oxide-based multicomponent catalysts for methanol synthesis from carbon dioxide and hydrogen. *Applied Catalysis A: General*, *138*(2), 311–318. https://doi.org/10.1016/0926-860X(95)00305-3.

Samei, E., Taghizadeh, M., & Bahmani, M. (2012). Enhancement of stability and activity of $Cu/ZnO/Al_2O_3$ catalysts by colloidal silica and metal oxides additives for methanol synthesis from a CO_2-rich feed. *Fuel Processing Technology*, *96*, 128–133. https://doi.org/10.1016/j.fuproc.2011.12.028.

Sharma, S., Hu, Z., Zhang, P., McFarland, E. W., & Metiu, H. (2011). CO_2 methanation on Ru-doped ceria. *Journal of Catalysis*, *278*(2), 297–309. https://doi.org/10.1016/j.jcat.2010.12.015.

Singha, R. K., Yadav, A., Agrawal, A., Shukla, A., Adak, S., Sasaki, T., & Bal, R. (2016). Synthesis of highly coke resistant Ni nanoparticles supported MgO/ZnO catalyst for reforming of methane with carbon dioxide. *Applied Catalysis B: Environmental*, *191*, 165–178. https://doi.org/10.1016/j.apcatb.2016.03.029.

Singha, R. K., Shukla, A., Yadav, A., Sivakumar Konathala, L. N., & Bal, R. (2017). Effect of metal-support interaction on activity and stability of Ni-CeO_2 catalyst for partial oxidation of methane. *Applied Catalysis B: Environmental*, *202*, 473–488. https://doi.org/10.1016/j.apcatb.2016.09.060.

Sreedhar, I., Varun, Y., Singh, S. A., Venugopal, A., & Reddy, B. M. (2019). Developmental trends in CO_2 methanation using various catalysts. *Catalysis Science and Technology*, *9*(17), 4478–4504. https://doi.org/10.1039/c9cy01234f.

Stangeland, K., Kalai, D., Li, H., & Yu, Z. (2017). CO_2 methanation: The effect of catalysts and reaction conditions. *Energy Procedia*, *105*, 2022–2027. https://doi.org/10.1016/j.egypro.2017.03.577.

Stangeland, K., Li, H., & Yu, Z. (2018). Thermodynamic analysis of chemical and phase equilibria in CO_2 hydrogenation to methanol, dimethyl ether, and higher alcohols. *Industrial and Engineering Chemistry Research*, *57*(11), 4081–4094. https://doi.org/10.1021/acs.iecr.7b04866.

Storsæter, S., Tøtdal, B., Walmsley, J. C., Tanem, B. S., & Holmen, A. (2005). Characterization of alumina-, silica-, and titania-supported cobalt Fischer-Tropsch catalysts. *Journal of Catalysis*, *236*(1), 139–152. https://doi.org/10.1016/j.jcat.2005.09.021.

Suib, S. L. (2013). In S. Suib (Ed.), *New and future developments in catalysis activation of carbon dioxide* (1st ed.). Elsevier. https://doi.org/10.1016/C2010-0-68570-1.

Sun, C., Beaunier, P., La Parola, V., Liotta, L. F., & Da Costa, P. (2020). Ni/CeO_2 nanoparticles promoted by yttrium doping as catalysts for CO_2 methanation. *ACS Applied Nano Materials*, *3*(12), 12355–12368. https://doi.org/10.1021/acsanm.0c02841.

Sun, C., Świrk, K., Wierzbicki, D., Motak, M., Grzybek, T., & Da Costa, P. (2020). On the effect of yttrium promotion on Ni-layered double hydroxides-derived catalysts for hydrogenation of CO_2 to methane. *International Journal of Hydrogen Energy*, *46*(22). https://doi.org/10.1016/j.ijhydene.2020.03.202.

Sun, C., Beaunier, P., & Da Costa, P. (2020). Effect of ceria promotion on the catalytic performance of Ni/SBA-16 catalysts for CO_2 methanation. *Catalysis Science and Technology*, *10*(18), 6330–6341. https://doi.org/10.1039/d0cy00922a.

Tada, S., Shimizu, T., Kameyama, H., Haneda, T., & Kikuchi, R. (2012). Ni/CeO_2 catalysts with high CO_2 methanation activity and high CH_4 selectivity at low temperatures. *International Journal of Hydrogen Energy*, *37*(7), 5527–5531. https://doi.org/10.1016/j.ijhydene.2011.12.122.

Tada, S., Shimizu, T., Kameyama, H., Haneda, T., & Kikuchi, R. (2012). Ni/CeO_2 catalysts with high CO_2 methanation activity and high CH_4 selectivity at low temperatures. *International Journal of Hydrogen Energy*, *37*(7), 5527–5531. https://doi.org/10.1016/j.ijhydene.2011.12.122.

Tada, S., Ochieng, O. J., Kikuchi, R., Haneda, T., & Kameyama, H. (2014). Promotion of CO_2 methanation activity and CH_4 selectivity at low temperatures over $Ru/CeO_2/Al_2O_3$ catalysts. *International Journal of Hydrogen Energy*, *39*(19), 10090–10100. https://doi.org/10.1016/j.ijhydene.2014.04.133.

Tada, S., Watanabe, F., Kiyota, K., Shimoda, N., Hayashi, R., Takahashi, M., … Satokawa, S. (2017). Ag addition to CuO-ZrO_2 catalysts promotes methanol synthesis via CO_2 hydrogenation. *Journal of Catalysis*, *351*, 107–118. https://doi.org/10.1016/j.jcat.2017.04.021.

Tahir, M., & Amin, N. S. (2013). Advances in visible light responsive titanium oxide-based photocatalysts for CO_2 conversion to hydrocarbon fuels. *Energy Conversion and Management*, *76*, 194–214. https://doi.org/10.1016/j.enconman.2013.07.046.

Tahir, B., Tahir, M., & Amin, N. S. (2015). Performance analysis of monolith photoreactor for CO_2 reduction with H_2. *Energy Conversion and Management*, *90*, 272–281. https://doi.org/10.1016/j.enconman.2014.11.018.

Tahir, M., Tahir, B., Amin, N. A. S., & Muhammad, A. (2016). Photocatalytic CO_2 methanation over NiO/In_2O_3 promoted TiO_2 nanocatalysts using H_2O and/or H_2

reductants. *Energy Conversion and Management, 119*, 368–378. https://doi.org/10.1016/j.enconman.2016.04.057.

Takano, H., Kirihata, Y., Izumiya, K., Kumagai, N., Habazaki, H., & Hashimoto, K. (2016). Highly active Ni/Y-doped ZrO_2 catalysts for CO_2 methanation. *Applied Surface Science, 388*, 653–663. https://doi.org/10.1016/j.apsusc.2015.11.187.

Thampi, K. R., Kiwi, J., & Grätzel, M. (1987). Methanation and photo-methanation of carbon dioxide at room temperature and atmospheric pressure. *Nature, 327*(6122), 506–508. https://doi.org/10.1038/327506a0.

Thema, M., Bauer, F., & Sterner, M. (2019). Power-to-gas: Electrolysis and methanation status review. *Renewable and Sustainable Energy Reviews, 112*, 775–787. https://doi.org/10.1016/j.rser.2019.06.030.

Tsoukalou, A., Abdala, P. M., Stoian, D., Huang, X., Willinger, M. G., Fedorov, A., & Müller, C. R. (2019). Structural evolution and dynamics of an In_2O_3 catalyst for CO_2 hydrogenation to methanol: An operando XAS-XRD and in situ TEM study. *Journal of the American Chemical Society, 141*(34), 13497–13505. https://doi.org/10.1021/jacs.9b04873.

Ulmer, U., Dingle, T., Duchesne, P. N., Morris, R. H., Tavasoli, A., Wood, T., & Ozin, G. A. (2019). Fundamentals and applications of photocatalytic CO_2 methanation. *Nature Communications, 10*(1). https://doi.org/10.1038/s41467-019-10996-2.

Varun, Y., Sreedhar, I., & Singh, S. A. (2020). Highly stable M/NiO–MgO (M = Co, Cu and Fe) catalysts towards CO_2 methanation. *International Journal of Hydrogen Energy, 45*(53), 28716–28731. https://doi.org/10.1016/j.ijhydene.2020.07.212.

Wang, W., & Gong, J. (2011). Methanation of carbon dioxide: An overview. *Frontiers of Chemical Engineering in China, 5*(1), 2–10. https://doi.org/10.1007/s11705-010-0528-3.

Wang, Z. Y., Chou, H. C., Wu, J. C. S., Ping Tsai, D., Mul, G., & Engineering, C. (2010). CO_2 photoreduction using NiO/$InTaO_4$ in optical-fiber reactor for renewable energy. *Applied Catalysis A: General, 380*(1–2), 172–177. https://doi.org/10.1016/j.apcata.2010.03.059.

Wang, W., Wang, S., Ma, X., & Gong, J. (2011). Recent advances in catalytic hydrogenation of carbon dioxide. *Chemical Society Reviews, 40*(7), 3703–3727. https://doi.org/10.1039/c1cs15008a.

Wang, S., Pan, Q., Peng, J., & Wang, S. (2014). In situ FTIR spectroscopic study of the CO_2 methanation mechanism on Ni/Ce0.5Zr0.5O2. *Catalysis Science and Technology, 4*(2), 502–509. https://doi.org/10.1039/c3cy00868a.

Wang, W. H., Himeda, Y., Muckerman, J. T., Manbeck, G. F., & Fujita, E. (2015). CO_2 hydrogenation to Formate and methanol as an alternative to photo- and electrochemical CO_2 reduction. *Chemical Reviews, 115*(23), 12936–12973. https://doi.org/10.1021/acs.chemrev.5b00197.

Wang, F., Li, C., Zhang, X., Wei, M., Evans, D. G., & Duan, X. (2015). Catalytic behavior of supported Ru nanoparticles on the {100}, {110}, and {111} facet of CeO_2. *Journal of Catalysis, 329*(1), 177–186. https://doi.org/10.1016/j.jcat.2015.05.014.

Wang, F., He, S., Chen, H., Wang, B., Zheng, L., Wei, M., … Duan, X. (2016). Active site dependent reaction mechanism over Ru/CeO_2 catalyst toward CO_2 methanation. *Journal of the American Chemical Society, 138*(19), 6298–6305. https://doi.org/10.1021/jacs.6b02762.

Wang, H. H., Hou, L., Li, Y. Z., Jiang, C. Y., Wang, Y. Y., & Zhu, Z. (2017). Porous MOF with highly efficient selectivity and chemical conversion for CO_2. *ACS Applied Materials and Interfaces, 9*(21), 17969–17976. https://doi.org/10.1021/acsami.7b03835.

Wang, J., Li, G., Li, Z., Tang, C., Feng, Z., An, H., … Li, C. (2017). A highly selective and stable ZnO-ZrO_2 solid solution catalyst for CO_2 hydrogenation to methanol. *Science Advances, 3*(10), e1701290. https://doi.org/10.1126/sciadv.1701290.

Wang, F., Han, B., Zhang, L., Xu, L., Yu, H., & Shi, W. (2018). CO2 reforming with methane over small-sized Ni@SiO2 catalysts with unique features of sintering-free and low carbon. *Applied Catalysis B: Environmental, 235*, 26–35. https://doi.org/10.1016/j.apcatb.2018.04.069.

Wang, L., Liu, H., Ye, H., Hu, R., Yang, S., Tang, G., … Yang, Y. (2018). Vacuum thermal treated Ni-CeO2/SBA-15 catalyst for CO_2 methanation. *Nanomaterials, 8*(10), 759. https://doi.org/10.3390/nano8100759.

Wang, J., Tang, C., Li, G., Han, Z., Li, Z., Liu, H., … Li, C. (2019). High-performance MaZrOx (Ma = Cd, Ga) solid-solution catalysts for CO_2 hydrogenation to methanol. *ACS Catalysis, 9*(11), 10253–10259. https://doi.org/10.1021/acscatal.9b03449.

Wang, X., Zhu, L., Zhuo, Y., Zhu, Y., & Wang, S. (2019). Enhancement of CO_2 methanation over La-modified Ni/SBA-15 catalysts prepared by different doping methods. *ACS Sustainable Chemistry & Engineering, 7*(17), 14647–14660. https://doi.org/10.1021/acssuschemeng.9b02563.

Wang, Y., Zhao, Q., Wang, Y., Hu, C., & Da Costa, P. (2020). One-step synthesis of highly active and stable Ni-ZrOxfor Dry reforming of methane. *Industrial and Engineering Chemistry Research, 59*(25), 11441–11452. https://doi.org/10.1021/acs.iecr.0c01416.

Wang, B., Mikhail, M., Galvez, M. E., Cavadias, S., Tatoulian, M., Da Costa, P., & Ognier, S. (2020). Coupling experiment and simulation analysis to investigate physical parameters of CO_2 methanation in a plasma-catalytic hybrid process. *Plasma Processes and Polymers, 17*(9). https://doi.org/10.1002/ppap.201900261.

Wang, B., Mikhail, M., Cavadias, S., Tatoulian, M., Da Costa, P., & Ognier, S. (2021). Improvement of the activity of CO_2 methanation in a hybrid plasma-catalytic process in varying catalyst particle size or under pressure. *Journal of CO_2 Utilization, 46*. https://doi.org/10.1016/j.jcou.2021.101471.

Westermann, A., Azambre, B., Bacariza, M. C., Graça, I., Ribeiro, M. F., Lopes, J. M., & Henriques, C. (2017). The promoting effect of Ce in the CO_2 methanation performances on NiUSY zeolite: A FTIR in situ/operando study. *Catalysis Today, 283*, 74–81. https://doi.org/10.1016/j.cattod.2016.02.031.

Wierzbicki, D., Debek, R., Motak, M., Grzybek, T., Gálvez, M. E., & Da Costa, P. (2016). Novel Ni-La-hydrotalcite derived catalysts for CO_2 methanation. *Catalysis Communications, 83*, 5–8. https://doi.org/10.1016/j.catcom.2016.04.021.

Wierzbicki, D., Baran, R., Dębek, R., Motak, M., Grzybek, T., Gálvez, M. E., & Da Costa, P. (2017). The influence of nickel content on the performance of hydrotalcite-derived catalysts in CO_2 methanation reaction. *International Journal of Hydrogen Energy, 42*(37), 23548–23555. https://doi.org/10.1016/j.ijhydene.2017.02.148.

Wierzbicki, D., Motak, M., Grzybek, T., Gálvez, M. E., & Da Costa, P. (2018). The influence of lanthanum incorporation method on the performance of nickel-containing hydrotalcite-derived catalysts in CO_2 methanation reaction. *Catalysis Today, 307*, 205–211. https://doi.org/10.1016/j.cattod.2017.04.020.

Wierzbicki, D., Baran, R., Dębek, R., Motak, M., Gálvez, M. E., Grzybek, T., ... Glatzel, P. (2018). Examination of the influence of La promotion on Ni state in hydrotalcite-derived catalysts under CO_2 methanation reaction conditions: Operando X-ray absorption and emission spectroscopy investigation. *Applied Catalysis B: Environmental, 232*, 409–419. https://doi.org/10.1016/j.apcatb.2018.03.089.

Wierzbicki, D., Moreno, M. V., Ognier, S., Motak, M., Grzybek, T., Da Costa, P., & Gálvez, M. E. (2020). Ni-Fe layered double hydroxide derived catalysts for non-plasma and DBD plasma-assisted CO_2 methanation. *International Journal of Hydrogen Energy, 45*(17), 10423–10432. https://doi.org/10.1016/j.ijhydene.2019.06.095.

Wu, H. C., Chang, Y. C., Wu, J. H., Lin, J. H., Lin, I. K., & Chen, C. S. (2015). Methanation of CO_2 and reverse water gas shift reactions on Ni/SiO_2 catalysts: The influence of particle size on selectivity and reaction pathway. *Catalysis Science and Technology, 5*(8), 4154–4163. https://doi.org/10.1039/c5cy00667h.

Yan, Y., Dai, Y., Yang, Y., & Lapkin, A. A. (2018). Improved stability of Y_2O_3 supported Ni catalysts for CO_2 methanation by precursor-determined metal-support interaction. *Applied Catalysis B: Environmental, 237*, 504–512. https://doi.org/10.1016/j.apcatb.2018.06.021.

Yan, X., Sun, W., Fan, L., Duchesne, P. N., Wang, W., Kübel, C., Wang, D., Kumar, S. G. H., Li, Y. F., Tavasoli, A., Wood, T. E., Hung, D. L. H., Wan, L., Wang, L., Song, R., Guo, J., Gourevich, I., Jelle, A. A., Lu, J., ... Ozin, G. A. (2019). Nickel@Siloxene catalytic nanosheets for high-performance CO_2 methanation. *Nature Communications, 10*(1). https://doi.org/10.1038/s41467-019-10464-x.

Yang, H., Zhang, C., Gao, P., Wang, H., Li, X., Zhong, L., ... Sun, Y. (2017). A review of the catalytic hydrogenation of carbon dioxide into value-added hydrocarbons. *Catalysis Science and Technology, 7*(20), 4580–4598. https://doi.org/10.1039/c7cy01403a.

Ye, R. P., Gong, W., Sun, Z., Sheng, Q., Shi, X., Wang, T., ... Yao, Y. G. (2019). Enhanced stability of Ni/SiO_2 catalyst for CO_2 methanation: Derived from nickel phyllosilicate with strong metal-support interactions. *Energy, 188*. https://doi.org/10.1016/j.energy.2019.116059, 116059.

Ye, R. P., Li, Q., Gong, W., Wang, T., Razink, J. J., Lin, L., ... Yao, Y. G. (2020). High-performance of nanostructured Ni/CeO_2 catalyst on CO_2 methanation. *Applied Catalysis B: Environmental, 268*. https://doi.org/10.1016/j.apcatb.2019.118474, 118474.

Yin, S., & Ge, Q. (2012). Selective CO_2 hydrogenation on the γ-Al_2O_3 supported bimetallic CO-Cu catalyst. *Catalysis Today, 194*(1), 30–37. https://doi.org/10.1016/j.cattod.2012.01.011.

Yoshihara, J., & Campbell, C. T. (1996). Methanol synthesis and reverse water-gas shift kinetics over Cu(110) model catalysts: Structural sensitivity. *Journal of Catalysis, 161*(2), 776–782. https://doi.org/10.1006/jcat.1996.0240.

Younas, M., Loong Kong, L., Bashir, M. J. K., Nadeem, H., Shehzad, A., & Sethupathi, S. (2016). Recent advancements, fundamental challenges, and opportunities in catalytic methanation of CO_2. *Energy and Fuels, 30*(11), 8815–8831. https://doi.org/10.1021/acs.energyfuels.6b01723.

Yu, Y., Guo, L., Cao, H., Lv, Y., Wang, E., & Cao, Y. (2015). A new $Ni/Ni_3(BO_3)_2/NiO$ heterostructured photocatalyst with efficient reduction of CO_2 into CH4. *Separation and Purification Technology, 142*, 14–17. https://doi.org/10.1016/j.seppur.2014.12.014.

Yuan, K., Yang, L., Du, X., & Yang, Y. (2014). Performance analysis of photocatalytic CO_2 reduction in optical fiber monolith reactor with multiple inverse lights. *Energy Conversion and Management, 81*, 98–105. https://doi.org/10.1016/j.enconman.2014.02.027.

Zağli, E., & Falconer, J. L. (1981). Carbon dioxide adsorption and methanation on ruthenium. *Journal of Catalysis, 69*(1), 1–8. https://doi.org/10.1016/0021-9517(81)90122-6.

Zeng, Y., & Tu, X. (2016). Plasma-catalytic CO_2 hydrogenation at low temperatures. *IEEE Transactions on Plasma Science*, *44*(4), 405–411. https://doi.org/10.1109/TPS.2015.2504549.

Zhang, T., & Liu, Q. (2020). Lanthanum-modified MCF-derived nickel phyllosilicate catalyst for enhanced CO_2 methanation: A comprehensive study. *ACS Applied Materials and Interfaces*, *12*(17), 19587–19600. https://doi.org/10.1021/acsami.0c03243.

Zhang, Y., Jacobs, G., Sparks, D. E., Dry, M. E., & Davis, B. H. (2002). CO and CO_2 hydrogenation study on supported cobalt Fischer-Tropsch synthesis catalysts. *Catalysis Today*, *71*(3–4), 411–418. https://doi.org/10.1016/S0920-5861(01)00468-0.

Zhang, S., Muratsugu, S., Ishiguro, N., & Tada, M. (2013). Ceria-doped Ni/SBA-16 catalysts for dry reforming of methane. *ACS Catalysis*, *3*(8), 1855–1864. https://doi.org/10.1021/cs400159w.

Zhao, K., Wang, W., & Li, Z. (2016). Highly efficient Ni/ZrO_2 catalysts prepared via combustion method for CO_2 methanation. *Journal of CO2 Utilization*, *16*, 236–244. https://doi.org/10.1016/j.jcou.2016.07.010.

Zhao, J., Li, Y., Zhu, Y., Wang, Y., & Wang, C. (2016). Enhanced CO_2 photoreduction activity of black TiO_2-coated cu nanoparticles under visible light irradiation: Role of metallic cu. *Applied Catalysis A: General*, *510*, 34–41. https://doi.org/10.1016/j.apcata.2015.11.001.

Zhao, K., Wang, W., & Li, Z. (2016). Highly efficient Ni/ZrO_2 catalysts prepared via combustion method for CO_2 methanation. *Journal of CO_2 Utilization*, *16*, 236–244. https://doi.org/10.1016/j.jcou.2016.07.010.

Zhou, G., Wu, T., Zhang, H., Xie, H., & Feng, Y. (2014). Carbon dioxide methanation on ordered mesoporous Co/kit-6 catalyst. *Chemical Engineering Communications*, *201*(2), 233–240. https://doi.org/10.1080/00986445.2013.766881.

Zhou, G., Liu, H., Cui, K., Xie, H., Jiao, Z., Zhang, G., … Zheng, X. (2017). Methanation of carbon dioxide over Ni/CeO_2 catalysts: Effects of support CeO_2 structure. *International Journal of Hydrogen Energy*, *42*(25), 16108–16117. https://doi.org/10.1016/j.ijhydene.2017.05.154.

CHAPTER

4

Sorption enhanced catalysis for CO_2 hydrogenation towards fuels and chemicals with focus on methanation

Liangyuan Wei[a], *Wim Haije*[b], *Henrik Grénman*[c], *and Wiebren de Jong*[a]

[a]Faculty 3mE, Department of Process and Energy, Large-Scale Energy Storage Section, Delft University of Technology, Delft, The Netherlands [b]Faculty of Applied Science, Department of Chemical Engineering, Materials for Energy Conversion and Storage Section, Delft University of Technology, Delft, The Netherlands [c]Faculty of Science and Engineering, Johan Gadolin Process Chemistry Centre, Laboratory in Industrial Chemistry and Reaction Engineering, Åbo Akademi University, Turku, Finland

4.1 Introduction

The combustion of fossil fuels is the major source of carbon emissions to the environment (Modak et al., 2020). To achieve the goal of reducing carbon emissions according to the Paris agreement, the use of fossil fuels should be radically diminished as a source of energy and chemicals (Rogelj et al., 2016; UNFCCC, 2015). This results in CO_2 and H_2 becoming increasingly important feedstock for the fuels and chemical industry, with hydrogen being produced by renewable electricity, e.g., solar or wind power (Lee et al., 2020; Rönsch et al., 2016).

Shifting from fossil to renewable resources means that a new industrial platform has to be developed to provide carbon-based fuels and large-scale base chemicals to replace the current petrochemical routes. The global demand cannot be met to a sufficient extent solely by the indirect use of carbon dioxide via biomass, necessitating the direct use from point sources or direct air capture. This increases the value of CO_2 from waste to a commodity chemical. The production of chemicals by hydrogenation of CO_2 (Fig. 4.1) is a promising way of CO_2 mitigation (Artz et al., 2018; Bailera et al., 2017; Rahman et al., 2017; Thomas & Harris, 2016), and also a possible solution for large-scale

Heterogeneous Catalysis
https://doi.org/10.1016/B978-0-323-85612-6.00004-8

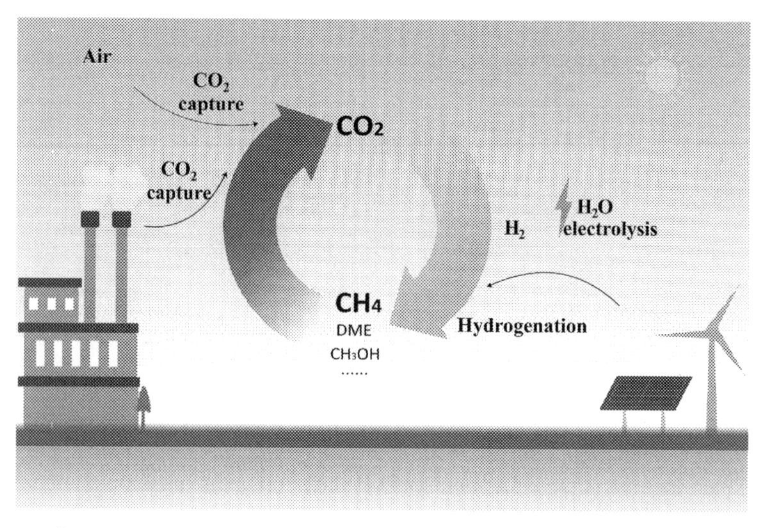

FIG. 4.1 A schematic of carbon cycle with CO_2 utilization by hydrogenation, producing methane.

energy storage to cope with the intermittent nature of wind and solar sources as well as the energy demand fluctuations (Bailera et al., 2017).

Several routes have been developed for CO_2 utilization through hydrogenation. Typical products include base and intermediary chemicals such as methane, methanol, formic acid, and dimethyl ether (Alvarez et al., 2017; Modak et al., 2020; Saeidi, Amin, & Rahimpour, 2014).

Methanol is an important and widely used liquid fuel and base chemical, around 95 million tons of methanol was produced in the world in 2019 (Bazaluk et al., 2020). The methanol synthesis from CO_2 hydrogenation is thermodynamically limited and more favorable at high pressure (Dang et al., 2019; Din et al., 2019; Xie et al., 2020). Around 700,000–800,000 tons of formic acid are produced per year in the world, CO_2 hydrogenation to synthesize formic acid is an important route for formic acid production (Bulushev & Ross, 2018). Dimethyl ether is an excellent fuel substituent for diesel and Liquefied Petroleum Gas (Roy, Cherevotan, & Peter, 2018) and the dimethyl ether global market was around 20 million tons in 2020 (Nakyai & Saebea, 2019). What is more, 3929.2 billion cubic metres methane (natural gas) was consumed in 2019 worldwide (BP p.l.c, 2020).

The CO_2 methanation reaction (Sabatier reaction (4.1)) was discovered in 1902 by Sabatier and Senderens (Sabatier & Senderens, 1902). CO_2 hydrogenation to produce methane (4.1) has great potential as an energy carrier (Thomas & Harris, 2016). This stems from the benefits of combining renewable hydrogen produced with wind or solar power, for instance, with CO_2 from traditional stack emissions such as power plants, biomass conversion processes (Li et al., 2017), or even air capture combined with the ease of distribution of the renewable methane in existing infrastructure. Moreover, CO_2 methanation has a higher energetic efficiency compared to producing, e.g., methanol from CO_2 for energy storage, the exergy efficiency of methanation and methanol synthesis being 30.1% to 18.2%, respectively (Uebbing, Rihko-Struckmann, & Sundmacher, 2019). Today, the methane, methanol, formic acid, and dimethyl ether are important chemicals used as fuels or raw materials in the industry (Alvarez et al., 2017).

$$CO_2 + 4H_2 \leftrightarrow CH_4 + 2H_2O;\ \Delta H^0_{298} = -165\,\text{kJ/mol} \quad (4.1)$$

Methanation is typically hampered by the thermodynamic conversion being far from 100% under currently viable reaction conditions (Fig. 4.2) (Alvarez et al., 2017; Ashok et al., 2020; Rönsch et al., 2016). According to Le Chatelier's principle (Carvill et al., 1996), the equilibrium can, however, be shifted by removing one of the reaction side products, which in many reactions between CO_2 and H_2 is water. This can be efficiently achieved by sorbents like zeolites (Borgschulte et al., 2013; Delmelle et al., 2016; Haije & Geerlings, 2011; Terreni et al., 2019; Van Berkel et al., 2020). This process is called a sorption enhanced reaction (Carvill et al., 1996). In addition to a high-purity reaction product, the sorption-enhanced CO_2 hydrogenation is also beneficial for high energy efficiency, lower temperature and pressure operation, and process simplification/intensification as less process steps and reactors are required (Carvill et al., 1996). Sorption enhanced CO_2 hydrogenation processes include sorption-enhanced CO_2 methanation (Borgschulte et al., 2013), methanol synthesis (Terreni et al., 2019) and dimethyl ether synthesis (Guffanti et al., 2021; van Kampen et al., 2020) to mention only a few. The current review mainly focuses on sorption enhanced CO_2 methanation as a case exemplifying the typical challenges encountered in CO_2 hydrogenation as well as some solutions for tackling them.

Sorption-enhanced CO_2 methanation (Fig. 4.3) was developed to obtain close to 100% conversion and yield in the otherwise thermodynamically limited conditions (Borgschulte et al., 2013; Walspurger et al., 2014). The water adsorbing capacity in the processes' operating window and catalytic performance are the critical material challenges in sorption enhanced CO_2 methanation. Even though novel catalysts, which are active at lower temperature than traditional catalysts are being developed, the water adsorbent still has to work at a relatively high

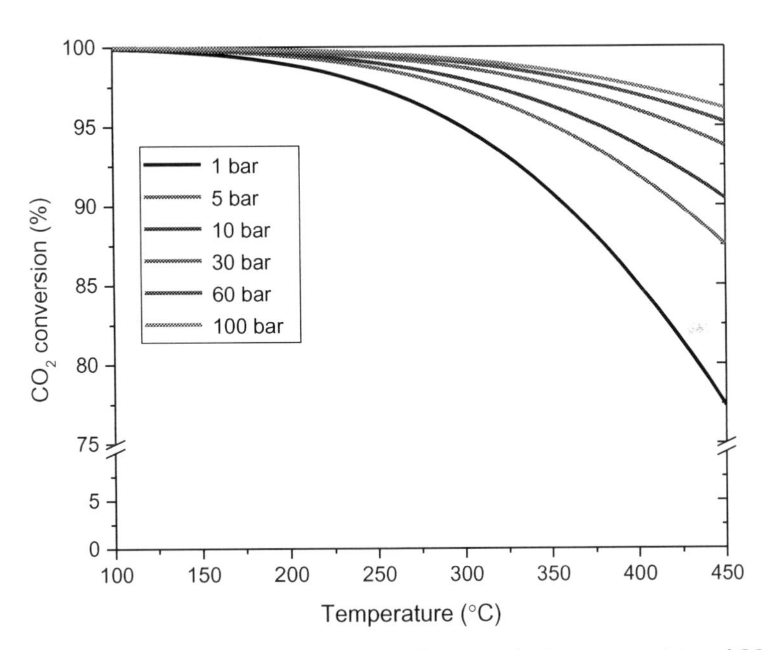

FIG. 4.2 Thermodynamic equilibrium conversion for the stoichiometric feed gas composition of CO_2 methanation.

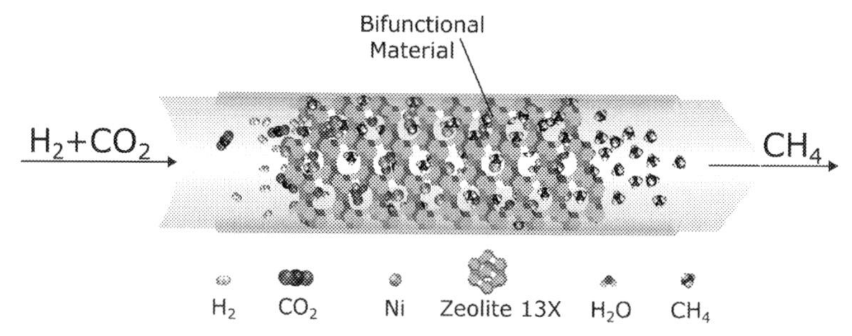

FIG. 4.3 Schematic of sorption enhanced CO_2 methanation.

temperature when thinking purely of adsorption, in order for reasonable reaction rates to be maintained. High temperature is not beneficial for water adsorption as high temperature rather enhances desorption. This emphasizes the importance of a proper material as the water adsorbent, which also acts as an efficient catalyst support (bifunctional material) in addition to developing low-temperature activity.

Despite the importance of the topic for developing efficient large-scale methanation processes, review publications on bifunctional materials synthesis and application for sorption enhanced CO_2 methanation are scarce. The current work presents the background and the state of the art in synthesis and application of these bifunctional materials and provides an outlook on future developments in sorption enhanced CO_2 methanation in particular and of sorption enhanced reactions in general.

4.2 Sorption enhanced catalysis for CO_2 methanation

Water sorbent

Sorbent choice

The requirement for a sorbent to be viable in sorption enhanced reactions is that the equilibrium vapor partial pressure is lower than that of the hydrogenation reaction. In the sorption enhanced CO_2 methanation, the water sorbent must be able to operate efficiently at a high temperature (>200°C) since the minimum working temperature of a conventional Ni-based catalysts, for example, is well over 300°C. Silica-gel cannot be used as its extremely low water capacity at the methanation reaction temperature (Goldsworthy, 2014; Wang & LeVan, 2009). The most promising sorbent class for CO_2 methanation is zeolites (van Kampen et al., 2019), which provides high water absorption capacity and are stable under the reactor and regeneration conditions.

During the last decades, around 70 scientific papers have been published on CO_2 methanation with the help of a zeolite, and the number of publications has increased significantly during the last 10 years (Fig. 4.4). The FAU (X and Y type) and LTA framework zeolites are the most studied ones. The FAU and LTA zeolite frameworks can be found in Fig. 4.5, more zeolite structure information can be found from reference (Newsam, 1986). Over 50% of the existing publications focused on utilizing FAU zeolite for CO_2 methanation. However, most publications mainly focus on the metal-zeolite catalysts preparation, characterization, and the catalytic performance in CO_2 methanation, where the zeolite is merely a support disregarding the sorption effect.

To investigate the sorption effect of Ni/5A zeolite catalysts in CO_2 methanation, Borgschulte et al. loaded Ni on 5A zeolite using the ion-exchange method. Their results show that

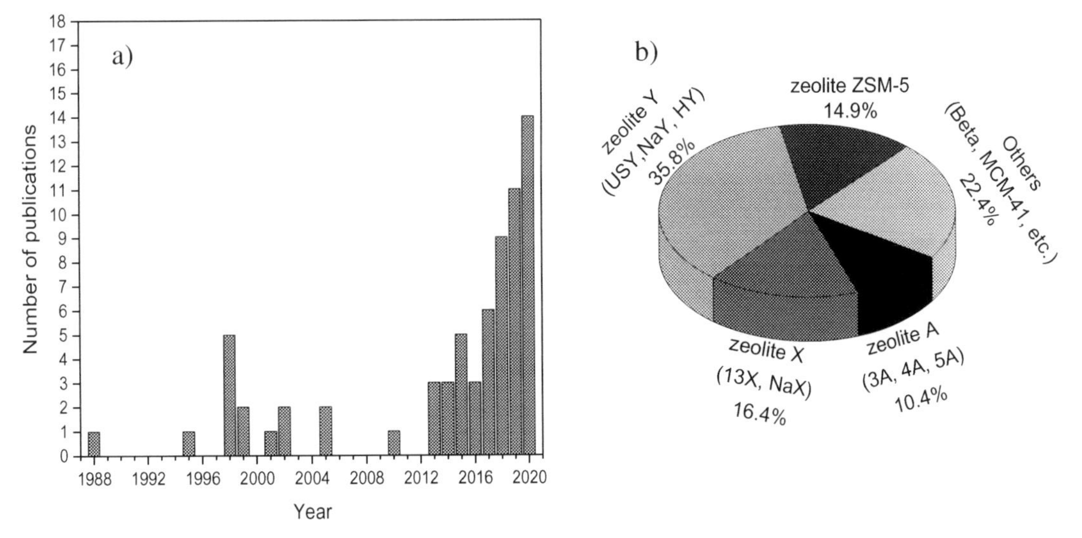

FIG. 4.4 Publications analysis (A) indexed topic as CO_2 methanation and zeolite/molecular sieve, (B) percentage of zeolite/molecular sieve used in CO_2 methanation. *Source: Web of Science, range 1900–2020, 3rd September.*

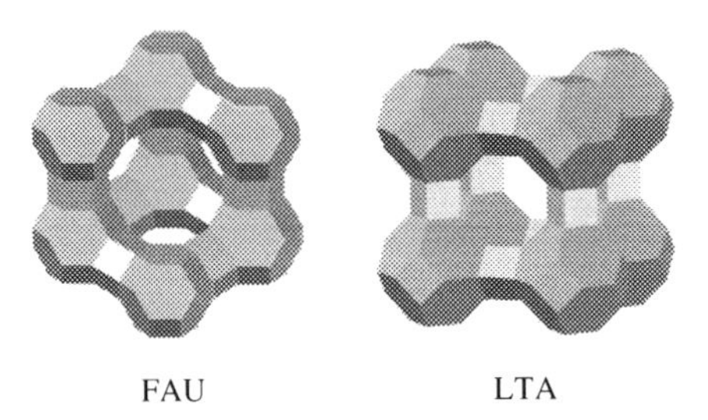

FIG. 4.5 The figurative construction of FAU and LTA zeolite frameworks that contain sodalite cages.

Ni/zeolite 5A can be used as an efficient bifunctional catalyst (Borgschulte et al., 2013). Similar to 5A zeolite, 4A zeolite has a high water sorption capacity. In 2014, Walspurger et al. reported results using 4A zeolite (physically mixed with commercial Ni/Al_2O_3 catalyst): around 100% CO_2 conversion could be obtained in the temperature range 250–350°C (Walspurger et al., 2014). However, by comparing the kinetics of Ni/5A and Ni/3A bifunctional material, Borgschulte et al. found that the CH_4 selectivity was greatly enhanced when the zeolite's pore size is larger than 5 Å (Borgschulte et al., 2015), 5 Å is large enough to allow the reactants (H_2 and CO_2) and the product (CH_4, CO) to enter and leave the zeolite.

A sorbent with a larger pore size should be beneficial for the rate diffusion of the reactants and products. Compared to LTA type zeolite (3A, 4A, and 5A), zeolite 13X (FAU) not only has a bigger pore size but it also presents a higher water sorption capacity (Ghodhbene et al., 2017; Tatlier, Munz, & Henninger, 2018). It was reported that 5%Ni/13X (5% is the weight percentage of Ni on the catalyst) displayed nearly threefold operation time compared to

5%Ni/5A in sorption enhanced CO_2 methanation, likely due to 13X having a higher water sorption capacity (Delmelle et al., 2016). Even though the CO_2 conversion levels with both 5A and 13X zeolite catalyst were similar (Delmelle et al., 2016). As can be seen from Fig. 4.4B, zeolite Y (FAU type) is also widely used as the catalyst support for CO_2 methanation. However, the water sorption effect of Y zeolite has not been discussed in the literature on CO_2 methanation (Graça et al., 2014; Quindimil et al., 2018; Westermann et al., 2015, 2017).

Regeneration of sorbents

The sorbent regeneration, i.e., the desorption of water is an essential step in the continuous operation of sorption enhanced CO_2 methanation. Sorbents have typically been regenerated under N_2, Air, or H_2 atmosphere at 300–500°C (Coppola et al., 2020; Delmelle et al., 2016; Walspurger et al., 2014). According to the Clausius-Clapeyron Eq. (4.2), the water sorption capacity decreases significantly when sorption temperature increases from 200 to 300°C as displayed in Fig. 4.6, while the regeneration temperature does not have a large effect on the subsequent water uptake for 4A and 3A zeolite (Coppola et al., 2020; Walspurger et al., 2014). Compared to H_2, air is a better carrier gas for the regeneration of the sorbent, since N_2 and O_2 have higher efficiency in carrying out water from the zeolite 5A and 13X due to oxygen and nitrogen molecules having sizes and weights that are comparable to the water molecule (Delmelle et al., 2016), but getting rid of the inert air constituents should be taken into consideration for the next cycle in practice operation.

$$\ln\frac{p}{p_0} = -\frac{\Delta H^0}{RT} + \frac{\Delta S^0}{R} \quad (4.2)$$

where p is the vapor partial pressure, p_0 is the reference pressure, T is adsorption temperature, R is the universal gas constant, ΔH^0 and ΔS^0 are the standard enthalpy and entropy change of the adsorption process, respectively (Bevers et al., 2007).

Walspurger et al. reported the zeolite 4A water sorption capacity at 200, 250, and 300°C under a water partial pressure of 0.039 bar. The

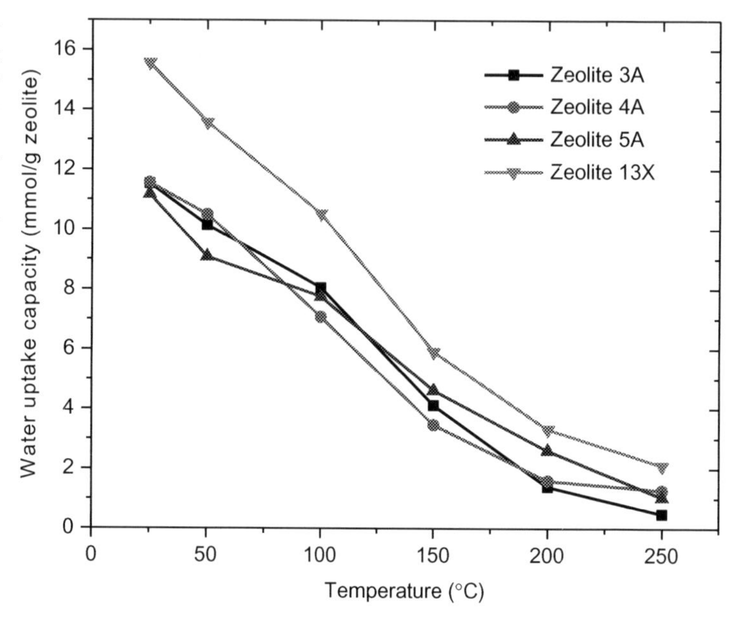

FIG. 4.6 Water mass uptake capacity under at 20°C saturated N_2 flow at different temperatures at 100 kPa total pressure for bead-shaped zeolites LTA-(3A, 4A, and 5A) and FAU-13X. *Data from Ghodhbene, M., et al. (2017). Hydrophilic zeolite sorbents for in-situ water removal in high temperature processes.* The Canadian Journal of Chemical Engineering, *95(10), 1842–1849.*

water capacity ranged from 0.98–2.00 mmol/g, clearly dependent on the sorption temperature as described by a Clausius-Clapeyron equation (Walspurger et al., 2014). Zeolite 3A showed similar water uptake capacity as zeolite 4A at sorption temperatures ranging 200–300°C (Coppola et al., 2020). The mass transfer rate during adsorption on zeolite 3A can be described by a linear driving force approximation, as the mass transfer resistance is predominantly determined by micropore resistance, due to the cage aperture (van Kampen, Boon, & van Sint Annaland, 2020). Zeolite 13X has a higher water uptake capacity compared to zeolite 3A, 4A, and 5A in the temperature range 20–250°C (Fig. 4.6) (Ghodhbene et al., 2017).

Other potential sorbents

Zeolites have been shown to be the most promising materials for in situ water removal in CO_2 methanation. The scientific articles published on CO_2 methanation with zeolites have mainly focused on FAU and LTA types zeolites, even though some other zeolites such as EMT and AFR could also be promising to be used for water removal (Ng & Mintova, 2008). Most work regarding the water sorption capacity for many typical zeolites has been performed at low temperature (<100°C) (Fig. 4.7) (Tatlier et al., 2018), however, the research on the water capacity of zeolites at high temperature (above 200°C) under different water partial pressures (>16.4 kPa) is scarce (Simo et al., 2009). Hardly any broad temperature and pressure range isosteres have been reported, even though the actual operational temperature in chemical processes is typically significantly above 100° C. Further work needs to be performed regarding the zeolites that could be used in sorption enhanced CO_2 hydrogenation.

Some new materials like MOF-74 and MOF-801 may also have the potential for water

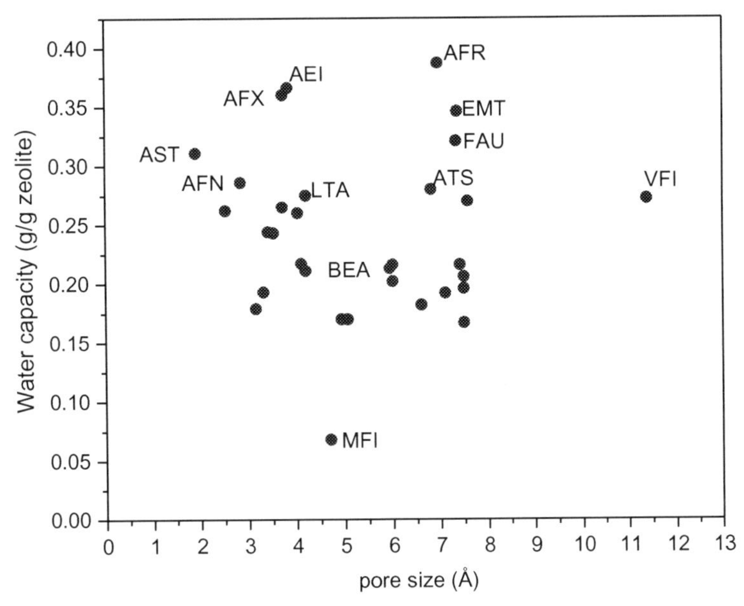

FIG. 4.7 Water adsorption capacity of zeolite with respect to its pore size. *Water capacity data from Tatlier, M., Munz, G., & Henninger S.K., (2018). Relation of water adsorption capacities of zeolites with their structural properties.* Microporous and Mesoporous Materials, 264, *70–75, pore size data from IZA-SC. Copyright © 2017 Structure Commission of the International Zeolite Association.*

removal since they have a large water uptake capacity at room temperature (Furukawa et al., 2014), however, their thermal stability at higher temperatures, >200°C, is questionable. Furthermore, one drawback is that their ecological footprint is substantial due to the organic precursors used in their manufacturing. Further study is needed for those materials.

Catalytic metals and promoters

Active metals such as Ni, Co, Ru, Rh, etc., have been studied for CO_2 methanation (Ashok et al., 2020; Lee et al., 2020; Nie et al., 2019; Sreedhar et al., 2019), and in some studies metals like W, La, and Ce, were used as promoters to enhance the catalytic metal dispersion increase, coke resistance, and anti-CO-poisoning ability (Sreedhar et al., 2019; Yan et al., 2016). The relevant active metals and promoters are displayed in Fig. 4.8.

The observed order of activity and selectivity for the respective metal catalysts on CO_2 methanation is shown below (Lee et al., 2020; Rönsch et al., 2016). However, this order is only a generalized trend and sometimes differs depending on, e.g., different metal-support interactions (Lee et al., 2020).

$$\text{Activity}: Ru > Fe > Ni > Co > Rh > Pd > Pt > Ir$$

$$\text{Selectivity}: Pd > Pt > Ir > Ni > Rh > Co > Fe > Ru$$

Catalytically active metals

Nickel (Ni) is the most widely used active metal for conventional (Aziz et al., 2015; Li et al., 2016; Rönsch et al., 2016; Walspurger, Haije, & Louis, 2014) and sorption enhanced CO_2 methanation (Bacariza et al., 2019a; Borgschulte et al., 2013; Delmelle et al., 2016; Walspurger et al., 2014), due to its rather high activity, CH_4 selectivity, and low cost, which makes Ni an interesting active metal from a commercial perspective (Fechete & Vedrine, 2015). 5%Ni/SiO_2 catalyst has an approximate apparent activation energy of 84 kJ/mol in CO_2 methanation (Aziz et al., 2014).

Cobalt (Co) catalysts exhibit a similar methanation activity and CH_4 selectivity comparable

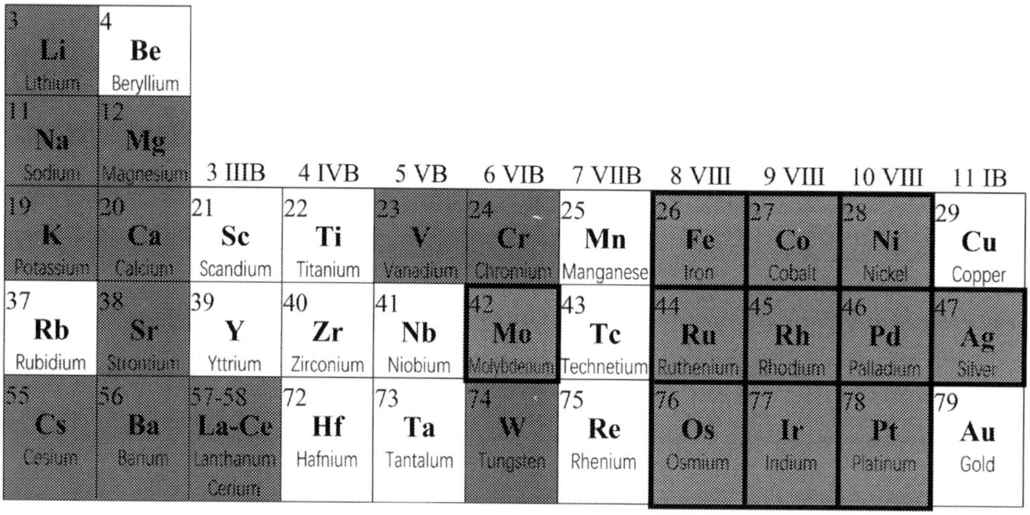

		3 IIIB	4 IVB	5 VB	6 VIB	7 VIIB	8 VIII	9 VIII	10 VIII	11 IB
3 Li Lithium	4 Be Beryllium									
11 Na Sodium	12 Mg Magnesium									
19 K Potassium	20 Ca Calcium	21 Sc Scandium	22 Ti Titanium	23 V Vanadium	24 Cr Chromium	25 Mn Manganese	26 Fe Iron	27 Co Cobalt	28 Ni Nickel	29 Cu Copper
37 Rb Rubidium	38 Sr Strontium	39 Y Yttrium	40 Zr Zirconium	41 Nb Niobium	42 Mo Molybdenum	43 Tc Technetium	44 Ru Ruthenium	45 Rh Rhodium	46 Pd Palladium	47 Ag Silver
55 Cs Cesium	56 Ba Barium	57-58 La-Ce Lanthanum Cerium	72 Hf Hafnium	73 Ta Tantalum	74 W Tungsten	75 Re Rhenium	76 Os Osmium	77 Ir Iridium	78 Pt Platinum	79 Au Gold

FIG. 4.8 Active metals *(marked in green; dark gray in print version, with thick borders)* and promotional metals *(marked in light blue; dark gray in print version)* for CO_2 methanation, excerpt from the periodic table of elements (Rönsch et al., 2016).

to Ni, the apparent activation energy of Co/MCF-17 in CO_2 methanation has been determined to be around 80 kJ/mol (Beaumont et al., 2014a). However, cobalt is not as widely used for commercial application since it is more expensive (Li et al., 2018; Sreedhar et al., 2019). Ashok et al. described that morphologies, surface orientations, catalyst supports, and cluster size of metals are the key factors in Co catalyst performance in CO_2 methanation (Ashok et al., 2020), like with other catalysts.

Iron (Fe) catalysts have been used for CO_2 reduction with H_2, and they exhibit a high activity, while showing low CH_4 selectivity, around 85% of the product gas is CO when using 5wt%Fe/13X at 350°C (Franken & Heel, 2020; Merkache et al., 2015; Wang et al., 2016), which hindered its use even though Fe is cheaper and less toxic than Ni and Co, and much cheaper than noble metals (Rh, Ru, Pd, Pt). 15%Fe/SiO_2 displayed a 134 kJ/mol of apparent activation energy for CO_2 methanation at 253 to 299°C (Weatherbee, 1984).

Molybdenum (Mo) has a low activity in CO_2 methanation and the number of publications on Mo catalysts for CO_2 methanation is low. A special feature of Mo is that it has the highest sulfur species tolerance (Rönsch et al., 2016). The reported apparent activation energy of Mo-based graphene catalysts for CO_2 methanation ranges between 62 and 115 kJ/mol (Primo et al., 2019).

As discussed above, the water adsorption capacity is higher at lower temperatures due to thermodynamics, therefore low-temperature catalytic activity of the bifunctional materials in CO_2 methanation would be highly beneficial for the sorption enhanced CO_2 methanation.

Ruthenium (Ru) is known as one of the most active metals for CO_2 methanation even at lower temperatures (Rönsch et al., 2016), and thus it is an attractive alternative (Ashok et al., 2020). An apparent activation energy of 41 kJ/mol was obtained by researchers for Ru/MgO catalyst in CO_2 methanation which was measured at 80 to 180°C (Mori et al., 1996). Ru on NaY and 5A zeolite catalysts have been observed to yield high CH_4 selectivity (Hastings et al., 2002), whereas 100% yield of CH_4 has been reported to be obtained at 160°C using Ru/TiO_2 catalyst (Abe et al., 2009). The dispersion of Ru on zeolite (13X and 5A) is important for the catalyst performance in CO_2 methanation (Wei et al., 2020) and high dispersion of Ru has been obtained on FAU zeolite with the ion-exchange method (Bando, Arakawa, & Ichikuni, 1999; Lee et al., 2019a).

Rhodium (Rh) has been reported to be an active metal for CO_2 methanation at low temperatures (Swalus et al., 2012), and its activation energy has been reported to be as low as 17.0 kJ/mol (3%Rh/TiO_2), and a turnover frequency of 0.524×10^2 (s^{-1}) in CO_2 methanation at 120°C (Karelovic & Ruiz, 2013a). Rh could thus be a promising active metal for sorption enhanced CO_2 methanation when loaded into zeolite, as the water uptake capacity of zeolites is increased dramatically at temperatures lower than 200°C (Fig. 4.6).

Platinum (Pt)-modified ZSM-5 zeolite catalyst has displayed high catalytic activity and CH_4 selectivity even with 0.5% loading of Pt (Sápi et al., 2019). Pt can promote the activity of Co for CO_2 methanation (Beaumont et al., 2014b). The apparent activation energy has been reported to be 94 kJ/mol for Pt/Al_2O_3 in CO_2 methanation (Kikkawa et al., 2019).

Luo et al. prepared a highly dispersed palladium (Pd)/Fe catalyst, and the results showed that Pd promoted and stabilized the catalyst significantly (Luo et al., 2020). 23.5 kJ/mol of apparent activation energy was measured at 250 to 350°C for Pd/γ-Al_2O_3 in CO_2 methanation (Karelovic & Ruiz, 2013b).

A summary of CO_2 methanation catalysts using different active metals can be found in Table 4.1.

Promoters

Promoters are used to improve the catalyst performance and their introduction may influence the catalyst properties such as the acidity,

TABLE 4.1 Summary of CO_2 methanation catalysts using different active metals.

Catalyst	m_{cat}[a] g	GHSV[b] mL/g/h	H_2: CO_2 –	p[c] (bar)	T[d] (°C)	X_{CO_2} (%)	S_{CH_4} (%)	E_a[e] (kJ/mol)	Reference
5%Ni/SiO_2	0.200	50,000	4:1	1	300	42.4	96.6	84	Aziz et al. (2014)
5.9%Ni/Al_2O_3	0.100	30,000	4:1	1	250	2	98.1	92	Kikkawa et al. (2019)
4.9%Co/MCF-17	0.050	60,000	3:1	6	250	5.1	58.8	80	Beaumont et al. (2014a)
15%Fe/SiO_2	0.39	1470[f]	4:1	1	253	7.7	12.9	134	Weatherbee (1984)
MoO_3-3/ graphene	0.020	12,000	3:1	10	400	21	100	79	Primo et al. (2019)
MoS_2-3/graphene	0.020	60,000	3:1	10	400	20	90	62	Primo et al. (2019)
0.5%Ru/γ-Al_2O_3	0.375	5000	4:1	1	350	82	99.5	68.1	Falbo et al. (2018)
3%Rh/TiO_2	0.200	6000	4:1	1	120	0.65	100	17	Karelovic and Ruiz (2013a)
2%Rh/γ-Al_2O_3	0.200	6000	4:1	1	200	6.2	100	16.4	Karelovic and Ruiz (2013b)
19.5%Pt/Al_2O_3	0.100	30,000	4:1	1	250	2	2	94	Beaumont et al. (2014a)
1.1%Pt/MCF-17	0.050	60,000	3:1	6	250	1	98	–	Kikkawa et al. (2019)
5%Pd/γ-Al_2O_3	0.200	6000	4:1	1	300	4.3	100	23.5	Karelovic and Ruiz (2013b)

[a] m_{cat}, *catalyst mass used in experiments.*
[b] GHSV, *gas hourly space velocity, mL/g_cat/h.*
[c] p, *reaction pressure.*
[d] T, *reaction temperature;* X_{CO2}, *CO_2 conversion;* S_{CH4}, *CH_4 selectivity*
[e] E_a, *apparent activation energy;* f, *unit is/h*

basicity, dispersion of the active metal, etc., and as a result affect the activity, selectivity, and the resistance to coke deposition.

In addition to promoting the catalyst activity and stability, adding W in the Ni-MgO_x catalyst can promote its anti-CO-poisoning ability and resistance against coke formation (Yan et al., 2016).

Na has been observed to have a negative effect on Na/Ni/CeO_2 catalysts performance on CO_2 methanation, even at low concentrations (0.1 wt%). The adding of Na was reported to decrease the amount of chemisorbed CO_2 on Na/Ni/CeO_2 catalysts, and the CO_2 methanation activity of Na/Ni/CeO_2 decreased with Na content. While a positive effect of Na was observed for CO_2 methanation over Na/Ni/SiO_2 catalysts in which the amount of chemisorbed CO_2 increases with Na content (Le et al., 2018). This was speculated to be related to the position on and interaction of Na with the support. The Na content in 13X zeolite is around 10 wt%, but Ni/13X catalyst has shown high activity in CO_2 methanation (Wei et al., 2021). The effect of alkali promoters (Li, Na, K, Cs) seems to depend also on the active metal, support, dispersion, and loading (Petala & Panagiotopoulou, 2018).

The addition of lanthanum (La) in Ni/BETA zeolite catalyst can promote CO_2 conversion since La enhances the formation of surface hydroxyl groups greatly, which can interact with CO_2 and thus promote CO_2 conversion (Chen et al., 2019). The increasing amount of CO_2 adsorbed by the catalyst as a result of the La_2O_3 was reported to lead to a higher CO_2 methanation activity and CH_4 selectivity (Quindimil et al., 2018). Quindimil et al. reported that the Ni particle growth was influenced by La during catalyst calcination (Quindimil et al., 2018). Cerium (Ce) is a typical metal used as a promoter. It has been reported that Ce can further improve the catalyst activity and selectivity, due to the ability to promote CO_2 dissociation (Graça et al., 2014; Westermann et al., 2017).

Preparation of the bifunctional material

In general, the catalyst activity and selectivity will be influenced by different preparation methods and different supports, since the active metal dispersion, particle size, and location (Graça et al., 2014), the acidity, and interaction with the Si-Al framework will be different. In order to remove the water produced during the CO_2 methanation, a zeolite should be used in preparing the bifunctional material, which sets certain requirements for the preparation. There are two types of routes for the preparation of materials combining both catalytic and high water uptake capabilities: (a) mixing and shaping (physical mixture) route; (b) catalytic metal loading (chemical) route as displayed in Fig. 4.9.

Mixing and shaping route

In the mixing and shaping route, the catalyst preparation process is separated from the synthesis of the sorbent. In this way, the bifunctional material preparation will not be limited by the sorbent structure and other properties, such as the pore size, the acid and basic sites, etc. It is an easier way to prepare the bifunctional material compared to loading the metals directly on the sorbent, as all possible catalyst preparation methods can be utilized, such as impregnation, ion-exchange, deposition precipitation, chemical vapor deposition, atomic layer deposition, and co-precipitation (Pierson, 1999; Schlögl, 2008). As long as a highly active catalyst can be obtained (even purchased from commercial companies), mixing the catalyst and sorbent can be performed. One of the few examples is from Walspurger et al.. They physically mixed

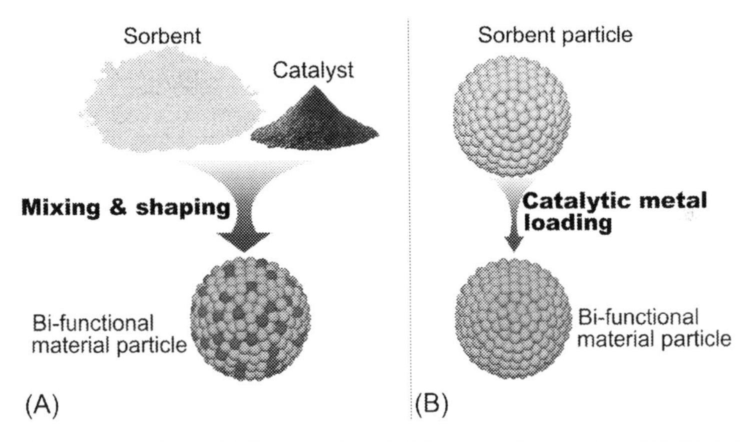

FIG. 4.9 A schematic representation of bifunctional material preparation routes. (A) Mixing and shaping route; (B) Catalytic metal loading route.

a commercial Ni catalyst with 4A zeolite and pelletized the mixture for sorption enhanced CO_2 methanation and, around 100% CO_2 conversion was obtained at 250–350°C (Walspurger et al., 2014). However, the pelletizing is not always straightforward as mechanical stability may pose an issue. Moreover, pore blockage during pelletizing can significantly decrease the surface area of the material. This again results in significantly lower performance both in the catalysis as well as in the absorptive capacity and kinetics.

Catalytic metal loading route

Although the mixing and shaping route is very flexible, the distance between the sorbing and catalytic site would be in a micro-meter scale in these bifunctional materials. Terreni et al. reported that nano-structuring sorption enhanced catalysts to shorten the diffusion pathway is superior over physical mixtures of macroscopic sorbents and other catalysts which result in longer diffusion path lengths (Terreni et al., 2018). In other words, in sorption enhanced CO_2 methanation, close proximity of sorption and catalytic sites is a prerequisite to conversion enhancement and process intensification. Materials prepared by loading the active metals directly into the zeolite would therefore be preferable.

Ion exchange is a typical method for catalyst preparation; the active metal, Ni for instance, can be loaded in the zeolite by exchanging it with the Na^+, K^+, Ca^{2+} from the zeolite framework (Schlögl, 2008). Recent publications have used ion exchange to prepare bifunctional materials with sorption enhancement (Borgschulte et al., 2013, 2015, 2016). In an ion-exchange process, a certain amount of metal salt (Ni precursor, e.g., nickel nitrate) is dissolved in distilled water. The sorbent, also acting as the catalyst support, is added into the solution and the mixture solution is stirred before filtering and washing with distilled water, after which the obtained solid sample is dried before calcination (Borgschulte et al., 2013; Schlögl, 2008). The metal loading is limited by the amount of exchangeable ions in the zeolite, however, the method results in highly dispersed catalyst when successful.

Impregnation is another conventional way for catalyst preparation, in which a certain amount of distilled water is used to dissolve the metal precursor. The water amount exceeds the pore volume of the support in wet impregnation, while it is equal to the pore volume of the support in incipient wetness impregnation. The sorbent is added to the solution, and the water in the mixture solution is removed by filtering or evaporation after several hours of stirring. The obtained solid is then dried further in an oven before calcination (Wei et al., 2020).

Other considerations for the bifunctional material preparation

Sub-nanometer or single-atom-based materials are the desired option, as they typically display high activity and they should largely retain their water-adsorption capacity after metal loading as low loadings can be used due to the high dispersion and pore blockage should then be avoided due to the small cluster size. Several strategies exist for synthesizing sub-nano/single-atom catalysts (Chen et al., 2018; Zhang et al., 2018a), of which mass-selected soft loading (Abbet et al., 2000) and atomic layer deposition (Peters et al., 2015; Yan et al., 2015) are limited by high cost and are complicated methods for large-scale catalyst production. In recent years, other synthesis strategies have emerged for single-atom catalyst preparation, such as defect engineering (Qiu et al., 2015; Zhang et al., 2018b), coordination pyrolysis (Chen et al., 2017; Han et al., 2017; Yin et al., 2016), and gas migration using volatile metal complexes (Qu et al., 2018). However, it is difficult to keep the material's structure intact since the high calcination or pyrolysis temperatures needed are prone to damage the structure of the support irreversibly. Furthermore, there is the possibility of sintering of the metal (nano) particles. Additionally, these preparation routes are difficult to scale up because of the complicated and expensive methods involved. Nevertheless, if solutions can be found for the drawbacks mentioned above, it will be feasible to prepare an extremely highly dispersed (maybe down to single-atom) catalyst and combined with a sorbent for sorption enhanced CO_2 methanation. However, physical mixtures are mainly the possible option in many of the cases.

It is possible to prepare highly dispersed zeolite catalysts by a novel strategy or by modifying conventional methods (Kistler et al., 2014; Lee et al., 2019a; Liu et al., 2019a). Liu et al. reported a general strategy to prepare a highly dispersed (even single atom) Pt, Pd, Ru, Rh, Co, Ni, and Cu on Y zeolite by adding ethylenediamine (EDA) as ligand to adjust the size the of precursors (Fig. 4.10) (Liu et al., 2019b).

Moreover, the metal precursor has also been observed to play an important role in the metal dispersion during the catalyst preparation, which influences the catalyst performance in CO_2 methanation. Wei et al. clearly showed how the selection of the precursor can influence the properties of a nickel-modified catalyst, in which nickel citrate precursor resulted in better dispersion compared to nitrate and acetate (Wei et al., 2021).

Characterization of the material

Similar to conventional catalysts, the bifunctional material has some important properties which relate to its performance in CO_2 methanation, such as the crystal structure and size, the dispersion of the active metal, the actual loading of active and promoter metal, the cluster size of metal, the acidity and basicity of the material, the pore size, volume and surface area, etc. Those properties can be characterized by conventional techniques such as X-ray powder diffraction (XRD) for crystal structure and size, scanning electron microscopy (SEM) with energy-dispersive X-ray spectroscopy (EDX) for material morphology and surface elemental content, transmission electron microscopy (TEM) for metal particle size on the support, scanning transmission electron microscopy equipped with energy-dispersive X-ray spectroscopy (STEM-EDX) for elemental distribution, X-ray photoelectron spectroscopy (XPS) for chemical valence states, hydrogen-temperature programmed reduction (H_2-TPR) for catalyst reduction behavior, CO_2-temperature programmed desorption (CO_2-TPD) for basicity, pyridine-FTIR for acidity distribution, and N_2 adsorption for pore size, volume and surface area, etc.

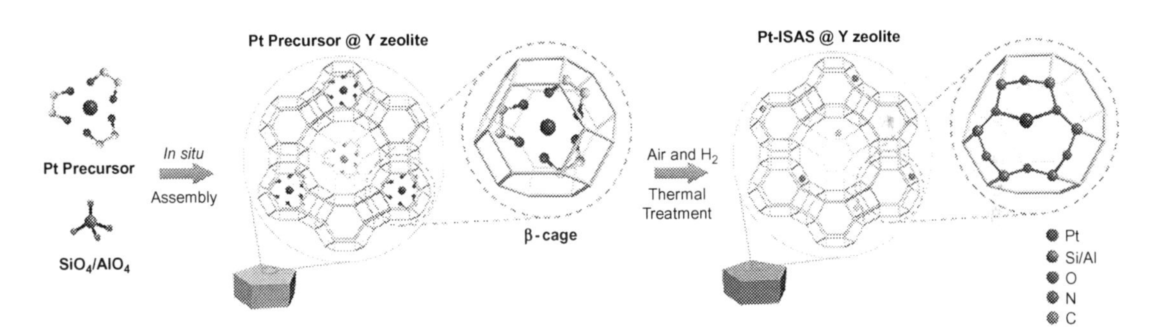

FIG. 4.10 A schematic illustration of the in-situ separation and confinement of platinum precursor in β-cage followed by thermal treatments. *Reprinted (adapted) with permission from Liu, Y., et al. (2019b). A general strategy for fabricating isolated single metal atomic site catalysts in Y zeolite.* Journal of the American Chemical Society, 141*(23), 9305–9311. Copyright (2019) American Chemical Society.*

Single-atom catalyst is a challenging topic in the analysis, HAADF-STEM (High-angle annular dark-field-scanning transmission electron microscope) has in some cases been successfully used to characterize the single atoms catalyst to get direct evidence of metal dispersion (Liu et al., 2019b).

For most single-atom catalyst, XANES (X-ray Absorption Near Edge Structure) and EXAFS (Extended X-Ray Absorption Fine Structure) are useful spectroscopic techniques to further determine the dispersion and coordination of the active metal in the zeolite (Liu et al., 2019b).

Performance of the material

Around 100% CO_2 conversion and 100% CH_4 selectivity have been obtained by using bifunctional material 6%Ni/5A in sorption-enhanced CO_2 methanation at temperatures lower than 200°C (Borgschulte et al., 2013), and 100% CH_4 selectivity was obtained using 5%Ni/5A and 5%Ni/13X under atmospheric pressure in fixed bed reactor (Delmelle et al., 2016). These materials are easily regenerated under H_2 or air atmosphere at elevated temperatures (Delmelle et al., 2016; Walspurger et al., 2014). The results of the studies are summarized in Table 4.2.

However, the number of publications on utilization of bifunctional materials for CO_2 hydrogenation is low so far, and the published papers are mainly focused on nickel-based LTA and FAU (13X) zeolite. Another zeolite, USY has been widely studied for CO_2 methanation, however, the sorption effect of the USY has been ignored. Results found in the literature are presented in Table 4.3. Other metals besides Ni have been explored in CO_2 methanation, with the aim of obtaining high low-temperature activity.

A study has shown that there are at least three possible reaction pathways for CO_2 methanation, (I) formate pathway; (II) carbide pathway; (III) carboxyl pathway (Fig. 4.11) (Vogt et al., 2019). The water removal might influence the reaction pathway, therefore the reaction rate and product distribution could be different. The reaction mechanisms of sorption enhanced CO_2 methanation are not yet exhaustively studied, including the bifunctional catalysts with or without promoters. The presence of promoters such as Ce could change the reaction pathway of CO_2 methanation, and there exist reports that the detected intermediates were different when a promoter was added (Westermann et al., 2017). However, this has not been extensively studied for sorption-enhanced CO_2 methanation.

TABLE 4.2 Performance of representative bifunctional materials for sorption-enhanced CO_2 methanation.

Bifunctional catalyst	Metal loading (wt%)	Pressure (bar)	Temp. (°C)	X_{CO2} (%)	S_{CH4} (%)	$T_{reg.}$[a] (°C)	Reference
Ni/5A	6	1.2	170	100	100	N.A.	Borgschulte et al. (2013)
Ni/Al_2O_3 mix 4A	N.A.	1	250–350	100	100	350–450	Walspurger et al. (2014)
Ni/5A	5	1	300	100	100	300	Delmelle et al. (2016)
Ni/13X	5	1	300	100	100	300	Delmelle et al. (2016)

[a] *T_{reg}, regeneration temperature of bifunctional material; X_{CO2}, CO_2 conversion; S_{CH4}, CH_4 selectivity.*

TABLE 4.3 Performance of representative zeolite catalysts for conventional CO_2 methanation.

Zeolite catalyst	Metal loading (wt %)	Prep. method[a]	Pressure (bar)	Temp. (°C)	X_{CO2} (%)	S_{CH4} (%)	Reference
Ru/5A	2–5	Impregnation	23	320	92	99.5	Hastings et al. (1988, 2002)
Rh/Y	6	Ion-exchange	30	150	5.9	99.8	Hastings et al. (1988)
Ru/HZSM-5	2	Impregnation	1	350	20	99	Scirè et al. (1998)
Ru/Y	3	Ion-exchange	30	150	12.4	96	Bando et al. (1999)
Ni/Beta	10	Impregnation	1	280	20	N.A.	Jwa et al. (2013)
Ni/HY	5	Impregnation	1	300	48.5	96.4	Aziz et al. (2014)
Ni/USY	5	IWI	1	400	24.7	61.4	Graça et al. (2014)
Ni/USY	14	Impregnation	1	300	8	36	Azzolina-Jury and Thibault-Starzyk (2017)
Ni/USY	4.8	IWI	1	300	10	95	Bacariza et al. (2017a)
Ni/HUSY	15	IWI	1	340	12	80	Bacariza et al. (2017b)
Ni/USY	5	IWI	1	350	12	72	Westermann et al. (2017)
Ni	15	IWI	1	337	12	80	Bacariza et al. (2018a)
Ni/Na-USY	15	IWI	1	305	14	95	Bacariza et al. (2018b)
Ni/USY	5	IWI	1	350	8	73	Bacariza et al. (2018c)
Ni/Na-Y	9.9	IWI	1	350	32	84	Quindimil et al. (2018)
Ni/Na-USY	15	IWI	1	360	62	96	Bacariza et al. (2019b)
Ni/ZSM-5	5	Impregnation	N.A.	450	44	84	Rasmussen et al. (2019)
Rh/HZSM-5	0.42	Seed-directing	10	300	20	100	Wang et al. (2019)
Ni/HZSM-5	10	IWI	1	400	68.4	94.8	Chen et al. (2020)
Ni/X	10	IWI	1	470	49	96	Czuma et al. (2020)
Ni/ITQ-2	5	IWI	1	250	6	97	da Costa-Serra, Cerdá-Moreno, and Chica (2020)
Fe/13X	5	Impregnation	1	350	13	11	Franken and Heel (2020)
Ru/13X	2.5	Impregnation	1	240	10	100	Wei et al. (2020)
Ni/13X	5	Impregnation	1	240	17	70	Wei et al. (2021)

[a] *IWI-incipient wetness impregnation.* X_{CO2}, *CO_2 conversion;* S_{CH4}, *CH_4 selectivity.*

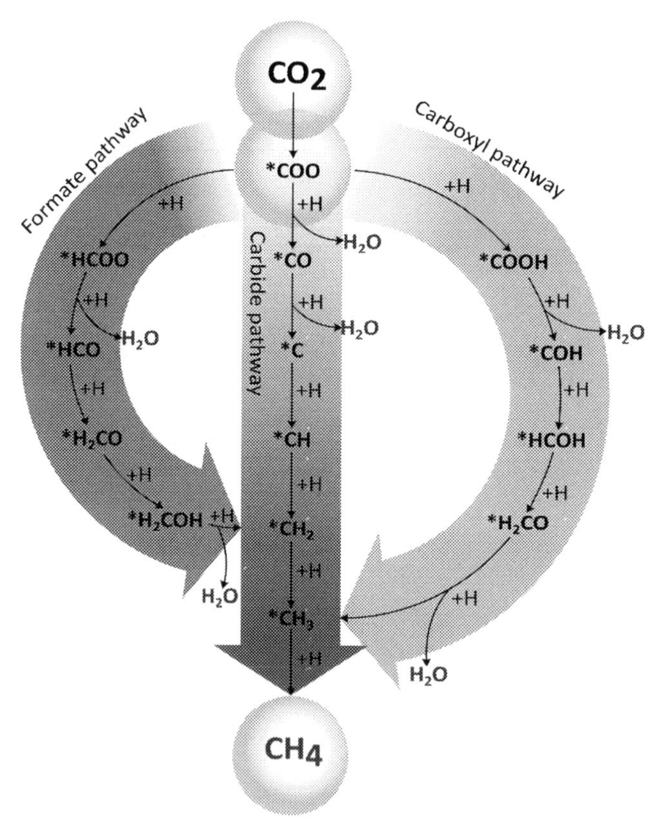

FIG. 4.11 Possible reaction pathways for the methanation of CO_2 (Vogt et al., 2019).

Stability of the material

The longevity and stability test of the catalyst/sorptive materials is extremely important for future commercial and large-scale projects on sorption-enhanced CO_2 methanation.

The factors which influence the stability and performance of the materials include two main parts: the deactivation of the catalytic metal and the change in water uptake capacity of the sorbent. Compared to the conventional CO_2 methanation processes, the bifunctional sorption catalyst can work at relatively mild reaction conditions. The calculated results of equilibrium in CO_2 methanation reaction confirm the extremely low carbon depositions present, although water removal in sorption enhanced CO_2 methanation can result in significant carbon generation at high temperature (higher than 400°C) (Catarina Faria, Miguel, & Madeira, 2018; Massa, Coppola, & Scala, 2020). On the other hand, carbon formation can be avoided by supplying H_2 in slight excess (Massa et al., 2020), however, a balance should be achieved in getting a high-quality product CH_4 with low H_2 concentration. Therefore, a good solution for avoiding carbon formation in bifunctional material is to operate the CO_2 methanation at low temperature. It has been shown that a CO_2 conversion of around 100% can be obtained at around 170°C using Ni-5A zeolite bifunctional material (Borgschulte et al., 2013), and it is also possible to obtain around full CO_2 conversion at 250–300°C using physical

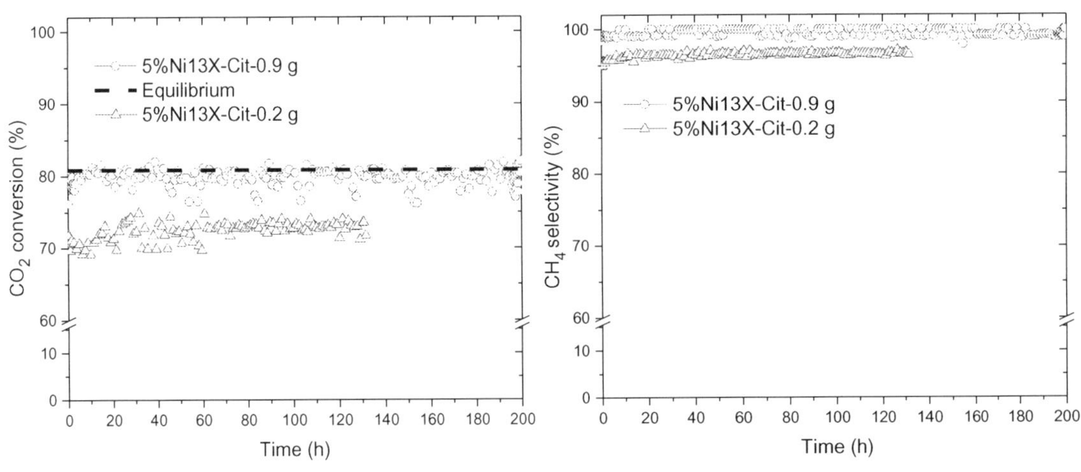

FIG. 4.12 CO_2 conversion *(left)* and CH_4 selectivity *(right)* of 5%Ni13X-Cit at 360°C with 150mL/min N_2, 40mL/min H_2, 10mL/min CO_2 (Wei et al., 2021).

mixtures (commercial Ni/Al_2O_3 catalyst with 4A zeolite) (Walspurger et al., 2014). In fact, even coking of catalyst was not found from the bifunctional material at around 100 to 200-h time scale operation (Fig. 4.12), it showed that Ni zeolite bifunctional materials are very stable for the CO_2 methanation (Delmelle et al., 2018; Walspurger et al., 2014; Wei et al., 2020, 2021).

Moreover, some promoter metals such as Ce and La might also be an option for decreasing the carbon formation on the bifunctional materials in sorption enhanced CO_2 methanation. It would be necessary to further study the catalyst poisoning in practical operation of sorption enhanced CO_2 methanation, depending on the source, because some poisoning gases, H_2S for instance in biomass-derived CO_2, would lead to the catalyst deactivation as in conventional catalysis.

A high water uptake capacity of the bifunctional material is vital in sorption enhanced CO_2 methanation. Delmelle et al. reported that no change in sorption performance was observed within six cycles of drying procedure for both 5Ni/5A and 5Ni/13X (Delmelle et al., 2016). However, they found a degradation mechanism for Ni/5A specific to the sorption catalysis under cyclic methanation/drying periods, which affects water diffusion kinetics in the zeolite support and showed a decrease of water-diffusion coefficient during multiple cycling (Delmelle et al., 2018). To understand the mechanism of decreased water sorption for different sorbents, further studies and many more operation cycles would be needed for further application of sorption enhanced CO_2 methanation.

Research systems and scale

Sorption enhanced CO_2 methanation is based on the conventional CO_2 methanation reaction (Sabatier reaction (4.1)). The CO_2 conversion values reported in the literature (around 80%) (Ashok et al., 2020; Rönsch et al., 2016) are typically far from 100% under currently viable reaction conditions because conventional CO_2 methanation is hampered by thermodynamic limitations (Alvarez et al., 2017; Ashok et al., 2020; Rönsch et al., 2016). The sorption enhanced CO_2 methanation has the potential to provide high purity of CH_4 even meeting the requirements of the gas grid (Walspurger et al., 2014).

To obtain high purity product gas in a conventional CO_2 methanation system (Fig. 4.13),

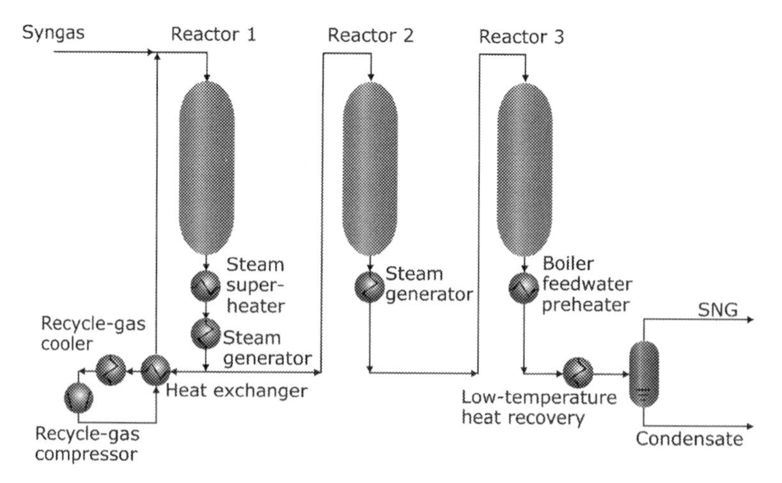

FIG. 4.13 Exemplary fixed-bed methanation process configuration with intermediate cooling and gas recycle. *Adapted from Rönsch, S., et al. (2016). Review on methanation—From fundamentals to current projects.* Fuel, 166, 276–296.

several consecutive reactors combined with water condensers are needed to be employed, and separation of CO_2 and CH_4 is typically also implemented on site (Fig. 4.14) (Uebbing et al., 2019). In conventional CO_2 methanation, a high operating pressure must be used for obtaining a high CO_2 conversion and CH_4 yield due to kinetics (Fig. 4.2) (Catarina Faria et al., 2018). The sorption enhanced CO_2 methanation route has the potential to produce high purity CH_4 with water removal in situ and it is operated at lower pressure as well as at relatively low temperatures. This provides an opportunity for process simplification and cost savings by decreasing the number of reactors needed and by elimination of the downstream separation steps (Carvill et al., 1996). The sorbent will be saturated after some time in operation, and needs to be regenerated. There are two basic options for regeneration in a continuous process, which are a double parallel fixed bed reactor system and a circulating fluidized reactor system. A system which combines the fixed bed reactor, circulating fluidized reactor, and integrates the heat of reaction utilization is a promising solution for getting a high system energetic efficiency.

Fixed bed reactor system

Sorption enhanced CO_2 methanation was first reported by Borgschulte et al. in 2013 (Haije & Geerlings, 2011) in a study where the experiments were performed in a lab-scale stainless steel tubular fixed bed reactor system. Some publications on sorption enhanced CO_2 methanation have been published since then of which all are lab-scale experimental research as commercial operation and review papers are scarce to date (van Kampen et al., 2019). The utilization of zeolites as supports for CO_2 methanation catalysts was reported by M. Carmen Bacariza et al. (Bacariza et al., 2019a) and, Walspurger et al. described preliminary results from sorption enhanced CO_2 methanation experiment performed in a quartz reactor (lab scale) system in 2014 (Walspurger et al., 2014). In 2015, Borgschulte et al. described the Sabatier reaction kinetics using Ni supported on zeolite 3A and 5A in a lab-scale fixed bed reactor system (Borgschulte et al., 2015).

A mechanistic study was performed by Borgschulte et al. in a lab-scale fixed bed reactor system. They used time-resolved neutron radiography on the reactor. Using this technique, they clearly showed that water accumulated in

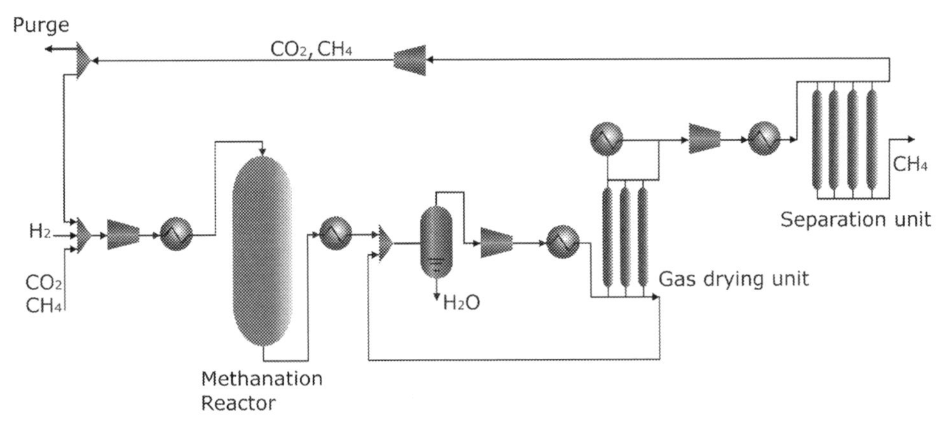

FIG. 4.14 Schematic figure of the process configuration for CO_2 methanation with separation unit. *Adapted from Uebbing, J., Rihko-Struckmann, L.K., Sundmacher, K. (2019). Exergetic assessment of CO_2 methanation processes for the chemical storage of renewable energies.* Applied Energy, *233–234, 271–282.*

the reactor during the sorption enhanced methanation, and was released from the reactor inventory gradually in the regeneration process (Borgschulte et al., 2016).

Fluidized bed reactor system

Compared to fixed bed reactor systems, fluidized bed reactor systems have advantages related to heat and mass transfer and they are favorable for integrating the sorbent regeneration involving large-scale operation of exothermic reactions (Rönsch et al., 2016). Compared to fixed beds, the application of fluidized bed reactors can also lead to energy-saving and favorable reaction conditions due to much higher efficiency in heat exchange as a result of the turbulent gas flow and rapid solids circulation flows (Lappas & Heracleous, 2016). However, the studies on sorption enhanced CO_2 methanation in fluidized bed reactor systems are scarce. Recently, Coppola et al. reported that they evaluated 3A zeolite and CaO for water removal in a lab-scale fluidized bed (40 mm inner diameter) system. They investigated the sorbents hydration in steam (balance air) at 200–300°C and dehydration in air at 350–450°C. The results show that zeolite 3A has a higher steam adsorption capacity compared to CaO, and that the capacity of CaO reduced as a result of the carbonation reaction (Coppola et al., 2020). However, the sorbents were not used in combination with CO_2 processing.

Membrane reactor system

Compared to conventional CO_2 methanation, besides the advantage on methane yield and achieving zero CO_2 and CO outlet concentration, the operating temperature, pressure, and CO_2/H_2 ratio range can also be extended significantly in sorption enhanced CO_2 methanation systems (Borgschulte et al., 2013; Catarina Faria et al., 2018; Massa et al., 2020).

A membrane reactor can be used in CO_2 methanation to enhance the conversion of CO_2 by removing H_2O from the product mixture (Ashok et al., 2020; Lee et al., 2019b). In other reaction setups, typically at least two reactors are needed for obtaining continuous operation, one is used for the sorption enhanced reaction, another is for the regeneration of the sorbent. However, the high cost of membrane reactors and operation costs due to the required high driving force (pressure) often outweigh the benefits of having a single reactor (Catarina Faria et al., 2018).

Other considerations and novel reactor system application

The Sabatier reaction (4.1) is highly exothermic, while the sorbent regeneration is endothermic, therefore the energy efficiency of the whole system can be improved by integrating the heat from the Sabatier reaction performed in a first stage reactor at relatively high temperature for sorbent regeneration. This would be beneficial for lowering the cost of a large-scale commercial application of sorption enhanced CO_2 methanation. It is also beneficial to remove heat efficiently from the methanation reactor to avoid hotspots in the catalyst bed. One option would be to combine a circulating fluidized bed reactor with a fixed bed reactor according to the scheme presented in Fig. 4.15. The system integrates the heat utilization for obtaining a high-system energetic efficiency.

In the proposed system, the conventional CO_2 methanation technologies can be used well for the fixed bed reactor. The operating conditions such as the input ratio of H_2/CO_2, temperature, and pressure of the fixed bed reactor could be regulated to optimize the output composition, which is then fed into the sorption enhanced fluidized bed reactor. For example, thermal oil could be cycled in the system for carrying the heat from the fixed bed reactor to the generator. The bifunctional material (sorptive catalyst) is cycled between the circulating fluidized bed and the regenerator. The removal of water from the methanation reactor results in the equilibrium shift for achieving high conversion of CO_2 and very pure CH_4.

The above-described sorption enhanced CO_2 methanation system is not only beneficial for achieving 100% purity for CH_4, but it is also beneficial for lower temperature and pressure operation and process simplification. In addition to the application on earth, it may also have the potential for undertaking the role of H_2O and carbon cycle in space (Fig. 4.16), e.g., planet Mars (Vogt et al., 2019).

4.3 Conclusions and outlook

In summary, sorption enhanced CO_2 methanation has a great potential in the utilization of CO_2 and H_2 transfer for energy storage. It has been proven, that loading Ni on a water sorbents (LTA or FAU zeolite) is a feasible method to enhance the CH_4 yield significantly during CO_2 methanation and that it is possible to obtain full CO_2 conversion and high purity of CH_4 with the sorption enhanced CO_2 methanation route. Moreover, the sorption enhanced CO_2 methanation is also beneficial for lower temperature and pressure operation and process simplification. In addition to the application on earth, it may

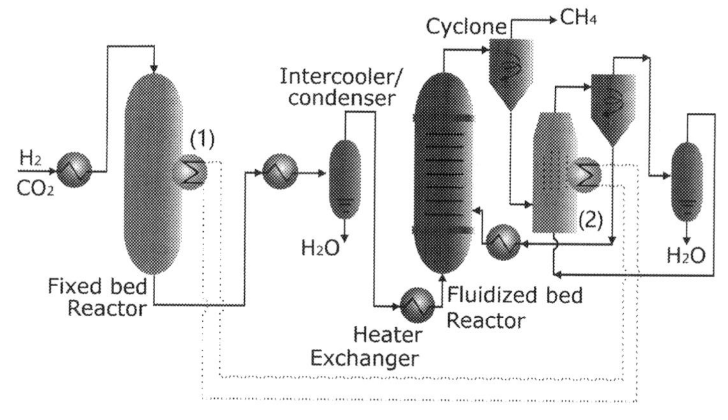

FIG. 4.15 A schematic diagram of sorption enhanced CO_2 methanation with circulating fluidized bed reactor setup and heat utilization for sorbents regeneration. (1) Heater exchanger in reactor or regenerator, (2) Fluidized bed regenerator.

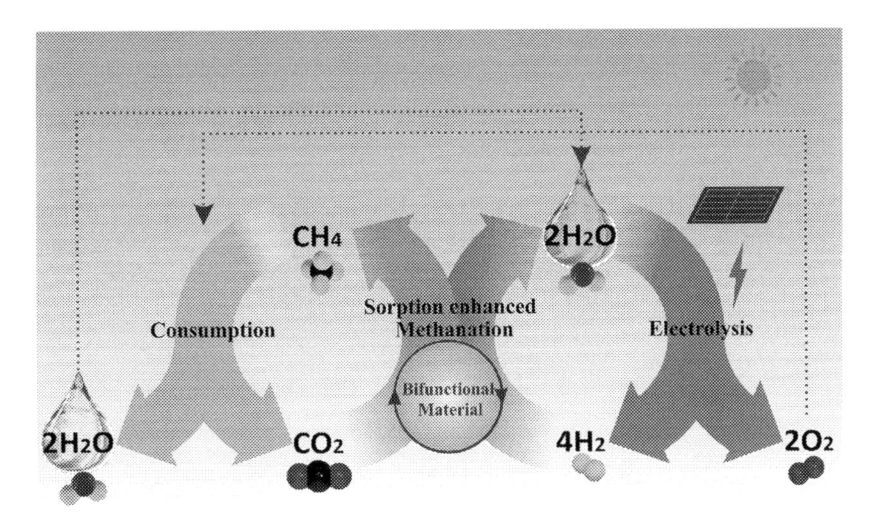

FIG. 4.16 A schematic diagram of sorption enhanced CO_2 methanation system with carbon and hydrogen cycle.

also have the potential for undertaking the role of H_2O and carbon cycle in space, e.g., in Mars.

However, there are still some unclear aspects in sorption enhanced CO_2 methanation. Even though the basic concept has been presented previously, there are still many fundamental and practical aspects unclear and unresolved in sorption enhanced CO_2 methanation. Before the commercial application of sorption enhanced CO_2 methanation, further studies are needed, which include the structure and properties (pore size, acidity, basicity, metal dispersion, etc.) tuning of sorbent and catalyst, the possibility for applying prolonged time and many regeneration cycles, bifunctional performance evaluation under practical operation conditions well as considering systematic scale-up for sorption enhanced CO_2 methanation by taking into account, e.g., mass and heat transfer phenomena.

To select a proper zeolite, or tune the zeolite property or synthesize a new zeolite for sorption enhanced CO_2 methanation is tedious. The zeolite water sorption capacity has a direct relationship to the sorption temperature, and a lower temperature is favored for a higher water sorption capacity. To develop a bifunctional catalyst which can work at a low temperature is regarded for a future sorption enhanced CO_2 methanation. This kind of bifunctional material should have a high dispersion of catalytic metal and activity. The loading of the catalytic metal should also have limited effect on the sorption capacity of the sorbent.

Further work regarding the mechanism of sorption enhanced CO_2 methanation will give a clearer image about the pathway of CH_4 formation under water sorption enhanced condition, which will reveal the selectivity of the bifunctional material.

It would be worth to study sorption enhanced CO_2 methanation with a systematic design which considers the heat utilization and mass cycle.

Some harsh conditions in the presence of poisoning gases such as CO and sulfur compounds are needed for further study in sorption enhanced CO_2 methanation, as it is an important issue to avoid the catalyst poisoning and water capacity decay for the bifunctional material.

Further study on more cycles for sorbent (bifunctional material) on sorption enhanced CO_2 methanation is needed before industrial application.

Acknowledgments

The research work is a part of the activities of the Process and Energy Department in Delft University of Technology and the Johan Gadolin Process Chemistry Centre, a center of excellence financed by Åbo Akademi University. The authors acknowledge the PhD scholarship awarded to Liangyuan Wei by the China Scholarship Council (CSC).

References

Abbet, S., et al. (2000). Acetylene cyclotrimerization on supported size-selected Pd n clusters ($1 \leq n \leq 30$): One atom is enough! *Journal of the American Chemical Society*, *122*(14), 3453–3457.

Abe, T., et al. (2009). CO_2 methanation property of Ru nanoparticle-loaded TiO_2 prepared by a polygonal barrel-sputtering method. *Energy & Environmental Science*, *2*(3), 315–321.

Alvarez, A., et al. (2017). Challenges in the greener production of formates/formic acid, methanol, and DME by heterogeneously catalyzed CO_2 hydrogenation processes. *Chemical Reviews*, *117*(14), 9804–9838.

Artz, J., et al. (2018). Sustainable conversion of carbon dioxide: An integrated review of catalysis and life cycle assessment. *Chemical Reviews*, *118*(2), 434–504.

Ashok, J., et al. (2020). A review of recent catalyst advances in CO_2 methanation processes. *Catalysis Today*, *356*, 471–489.

Aziz, M. A. A., et al. (2014). Highly active Ni-promoted mesostructured silica nanoparticles for CO_2 methanation. *Applied Catalysis B: Environmental*, *147*, 359–368.

Aziz, M. A. A., et al. (2015). CO_2 methanation over heterogeneous catalysts: Recent progress and future prospects. *Green Chemistry*, *17*(5), 2647–2663.

Azzolina-Jury, F., & Thibault-Starzyk, F. (2017). Mechanism of low pressure plasma-assisted CO_2 hydrogenation over Ni-USY by microsecond time-resolved FTIR spectroscopy. *Topics in Catalysis*, *60*(19–20), 1709–1721.

Bacariza, M. C., et al. (2017a). Magnesium as promoter of CO_2 Methanation on Ni-based USY zeolites. *Energy & Fuels*, *31*(9), 9776–9789.

Bacariza, M. C., et al. (2017b). The effect of the compensating cation on the catalytic performances of Ni/USY zeolites towards CO_2 methanation. *Journal of CO_2 Utilization*, *21*, 280–291.

Bacariza, M. C., et al. (2018a). Enhanced activity of CO_2 hydrogenation to CH_4 over Ni based zeolites through the optimization of the Si/Al ratio. *Microporous and Mesoporous Materials*, *267*, 9–19.

Bacariza, M. C., et al. (2018b). Micro- and mesoporous supports for CO_2 methanation catalysts: A comparison between SBA-15, MCM-41 and USY zeolite. *Chemical Engineering Science*, *175*, 72–83.

Bacariza, M. C., et al. (2018c). Ni-Ce/zeolites for CO_2 hydrogenation to CH_4: Effect of the metal incorporation order. *ChemCatChem*, *10*(13), 2773–2781.

Bacariza, M. C., et al. (2019a). Tuning zeolite properties towards CO_2 methanation: An overview. *ChemCatChem*, *11*(10), 2388–2400.

Bacariza, M. C., et al. (2019b). Power-to-methane over Ni/zeolites: Influence of the framework type. *Microporous and Mesoporous Materials*, *274*, 102–112.

Bailera, M., et al. (2017). Power to gas projects review: Lab, pilot and demo plants for storing renewable energy and CO_2. *Renewable and Sustainable Energy Reviews*, *69*, 292–312.

Bando, K. K., Arakawa, H., & Ichikuni, N. (1999). CO_2 hydrogenation over micro- and mesoporous oxides supported Ru catalysts. *Catalysis Letters*, *60*(3), 125–132.

Bazaluk, O., et al. (2020). Assessment of green methanol production potential and related economic and environmental benefits: The case of China. *Energies*, *13*(12), 3113.

Beaumont, S. K., et al. (2014a). A nanoscale demonstration of hydrogen atom spillover and surface diffusion across silica using the kinetics of CO_2 methanation catalyzed on spatially separate Pt and Co nanoparticles. *Nano Letters*, *14*(8), 4792–4796.

Beaumont, S. K., et al. (2014b). Combining in situ NEXAFS spectroscopy and CO(2) methanation kinetics to study Pt and co nanoparticle catalysts reveals key insights into the role of platinum in promoted cobalt catalysis. *Journal of the American Chemical Society*, *136*(28), 9898–9901.

Bevers, E., et al. (2007). Investigation of thermodynamic properties of magnesium chloride amines by HPDSC and TG: For application in a high-lift high-temperature chemical heat pump. *Journal of Thermal Analysis and Calorimetry*, *90*(3), 923–929.

Borgschulte, A., et al. (2013). Sorption enhanced CO_2 methanation. *Physical Chemistry Chemical Physics*, *15*(24), 9620–9625.

Borgschulte, A., et al. (2015). Manipulating the reaction path of the CO_2 hydrogenation reaction in molecular sieves. *Catalysis Science & Technology*, *5*(9), 4613–4621.

Borgschulte, A., et al. (2016). Water distribution in a sorption enhanced methanation reactor by time resolved neutron imaging. *Physical Chemistry Chemical Physics*, *18*(26), 17217–17223.

BP p.l.c. (2020). *Statistical review of world energy*. June.

Bulushev, D. A., & Ross, J. R. H. (2018). Heterogeneous catalysts for hydrogenation of CO_2 and bicarbonates to formic acid and formates. *Catalysis Reviews*, *60*(4), 566–593.

Carvill, B. T., et al. (1996). Sorption-enhanced reaction process. *AICHE Journal*, *42*(10), 2765–2772.

Catarina Faria, A., Miguel, C. V., & Madeira, L. M. (2018). Thermodynamic analysis of the CO_2 methanation reaction with in situ water removal for biogas upgrading. *Journal of CO_2 Utilization*, *26*, 271–280.

Chen, H., et al. (2019). Coupling non-thermal plasma with Ni catalysts supported on BETA zeolite for catalytic CO_2 methanation. *Catalysis Science & Technology*, *9*(15), 4135–4145.

Chen, Y., et al. (2017). Isolated single Iron atoms anchored on N-doped porous carbon as an efficient electrocatalyst for the oxygen reduction reaction. *Angewandte Chemie. International Edition*, *56*(24), 6937–6941.

Chen, Y., et al. (2018). Single-atom catalysts: Synthetic strategies and electrochemical applications. *Joule*, *2*(7), 1242–1264.

Chen, Y., et al. (2020). An active and stable nickel-based catalyst with embedment structure for CO_2 methanation. *Applied Catalysis B: Environmental*, *269*.

Coppola, A., et al. (2020). Evaluation of two sorbents for the sorption-enhanced methanation in a dual fluidized bed system. *Biomass Conversion and Biorefinery*, *11*(1), 111–119.

Czuma, N., et al. (2020). Ni/zeolite X derived from fly ash as catalysts for CO_2 methanation. *Fuel*, *267*, 117139.

da Costa-Serra, J. F., Cerdá-Moreno, C., & Chica, A. (2020). Zeolite-supported Ni catalysts for CO_2 methanation: Effect of zeolite structure and Si/Al ratio. *Applied Sciences*, *10*(15).

Dang, S., et al. (2019). A review of research progress on heterogeneous catalysts for methanol synthesis from carbon dioxide hydrogenation. *Catalysis Today*, *330*, 61–75.

Delmelle, R., et al. (2016). Development of improved nickel catalysts for sorption enhanced CO_2 methanation. *International Journal of Hydrogen Energy*, *41*(44), 20185–20191.

Delmelle, R., et al. (2018). Evolution of water diffusion in a sorption-enhanced methanation catalyst. *Catalysts*, *8*(9).

Din, I. U., et al. (2019). Recent developments on heterogeneous catalytic CO_2 reduction to methanol. *Journal of CO_2 Utilization*, *34*, 20–33.

Falbo, L., et al. (2018). Kinetics of CO_2 methanation on a Ru-based catalyst at process conditions relevant for power-to-gas applications. *Applied Catalysis B: Environmental*, *225*, 354–363.

Fechete, I., & Vedrine, J. C. (2015). Nanoporous materials as new engineered catalysts for the synthesis of green fuels. *Molecules*, *20*(4), 5638–5666.

Franken, T., & Heel, A. (2020). Are Fe based catalysts an upcoming alternative to Ni in CO_2 methanation at elevated pressure? *Journal of CO_2 Utilization*, *39*, 101175.

Furukawa, H., et al. (2014). Water adsorption in porous metal-organic frameworks and related materials. *Journal of the American Chemical Society*, *136*(11), 4369–4381.

Ghodhbene, M., et al. (2017). Hydrophilic zeolite sorbents for in-situ water removal in high temperature processes. *The Canadian Journal of Chemical Engineering*, *95*(10), 1842–1849.

Goldsworthy, M. J. (2014). Measurements of water vapour sorption isotherms for RD silica gel, AQSOA-Z01, AQSOA-Z02, AQSOA-Z05 and CECA zeolite 3A. *Microporous and Mesoporous Materials*, *196*, 59–67.

Graça, I., et al. (2014). CO_2 hydrogenation into CH_4 on NiH-NaUSY zeolites. *Applied Catalysis B: Environmental*, *147*-(Suppl C), 101–110.

Guffanti, S., et al. (2021). Reactor modelling and design for sorption enhanced dimethyl ether synthesis. *Chemical Engineering Journal*, *404*, 126573.

Haije, W., & Geerlings, H. (2011). Efficient production of solar fuel using existing large scale production technologies. *Environmental Science & Technology*, *45*(20), 8609–8610.

Han, Y., et al. (2017). Hollow N-doped carbon spheres with isolated cobalt single atomic sites: Superior electrocatalysts for oxygen reduction. *Journal of the American Chemical Society*, *139*(48), 17269–17272.

Hastings, W. R., et al. (1988). Carbon monoxide and carbon dioxide hydrogenation catalyzed by supported ruthenium carbonyl clusters. A novel procedure for encapsulating triruthenium dodecacarbonyl within the pores of Na-Y zeolite. *Inorganic Chemistry*, *27*(17), 3024–3028.

Hastings, W. R., et al. (2002). Carbon monoxide and carbon dioxide hydrogenation catalyzed by supported ruthenium carbonyl clusters. A novel procedure for encapsulating triruthenium dodecacarbonyl within the pores of Na-Y zeolite. *Inorganic Chemistry*, *27*(17), 3024–3028.

Jwa, E., et al. (2013). Plasma-assisted catalytic methanation of CO and CO_2 over Ni–zeolite catalysts. *Fuel Processing Technology*, *108*, 89–93.

Karelovic, A., & Ruiz, P. (2013a). Mechanistic study of low temperature CO_2 methanation over Rh/TiO_2 catalysts. *Journal of Catalysis*, *301*, 141–153.

Karelovic, A., & Ruiz, P. (2013b). Improving the hydrogenation function of $Pd/\gamma\text{-}Al_2O_3$ catalyst by Rh/γ-Al2O3 addition in CO_2 methanation at low temperature. *ACS Catalysis*, *3*(12), 2799–2812.

Kikkawa, S., et al. (2019). Isolated platinum atoms in $Ni/\gamma\text{-}Al_2O_3$ for selective hydrogenation of CO_2 toward CH_4. *The Journal of Physical Chemistry C*, *123*(38), 23446–23454.

Kistler, J. D., et al. (2014). A single-site platinum CO oxidation catalyst in zeolite KLTL: Microscopic and spectroscopic determination of the locations of the platinum atoms. *Angewandte Chemie (International Ed. in English)*, *53*(34), 8904–8907.

Lappas, A., & Heracleous, E. (2016). Production of biofuels via Fischer–Tropsch synthesis: biomass-to-liquids. In *Handbook of biofuels production* (pp. 549–593). Elsevier.

Le, T. A., et al. (2018). Effects of Na content in $Na/Ni/SiO_2$ and $Na/Ni/CeO_2$ catalysts for CO and CO_2 methanation. *Catalysis Today*, *303*, 159–167.

Lee, B., et al. (2019b). Quantification of economic uncertainty for synthetic natural gas production in a H_2O permeable membrane reactor as simultaneous power-to-gas and CO_2 utilization technologies. *Energy*, *182*, 1058–1068.

Lee, W. J., et al. (2020). Recent trend in thermal catalytic low temperature CO_2 methanation: A critical review. *Catalysis Today*, *368*, 2–19.

Lee, W. T., et al. (2019a). Indirect CO_2 methanation: hydrogenolysis of cyclic carbonates catalyzed by Ru-modified zeolite produces methane and diols. *Angewandte Chemie (International Ed. in English), 58*(2), 557–560.

Li, B., et al. (2017). Influence of addition of a high amount of calcium oxide on the yields of pyrolysis products and noncondensable gas evolving during corn stalk pyrolysis. *Energy & Fuels, 31*(12), 13705–13712.

Li, W., et al. (2018). ZrO_2 support imparts superior activity and stability of Co catalysts for CO_2 methanation. *Applied Catalysis B: Environmental, 220*, 397–408.

Li, Y., et al. (2016). Metal-foam-structured Ni–Al_2O_3 catalysts: Wet chemical etching preparation and syngas methanation performance. *Applied Catalysis A: General, 510*, 216–226.

Liu, L., et al. (2019a). Regioselective generation and reactivity control of subnanometric platinum clusters in zeolites for high-temperature catalysis. *Nature Materials, 18*(8), 866–873.

Liu, Y., et al. (2019b). A general strategy for fabricating isolated single metal atomic site catalysts in Y zeolite. *Journal of the American Chemical Society, 141*(23), 9305–9311.

Luo, L., et al. (2020). Surface iron species in palladium-iron intermetallic nanocrystals that promote and stabilize CO_2 methanation. *Angewandte Chemie (International Ed. in English), 59*(34), 14434–14442.

Massa, F., Coppola, A., & Scala, F. (2020). A thermodynamic study of sorption-enhanced CO_2 methanation at low pressure. *Journal of CO2 Utilization, 35*, 176–184.

Merkache, R., et al. (2015). 3D ordered mesoporous Fe-KIT-6 catalysts for methylcyclopentane (MCP) conversion and carbon dioxide (CO_2) hydrogenation for energy and environmental applications. *Applied Catalysis A: General, 504*, 672–681.

Modak, A., et al. (2020). Catalytic reduction of CO_2 into fuels and fine chemicals. *Green Chemistry, 22*(13), 4002–4033.

Mori, S., et al. (1996). Mechanochemical activation of catalysts for CO_2 methanation. *Applied Catalysis A: General, 137*(2), 255–268.

Nakyai, T., & Saebea, D. (2019). Exergoeconomic comparison of syngas production from biomass, coal, and natural gas for dimethyl ether synthesis in single-step and two-step processes. *Journal of Cleaner Production, 241*, 118334.

Newsam, J. M. (1986). The zeolite cage structure. *Science, 231*(4742), 1093–1099.

Ng, E.-P., & Mintova, S. (2008). Nanoporous materials with enhanced hydrophilicity and high water sorption capacity. *Microporous and Mesoporous Materials, 114*(1–3), 1–26.

Nie, X., et al. (2019). Recent advances in catalytic CO_2 hydrogenation to alcohols and hydrocarbons. In C. Song (Ed.), *Vol. 65. Advances in catalysis* (pp. 121–233).

Petala, A., & Panagiotopoulou, P. (2018). Methanation of CO_2 over alkali-promoted Ru/TiO_2 catalysts: I. Effect of alkali additives on catalytic activity and selectivity. *Applied Catalysis B: Environmental, 224*, 919–927.

Peters, A. W., et al. (2015). Atomically precise growth of catalytically active cobalt sulfide on flat surfaces and within a metal–organic framework via atomic layer deposition. *ACS Nano, 9*(8), 8484–8490.

Pierson, H. O. (1999). *Handbook of chemical vapor deposition (CVD)—Principles, technology and applications* (2nd Ed.). William Andrew Publishing/Noyes.

Primo, A., et al. (2019). CO_2 methanation catalyzed by oriented MoS_2 nanoplatelets supported on few layers graphene. *Applied Catalysis B: Environmental, 245*, 351–359.

Qiu, H. J., et al. (2015). Nanoporous graphene with single-atom nickel dopants: An efficient and stable catalyst for electrochemical hydrogen production. *Angewandte Chemie. International Edition, 54*(47), 14031–14035.

Qu, Y., et al. (2018). Direct transformation of bulk copper into copper single sites via emitting and trapping of atoms. *Nature Catalysis, 1*(10), 781–786.

Quindimil, A., et al. (2018). Ni catalysts with La as promoter supported over Y- and BETA- zeolites for CO_2 methanation. *Applied Catalysis B: Environmental, 238*, 393–403.

Rahman, F. A., et al. (2017). Pollution to solution: Capture and sequestration of carbon dioxide (CO2) and its utilization as a renewable energy source for a sustainable future. *Renewable and Sustainable Energy Reviews, 71*, 112–126.

Rasmussen, K. H., et al. (2019). Stabilization of metal nanoparticle catalysts via encapsulation in mesoporous zeolites by steam-assisted recrystallization. *ACS Applied Nano Materials, 2*(12), 8083–8091.

Rogelj, J., et al. (2016). Paris agreement climate proposals need a boost to keep warming well below 2 degrees C. *Nature, 534*(7609), 631–639.

Rönsch, S., et al. (2016). Review on methanation—From fundamentals to current projects. *Fuel, 166*, 276–296.

Roy, S., Cherevotan, A., & Peter, S. C. (2018). Thermochemical CO_2 hydrogenation to single carbon products: Scientific and technological challenges. *ACS Energy Letters, 3*(8), 1938–1966.

Sabatier, P., & Senderens, J. J. P. I. G.-V. (1902). *Comptes Rendus Des Séances De L'Académie Des Sciences, Section VI–Chimie.*

Saeidi, S., Amin, N. A. S., & Rahimpour, M. R. (2014). Hydrogenation of CO_2 to value-added products—A review and potential future developments. *Journal of CO_2 Utilization, 5*, 66–81.

Sápi, A., et al. (2019). Synergetic of Pt nanoparticles and H-ZSM-5 zeolites for efficient CO_2 activation: Role of interfacial sites in high activity. *Frontiers in Materials, 6*. https://doi.org/10.3389/fmats.2019.00127.

Schlögl, R. (2008). *Handbook of heterogeneous catalysis*. Wiley-VCH Verlag GmbH & Co. KGaA.

Scirè, S., et al. (1998). Influence of the support on CO_2 methanation over Ru catalysts: An FT-IR study. *Catalysis Letters, 51*(1/2), 41–45.

Simo, M., et al. (2009). Adsorption/desorption of water and ethanol on 3A zeolite in near-adiabatic fixed bed. *Industrial & Engineering Chemistry Research*, *48*(20), 9247–9260.

Sreedhar, I., et al. (2019). Developmental trends in CO_2 methanation using various catalysts. *Catalysis Science & Technology*, *9*(17), 4478–4504.

Swalus, C., et al. (2012). CO_2 methanation on Rh/γ-Al_2O_3 catalyst at low temperature: "In situ" supply of hydrogen by Ni/activated carbon catalyst. *Applied Catalysis B: Environmental*, *125*, 41–50.

Tatlier, M., Munz, G., & Henninger, S. K. (2018). Relation of water adsorption capacities of zeolites with their structural properties. *Microporous and Mesoporous Materials*, *264*, 70–75.

Terreni, J., et al. (2018). Observing chemical reactions by time-resolved high-resolution neutron imaging. *The Journal of Physical Chemistry C*, *122*(41), 23574–23581.

Terreni, J., et al. (2019). Sorption-enhanced methanol synthesis. *Energy Technology*, *7*(4), 9.

Thomas, J. M., & Harris, K. D. M. (2016). Some of tomorrow's catalysts for processing renewable and non-renewable feedstocks, diminishing anthropogenic carbon dioxide and increasing the production of energy. *Energy & Environmental Science*, *9*(3), 687–708.

Uebbing, J., Rihko-Struckmann, L. K., & Sundmacher, K. (2019). Exergetic assessment of CO_2 methanation processes for the chemical storage of renewable energies. *Applied Energy*, *233–234*, 271–282.

UNFCCC. (2015). *Adoption of the Paris agreement*.

Van Berkel, F. P. F., et al. (2020). *Process and system for producing dimethyl ether*. Google Patents.

van Kampen, J., Boon, J., & van Sint Annaland, M. (2020). Steam adsorption on molecular sieve 3A for sorption enhanced reaction processes. *Adsorption*, *27*(4), 577–589.

van Kampen, J., et al. (2019). Steam separation enhanced reactions: Review and outlook. *Chemical Engineering Journal*, *374*, 1286–1303.

van Kampen, J., et al. (2020). Sorption enhanced dimethyl ether synthesis for high efficiency carbon conversion: Modelling and cycle design. *Journal of CO_2 Utilization*, *37*, 295–308.

Vogt, C., et al. (2019). The renaissance of the Sabatier reaction and its applications on earth and in space. *Nature Catalysis*, *2*(3), 188–197.

Walspurger, S., Haije, W. G., & Louis, B. (2014). CO_2 reduction to substitute natural gas: Toward a global low carbon energy system. *Israel Journal of Chemistry*, *54*(10), 1432–1442.

Walspurger, S., et al. (2014). Sorption enhanced methanation for substitute natural gas production: Experimental results and thermodynamic considerations. *Chemical Engineering Journal*, *242*, 379–386.

Wang, C., et al. (2019). Product selectivity controlled by nanoporous environments in zeolite crystals enveloping rhodium nanoparticle catalysts for CO_2 hydrogenation. *Journal of the American Chemical Society*, *141*(21), 8482–8488.

Wang, X., et al. (2016). Synthesis of isoalkanes over a core (Fe-Zn-Zr)-shell (zeolite) catalyst by CO_2 hydrogenation. *Chemical Communications (Cambridge)*, *52*(46), 7352–7355.

Wang, Y., & LeVan, M. D. (2009). Adsorption equilibrium of carbon dioxide and water vapor on zeolites 5A and 13X and silica gel: Pure components. *Journal of Chemical & Engineering Data*, *54*(10), 2839–2844.

Weatherbee, G. (1984). Hydrogenation of CO_2 on group VIII metals IV. Specific activities and selectivities of silica-supported Co, Fe, and Ru. *Journal of Catalysis*, *87*(2), 352–362.

Wei, L., et al. (2020). Can bi-functional nickel modified 13X and 5A zeolite catalysts for CO_2 methanation be improved by introducing ruthenium? *Molecular Catalysis*, *494*, 111115.

Wei, L., et al. (2021). Influence of nickel precursors on the properties and performance of Ni impregnated zeolite 5A and 13X catalysts in CO_2 methanation. *Catalysis Today*, *362*, 35–46.

Westermann, A., et al. (2015). Insight into CO_2 methanation mechanism over NiUSY zeolites: An operando IR study. *Applied Catalysis B: Environmental*, *174-175*, 120–125.

Westermann, A., et al. (2017). The promoting effect of Ce in the CO_2 methanation performances on NiUSY zeolite: A FTIR in situ/operando study. *Catalysis Today*, *283*, 74–81.

Xie, S., et al. (2020). CO_2 reduction to methanol in the liquid phase: A review. *ChemSusChem*, *13*(23), 6141–6159.

Yan, H., et al. (2015). Single-atom Pd1/graphene catalyst achieved by atomic layer deposition: Remarkable performance in selective hydrogenation of 1, 3-butadiene. *Journal of the American Chemical Society*, *137*(33), 10484–10487.

Yan, Y., et al. (2016). A novel W-doped Ni-mg mixed oxide catalyst for CO_2 methanation. *Applied Catalysis B: Environmental*, *196*, 108–116.

Yin, P., et al. (2016). Single cobalt atoms with precise N-coordination as superior oxygen reduction reaction catalysts. *Angewandte Chemie. International Edition*, *55*(36), 10800–10805.

Zhang, J., et al. (2018b). Cation vacancy stabilization of single-atomic-site Pt 1/Ni (OH)x catalyst for diboration of alkynes and alkenes. *Nature Communications*, *9*(1), 1002.

Zhang, L., et al. (2018a). Single-atom catalyst: A rising star for green synthesis of fine chemicals. *National Science Review*, *5*(5), 653–672.

CHAPTER

5

Hydrogenation of CO_2 by photocatalysis: An overview

T. Vijayaraghavan and Anuradha M. Ashok

PSG Institute of Advanced Studies, Coimbatore, India

5.1 Introduction

Extensive utilization of fossil fuels for energy generation and various other industrial and domestic activities not only lead to their faster depletion but also increase CO_2 level in the atmosphere. Extensive efforts are under progress to minimize the environmental damage due to increased level of CO_2 in the atmosphere. Most of the methods proposed on CO_2 reduction from fossil fuels such as thermal reforming, hydrothermal conversion, electrochemical reduction, chemical transformation, etc., consume a large amount of thermal/electrical energy for the breakage of C=O bond. The hydrogenation of CO_2 by using photocatalysis is one of the sustainable methods which is derived from nature. This will generate several useful products from CO_2 such as formic acid, carbon monoxide, formaldehyde, methanol, and methane, based on their combustion-free energy. The detailed mechanism of photocatalytic hydrogenation of CO_2, existing photocatalysts, their advantages/limitations and recent developments reported on such photocatalysts are discussed in this chapter.

5.2 Photocatalytic hydrogenation of CO_2

Due to the extensive usage of fossil fuels for the energy needs of various industrial and domestic activities humankind has been facing serious problems such as faster depletion of fossil fuels and increasing CO_2 level in the atmosphere, etc. IPCC (Intergovernmental Panel on Climate Change) predicted that the CO_2 level will increase up to 590 ppm and the global average temperature may increase up to 1.9°C by the year 2100 (Li, Peng, & Peng, 2016). Reducing CO_2 into environment-friendly nontoxic products is one of the solutions to this problem. CO_2 is one of the most thermodynamically stable carbon-based compound. The methods proposed in recent research articles published on CO_2 reduction from fossil fuels such as thermal reforming (Hu, 2009), hydrothermal conversion (Huo, Hu, Zeng, Yun, & Jin, 2012), electrochemical reduction (Garg et al., 2020), chemical transformation (Yu & He, 2015), etc., consume a large amount of thermal/electrical energy for the breakage of C=O bond. The bond energy of C=O (750 kJ/mol) is higher than C—H (411 kJ/mol) and C—C (336 kJ/mol).

Heterogeneous Catalysis
https://doi.org/10.1016/B978-0-323-85612-6.00005-X

Therefore it is necessary to combine thermal energy with other eco-friendly methods for the efficient hydrogenation of CO_2. Photothermocatalysis under sunlight irradiation is one of the sustainable methods (Li et al., 2016). In this case, the photocatalyst developed for CO_2 reduction produces heat energy locally within itself thus enhancing the overall efficiency of photoconversion.

The reduction of CO_2 by using semiconductor photocatalysis under sunlight irradiation is one of the sustainable methods derived from mother nature (Yi, Li, Feng, & Xie, 2015). The reduction of CO_2 by this method will lead to several products which are valuable for mankind such as formic acid, carbon monoxide, formaldehyde, methanol, and methane, based on their combustion-free energy. Fig. 5.1 shows the useful by-products derived from CO_2 and the number of proton-coupled electrons required for the chemical reaction to generate them (Neatu, Antonio, Agullo, & Garcia, 2014).

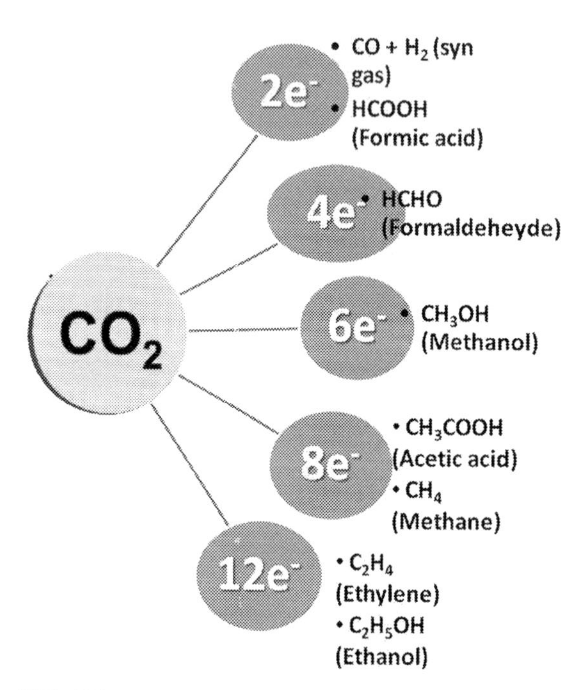

FIG. 5.1 By-products derived from CO_2 and number of proton-coupled electrons required for each of them *(circled in green color; dark gray in print version)*.

5.3 Mechanism for photocatalytic hydrogenation of CO_2

The mechanism for photocatalytic CO_2 reduction is more complicated than other similar processes such as hydrogen generation through water splitting due to several factors such as solubility, by-product formation, etc.

CO_2 reduction process involves three steps; (1) adsorption of light and creation of charge, (2) separation of charges and (3) redox reactions at the surface of the catalyst. Fig. 5.2 shows the schematic representation of photocatalytic CO_2 conversion mechanism. When photocatalysts are exposed to photon energy which is higher than their bandgap, charge carriers such as electrons and holes will be produced in the conduction band and valence band. The electrons in the conduction band are responsible for the reduction and conversion of adsorbed CO_2 to various by-products such as CO, HCOOH, CH_3OH, CH_4, etc. The type of the by-product depends on the number of electrons available for the reactions (as mentioned in Fig. 5.1). Holes in the valence band will undergo oxidation reaction with water and CO_2. The type of by-product formed during the reduction reaction of CO_2can be controlled by the type of co-catalysts used for reduction reaction. Unlike water-splitting reactions, reduction reaction will contribute more for CO_2 conversion.

The reduction potentials of various by-products are shown in Eqs. (5.1)–(5.8). The one electron reduction from CO_2 to $CO_2^{\bullet -}$ is not possible due to the large negative reduction potential (i.e., -1.85 V vs SHE (standard hydrogen electrode)) relative to the conduction band potential of many of the semiconductors. This highly negative reduction potential arises from the change in hybridization of carbon from sp^2

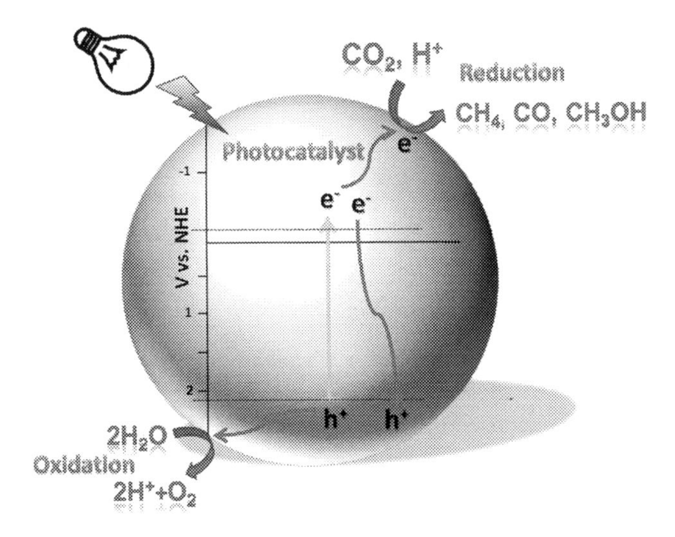

FIG. 5.2 Mechanism of photocatalytic carbon dioxide reduction.

to sp^3. The reduction potential for the formation of by-products such as HCOOH (−0.665 V), HCHO (−0.485 V), CH_3OH (−0.399 V) and CH_4 (−0.246 V) are smaller and lie toward the positive of the CB (conduction band) edge potentials of many semiconductors. Therefore it is preferable to undergo proton-coupled electron transfer, which is electron transfer to CO_2 associated with proton transfer. The formation energy (i.e., ΔG) of the by-products are in the order $HCOOH < CO < HCHO < CH_3OH < CH_4$.

$$CO_2 + e^- \rightarrow CO_2^{\cdot-}\ (-1.85\ V\ vs\ SHE) \quad (5.1)$$

$$CO_2 + H_2O + 2e^- \rightarrow HCOO^- + OH^-\ (-0.665\ V\ vs\ SHE) \quad (5.2)$$

$$CO_2 + H_2O + 2e^- \rightarrow CO + 2OH^-\ (-0.521\ V\ vs\ SHE) \quad (5.3)$$

$$CO_2 + 3H_2O + 4e^- \rightarrow HCHO + 4OH^-\ (-0.485\ V\ vs\ SHE) \quad (5.4)$$

$$CO_2 + 5H_2O + 6e^- \rightarrow CH_3OH + 6OH^-\ (-0.399\ V\ vs\ SHE) \quad (5.5)$$

$$CO_2 + 6H_2O + 8e^- \rightarrow CO_4 + 8OH^-\ (-0.246\ V\ vs\ SHE) \quad (5.6)$$

$$2H_2O + 2e^- \rightarrow H_2 + 2OH^-\ (-0.414\ V\ vs\ SHE) \quad (5.7)$$

$$2H_2O \rightarrow O_2 + 4H^+ + 4e^-\ (+0.816\ V\ vs\ SHE) \quad (5.8)$$

5.4 Important criteria for photocatalytic hydrogenation of CO_2

A photocatalyst should possess the following characteristics to be used for CO_2 reduction.

1. Suitable bandgap (The bandgap should be in between 1.7 and 3.2 eV to be visible light active).
2. Charge separation (generated electron-hole pairs should be effectively separated to avoid recombination).

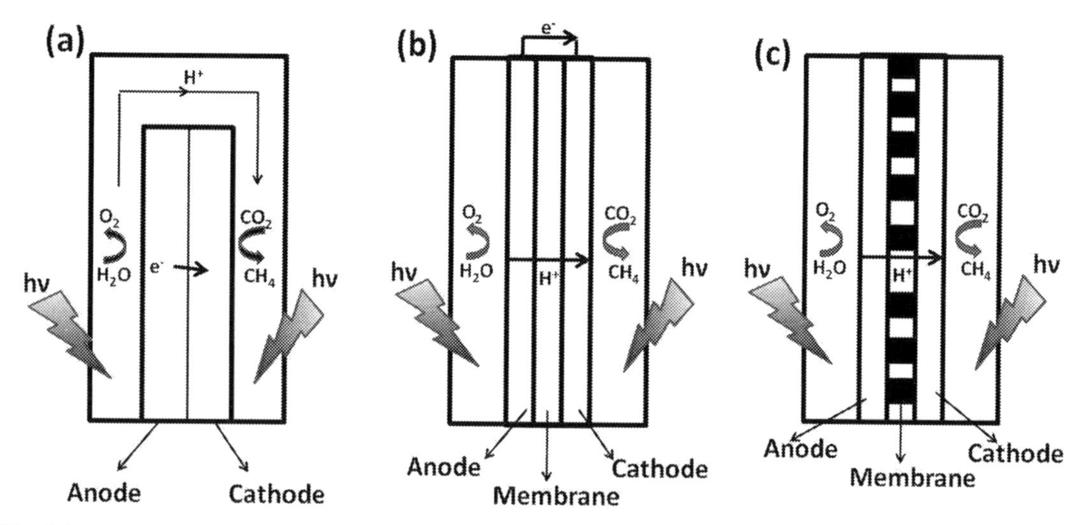

FIG. 5.3 Hydrogeneration of CO_2 by photoelectrochemical cells (a) Monolithic device, (b) Wired device, and (c) Photoelectrochemical device.

3. Redox reduction and product formation (should have more negative conduction band potential). Fig. 5.3 shows some of the devices for hydrogenation of CO_2 using semiconductor photocatalysts with suitable redox potentials.
4. Desorption of products (from the surface of the photocatalysts so that it is free for the continuation of reaction for longer duration at a higher temperature)

5.5 Types of photocatalytic hydrogenation of CO_2

Photocatalytic hydrogenation of CO_2 is broadly classified into (1) Photothermal, (2) Bio-photocatalytic, (3) Homogeneous, and (4) Heterogeneous processes.

Photothermal hydrogenation of CO_2

The conventional photocatalysis utilizes light energy to overcome the bandgap energy difference of a semiconductor, subsequently generating active electrons and holes that are capable of the redox reactions, depending on the positions of the conduction/valence bands. Although various semiconductor-based photocatalysts have been reported for photocatalytic hydrogenation of CO_2, the efficiency of conversion is very low due to environmental conditions. The hydrogenation of CO_2 requires higher temperature and pressure (i.e., 240–450°C and 2–5 MPa) for better efficiency (Porosoff, Yan, & Chen, 2016). So, combining both photo and thermo catalysis is one of the novel approaches (Jaiswal, Landge, Jagadeesan, & Balaraman, 2017; Vernekar, Sakate, Rode, & Jagadeesan, 2019).

The photothermal catalysis begins with the excitation of electrons by light absorption. The excited electrons will involve in interband, intraband, or plasmonic excitations. Interband excitation will occur between valence band to conduction band, intraband excitation will occur between defect states within the band and plasmonic excitation involves the collective excitation of conduction band electrons.

It is very important to design the photothermal catalysts in such a way that the generated heat should be localized to the catalytic sites. Many of the catalysts will employ supply material to enhance the catalytic activity, stability, electrical and thermal conductivities. These materials act as thermal energy carriers to scatter into active catalyst material, which can give more effective localized heat and thus increase photothermal effect. The photothermal catalysts also initiate surface chemical reactions by hot electron injection, where the energy of localized surface plasmons excite charge carriers (i.e., electrons and holes) on the metal surface. The charge carriers are transferred to reactant or intermediate phase and create excited states that facilitate chemical transformation of the adsorbed species.

There are many metals which are used as co-catalysts (Ni, Ru, Rh, Fe, Au, and Pd), due to their strong broadband optical absorption. These co-catalysts are supported on many photocatalysts such as TiO_2, Al_2O_3, ZnO, Nb_2O_5, Si, and metal-organic frameworks (MOFs) to enhance their dispersion and stability. Generally, the corners and edges of the photocatalysts will enhance the local electric fields generated by oscillating plasmonic electrons, so that light absorption is also improved. Additionally, the lower coordination of atoms found at edge and corner sites, in combination with these enhanced electric fields will increase the injection of electrons into the adsorbed reactants or intermediates.

Arai, Sato, and Morikawa (2015) developed table sized device for photoreduction of CO_2 to formic acid under solar light irradiation. The device consists of IrO_2-coated ITO (indium doped tin oxide) plate with Si-Ge junction, carbon cloth with stainless steel and Ruthenium complex polymer catalyst. It reduces CO_2 to formic acid of 50 μmol/g within 180 min under solar irradiation. Wu et al. (2019) prepared tungstates for conversion of CO_2 to methanol under Xenon lamp irradiation. Recently, Guangbo Chen et al. reported novel alumina based nanocomposites for photothermal catalytic hydrogenation of CO_2 to methanol (Chen et al., 2017). They claimed that the LDH (layered double hydroxide) is capable of absorbing more CO_2 at higher temperatures due to the formation of amorphous Al_2O_3 at higher temperatures. Formation of CoFe alloy helps to increase the photocatalytic hydrogenation of CO_2 to methanol. They also found that the conversion efficiency randomly increases with increasing operating temperature from 240 to 320°C. Jia et al. (2016) showed in-situ diffuse reflectance infrared spectroscopic studies on Pd@Nb_2O_5 nanorods to find the reaction intermediates formed during hydrogenation of CO_2 by photothermal catalysis method. Srinivas et al. (2011) prepared Cu-TiO_2 by chemical method and showed oxalic acid production using photocatalytic CO_2 conversion. Singhal, Goyal, and Kumar (2017) prepared $InTaO_4$ by sol-gel method and irradiated under 20 W LED lamp. The material exhibited 200 μmol of CO_2 to methanol conversion after 4 h irradiation. Most of these photocatalysts are tested in the liquid phase except reference (Kandy & Gaikar, 2019; Sharma et al., 2019). Though there are several photocatalysts reported in reputed journals, the process of utilizing these photocatalysts for making device is still under progress.

There are many challenges associated with the design of photothermal catalyst for hydrogenation of CO_2. They are (1) Correlating morphology, size, and composition of the photothermal catalyst/support system with its light-harvesting and catalytic properties, (2) Identifying effects of light intensity and spectral distribution on the light-harvesting properties, quantum efficiency, catalytic rate/selectivity, temperature evolution and (3) Distinguishing between thermal and photocatalytic effects on catalytic activity and selectivity. There are few advantages over other type of photocatalytic conversion of CO_2. First, the high local temperature can be generated very close to the catalytic site, thereby reducing heat transfer distances as compared to traditional heating methods and enable the faster start-up times, improved load flexibility. Second, the introduction of light into the hydrogenation reactor could increase the effective reaction rate relative to the thermocatalytic reaction through plasmonic effects. So this will increase the required throughput of reactants for a given catalyst mass or enable a smaller reactor to achieve the same throughput. Third, though the presence of light is not sufficient for the reaction, the plasmonic photocatalysts can be used to adjust the caloric value of synthetic natural gases by producing higher value chemicals such as ethane or other higher hydrocarbons. The photocatalytic dehydrogenative coupling also can be applied to convert CH_4 into higher hydrocarbons.

Recently, Au nanoparticles deposited ZnO nanosheets (Meng et al., 2018) showed quantum efficiency of 0.08% which is comparable to natural photosynthesis. The experimental and theoretical studies showed that the reaction is induced by the electron transfer between the photoexcited ZnO nanosheets and a surface-adsorbed CH_4 molecule. So this plasmonic photocatalysis represents a convincing method of tuning product selectivity and forming high-value products at milder temperatures than those used in common thermocatalytic reactions.

Biophotocatalytic hydrogenation of CO_2

In this method, hydrogenated microbes, or the hybrid systems composed of inorganic materials coupled with microbes convert CO_2 and H_2O into CH_4. The inorganic materials produce molecular H_2, which is fed to the microorganisms to fuel the CO_2 reduction reaction. Microbes live in anaerobic environments on earth and release around one billion tons of gas every year (Welte & Deppenmeier, 2014). The reaction of H_2 with CO_2 is an exergonic process, it can be utilized by the microbes as a source of energy, which is stored in molecules such as ATP. Though photosynthesis is not required for this form of CO_2 metabolism, the generation of dihydrogen, or other reducing equivalents, using solar energy can be linked to CH_4 formation through the use of bioreactors containing microbes.

The bio-inorganic hybrid systems place the archaea in the cathode compartment of a two-compartment electrolysis cell have recently been described (Nichols et al., 2015). First demonstration of the feasibility of hybrid bio/inorganic systems for light-driven CO_2 methanation uses water as the hydrogen source. The 110 mL of methane was obtained using platinum cathode for 7 days of duration with a faradaic efficiency of up to 86% at low overpotential of 360 mV. The system could be made completely solar-powered by replacing the platinum cathode with platinum-coated photoactive *p*-InP cathode to feed the archaea with H_2 and photoactive *n*-TiO_2 anode in the second compartment to provide the electrons for H_2 production from water oxidation. This solar-powered system produced 1.8 mL CH_4 after 3 days and required an anion exchange membrane between the compartments to minimize pH changes. The observed faradaic efficiency of this system was higher up to 74%, when blue light was filtered out from the cathode compartment, due to the sensitivity of the microbes to these wavelengths.

The above-mentioned hybrid system was reported to have a solar-to-chemical efficiency of 52%. The scaling up of this bioreactor based process needs more attention unlike conventional heterogeneous and homogeneous catalytic processes.

Photoredox hydrogenation of CO_2

In photoredox catalysis the catalyst will be in an electronically excited state and therefore can be more easily reduced or oxidized than in its ground state. These photocatalysts are used to reduce CO_2 to CH_4 by light-induced redox reactions. This photoredox catalysis can be classified as homogeneous and heterogeneous photoredox reactions. In homogeneous photoredox catalysis, the light absorption, reduction, and oxidation occur through the formation of several complexes in solution. In heterogenous photoredox catalysis, semiconductor materials generate excited electronic states to drive reduction and oxidation reactions. A photon of sufficient energy must excite an electron from the valence band of a semiconductor to its conduction band to create an electron-hole pair. The photogenerated hole initiates the oxidation of water to O_2 and H^+, and the photoexcited electron activates the CO_2 reduction reaction.

The oxidation and reduction reactions occur on the surface of the semiconductor photocatalyst. But it is difficult to achieve both reactions on the surface of a single semiconductor. This is due to the complex, multi-step reaction mechanism that requires eight protons and eight electrons, which must be constantly supplied to specific sites and intermediate stages of the reactions. So far no single semiconductor photocatalyst has been reported for hydrogenation of CO_2. So many researchers are working on developing composite materials to enhance the efficiency of hydrogenation of CO_2.

The co-catalysts can be used to tune the product selectivity, reduce the activation barrier for the redox reactions and facilitate the separation of charge carriers. Generally, the composite heterogeneous photoredox catalysts composed of metals (e.g., Cu, Ru, or Re) coupled with semiconductors (e.g., P-Si, GaP, GaAs, GaN or TiO_2) or other metal co-catalysts like Au or Pt. The lead-halide perovskite quantum dots supported on graphene oxide have also been shown to photocatalytically reduce CO_2 to CH_4 and other products. Sorcar, Hwang, Grimes, and In (2017) have demonstrated that the addition of Pt and Cu onto TiO_2 photocatalysts yielded CH_4 under solar irradiation. The addition of metal oxide co-catalysts, such as NiO and In_2O_3 on semiconductors (TiO_2) has also increased the hydrogenation of CO_2 to CH_4.

The common device architecture for hydrogenation of CO_2 photoelectrochemical cells are shown in Fig. 5.3a and b. Fig. 5.3a is the sketch of a monolithic device, in which protons and electrons are transferred from the anode to the cathode through an electrolyte or via conduction, respectively. Fig. 5.3b ahows a wired device, in which protons and electrons are transferred from anode to cathode via a membrane and an external circuit, respectively and, Fig. 5.3c depicts a photoelectrochemical cell, in which the anode and cathode are separated by a proton-conducting membrane with integrated electron-conducting material. A number of layouts of photoelectron chemical cells with anode and cathode are reported where the oxidation and reduction half-cell reactions will occur. These two are separated from each other by electron ($e^{-\cdot}$)

These complex architectures are also called as "artificial leaf", to mimic the natural leaf and also have spatially separated photoredox reactions, micron charge-carrier transport distances for both electrodes. Zhou, Guo, Li, Fan, and Zhang (2013) reported 3D leaf-like titanate-based architecture for the direct hydrogenation of CO_2. The electrochemical potentials of various CO_2 reduction products are very close to that of the hydrogenation or methanation half-reaction (Fig. 5.4). It means that the multiple reactions are likely to occur in parallel, resulting in lower CH_4 selectivity. So, this makes it rather challenging to achieve high faradaic efficiency (i.e.) the percentage of charge carriers consumed in a particular reaction.

The cell level efficiency losses occur due to limitations on ionic conduction between electrodes. Nafion is one of the best and most commonly used proton conductors in electrochemical systems. The other media such as anion exchange membranes, ionic liquids, and bipolar membranes also offer interesting approaches to improving photoelectrochemical systems.

The hydrogenation of homogeneous photoredox system involves the use of several molecular components. The photo-sensitizer, usually but not necessarily a metal complex, acts as the light-absorbing agent. The absorbed energy is

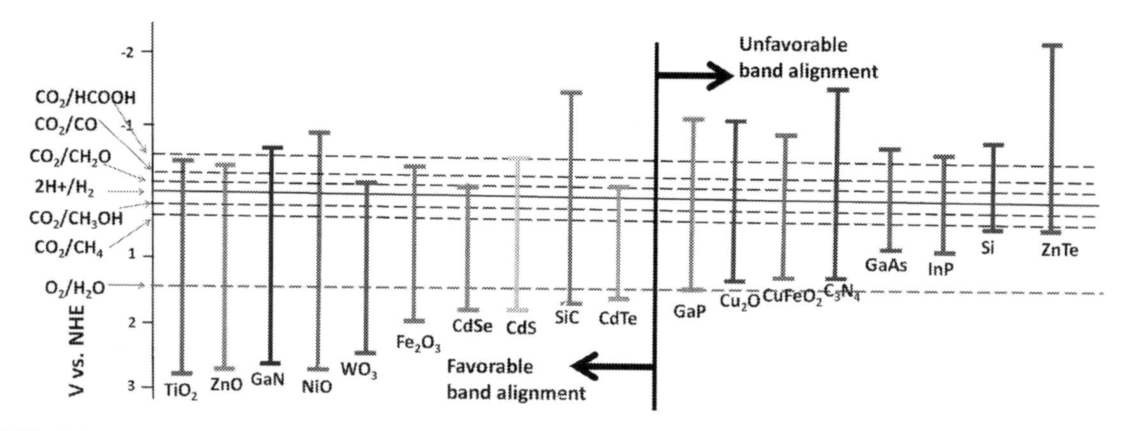

FIG. 5.4 Band edge potential diagram of semiconductors used for hydrogenation of carbon dioxide.

then either transferred to another metal ion complex called the photocatalyst, where the CO_2 reduction reaction will take place, or it reacts with a sacrificial electron donor, whose electrons are transferred to the photocatalyst to perform the CO_2 reduction reaction. These two complexes work synergistically to absorb light and convert dissolved CO_2 into CH_4 and other products. However, very few reported catalyst complexes include iron, copper, and cobalt metal active centres. In each case, a low valent metal complex (Co^+, Cu^+, Fe^0) bound to CO_2 is postulated as part of the catalytic cycle. The cobalt and copper systems continue to convert CO to further reduced products, including the eight electron product, methane. There are some electrode-free photocatalytic systems that utilize a photosensitizer to strip electrons from a sacrificial donor, such as triethylamine (TEA) to enable the reduction process.

Recently, an iron tetraphenylporphyrin complex functionalized with triethylammonio groups was incorporated into an electrode-free photocatalytic system utilizing visible light for the hydrogenation of CO_2 at ambient pressure and temperature (Rao, Lim, Bonin, Miyake, & Robert, 2018). In this system, TEA is oxidized to an iminium radical that then decomposes to provide electrons at a negative electrochemical potential (less than −1.5 V vs NHE). These electrons are extracted from TEA by an iridium photosensitizer $[Ir(ppy)_3]^+/[Ir(ppy)_3]$ (E0 = −1.7 V vs NHE, ppy = cyclometallated phenylpyridine) which then shuttles them to the CO_2 activating iron center. The first hydrogenation step is the conversion of CO_2 to CO. Under optimized conditions, and in the presence of added trifluoroethanol, the photocatalytic oxidation of additional TEA molecules ([TEA]/[Fe] = 25,000) supplies two electrons to produce CO and another six electrons to reduce CO to CH_4 with 82% selectivity, an estimated quantum yield of 0.18% and a turn over a number of up to 159 for CH_4 production along with up to 34 equivalents of H_2 produced in side reactions.

The iron-bound intermediates in this process remain unidentified, but the existence of an iron formyl (Fe-CHO) intermediate was postulated. It is worth pointing out that the alpha-/beta-elimination and hydride formation reactions leading to H_2 evolution are inhibited by saturating the iron coordination sites, located CIS to the reduced carbon species, with the four nitrogen atoms of the porphyrin ring. The iridium complex can also be replaced by a metal-free phenoxazine photosensitizer (Rao et al., 2018). Several other metal-complex catalysts have been described in a review, including those based on

Mn^+, Re^+, Fe^{3+}, Ru^{2+}, Co^+, Ir^{3+}, and Ni^{2+} for the photocatalytic reduction of CO_2 to CO or formate, with triethanolamine or TEA used as the sacrificial electron donor (Liu, Inagaki, & Gong, 2016). The advantage of homogeneous photoredox systems is their high product selectivity, while heterogeneous photoredox systems exhibit higher optical efficiencies and tunable optoelectronic properties as compared to homogeneous systems. Some novel approaches to CO_2 photoredox catalysis aim to combine catalyst design principles from both heterogeneous and homogeneous photoredox systems, with the goal of developing photoredox catalysts that exhibit both the high optical efficiencies of heterogeneous semiconductor/plasmon-based light-harvesting systems with the superior product selectivity of homogeneous catalysts.

Metal Organic Frameworks (MOF) are solid-state compounds consisting of metal ions or clusters coordinated to organic ligands, and are demonstrably suitable for such catalytic reactions (Diercks, Liu, Cordova, & Yaghi, 2018). Initial reports have shown the successful integration of known homogeneous CO_2 photoreduction catalysts into the backbone of a solid-state framework and demonstrated the photocatalytic CO_2 reduction activity of the obtained compounds. This concept has been expanded to integrate plasmonic metal clusters, metal oxides, and photosensitizers into MOF architectures, thereby yielding new, high surface area, porous materials with tunable optoelectronic properties and catalytic activity. Reaction products were primarily formic acid, formate, and CO; however, this concept could potentially be applied to design photoredox catalysts with high selectivity for CH_4.

The various aspects of hydrogenation of photoredox reactions make it advantageous for commercialization. First, the opportunity to avoid hydrogen production and storage steps makes photoredox systems a compelling way to increase the economic viability of industrial CO_2 reduction. While the heterogeneous photoredox systems described in this review have effectively achieved the desired direct conversion of CO_2 and H_2O into CH_4, their homogeneous counterparts require expensive sacrificial electron donors that limit their large-scale utility. Further research is required in order to identify homogeneous catalysts capable of producing hydrocarbons directly from CO_2 and H_2O. Second, photoredox systems operate at near-ambient pressure and temperature conditions, which are milder than the high temperature, high-pressure conditions experienced in state-of the-art heterogeneous catalytic hydrogenation. In addition to advances in materials development, further research is necessary to improve the designs of photoelectrochemical devices intended to reach technologically significant scales and improve solar-to-CH_4 efficiencies. A solar-to-fuel conversion efficiency of 10% is regarded as the minimum value that PEC systems must achieve to exhibit similar or improved efficiency compared with state-of-the-art Pt-G systems. The highest reported solar-to-CH_4 efficiency is 2.41% (Wang, Thompson, Baltrus, & Matranga, 2010) and most PEC systems exhibit efficiencies of at least one order of magnitude lower than this.

5.6 Reported heterogeneous photocatalysts for hydrogenation of CO_2

TiO_2 based photocatalysts

After Fujishima's discovery, TiO_2 with several modifications has been extensively used as a photocatalyst for CO_2 reduction (Inoue, Fujishima, Konishi, & Honda, 1979). Among different types of oxide semiconductors, TiO_2 is the best and most studied photocatalyst for the hydrogenation of CO_2, because of its chemically and biologically nontoxic nature, high stability, low cost, and abundance. TiO_2 is a typical d^0

metal oxide with suitable conduction band electrons (−0.50 V vs NHE) to drive the photocatalytic hydrogenation of CO_2. Several efforts have been made to increase its visible-light absorption and enhance the quantum efficiency. The doping of TiO_2 with nonmetals such as nitrogen, sulfur, and flourine could increase the narrowed bandgap and shift in absorption band of TiO_2 towards the visible light region. (Li, Liu, Mao, Xing, & Zhang, 2015) reported that the nitrogen-doped mesoporous TiO_2 photocatalyst showed a good visible-light absorption and enhanced activity for the hydrogenation of CO_2 to CH_4 under visible light irradiation. (Ong, Tan, Chai, Yong, & Mohamed, 2014) synthesized anatase nitrogen-doped TiO_2 nanoparticles with exposed (001) facets, showing photocatalytically active under visible light for the hydrogenation of CO_2 to CH_4 with an activity of 11-fold higher than that of pure TiO_2 under visible light irradiation. (Zhang, Li, Ackerman, Gajdardziska-Josifovska, & Li, 2012) successfully prepared 10% iodine-doped TiO_2 samples with a mixture of anatase and brookite phases exhibiting a high CO_2 hydrogenation activity (2.4 $\mu mol\,g^{-1}\,h^{-1}$) under visible light irradiation. The doping of transient metal ions such as V, Cr, Mn, Fe, and Ni is another effective strategy to extend the light adsorption range of TiO_2. The incorporation of an optimal weight percentage (1.5 wt%) Ni^{3+} into TiO_2 matrix not only extended optical absorption in the visible light range, but also enhanced the hydrogenation of CO_2 (Ola & Mercedes Maroto-Valer, 2014). The different concentration of Cr, V, and Co, metals-doped TiO_2 shows obvious red shifts of absorption edges and increased light absorption in the visible light region compared to pure TiO_2 (Ola & Mercedes Maroto-Valer, 2015). The 0.5 wt % V doped TiO_2 sample showed a maximum acetaldehyde rate of 11.13 $\mu mol\,g^{-1}\,h^{-1}$ in vapor phase hydrogenation of CO_2 under visible light irradiation. The metal and nonmetal nonmetal co-doped TiO_2 was also found to be efficient in hydrogenation of CO_2 under visible light irradiation (Ola & Mercedes Maroto-Valer, 2016). The effects of N and Ni^{2+} doped on TiO_2 in hydrogenation of CO_2 photocatalytic reduction showed an optimal methanol yield of 3.59 $\mu mol\,g^{-1}\,h^{-1}$ under visible light irradiation (Fan et al., 2011). The doped N and Ni^{2+} could enhance the light response of TiO_2 and improve the photocatalytic activity of TiO_2 by acting as a probable electron trapper to ensure a good separation of electron-hole pairs, respectively.

Plasmonic metal nanoparticles such as Au and Ag nanoparticles deposited on semiconductors are expected to exhibit some effects such as (1) enhancing the electron-hole separation for the formation of Schottky barriers, (2) extending the light absorption into the visible light range and enhancing the surface electron excitation by their surface plasmon resonance effects. Recently several studies have been reported the use of plasmonic Au or Ag nanoparticles decorated TiO_2 materials for photocatalytic hydrogenation of CO_2. Hou et al. (2011) reported that TiO_2-Au material catalyzed reduction of CO_2 and water vapor over a wide range of wavelengths. When they used 532 nm light to catalyze the photoreduction of aqueous CO_2, methane was the only product and a 24-fold enhancement was observed because of the strong electric fields created by the surface plasmon resonance of the Au nanoparticles, which excited electron-hole pairs locally in the TiO_2 at a rate several orders of magnitude higher than the normal incident light. When the photon energy was higher to excite d-band electronic transitions of Au nanoparticles in the UV range, different products including C_2H_6, CH_3OH, and HCHO were formed. Plasmonic Ag nanoparticles deposited on TiO_2 nanowire films were used for photocatalytic hydrogenation of CO_2 to CH_3OH under visible light irradiation (Liu et al., 2015). The results indicated that the CH_3OH yield reached 8.3 $\mu mol\,cm^{-2}$ over the optimal TiO_2-Ag sample due to the charge transfer property of Ag nanoparticles and the efficient light utilization based on the overlapped visible light-harvesting of Ag nanoparticles and TiO_2 films. Liu et al. (2014) prepared 2.5wt% Ag-TiO_2 composite which exhibited a methanol production rate of 135.1 $\mu mol\,g^{-1}\,h^{-1}$ under UV and visible light irradiation. The rate was 9.4 times higher than the pure TiO_2. Compared with Au or Ag plasmonic nanoparticles, Cu nanoparticles are reported to exhibit the surface plasmon resonance effect for visible light harvesting property. Liu et al. (2015) used plasmonic Cu nanoparticles modified TiO_2 nanoflower films for hydrogenation of CO_2 to CH_3OH. The CH_3OH production rate was 1.8 $\mu mol\,g^{-1}\,h^{-1}$ over 0.5wt% Cu-TiO_2 under UV and visible light irradiation., which is 6 times higher rate than the pure TiO_2 film. The commercial P25-TiO_2 decorated with Au-Cu alloy nanoparticles (Au:Cu = 1.2) exhibited a CH_4 production rate of 2000 $\mu mol\,g^{-1}\,h^{-1}$ under-stimulated sunlight (Neaţu, Maciá-Agulló, Concepción, & Garcia, 2014). The combination of TiO_2 with graphene proves to be a promising way to improve the photocatalytic reduction of CO_2. When combining TiO_2 with graphene the excited electrons of TiO_2 transferring from the conduction band graphene react with adsorbed CO_2 and H_2O to produce CH_3OH and O_2 respectively (Hsu et al., 2013). Solvent exfoliated graphene showed enhanced photocatalytic activity than conventional graphene for the photocatalytic reduction of CO_2 to CH_4, with up to sevenfold improvement compared to pure TiO_2 under visible light irradiation. This is due to its superior electrical mobility, which facilitates the diffusion of photoinduced electrons to reactive sites (Liang, Vijayan, Gray, & Hersam, 2011). Liang, Vijayan, Lyandres, Gray, and Hersam (2012) found that carbon nanomaterial dimensionality was a key factor in determining the spectral response and reaction specificity of graphene titania nanosheet composite photocatalysts. The 2D-2D graphene titania nanosheet composites and superior electronic coupling compared to the 1D-2D carbon nanotube-titania nanosheet composites, leading to greater enhancement factors for CO_2 photoreduction under UV irradiation. Cu-Pt bimetallic doped on TiO_2 nanotubes showed 4 times enhanced CO_2 reduction to CH_4, C_2H_4, and C_2H_6 in presence of H_2O (Shiraishi, Sakamoto, Sugano, Ichikawa, & Hirai, 2013). They have been attempting to mimick natural photosynthesis by making three-dimensional leaf-like architectures using titanates (Zhou et al., 2013) and reported reasonable 3 $\mu mol/g$ of CO and 0.3 $\mu mol/g$ of CH_4 conversion from CO_2. Shan et al. (2017) showed that the boron-doped $SrTiO_3$ exhibited four times increased conversion of CO_2 to useful products such as CO, CH_4, and oxygen. Bi_2WO_6 with oxygen vacancies also showed photoreduction of CO_2 to CO (Li et al., 2020). List of reported complex oxide photocatalysts are shown in Table 5.1. Tahir (2020) showed that the $ZnFe_2O_4$/Ag/TiO_2 nanocomposite exhibited excellent formation of various byproducts such as CO, C_2H_6, CH_3OH, and CH_4 using his indigenous photocatalytic reactor. He also developed monolith photoreactor (Tahir, 2020) that enhances CO_2 reduction than the liquid phase commercial reactors.

Quantum dot based photocatalysts

Quantum dots (QDs) are exceptional candidates for hydrogenation of CO_2 photocatalyst due to their small band gap and ability to undergo efficient carrier multiplication under visible light irradiation. Wang, Thompson, Ohodnicki, Baltrus, and Matranga (2011) modified the surface of TiO_2 with PbS QDs and showed that the band alignment between PbS and TiO_2 favored charge separation across the interface, eliminating inefficiencies associated with direct carrier recombination. Wang, Thompson, Baltrus, and Matranga (2010) synthesized a series of CdSe QD sensitized TiO_2 heterostructures for CO_2 photoreduction in the

TABLE 5.1 List of complex oxide photocatalysts for CO_2 reduction along with their by-products and yield.

Material	Preparation method	Reaction conditions	By product and yield (μmol/g h)
Boron doped $SrTiO_3$ (Shan et al., 2017)	Ball milling	Liquid phase reaction under 500 W Xenon lamp	CO: 21 μmol, CH_4: 14 μmol & O_2: 35 μmol
$Ca_xTi_yO_3$ (Lin et al., 2019)	Hydrothermal	Liquid phase reaction under 6 W UV lamp	CH_4: 17 μmol/g & O_2: 34 μmol.
3D leaf like $SrTiO_3$ (Zhou et al., 2013)	Chemical method	Liquid phase reaction under 300 W Xenon lamp	CO: 3 μmol/g & CH_4: 0.29 μmol/g
$LaFeO_3$-TiO_2 (Humayun et al., 2016)	Chemical method	Liquid phase reaction under 300 W Xenon lamp	CO: 18.75 μmol/g & CH_4: 13.75 μmol/g
CuO loaded $SrTiO_3$ nanorod thin film (Shoji et al., 2016)	Hydrothermal method	Liquid phase reaction under UV lamp	CO: 1.4 μmol/g
$ZnFe_2O_4$/Ag/TiO_2 (Tahir, 2020)	Solvothermal and Physical mixing methods	Gas phase reaction under 200 W Hg lamp	CO: 1025 μmol/g, CH_4: 132 μmol/g, CH_3OH: 31 μmol/g & C_2H_6: 19 μmol/g
$ZnFe_2O_4$/TiO_2 (Iqbal et al., 2020)	Chemical method	Liquid phase reaction under 500 W Xenon lamp	CH_3OH: 141.22 μmol/g
$M_{0.33}WO_3$ (M = K, Rb, Cs) (Wu et al., 2019)	Solvothermal method	Liquid phase reaction under 500 W Xenon lamp	CH_3OH: 17.5 μmol/g h
Vacancy Bi_2WO_6 (Kong, Lee, Mohamed, & Chai, 2019)	Hydrothermal	Liquid phase reaction under 500 W Xenon lamp	CO: 40 μmol/g h
Cu-TiO_2 (Li et al., 2010)	Chemical method	Liquid phase reaction under 250 W Xenon lamp	Oxalic acid: 65.6 μmol/g, Acetic acid: 12 μmol/g & Methanol: 0.8 μmol/g
Ni doped $InTaO_4$ (Tsai, Chen, Liu, Asakura, & Chan, 2011)	Sol-gel method	Liquid phase reaction under 20 W white LED	Methanol: 200 μmol/g
P-25 TiO_2 (Razzaq & In, 2019)	Sol-gel method	Liquid phase reaction under 250 W Hg lamp	Methanol: 914 μmol/g & Methane: 0.7 μmol/g
Co(II)Phathalocyanine with Ni/NiO core shell (Prajapati et al., 2019)	Chemical method	Liquid phase reaction under 20 W white LED	Methanol: 3641.2 μmol/g
Oxygen vacancy with CeO_2 (Wang et al., 2019)	Solution combustion method	Liquid phase reaction under 300 W Xenon lamp	Methanol: 0.702 μmol/g h
rGO/$InVO_4$/Fe_2O_3 (Kumar, Prajapati, Pal, & Jain, 2018)	Chemical method	Liquid phase reaction under 20 W white LED	Methanol: 16.9 mmol/g

TABLE 5.1 List of complex oxide photocatalysts for CO_2 reduction along with their by-products and yield—cont'd

Material	Preparation method	Reaction conditions	By product and yield (μmol/gh)
g-C_3N_4/$FeWO_4$ (Bhosale, Jain, Vinod, Kumar, & Ogale, 2019)	Hydrothermal method	Liquid phase reaction under 300 W Xenon lamp	CO: 6 μmol/gh
rGO-CdS on porous alumina support (Kandy and Gaikar, 2019)	Chemical method	Gas phase reaction under sunlight irradiation	Methanol: 153.8 μmol/gh
Au-S-TiO_2 (Awate, Deshpande, Rakesh, Dhanasekarana, & Gupta, 2011)	Chemical method	Liquid phase reaction under 20 W white LED	Methanol: 74.3 μL/g & H_2: 57.01 μmol/gh
$InVO_4$ (Awate, Deshpande, Rakesh, Dhanasekaran, & Gupta, 2011)	Chemical method	Liquid phase reaction under 20 W white LED	CO: 520 ppm & H_2: 236.41 ppm
Cu_3SnS_4 (Sharma et al., 2019)	Chemical method	Gas phase reaction under solar simulator irradiation	Methanol: 14 μmol/gh
g-C_3N_4/NiAl-LDH (Tonda, Kumar, Bhardwaj, Yadav, & Ogale, 2018)	Chemical method	Liquid phase reaction under 300 W Xenon lamp	Methanol: 2 μmol/g & H_2: 0.45 μmol/g, O_2: 1.2 μmol/g
CNT-TiO_2 (Rodríguez, Camarillo, Martínez, Jiménez, & Rincón, 2020)	Chemical method	Liquid phase reaction under 20 W white LED	CH_4: 2360 μmol/gh, H_2: 3246 μmol/gh & HCOOH: 1520 μmol/gh

presence of H_2O. This composite showed an efficiency rate of 3.3 ppm $g^{-1} h^{-1}$ for CH_3OH and 48 μmol $g^{-1} h^{-1}$ under visible light irradiation. CdS is one of the best visible light-driven photocatalysts due to its narrower bandgap. Park, Ou, Kang, Choi, and Hoffmann (2016) reported that TiO_2 coupled CdS materials exhibited exceptional visible-light photocatalytic activities for CO_2 conversion to CH_4. Compared to bare TiO_2, TiO_2-CdS composite increased the production of CH_4 (48 μmol $g^{-1} h^{-1}$).

Cu-based photocatalysts

Cu-based semiconductors including copper oxide (CuO), cuprous oxide (Cu_2O) and cuprous sulfide (Cu_2S) with band gaps of 1.7 eV, 2.2 eV, and 1.2 eV, respectively, are of particular interest due to their efficient visible light-harvesting, suitable conduction band states and good selectivity towards hydrogenation of CO_2. But the narrow bandgap leads to poor photocatalytic activity because of the fast recombination of photoinduced carriers. So, incorporating suitable n-type semiconductor with Cu-semiconductor to form p-n junction, can effectively increase the lifetime of the charge carriers by establishing an electric field at the interface with a potential difference between the two sides, is considered to be an effective strategy to promote the performance of hydrogenation of CO_2. The iron oxide cuprous oxide nanocomposites showed CO yield of 1.67 μmol $g^{-1} h^{-1}$ under visible light irradiation in presence of water vapor. The ZnO-CuO nanocomposites showed highest CO yields of 1.98 m mole $g^{-1} h^{-1}$ under UV-visible irradiation

(Wang et al., 2015). In this composite, the CuO was surface engineered with ZnO islands using a few pulsed cycles of atomic layer deposition. The detailed investigations revealed that the formation of the p-n heterojunction as the contact surface, which significantly influenced the migration behavior of charge carriers. The PL studies also confirmed a very high density of defects on these ZnO islands, which could enhance the lifetime of photogenerated electrons. Gusain, Kumar, Sharma, Jain, and Khatri (2016) synthesized CuO nanoparticles covered with rGO nanosheets via chemical synthesis. This composite showed excellent photocatalytic activity and stability for transforming CO_2 into CH_3OH at the rate of 1228 which is seven times higher than that of the pure CuO sample under visible light irradiation. The yield of CH_3OH remained almost the same after six cycles of reaction, which ruled out the possibility of surface carbon contamination. The improved photoactivity of rGO-CuO was explained by the strong interaction between rGO and CuO, which facilitated rGO to efficiently capture and transfer photoelectrons generated from CuO to the surface catalytic sites for reduction of CO_2.

Bismuth based photocatalysts

The bismuth-based semiconductors such as $BiVO_4$, Bi_2MoO_6, and $BiWO_6$ have been reported due to their narrower, visible light active bandgap, and its electronic structures. The Bi_2WO_6 nanoplates with (001) facets exhibited a CH_4 production of 1.09 micromole $g^{-1}\,h^{-1}$ when used for visible light hydrogenation of CO_2 (Zhou et al., 2011). The ultrathin geometry of nanoplates allowed charge carriers to move rapidly from the interior to the surface in order to participate in photoreduction reaction and also improved the separation of photo-generated electrons and holes. The BiOCl nanosheets with (001) facets showed CO_2 to CH_4 conversion rate of 41.48 micromole $g^{-1}\,h^{-1}$ in 8h under visible light irradiation (Jin, Wang, & He, 2015).

WO_3-based photocatalysts

WO_3 is one of the narrower bandgap material around 2.5–2.8eV and shows good light absorption up to NIR region (Yan et al., 2015). It also has a good electron transport mobility than TiO_2. The crystal structure of WO_3 also affects its charge separation, optical absorption, redox capability, and electrical conductivity. Huang et al. (2015) single-crystal WO_2 nanosheet of 4–5nm thickness showed enhanced CO_2 to CH_4 conversion rate of 1.14 micromole $g^{-1}\,h^{-1}$. Xie, Liu, Yin, and Cheng (2012) synthesized rectangular sheet-like WO_3 with predominant (002) facets by controlling acid and showed CO_2 to CH_4 conversion rate of 0.34 micromole $g^{-1}\,h^{-1}$.

C_3N_4 based photocatalysts

Graphitic carbon nitride is a two-dimensional conjugated polymer, which is a promising metal-free, visible light responsive photocatalyst. The conduction band potential of C_3N_4 is located at 1.3V which is high enough to drive the multi-electron transfer reactions to produce CO, HCOOH, CH_3OH, and other hydrocarbon fuels. At the same time, H_2O can act as both H^+ source and hole consumption agent due to its higher oxidation potential than the g-C_3N_4 valence band (+1.4V). This material can be easily synthesized by thermal polymerization of abundant nitrogen-rich precursors, such as urea, melamine, dicyanide, cyanamide, thiourea, ammonium thiocyanate, which provides a great possibility for practical application of g-C_3N_4. The heterojunction of g-C_3N_4/Fe_2O_3 hybrid showed enhanced photoreduction of CO_2 to CO at the rate of 27.2 micromole $g^{-1}\,h^{-1}$ without cocatalyst and sacrificial reagent which was 2.2 times higher than that bare g-C_3N_4 (10.3 micromole $g^{-1}\,h^{-1}$) (Jiang et al., 2018). Li et al. (2016) hybridized mesoporous structured CeO_2 with g-C_3N_4 to form heterojunction photocatalyst for CO_2 reduction. The optimal CeO_2-g-C_3N_4 with 3 wt% CeO_2 showed the highest rate of 13.88 $\mu mol\,g^{-1}\,h^{-1}$ CH_4. When

$g-C_3N_4$ was combined with 42 wt% SnO_{2-x} the CO_2 reduction rate reached $22.7\,\mu mol\,g^{-1}\,h^{-1}$ which was 4 and 5 times higher than those of $g-C_3N_4$ and P25 respectively. This is due to the formation of a heterojunction structure between two components, which efficiently promote the separation of electron-hole pairs by a direct Z scheme mechanism to enhance the photocatalytic activity. The Boron and phosphorus co-doped $g-C_3N_4$ coupled with SnO_2 exhibited an enhanced visible light activity for CO_2 conversion to CH_4 with the yield of $37.5\,\mu mol\,g^{-1}\,h^{-1}$ from CO_2 containing water (Raziq et al., 2017). Zhou et al. (2014) reported the coupling of $g-C_3N_4$ with $n-TiO_2$ and showed enhanced CO conversion rate of $12.2\,\mu mol\,g^{-1}\,h^{-1}$. Wang et al. (2017) synthesized novel 2D MnO_2-$g-C_3N_4$ heterojunction photocatalyst for enhanced CO_2 to CO conversion efficiency. Metal oxide frameworks have attracted intense attention due to their extremely high surface area, and availability in both inorganic and organic nature. $g-C_3N_4$ nanotubes integrated ZIF-8-Zn showed enhanced CH_3OH production efficiency by a factor of more than three times than that of the bare $g-C_3N_4$ (Shi, Wang, Zhang, Chang, & Ye, 2015). $g-C_3N_4$ decorated with carbon materials such as carbon dots, reduced graphene oxides has attracted great attention. Ong, Tan, Chai, Yong, and Mohamed (2015) reported that the enhanced CH_4 conversion rate of $29.23\,\mu mol\,g^{-1}\,h^{-1}$ after 10 h of irradiation under visible light, which were 3.6 times higher than the bare $g-C_3N_4$. Ong et al. (2017) constructed 2D heterojunction nanocomposite containing $g-C_3N_4$-rGO with different ratio of rGO contents exhibit remarkably enhanced activity than the bare $g-C_3N_4$. The optimized sample ($g-C_3N_4$-15 wt% rGO) showed enhanced CH_4 evolution of $13.93\,\mu mol\,g^{-1}\,h^{-1}$ in 10 h with a quantum efficiency of 0.56% and high stability. They explained the reason that the charge modification at the surface of $g-C_3N_4$ extended the 2D/2D interlamination region and facilitated the separation of photoinduced charge carriers, thus improving photoreduction performance

Complex oxide-based photocatalysts

There are a few iron-based perovskites such as $LaSrCoFeO_3$, $LaCoFeO_3$, and $LaNiFeO_3$ that have been reported for hydrogenation of CO_2 by photothermal catalysis. Ha, Lu, Liu, Wang, and Zhao (2016) developed three-dimensionally-ordered macroporous $LaSrCoFeO_3$ and it showed excellent thermal stability, higher surface area, and good catalytic conversion rate of CH_4 from CO_2. Zheng et al. (2019) also prepared $LaNiFeO_3$ and reported 4 times increased methanol conversion from CO_2 than the $LaFeO_3$ & $LaNiO_3$. These experiments were carried out at 350°C under visible light irradiations.

5.7 Methods of reduction or hydrogenation of CO_2

The reduction of CO_2 can be carried out in the liquid phase and gas phase. In the liquid phase, the reduction can be done in saturated aqueous solution of CO_2. But the poor solubility rate of CO_2 in water is one of the major drawbacks. The solubility rate of CO_2 in water can be improved by adding additives such as NaOH, $NaHCO_3$, Na_2CO_3. Another issue is the surface adsorption of H_2O is more preferable over CO_2 in the liquid phase, so the reduction of water is favorable. The efficiency of CO_2 reduction in the liquid phase depends on pH of the reaction medium, surface hydroxyl groups, solvent, and additives. The increase in the pH increases the rate of reaction. So it forms different concentrations of species such as CO_3^{2-}, HCO_3^- and CO_2 at different values of pH. These chemical species having different reduction potentials get adsorbed on the catalyst surface. Addition of NaOH can improve the dissolution of CO_2 and increase the efficiency of photoreduction of CO_2 on TiO_2 supported Cu catalysts. The gas-phase reactions can be carried out with humidified CO_2. Xie, Wang, Zhang, Deng, and Wang (2014) demonstrated both liquid and gas phase reactions with TiO_2 and Pt-TiO_2. He found that the CH_4 production is nearly three times more and hydrogen production is less in the gas phase compared to the liquid phase. Consequently three times more selectivity in CO_2 reduction over water reduction was observed in gas phase. The efficiency of CO_2 reduction in gas phase reactions depends on surface properties of photocatalysts, CO_2-H_2O mixture ratio, feed pressure, temperature, etc.

5.8 Limitations and important aspects for future perspectives

The photocatalytic reduction of CO_2 not only depends on the bandgap (i.e.) visible light activity of photocatalysts, also depends on particle size, surface area, crystallinity, carrier mobility, and defects.

Discovery of novel materials is important, in order to satisfy all the above factors and also it should be stable under different environmental conditions, eco-friendly and low cost for the preparation.

Suitable novel, cost-effective cocatalysts, heterostructure with easily available materials will improve the charge separation as well as quantum efficiency during CO_2 reduction reaction.

Proper reactor design is very important for CO_2 reduction under different conditions. Gas-phase reactors are promising new set ups that reported higher quantum yields.

In-situ techniques should be investigated more to understand the mechanism during reduction reactions.

5.9 Conclusion

The hydrogenation of CO_2 is one of the promising green energy technologies. The present techniques used for this process are not suitable for the industrial level of progress due to several reasons such as lower conversion efficiency, long reaction time, complications in experimental methods, etc. It is expected that the material should give at least 10% of quantum efficiency in the visible region of solar spectrum. The only way to improve its efficiency is through globalized collaboration in development of novel materials and experimental procedures. This would initiate efficient conversion of the excess CO_2 into useful products and make earth safer for living.

References

Arai, T., Sato, S., & Morikawa, T. (2015). A monolithic device for CO_2 photoreduction to generate liquid organic substances in a single-compartment reactor. *Energy & Environmental Science, 8*, 1998–2002. https://doi.org/10.1039/C5EE01314C.

Awate, S. V., Deshpande, S. S., Rakesh, K., Dhanasekaran, P., & Gupta, N. M. (2011). Role of micro-structure and interfacial properties in the higher photocatalytic activity of TiO_2-supported nanogold for methanol-assisted visible-light-induced splitting of water. *Physical Chemistry Chemical Physics, 13*, 11329–11339.

Awate, S. V., Deshpande, S. S., Rakesh, K., Dhanasekarana, P., & Gupta, N. M. (2011). Role of micro-structure and interfacial properties in the higher photocatalytic activity of TiO_2-supported nanogold for methanol-assisted visible-light-induced splitting of water. *Physical Chemistry Chemical Physics, 13*, 11329–11339.

Bhosale, R., Jain, S., Vinod, C. P., Kumar, S., & Ogale, S. (2019). Direct Z-scheme g-C_3N_4/$FeWO_4$ nanocomposite for enhanced and selective photocatalytic CO_2 reduction under visible light. *ACS Applied Materials & Interfaces, 11*(6), 6174–6183.

Chen, G., Gao, R., Zhao, Y., Li, Z., Waterhouse, G. I. N., Shi, R., et al. (2017). Alumina-supported CoFe alloy catalysts derived from layered-double-hydroxide nanosheets for efficient photothermal CO_2 hydrogenation to hydrocarbons. *Advanced Materials, 30*, 1704663.

Diercks, C. S., Liu, Y., Cordova, K. E., & Yaghi, O. M. (2018). The role of reticular chemistry in the design of CO_2 reduction catalysts. *Nature Materials, 17*(4), 301–307. https://doi.org/10.1038/s41563-018-0033-5.

Fan, J., Liu, E.-z., Tian, L., Hu, X.-y., He, Q., & Sun, T. (2011). Synergistic effect of N and Ni^{2+} on nanotitania in photocatalytic reduction of CO_2. *Journal of Environmental Engineering, 137*, 171–176.

Garg, S., Li, M., Weber, A. Z., Ge, L., Li, L., Rudolph, V., et al. (2020). Advances and challenges in electrochemical CO_2 reduction processes: An engineering and design perspective looking beyond new catalyst materials. *Journal of Materials Chemistry A, 8*(4), 1511–1544. https://doi.org/10.1039/c9ta13298h.

Gusain, R., Kumar, P., Sharma, O. P., Jain, S. L., & Khatri, O. P. (2016). Reduced graphene oxide–CuO nanocomposites for photocatalytic conversion of CO_2 into methanol under visible light irradiation. *Applied Catalysis B: Environmental, 181*, 352–362.

Ha, M. N., Lu, G., Liu, Z., Wang, L., & Zhao, Z. (2016). 3DOM-LaSrCoFeO $6-\delta$ as a highly active catalyst for the thermal and photothermal reduction of CO_2 with H_2O to CH_4. *Journal of Materials Chemistry A, 4*, 13155–13165. https://doi.org/10.1039/C6TA05402A.

Hou, W., Hung, W. H., Pavaskar, P., Goeppert, A., Aykol, M., & Cronin, S. B. (2011). Photocatalytic conversion of CO_2 to hydrocarbon fuels via plasmon-enhanced absorption and metallic interband transitions. *ACS Catalysis, 1*, 929–936.

Hsu, H.-C., Shown, I., Wei, H.-Y., Chang, Y.-C., Du, H.-Y., Lin, Y.-G., … Chen, K.-H. (2013). Graphene oxide as a promising photocatalyst for CO_2 to methanol conversion. *Nanoscale, 5*, 262–268.

Hu, Y. H. (2009). Solid-solution catalysts for CO_2 reforming of methane. *Catalysis Today, 148*(3–4), 206–211. https://doi.org/10.1016/j.cattod.2009.07.076.

Huang, Z. F., Song, J., Pan, L., Zhang, X., Wang, L., & Zou, J.-J. (2015). Tungsten oxides for photocatalysis, electrochemistry, and phototherapy. *Advanced Materials, 27*, 5309–5327.

Humayun, M., Qu, Y., Raziq, F., Yan, R., Li, Z., Zhang, X., & Jing, L. (2016). Exceptional visible-light activities of TiO_2-coupled N-doped porous perovskite $LaFeO_3$ for 2,4-dichlorophenol decomposition and CO_2 conversion. *Environmental Science & Technology, 50*(24), 13600–13610.

Huo, Z., Hu, M., Zeng, X., Yun, J., & Jin, F. (2012). Catalytic reduction of carbon dioxide into methanol over copper under hydrothermal conditions. *Catalysis Today, 194*(1), 25–29. https://doi.org/10.1016/j.cattod.2012.06.013.

Inoue, T., Fujishima, A., Konishi, S., & Honda, K. (1979). Photoelectrocatalytic reduction of carbon dioxide in aqueous suspensions of semiconductor powders. *Nature, 277*, 637–638.

Iqbal, F., Mumtaz, A., Shahabuddin, S., Abd Mutalib, M. I., Shaharun, M. S., Nguyen, T. D., … Abdullah, B. (2020). Photocatalytic reduction of CO_2 to methanol over $ZnFe_2O_4/TiO_2$ (p–n) heterojunctions under visible light irradiation. *Journal of Chemical Technology and Biotechnology, 95*(8), 2208–2221.

Jaiswal, G., Landge, V. G., Jagadeesan, D., & Balaraman, E. (2017). Iron-based nanocatalyst for the acceptorless dehydrogenation reactions. *Nature Communications, 8*(1). https://doi.org/10.1038/s41467-017-01603-3.

Jia, J., O'Brien, P. G., He, L., Qiao, Q., Fei, T., Reyes, L. M., et al. (2016). Visible and near-infrared photothermal catalyzed hydrogenation of gaseous CO_2 over nanostructured Pd@ Nb2O5. *Advancement of Science, 3*, 1600189.

Jiang, Z., Wan, W., Li, H., Yuan, S., Zhao, S., & Wong, P. K. (2018). A hierarchical Z-scheme α-Fe_2O_3/g-C_3N_4 hybrid for enhanced photocatalytic CO_2 reduction. *Advanced Materials, 30*, 1706108.

Jin, J., Wang, Y., & He, T. (2015). Preparation of thickness-tunable BiOCl nanosheets with high photocatalytic activity for photoreduction of CO_2. *RSC Advances, 5*, 100244–100250.

Kandy, M. M., & Gaikar, V. G. (2019). Continuous photocatalytic reduction of CO_2 using nanoporous reduced graphene oxide (RGO)/cadmium sulfide (CdS) as catalyst on porous anodic alumina (PAA)/aluminum support. *Journal of Nanoscience and Nanotechnology, 19*(8), 5323–5331. https://doi.org/10.1166/jnn.2019.16817.

Kong, X. Y., Lee, W. Q., Mohamed, A. R., & Chai, S.-P. (2019). Effective steering of charge flow through synergistic inducing oxygen vacancy defects and p-n heterojunctions in 2D/2D surface-engineered Bi_2WO_6/BiOI cascade: Towards superior photocatalytic CO_2 reduction activity. *Chemical Engineering Journal, 372*, 1183–1193.

Kumar, A., Prajapati, P. K., Pal, U., & Jain, S. L. (2018). Ternary rGO/$InVO_4$/Fe_2O_3 Z-scheme heterostructured photocatalyst for CO_2 reduction under visible light irradiation. *ACS Sustainable Chemistry & Engineering, 6*(7), 8201–8211.

Li, K., Peng, B., & Peng, T. (2016). Recent advances in heterogeneous photocatalytic CO_2 conversion to solar fuels. *ACS Catalysis, 6*(11), 7485–7527. https://doi.org/10.1021/acscatal.6b02089.

Li, Q., Zhu, X., Yang, J., Yu, Q., Zhu, X., Chu, J., … Xu, H. (2020). Plasma treated Bi_2WO_6 ultrathin nanosheets with oxygen vacancies for improved photocatalytic CO_2 reduction. *Inorganic Chemistry Frontiers, 7*, 597–602.

Li, X., Liu, P., Mao, Y., Xing, M., & Zhang, J. (2015). Preparation of homogeneous nitrogen-doped mesoporous TiO_2 spheres with enhanced visible-light photocatalysis. *Applied Catalysis B: Environmental, 164*, 352–359.

Li, Y., Wang, W.-N., Zhan, Z., Woo, M.-H., Wu, C.-Y., & Biswas, P. (2010). Photocatalytic reduction of CO_2 with H_2O on mesoporous silica supported Cu/TiO_2 catalysts. *Applied Catalysis B: Environmental, 100*(1–2), 386–392.

Liang, Y. T., Vijayan, B. K., Gray, K. A., & Hersam, M. C. (2011). Minimizing graphene defects enhances titania nanocomposite-based photocatalytic reduction of CO_2 for improved solar fuel production. *Nano Letters, 11*, 2865–2870.

Liang, Y. T., Vijayan, B. K., Lyandres, O., Gray, K. A., & Hersam, M. C. (2012). Effect of dimensionality on the photocatalytic behavior of carbon–titania nanosheet composites: Charge transfer at nanomaterial interfaces. *The Journal of Physical Chemistry Letters, 3*, 1760–1765.

Lin, J., Hu, J., Qiu, C., Huang, H., Chen, L., Xie, Y., … Wang, X. (2019). In situ hydrothermal etching fabrication of $CaTiO_3$ on TiO_2 nanosheets with heterojunction effects to enhance CO_2 adsorption and photocatalytic reduction. *Catalysis Science & Technology, 9*, 336–346.

Liu, E., Hu, Y., Li, H., Tang, C., Hu, X., Fan, J., ... Bian, J. (2015). Photoconversion of CO_2 to methanol over plasmonic Ag/TiO_2 nano-wire films enhanced by overlapped visible-light-harvesting nanostructures. *Ceramics International, 41*, 1049–1057.

Liu, E., Kang, L., Wu, F., Sun, T., Hu, X., Yang, Y., ... Fan, J. (2014). Photocatalytic reduction of CO_2 into methanol over Ag/TiO_2 nanocomposites enhanced by surface plasmon resonance. *Plasmonics, 9*, 61–70.

Liu, E., Qi, L., Bian, J., Chen, Y., Hu, X., Fan, J., ... Wang, Q. (2015). A facile strategy to fabricate plasmonic Cu modified TiO_2 nano-flower films for photocatalytic reduction of CO_2 to methanol. *Materials Research Bulletin, 68*, 203–209.

Liu, X., Inagaki, S., & Gong, J. (2016). Heterogeneous molecular systems for photocatalytic CO_2 reduction with water oxidation. *Angewandte Chemie International Edition in English, 55*(48), 14924–14950.

Meng, L., Chen, Z., Ma, Z., He, S., Hou, Y., Li, H. H., et al. (2018). Gold plasmon-induced photocatalytic dehydrogenative coupling of methane to ethane on polar oxide surfaces. *Energy and Environmental Science, 11*(2), 294–298. https://doi.org/10.1039/c7ee02951a.

Neatu, S., Antonio, J., Agullo, M., & Garcia, H. (2014). Solar light photocatalytic CO_2 reduction: General considerations and selected bench-mark photocatalysts. *International Journal of Molecular Sciences, 15*, 5246–5262.

Neaţu, S., Maciá-Agulló, J. A., Concepción, P., & Garcia, H. (2014). Gold–copper nanoalloys supported on TiO_2 as photocatalysts for CO_2 reduction by water. *Journal of American Chemical Society, 136*, 15969–15976.

Nichols, E. M., Gallagher, J. J., Liu, C., Su, Y., Resasco, J., Yu, Y., et al. (2015). Hybrid bioinorganic approach to solar-to chemical conversion. *Proceedings of the National Academy of Sciences of the United States of America, 112*(37), 11461–11466. https://doi.org/10.1073/pnas.1508075112.

Ola, O., & Mercedes Maroto-Valer, M. (2014). Role of catalyst carriers in CO_2 photoreduction over nanocrystalline nickel loaded TiO_2-based photocatalysts. *Journal of Catalysis, 309*, 300–308.

Ola, O., & Mercedes Maroto-Valer, M. (2015). Transition metal oxide based TiO_2 nanoparticles for visible light induced CO_2 photoreduction. *Applied Catalysis A: General, 502*, 114–121.

Ola, O., & Mercedes Maroto-Valer, M. (2016). Synthesis, characterization and visible light photocatalytic activity of metal based TiO_2 monoliths for CO_2 reduction. *Chemical Engineering Journal, 283*, 1244–1253.

Ong, W. J., Putri, L. K., Tan, Y.-C., Tan, L.-L., Li, N., Ng, Y. H., ... Chai, S.-P. (2017). Unravelling charge carrier dynamics in protonated g-C_3N_4 interfaced with carbon nanodots as co-catalysts toward enhanced photocatalytic CO_2 reduction: A combined experimental and first-principles DFT study. *Nano Research, 10*, 1673–1696.

Ong, W.-J., Tan, L.-L., Chai, S.-P., Yong, S.-T., & Mohamed, A. R. (2014). Self-assembly of nitrogen-doped TiO_2 with exposed {001} facets on a graphene scaffold as photo-active hybrid nanostructures for reduction of carbon dioxide to methane. *Nano Research, 7*, 1528–1547.

Ong, W. J., Tan, L.-L., Chai, S.-P., Yong, S.-T., & Mohamed, A. R. (2015). Surface charge modification via protonation of graphitic carbon nitride (g-C_3N_4) for electrostatic self-assembly construction of 2D/2D reduced graphene oxide (rGO)/g-C_3N_4 nanostructures toward enhanced photocatalytic reduction of carbon dioxide to methane. *Nano Energy, 13*, 757–770.

Park, H., Ou, H.-H., Kang, U., Choi, J., & Hoffmann, M. R. (2016). Photocatalytic conversion of carbon dioxide to methane on TiO_2/CdS in aqueous isopropanol solution. *Catalysis Today, 266*, 153–159.

Porosoff, M. D., Yan, B., & Chen, J. G. (2016). Catalytic reduction of CO_2 by H_2 for synthesis of CO, methanol and hydrocarbons: Challenges and opportunities. *Energy and Environmental Science, 9*(1), 62–73. https://doi.org/10.1039/c5ee02657a.

Prajapati, P. K., Singh, H., Yadav, R., Sinha, A. K., Szunerits, S., Boukherroub, R., & Jain, S. L. (2019). Core-shell Ni/NiO grafted cobalt (II) complex: An efficient inorganic nanocomposite for photocatalytic reduction of CO_2 under visible light irradiation. *Applied Surface Science, 467–468*, 370–381.

Rao, H., Lim, C.-H., Bonin, J., Miyake, G. M., & Robert, M. (2018). Visible-light-driven conversion of CO_2 to CH_4 with an organic sensitizer and an iron porphyrin catalyst. *Journal of American Chemical Society, 140*(51), 17830–17834.

Raziq, F., Qu, Y., Humayun, M., Zada, A., Yu, H., & Jing, L. (2017). Synthesis of SnO_2/B-P codoped g-C_3N_4 nanocomposites as efficient cocatalyst-free visible-light photocatalysts for CO_2 conversion and pollutant degradation. *Applied Catalysis B: Environmental, 201*, 486–494.

Razzaq, A., & In, S. I. (2019). TiO_2 based nanostructures for photocatalytic CO_2 conversion to valuable chemicals. *Micromachines, 10*, 326.

Rodríguez, V., Camarillo, R., Martínez, F., Jiménez, C., & Rincón, J. (2020). CO_2 photocatalytic reduction with CNT/TiO_2 based nanocomposites prepared by high-pressure technology. *The Journal of Supercritical Fluids, 163*, 104876.

Shan, J., Raziq, F., Humayun, M., Zhou, W., Qu, Y., Wang, G., et al. (2017). Improved charge separation and surface activation via boron-doped layered polyhedron $SrTiO_3$ for CO-catalyst free photocatalytic CO_2 conversion. *Applied Catalysis B: Environmental, 219*, 10–17. https://doi.org/10.1016/j.apcatb.2017.07.024.

Shan, J., Raziq, F., Humayun, M., Zhou, W., Qu, Y., Wang, G., & Li, Y. (2017). Improved charge separation and surface activation via boron-doped layered polyhedron $SrTiO_3$ for co-catalyst free photocatalytic CO_2 conversion. *Applied Catalysis B: Environmental, 219*, 10–17.

Sharma, N., Das, T., Kumar, S., Bhosale, R., Kabir, M., & Ogale, S. (2019). Photocatalytic activation and reduction of CO_2 to CH_4 over single phase nano Cu_3SnS_4: A combined experimental and theoretical study. *ACS Applied Energy Materials*, 2(8), 5677–5685.

Sharma, N., Das, T., Kumar, S., Bhosale, R., Kabir, M., & Ogale, S. (2019). Photocatalytic activation and reduction of CO_2 to CH_4 over single phase nano Cu_3 SnS_4: A combined experimental and theoretical study. *ACS Applied Energy Materials*, 2(8), 5677–5685. https://doi.org/10.1021/acsaem.9b00813.

Shi, L., Wang, T., Zhang, H., Chang, K., & Ye, J. (2015). Electrostatic self-assembly of nanosized carbon nitride nanosheet onto a zirconium metal–organic framework for enhanced photocatalytic CO_2 reduction. *Advanced Functional Materials, 25*, 5360–5367.

Shiraishi, Y., Sakamoto, H., Sugano, Y., Ichikawa, S., & Hirai, T. (2013). Pt–Cu bimetallic alloy nanoparticles supported on anatase TiO_2: Highly active catalysts for aerobic oxidation driven by visible light. *ACS Nano*, 7(10), 9287–9297.

Shoji, S., Yin, G., Nishikawa, M., Atarashi, D., Sakaia, E., & Miyauchi, M. (2016). Photocatalytic reduction of CO_2 by Cu_xO nanocluster loaded $SrTiO_3$ nanorod thin film. *Chemical Physics Letters, 658*, 309–314.

Singhal, N., Goyal, R., & Kumar, U. (2017). Visible-light-assisted photocatalytic CO_2 reduction over $InTaO_4$: Selective methanol formation. *Energy & Fuels*, 31(11), 12434–12438. https://doi.org/10.1021/acs.energyfuels.7b02123.

Sorcar, S., Hwang, Y., Grimes, C. A., & In, S.-I. (2017). Highly enhanced and stable activity of defect-induced titania nanoparticles for solar light-driven CO_2 reduction into CH_4. *Materials Today, 20*, 507–515.

Srinivas, B., Shubhamangala, B., Lalitha, K., Anil Kumar Reddy, P., Durga Kumari, V., Subrahmanyam, M., et al. (2011). Photocatalytic reduction of CO_2 over Cu-TiO_2/molecular sieve 5A composite. *Photochemistry and Photobiology*, 87(5), 995–1001. https://doi.org/10.1111/j.1751-1097.2011.00946.x.

Tahir, M. (2020). Well-designed $ZnFe_2O_4$/Ag/TiO_2 nanorods heterojunction with Ag as electron mediator for photocatalytic CO_2 reduction to fuels under UV/visible light. *Journal of CO_2 Utilization, 37*, 134–146. https://doi.org/10.1016/j.jcou.2019.12.004.

Tonda, S., Kumar, S., Bhardwaj, M., Yadav, P., & Ogale, S. (2018). g-C_3N_4/NiAl-LDH 2D/2D hybrid heterojunction for high-performance photocatalytic reduction of CO_2 into renewable fuels. *ACS Applied Materials & Interfaces*, 10(3), 2667–2678.

Tsai, C.-W., Chen, H. M., Liu, R.-S., Asakura, K., & Chan, T.-S. (2011). Ni@NiO core–shell structure-modified nitrogen-doped $InTaO_4$ for solar-driven highly efficient CO_2 reduction to methanol. *The Journal of Physical Chemistry C, 115*, 10180–10186.

Vernekar, D., Sakate, S. S., Rode, C. V., & Jagadeesan, D. (2019). Water-promoted surface basicity in FeO(OH) for the synthesis of pseudoionones (PS) and their analogues. *Journal of Catalysis*, 80–89. https://doi.org/10.1016/j.jcat.2019.08.026.

Wang, C., Thompson, R. L., Baltrus, J., & Matranga, C. (2010). Visible light photoreduction of CO_2 using CdSe/Pt/TiO_2 heterostructured catalysts. *The Journal of Physical Chemistry Letters, 1*, 48–53.

Wang, C., Thompson, R. L., Baltrus, J., & Matranga, C. (2010). Visible light photoreduction of CO_2 using CdSe/Pt/TiO_2 heterostructured catalysts. *Journal of Physical Chemistry Letters, 1*, 48–53. https://doi.org/10.1021/jz9000032.

Wang, C., Thompson, R. L., Ohodnicki, P., Baltrus, J., & Matranga, C. (2011). Size-dependent photocatalytic reduction of CO_2 with PbS quantum dot sensitized TiO_2 heterostructured photocatalysts. *Journal of Materials Chemistry, 21*, 13452–13457. https://doi.org/10.1039/C1JM12367J.

Wang, M., Shen, M., Jin, X., Tian, J., Li, M., Zhou, Y., … Shi, J. (2019). Oxygen vacancy generation and stabilization in CeO_{2-x} by Cu introduction with improved CO_2 photocatalytic reduction activity. *ACS Catalysis*, 9(5), 4573–4581.

Wang, M., Shen, M., Zhang, L., Tian, J., Jin, X., Zhou, Y., & Shi, J. (2017). 2D-2D MnO_2/g-C_3N_4 heterojunction photocatalyst: In-situ synthesis and enhanced CO_2 reduction activity. *Carbon, 120*, 23–31.

Wang, W. N., Wu, F., Myung, Y., Niedzwiedzki, D. M., Im, H. S., Park, J., … Biswas, P. (2015). Surface engineered CuO nanowires with ZnO islands for CO_2 photoreduction. *ACS Applied Materials & Interfaces, 7*, 5685–5692.

Welte, C., & Deppenmeier, U. (2014). Bioenergetics and anaerobic respiratory chains of aceticlastic methanogens. *Biochimica et Biophysica Acta, 1837*, 1130–1147.

Wu, X., Li, Y., Zhang, G., Chen, H., Li, J., Wang, K., et al. (2019). Photocatalytic CO_2 conversion of M0. 33WO3 directly from the air with high selectivity: Insight into full spectrum-induced reaction mechanism. *Journal of the American Chemical Society, 141*, 5267–5274.

Wu, X., Li, Y., Zhang, G., Chen, H., Li, J., Wang, K., … Xie, Y. (2019). Photocatalytic CO_2 conversion of $M_{0.33}WO_3$ directly from the air with high selectivity: Insight into full spectrum-induced reaction mechanism. *Journal of the American Chemical Society*, 141(13), 5267–5274.

Xie, S., Wang, Y., Zhang, Q., Deng, W., & Wang, Y. (2014). MgO- and Pt-promoted TiO_2 as an efficient photocatalyst for the preferential reduction of carbon dioxide in the presence of water. *ACS Catalysis*, *4*(10), 3644–3653. https://doi.org/10.1021/cs500648p.

Xie, Y. P., Liu, G., Yin, L., & Cheng, H.-M. (2012). Crystal facet-dependent photocatalytic oxidation and reduction reactivity of monoclinic WO_3 for solar energy conversion. *Journal of Materials Chemistry*, *22*, 6746–6752.

Yan, J., Wang, T., Wu, G., Dai, W., Guan, N., Li, L., & Gong, J. (2015). Tungsten oxide single crystal nanosheets for enhanced multichannel solar light harvesting. *Advanced Materials*, *27*, 1579.

Yi, Q., Li, W., Feng, J., & Xie, K. (2015). Carbon cycle in advanced coal chemical engineering. *Chemical Society Reviews*, *44*, 5409–5445.

Yu, B., & He, L. N. (2015). Upgrading carbon dioxide by incorporation into heterocycles. *ChemSusChem*, *8*(1), 52–62. https://doi.org/10.1002/cssc.201402837.

Zhang, Q., Li, Y., Ackerman, E. A., Gajdardziska-Josifovska, M., & Li, H. (2012). Visible light responsive iodine-doped TiO_2 for photocatalytic reduction of CO_2 to fuels. *Applied Catalysis A: General*, *400*, 195–202.

Zheng, D., Wei, G., Xu, L., Guo, Q., Hu, J., Sha, N., et al. (2019). LaNi Fe1-O3 ($0 \leq x \leq 1$) as photothermal catalysts for hydrocarbon fuels production from CO_2 and H_2O. *Journal of Photochemistry and Photobiology A: Chemistry*, *377*, 182–189. https://doi.org/10.1016/j.jphotochem.2019.03.045.

Zhou, H., Guo, J., Li, P., Fan, T., & Zhang, D. (2013). Leaf-architectured 3D hierarchical artificial photosynthetic system of perovskite titanates towards CO_2 photoreduction into hydrocarbon fuels. *Scientific Reports*, *3*, 1–9.

Zhou, S., Liu, Y., Li, J., Wang, Y., Jiang, G., Zhao, Z., … Wei, Y. (2014). Facile in situ synthesis of graphitic carbon nitride (g-C_3N_4)-N-TiO_2 heterojunction as an efficient photocatalyst for the selective photoreduction of CO_2 to CO. *Applied Catalysis B: Environmental*, *158–159*, 20–29.

Zhou, Y., Tian, Z., Zhao, Z., Liu, Q., Kou, J., Chen, X., … Zou, Z. (2011). High-yield synthesis of ultrathin and uniform Bi_2WO_6 square nanoplates benefitting from photocatalytic reduction of CO_2 into renewable hydrocarbon fuel under visible light. *ACS Applied Materials & Interfaces*, *3*, 3594–3601.

Further reading

Gao, Y., Qian, K., Xu, B., Li, Z., Zheng, J., Zhao, S., … Xu, Z. (2020). Recent advances in visible-light-driven conversion of CO_2 by photocatalysts into fuels or value-added chemicals. *Carbon Resources Conversion*, *3*, 46–59.

Kong, T., Jiang, Y., & Xiong, Y. (2020). Photocatalytic CO_2 conversion: What can we learn from conventional COx hydrogenation? *Chemical Society Reviews*, *49*, 6579–6591.

Shyamal, S., & Pradhan, N. (2020). Halide perovskite nanocrystals photocatalysts for CO_2 reduction: Success and challenges. *Journal of Physical Chemistry Letters*, *11*, 6921–6934.

Ulmer, U., Dingle, T., Duchesne, P. N., Morris, R. H., Tavasoli, A., Wood, T., & Ozin, G. A. (2019). Fundamentals and applications of photocatalytic CO_2 methanation. *Nature Communications*, *10*, 3169.

CHAPTER

6

The role of CO_2 sorbents materials in SESMR for hydrogen production

Thamyscira H.S. da Silva[a,b], *Muriel Chaghouri*[a], *Moisés R. Cesário*[b], *Haingomalala Lucette Tidahy*[a], *Cédric Gennequin*[a], *Daniel A. Macedo*[b], *and Edmond Abi-Aad*[a]

[a]Environmental Chemistry and Life Interactions Unit (UCEIV), University of the Littoral Opal Coast, UR 4492, SFR Condorcet—FR CNRS 3417, Dunkerque, France [b]Materials Science and Engineering Postgraduate Program (PPCEM), Federal University of Paraíba (UFPB), João Pessoa, Brazil

6.1 Introduction

In December 2015, the first universal agreement was signed between 195 countries and the European Union, in the city of Paris, in which public officials recognized the urgency of taking measures to contain climate change. The Paris Agreement (Methane tracker 2020, 2020) was an important milestone in determining public policies aimed at reducing the anthropogenic impact on global emissions.

More than recognizing the impending problem, the Paris agreement brought to the political and economic debate the need for measures to drastically reduce the global emissions of greenhouse gases that were discussed in the scientific sphere.

The agreement has a specific objective: the search for new technologies that assist in mitigating global emissions and that keep the global average temperature rise below 2°C above preindustrial levels, striving to maintain the temperature rise at 1.5°C above preindustrial levels (Methane tracker 2020, 2020).

The goal of maintaining global warming to 1.5°C above preindustrial levels until 2100 is very ambitious. To achieve this goal, environmental policies for international cooperation are needed to transform the supply and demand chains of energy, agriculture, and services production systems (Rogelj, 2018).

Heterogeneous Catalysis
https://doi.org/10.1016/B978-0-323-85612-6.00006-1

Taking into account only the anthropological factors that lead to global warming, the projections of the report produced by the IPCC (Intergovernmental Panel on Climate Change) show that to achieve the ambitious goal of 1.5°C of global warming it would be necessary to achieve zero CO_2 emissions in less than 15 years, as well as maintaining greenhouse gas (GHG) emissions in 2030 between 25 and 30 Gt CO_2 e/year. However, with current technologies for mitigating GHG emissions, it is estimated that in 2030 emissions will be between 52 and 58 Gt CO_2 e/year (Rogelj, 2018).

To change this scenario, the alternatives would be a reduction in the energy demand accompanied by decarbonization in the production of electricity and fuels, a reduction in emissions caused by the agricultural sector, large-scale adoption of technologies that sequester carbon dioxide, and the imposition of regulatory policies that price the carbon emitted by countries. The change process is challenging and complex, as it involves making political and economic decisions and a change in society's lifestyle.

In 2019, the expectation was that carbon emissions related to energy production would continue to follow an upward trend as in previous years, however, in the report presented by the IEA (IEA – Global CO_2 emissions in 2019, 2019) the trend was for a reduction in CO_2 emissions even with a 2.9% growing economy. The study points out that the reduction was more pronounced in developed countries and was related to the means of producing electricity from renewable sources of energy, mainly solar and wind, as well as a change of fuel from coal to natural gas, in addition to the greater use of nuclear energy.

The report on methane emissions in 2020 produced by the International Energy Agency (IEA) (Methane tracker 2020, 2020), points out that methane has an important role in transitional energy policies in the coming decades, as a substitute for polluting fuels in a scenario of sustainable development. In the IEA's projections for 2030, natural gas will become the main fossil fuel in developing countries.

Natural gas emits fewer greenhouse gases when compared to coal and is composed mostly of CH_4, the second greenhouse gas that most impacts global warming. The methane present in natural gas can be used in processes that produce hydrogen, considered a promising fuel for more sustainable energy systems.

Hydrogen is the most abundant element in the earth's crust, but it is not available naturally in the form of H_2. Hydrogen can be produced through water electrolysis, however, changes in production processes are necessary to increase efficiency, making the application in industrial sectors still economically unviable (Kalamaras & Efstathiou, 2013).

While water electrolysis is not yet economically interesting, H_2 can be obtained through fossil fuels, more precisely by methane conversion techniques, such as steam reform, dry reform, partial oxidation, and autothermal reform of methane. In this way, it is possible to use the existing structure, reduce greenhouse gas emissions, and proportionate a more economical H_2 production (Carapellucci & Giordano, 2020).

6.2 The steam methane reforming process

Steam methane reforming of (SMR) is a commercially consolidated technology and the most used to obtain hydrogen and synthesis gas, which is responsible for 48% of the total production of hydrogen in the world Noh, Lee and Moon, 2019. Pure hydrogen can be applied to fuel cells while synthesis gas (syngas), a mixture of hydrogen and carbon monoxide, can be used in several processes to obtain other products, such as ammonia, methanol, and larger chain hydrocarbons (Balat, 2008; Nieva, Villaverde, Monzón, Garetto, & Marchi, 2014).

The SMR process consists of the catalytic conversion of methane (in the presence of water vapor) to hydrogen and syngas (Eq. 6.1) (Noh, Lee, & Moon, 2019).

$$CH_{4(g)} + H_2O_{(g)} \leftrightarrow CO_{(g)} + 3H_{2(g)} \quad \Delta H_{298\,K} = 206.2\,kJ\,mol^{-1} \tag{6.1}$$

This catalytic reaction occurs under severe conditions of high temperatures ($T \geq 800°C$) and pressures between 8 and 35 bar in tubular reactors positioned in an oven, where the catalyst is usually nickel-based. An excess of steam is used to favor the reduction of carbon deposits, in the form of CO or CO_2, resulting from the dissociation of methane (Carapellucci & Giordano, 2020; Noh, Lee, & Moon, 2019).

The breaking of the C–H bond requires high energy, being a very endothermic reaction, and for this reason, it makes the process expensive in which the heating comes from the use of natural gas (Hufton, Mayorga, & Sircar, 1999). The decomposition of methane is favorable at $T > 500°C$ (Eq. 6.1), the Boudouard reaction (Eq. 6.2) favorable at $T < 750°C$, and the reverse carbon gasification favorable at $T < 750°C$, they are secondary reactions responsible for the formation of carbon that can occur during steam methane reform (Nieva, Villaverde, Monzón, Garetto, & Marchi, 2014). This carbon deposit is the main cause of the deactivation of the catalysts (Rostrup-Nielsen, 1997).

$$CH_{4(g)} \leftrightarrow C_{(s)} + 2H_{2(g)} \quad \Delta H_{298\,K} = 75.0\,kJ\,mol^{-1} \tag{6.2}$$

$$2CO_{(g)} \leftrightarrow C_{(s)} + CO_{2(g)} \quad \Delta H_{298\,K} = -172.52\,kJ\,mol^{-1} \tag{6.3}$$

$$CO_{(g)} + H_{2(g)} \leftrightarrow C_{(s)} + H_2O_{(g)} \quad \Delta H_{298\,K} = -131.34\,kJ\,mol^{-1} \tag{6.4}$$

Nickel-based catalysts are the most widely used and industrially consolidated in the SMR process and generally undergo deactivation due to carbon deposits that can be formed according to reactions (Eqs. 6.2–6.4). In addition to catalyst deactivation by carbon deposition, the formation of carbon can obstruct the transit of the gases, and consequently, a drop in the pressure used. The nickel catalyst can be deactivated due to the sintering effect and by sulfur poisoning (Hashemnejad & Parvari, 2011).

After the methane catalytic reform step, the resulting gases contain 76% H_2 (mol%), 13% CH_4, 12% CO, and 10% CO_2 on a dry basis (Barelli et al., 2008). However, the steam reforming reaction (Eq. 6.1) is always accompanied by the water gas shift reaction (WGS) (Eq. 6.5), an exothermic reaction that is favored at low temperatures (Carapellucci & Giordano, 2020).

$$CO_{(g)} + H_2O_{(g)} \leftrightarrow CO_{2(g)} + H_{2(g)} \quad \Delta H_{298\,K} = -41.2\,kJ\,mol^{-1} \tag{6.5}$$

A dedicated reactor performs the WGS reaction in which the syngas produced by SMR is introduced at temperatures between 300°C and 400°C, where the mole fraction of CO is reduced and the rate of H_2 production increases (Carapellucci & Giordano, 2020).

There are three basic steps to produce H_2 by the SMR process. A simplified process of SMR is shown in Fig. 6.1. The first step is addressed to reforming step at elevated temperature and pressure, the second step is dedicated to increasing the H_2 production by using the WGS reaction at a lower temperature, and the final step consists of a mixture gases purification process to reduce the CO_2 produced by absorption after the WGS reaction step.

The WGS reaction leads to a higher formation of CO_2, an undesirable gas for the environment. It is interesting to reduce or eliminate this CO_2 formation considering the CO_2 emission

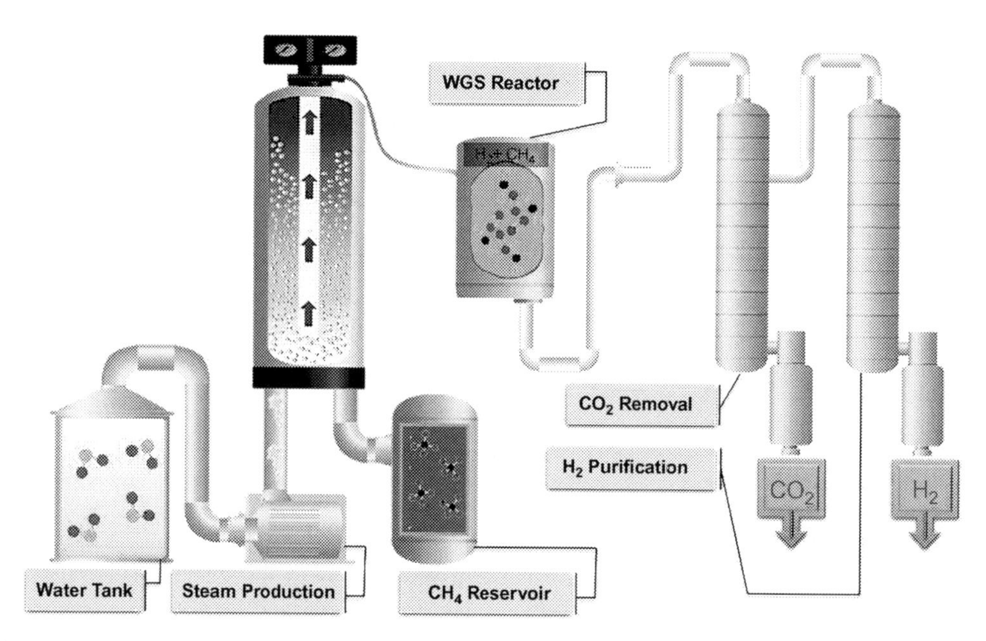

FIG. 6.1 Scheme for a conventional SMR process. In the schematic design for the SMR process it is possible to understand what the basic units for H_2 production are, such as, steam production unit, reactor with a catalyst for the methane reforming reaction, reactor for the WGS reaction, CO_2 removal unit, and H_2 purification unit.

reduction targets set out in the Paris agreement. Configurations of new SMR process systems that enable the improvement of methane conversion, assisted by CO_2 capture, and consequently, favoring an increase in the production of pure H_2 have been extensively studied (Anderson, Yun, Kottke, & Fedorov, 2017; Lu & Xie, 2016).

One of the studied solutions involves the use of membranes that allow the selective permeation of specific molecules. An oxygen transport membrane (OTM) (Cai, Wu, Zhu, Ghoniem, & Yang, 2020; Park, Lee, Dorris, Lu, & Balachandran, 2011; Rosen et al., 2011) and a hydrogen transport membrane (HTM) (Buxbaum & Kinney, 1996; Phair & Badwal, 2006) are positioned in the reactor to create three compartments that promote a different reaction: syngas from natural gas, promotion of the WGS reaction, and hydrogen separation. They have generally composed of conductive ceramics, and the operating temperature is between 800°C and 1000°C.

In this technique, low-temperature air is injected from the cathodic side of the OTM membrane to promote partial oxidation of methane and hydrogen formation, according to Eq. (6.6). On the anode side of the OTM membrane, a mixture of natural gas and steam is added and the steam reforming reaction takes place in the presence of a catalyst. The HTM membrane exposed to the syngas side allows the removal of H_2 due to a difference in partial pressure, which allows the formation of more H_2 due to the reaction of WGS until there is no more difference in partial pressure between the reaction compartment and the HTM membrane.

$$CH_{4(g)} + \frac{1}{2}O_{2(g)} \leftrightarrow CO_{(g)} + 2H_{2(g)} \qquad (6.6)$$

Despite the benefits of membranes, problems due to pore blockage, low thermal and mechanical stability, and dilution of the products obtained caused by the use of drag gases can limit their application to CO_2 capture (Barelli et al., 2008).

Another approach to capture CO_2 widely studied is the sorption enhanced reform process (SERP) of natural gas, or more specifically the sorption enhanced steam methane reforming (SESMR), using specific sorbents that selectively capture CO_2 and allow the regeneration of the absorber after saturation (Balasubramanian, Ortiz, Kaytakoglu, & Harrison, 1999; Harrison, 2009; Martavaltzi, Pampaka, Korkakaki, & Lemonidou, 2010; Ortiz & Harrison, 2001).

When sorbents are applied in SMR processes they can remove CO_2 from the gas mixture during the reaction process, favoring the reaction for the hydrogen production side. In this way, the production capacity of pure hydrogen increases (>90%) even at lower reaction temperatures (Martavaltzi et al., 2010), between 400°C and 650°C, compared to the conventional SMR process without the need for profound changes in the configuration of the reactors used, making the costs for hydrogen production decrease (Ochoa-Fernández et al., 2007).

This chapter proposes to provide an overview of sorption enhanced steam methane reforming, addressing the types of sorbents most used and the type of configuration of the most studied reactors for industrial applications. Studies on the carbonation-calcination cycles and the lifetime of the sorbents for SESMR will be addressed. The literature in this field is vast, and therefore only the most promising materials and methods will be covered.

6.3 The SESMR process

The SMR process is the most used for hydrogen production and consists of the reforming of methane in the presence of a catalyst, generally nickel-based, in which the reforming reaction is highly endothermic and is always accompanied by the WGS reaction. To increase the production of H_2 by SMR, additional steps are used with dedicated reactors for the reaction of WGS and steps to purify the gases produced to obtain greater amounts of H_2 and reduce CO_2 emissions. These additional steps are equivalent to 30% of the total cost of H_2 production (Hufton, Mayorga, & Sircar, 1999; Myers, Mastin, Bjørnebøle, Ryberg, & Eldrup, 2011).

To reduce the energy costs of H_2 production by SMR, the process of sorption enhanced steam methane reform (SESMR) (Carvill, Hufton, Anand, & Sircar, 1996; Ding & Alpay, 2000) was proposed, which consists of combining the catalytic reactions of the steam reform of methane with the adsorption separation of one of the products in the same reactor unit equipped with a system for regenerating the material used as a sorbent.

The SESMR technique is shown as an alternative to increase the efficiency of the industrially consolidated SMR technique without the need for reactors dedicated to other process steps, and, consequently, to reduce the thermodynamic losses associated with the WGS reaction steps and H_2 purification (Balasubramanian et al., 1999; Harrison, 2008).

The SESMR process is based on the Le Chatelier principle (Carvill et al., 1996; Chen, Yu, & Chen, 2021) in which the selective removal of one of the unwanted products (CO_2) from the SMR reaction, through in situ capture using adsorbent materials, shifts the equilibrium of the reaction of reform to favor greater production of the desired product (H_2) in just a single step, with a conversion to hydrogen that can reach values greater than 90% (dry basis) (Balasubramanian et al., 1999; Chen, Yu, & Chen, 2021).

The materials used for CO_2 absorption consist mostly of metal oxides (MeO) that can be from natural or synthesized sources and that will selectively remove CO_2 according to the exothermic carbonation reaction (Eq. 6.7):

$$MeO_{(s)} + CO_{2(g)} \leftrightarrow MeCO_{3(s)} \tag{6.7}$$

With reactions (Eqs. 6.1–6.5) being able to occur simultaneously, it can be said that the total reaction for the SESMR process is according to reaction (Eq. 6.8).

$$CH_{4(g)} + 2H_2O_{(g)} + MeO_{(s)} + \leftrightarrow MeCO_{3(s)} + 4H_{2(g)} \quad (6.8)$$

Several materials are potential candidates for in situ capture of CO_2 for the SESMR process, however, the selection of these materials must take into account some conditions.

In general, sorbent materials must have a high CO_2 absorption capacity, have high thermal and performance stability due to cyclical processes of carbonation and calcination, in addition to adequate reaction kinetics at the usual operating temperatures for both capture and regeneration processes (Barelli et al., 2008; Ochoa-Fernández et al., 2007). In the case of applications in fluidized bed reactors, it is interesting that the sorbents have mechanical stability due to the flow of particles generated. When meeting the criteria for application as sorbents, the material, and its production must be low cost so as not to lead to an increase in hydrogen production costs.

One of the metal oxides most used in applications as sorbents for SESMR is calcium oxide, which has a high CO_2 capture capacity and good kinetics, however, its disadvantage is the loss of the absorption capacity over the carbonation/calcination cycles (Albrecht, Wagenbach, Satrio, Shanks, & Wheelock, 2008). To explain the potential of the SESMR process through the kinetics of the reactions involved, a CaO absorbent will be used as an example. In addition to the methane (Eq. 6.1) and WGS (Eq. 6.5) reform reactions that occur in the conventional SMR process, reaction (Eq. 6.9) represents the in situ capture of CO_2 that occurs when adding the CaO adsorbent to the system:

$$CaO_{(s)} + CO_{2(g)} \leftrightarrow CaCO_{3(s)} \quad \Delta H_{298\,K} = -178\,kJ\,mol^{-1} \quad (6.9)$$

Reform reaction (Eq. 6.1) needs a lot of energy to promote the disruption of C–H bonds, making the SMR process considered highly endothermic. The SESMR process is considered thermodynamically neutral according to reaction (Eq. 6.10) because the CaO carbonation reaction (Eq. 6.9) is highly exothermic, reducing the need for high temperatures to promote the conversion of methane to hydrogen due to the in situ energy supply supplement to the methane conversion process (Zhu, Li, & Fan, 2015), and therefore, problems with sintering and coke formation of the catalyst are reduced. Although the regeneration process of the sorbent is endothermic (reverse of reaction, Eq. 6.9), the total energy demand is 20%–25% less than the energy required for the SMR process (Balasubramanian et al., 1999; Barelli et al., 2008; Harrison, 2009).

$$CH_{4(g)} + 2H_2O_{(g)} + CaO_{(s)} + \leftrightarrow CaCO_{3(s)} + 4H_{2(g)} \quad \Delta H_{298\,K} = 1.4\,kJ\,mol^{-1} \quad (6.10)$$

The lower energy demand for hydrogen production is demonstrated in Fig. 6.2 of the study of the thermodynamic balance carried out by Balasubramanian et al. (1999), which compares the formation of H_2 in the presence or not of a CaO adsorbent in relation to temperature, at a pressure of 15 atm and a steam-to-methane ratio of 4.

The maximum production of H_2 without the presence of the absorbent occurs at 850°C and represents 76% of the gas produced on a dry basis, in which the production rate increases with temperature. In the presence of an absorbent, the maximum production of H_2 reaches 96% at 650°C and remains close to that level up to 750°C, even considering that $Ca(OH)_2$ can form during the process, a product that hinders the carbonation of CaO for the solid $CaCO_3$ product and decomposes at around 600°C. At low temperatures, unreacted methane is the main impurity of the product and at higher

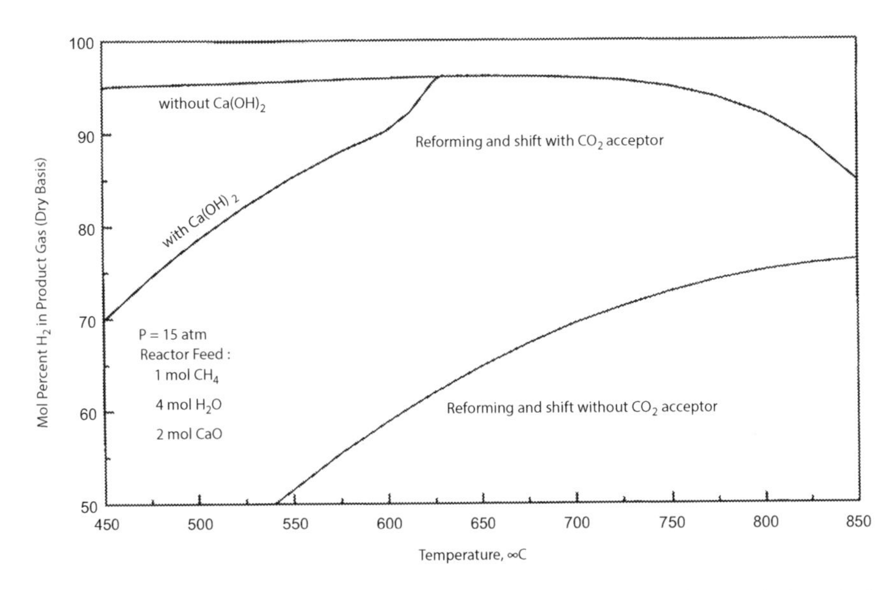

FIG. 6.2 Hydrogen content with and without adsorbent material. From the thermodynamic balance in the presence of a CaO adsorbent, the H_2 production reached a maximum of 96% at a lower temperature (650°C) when compared to the sample without the addition of the adsorbent, which reached a maximum of 76% H_2 at 850°C (Balasubramanian, Ortiz, Kaytakoglu, & Harrison, 1999).

temperatures, CO and CO_2 not captured become the main impurity.

The absorption process by a solid absorbent can be better understood through the gas-solid interaction dynamics between CO_2-CaO (Barelli et al., 2008; Ding & Alpay, 2000; Gupta & Fan, 2002). The process can be divided into two stages that take place at different rates. In the first stage, the kinetics of the absorption reaction occurs quickly due to the extensive surface area available of the absorbent. A layer of the $CaCO_3$ product begins to form on the surface of the CaO from clusters that grow as the absorption process develops. The process continues with high reaction kinetics until the CaO surface is almost completely transformed into a $CaCO_3$ shell, then the kinetics is governed by ion diffusion processes of the CO_2 molecules to reach the CaO contained inside. In this kinetic model, only 70%–90% of CaO is converted to $CaCO_3$ (Gupta & Fan, 2002). $CaCO_3$ can be regenerated through calcination, which in turn has fast kinetics and due to the release of CO_2, allows the regenerated CaO to have a porous structure.

Fig. 6.3 shows the evolution of a SESMR test over time, at a fixed temperature of 650°C, pressure equivalent to 15 atm, and steam-CH_4 ratio equal to 4, where the ratio between absorbent and catalyst is 1:1 (Balasubramanian et al., 1999). The initial phase corresponds to the moment when the reactant gases are added to the reactor and the NiO is reduced to metallic Ni (active phase of the catalyst). A first period called prebreakthrough the CH_4, WGS, and CO_2 capture reactions occur simultaneously with maximum efficiency, causing the H_2 produced to reach its maximum value, while unreacted CH_4, CO, and CO_2 remain at their minimum values. The molar fractions are close to the expected balance, where the horizontal lines correspond to the equilibrium value indicated for each product.

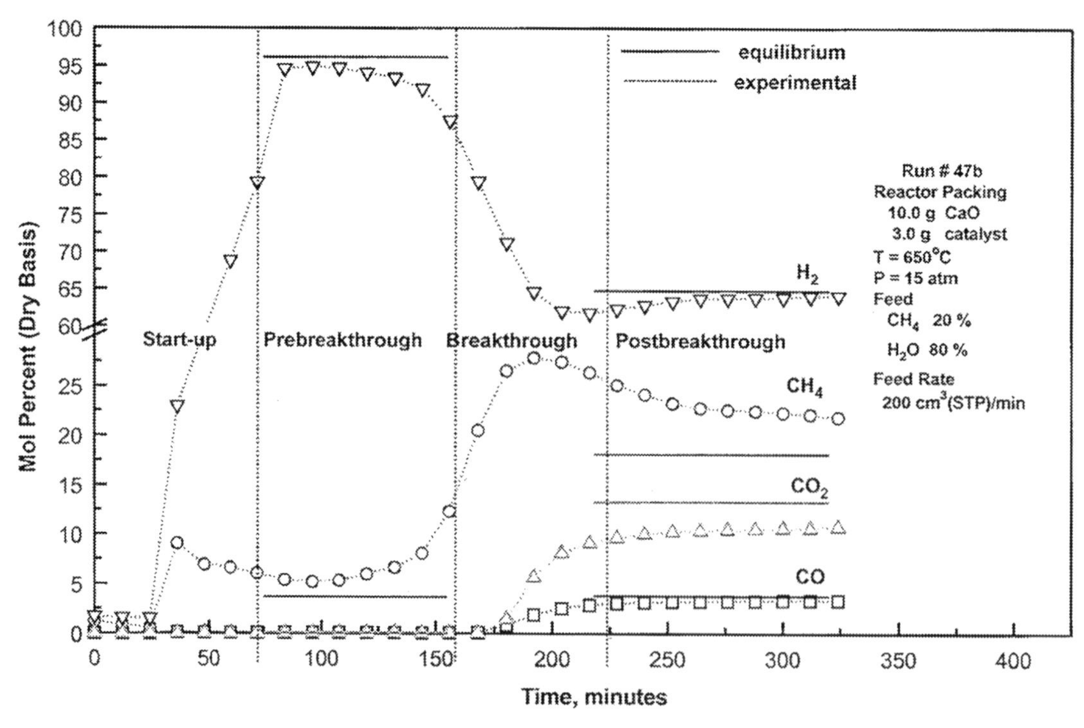

FIG. 6.3 Evolution of a SESMR test over time. When the adsorbent is active the H_2 production reaches its maximum and without large amounts of CO_2 and CO present, in which phase it is known as prebreakthrough. The breakthrough phase indicates that the adsorbent begins its loss of capture capacity. And finally, the postbreakthrough phase is related to the simple SMR process (Balasubramanian, Ortiz, Kaytakoglu, & Harrison, 1999).

The period corresponding to the breakthrough is associated with a reduction in the CO_2 absorption capacity by the absorbent, leading to a reduction in the H_2 produced and an increase in the concentrations of CH_4, CO, and CO_2. After the saturation of the absorbent, the last period called postbreakthrough, the absorption corresponds to zero and the CH_4 reform reaction and the WGS reaction are the only active ones and the concentration of the gases produced are similar to the SMR process (Balasubramanian et al., 1999).

For greater efficiency of the SESMR process, it is interesting that the prebreakthrough period is as long as possible to delay the need for regeneration of the absorbent.

In summary, the SESMR process is thermodynamically advantageous for the following reasons (Carvill et al., 1996; Harrison, 2008):

- The process is simpler as it provides for the separation of CO_2 in a single reaction step;
- The purification steps can be eliminated to obtain large amounts of pure H_2, as it performs the direct production of the desired product with high purity;
- Does not need dedicated reactors or catalysts for the reaction of WGS;
- Reduces the need for special alloys for reactors that operate at high temperature, reducing the construction costs of the reactors;
- Reduces (or can eliminate) the formation of carbon deposits;

- Allows operation at lower temperatures to achieve the same conversion rates when compared to the use of catalysts only;
- Lower operating temperatures reduce losses due to heat exchange;
- Higher methane conversion rates are obtained;
- Although the sorbent regeneration process through its calcination requires high energy, it is possible to recycle it simply and the total energy consumption is less than for the SMR process;
- Reduces the need to move solid material between reactors, reducing friction problems, dust formation, and eliminating the need for a pneumatic conveying system.

Deactivation

As mentioned previously, the catalyst involved in the sorption enhanced steam reforming of methane reaction needs to be highly active, selective to the specified reaction as opposed to secondary reaction, chemically and thermally stable for several cycles of catalysis-regeneration. In this perspective, it is important to detail the different types of deactivation that the catalyst can go through.

Influence of the temperature

Because of the endothermic nature of the studied reaction, temperatures of 600°C and above are used in general. This high-temperature influences two highly frequent types of deactivation: coking and sintering. Thermodynamically, methane cracking reaction (Eq. 6.2) is favored at temperatures of 500°C and above whereas Boudouard reaction (Eq. 6.3) is favored at temperatures below 750°C. Reverse gasification reaction (Eq. 6.4) also takes place in the same temperature range of SESMR, methane cracking, and Boudouard reaction.

The three mentioned reactions produce carbon as a secondary product which can cause the deactivation of the catalytic material. During the test, an accumulation of carbon on the surface of the catalyst can cause the blockage of the active sites and provoke a progressive or a sudden decrease in activity (Chen, Yu, & Chen, 2021). This type of deactivation is very common but is not irreversible. Catalysts are usually recovered after calcination at high temperature in order to eliminate all of the carbon by reoxidation. Even though the activity can be redeemed, it is usually not completely restored.

In addition, sintering is a physical phenomenon that can also cause partial deactivation of the catalysts. This type of thermal degradation usually takes place after extreme exposure to high temperatures. In this case, the original crystallite structure of the catalyst can be modified which causes a permanent loss of activity. In general, this phenomenon is due to a loss of surface area due to support or pore collapse. Particle size is consequently increased which raises the risk of carbon formation (Zhu, Li, & Fan, 2015).

Influence of the gas composition

In most cases, the gas mixture used for SESMR studies contains steam and methane. However, in more realistic conditions, gas could contain impurities such as sulfur and halogen compounds (Guo, 2011; Hashemnejad & Parvari, 2011). Upon contact with the catalyst, these molecules can cause a rapid deactivation by pore blockage. As a result, the reactants can no more interact with the active sites of the catalyst. This is caused when the affinity of the impurity to certain catalytic sites is higher than the affinity of the reactants. In general, the poisoning phenomenon is not irreversible and catalysts can be recovered after thermal treatment at high temperature in the presence of an excess of steam.

Influence of the H_2O and CO_2

Water and carbon dioxide are two of the main molecules involved in the SESMR process. H_2O is one of the reactants of both the steam reforming of methane reaction and the water gas shift reaction. CO_2 on the other hand, is the product of the water-gas shift reaction and the reactant of the adsorption reaction on the adsorbent compound such as CaO. In general, these molecules lead to a modification in the adsorption properties of the catalysts by reacting with the CaO sites according to reactions (Eqs. 6.9 and 6.11).

$$CaO_{(s)} + H_2O_{(g)} \leftrightarrow Ca(OH)_{2_{(s)}} \quad \Delta H_{298K} = -109\,kJ\,mol^{-1} \tag{6.11}$$

The occurrence of both of these reactions is common during the SESMR process, especially on CaO based materials. The hydration reaction (6.11) and the carbonation reaction (6.9) lead to a decrease in the production of hydrogen due to the Le Chatelier's principle. While the hydration reaction is taking place, H_2O is reacting with both methane and CaO. This means that the ratio H_2O/CH_4 is lower than expected which slows down both the SMR reaction and WGS reaction (Ochoa-Fernández et al., 2007). On the other hand, the carbonation reaction causes the saturation of all the CaO sites which drastically reduces the adsorption parameter of the material. This leads to an increase of CO_2 in the environment and results in favoring the reverse water gas shift reaction which reduces the production of hydrogen. In this case, a regeneration step is necessary to recover the full efficiency of the catalyst (García-Lario, Aznar, Martinez, Grasa, & Murillo, 2015).

Regeneration

As mentioned previously, catalysts can be completely and partially deactivated through different mechanisms such as sintering, carbon formation, poisoning, and especially carbonation reaction. Sintering is an irreversible mechanism whereas catalysts deactivated by carbon deposition and poisoning can usually be recovered with a thermal treatment in an oxidative environment. In order to recover the activity of a catalyst that underwent hydration, a thermal treatment in a reductive atmosphere should be performed (Yancheshmeh, Radfarnia, & Iliuta, 2016).

Carbonation is the main cause in the loss of activity during SESMR because it is specific for the adsorption step of the process. It is known that all the materials used in SESMR undergo this phenomenon, which translates into a complete saturation of the adsorbent with CO_2 through the formation of carbonates. For a continuous process, regeneration becomes a necessary step built in the SESMR process. In general, the reaction followed for the regeneration step is the reverse carbonation reaction, as exemplified for calcium oxide-based sorbents in reaction (Eq. 6.9).

In a more generalized way, this reverse carbonation reaction can be summarized in Eq. (6.12).

$$MeCO_{3(s)} \leftrightarrow MeO_{(s)} + CO_{2(g)} \quad \Delta H_{298\,K} > 0\,kJ\,mol^{-1} \tag{6.12}$$

With *Me* being the sorbent. This reaction needs a high amount of energy in order to take place.

In general, regeneration of a catalyst that underwent full or partial carbonation is done through thermal treatment. This usually takes place via temperature swing, pressure swing or a combination of both (Harrison, 2008). Temperature swing is the most used process for the regeneration of CaO-based sorbents. In this case, temperature is increased which favors the dissociation of $CaCO_3$ and the regeneration of the CaO sites. This technique is mostly used for calcium oxide-based materials. In most cases, its efficiency leads to the desorption of nearly 100% of the adsorbed CO_2. The disadvantage of the temperature swing regeneration is the need for extreme thermal conditions which leads to the sintering of the catalyst and the loss of activity throughout the regeneration cycles. Another disadvantage for this technique is the need for a longer waiting time for the temperature to decrease from above 800°C (regeneration temperature) to less than 650°C (reforming temperature).

Vacuum swing adsorption is mainly used for fixed bed reactors configuration. It is also more suitable for industrial processes reducing flue gases for high CO_2 partial pressure. In this technique, the pressure is modified in order to favor the desorption of CO_2. Working at lower pressure leads to the dissociation of the carbonates resulting in CO_2 release. A study showed that even though this pressure swing adsorption may protect the material from sintering and is an overall shorter process, it allows the recovery of only 85% of CO_2. On the other hand, the recovered CO_2 is in general very pure (Tlili, Grévillot, & Vallières, 2009).

In some cases, a combination of pressure and temperature swing can be used for the regeneration of the sorbent. In this case, the reduction of the pressure lowers the input energy required for the reverse carbonation reaction to take place. Several authors have reported using this hybrid technique and noticing a high CO_2 recovery. Moreover, this method also decreases the risk of sintering.

In addition to the pressure and temperature conditions of the regeneration process, different gas compositions can be used for this step. Authors have used gas mixtures containing O_2, H_2, N_2, and H_2O for the regeneration step. Air, pure O_2, or a mixture containing O_2 is still the most used oxidant for regeneration (Antzaras, Heracleous, & Lemonidou, 2020; García-Lario, Aznar, Martinez, Grasa, & Murillo, 2015; Hildenbrand, Readman, Dahl, & Blom, 2005). Hildendrand et al. and Antzaras et al. have both used a mixture of O_2 diluted in N_2 or Ar for regeneration at 800°C. Both studies have shown that 100% of CO_2 was eliminated and the sorbent capacity was fully recovered during the first few cycles (Antzaras et al., 2020; Hildenbrand et al., 2005).

Even though O_2 is the main gas used for regeneration of sorbents, other compounds have also shown interesting results for CO_2 recovery. Acharya et al. (Acharya, Dutta, & Basu, 2012) studied the effect of the gas composition of the efficiency of the calcination step (reverse carbonation reaction). For this study, they used CO_2, H_2O, and N_2. A total regeneration could be obtained in the presence of H_2O at 900°C and N_2 at 1000°C. It was concluded that steam could be the best gas for regeneration of CaO.

Several authors have confirmed the efficient use of steam for CO_2 desorption (Arstad, Prostak, & Blom, 2012; Dewoolkar & Vaidya, 2017; Xie, Zhou, Qi, Cheng, & Yuan, 2012). Cobden et al. have used a mixture of 30% H_2O diluted in N_2 to perform the regeneration step (Cobden et al., 2007). However, the downside of this technique is the need to remove H_2O after CO_2 capture and before its storage. It is important to mention that, in the case SESMR, a reductive pretreatment enhances the activation of certain materials and increases their activity. However, both O_2 and H_2O are oxidative molecules, which can provoke the oxidation of the material. Under those circumstances, a reactivation step under reductive atmosphere may be required. For this reason, several authors have opted for the use of an inert gas such as N_2, Ar, and He or a reductive mixture (e.g., with H_2). Even though this gas mixture necessitates higher temperatures, their use eliminates a potential reactivation step.

Antzaras et al. (2020) showed that the regeneration is not only a function of the composition of the gas but also the GHSV used. This study proved that the time for full CO_2 desorption from the catalyst decreases with the increase of GHSV (increase in the flow rate).

In summary, all of these parameters highly influence the cyclic stability and the multicycle ability of the materials. It is important for industrial application that the catalytic cycle shows a high thermal stability which can result in a stable activity even after multiple reforming/regeneration cycles.

In the next sessions, the materials most used as absorbents will be discussed.

6.4 Adsorbents

Adsorption is a process in which molecules of a gas are captured by a solid material that we call adsorbent, being considered a promising technique for capturing CO_2, in which the properties of both the gas mixture and the solid will dictate the nature and extent of the interaction between them (Bhatta, Subramanyam, Chengala, Olivera, & Venkatesh, 2015; Drage, Smith, Pevida, Arenillas, & Snape, 2009). There is a vast amount of solid adsorbent available for the capture of CO_2, however, the choice depends on factors such as the type of application, reactor design, desired performance for the process, and configuration of the adsorbent regeneration system after its saturation (Fiaschi & Lombardi, 2002).

Adsorbents can be classified to the nature of interaction in physisorbents and chemisorbants, or the adsorption/desorption temperature at low temperature (below 200°C), intermediate temperature (between 200°C and 400°C), and high-temperature adsorbent (above 400°C) (Wang, Luo, Zhong, & Borgna, 2011). In this work, we will address only high-temperature adsorbents due to the operating temperature used in SESMR.

For industrial applications, the absorbents must have some general characteristics (Abanades, Rubin, & Anthony, 2004; Krishna & Van Baten, 2012; Maring & Webley, 2013; Samanta, Zhao, Shimizu, Sarkar, & Gupta, 2012; Sayari, Belmabkhout, & Serna-Guerrero, 2011):

- Present good performance in the conditions of temperature and pressure of the SESMR process and rapid reaction kinetics, in addition to low adsorption hysteresis;
- Have a large work capacity, that is, a small amount of material must be able to capture large amounts of CO_2, as it directly implies reducing the size of equipment, costs, and energy demand;
- A high selectivity for CO_2 is desired, because, in addition to increasing the purity of the H_2 produced, after the saturation of the absorbent followed by its calcination, the released CO_2 will have high purity, being able to be used for other industrial processes, such as methanation, production of ammonia or light chain hydrocarbons;
- Present a long period of prebreakthrough, that is, remain with the absorption reactions for a long time to avoid as much as possible the regeneration of the absorbent;
- Energy for regeneration of the absorbent must not be so high as to cause a loss in the energy efficiency of the process, and cause damage to the absorbent material such as loss of performance, stability, reduction of the useful life, or change in the composition of the sorbent;
- The sorbent must have good mechanical resistance, good microstructural and morphological stability to maintain the working capacity and high kinetics, due to the severe working conditions that can cause friction, mainly in fluidized bed reactors;
- Must present high tolerance to the impurities present in the feed gas because such impurities can reduce the CO_2 capture capacity and degrade the crystalline structure of the absorbent;
- The costs of the absorbent must be minimal, around $\$10\,kg^{-1}$ or less;
- The adsorbents and catalysts must have a similar useful life to prevent a catalyst that is still active from being discarded or to avoid the need to separate the catalyst and absorbent.

Only one type of absorbent is not able to meet all the requirements for application in SESMR, therefore, in the next session, we will present the most used materials as an adsorbent for the SESMR process.

CaO-based

The use of CaO as a sorbent is not recent, and its application for this purpose was registered in a patent dating from 1933, by Roger Williams (Harrison, 2008), in which the steam methane reforming in the presence of a nickel catalyst and CaO as sorbent material is proposed.

Currently, CaO-based sorbents are the most promising among metal oxides for capturing CO_2 under the working conditions of the steam methane reforming process, due to their thermodynamic characteristics and their ability to capture low concentrations of CO_2 even at low temperatures (450–750°C) and atmospheric pressure (Balasubramanian et al., 1999; Harrison, 2009; Martavaltzi et al., 2010; Ortiz & Harrison, 2001).

As seen previously, CO_2 capture occurs according to reaction (Eq. 6.9), being an exothermic reaction that when combined with reactions (Eqs. 6.1 and 6.5), makes the SESMR process thermodynamically neutral, in which the required supplementary energy is dedicated only to the regeneration of the solid sorbent.

The thermodynamic study and experimental analysis carried out by Balasubramanian et al. (1999) discussed earlier, demonstrate the potential of the SESMR process for greater production of H_2 (>95% on a dry basis), with the formation of pure $CaCO_3$, occurring in a single reactor containing catalyst and sorbent, in which the values obtained are very close to the equilibrium concentrations. For industrial applications, it is necessary to consider the carbonation/calcination cycles and the effectiveness of CO_2 adsorption over the cycles to guarantee the economic and commercial viability of the SESMR process.

Fig. 6.4 shows the equilibrium pressure ratio of CO_2 versus temperature used to select the experimental conditions for regeneration. The formation of $CaCO_3$ (carbonation process) is favored under conditions of temperature and pressure to the left and above the equilibrium line in the graph of, while the decomposition of carbonate is favored under conditions below and to the right of the line (Ortiz & Harrison, 2001). To obtain pure CO_2, at atmospheric pressure of 1 atm, the minimum temperature required is 900°C.

The biggest problem associated with calcium-based sorbents is the high temperature for regeneration, which can lead to the sintering of particles and loss of absorption activity over the multiple cycles of operation (Kuramoto et al., 2003; Li, King, Nie, & Howard, 2009; Maya, Chejne, Gómez, & Bhatia, 2018). The sintering process consists of the progressive agglomeration of the particles due to the molten material that promotes the shrinkage and pore-filling phenomena (Lysikov, Salanov, & Okunev, 2007; Yancheshmeh, Radfarnia, & Iliuta, 2016). Also, the volume of the absorbent due to the formation of $CaCO_3$ increases substantially, implying macro- and microstructural changes that affect the properties of the sorbent, such as pore blockage, reduction of the surface area, and alteration of the CaO surface crystallographic planes (Abanades & Alvarez, 2003; Bazaikin, Malkovich, Prokhorov, & Derevschikov, 2021; Sun, Grace, Lim, & Anthony, 2007).

The deactivation of the sorbent begins during the first carbonation cycle, in which a layer of $CaCO_3$ is formed on the outside of the CaO particles that contain a large amount of micropores (<220 nm). The first calcination cycle provides highly dispersed and reactive CaO particles in nanosize (Kierzkowska, Pacciani, & Müller, 2013; Scaltsoyiannes, Antzaras, Koilaridis, & Lemonidou, 2021; Sun et al., 2007). During subsequent cycles, the CaO available for absorption is not fully reacted due to the slow diffusion process inside the grain, giving rise to an increasing

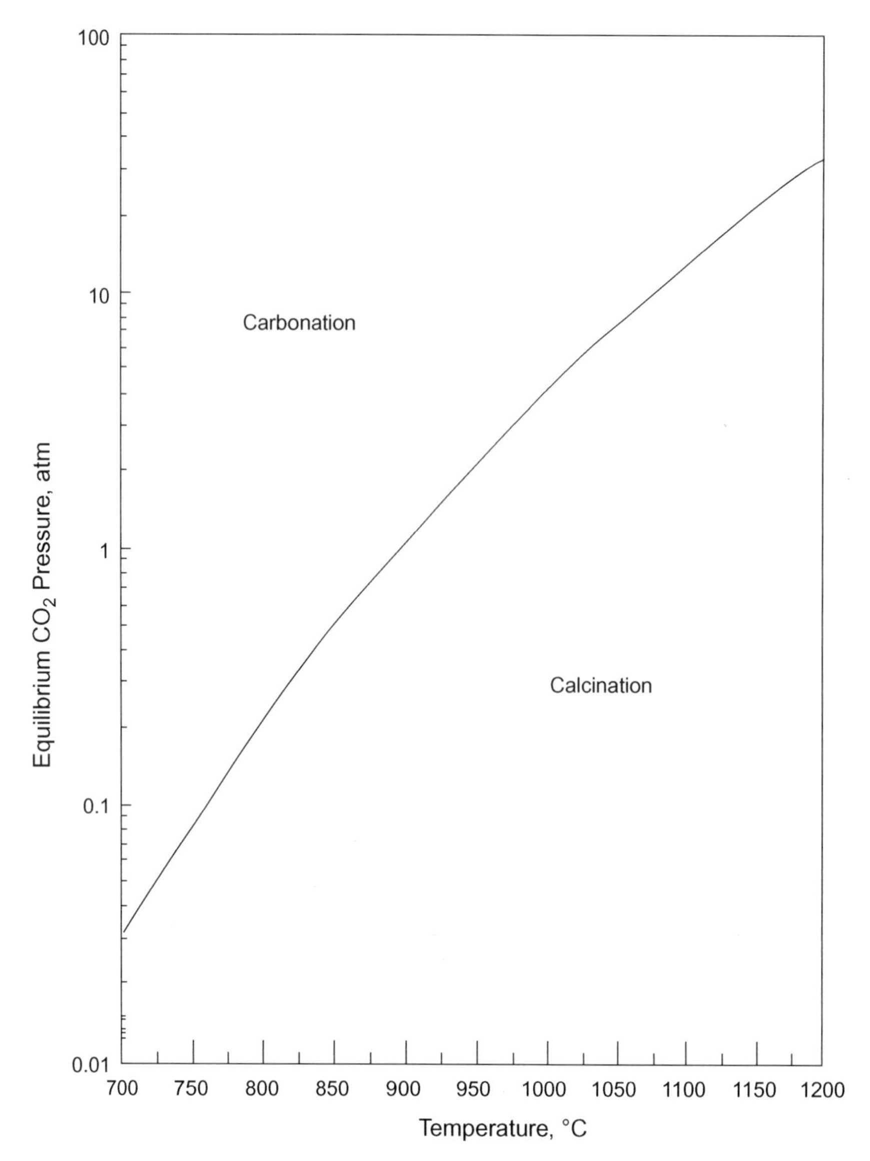

FIG. 6.4 Equilibrium CO_2 pressure over temperature. According to the CO_2 equilibrium pressure and temperature used, the calcination or carbonation processes take place, i.e., to the left and above the equilibrium curve, the carbonation step is favored, while to the right and below, the calcination step is favored (Ortiz and Harrison, 2001). *Reprinted (adapted) with permission from Ortiz, A. L., & Harrison, D. P. (2001). Hydrogen production using sorption-enhanced reaction.* Industrial and Engineering Chemistry Research, *40(23), 5102–5109. https://doi.org/10.1021/ie001009c. Copyright 2001 American Chemical Society.*

amount of unreacted CaO particles, forming a CaO skeleton that works as refractory support for reactive CaO particles (Lysikov et al., 2007). Changes in the microstructure caused by the sintering of the particles due to high temperature and macropore formations (>220nm) inside the CaO particles resulting from the release of CO_2 will affect the absorption capacity over successive sorbent regeneration cycles (Sun et al., 2007). The micropores inside the CaO particles are mainly responsible for the CO_2 absorption process due to their larger surface area, while the larger pores contribute less to the carbonation process (Abanades & Alvarez, 2003; Dennis & Pacciani, 2009).

To increase the durability of the CaO-based absorbers, investigations on the type of precursor, use of nanoparticles, and incorporation of inert or doping materials have been addressed in the literature. In addition to the properties necessary to obtain a stable sorbent, it is important to consider the preparation methods and the associated costs when choosing the precursor that will give rise to CaO, in which the improvement in the absorption capacity must be sufficiently advantageous to compensate for the costs of the process obtaining the absorbent.

In this sense, precursors of natural origin, such as limestone ($CaCO_3$) and dolomite ($CaMg(CO_3)_2$), have a low cost and their process of obtaining is simple, however, they lose efficiency in the adsorption capacity along the multicycles of carbonation/calcination. Limestone has the advantage of wide availability and easy extraction, and a capture capacity of 0.79 (ratio of the mass of chemisorbed CO_2 to the mass of the adsorbent), while dolomite has a capture capacity of less than 0.46 (Valverde, Sanchez-Jimenez, & Perez-Maqueda, 2015). If in the first cycles the adsorption capacity of the limestone is greater than that of the dolomite, in the following cycles the trend is a progressive reduction in the absorption capacity for the limestone while the dolomite is deactivated at a lower rate, as can be seen in Fig. 6.5 where the adsorbents were regenerated both in an air atmosphere at 850°C (Fig. 6.5A) and in an atmosphere with 70% CO_2 at 950°C (Fig. 6.5B), real calcination conditions for capture in postcombustion processes, demonstrating that the absorption capacity for dolomite in these conditions it is twice that of the limestone even after 20 carbonation/calcination cycles.

The stability in the adsorption capacity for CaO from dolomite is due to the sintering

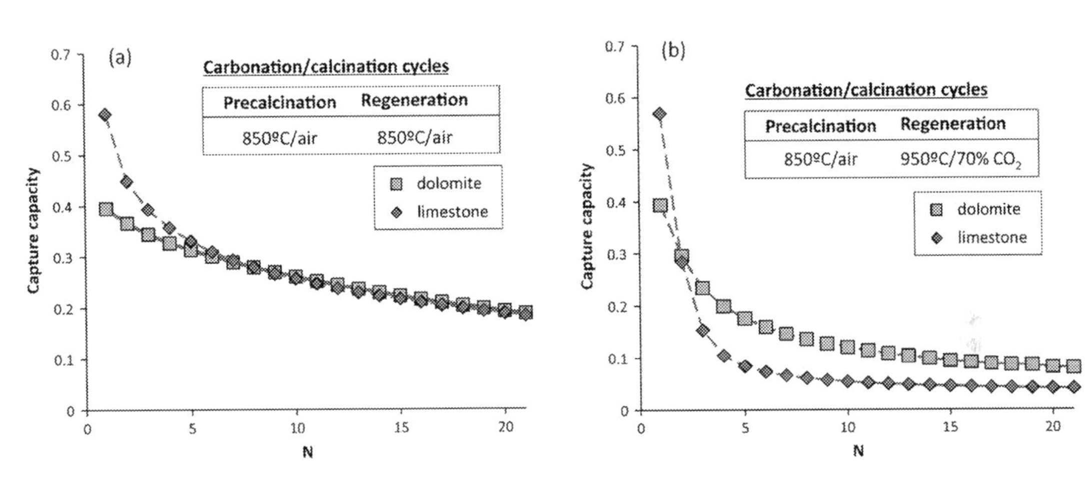

FIG. 6.5 Capture capacity of dolomite and limestone with regeneration in an air atmosphere at 850°C (A) and in an atmosphere with 70% CO_2 at 950°C (B). The stability of the sorbent can be evaluated through the CO_2 capture capacity after carbonation/calcination cycles. It can be seen that the dolomite, under conditions closer to those used in industrial processes (B), has a higher capture capacity than the limestone sample (Valverde, Sanchez-Jimenez, & Perez-Maqueda, 2015).

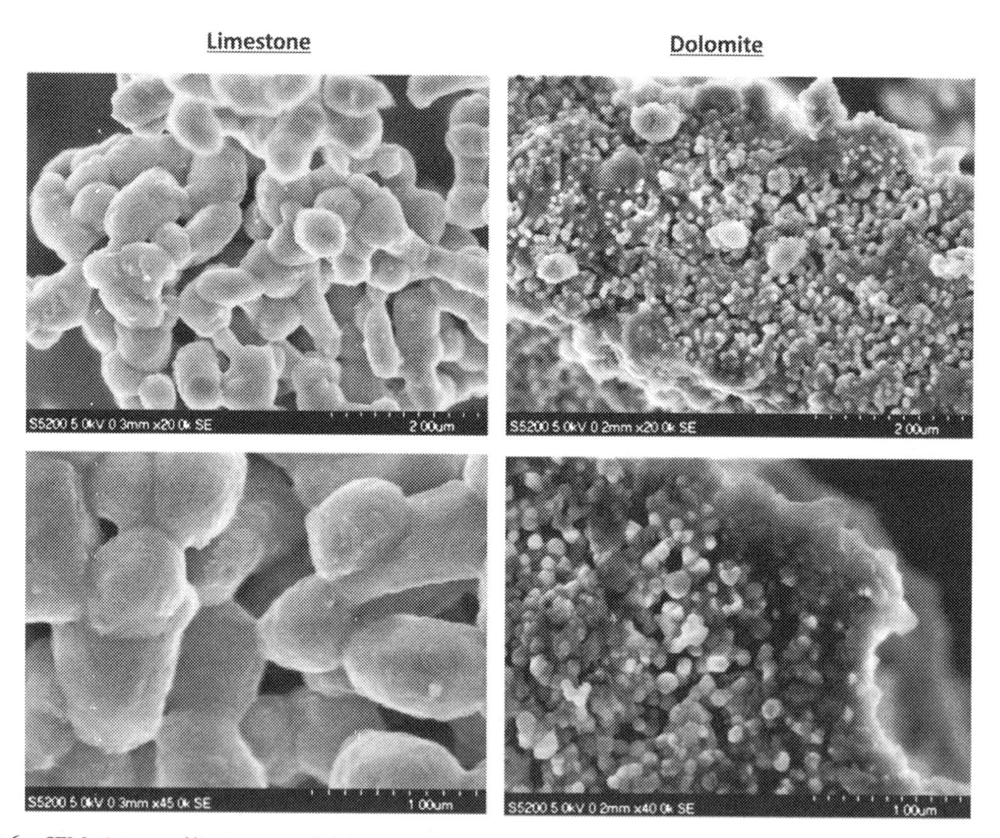

FIG. 6.6 SEM pictures of limestone and dolomite after precalcination and regenerated in an atmosphere of 70% of CO_2 at 950°C. After a carbonation/calcination cycle and precalcination step, the morphology of the limestone sample is composed of agglomerated and large-sized grains, while the dolomite sample has smaller particles and high porosity. From the morphology, it is possible to imagine that the limestone sample is more susceptible to the effects of siterization than the dolomite (Valverde, Sanchez-Jimenez, & Perez-Maqueda, 2015).

resistance of the particles due to the presence of MgO. Fig. 6.6 presents SEM images for adsorbents from limestone and dolomite precalcined in an air atmosphere at 850°C and regenerated by calcination in an atmosphere of 70% CO_2 at 950°C. The particles of the absorbent produced from $CaCO_3$ present coalescence and large size, which are indications of a sintering process, while the particles of the adsorbent produced from $CaMg(CO_3)_2$ present particles of smaller size and higher porosity, providing, thus, a larger surface area for subsequent carbonation cycles.

The reactivity of CaO from synthetic precursors has been investigated as an alternative to the rapid deactivation of sorbents from natural precursors and a comparison is presented in Table 6.1. CaO derived from synthetic precursors has structural characteristics such as high surface area and large pore volume that facilitate the diffusion of CO_2 gas molecules.

Although CaO sorbents prepared from synthetic precursors have a greater adsorption capacity in adsorption tests of only one cycle when applied in real conditions of SESMR, the capture capacity is similar to sorbents from

TABLE 6.1 Comparison of CO_2 capture capacity for different types of precursors.

Precursors	Reaction conditions		CO_2 capture capacity (wt %)	Specific surface area (m^2/g)	Poro Vol. (cm^3/g)	References
	Adsorption	Regeneration				
$CaCO_3$	650°C 60 min	700°C	49.0	17.8	0.078	Gupta and Fan (2002)
PCC	650°C 60 min	700°C	90.0	12.8	0.027	Gupta and Fan (2002)
$CaCO_3$	600 °C 300 min	750 °C 30 min	66.0	5.3	0.08	Lu, Reddy, and Smirniotis (2006)
CaO	600°C 300 min	750°C 30 min	25.0	4.2	0.02	Lu et al. (2006)
$Ca(OH)_2$	600°C 300 min	750°C 30 min	63.0	13.9	0.15	Lu et al. (2006)
$Ca(NO_3)_2$	600°C 300 min	750°C 30 min	2.5	–[a]	–[a]	Lu et al. (2006)
$Ca(CH_3COO)_2$	600°C 300 min	750°C 30 min	97.0	20.2	0.23	Lu et al. (2006)
$Ca(C_2H_5COO)_2$	700°C 300 min	750°C 30 min	75.0	15.0	0.18	Lu, Khan, and Smirniotis (2008)
$Ca(CH_3COO)_2$	700°C 300 min	750°C 30 min	70.7	20.0	0.22	Lu et al. (2008)
$Ca(CH_3COCHCOCH_3)_2$	700°C 300 min	750°C 30 min	60.5	12.0	0.09	Lu et al. (2008)
$Ca(COO)_2$	700°C 300 min	750°C 30 min	39.3	5.9	0.02	Lu et al. (2008)
$Ca(C_7H_{15}COO)_2$	700°C 300 min	750°C 30 min	19.6	9.3	0.015	Lu et al. (2008)

[a] *Under the conditions of BET measurements, such properties had values so low that it was not possible to measure.*

natural sources, and therefore, an increase in cost when using synthetic precursors would not be justified for applications in real reaction conditions (Yancheshmeh, Radfarnia, & Iliuta, 2016).

Another approach is to insert inert, doping, or refractory materials that assist in preventing CaO-$CaCO_3$ sintering and that allow greater sorbent recyclability. To provide greater stability to the sorbent, the material that will function as support must not react with CO_2 under operating conditions, must have a high sintering temperature (known as Tammann temperature), present a good dispersion, and have a high specific area. Hu, Lu, Liu, Yang, and Li (2020), in a recent review, discuss the synthesis methods for incorporating inert materials in CaO particles and compares studies of various types of supports to prevent the sintering of CaO particles. The materials can be divided into two large groups: metal oxides as a support and calcium metal oxides.

Among metal oxides, manganese oxide (MgO) is one of the most used, as it has a sintering temperature of 1289°C. Through the formation of a framework that circulates the calcium particles, acting as a physical barrier against their agglomeration, MgO provides greater resistance to sintering, mainly of $CaCO_3$ particles (Hu et al., 2020). However, the disadvantage of inserting large amounts of MgO is in the reduction of the sorbent's absorption capacity, in which the ideal values to maintain a good absorption with good resistance to sintering is between 11 and 25 wt% MgO (Guo, 2011; Naeem et al., 2018). When 20 wt% of Y_2O_3 nanoparticles are uniformly dispersed over CaO, an improvement is observed in the cyclic capacity as well as in the sorbent absorption stability. Additions of Y_2O_3 can improve the carbonation rate of CaO due to the better surface area and pore volume, and due to the high sintering temperature of 1083°C that prevents the agglomeration of particles at high temperatures (Zhang, Li, et al., 2014). CeO_2 is another metal oxide with a high sintering temperature (1064°C) that can be incorporated as an inert material into the CaO particles, promoting a greater capture capacity and being more tolerant to the presence of SO_2 than the CaO sorbents supported in other types of materials such as Mn_2O_3, Cr_2O_3, CoO, and CuO. The best performance of the CeO_2 supported sorbent is due both to the reticulated morphology that favors the contact between CaO and CO_2, as well as its performance as a protective barrier against the growth and sintering of CaO crystallites, and also, the CaO carbonation is improved due to the presence of oxygen vacancies that reduces the activation energy when the Ce/Ca ratio increases (Wang et al., 2011).

Calcium aluminates are another group of materials that have great potential as a support for CaO. Reactions between CaO and Al_2O_3 followed by calcinations at high temperatures produce calcium aluminates with different crystalline phases such as $Ca_3Al_2O_6$, $Ca_9Al_6O_{18}$, and $Ca_{12}Al_{14}O_{33}$ (Cesário, Barros, Courson, Melo, & Kiennemann, 2015). The calcination temperature of the mixtures of calcium and alumina precursors produce different phases, in which the formation of a pure $Ca_3Al_2O_6$ phase is obtained at 1200°C and the $Ca_{12}Al_{14}O_{33}$ phase preferably forms below 1000°C. Radfarnia and Sayari (2015) obtained $Ca_9Al_6O_{18}$ after 2 h at a calcination temperature of 900°C. Calcium aluminate supports to prevent the sintering of CaO sorbents by increasing their life cycle due to their thermal stability and resistance to sintering. In tests with more than 150 cycles (Martavaltzi & Lemonidou, 2008), a CaO sorbent with 25% $Ca_{12}Al_{14}O_{33}$ showed an increasing and continuous absorption capacity, including for CaO prepared with eggshell precursors and 37% Al_2O_3 from bauxite tailings (Shan et al., 2016). The SiO_2 present in the bauxite tailings promotes the formation of Ca_2SiO_4 in addition to $Ca_{12}Al_{14}O_{33}$, forming a multisupport configuration for the sorbent. The addition of bauxite helps to stabilize the structure by improving ionic diffusion as $CaCO_3$ decomposes into CaO and CO_2, increasing the carbonation conversion up to 55% even after 40 cycles. For the sorbent containing 25% of the $Ca_3Al_2O_6$ phase, the CaO absorption capacity increased from 0.30 to 0.50 gCO_2/g of sorbent in a test with 15 cycles. The intermediate phase $Ca_9Al_6O_{18}$ has an absorption capacity of 0.57 gCO_2/g of sorbent without any deactivation during 31 cycles (Radfarnia & Sayari, 2015), due to its favorable structural properties and fine dispersion over the CaO matrix. Guo (2011) attributed the better performance of CaO-$Ca_9Al_6O_{18}$ sorbent (75/25 wt%), compared to limestone, dolomite, and CaO/MgO, to the CO_2 inert nature of calcium aluminate, to the reduction of the crystallites of CaO and the prevention of pore blockage.

Calcium metal oxides are also the focus of investigations regarding resistance to CaO sintering. The reaction between ZrO_2 and CaO forms $CaZrO_3$ nanoparticles that have excellent resistance to sintering at high temperatures under tests of 1200 cycles under severe conditions when 73 wt% of $CaZrO_3$ is used (Koirala, Gunugunuri, & Smirniotis, 2011). Studies (Krohn & Krohn, 2001; Wu & Zhu, 2010) with $CaTiO_3$ attempt to introduce an inert framework capable of preventing CaO sintering and maintaining its reactivity. The results showed that coating CaO nanoparticles with $CaTiO_3$ lead to improved stability in CO_2 absorption cycles by avoiding contact between nanoparticles at high temperatures (Wu & Zhu, 2010), when compared to $CaZrO_3$ and Nd_2O_3 supports (Zhang et al., 2018), for example. For Ca_2MnO_4 supports the sintering resistance is higher than La_2O_3 supports but less than MgO and $Ca_{12}Al_{14}O_{33}$ support (Luo et al., 2013). Using a natural precursor of manganocalcite ($(Ca/Mn)CO_3$) to prepare CaO sorbents supported on Ca_2MnO_4 by calcination, the results showed greater stability over the cycles than the calcined limestone, however, the sulfur tolerance is reduced (Hu, Lu, Liu, Yang, & Li, 2020).

Lanthanum oxide (La_2O_3), unlike the supports presented so far, is not inert in the presence of CO_2, being able to capture around 0.1 mmol/g after four cycles (Albrecht, Wagenbach, Satrio, Shanks, & Wheelock, 2008), however, when used as a dopant in sorbents of CaO, La_2O_3 improves stability and increases CaO conversion. Hexagonal mesoporous silica, SBA-15 (Zhao et al., 1998), is also considered to be a good dopant for CaO sorbents due to ease of synthesis, adjustable pore size, and excellent mechanical and thermal stability. The CaO/SBA-15 sorbent shows 80% conversion even at high temperatures and after 40 cycles of carbonation/calcination, due to the high surface area, good diameter, and pore volume and mesostructure attributed to SBA-15 (Huang, Chang, Yu, Chiang, & Wang, 2010).

Another approach to improve the performance of CaO sorbents is to control their structure and size by increasing the surface area and increasing the reaction kinetics in the first stage of CaO to $CaCO_3$ conversion. When using CaO nanoparticles, it is possible to cover the entire surface of the sorbent, restricting the carbonation reaction to only the rapid conversion phase, that is, the slowest stage of the conversion reaction that is controlled by the diffusion of ions through the $CaCO_3$ layer it does not occur. Nanosize sorbents have more active sites for the adsorption process and have a larger surface area, improving the carbonation rate and CaO conversion (Abanades & Alvarez, 2003; Bazaikin, Malkovich, Prokhorov, & Derevschikov, 2021; Sun et al., 2007). When comparing $CaCO_3/Al_2O_3$ sorbents in nano and micro sizes, nanosorbents require a lower decarbonation temperature and greater stability in the adsorption capacity (68.3%) after 50 cycles (Wu, Li, Kim, & Yi, 2008). However, in practice, the greater reactivity of nanosorbents implies a greater susceptibility to sintering in addition to the need for complex and expensive methods to produce nanostructures of CaO sorbents.

Alkaline materials

Calcium-based adsorbents have been extensively studied due to their good characteristics such as good CO_2 absorption capacity, long-term durability, good absorption/regeneration kinetics, good mechanical properties, wide availability, and low cost. However, calcium-based adsorbents suffer from problems of reduced CO_2 capture capacity over multiple carbonation/calcination cycles due to the sintering phenomenon ($T > 900°C$) during the adsorbent regeneration process, which leads to a reduction in the surface area of the material (Abanades & Alvarez, 2003; Harrison, 2009).

To circumvent this problem, alkaline ceramic materials are gaining special interest for CO_2 capture applications, especially for SESMR processes. The alkaline ceramics are materials capable of performing CO_2 capture at temperatures above 500°C with good kinetics and selectivity for absorption, and present good cyclic and thermal stability, without the need for major modifications through the incorporation of inert materials to stabilize their framework and avoid sintering of the adsorbent. However, when compared to CaO-based adsorbents, the theoretical CO_2 absorption capacity of alkaline ceramics is lower (Wang, Memon, et al., 2021).

In general, the model of the CO_2 capture mechanism performed by alkaline ceramics is divided into two steps with different absorption rates that correspond to chemisorption and diffusion processes. In the step corresponding to chemisorption, CO_2 is captured on the surface of the adsorbent at a much higher absorption rate than in the diffusion step, and this step is the limiting one for CO_2 capture of these materials (Rodríguez & Pfeiffer, 2008). After forming a layer of alkaline carbonate on the surface of the adsorbent particle, the diffusion step comes into action limiting the conversion of CO_2, therefore, to reduce this obstacle it is necessary to adopt strategies that favor the chemisorption step, such as reducing the size of the adsorbent particles, for example, or reducing the resistance to the diffusion processes of the adsorbent (Mejía-Trejo, Fregoso-Israel, & Pfeiffer, 2008).

Li-based adsorbents

Lithium-based materials have been studied as sorbents for SESMR due to the temperature for their regeneration being lower than the temperature required to regenerate calcium-based sorbents. Materials such as Li_2ZrO_3, $LiFeO_2$, Li_4TiO_4, Li_5AlO_4, and especially Li_4SiO_4 have been investigated as good candidates to capture CO_2 in situ (Bhatta et al., 2015).

Li_2ZrO_3 ceramics have good thermal stability and low volumetric change coupled (Nakagawa & Ohashi, 1998) with the ability to absorb large amounts of CO_2 (4.5 mol/kg) at high temperatures (400–700°C) and low CO_2 concentrations. Regeneration of the adsorbent can be carried out at a lower temperature (700°C) when compared to CaO adsorbents (900–950°C) (Ida, Xiong, & Lin, 2004).

The adsorption of Li_2ZrO_3 occurs according to Eq. (6.13) and can be divided into two steps. The first consists of the reaction that occurs on the surface of the adsorbent forming a solid shell of Li_2CO_3 and ZrO_2 and the second step consist of the diffusion of CO_2 through the Li_2CO_3 layer formed, while Li^+ and O^{2-} diffuse externally through the ZrO_2 layer so that CO_2 is absorbed in the interface of the two layers forming the mechanism known as double-shell model. A description of this model can be seen in Fig. 6.7 (Wang, Memon, et al., 2021).

$$Li_2ZrO_{3(s)} + CO_{2(g)} \leftrightarrow Li_2CO_{3(s)} + ZrO_{2(s)} \quad \Delta H_{298\,K} = -160\,kJ\,mol^{-1} \tag{6.13}$$

However, Li_2ZrO_3 has the limitation of low reaction rate which can be improved by doping with potassium. Doping with K salts can increase the CO_2 adsorption rate compared to the undoped adsorbent due to the formation of a eutectic layer of molten Li_2CO_3 and K_2CO_3 carbonate above 500°C that facilitates the diffusion of CO_2 molecules (Ida et al., 2004). However, in another study conducted by Ochoa-Fernández, Rønning, Yu, Grande, and Chen (2008), showed that despite the improvement in the absorption reaction kinetics, the cyclic stability and CO_2 absorption capacity was reduced.

Li_2ZrO_3 adsorbents are usually prepared by solid-state reaction, in which the material produced has a large particle size and low porosity, morphology that confers low reaction kinetics and small CO_2 capture capacity. The use of ZrO_2 and Li_2CO_3 precursors with smaller particle sizes can reduce the particle size of Li_2ZrO_3 (Xiong, Ida, & Lin, 2003), but they promote sintering of the final product due to the high temperatures employed to produce the adsorbent (>900°C). A smaller Li_2ZrO_3 particle size increases the surface area of the adsorbent

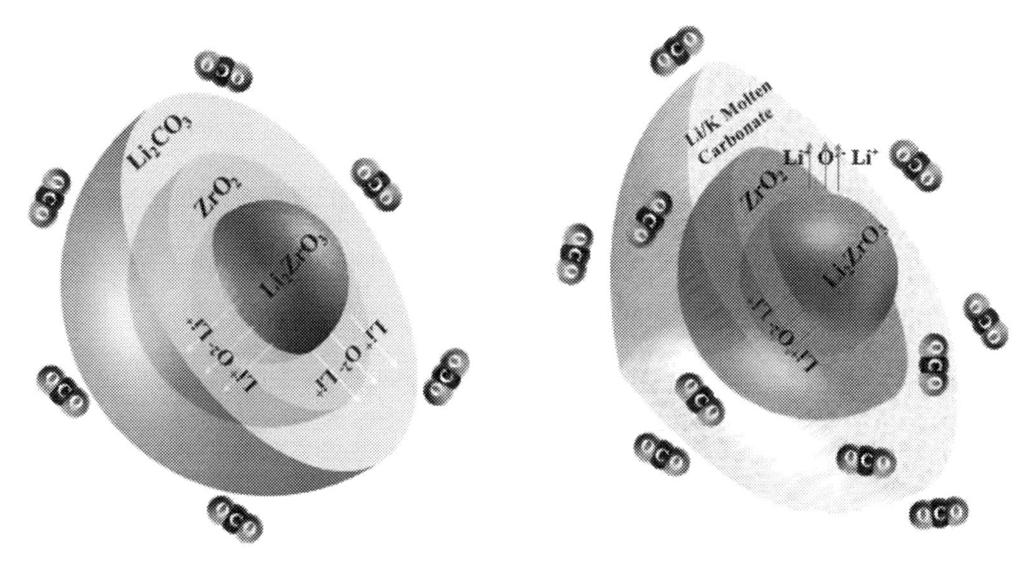

FIG. 6.7 Adsorption by Li_2ZrO_3 undoped and doped with potassium. Double-shell model for CO_2 absorption. When the sample is doped with K, a molten eutectic layer forms and reduces the diffusion resistance (Wang et al., 2021). *From Wang, Y., Memon, M. Z., Seelro, M. A., Fu, W., Gao, Y., Dong, Y., & Ji, G. (2021). A review of CO_2 sorbents for promoting hydrogen production in the sorption-enhanced steam reforming process.* International Journal of Hydrogen Energy, *46(45), 23358–23379. https://doi.org/10.1016/j.ijhydene.2021.01.206.*

(Ida et al., 2004), and consequently, more active sites for CO_2 absorption, and also forms a thinner ZrO_2 layer during carbonation, facilitating the CO_2 diffusion mechanism.

Lithium orthosilicate is lighter and less expensive, exhibiting a 30 times faster absorption rate and a higher absorption capacity when compared to Li_2ZrO_3 sorbent, and thus is considered a promising material for CO_2 capture (Kato, Yoshikawa, & Nakagawa, 2002). Furthermore, when compared to Ca-based sorbents, the regeneration temperature is lower. However, for SESMR applications, Li_4SiO_4 sorbents have low reaction kinetics at low CO_2 concentrations (between 4 and 20 vol%) (Kuramoto et al., 2003; Li et al., 2009; Maya et al., 2018), mainly due to their low surface area that limits chemisorption processes (Ida et al., 2004).

The CO_2 uptake is given by Eq. (6.14), in which the theoretical uptake capacity is 36.7 wt %, higher than that of Li_2ZrO_3-based adsorbents (28.7 wt%), and can occur under typical conditions of sorption enhanced steam methane reforming, i.e., over a large temperature range (400°C and 600°C) and low CO_2 concentrations (Nair, Burwood, Goh, Nakagawa, & Yamaguchi, 2009; Wang, Memon, et al., 2021).

$$Li_4SiO_{4(s)} + CO_{2(g)} \leftrightarrow Li_2CO_{3(s)} + Li_2SiO_{3(s)} \quad \Delta H_{298\,K} = -142\,\text{kJ}\,\text{mol}^{-1} \tag{6.14}$$

Li_4SiO_4 shows a considerably higher absorption rate when compared to Li_2ZrO_3. In an atmosphere containing 20% CO_2, Li_4SiO_4 was able to absorb CO_2 at a rate greater than 50 mg/g min at 500°C, 30 times faster than for the Li_2ZrO_3 sample. By reducing the CO_2 concentration to 2% in the feed gas mixture at a temperature of 500°C, the CO_2 absorption rate was 7.5 mg g^{-1} min^{-1} and 25 wt% absorption was achieved after 50 min for Li_4SiO_4, while for Li_2ZrO_3 only 1 wt% absorption was achieved during the same period and at an extremely slow reaction rate (Kato et al., 2002).

The CO_2 adsorption mechanism for Li_4SiO_4 sorbents can be explained through the Avrami-Erofeev and double-shell model, in which CO_2 molecules come into contact with the surface of the sorbent and react rapidly forming Li_2CO_3 and Li_2SiO_3 cores that will give rise to a double-layered structure (Qi, Daying, Yang, Qian, & Zibin, 2013). CO_2 passes through these layers through a diffusional process with a slow reaction rate. The Li_2SiO_3 layer, as well as the presence of water vapor, facilitates the diffusion of Li^- and O^{2+} ions, increasing the speed of the diffusional processes when compared to the ZrO_2 layer formed on the Li_2ZrO_3 adsorbent (Ochoa-Fernández, Zhao, Rønning, & Chen, 2009). As the adsorption occurs, the thickness of the products layer increases, further reducing the reaction speed due to the increased diffusion resistance through the products layer (Qi et al., 2013).

Pure Li_4SiO_4 adsorbents are usually prepared by solid-state reaction (Pfeiffer, Bosch, & Bulbulian, 1998; Venegas, Fregoso-Israel, Escamilla, & Pfeiffer, 2007), a simple technique that consists of mixing Li_2CO_3 and SiO_2 powders to form a solid solution, followed by high-temperature calcination for several hours to obtain the desired Li_4SiO_4 product (Eq. 6.15). Due to the high calcination temperatures (>900°C) and several hours of residence time (4h), the resulting product does not exhibit a large surface area or excellent porosity that promotes good performance for CO_2 capture (Izquierdo, Turan, García, & Maroto-Valer, 2018). Problems such as contamination and particle agglomeration are possible to occur with this preparation method.

$$Li_2CO_{3(s)} + SiO_{2(s)} \rightarrow Li_4SiO_{4(s)} + CO_{2(g)} \quad (6.15)$$

To improve the surface characteristics of the adsorbent and improve the reaction kinetics for CO_2 capture, modifications in the morphology of sorbents prepared by solid-state reaction have been tested. Modifications by dry ball-milling (Kanki, Maki, & Mizuhata, 2016; Romero-Ibarra, Ortiz-Landeros, & Pfeiffer, 2013), have been shown to increase the surface area up to 10 times (from 0.4 to 4.9 m^2/g) and to reduce the crystal size from >500 to 175 Å, besides reducing the activation energy for the adsorption reaction. The modification by hydration-calcination technique (Yin, Wang, Zhao, & Tang, 2016) consists in adding water to a product already prepared by solid-state reaction, under constant stirring, followed by second calcination at 800°C for 4h. The results of this study showed that the hydrated sample has a larger surface area, smaller particle size, and a nondense structure of loose particles that favors the increase of absorption rates and capacity, maintaining a high absorption-desorption performance over several cycles.

Materials with controlled particle size, uniform size distribution, good homogeneity, and high surface area can be obtained by aqueous synthesis methods, such as those involving sol-gel processes. For sol-gel methods, the synthesis route involves mixing lithium and silica precursors in aqueous solutions until they form a stable solution, followed by a slow polymerization that forms a three-dimensional gel structure, whereas the reactions occur in the solution, the solvent loses its fluidity. Finally, the resulting gel undergoes a drying step and subsequently calcination (at a lower temperature than the solid-state reaction method) to form the desired end product (Hu, Liu, Yang, Qu, & Li, 2019).

Compared with the samples prepared by solid-state reaction (CO_2 adsorption rate of 13.3 $mg\,g^{-1}\,min^{-1}$), the adsorbent produced by sol-gel showed an adsorption rate of 22.5 $mg\,g^{-1}\,min^{-1}$ and adsorption capacity of 350 $mg\,g^{-1}$, and good stability after five cycles of adsorption-desorption (Subha et al., 2014). Due to the nanoparticles, good porosity, and larger surface area, Li_4SiO_4 adsorbents prepared by sol-gel methods show a fast adsorption rate, accompanied by a higher uptake capacity and higher cyclic stability, without profound changes in the adsorption mechanism, i.e., the chemical adsorption process followed by the diffusion processes of lithium (Subha et al., 2016; Wang, Wang, Zhao, & Guo, 2014).

Other methods, such as impregnated suspension method (Bretado, Guzmán Velderrain, Lardizábal Gutiérrez, Collins-Martínez, & Ortiz, 2005; Hu, Liu, Zhou, & Yang, 2018; López Ortiz et al., 2014), carbon-templates (Li, Qu, & Hu, 2020; Ma, Qin, Pi, & Cui, 2020), combustion (Choudhary, Sahu, Mazumder, Bhattacharyya, & Chaudhuri, 2014; Rao, Mazumder, Bhattacharyya, & Chaudhuri, 2017), solvo-plasma (Nambo et al., 2017), spray-drying (Carella & Hernandez, 2014) , etc., have also been researched to produce sorbents with smaller particle size, larger surface area, and good porosity. In addition, the need for lower calcination temperature, as well as, shorter plateau duration can prevent lithium sublimation and increase control over the morphology of the sorbent. However, from an economic perspective, the solid-state reaction method is still more widely used due to the cost-effectiveness of starting materials and the ease of synthesis when compared to the other methods, even if the properties for adsorption prove to be inferior (Hu et al., 2019). A comparison between the morphology of the particles produced by different methods can be seen in Fig. 6.8.

In addition to morphology control through preparation methods, doping with alkali salts and heteroatoms can improve the performance of the adsorbents by forming low-temperature molten shell eutectic layers that facilitate CO_2 diffusion, or by creating defects or vacancies that increase the diffusion rate of Li^+ and O^{2-} ions.

The addition of different alkali carbonates has been widely investigated as promoters for CO_2 diffusion due to the formation, at temperatures greater than 500°C, of a eutectic mixture with the formed Li_2CO_3, which reduces the diffusion resistance of CO_2 (Yang et al., 2016; Zhang, Zhang, et al., 2014). A study (Seggiani, Puccini, & Vitolo, 2013) compared the absorption ability of adsorbents without incorporation and with incorporations of Na and K carbonates, and it was shown that both can increase the absorption rates and improve CO_2 conversion when compared to pure Li_4SiO_4 adsorbents. However, the Na-modified adsorbents showed sintering of the particles that reduce the surface area, and consequently, the absorption capacity over multiple sorption/desorption cycles, while the K_2CO_3-modified adsorbents remain stable even after 25 cycles of sorption/desorption.

Physical doping and eutectic doping methods are the most widely used methods to make the incorporation of K_2CO_3 and Na_2CO_3 into Li_4SiO_4 adsorbents (Zhou et al., 2017). In physical doping, the alkali salts are mixed with the already prepared Li_4SiO_4 adsorbent, while in eutectic doping, the alkali salts are directly mixed with the lithium and silica precursors, followed by high-temperature calcination. Although both methods provide adsorbents with higher adsorption capacity when compared to pure adsorbents, the adsorbents prepared by physical doping show higher adsorption capacity than the adsorbents prepared by eutectic doping (Seggiani, Puccini, & Vitolo, 2011).

According to the double-shell model, the diffusion of Li^+ and O^{2-} ions is considered as a limiting step for CO_2 adsorption processes. Thus, the performance of adsorbents can be improved by doping with heteroatoms such as Al and Fe, for example, in order to create defects or vacancies in the crystal structure of Li_4SiO_4, and consequently facilitate the mobility of Li^+ and O^{2-} ions (Zhang, Zhang, et al., 2014). A substitution of Si^{4+} for Al^{3+} creates interstitial Li^+, while a vacancy can be originated by substituting Li^+ for Al^{3+}, with oxygen vacancies being responsible for a greater

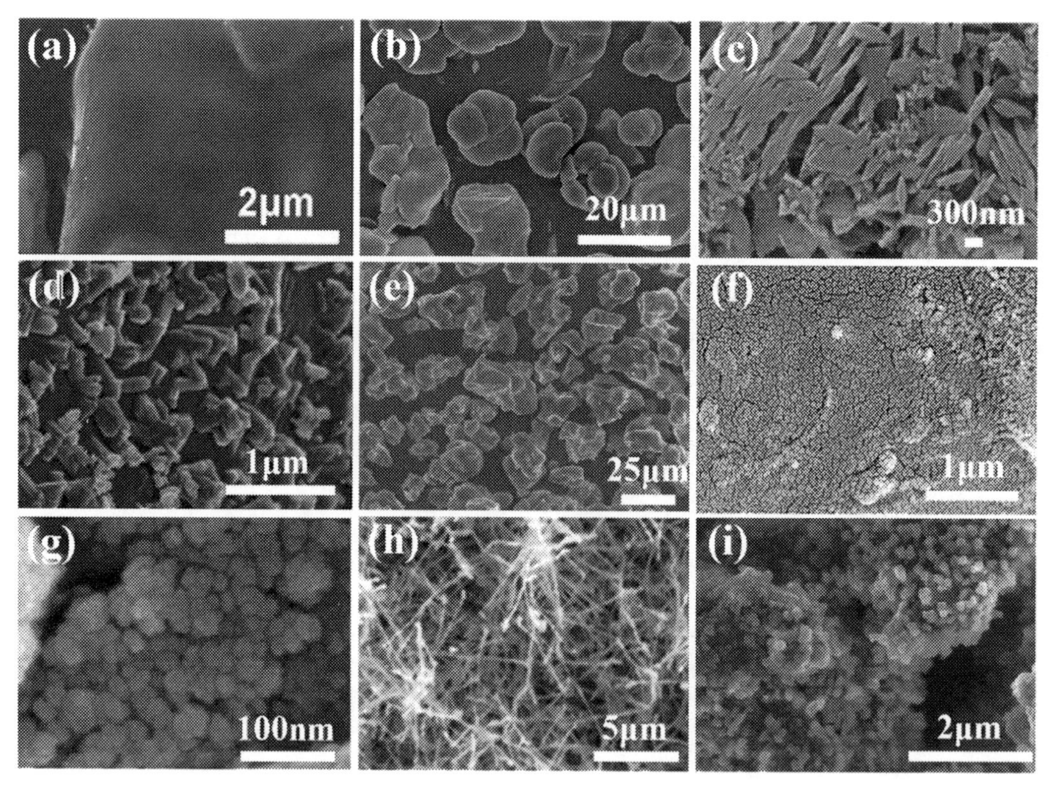

FIG. 6.8 SEM images of Li_4SiO_4 produced by different methods. The methods are (A) solid-state reaction method; (B)solid-state method between rice husk and Si source; (C) sol-gel method; (D) water-based sol-gel method; (E) precipitation method; (F) precipitation method with organic Li precursor; (G) glycine-nitrate solution combustion method; (H) solvo-plasma method; (I) spray-drying technique (Hu, Liu, Yang, Qu and Li, 2019).

response in the reactivity of Li_4SiO_4 due to the size of the Li^+ ion being much smaller than O^{2-}, as evidenced by the higher reactivity of $Li_{3.7}Al_{0.1}SiO_4$ adsorbents compared to $Li_{4.1}Al_{0.1}Si_{0.9}O_4$ adsorbents. In the case of the iron-doped sample, in addition to the higher absorption rate at 500°C, CO_2 is more easily desorbed and the sample also exhibits catalytic activity, which is advantageous for SESMR processes (Gauer & Heschel, 2006).

Some disadvantages associated with doping are due to a large amount of dopant used which can reduce the theoretical absorption capacity, cyclic stability can be reduced due to loss of porosity and grain aggregation, and occurrence of particle sintering problems due to the formation of the fused eutectic layer (Cui et al., 2020; Seggiani et al., 2013; Wang, Gu, et al., 2021).

Furthermore, the effective costs associated with the production of lithium-based adsorbents are high and almost prohibitive for large-scale SESMR applications when compared to Ca-based adsorbents. Especially the costs of Li_2ZrO_3 adsorbents due to the cost of precursors, while the costs for preparing Li_4SiO_4 can be reduced by using cheaper silica sources, such as using rice husk ash and fly ash waste, for example (Choudhary et al., 2014).

Na_2ZrO_3 adsorbents

The most commonly used adsorbents for obtaining hydrogen by SESMR are calcium-based, which suffer from the loss of CO_2 capture capacity. To overcome this problem, Li-based adsorbents have been studied for the same application, however, the disadvantage is the sintering problems of the particle and high costs for its preparation. In this context, sodium-based adsorbents, particularly Na_2ZrO_3 adsorbents, have been shown as an alternative for CO_2 capture due to the lower cost of the precursors, ease of preparation, and stability over carbonation/regeneration cycles.

Na_2ZrO_3 adsorbent can be obtained by a solid-state reaction method from Na_2CO_3 and ZrO_2 according to Eq. (6.16). The crystal structure of Na_2ZrO_3 can be crystallized in monoclinic or hexagonal phase and depending on the predominant phase, the absorption rate, CO_2 capture capacity, and cyclic stability can be improved according to the higher crystallinity and smaller average crystal size (Munro, Åhlén, Cheung, & Sanna, 2020; Zhao et al., 1998).

$$Na_2CO_{3(s)} + ZrO_{2(s)} \leftrightarrow Na_2ZrO_{3(s)} + CO_{2(g)} \tag{6.16}$$

In comparison to lithium-based adsorbents, Na_2ZrO_3 shows lower theoretical absorption capacity, between 18 and 23.8 wt%, compared to 28.7 wt% for Li_2ZrO_3 and 36.7 wt% for Li_4SiO_4 (Ji, Memon, Zhuo, & Zhao, 2017). The CO_2 sorption process occurs between room temperature to 600°C and at low CO_2 concentrations according to the carbonation reaction (Eq. 6.17). Among the sorbents from alkaline materials, it is the one with the highest reaction kinetics and its chemisorption properties remain stable even under conditions of high regeneration temperatures and in multicycle carbonation/calcination (Memon, Ji, Li, & Zhao, 2017).

$$Na_2ZrO_{3(s)} + CO_{2(g)} \leftrightarrow Na_2CO_{3(s)} + ZrO_{2(s)} \quad \Delta H_{298\,K} = -149\,kJ\,mol^{-1} \tag{6.17}$$

The mechanism of CO_2 capture is divided into two steps. The first step named CO_2 diffusion occurs at low temperatures (400°C) in which CO_2 diffuses through the vacancies in the Na_2CO_3 and ZrO_2 layers until it reaches the interior of the adsorbent composed of Na_2ZrO_3. In the second step, called Na diffusion, Na^+ ions diffuse from the inner layer to the outer shell forming Na_2CO_3. Unlike the step governed by CO_2 diffusion, the Na^+ diffusion mechanism does not suffer effects due to sintering and presents a higher rate of CO_2 capture, however, it presents problems in the desorption step, which can decrease the efficiency of CO_2 capture (Alcérreca-Corte, Fregoso-Israel, & Pfeiffer, 2008).

A study conducted by López-Ortiz's team (López-Ortiz, Rivera, Rojas, & Gutierrez, 2004), evaluated the sorption activity of Na_2ZrO_3, Na_2TiO_3, and Na_3SbO_4, prepared by solid-state reaction and compared it with Li_2ZrO_3 and Li_4SiO_4 sorbents. The reaction rates were higher for the Na_2ZrO_3 samples containing an excess of 10% Na (of nomenclature Na_2ZrO_3-F61) when compared to the other samples, however, the amount of CO_2 absorbed was higher for the Li_4SiO_4 sample (30%) while the Na_2ZrO_3-F61 sample showed an absorption capacity of 25.77%. Considering sorbent regeneration, the samples based on Na have a low regeneration rate, which might hinder its cyclical activity.

6.5 Reactors

In some cases, SESMR is performed in a one-stage multifunctional reactor in which the reforming of methane reaction and CO_2 sorption takes place. As mentioned previously, these processes can take place on two different materials (catalyst and sorbent) or on the same material (bifunctional material).

In other cases, a different strategy is proposed where several reactors are implemented in a successive matter in order to better control the temperature and the catalyst/adsorbent ratio for each of the SESMR/calci steps. This is called the subsection controlling strategy (Xiu & Rodrigues, 2004). Several types of reactors can be used for each of the strategies (Cherbanski and Molga, 2018; Dhoke, Zaabout, Cloete, & Amini, 2021).

1. **Adsorptive reactors (cyclic behavior):** Adsorptive reactors (Fig. 6.9) are the most used for CO_2 capture. These reactors can be one stage, multistage, or have a transient configuration. The main three types of adsorptive reactors are discussed in the following sections.
   (a) Fixed bed reactors: In this type of reactor, the catalyst/sorbent material is immobilized on a fixed bed. The SESMR and the calcination step are done consecutively. First, the reforming step takes place in the presence of methane and steam at a temperature between 400°C and 600°C. Then, when the material is almost or completely saturated with CO_2, the gas mixture is changed to $H_2O/O_2/CO_2$ or an inert gas such as He or N_2. This step takes place at higher temperatures (800–1000°C). After that, the cycle of reforming calcination starts over (Oliveira, Grande, & Rodrigues, 2008; Xiu, Li, & Rodrigues, 2002; Zou & Rodrigues, 2001).
   (b) Circulating fluidized reactor: In this type of reactor, two subunits are included and placed simultaneously and in parallel. The first subunit is responsible for the SESMR step (400–600°C, H_2O+CH_4) whereas the second represents the calcination step (800–1000°C, inert gas). In this step, the material is moving from a unit to the other, all while respecting the state of the compounds. To be more specific, the saturated portion moves from the reformer to the regenerator while the regenerated material moves from the regenerator to the reformer (Arstad, Prostak, & Blom, 2012; Lindborg & Jakobsen, 2009).
   (c) Gas-solid-solid trickle flow reactor (GSSTFR): In this step-up, two subunits are used: a reformer unit (400–800°C; H_2O+CH_4) and a regenerator unit (800–000°C, gas). The difference between this type of reactor and the circulating fluidized bed reactor is the state of the material. In this case, the active phase of the catalyst is

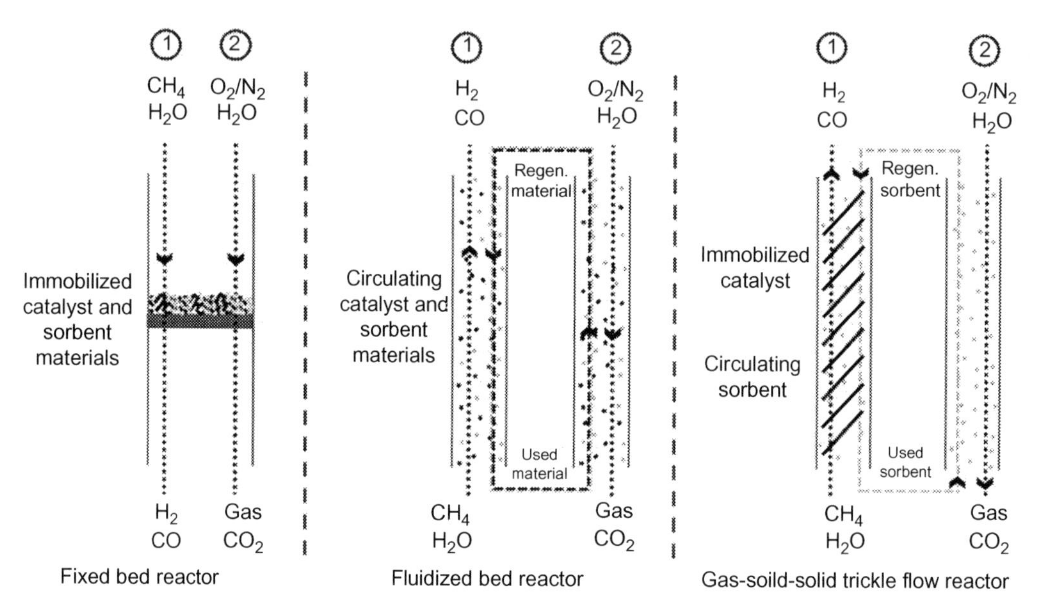

FIG. 6.9 Adsorptive reactors for CO_2 capture. In the fixed bed reactor both catalyst and sorbent are immobile in the reaction medium. In the fluidized bed reactor configuration, both catalyst and sorbent are circulating. In the gas-solid trickle flow reactor configuration, the catalyst is immobile and the sorbent is circulating in the reaction medium.

immobilized inside of the reformer, while the sorbent is circulating between the reformer and the regenerator. For better performance, the gas mixture can be flown either co-currently or counter-currently with the direction of the sorbent (Cherbański & Molga, 2018; Dallos, Kafarov, & Filho, 2007).

2. Membrane reactors: Membrane reactors for SESMR reaction is a relatively new application concept (Lu & Xie, 2016). The particularity of this model is the use of a membrane that is H_2 permeable (Anderson, Yun, Kottke, & Fedorov, 2017). In this case, the produced H_2 from the steam reforming of methane and the water gas shift reaction passes through the membrane (Barelli et al., 2008). Consequently, the concentration of H_2 in the gas phase drops and both of these reactions are favored following the Le Chatelier principle. Membrane reactor combines both CO_2 and H_2 selective removal in order to shift the equilibrium of both reactions to lower operating temperatures. A 90% conversion fuel conversion can be obtained in this type of reactor at 400°C and with a steam to carbon ratio of 2:1 (Anderson, Yun, Kottke, & Fedorov, 2017).

6.6 Conclusions

Hydrogen can be considered as a low carbon energy vector that can be applied in industrial, transportation, and residential sectors, but it is not available in H_2 form in nature. Among the techniques for obtaining H_2, the most used today is SMR that transforms CH_4, in the presence of water vapor and with the aid of a catalyst, into H_2 and CO, in which, in addition to the methane reform reaction, the reaction WGS takes over, producing CO_2 as a by-product. The formation of carbon during the process can cause the catalyst to be deactivated, usually nickel, and also makes it necessary to use an H_2 purification system produced. In the SESMR technique, adsorbing materials are applied in the reaction medium together with the catalyst, to capture in situ the CO_2 produced during the SMR process, increasing the production and purity of H_2, reducing the temperature required for operation, and increasing efficiency energy through the elimination of purification steps. Calcium-based sorbents are the most studied for applications in SESMR due to their thermodynamic characteristics, their high capacity to capture low concentrations of CO_2, and their low cost. However, the sintering of the particles and the change in the CaO microstructure are responsible for decreasing the CO_2 capture capacity, and consequently, reducing the useful life of this material. To reduce the deactivation of CaO sorbent, the use of different types of precursors, the addition of inert materials and/or dopants, and the use of nanomaterials have been intensively investigated to produce a material capable of increasing the number of cycles of carbonation/calcination at levels that are feasible for applications in the industrial sector. Alkaline ceramic-based sorbents are part of another group of widely investigated materials for SESMR applications. Among them, lithium orthosilicate is one of the most studied due to its thermal and microstructural stability over multicycle calcination/carbonation, good reaction kinetics, and lower temperature requirement for adsorbent regeneration. However, the CO_2 absorption capacity when compared to CaO adsorbents is lower, a characteristic that can be compensated by the thermal stability, and consequently, in the CO_2 capture capacity over multicycles. For the production of H_2 by SESMR, the adsorbent materials and catalysts must be present in the same reaction medium, thus, the configuration of the reactors is an important characteristic to control the temperature for the reform reaction, as well as the proportion between adsorbent and catalyst, and finally to control the regeneration processes of the adsorbent and prolong the life of the materials.

Acknowledgments

The authors would like to thank FAPESQ/CAPES/BRAZIL for their support and CNPq, Conselho Nacional de Desenvolvimento Científico e Tecnológico—Brazil.

Conflict of interest

The authors declare that they have no known competing financial interests or personal relationships that could have appeared to influence the work reported in this chapter.

References

Abanades, J. C., & Alvarez, D. (2003). Conversion limits in the reaction of CO_2 with lime. *Energy and Fuels*, *17*(2), 308–315. https://doi.org/10.1021/ef020152a.

Abanades, J. C., Rubin, E. S., & Anthony, E. J. (2004). Sorbent cost and performance in CO_2 capture systems. *Industrial and Engineering Chemistry Research*, *43*(13), 3462–3466. https://doi.org/10.1021/ie049962v.

Acharya, B., Dutta, A., & Basu, P. (2012). Circulating-fluidized-bed-based calcium-looping gasifier: Experimental studies on the calcination–carbonation cycle. *Industrial & Engineering Chemistry Research*, *51*, 8652–8660. https://doi.org/10.1021/ie300629a.

Albrecht, K. O., Wagenbach, K. S., Satrio, J. A., Shanks, B. H., & Wheelock, T. D. (2008). Development of a CaO-based CO_2 sorbent with improved cyclic stability. *Industrial and Engineering Chemistry Research*, *47*(20), 7841–7848. https://doi.org/10.1021/ie8007743.

Alcérreca-Corte, I., Fregoso-Israel, E., & Pfeiffer, H. (2008). CO_2 absorption on Na_2ZrO_3: A kinetic analysis of the chemisorption and diffusion processes. *Journal of Physical Chemistry C*, *112*(16), 6520–6525. https://doi.org/10.1021/jp710475g.

Anderson, D. M., Yun, T. M., Kottke, P. A., & Fedorov, A. G. (2017). Comprehensive analysis of sorption enhanced steam methane reforming in a variable volume membrane reactor. *Industrial and Engineering Chemistry Research*, *56*(7), 1758–1771. https://doi.org/10.1021/acs.iecr.6b04392.

Antzaras, A. N., Heracleous, E., & Lemonidou, A. A. (2020). Sorption enhanced–chemical looping steam methane reforming: Optimizing the thermal coupling of regeneration in a fixed bed reactor. *Fuel Processing Technology*, *208*, 106513.

Arstad, B., Prostak, J., & Blom, R. (2012). Continuous hydrogen production by sorption enhanced steam methane reforming (SE-SMR) in a circulating fluidized bed reactor: Sorbent to catalyst ratio dependencies. *Chemical Engineering Journal*, *189–190*, 413–421. https://doi.org/10.1016/j.cej.2012.02.057.

Balasubramanian, B., Ortiz, A. L., Kaytakoglu, S., & Harrison, D. P. (1999). Hydrogen from methane in a single-step process. *Chemical Engineering Science*, *54*(15–16), 3543–3552. https://doi.org/10.1016/S0009-2509(98)00425-4.

Balat, M. (2008). Potential importance of hydrogen as a future solution to environmental and transportation problems. *International Journal of Hydrogen Energy*, *33*, 4013–4029. https://doi.org/10.1016/j.ijhydene.2008.05.047.

Barelli, L., Bidini, G., Gallorini, F., & Servili, S. (2008). Hydrogen production through sorption-enhanced steam methane reforming and membrane technology: A review. *Energy*, *33*(4), 554–570. https://doi.org/10.1016/j.energy.2007.10.018.

Bazaikin, Y. V., Malkovich, E. G., Prokhorov, D. I, & Derevschikov, V. S. (2021). Detailed modeling of sorptive and textural properties of CaO-based sorbents with various porous structures. *Separation and Purification Technology*, *255*(15), 1–8. https://doi.org/10.1016/j.seppur.2020.117746.

Bhatta, L. K. G., Subramanyam, S., Chengala, M. D., Olivera, S., & Venkatesh, K. (2015). Progress in hydrotalcite like compounds and metal-based oxides for CO_2 capture: A review. *Journal of Cleaner Production*, *103*, 171–196. https://doi.org/10.1016/j.jclepro.2014.12.059.

Bretado, M. E., Guzmán Velderrain, V., Lardizábal Gutiérrez, D., Collins-Martínez, V., & Ortiz, A. L. (2005). A new synthesis route to Li_4SiO_4 as CO_2 catalytic/sorbent. *Catalysis Today*, *107–108*, 863–867. https://doi.org/10.1016/j.cattod.2005.07.098.

Buxbaum, R. E., & Kinney, A. B. (1996). Hydrogen transport through tubular membranes of palladium-coated tantalum and niobium. *Industrial and Engineering Chemistry Research*, *35*(2), 530–537. https://doi.org/10.1021/ie950105o.

Cai, L., Wu, X. Y., Zhu, X., Ghoniem, A. F., & Yang, W. (2020). High-performance oxygen transport membrane reactors integrated with IGCC for carbon capture. *AICHE Journal*, *66*(7). https://doi.org/10.1002/aic.16247.

Carapellucci, R., & Giordano, L. (2020). Steam, dry and autothermal methane reforming for hydrogen production: A thermodynamic equilibrium analysis. *Journal of Power Sources*, *469*. https://doi.org/10.1016/j.jpowsour.2020.228391.

Carvill, B. T., Hufton, J. R., Anand, M., & Sircar, S. (1996). Sorption-enhanced reaction process. *AICHE Journal*, *42*(10), 2765–2772. https://doi.org/10.1002/aic.690421008.

Chen, C. H., Yu, C. T., & Chen, W. H. (2021). Improvement of steam methane reforming via in-situ CO_2 sorption over a nickel-calcium composite catalyst. *International Journal of Hydrogen Energy*, *46*(31), 1–12. https://doi.org/10.1016/j.ijhydene.2020.08.284.

Cherbanski, R., & Molga, E. (2018). Sorption-enhanced steam methane reforming (SE-SMR) – A review: Reactor types, catalyst and sorbent characterization, process modeling. *Chemical and Process Engineering, 39*(4), 427–448. https://doi.org/10.24425/122961.

Cherbański, R., & Molga, E. (2018). Sorption-enhanced steam-methane reforming with simultaneous sequestration of CO_2 on fly ashes – Proof of concept and simulations for gas-solid-solid trickle flow reactor. *Chemical Engineering and Processing: Process Intensification, 124*, 37–49. https://doi.org/10.1016/j.cep.2017.11.010.

Choudhary, A., Sahu, B. S., Mazumder, R., Bhattacharyya, S., & Chaudhuri, P. (2014). Synthesis and sintering of Li_4SiO_4 powder from rice husk ash by solution combustion method and its comparison with solid state method. *Journal of Alloys and Compounds, 590*, 440–445. https://doi.org/10.1016/j.jallcom.2013.12.084.

Cobden, P. D., van Beurden, P., Reijers, H. T. J., Elzinga, G. D., Kluiters, S. C. A., Dijkstra, J. W., et al. (2007). Sorption-enhanced hydrogen production for pre-combustion CO_2 capture: Thermodynamic analysis and experimental results. *International Journal of Greenhouse Gas Control, 1*(2), 170–179. https://doi.org/10.1016/S1750-5836(07)00021-7.

Cui, H., Li, X., Chen, H., Gu, X., Cheng, Z., & Zhou, Z. (2020). Sol-gel derived, Na/K-doped Li_4SiO_4-based CO_2 sorbents with fast kinetics at high temperature. *Chemical Engineering Journal, 382*. https://doi.org/10.1016/j.cej.2019.122807.

Dallos, C. G., Kafarov, V., & Filho, R. M. (2007). A two dimensional steady-state model of the gas-solid-solid reactor. Example of the partial oxidation of methane to methanol. *Chemical Engineering Journal, 134*(1–3), 209–217. https://doi.org/10.1016/j.cej.2007.03.044.

Dennis, J. S., & Pacciani, R. (2009). The rate and extent of uptake of CO_2 by a synthetic, CaO-containing sorbent. *Chemical Engineering Science, 64*(9), 2147–2157. https://doi.org/10.1016/j.ces.2009.01.051.

Dewoolkar, K. D., & Vaidya, P. D. (2017). Tailored Ce- and Zr-doped Ni/hydrotalcite materials for superior sorption-enhanced steam methane reforming. *International Journal of Hydrogen Energy, 42*(34), 21762–21774. https://doi.org/10.1016/j.ijhydene.2017.06.235.

Dhoke, C., Zaabout, A., Cloete, S., & Amini, S. (2021). Review on reactor configurations for adsorption-based CO_2 capture. *Industrial and Engineering Chemistry Research, 60*(10), 3779–3798. https://doi.org/10.1021/acs.iecr.0c04547.

Ding, Y., & Alpay, E. (2000). Adsorption-enhanced steam-methane reforming. *Chemical Engineering Science, 55*(18), 3929–3940. https://doi.org/10.1016/S0009-2509(99)00597-7.

Drage, T. C., Smith, K. M., Pevida, C., Arenillas, A., & Snape, C. E. (2009). Development of adsorbent technologies for post-combustion CO_2 capture. *Energy Procedia, 1*(1), 881–884. https://doi.org/10.1016/j.egypro.2009.01.117.

Fiaschi, D., & Lombardi, L. (2002). Integrated gasifier combined cycle plant with integrated CO_2—H_2S removal: Performance analysis, life cycle assessment and exergetic life cycle assessment. *International Journal of Applied Thermodynamics, 5*(1), 13–24.

García-Lario, A. L., Aznar, M., Martinez, I., Grasa, G. S., & Murillo, R. (2015). Experimental study of the application of a $NiO/NiAl_2O_4$ catalyst and a CaO-based synthetic sorbent on the Sorption Enhanced Reforming process. *International Journal of Hydrogen Energy, 40*(1), 219–232. https://doi.org/10.1016/j.ijhydene.2014.10.033.

Gauer, C., & Heschel, W. (2006). Doped lithium orthosilicate for absorption of carbon dioxide. *Journal of Materials Science, 41*(8), 2405–2409. https://doi.org/10.1007/s10853-006-7070-1.

Guo, M. (2011). Removal of CO_2 by CaO/MgO and CaO/$Ca_9Al_6O_{18}$ in the presence of SO_2. *Energy and Fuels, 25*, 5514–5520.

Gupta, H., & Fan, L. S. (2002). Carbonation-calcination cycle using high reactivity calcium oxide for carbon dioxide separation from flue gas. *Industrial and Engineering Chemistry Research, 41*(16), 4035–4042. https://doi.org/10.1021/ie010867l.

Harrison, D. P. (2008). Sorption-enhanced hydrogen production: A review. *Industrial and Engineering Chemistry Research, 47*(17), 6486–6501. https://doi.org/10.1021/ie800298z.

Harrison, D. P. (2009). Calcium enhanced hydrogen production with CO_2 capture. *Energy Procedia, 1*(1), 675–681. https://doi.org/10.1016/j.egypro.2009.01.089.

Hashemnejad, S. M., & Parvari, M. (2011). Deactivation and regeneration of nickel-based catalysts for steam-methane reforming. *Chinese Journal of Catalysis, 32*(1), 273–279. https://doi.org/10.1016/S1872-2067(10)60175-1.

Hildenbrand, N., Readman, J., Dahl, I. M., & Blom, R. (2005). Sorbent enhanced steam reforming (SESR) of methane using dolomite as internal carbon dioxide absorbent: Limitations due to $Ca(OH)_2$ formation. *Applied Catalysis A: General, 303*(1), 131–137. https://doi.org/10.1016/j.apcata.2006.02.015.

Hu, Y., Liu, W., Yang, Y., Qu, M., & Li, H. (2019). CO_2 capture by Li_4SiO_4 sorbents and their applications: Current developments and new trends. *Chemical Engineering Journal, 359*, 604–625. https://doi.org/10.1016/j.cej.2018.11.128.

Hu, Y., Liu, W., Zhou, Z., & Yang, Y. (2018). Preparation of Li_4SiO_4 sorbents for carbon dioxide capture via a spray-drying technique. *Energy and Fuels, 32*(4), 4521–4527. https://doi.org/10.1021/acs.energyfuels.7b03051.

Hu, Y., Lu, H., Liu, W., Yang, Y., & Li, H. (2020). Incorporation of CaO into inert supports for enhanced CO_2 capture: A review. *Chemical Engineering Journal, 396*. https://doi.org/10.1016/j.cej.2020.125253.

Huang, C. H., Chang, K. P., Yu, C. T., Chiang, P. C., & Wang, C. F. (2010). Development of high-temperature

CO_2 sorbents made of CaO-based mesoporous silica. *Chemical Engineering Journal*, *161*(1–2), 129–135. https://doi.org/10.1016/j.cej.2010.04.045.

Hufton, J. R., Mayorga, S., & Sircar, S. (1999). Sorption-enhanced reaction process for hydrogen production. *AICHE Journal*, *45*(2), 248–256. https://doi.org/10.1002/aic.690450205.

Ida, J. I., Xiong, R., & Lin, Y. S. (2004). Synthesis and CO_2 sorption properties of pure and modified lithium zirconate. *Separation and Purification Technology*, *36*(1), 41–51. https://doi.org/10.1016/S1383-5866(03)00151-5.

(2019). *IEA – Global CO_2 emissions in 2019*. https://www.iea.org/articles/global-co2-emissions-in-2019.

Izquierdo, M. T., Turan, A., García, S., & Maroto-Valer, M. M. (2018). Optimization of Li_4SiO_4 synthesis conditions by a solid state method for maximum CO_2 capture at high temperature. *Journal of Materials Chemistry A*, *6*(7), 3249–3257. https://doi.org/10.1039/c7ta08738a.

Ji, G., Memon, M. Z., Zhuo, H., & Zhao, M. (2017). Experimental study on CO_2 capture mechanisms using Na_2ZrO_3 sorbents synthesized by soft chemistry method. *Chemical Engineering Journal*, *313*, 646–654. https://doi.org/10.1016/j.cej.2016.12.103.

Kalamaras, C. M., & Efstathiou, A. M. (2013). Hydrogen production technologies: Current state and future developments. *Conference Papers in Energy*, 1–9. https://doi.org/10.1155/2013/690627.

Kanki, K., Maki, H., & Mizuhata, M. (2016). Carbon dioxide absorption behavior of surface-modified lithium orthosilicate/potassium carbonate prepared by ball milling. *International Journal of Hydrogen Energy*, *41*(41), 18893–18899. https://doi.org/10.1016/j.ijhydene.2016.06.158.

Kato, M., Yoshikawa, S., & Nakagawa, K. (2002). Carbon dioxide absorption by lithium orthosilicate in a wide range of temperature and carbon dioxide concentrations. *Journal of Materials Science Letters*, *21*(6), 485–487. https://doi.org/10.1023/A:1015338808533.

Kierzkowska, A. M., Pacciani, R., & Müller, C. R. (2013). CaO-based CO_2 sorbents: From fundamentals to the development of new, highly effective materials. *ChemSusChem*, *6*(7), 1130–1148. https://doi.org/10.1002/cssc.201300178.

Koirala, R., Gunugunuri, K. R., & Smirniotis, P. (2011). Effect of zirconia doping on calcium oxide on stability and performance during the extended operating cycles. In *11AIChE – 2011 AIChE annual meeting, conference proceedings*.

Krishna, R., & Van Baten, J. M. (2012). A comparison of the CO_2 capture characteristics of zeolites and metal-organic frameworks. *Separation and Purification Technology*, *87*, 120–126. https://doi.org/10.1016/j.seppur.2011.11.031.

Krohn, C. M., & Krohn, C. G. (2001). Development of porous solid reactant for thermal-energy storage and temperature upgrade using carbonation/decarbonation reaction. *Applied Energy*, *69*(3), 225–238. https://doi.org/10.1016/S0306-2619(00)00072-6.

Kuramoto, K., Fujimoto, S., Morita, A., Shibano, S., Suzuki, Y., Hatano, H., et al. (2003). Repetitive carbonation-calcination reactions of Ca-based sorbents for efficient CO_2 sorption at elevated temperatures and pressures. *Industrial and Engineering Chemistry Research*, *42*(5), 975–981. https://doi.org/10.1021/ie0207111.

Li, L., King, D. L., Nie, Z., & Howard, C. (2009). Magnesia-stabilized calcium oxide absorbents with improved durability for high temperature CO_2 capture. *Industrial and Engineering Chemistry Research*, *48*(23), 10604–10613. https://doi.org/10.1021/ie901166b.

Li, H., Qu, M., & Hu, Y. (2020). Preparation of spherical Li_4SiO_4 pellets by novel agar method for high-temperature CO_2 capture. *Chemical Engineering Journal*, *380*. https://doi.org/10.1016/j.cej.2019.122538.

Lindborg, H., & Jakobsen, H. A. (2009). Sorption enhanced steam methane reforming process performance and bubbling fluidized bed reactor design analysis by use of a two-fluid model. *Industrial and Engineering Chemistry Research*, *48*(3), 1332–1342. https://doi.org/10.1021/ie800522p.

López Ortiz, A., Escobedo Bretado, M. A., Guzmán Velderrain, V., Meléndez Zaragoza, M., Salinas Gutiérrez, J., Lardizábal Gutiérrez, D., et al. (2014). Experimental and modeling kinetic study of the CO_2 absorption by Li_4SiO_4. *International Journal of Hydrogen Energy*, *39*(29), 16656–16666. https://doi.org/10.1016/j.ijhydene.2014.05.015.

López-Ortiz, A., Rivera, N. G. P., Rojas, A. R., & Gutierrez, D. L. (2004). Novel carbon dioxide solid acceptors using sodium containing oxides. *Separation Science and Technology*, *39*(15), 3559–3572. https://doi.org/10.1081/SS-200036766.

Lu, H., Khan, A., & Smirniotis, P. G. (2008). Relationship between structural properties and CO_2 capture performance of CaO-based sorbents obtained from different organometallic precursors. *Industrial & Engineering Chemistry Research*, *47*(16), 6216–6220. https://doi.org/10.1021/ie8002182.

Lu, H., Reddy, E. P., & Smirniotis, P. G. (2006). Calcium oxide based sorbents for capture of carbon dioxide at high temperatures. *Industrial and Engineering Chemistry Research*, *45*(11), 3944–3949. https://doi.org/10.1021/ie051325x.

Lu, N., & Xie, D. (2016). Novel membrane reactor concepts for hydrogen production from hydrocarbons: A review. *International Journal of Chemical Reactor Engineering*, *14*(1), 1–31. https://doi.org/10.1515/ijcre-2015-0050.

Luo, C., Zheng, Y., Yin, J., Qin, C., Ding, N., Zheng, C., et al. (2013). Effect of support material on carbonation and sulfation of synthetic CaO-based sorbents in calcium looping cycle. *Energy and Fuels*, *27*(8), 4824–4831. https://doi.org/10.1021/ef400564j.

Lysikov, A. I., Salanov, A. N., & Okunev, A. G. (2007). Change of CO_2 carrying capacity of CaO in isothermal recarbonation-decomposition cycles. *Industrial and Engineering Chemistry Research, 46*(13), 4633–4638. https://doi.org/10.1021/ie0702328.

Ma, L., Qin, C., Pi, S., & Cui, H. (2020). Fabrication of efficient and stable Li_4SiO_4-based sorbent pellets via extrusion-spheronization for cyclic CO_2 capture. *Chemical Engineering Journal, 379*. https://doi.org/10.1016/j.cej.2019.122385.

Maring, B. J., & Webley, P. A. (2013). A new simplified pressure/vacuum swing adsorption model for rapid adsorbent screening for CO_2 capture applications. *International Journal of Greenhouse Gas Control, 15*, 16–31. https://doi.org/10.1016/j.ijggc.2013.01.009.

Martavaltzi, C. S., & Lemonidou, A. A. (2008). Parametric study of the CaO-$Ca_{12}Al_{14}O_{33}$ synthesis with respect to high CO_2 sorption capacity and stability on multicycle operation. *Industrial and Engineering Chemistry Research, 47*(23), 9537–9543. https://doi.org/10.1021/ie800882d.

Martavaltzi, C. S., Pampaka, E. P., Korkakaki, E. S., & Lemonidou, A. A. (2010). Hydrogen production via steam reforming of methane with simultaneous CO_2 capture over CaO-$Ca_{12}Al_{14}O_{33}$. *Energy and Fuels, 24*(4), 2589–2595. https://doi.org/10.1021/ef9014058.

Maya, J. C., Chejne, F., Gómez, C. A., & Bhatia, S. K. (2018). Effect of the CaO sintering on the calcination rate of $CaCO_3$ under atmospheres containing CO_2. *AICHE Journal, 64*(10), 3638–3648. https://doi.org/10.1002/aic.16326.

Mejía-Trejo, V. L., Fregoso-Israel, E., & Pfeiffer, H. (2008). Textural, structural, and CO_2 chemisorption effects produced on the lithium orthosilicate by its doping with sodium Li_{4-x}, Na_xSiO_4. *Chemistry of Materials, 20*(22), 7171–7176. https://doi.org/10.1021/cm802132t.

Memon, M. Z., Ji, G., Li, J., & Zhao, M. (2017). Na_2ZrO_3 as an effective bifunctional catalyst-sorbent during cellulose pyrolysis. *Industrial and Engineering Chemistry Research, 56*(12), 3223–3230. https://doi.org/10.1021/acs.iecr.7b00309.

(2020). *Methane tracker 2020*. https://www.iea.org/reports/methane-tracker-2020.

Myers, J., Mastin, J., Bjørnebøle, T., Ryberg, T., & Eldrup, N. (2011). Techno-economical study of the Zero Emission Gas power concept. *Energy Procedia, 4*, 1949–1956. https://doi.org/10.1016/j.egypro.2011.02.075.

Naeem, M. A., Armutlulu, A., Imtiaz, Q., Donat, F., Schäublin, R., Kierzkowska, A., et al. (2018). Optimization of the structural characteristics of CaO and its effective stabilization yield high-capacity CO_2 sorbents. *Nature Communications, 9*(1). https://doi.org/10.1038/s41467-018-04794-5.

Nair, B. N., Burwood, R. P., Goh, V. J., Nakagawa, K., & Yamaguchi, T. (2009). Lithium based ceramic materials and membranes for high temperature CO_2 separation. *Progress in Materials Science, 54*(5), 511–541. https://doi.org/10.1016/j.pmatsci.2009.01.002.

Nakagawa, K., & Ohashi, T. (1998). A novel method of CO_2 capture from high temperature gases. *Journal of the Electrochemical Society, 145*(4), 1344–1346. https://doi.org/10.1149/1.1838462.

Nambo, A., He, J., Nguyen, T. Q., Atla, V., Druffel, T., & Sunkara, M. (2017). Ultrafast carbon dioxide sorption kinetics using lithium silicate nanowires. *Nano Letters, 17*(6), 3327–3333. https://doi.org/10.1021/acs.nanolett.6b04013.

Nieva, M. A., Villaverde, M. M., Monzón, A., Garetto, T. F., & Marchi, A. J. (2014). Steam-methane reforming at low temperature on nickel-based catalysts. *Chemical Engineering Journal, 235*, 158–166. https://doi.org/10.1016/j.cej.2013.09.030.

Noh, Y. S., Lee, K. Y., & Moon, D. J. (2019). Hydrogen production by steam reforming of methane over nickel based structured catalysts supported on calcium aluminate modified SiC. *International Journal of Hydrogen Energy, 44*(38), 21010–21019. https://doi.org/10.1016/j.ijhydene.2019.04.287.

Ochoa-Fernández, E., Haugen, G., Zhao, T., Rønning, M., Aartun, I., Børresen, B., et al. (2007). Process design simulation of H_2 production by sorption enhanced steam methane reforming: Evaluation of potential CO_2 acceptors. *Green Chemistry, 9*(6), 654–666. https://doi.org/10.1039/b614270b.

Ochoa-Fernández, E., Rønning, M., Yu, X., Grande, T., & Chen, D. (2008). Compositional effects of nanocrystalline lithium zirconate on its CO_2 capture properties. *Industrial and Engineering Chemistry Research, 47*(2), 434–442. https://doi.org/10.1021/ie0705150.

Ochoa-Fernández, E., Zhao, T., Rønning, M., & Chen, D. (2009). Effects of steam addition on the properties of high temperature ceramic CO_2 acceptors. *Journal of Environmental Engineering, 135*(6), 397–403. https://doi.org/10.1061/(ASCE)EE.1943-7870.0000006.

Oliveira, E. L. G., Grande, C. A., & Rodrigues, A. E. (2008). CO_2 sorption on hydrotalcite and alkali-modified (K and Cs) hydrotalcites at high temperatures. *Separation and Purification Technology, 62*(1), 137–147. https://doi.org/10.1016/j.seppur.2008.01.011.

Ortiz, A. L., & Harrison, D. P. (2001). Hydrogen production using sorption-enhanced reaction. *Industrial and Engineering Chemistry Research, 40*(23), 5102–5109. https://doi.org/10.1021/ie001009c.

Park, C. Y., Lee, T. H., Dorris, S. E., Lu, Y., & Balachandran, U. (2011). Oxygen permeation and coal-gas-assisted hydrogen production using oxygen transport membranes.

International Journal of Hydrogen Energy, *36*(15), 9345–9354. https://doi.org/10.1016/j.ijhydene.2011.04.090.

Pfeiffer, H., Bosch, P., & Bulbulian, S. (1998). Synthesis of lithium silicates. *Journal of Nuclear Materials*, *257*(3), 309–317. https://doi.org/10.1016/S0022-3115(98)00449-8.

Phair, J. W., & Badwal, S. P. S. (2006). Review of proton conductors for hydrogen separation. *Ionics*, *12*(2), 103–115. https://doi.org/10.1007/s11581-006-0016-4.

Qi, Z., Daying, H., Yang, L., Qian, Y., & Zibin, Z. (2013). Analysis of CO_2 sorption/desorption kinetic behaviors and reaction mechanisms on Li_4SiO_4. *AICHE Journal*, *59*(3), 901–911. https://doi.org/10.1002/aic.13861.

Radfarnia, H. R., & Sayari, A. (2015). A highly efficient CaO-based CO_2 sorbent prepared by a citrate-assisted sol-gel technique. *Chemical Engineering Journal*, *262*, 913–920. https://doi.org/10.1016/j.cej.2014.09.074.

Rao, G. J., Mazumder, R., Bhattacharyya, S., & Chaudhuri, P. (2017). Synthesis, CO_2 absorption property and densification of Li_4SiO_4 powder by glycine-nitrate solution combustion method and its comparison with solid state method. *Journal of Alloys and Compounds*, *725*, 461–471. https://doi.org/10.1016/j.jallcom.2017.07.163.

Rodríguez, M. T., & Pfeiffer, H. (2008). Sodium metasilicate (Na_2SiO_3): A thermo-kinetic analysis of its CO_2 chemical sorption. *Thermochimica Acta*, *473*(1–2), 92–95. https://doi.org/10.1016/j.tca.2008.04.022.

Rogelj, J. (2018). Mitigation pathways compatible with 1.5°C in the context of sustainable development. In *Global Warming of 1.5°C. An IPCC special report on the impacts of global warming of 1.5°C above pre-industrial levels and related global greenhouse gas emission pathw. IPCC special report Global Warming of 1.5°C.*

Romero-Ibarra, I. C., Ortiz-Landeros, J., & Pfeiffer, H. (2013). Microstructural and CO_2 chemisorption analyses of Li_4SiO_4: Effect of surface modification by the ball milling process. *Thermochimica Acta*, *567*, 118–124. https://doi.org/10.1016/j.tca.2012.11.018.

Rosen, L., Degenstein, N., Shah, M., Wilson, J., Kellya, S., Peck, J., et al. (2011). Development of oxygen transport membranes for coal-based power generation. *Energy Procedia*, *4*, 750–755. https://doi.org/10.1016/j.egypro.2011.01.115.

Rostrup-Nielsen, J. R. (1997). Industrial relevance of coking. *Catalysis Today*, *37*(3), 225–232. https://doi.org/10.1016/S0920-5861(97)00016-3.

Samanta, A., Zhao, A., Shimizu, G. K. H., Sarkar, P., & Gupta, R. (2012). Post-combustion CO_2 capture using solid sorbents: A review. *Industrial and Engineering Chemistry Research*, *51*(4), 1438–1463. https://doi.org/10.1021/ie200686q.

Sayari, A., Belmabkhout, Y., & Serna-Guerrero, R. (2011). Flue gas treatment via CO_2 adsorption. *Chemical Engineering Journal*, *171*(3), 760–774. https://doi.org/10.1016/j.cej.2011.02.007.

Scaltsoyiannes, A., Antzaras, A., Koilaridis, G., & Lemonidou, A. (2021). Towards a generalized carbonation kinetic model for CaO-based materials using a modified random pore model. *Chemical Engineering Journal*, *407*. https://doi.org/10.1016/j.cej.2020.127207.

Seggiani, M., Puccini, M., & Vitolo, S. (2011). High-temperature and low concentration CO_2 sorption on Li_4SiO_4 based sorbents: Study of the used silica and doping method effects. *International Journal of Greenhouse Gas Control*, *5*(4), 741–748. https://doi.org/10.1016/j.ijggc.2011.03.003.

Seggiani, M., Puccini, M., & Vitolo, S. (2013). Alkali promoted lithium orthosilicate for CO_2 capture at high temperature and low concentration. *International Journal of Greenhouse Gas Control*, *17*, 25–31. https://doi.org/10.1016/j.ijggc.2013.04.009.

Shan, S. Y., Ma, A. H., Hu, Y. C., Jia, Q. M., Wang, Y. M., & Peng, J. H. (2016). Development of sintering-resistant CaO-based sorbent derived from eggshells and bauxite tailings for cyclic CO_2 capture. *Environmental Pollution*, *208*, 546–552. https://doi.org/10.1016/j.envpol.2015.10.028.

Subha, P. V., Nair, B. N., Hareesh, P., Mohamed, A. P., Yamaguchi, T., Warrier, K. G. K., et al. (2014). Enhanced CO_2 absorption kinetics in lithium silicate platelets synthesized by a sol-gel approach. *Journal of Materials Chemistry A*, *2*(32), 12792–12798. https://doi.org/10.1039/c4ta01976h.

Subha, P. V., Nair, B. N., Mohamed, A. P., Anilkumar, G. M., Warrier, K. G. K., Yamaguchi, T., et al. (2016). Morphologically and compositionally tuned lithium silicate nanorods as high-performance carbon dioxide sorbents. *Journal of Materials Chemistry A*, *4*(43), 16928–16935. https://doi.org/10.1039/c6ta06133h.

Sun, P., Grace, J. R., Lim, C. J., & Anthony, E. J. (2007). The effect of CaO sintering on cyclic CO_2 capture in energy systems. *AICHE Journal*, *53*(9), 2432–2442. https://doi.org/10.1002/aic.11251.

Tlili, N., Grévillot, G., & Vallières, C. (2009). Carbon dioxide capture and recovery by means of TSA and/or VSA. *International Journal of Greenhouse Gas Control*, *3*(5), 519–527. https://doi.org/10.1016/j.ijggc.2009.04.005.

Valverde, J. M., Sanchez-Jimenez, P. E., & Perez-Maqueda, L. A. (2015). Ca-looping for postcombustion CO_2 capture: A comparative analysis on the performances of dolomite and limestone. *Applied Energy*, *138*, 202–215. https://doi.org/10.1016/j.apenergy.2014.10.087.

Venegas, M. J., Fregoso-Israel, E., Escamilla, R., & Pfeiffer, H. (2007). Kinetic and reaction mechanism of CO_2 sorption on Li_4SiO_4: Study of the particle size effect. *Industrial and Engineering Chemistry Research*, *46*(8), 2407–2412. https://doi.org/10.1021/ie061259e.

Wang, K., Gu, F., Clough, P. T., Zhao, Y., Zhao, P., & Anthony, E. J. (2021). Molten shell-activated, high-performance, un-doped Li_4SiO_4 for high-temperature CO_2 capture at low CO_2 concentrations. *Chemical Engineering Journal, 408*. https://doi.org/10.1016/j.cej.2020.127353.

Wang, Q., Luo, J., Zhong, Z., & Borgna, A. (2011). CO_2 capture by solid adsorbents and their applications: Current status and new trends. *Energy and Environmental Science, 4*(1), 42–55. https://doi.org/10.1039/c0ee00064g.

Wang, Y., Memon, M. Z., Seelro, M. A., Fu, W., Gao, Y., Dong, Y., et al. (2021). A review of CO_2 sorbents for promoting hydrogen production in the sorption-enhanced steam reforming process. *International Journal of Hydrogen Energy, 46*(45), 23358–23379. https://doi.org/10.1016/j.ijhydene.2021.01.206.

Wang, K., Wang, X., Zhao, P., & Guo, X. (2014). High-temperature capture of CO_2 on lithium-based sorbents prepared by a water-based sol-gel technique. *Chemical Engineering and Technology, 37*(9), 1552–1558. https://doi.org/10.1002/ceat.201300584.

Wu, S. F., Li, Q. H., Kim, J. N., & Yi, K. B. (2008). Properties of a nano CaO/Al_2O_3 CO_2 sorbent. *Industrial and Engineering Chemistry Research, 47*(1), 180–184. https://doi.org/10.1021/ie0704748.

Wu, S. F., & Zhu, Y. Q. (2010). Behavior of $CaTiO_3$/nano-CaO as a CO_2 reactive adsorbent. *Industrial and Engineering Chemistry Research, 49*(6), 2701–2706. https://doi.org/10.1021/ie900900r.

Xie, M., Zhou, Z., Qi, Y., Cheng, Z., & Yuan, W. (2012). Sorption-enhanced steam methane reforming by in situ CO_2 capture on a CaO-$Ca_9Al_6O_{18}$ sorbent. *Chemical Engineering Journal, 207–208*, 142–150. https://doi.org/10.1016/j.cej.2012.06.032.

Xiong, R., Ida, J., & Lin, Y. S. (2003). Kinetics of carbon dioxide sorption on potassium-doped lithium zirconate. *Chemical Engineering Science, 58*(19), 4377–4385. https://doi.org/10.1016/S0009-2509(03)00319-1.

Xiu, G. H., Li, P., & Rodrigues, A. E. (2002). Sorption-enhanced reaction process with reactive regeneration. *Chemical Engineering Science, 57*(18), 3893–3908. https://doi.org/10.1016/S0009-2509(02)00245-2.

Xiu, G. H., & Rodrigues, A. E. (2004). Subsection-controlling strategy for improving sorption-enhanced reaction process. *Chemical Engineering Research and Design, 82*(2), 192–202. https://doi.org/10.1205/026387604772992765.

Yancheshmeh, M. S., Radfarnia, H. R., & Iliuta, M. C. (2016). High temperature CO_2 sorbents and their application for hydrogen production by sorption enhanced steam reforming process. *Chemical Engineering Journal, 283*, 420–444. https://doi.org/10.1016/j.cej.2015.06.060.

Yang, X., Liu, W., Sun, J., Hu, Y., Wang, W., Chen, H., et al. (2016). Alkali-doped lithium orthosilicate sorbents for carbon dioxide capture. *ChemSusChem, 9*(17), 2480–2487. https://doi.org/10.1002/cssc.201600737.

Yin, Z., Wang, K., Zhao, P., & Tang, X. (2016). Enhanced CO_2 chemisorption properties of Li_4SO_4, using a water hydration-calcination technique. *Industrial and Engineering Chemistry Research, 55*(4), 1142–1146. https://doi.org/10.1021/acs.iecr.5b03746.

Zhang, X., Li, Z., Peng, Y., Su, W., Sun, X., & Li, J. (2014). Investigation on a novel CaO-Y_2O_3 sorbent for efficient CO_2 mitigation. *Chemical Engineering Journal, 243*, 297–304. https://doi.org/10.1016/j.cej.2014.01.017.

Zhang, S., Zhang, Q., Wang, H., Ni, Y., & Zhu, Z. (2014). Absorption behaviors study on doped Li_4SiO_4 under a humidified atmosphere with low CO_2 concentration. *International Journal of Hydrogen Energy, 39*(31), 17913–17920. https://doi.org/10.1016/j.ijhydene.2014.07.011.

Zhao, D., Feng, J., Huo, Q., Melosh, N., Fredrickson, G. H., Chmelka, B. F., et al. (1998). Triblock copolymer syntheses of mesoporous silica with periodic 50 to 300 angstrom pores. *Science, 279*(5350), 548–552. https://doi.org/10.1126/science.279.5350.548.

Zhou, Z., Wang, K., Yin, Z., Zhao, P., Su, Z., & Sun, J. (2017). Molten K_2CO_3-promoted high-performance Li_4SiO_4 sorbents at low CO_2 concentrations. *Thermochimica Acta, 655*, 284–291. https://doi.org/10.1016/j.tca.2017.07.014.

Zhu, L., Li, L., & Fan, J. (2015). A modified process for overcoming the drawbacks of conventional steam methane reforming for hydrogen production: Thermodynamic investigation. *Chemical Engineering Research and Design, 104*, 792–806. https://doi.org/10.1016/j.cherd.2015.10.022.

Zou, Y., & Rodrigues, A. E. (2001). The separation enhanced reaction process (SERP) in the production of hydrogen from methane steam reforming. *Adsorption Science and Technology, 19*(8), 655–672. https://doi.org/10.1260/0263617011494475.

Further reading

Carella, E., & Hernandez, M. T. (2014). High lithium content silicates: A comparative study between four routes of synthesis. *Ceramics International, 40*(7), 9499–9508. https://doi.org/10.1016/j.ceramint.2014.02.023.

Cesário, M. R., Barros, B. S., Courson, C., Melo, D. M. A., & Kiennemann, A. (2015). Catalytic performances of Ni-CaO-mayenite in CO_2 sorption enhanced steam methane reforming. *Fuel Processing Technology, 131*, 247–253. https://doi.org/10.1016/j.fuproc.2014.11.028.

Lu, H., Khan, A., & Smirniotis, P. G. (2008). Relationship between structural properties and CO_2 capture performance of CaO-based sorbents obtained from different organometallic precursors. *Industrial and Engineering Chemistry Research, 47*(16), 6216–6220. https://doi.org/10.1021/ie8002182.

Lu, H., Reddy, E. P., & Smirniotis, P. (2005). Calcium oxide based sorbents for adsorption of CO_2 at high temperatures. In *AIChE annual meeting, conference proceedings* (p. 13179).

Lu, H., & Smirniotis, P. G. (2009). Calcium oxide doped sorbents for CO_2 uptake in the presence of SO_2 at high temperatures. *Industrial and Engineering Chemistry Research, 48*(11), 5454–5459. https://doi.org/10.1021/ie900162k.

Munro, S., Åhlén, M., Cheung, O., & Sanna, A. (2020). Tuning Na_2ZrO_3 for fast and stable CO_2 adsorption by solid state synthesis. *Chemical Engineering Journal, 388*. https://doi.org/10.1016/j.cej.2020.124284.

Sun, Z., Luo, S., Qi, P., & Fan, L. S. (2012). Ionic diffusion through calcite ($CaCO_3$) layer during the reaction of CaO and CO_2. *Chemical Engineering Science, 81*, 164–168. https://doi.org/10.1016/j.ces.2012.05.042.

Wang, S., Fan, S., Fan, L., Zhao, Y., & Ma, X. (2015). Effect of cerium oxide doping on the performance of CaO-based sorbents during calcium looping cycles. *Environmental Science and Technology, 49*(8), 5021–5027. https://doi.org/10.1021/es5052843.

Yanase, I., Maeda, T., & Kobayashi, H. (2017). The effect of addition of a large amount of CeO_2 on the CO_2 adsorption properties of CaO powder. *Chemical Engineering Journal, 327*, 548–554. https://doi.org/10.1016/j.cej.2017.06.140.

Zhang, Yang, Gong, Xun, Chen, Xuanlong, Yin, Li, Zhang, Jian, & Liu, Wenqiang. (2018). Performance of synthetic CaO-based sorbent pellets for CO_2 capture and kinetic analysis. *Fuel, 232*(15), 205–214. https://doi.org/10.1016/j.fuel.2018.05.143.

Zhao, T., Ochoa-Fernández, E., Rønning, M., & Chen, D. (2007). Preparation and high-temperature CO_2 capture properties of nanocrystalline Na_2ZrO_3. *Chemistry of Materials, 19*(13), 3294–3301. https://doi.org/10.1021/cm062732h.

CHAPTER

7

Catalysts for syngas production by dry reforming of methane

Jorge Álef Estevam Lau Bomfim[a,c], José Faustino Souza Carvalho Filho[b], Tércia Diniz Bezerra[c], Fernando Césário Rangel[a,c], Thiago Araujo Simões[a,d,e], Pedro Nothaft Romano[f], and Rosenira Serpa da Cruz[a,c]

[a]Materials Science and Engineering Postgraduate Program (PPCEM), Federal University of Paraíba (UFPB), João Pessoa, Brazil [b]Escola de Química, Universidade Federal do Rio de Janeiro—UFRJ, Rio de Janeiro, RJ, Brazil [c]Grupo Bioenergia e Meio Ambiente, Universidade Estadual de Santa Cruz—UESC, Ilhéus, BA, Brazil [d]Postgraduate Program in Science, Innovation and Modeling in Materials (PROCIMM), State University of Santa Cruz—UESC, Ilhéus, BA, Brazil [e]Center for Science and Technology in Energy and Sustainability (CETENS), Federal University of the Recôncavo of Bahia (UFRB), Feira de Santana, Brazil [f]Universidade Federal do Rio de Janeiro—UFRJ, Campus Duque de Caxias, Rio de Janeiro, RJ, Brazil

7.1 Introduction

In a global context where the concentration of greenhouse gases (GHG) in the atmosphere is increasing mainly due to human interference, causing ecosystem imbalances and global warming (Al-Fatesh et al., 2021; Araiza, González-Vigi, Gómez-Cortés, Arenas-Alatorre, and Díaz, 2021; Beheshti Askari et al., 2020; Saelee et al., 2021; Shanshan, Huanhao, Christopher, & Xiaolei, 2021), dry reforming of methane (DRM) stands out as it converts two major greenhouse gases into syngas (H_2+CO), as shown in Eq. (7.1); a key component for many industrially relevant processes such as the production of methanol, liquid fuels and chemicals (Angeli et al., 2021; Araiza, González-Vigi, et al., 2021; Shahnazi & Firoozi, 2020; Yusuf, Farooqi, Keong, Hellgardt, & Abdullah, 2021). It is worth mentioning that the syngas, with molar ratio $H_2/CO=1$, produced in the DRM is ideal for the production of long-chain hydrocarbons via Fischer-Tropsch processes, highlighting the economic and environmental advantages of DRM (Al-

Heterogeneous Catalysis
https://doi.org/10.1016/B978-0-323-85612-6.00007-3

Fatesh et al., 2021; Araiza, González-Vigi, et al., 2021; Beheshti Askari et al., 2020; de Dios García, Stankiewicz, & Nigar, 2021; Saelee et al., 2021; Shahnazi & Firoozi, 2020; Yusuf et al., 2021) and its contribution to improving global energy efficiency (Shanshan et al., 2021). Regarding the dry reforming of methane reaction thermodynamics, it is a well-known fact that the DRM reaction is highly endothermic (Eq. 7.1). In this sense, high reaction temperature values (973.15–1273.15K) and low pressures (around 0.1MPa) favor the reaction thermodynamics and kinetics, leading to higher CO_2 and CH_4 conversion values (Araiza, González-Vigi, et al., 2021; Beheshti Askari et al., 2020; Cao, Adegbite, Zhao, Lester, & Wu, 2018;de Dios García et al., 2021 ; Shahnazi & Firoozi, 2020 ; Yusuf et al., 2021).

$$CH_4 + CO_2 \rightarrow 2CO + 2H_2 \quad \Delta H^{\circ}_{298\,k} = 247\,kJ/mol \tag{7.1}$$

As defined by Mark et al., the dry reforming of methane itself is not a stoichiometrically independent reaction, but belongs to a rather complex set of reactions (Mark, Mark, & Maier, 1997):

$$H_2 + CO_2 \leftrightarrow H_2O + CO \quad \Delta H^{\circ}_{298\,k} = 41\,kJ/mol \tag{7.2}$$

$$CH_4 \leftrightarrow C + 2H_2 \quad \Delta H^{\circ}_{298\,k} = 75\,kJ/mol \tag{7.3}$$

$$2CO \leftrightarrow C + CO_2 \quad \Delta H^{\circ}_{298\,k} = -172\,kJ/mol \tag{7.4}$$

$$CO_2 + 2H_2 \leftrightarrow C + 2H_2O \quad \Delta H^{\circ}_{298\,k} = -90\,kJ/mol \tag{7.5}$$

$$H_2 + CO \leftrightarrow C + H_2O \quad \Delta H^{\circ}_{298\,k} = -131\,kJ/mol \tag{7.6}$$

However, following the DRM reaction, the reaction between H_2 and CO_2 consumes the CO_2 reactant and generates H_2O through the reverse water-gas shift reaction (Eq. 7.2) (de Dios García et al., 2021; Shahnazi & Firoozi, 2020; Yusuf et al., 2021), further, the H_2/CO ratio is found to be less than 1. Along those lines, the deactivation process carried out in DRM reaction is constantly referred on the literature as guided by five distinct reactions: methane cracking reaction (Eq. 7.3), carbon monoxide disproportionation reaction (Boudouard reaction) (Eq. 7.4), carbon dioxide hydrogenation reaction (Eq. 7.5), and carbon monoxide hydrogenation (Eq. 7.6) (Al-Fatesh et al., 2021; Angeli et al., 2021; Araiza, González-Vigi, et al., 2021; de Dios García et al., 2021; Shahnazi & Firoozi, 2020; Yusuf et al., 2021).

Another side reaction that we can mention in the DRM context is the steam reforming of methane (Eq. 7.7) (Mark et al., 1997), which produces syngas with H_2/CO ratio equals to 3, which is not ideal for Fischer-Tropsch, but is clearly a less detrimental parallel reaction.

$$CH_4 + H_2O \leftrightarrow CO + 3H_2 \quad \Delta H^{\circ}_{298\,k} = 206\,kJ/mol \tag{7.7}$$

Therefore, by analyzing the productivity in DRM, we can separate the reactions between those that contribute to synthesis gas formation, and those that contribute to reduced reaction productivity by forming CH_4 and CO_2, and to decreased catalytic activity and/or stability due to the formation of by-products such as H_2O and C. This classification is shown in Fig. 7.1.

Shanshan et al. explain that thermodynamically, DRM cannot occur at low temperatures due to its Gibbs free energy being negative, which is a consequence of the high thermodynamic stability and ionization potential of the linear CO_2 molecule, whose binding enthalpy is $805\,kJ\,mol^{-1}$, while the activation of CH_4 molecules also requires a lot of energy, occurring at temperatures between 873.15 and 1373.15K, due to its stable C—H bonds with dissociation energy of 4.5eV (Shanshan et al., 2021). Therefore, observing from the thermodynamic point of view, the parallel reactions represented by Eqs. (7.4)–(7.6) are favored at temperatures below 1273.15K (Angeli et al., 2021), leading to the inevitable formation of carbon deposits (Al-Fatesh et al., 2021), something

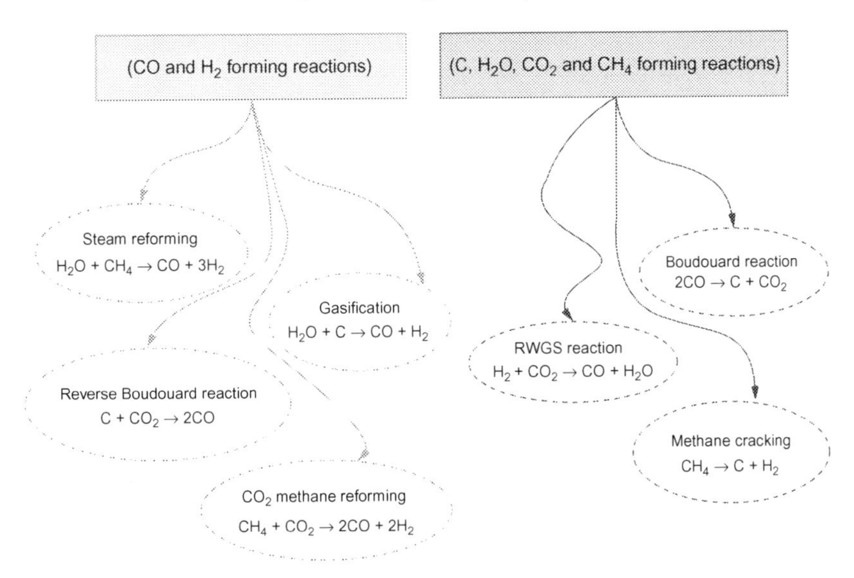

FIG. 7.1 Scheme of possible parallel reactions in DRM.

extremely undesirable industrially since these deposits besides encapsulating and deactivating the active site (Beheshti Askari et al., 2020), accumulate in the reactors causing pressure drop and even clogging (Angeli et al., 2021).

If on the one hand, performing DRM reactions above 1273.15K ensures lower coke deposition and higher CH_4 and CO_2 conversions; on the other hand, harsh reaction conditions end up being crucial for the agglomeration and growth of catalyst active metals, something that decreases the catalytic activity by hindering the adsorption of reagents in the active sites and consequently contributes to catalytic deactivation and reduction of DRM productivity. This process is called sintering (Araiza, González-Vigi, et al., 2021; Shahnazi & Firoozi, 2020).

Because there is no possibility of fully eliminating the problem of coke deposition in DRM, it is essential to think both about its minimization and in the reduction of sintering (Al-Fatesh et al., 2021). Therefore, the catalyst design appears as a crucial variable for improving DRM productivity since it provides a reaction path with lower activation energy, reduces the reaction temperature and sintering in the process (Yusuf et al., 2021). Also, by changing the reaction rates, it can favor the production of synthesis gas over its consumption, suppressing coke deposition and increasing the yield of the process (Yusuf et al., 2021). Another challenge is to keep nickel in its metallic form on the support, as its oxidized form reduces its activity (Abdelsadek et al., 2021).

In summary, an important problem related to the dry reforming reaction of methane is the coke generation, therefore, from the industrial point of view, the development of an effective and economic catalyst to prevent these with high activity and selectivity as well as long-term stability is a critical gap for the industry-wide application of DRM (Ranjekar & Yadav, 2021). Additionality, operating DRM processes at higher temperatures to reduce carbon formation, has the disadvantage of sintering nickel particles, which eventually reduces the activity over time.

7.2 DRM heterogeneous catalysts

In order to try to overcome the limitations related to the deactivation of the catalysts used for DRM, different strategies can be followed combined or not: the design of reactors, the optimization of reaction conditions, and the development of active and stable, and preferably, low-cost catalysts. The latter is considered by many authors (Abbasi, Abbasi, Tabkhi, & Akhlaghi, 2020; Mourhly, Kacimi, Halim, & Arsalane, 2020; Ray, Bhardwaj, Singh, & Deo, 2018) to be the focal point for DRM to become a consolidated industrial process, with resistance to carbon deposits being the main bottleneck for the development of DRM (Marinho et al., 2021). Fig. 7.2 highlights the main challenges and the relevant parameters and strategies for developing active and stable catalysts for DRM that will be discussed throughout this chapter.

In general, catalysts applied to DRM have the noble metals Ru, Rh, Ir, Pd, and Pt as active sites and/or the nonnoble transition metals Ni, Co, Cu, and Fe. Those containing noble metals have high activity and selectivity for DRM in addition to resistance to carbon deposition. However, the high cost and low availability of these make them economically uncompetitive compared to other materials based on nonnoble transition metals. Among the latter, Ni-based catalysts are the most active, but also highly prone to carbon formation, since, together with the ability to form the CH bond, Ni has high interaction with carbon (Zhenghong & Fei, 2018). In addition, these catalysts are also prone to the sintering of Ni particles and their oxidation by surface adsorbed oxygen species during the catalytic process (Zhang, Zhang, et al., 2020; Zhang, Liu, Xu, & Sun, 2018). It is worth mentioning that, depending on the methane source, the presence of sulfur is another determining factor in deactivating the catalyst.

Different methods of synthesizing catalysts, pretreatments (activation), adding dopants, or promoters have been studied to achieve better performances concerning the activity and stability of nickel-based catalysts supported on different materials (Li, Yuan, Li, & Wang, 2020). Among these strategies we can mention the following (Abdulrasheed et al., 2019; Zhang et al., 2018):

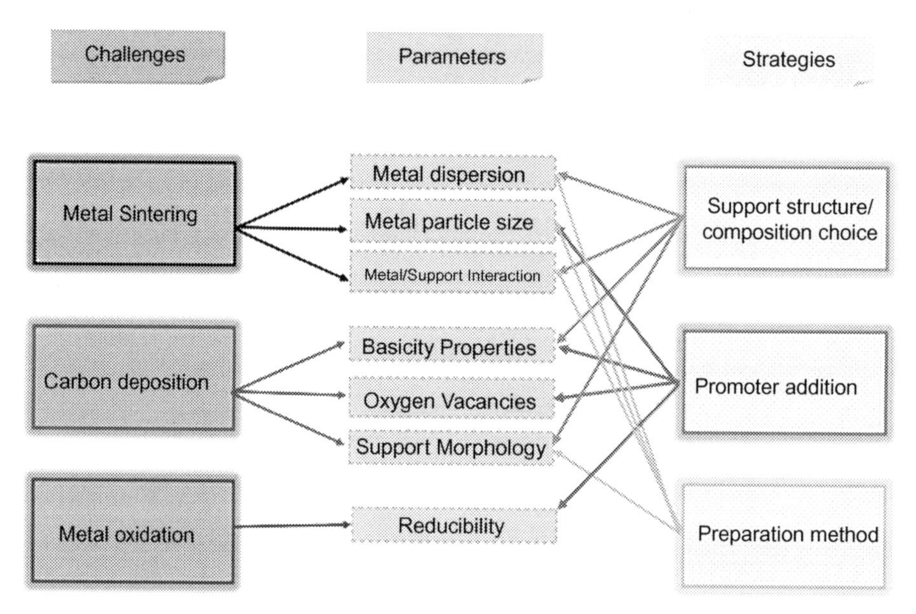

FIG. 7.2 Main challenges and the relevant parameters and strategies for developing active and stable catalysts for DRM.

The addition of a second metal, such as Co, Cu, Fe, and Sn, results in the formation of alloys less sensitive to carbon formation;
Addition of Lewis bases, which adsorb CO_2, thus reducing the formation of carbon via the reaction of C to form CO;
The addition of oxides, which have the capacity to store and release oxygen, leading to the removal of carbon through the reaction between the deposited carbon and the oxygen in the network formed in these oxides;
Preparation and activation methods lead to greater dispersion and smaller size of Ni particles.

Different supports, with the most diverse morphologies and topologies, have been evaluated to understand how the synergistic effect between active species, promoters, dopants, and support affects basicity, redox properties, oxygen mobility, particle size, reducibility, and mass transfer limitations factors that influence the activity and stability of Ni-based catalysts in DRM (Aziz, Jalil, Wongsakulphasatch, & Vo, 2020; Zhang et al., 2018). Table 7.1 shows some recent studies that have evaluated different catalysts with emphasis on the active site, nature of the support, type of promoter, and preparation method for DRM.

In the following sections, some relevant topics for the development of catalysts for application in DRM with an emphasis on nickel-based materials will be discussed.

Active sites and methods of synthesis

DRM occurs via the bifunctional active sites approach such that the active sites for CO_2 are on supports with adequate acid-basic properties while metals are active sites for CH_4 dissociation. Catalysts with inert support follow a monofunctional mechanism where both CO_2 and CH_4 are activated on active metals, as shown in Fig. 7.3. The metals in groups 8, 9, and 10, except osmium, are the most studied in DRM. Noble metals stand out for their greater activity and stability, probably due to the greater ability to dissociate CO_2 resulting in the adsorbed species CO and O which, in turn, promote the dissociation of methane in the adsorbed CH_x and H species thus

TABLE 7.1 Recent catalysts used in DRM.

Active site	Support	Promoter	Preparation method	$\%CO_2/\%CH_4$ Conversion rate[a]	References
Ni	$La_2O_3 + ZrO_2$	Ce	Wet impregnation method	(91/86)	Kasim et al. (2020)
Ni	Al_2O_3	Ca	Solution combustion method	(96.7/52)	Wang et al. (2020)
Ni	γ-Al_2O_3	–	Sol-gel method followed by impregnation	(81.2/69.7)	Lyu, Han, et al. (2020)
Ni	$Ca_xZr_yO_z$	Ce	Combined nanocoating and co-impregnation method	(72/80)	Hu et al. (2021)
Ni	K_2TiO_3	–	Solvo-plasma technique followed by incipient wetness impregnation	(90/-)	Nambo et al. (2021)
Ni	$MgAl_2O_4$	K	Co-precipitation method followed by wet impregnation	(90/72)	Azancot et al. (2021)

Continued

TABLE 7.1 Recent catalysts used in DRM—cont'd

Active site	Support	Promoter	Preparation method	$\%CO_2/\%CH_4$ Conversion rate[a]	References
Ni	Al_2O_3	–	Sol-gel method followed by incipient impregnation	(85.4/77.6)	Bian et al. (2021)
Ni	Hydrochar	–	Hydrothermal carbonization followed by ultrasound-assisted impregnation	(80/50)	Li, Wang, Zhang, Liu, and Yang (2020)
Ni	MCM-41	–	Sol-gel one-pot microwave-assisted method	(~80/~65)	Salazar Hoyos, Faroldi, and Cornaglia (2020)
Ni	SBA-15	–	Sol-gel method followed by "two solvents" impregnation technique	(~94/~92)	Kaydouh, El Hassan, Davidson, and Massiani (2021)
Ni	MgO	Mo	Autothermal combustion of Mg chips under CO_2 flow followed by dispersion in nitrates and hydrazine reduction in presence of polyvinylpyrrolidone	(80/75)	Hu and Ruckenstein (2020)
Ni	$La_{1.8}Ba_{0.2}Ni_{0.9}Cu_{0.1}O_4$	Ba, Cu	Autoignition method	(~80%/~80%)	Bekheet et al. (2021)
Co	$La_{0.8}Sr_{0.2}Co_{0.9}M_{0.1}O_3$	Sr, Mn	Thermal treatment of stoichiometric mixture of standard powders	(90/83)	Marin et al. (2021)
Rh	γ-Al_2O_3	–	Incipient wetness impregnation	(64/53)	de Araujo Moreira et al. (2020)
Ni-Fe	Mg(*Al*)O (LDH derived)	–	Co-precipitation method	(96.5/93.4)	Wan et al. (2020)
Ni-Pd-Pt	$Mg_{0.85}Ce_{0.15}O$	Ce	Co-precipitation followed by impregnation	(90/78)	Al-Doghachi, Jassim, and Taufiq-Yap (2020)
Pt-Pd-Ni	$Mg_{0.85}La_{0.15}O$	La	Co-precipitation followed by impregnation	(98.97/85.01)	Al-Najar, Al-Doghachi, Al-Riyahee, and Taufiq-Yap (2020)
Ni	ZSM-5	Ta	Microemulsion coupled with zeolite seed crystallization technique followed by incipient wetness impregnation	(90.8/96.6)	Hambali et al. (2021)
Ni	La_2O_3/SiO_2	–	One-pot colloidal solution combustion method	(90.2/84.4)	Wang et al. (2019)

TABLE 7.1 Recent catalysts used in DRM—cont'd

Active site	Support	Promoter	Preparation method	$\%CO_2/\%CH_4$ Conversion rate[a]	References
Ni-Co	CMOF-74	–	Template-directed synthesis with DHTA	(65/57)	Khan, Ramirez, Shterk, Garzón-Tovar, and Gascon (2020)
Ni-Rh	$MgAl_2O_4$	–	Co-precipitation followed by incipient wetness impregnation	(70/90)	Schiaroli et al. (2020)
Ni	$ZrAl_3O_x$	–	Modified Pechini sol-gel method followed by impregnation	(~20/~15)	Shin et al. (2020)
Co-Ce	Nitrogen-doped activated carbon	Ca	Melamine doping and co-impregnation method	(80.2/67.8)	Sun, Zhang, Liu, Xu, and Lv (2020) and Sun et al. (2020)
Ni-Co	γ-Al_2O_3	–	Capillary simultaneous impregnation	(~87/~89)	Myltykbayeva et al. (2020)
Rh-Ni	Al_2O_3	–	Sol-gel hydrothermal method	(85/70)	Mozammel et al. (2020)
Ni-Mo	SiO_2	–	Incipient wetness co-impregnation	(90/90)	Liu et al. (2021)
Ni-Pt	CeO_2	–	Precipitation followed by incipient wetness co-impregnation	(~90/~80)	Araiza, Arcos, Gómez-Cortés, and Díaz (2021)

[a] *The best results reported.*

enabling a higher rate of conversion of methane (Abdulrasheed et al., 2019; Ranjekar & Yadav, 2021; Singh, Nguyen, et al., 2020). Among the non-noble metals, nickel, which is already used as an active site in industrial processes of steam reforming of methane, is the most studied. However, nickel-based catalysts are quite prone to deactivation by carbon deposits, sintering, and oxidation of nickel (Xingyuan, Jangam, and Sibudjing, 2020; Zhang et al., 2018).

Araiza et al. highlight the following three crucial factors for good catalytic performance in DRM (Al-Fatesh et al., 2021; Araiza, Arcos, et al., 2021; Beheshti Askari et al., 2020; Saelee et al., 2021; Shanshan et al., 2021):

- High dispersion of the active site;
- Adequate choice of support;
- Degree of interaction between the metal and the support.

The optimization of these factors goes through four important aspects as suggested by Xiangyuan et al. (Xingyuan et al., 2020; Zhang et al., 2018):

- Surface regulation of the surface acidity and basicity;
- Oxygen defects including lattice and surface oxygen and oxygen vacancies;
- Interfacial engineering involves metal-metal synergy or metal-support interaction and structural optimization of support, such as

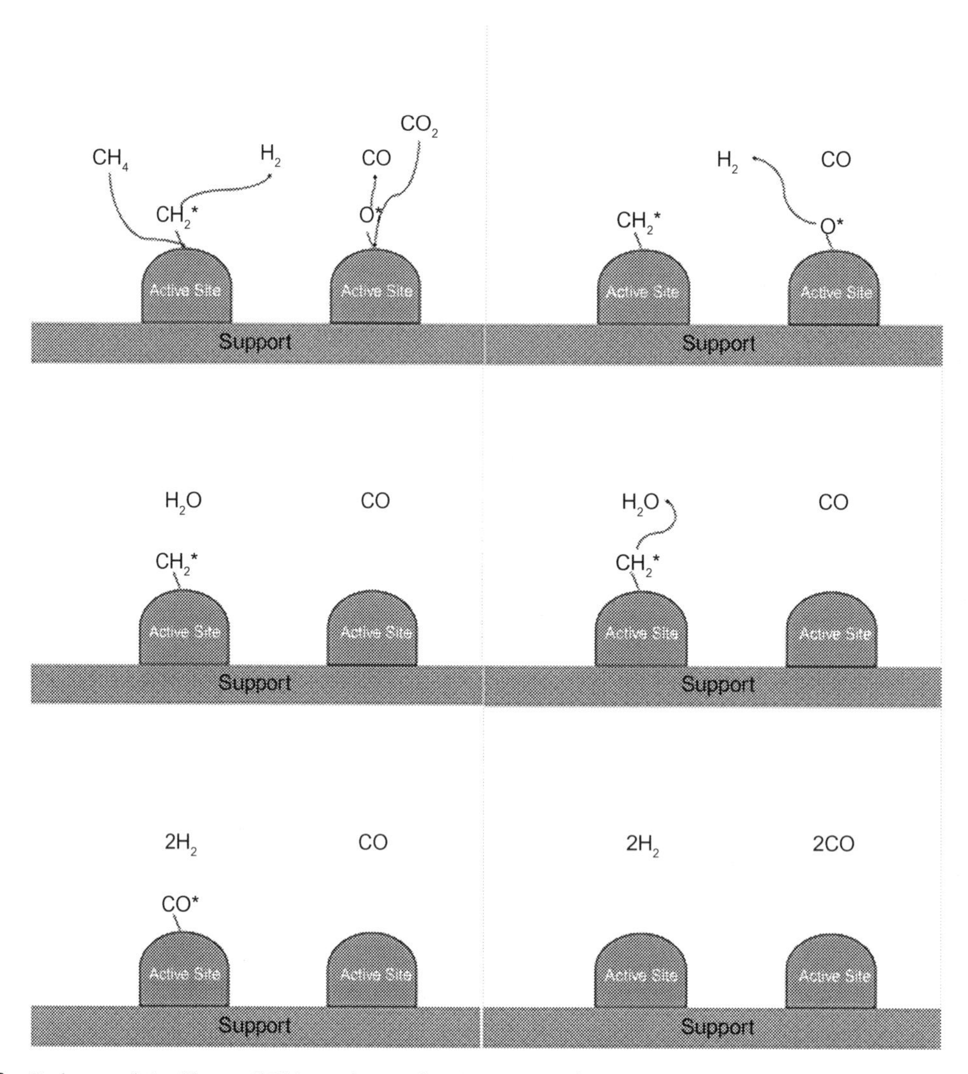

FIG. 7.3 Bodrov and Apel'baum DRM reaction mechanism proposed in 1967.

porous materials, core-shell, yolk-shell, embedded, and hollow structures among others.

Therefore, the major challenge for the development of active, selective, stable, and low-cost catalysts for RSM is to develop synthesis methods that allow for adequate size and dispersion of Ni particles and good metal-support interaction to suppress the carbon deposit and the sintering of nickel particles. Moreira et al. evaluated catalysts 0.1% Rh/Al_2O_3 and 10% $Ni/Ni/Al_2O_3$ prepared by the incipient wet impregnation method, to compare the catalytic efficiency of Rh even in a significantly lower amount than Ni. The physical-chemical properties obtained by different techniques showed that the catalyst with 0.1% Rh had a larger surface area, significantly smaller particle diameter (more than 24× smaller), and greater metallic dispersion (more than 23× larger) than the catalyst with 10% Ni, which naturally influenced the performance of the reform reaction (Al-Doghachi et al., 2020; Al-Najar et al., 2020; Araiza, Arcos, et al., 2021; Azancot, Bobadilla, Centeno, & Odriozola, 2020; Bekheet et al., 2021; Bian, Zhong, et al., 2021; de Araujo Moreira et al., 2020; Hambali et al., 2021; Hoyos, Faroldi, & Cornaglia, 2019; Hu, Hongmanorom, Galvita, Li, & Kawi, 2020; Hu & Ruckenstein, 2020; Kasim et al., 2020; Kaydouh et al., 2021; Khan et al., 2020; Li, Wang, et al., 2020; Liu et al., 2020; Lyu, Han, et al., 2020; Marin et al., 2020; Mozammel et al., 2020; Myltykbayeva et al., 2020; Nambo et al., 2021; Schiaroli et al., 2020; Shin et al., 2020; Sun, Zhang, Cheng, Liu, & Li, 2021; Wan et al., 2020; Wang et al., 2019, 2020).

The physicochemical properties of the catalyst and its performance are determined by the preparation and activation method of the catalyst. Different preparation methods influence the performance of the catalyst since it determines properties such as dispersion, morphology, metal-support interaction, specific area, etc. The sol-gel, co-precipitation, and impregnation methods are the most reported in the literature, however, the latter hardly allows achieving high dispersions of the metallic particles, and generally, the metal-support interaction is not strong enough to guarantee the desirable requirements for DRM. Hassani Rad et al. prepared Ni/Al_2O_3--CeO_2 and Ni/Al_2O_3-MgO nanocatalysts by impregnation and sol-gel methods to clarify the influence of preparation method and promoter loading on the catalytic performance. The results confirm the reinforced impact of the sol-gel method on NiO dispersion in the presence of support promoters. Besides, it provides the potential of the sol-gel method in the enhancement of metal-support interaction via $NiAl_2O_4$ spinel formation. Sol-gel synthesized Ni/Al_2O_3-CeO_2 nanocatalyst showing the H_2/CO molar ratio close to the unit, the best H_2 yield 94% and stable performance during 1440 min stability test at 1123.15 K. This result revealed that better catalytic performance is obtained by the employment of the sol-gel method and support promotion (Hassani Rad, Haghighi, Alizadeh Eslami, Rahmani, & Rahemi, 2016).

Ni@SBA-15 monoliths with up to 5 wt% of Ni were synthesized by means of easy one-pot sol-gel method by Daoura et al. and characterized by different techniques. After H_2-reduction, those solids exhibited small Ni particles (between 1 and 3 nm) highly dispersed in strong interaction with the silica support. Such confined Ni nanoparticles were shown to be very selective and stable in the dry reforming of methane with much more limited coke formation and sintering (Daoura et al., 2021).

Developing advanced synthesis approaches, such as atomic layer deposition plasma methods, nanocasting, evaporation induced self-assembly, equilibrium deposition filtration, etc., is interesting for the preparation of efficient and active catalysts containing nano-size Ni with well-defined structures or entrapped in different supports (Goula et al., 2015; Joo et al., 2020; Seo et al., 2021; Shah, 2021).

Liang et al. prepared by ALD a stable and active nickel (Ni) nanoparticle catalyst, supported on porous γ-Al_2O_3 particles (Liang, Chen, & Tsai, 2020). The catalyst showed high catalytic activity and long-term stability over 300 h DRM reaction at a temperature range from 973.15 to 1123.15 K. Its long-term stability is attributed to the formation of $NiAl_2O_4$ spinel and high activity was assigned to the effective and uniform dispersion of Ni crystallites by ALD and the reduction of $NiAl_2O_4$ spinel to Ni during the DRM reaction at 1123.15 K (Li et al., 2017). Marinho et al. synthesized a series of Ni-based mesoporous mixed CeO_2-Al_2O_3 oxide catalysts by the one-pot EISA method. The calcined catalysts present mesoporous structures with highly dispersed Ni in the form of $NiAl_2O_4$ spinel clusters with inhibits the sintering process. Additionally, the authors suggested that the presence of Ce interacting strongly with Al_2O_3 promotes the oxygen mobility in the material, responsible for the carbon removal mechanism, and enhances the activity of the catalyst (Marinho et al., 2021).

As can be seen in Table 7.1, many studies show that the addition of one or more metals to nickel-based catalysts has shown good results when compared to monometallic, probably due to the synergistic effect between metals in alloys they could tune the electronic and chemical properties from monometallic ones. Ni is also the most widely used element in the transition metal-based catalysts and has an excellent ability to form bi- and trimetallic systems with other metals.

Gao et al. investigated the catalytic performance of bimetallic Ni-Co catalyst over silica support in DRM. The addition of Co into Ni catalyst was believed to enable the controlling of Ni metal size and increase the interaction of Ni-silica support as well as inhibit the Ni sintering and oxidation through the electron transfer of Co. They concluded that the catalytic activity of this catalyst remained the same over 30h without any carbon deposition or metal sintering and Co takes part in the oxidation of surface carbon (Gao, Tan, Hidajat, & Kawi, 2017). Song et al. utilized the method of evaporation induced self-assembly to synthesize $NiFe/Al_2O_3$ catalysts with high surface area and spinel structure, which resulted in a stable catalyst, whose NiFe alloy, where Fe has a greater affinity for O than Ni, stabilized the Ni^0 metal through the oxidation of Fe^0 to FeO_x, and the FeO reaction with dissociated carbons balanced the deposition and gasification of carbon species during DRM at low temperature (Song et al., 2020).

Ergazieva et al. synthesized $Ni\text{-}Co/\gamma\text{-}Al_2O_3$ bimetallic catalysts were synthesized by capillary impregnation (CI) and solution combustion methods (SC). Synthesis of $Ni\text{-}Co/\gamma\text{-}Al_2O_3$ by SC method leads to the formation of a solid solution of $NiCo_2O_4$, increases the proportion of metal ions (Ni, Co) reduced at temperatures of 673.15–973.15K, uniform distribution of nanosized particles of active phase on the carrier surface. SC provides an efficient way to develop an active catalyst for DRM at low temperatures (873.15–1023.15K) (Ergazieva et al., 2021). This study shows that the preparation method plays an important role in regulating the textural and morphological properties of catalysts and provides a different performance of their activity.

Mozammel et al. have demonstrated that the addition of rhodium and cobalt to the nickel phase supported on mesoporous alumina (MAl) to form bimetallic and trimetallic RhNi/MAl, Ni-Co/MAl, and NiCoRh/MAl catalysts and alloying of Ni was found to enhance the catalyst activity and high resistance toward coking. The high conversions of CH_4 and the nature of carbon formation on these bimetallic catalysts were found to be different from those of the monometallic catalysts which were mainly due to the nature of alloy formation (Mozammel et al., 2020).

As seen above, several researchers reported that the major factors that can contribute to improving the stability of supported nickel catalysts are uniform metal dispersion, such as small size particles, high metal-support interaction, or the addition of a second metal to nickel and base promoted support by making nickel catalysts that are having kinetically higher resistance toward coking, preventing the deposition of inactive carbon and high oxidation tolerance (Mozammel et al., 2020). The effect of the addition of the promoters was discussed in the next section.

Role of promoters

As shown in Table 7.1 it is common to add a small amount of promoting or dopants elements to the catalyst design to improve the catalytic performance and stability of the nonnoble metal catalyst in DRM, particularly the carbon resistance, particles sintering, and catalytic performance. Promoters are responsible for enhancing the chemical properties of the catalyst, providing new active sites of the catalyst, or change textural proprieties. There are two types of promoters, chemical (electronic) and textural (structural) (Gholami, Tišler, & Rubáš, 2020). These promoters have different functions whenever they are used in reactions. Textural promoters are used to modify the textural properties to delay coke formation (Ranjekar & Yadav, 2021; Singh, Nguyen, et al., 2020).

Depending on their properties, the promoters can act, synergistically or not, in different ways impacting the performance of the catalyst (Li, Wang, Zhao, & Hu, 2021; Liqing, Xiangjuan, Hailian, & Xingyuan, 2020; Zhang et al., 2018) by:

- Neutralizing the surface acidity of the catalyst;
- Reducing the activity of methane dehydrogenation;
- Increasing the adsorption amount of CO_2 on the catalyst surface;
- Improving the carbon elimination effect of CO_2;
- Enhancing the oxygen adsorption and dissociation ability;
- Enhancing the interaction between the metal and the support;
- Regulating the electron density of the metal atom;
- Allowing to obtain smaller nickel particles and improving the dispersion of nickel particles.

Alkali metal and alkaline earth metal oxides, rare earth metals oxides, and others like Bi, Fe, Sn, Zr, Mo are commonly used as promoters for DRM catalysts. The lanthanide group metals are utilized on Ni-based catalysts on account of their alkaline properties that favor CO_2 adsorption on the support and great capability of oxygen storage/release, contributing to lower carbon deposition (Liqing et al., 2020). Different nickel-based catalysts deposited on $MgAl_2O_4$ spinel were prepared by Akiki et al. using γ-alumina supplied from different manufacturers (Sigma Aldrich, Alfa Aesar, and Degussa). The authors evaluated, among other factors, the effect of doping nickel with cerium and lanthanum on the performance and stability of materials. The results showed that the catalysts 1.5Ce-$Ni_5/MgAl_2O_4$, synthesized with alumina from Alfa Aesar, exhibited the best catalytic activity and stability for DRM probably due to an effective active oxygen transfer due to the redox properties of CeO_2, leading to the formation of oxygen vacancies (Akiki et al., 2020). The role of oxygen vacancies in DRM reaction will be discussed later in this chapter.

Luisetto et al. investigated nickel supported on metal-doped ceria (Me-DC) catalysts with Me = Zr^{4+}, La^{3+} or Sm^{3+} synthesized by a one-step citric acid method. Zr-doped catalysts that showed small Ni particles gave the highest activity, while Sm and La-doped catalysts with big Ni particles produced the lowest coke formation. These results excluded the dependence of coking on particle size, in this case. Additionally, La-doped CeO_2 had a much higher amount of surface basic sites compared to Sm doped CeO_2 or pure CeO_2, suggesting that the surface basicity by itself cannot explain the differences in carbon formation/removal. Their finds also revealed that extra oxygen vacancies introduced by La^{3+} and Sm^{3+} were not capable to split CO_2. In summary, they suggest the ceria dopants can change the interaction of Ni with the support to explain the formation/removal of carbon on these catalysts (Luisetto, Tuti, Romano, Boaro, & Di Bartolomeo, 2019).

The influences of promoter (Ce, La, Y, and Sm) addition toward physiochemical properties and catalytic performances of Co/mesoporous alumina (MA) catalysts were investigated by Bahari et al. The physiochemical characterization revealed that the large crystallite size resulted from the addition of Ce and La promoters weakened the interaction between Co and MA support while the small crystallite size of the Y and Sm promoters increased Co-MA interaction. The reactant conversions were in the order of YCo/MA > LaCo/MA > CeCo/MA > SmCo/MA > Co/MA, with YCo/MA achieving the highest conversion values at about 85.8% (CH_4) and 92.2% (CO_2). The authors attributed the best performance in DRM of the YCo/MA due to great Co dispersion, small Co particle size with strong Co-MA interaction, and higher oxygen storage capacity compared to other promoted catalysts (Bahari et al., 2020).

Transition metal oxides or mixed oxides are the oxides where the *d* layer is not filled and that have a variety of oxidation states. In the catalytic reaction process, transition metal oxides can produce lattice oxygen, reacting with the deposited carbon, so the nickel-based catalyst doped with transition metal oxides exhibits a good coke resistance (Abdullah, Ainirazali, & Ellapan, 2020; Liqing et al., 2020; Okutan, Arbag, Yasyerli, & Yasyerli, 2020). Additionally, transitions metal oxide can allow to obtain smaller nickel particles and improve the dispersion of nickel particles on the support.

Abdullah et al. synthesized 5 wt% of Ni/SBA-15 catalysis supported with different Zr contents (from 1 to 7 wt%) using a sol-gel process. The catalytic activity test showed that the optimum Zr loading was 1 wt% at which CH_4 and CO_2 conversions were 87.07% and 4.01%. Higher production of H_2 over 1Zr/5Ni-SBA-15 was strongly related to small and well-dispersed Ni species deposited on the SBA-15 surfaces that formed strong Ni active sites interaction with SBA-15 support (Abdullah et al., 2020). Okutan et al. evaluated transitions metal (Zr e Ti) in addition to the Al as promoters. They prepared by a simultaneous impregnation method SBA-15 supported Ni-Al, Ni-Zr, and Ni-Ti catalysts. In addition, Al incorporated SBA-15 (Al-SBA-15) materials were synthesized following a one-pot hydrothermal route in three different conditions: in the presence of only HCl, only NaCl, and both HCl and NaCl. According to activity test results, Al-SBA-15 prepared in presence of HCl e NaCl showed the highest activity in dry reforming of methane. The authors noted that is important to incorporate low concentrations of Al (0.11 wt%) to maintain a low acidity of the support (Okutan et al., 2020).

The addition of alkali metal (Na_2O and K_2O) and alkaline earth metal oxides (MgO and CaO) can neutralize the surface acidity of the catalyst, reduce the activity of methane dehydrogenation, increase the adsorption amount of CO_2 on the catalyst surface, and improve the carbon elimination effect of CO_2 (Al-Fatesh et al., 2020; Aziz et al., 2020; Fertout et al., 2020; Zhang et al., 2018). Al-Fatesh et al. prepared a series of magnesium-promoted NiO supported on mesoporous zirconia, 5Ni/*x*Mg–ZrO_2 ($x=0$, 3, 5, 7 wt%) by wet impregnation method. 5Ni/3Mg–ZrO_2 catalyst showed the highest catalyst stability attributed to the ability of CO_2 to partially oxidize the carbon deposit over the surface of the catalyst. They affirm that NiO–MgO solid solution played a significant role during the DRM (Al-Fatesh et al., 2020). Fertout et al. investigated the effect of using MgO, CaO, and SrO as promoters of Ni/Al_2O_3–La_2O_3 catalysts in DRM. Although the addition of these promoters decreased the surface area, it improved catalytic activity probably because of faster activation of CO_2. The Ni/SrLaAl catalyst presented the best activity and stability, which is explained by the strong basicity of strontium (Fertout et al., 2020). The influence of acidity and basicity on catalyst performance will be discussed in the next section.

As could be seen above, different studies have reported the role of a given promoter and there are still important discrepancies between these studies. This can be attributed to the fact that the influence of a given promoter, is extremely dependent on the active metal species and support involved. These finds revealed the dominant influence of metal and support selection on the overall catalytic performance (Aramouni, Touma, Tarboush, Zeaiter, & Ahmad, 2018).

Acidity and basicity influence

The acidity/basicity of catalyst surface has been found to play a decisive role in the performance of catalysts for DRM and is controlled by the structure and nature of the catalyst, support and promoters used. In DRM, CO_2 and CH_4 are converted to H_2 and CO by adsorption, dissociation, and recombination on a catalytic surface (Scaccia, Della Seta, Mirabile Gattia, & Vanga, 2021; Xingyuan et al., 2020). Within this context, modifying the properties of a catalytic surface can have significant effects on catalytic processes and performance since the acidity or basicity of such a surface affects the electronic density of the active site, and consequently, its ability to adsorb reactants (Scaccia et al., 2021; Xingyuan et al., 2020).

The presence of Lewis acid sites on a catalytic surface reduces the electronic density of the active site and balances the activation of CH_4 (Aziz et al., 2020; Xingyuan et al., 2020), which decreases coke deposition and provides greater stability to the catalyst (Xingyuan et al., 2020). However, carbon deposition from CH_4 decomposition is favored by the acidic sites of support surfaces (Jang, Shim, Kim, Yoo, & Roh, 2019). The basic Lewis sites, on the other hand, gain prominence in the DRM by exerting great influence on catalytic conversions and also in the resistance against coke deposition (Alabi, Wang, Adesanmi, Shakouri, & Hu, 2021; Azancot et al., 2020; Scaccia et al., 2021; Xingyuan et al., 2020).

Abdullah et al. concluded that the basic nature of catalyst enhances the catalytic activity by bringing down the activation energy of reactants. Methane being more stable requires more activation energy than carbon dioxide. Thus, the property of basicity of catalyst alone or synergistically of promoter and support can be used advantageously for overcoming the activation barrier (Abdullah et al., 2020).

Hambali et al. performed a comparative study of the role of different promoters (Ca, Mg, Ta e Ga) in the enhancement of catalytic performance to identify the possible interplay between surface acidity-basicity and coking resistance in DRM. They synthesized fibrous spherical Ni-M/ZSM-5 catalysts by microemulsion method. In conformity with BET, TPR, TPD, and FTIR results, it can be concluded that inherent fibrosity and bimetallic Ni-Ta synergism play a significant role in modifying the catalyst system. The activity follows the order: Ni/ZSM-5 < Ni/FZSM-5 < Ni-Ga/FZSM-5 < Ni-Ca/FZSM5 < Ni-Mg/FZSM-5 < Ni-Ta/FZSM-5. The presence of medium-weak acid sites and bimetallic Ni-Ta synergism amplified the interaction of catalyst components, resulting in improved interaction with the reactants, thus impeding metal sintering and coke deposition. On the other hand, Ca e Mg promoted catalysts exhibited a high concentration of basic sites, leading to strong CO_2 adsorption. Basically, the high basicity of Ca promoted catalyst had a negative influence on activity, leading to deactivation via RWGS side reaction (Hambali et al., 2020).

Looking at the dry reforming reaction of methane as an acid-base interaction followed by catalytic steps based on gas-solid interactions, one will realize that in this context the CO_2 molecules act as acids with respect to the catalyst, and due to this fact, basic properties have been increasingly desired for catalysts used in this reaction (Aziz et al., 2020). Increasing the basicity of supports enhances the adsorption of CO_2, which in turn dissociates into CO and oxygen atoms that react with CH_x intermediates to produce more CO (Azancot et al., 2020; Hambali et al., 2020; Scaccia et al., 2021; Xingyuan et al., 2020). Thus, it is evident that the function of the basic sites is to enhance the activation of CO_2 on the catalyst surface and inhibit carbon deposition on the catalyst (Jang et al., 2019; Scaccia et al., 2021), which in turn also improves catalytic stability (Alabi et al., 2021; Aziz et al., 2020). As options for providing basic sites in DRM, several compounds are used as supports, including monometallic oxides (Aziz et al., 2020; Xingyuan et al., 2020), supported alkali metals (Azancot et al., 2020; Aziz et al., 2020; Jang et al., 2019; Xingyuan et al., 2020), mixed metal oxides (Aziz et al., 2020; Xingyuan et al., 2020), rare earth metal oxides (Xingyuan et al., 2020), zeolites (Aziz et al., 2020), hydrotalcite-type anionic clays (Aziz et al., 2020; Świrk, Rønning, Motak, Grzybek, & Costa, 2020), carbon-based materials, and ion exchange resins (Aziz et al., 2020).

Strategies to increase the basicity of a solid catalyst, the main options are the application of promoters and exposure to pretreatment with H_2. Within this context, the literature mentions promoters such as gallium and boron (Xingyuan et al., 2020), cerium, zirconium, neodymium, and some alkaline earth metals (Azancot et al., 2020; Aziz et al., 2020; Jang et al., 2019; Scaccia et al., 2021), with some emphasis on the application of MgO (Alabi et al., 2021; Azancot et al., 2020; Scaccia et al., 2021; Xingyuan et al., 2020; Yu, Hu, Cui, Cheng, & Zhou, 2021), CaO (Scaccia et al., 2021), and La_2O_3 (Aziz et al., 2020). The basic promoters, in addition to providing more basic sites on the catalytic surface, also act by allowing greater mobility of oxygen molecules on the support (Aziz et al., 2020) in order to contribute directly to the minimization of coke (Scaccia et al., 2021). However, in certain situations, if the basicity of the support is too strong, the adsorption of CO_2 instead of inhibiting it, can lead to coke deposition (Xingyuan et al., 2020), or in other words, it is the medium and weak sites that have a significant contribution to the activation of CO_2 on the catalyst (Aziz et al., 2020). Therefore, to select the best promoter to promote the basicity of the catalytic surface, it is important to note the contribution of each promoter species, because since oxides such as MgO and La_2O_3 bring strong sites, promoters such as Y_2O_3 serve to introduce more weak and medium sites on the surface (Xingyuan et al., 2020). Even CeO_2, a rare earth metal oxide, can decrease the overall basicity of the catalyst. Meanwhile, the medium and strong sites can be enhanced, improving CO_2 adsorption and modifying the catalytic activity (Xingyuan et al., 2020).

Still dealing with strategies to increase the basicity of the catalyst, it is common to be reported in the literature the use of reduction pretreatments (Scaccia et al., 2021). In this type of treatment, the catalyst bed is exposed to the flow of pure H_2 under elevated temperatures, the basic character of the solid catalyst is enhanced through the desorption of water and CO_2 from its surface, and the metal oxide is reduced to metal, which improves the decomposition rate of methane and allows the creation of oxygen vacancies (Aziz et al., 2020). Such oxygen vacancies play a very important role within the DRM, as already mentioned in this chapter.

One of the parallel reactions of the RSM that deserves attention within the context of catalyst basicity is the reverse Boudouard reaction (Eq. 7.8). In this reaction, we have the so-called gasification of the coke deposited on the catalytic surface, as CO_2 reacts with C forming 2CO (Jang et al., 2019), and as expected, such reaction is highly endothermic and only favored above 946.15 K (Aziz et al., 2020). Therefore, to increase the rate of this reaction and favor it kinetically, a great artifice is the basic nature of the catalyst that will induce a higher degree of CO_2 adsorption on its surface (Alabi et al., 2021; Azancot et al., 2020; Aziz et al., 2020; Yu et al., 2021) collaborating also to minimize CH_4 decomposition and/or CO disproportionation (Scaccia et al., 2021). It is worth reinforcing that besides increasing the catalyst lifetime by reducing the amount of coke deposited on its surface (Jang et al., 2019), the favoring of the reverse Boudouard reaction also promotes the production of CO and is, therefore, a key artifice for maintaining the $H_2/CO = 1$ ratio of the final product.

$$C + CO_2 \rightarrow 2CO \quad \Delta H^{\circ}_{298\,k} = 172\,kJ/mol \qquad (7.8)$$

Ibrahim et al. developed a catalyst with 10 wt%. Ni supported on $8\%PO_4 + ZrO_2$, through incipient wet impregnation. Aiming to observe the influence of Ce on the basicity of the catalyst, the addition of Ce at concentrations 1%, 1.5%, 2%, 2.5%, 3%, and 5% were tested, but the

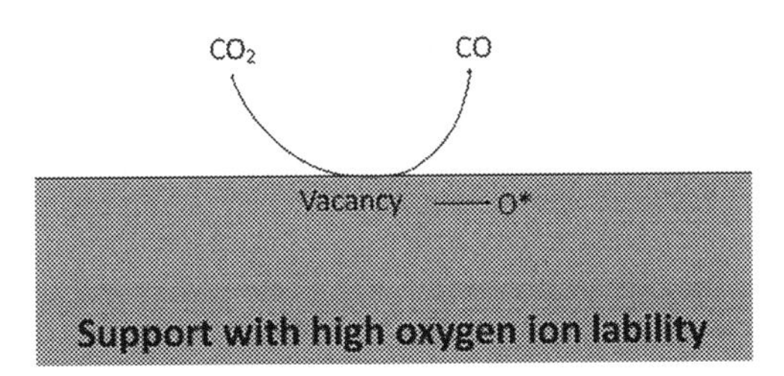

FIG. 7.4 Generation of CO promoted by oxygen vacancy.

concentration with the greatest influence on the basicity of the catalyst was 3% Ce, which allowed them to increase the total basicity of their catalyst by 105.17%, being an addition of 71.30% weak basic sites and 135.76% medium basic sites, without observing strong basic sites. In a DRM reaction at 1073.15K and 1bar, the catalyst with the highest basicity, $10Ni+3\%Ce/8\%PO_4+ZrO_2$, was the one that obtained the highest conversion values of CO_2 (96%) and CH_4 (95%), which allows us to observe the existing synergy between the weak and medium basic sites and the catalytic efficiency (Ibrahim et al., 2020).

In summary, it is important to balance adequately on moderate acid-base sites providing minimum equivalent activation energy for CH_4 and CO_2 toward dry reforming at the least chance of catalyst deactivation. In order to achieve this balance, it is crucial to assess the nature of the active site and the support, types of promoters as well as methods for synthesis and activation of catalysts.

Oxygen vacancies

As previously discussed, carbon deposition is one of the major problems associated with DRM because it reduces the useful life of catalysts and makes the process less efficient. According to Al-Fatesh et al., there are three main ways to combat the deposition of coke in the DRM (Al-Fatesh et al., 2021), these being:

- Injecting O_2 into the reactor supply chain;
- Regenerate the deactivated catalysts in the reactor bed, extra-situ, after the reaction cycle;
- Develop a catalyst that is able to gasify, in situ, the accumulated carbon simultaneously with its appearance.

As we are dealing with the development of catalysts in this topic, we will limit ourselves to the third option, which leads us to talk about oxygen vacancies, a crucial device in improving the catalytic performance in DRM reactions.

Oxygen vacancies, in addition to being essential for the removal of carbon deposits, which leads to the generation of CO (Fig. 7.4) (Yentekakis et al., 2019), also play a vital role in the activation of reactive gases (Sun, Zhang, et al., 2020; Sun et al., 2020), which kinetically favors DRM, and the activation of hydrogen (Bhattar, Abedin, Kanitkar, & Spivey, 2020; Bhattar, Abedin, Shekhawat, Haynes, & Spivey, 2020), which contributes to the reduction of oxidized metals present on the catalytic surface, providing activity and stability to the catalyst (Al-Fatesh et al., 2021). Metal oxides such as Al_2O_3, SiO_2, CaO, and MgO, have good oxygen mobility and redox attributes (Chong, Cheng, et al., 2020; Chong, Setiabudi, & Jalil, 2020). Akiki et al. emphasize the formation of oxygen vacancies when using $Ni/MgAl_2O_4$ due to the incorporation of nickel in the spinel structure (Akiki et al., 2020). It is also reported a great prominence for metals in the group of lanthanides such as La, Ce, Y, and Sm that are extensively used in nickel-based catalysts due to their ability to store/release oxygen (Bahari et al., 2020).

Several works (Bahari et al., 2020; Boaro, Colussi, & Trovarelli, 2019; Gomez, Lopez-

Martin, Ramire, Gascon, & Caballero, 2020; Guo et al., 2020; Hu et al., 2020; Kasim et al., 2020) bring the functionality added by the cerium within the heterogeneous catalysis of DRM, either as a support or as a promoter, where, it is possible to highlight the versatility made possible by the design of the catalyst from the presence of redox and acid-base supplied by this rare earth metal. In DRM, the addition of cerium to nickel-based catalysts significantly increases the reaction rate due to the high oxygen storage capacity and mutual conversion of Ni^{2+}/Ni^{0} e Ce^{3+}/Ce^{4+} species (Al-Fatesh et al., 2021; Boaro et al., 2019; Sun, Zhang, et al., 2020; Sun et al., 2020; Yentekakis et al., 2019). Thus, on the catalytic surface, the reduced species favor the adsorption of CO_2 and its decomposition (Bhattar, Abedin, Kanitkar, et al., 2020; Bhattar, Abedin, Shekhawat, et al., 2020; Yentekakis et al., 2019), while the presence of oxygen vacancies promotes the formation and dissociation of reaction intermediates, also helping in the process of decoking and prevention sintering of the active site (Sun, Zhang, et al., 2020; Sun et al., 2020). This process of activating CO_2 on the catalytic surface is addressed in the literature taking into account two main routes, the formate route ($HCOO^-$), which applies to acid supports such as Al_2O_3, and the carbonate route (CO_3^{2-}), which it applies to basic supports such as CeO_2 and La_2O_3 (Boaro et al., 2019; Singh, Nguyen, et al., 2020).

Taking into account that oxygen vacancies are not favored by the high temperatures associated with DRM, and therefore, carbon deposition rates can be even higher making the gasification of coke an even greater challenge, it is interesting that there is a sequential generation of oxygen vacancies, contributing to a longer period of the useful life of the catalyst (Al-Fatesh et al., 2021). Within this context, ZrO_2 appears as an extremely important oxide in DRM, having excellent thermal stability, medium basicity and acidity, and oxygen mobility (Ibrahim et al., 2020). The reason for such importance attributed to this oxide is also due to the improvement of the properties of CeO_2 since the addition of ZrO_2 in CeO_2 improves redox properties, oxygen storage capacity, and especially, thermal resistance (Al-Fatesh et al., 2021; Hongmanorom et al., 2020). Due to this synergy, the combined use of Ce and Zr in DRM is extensively reported in the literature (Al-Doghachi et al., 2020; Al-Najar et al., 2020; Araiza, Arcos, et al., 2021; Azancot et al., 2020; Bekheet et al., 2021; Bian, Zhong, et al., 2021; Hambali et al., 2021; Hoyos et al., 2019; Hu et al., 2020; Hu & Ruckenstein, 2020; Kasim et al., 2020; Kaydouh et al., 2021; Khan et al., 2020; Li, Wang, et al., 2020; Liu et al., 2020; Lyu, Han, et al., 2020; Marin et al., 2020; de Araujo Moreira et al., 2020; Mozammel et al., 2020; Myltykbayeva et al., 2020; Nambo et al., 2021; Schiaroli et al., 2020; Shin et al., 2020; Sun et al., 2021; Wan et al., 2020; Wang et al., 2019, 2020). Al Fatesh et al. presents a very interesting application, where the addition of ZrO_2 to CeO_2 yielded excellent thermal stability and strong metal-support interaction through the formation of solid fluorite solutions (CeO_2-ZrO_2), still allowing the addition of tungsten to the catalyst that provided additional active sites, thus forming a thermally stable, bifunctional Ni-Ce/W-Zr catalyst and rich in oxygen vacancies (Al-Fatesh et al., 2021).

Cases, where the addition of promoters served to assist the existing synergy between Ce and Zr in DRM, are also reported by Simonov et al. and Dan Guo et al. (Guo et al., 2020; Simonov et al., 2020). Simonov et al. observed that when incorporating Nb and Ti to the Ceria-Zirconia catalyst supporting Ni, the amount of oxygen vacancies in the catalytic structure was increased, which also increased stability since greater oxygen mobility was crucial for the coke gasification. Differently from Simonov et al., Dan Guo et al. report the role of promoters in the formation of oxygen vacancies by causing extrinsic defects by incorporating lower valence ions such as Ni^{2+}, Pr^{3+}, Sm^{3+}, and Gd^{3+} or smaller ions such as Si^{4+} and Zr^{4+} to the Si^{4+} and Zr^{4+} crystalline network.

Supports

The supported catalysts have different characteristics and properties, having, above all, different performances in DRM with regard to activity and stability. Such difference can be attributed to the different characteristics of the support and its adsorption/desorption capacity of the intermediate species, which is closely associated with the breaking or not of the energy barriers existing in the main elementary stages of the dry methane reform reaction (Chen, Yin, et al., 2020; Chen, Zaffran, & Yang, 2020).

Marinho et al. point out that mesoporous materials, which have a high surface area and well-defined pores, have shown good results with regard to the dispersion of metals, which ends up being interesting for the dry reforming of methane given the recurrent problems associated with the sintering of active sites already mentioned above (Marinho et al., 2021). Shi et al. also emphasize that ordered mesoporous materials are able to control nanoparticles very well through the effect of pore confinement, another attractive factor for DRM (Shi et al., 2021). In this context, a mesoporous material whose use is widely reported in the literature is SBA-15 (Abdulrasheed et al., 2019; Ranjekar & Yadav, 2021; Singh, Dhir, Mohapatra, & Mahla, 2020). SBA-15 is mesoporous silica, which has a high surface area (600–1000 m^2g^{-1}), 2d molecular sieves with hexagonal symmetry P6mm (Singh, Kumar, Setiabudi, Nanda, & Vo, 2018), combining micropores (0.5–3 nm) and mesopores (4–14 nm) (Bian, Zhong, et al., 2021)in hexagonal cylindrical channels (5–10 nm) (Singh et al., 2018) made up of thick silica walls (3–6 nm) (Meynen, Cool, & Vansant, 2009; Singh et al., 2018) that give this material good hydrothermal stability (Meynen et al., 2009; Singh et al., 2018; Tian, Ma, Bu, Yuan, & Zhao, 2018) and the ability to encapsulate metal nanoparticles (Singh et al., 2018; Tian et al., 2018). However, for the active site to be placed inside the mesoporous channels, the existing techniques are based on impregnation, precipitation, and hydrothermal methods, making the synthesis process end up accumulating a series of steps involving calcination, which implies a better distribution of the nanoparticles on the external and internal surfaces of the support, than inside the mesoporous channels (Tian et al., 2018). Singh et al. also highlight that the rational design of SBA-15 supports may be the future of reform reactions, since by having micropores in the silica walls, the support channels are interconnected, favoring mass transfer with minimal limitations and exercising kinetic control over the activity and stability of the catalyst, which is crucial for reform reactions (Singh et al., 2018).

In addition to SBA-15, another mesoporous silica-based structure that can serve as catalytic support in reform reactions is MCM-41 (Singh et al., 2018), although the number of publications reporting its use is limited (Cakiryilmaz, Arbag, Oktar, Dogu, & Dogu, 2019). In comparison with MCM-41, SBA-15 has thicker walls, wider mesopores, and a greater variety of pore types (Singh et al., 2018), giving more interesting properties that justify its greater use in DRM. However, it is worth noting that both are usable because they have parallel mesopores with uniform diameters (Arbag, Yasyerli, Yasyerli, & Dogu, 2010), which gives them greater resistance against the undesirable deactivation by coke deposition than conventional microporous materials (Cakiryilmaz et al., 2019; Yasyerli, Filizgok, Arbag, Yasyerli, & Dogu, 2011). In fact, Cakiryilmaz et al. emphasize that the discovery of such mesoporous silicas opened space for the development of a range of new catalysts, capable of showing less resistance to the transport of reagents to active sites (Cakiryilmaz et al., 2019).

MCM-41 is considered a simple and easy to prepare nanomaterial, consisting of an amorphous aluminum/metallosilicate structure with hexagonal pores (Meynen et al., 2009), with a high surface area (1200 $m^2 g^{-1}$), bulky pores, uniform (Aguiar et al., 2019; Meynen et al., 2009), diameters between 1.5 and 20 nm, and thin walls with thicknesses between 1 and 1.5 nm, which ends up giving less chemical and hydrothermal stability to this silica mesoporous (Meynen et al., 2009) which on the other hand also has good adsorption capacity (Aguiar et al., 2019). Aguiar et al. still highlight Si-MCM-41 as an attractive material as catalytic support in dry methane reforming also since it does not have acidic characteristics, which makes it less crucial to add promoters to increase basic sites, something that is recurrent in catalyst design, as we saw in previous topics (Aguiar et al., 2019).

Wang et al. compared the use of SBA-15 and MCM-41, both supporting $LaNiO_3$. $LaNiO_3$/MCM-41 showed a higher initial catalytic activity, due to the high dispersion of Ni, while $LaNiO_3$/SBA-15 obtained superior stability, due to the strong confinement effect of the support, avoiding agglomeration of Ni particles (Wang, Yu, Wang, Chu, & Liu, 2013). Regarding the resistance of the supports during the DRM reaction, Wang et al. found that the hexagonal pores of $LaNiO_3$/SBA-15 remained intact after the reaction cycle, while the mesoporous structure of $LaNiO_3$/MCM-41 collapsed during the reaction, resulting in the conjugation of metallic particles. As for the conversion of CH_4 and CO_2, both catalysts showed conversions that were severely similar, around 70% and 75%, respectively (Wang et al., 2013).

Alumina (Al_2O_3) is a support widely used industrially and of great applicability in reform reactions (Al-Doghachi et al., 2020; Al-Najar et al., 2020; Araiza, Arcos, et al., 2021; Azancot et al., 2020; Bekheet et al., 2021; Bian, Zhong, et al., 2021; Hambali et al., 2021; Hoyos et al., 2019; Hu et al., 2020; Hu & Ruckenstein, 2020; Kasim et al., 2020; Kaydouh et al., 2021; Khan et al., 2020; Li, Wang, et al., 2020; Liu et al., 2020; Lyu, Han, et al., 2020; Marin et al., 2020; de Araujo Moreira et al., 2020; Mozammel et al., 2020; Myltykbayeva et al., 2020; Nambo et al., 2021; Schiaroli et al., 2020; Shin et al., 2020; Sun et al., 2021; Wan et al., 2020; Wang et al., 2019, 2020), given its large surface area (Li, Pei, et al., 2020; Singh, Dhir, et al., 2020; Song et al., 2020), thermal stability (de la Cruz-Flores, Martinez-Hernandez, & Gracia-Pinilla, 2020; Li, Pei, et al., 2020), porosity (Song et al., 2020), strong metal-support interaction (Li, Pei, et al., 2020), high dispersion of the active site (de la Cruz-Flores et al., 2020; Li, Pei, et al., 2020; Marinho et al., 2021) and high catalytic activity (Li, Pei, et al., 2020). Liang et al. also point out that Al_2O_3 has the ability to disperse Ni crystallites evenly at the nanoscale, which contributes to the prevention of sintering and helps in the removal of moss-like carbon from the surface of the active site (Liang et al., 2020). Despite the great advantages attributed to this support, due to the fact that it has an acid character (de la Cruz-Flores et al., 2020), it cannot provide basic sites for the activation of CO_2 (Li, Pei, et al., 2020), which ends up being crucial for the performance in DRM and for the useful life of the catalyst, as stated in previous topics.

The acid character of alumina is explained by Aguiar et al. that attribute this characteristic to the presence of hydroxyl groups (–OH) or water molecules on the surface in coordination with aluminum ions and oxygen, constituting the behavior of Brønsted acids (Aguiar et al., 2019). Aguiar et al. also explain that when –OH groups are desorbed, Lewis acids are formed, responsible for increasing the catalytic activity of alumina, and, especially in DRM, this favors adverse reactions, which implies the formation of coke (Aguiar et al., 2019). Therefore, in addition to the addition of alkaline earth promoters such as MgO (Singh, Dhir, et al., 2020) and rare earth metals such as La_2O_3 and CeO_2 (Li, Pei, et al., 2020) to increase basicity (Aguiar et al., 2019), the stability of alumina against deposition of coke also depends on the structure (Singh, Dhir, et al., 2020), calcination parameters (Li, Pei, et al., 2020; Singh, Dhir, et al., 2020) and synthesis methods (Singh, Dhir, et al., 2020).

The spinel oxides have an $A^{2+}B_2^{3+}O_4^{2-}$ structure which can be, for example, $MgAl_2O_4$, $NiFe_2O_4$, or $NiAl_2O_4$ (Azancot et al., 2020; Schiaroli et al., 2020), having excellent stability at high temperatures and other very interesting characteristics for DRM (Khalighi, Bahadoran, Panjeshahi, Zamaniyan, & Tahouni, 2020; Shi et al., 2021; Song et al., 2020). Singh et al. noted that the appropriate addition of magnesium to alumina, resulted in the formation of spinel $MgAl_2O_4$, and this led to an increase in catalytic activity accompanied by increased basicity and improved resistance against coke deposition (Singh, Dhir, et al., 2020). Khalighi et al. define cobalt aluminate spinel, $CoAl_2O_4$ as a promising catalyst in DRM for having a very high melting point, good thermal resistance, and high mechanical resistance at high temperatures, in addition to providing high surface area, smaller crystallite sizes, high porosity, and cobalt, an active site comparable to nickel, composing its structure in a dispersed way (Khalighi et al., 2020). In fact, a very important parameter in the design of catalysts for DRM, and already mentioned earlier in this chapter, is the guarantee of a good dispersion of the active site to avoid sintering it. In this context, the possibility of maintaining a transition metal such as Ni, Co, or Cu inside a structure in a dispersed form, even under exposure to severe oxidation conditions and high temperatures, is an advantage presented by structured oxides, as is the case of perovskite oxides, spinel-type oxides, and oxides derived from hydrotalcite (Khalighi et al., 2020).

Perovskite oxides are a class of crystalline oxides that receive this name in reference to the mineral $CaTiO_3$ (Bhattar, Abedin, Shekhawat, et al., 2020). These oxides have a structure of type $A^{2+}B^{3+}O_3^{2-}$ (Al-Doghachi et al., 2020; Al-Najar et al., 2020; Araiza, Arcos, et al., 2021; Azancot et al., 2020; Bekheet et al., 2021; Bian, Zhong, et al., 2021; Hambali et al., 2021; Hoyos et al., 2019; Hu et al., 2020; Hu & Ruckenstein, 2020; Kasim et al., 2020; Kaydouh et al., 2021; Khan et al., 2020; Li, Wang, et al., 2020; Liu et al., 2020; Lyu, Han, et al., 2020; Marin et al., 2020; de Araujo Moreira et al., 2020; Mozammel et al., 2020; Myltykbayeva et al., 2020; Nambo et al., 2021; Schiaroli et al., 2020; Shin et al., 2020; Sun et al., 2021; Wan et al., 2020; Wang et al., 2019, 2020), which gives this type of oxide a remarkable dispersion of the active metal (Shi et al., 2021). Position A can be occupied by a rare earth metal (Bhattar, Abedin, Kanitkar, et al., 2020; Kryuchkova et al., 2020; Mousavi, Nakhaei Pour, Gholizadeh, Mohammadi, & KamaliShahri, 2020; Xianglei, Shen, Baoyi, & Laihong, 2021), alkaline or alkaline earth (Bhattar, Abedin, Kanitkar, et al., 2020; Mousavi et al., 2020; Xianglei et al., 2021), while position B is occupied by a transition metal (Bhattar, Abedin, Kanitkar, et al., 2020; Kryuchkova et al., 2020; Mousavi et al., 2020; Xianglei et al., 2021). Bhattar et al. emphasize that there are still perovskite-like oxides, which have $A_2^{2+}B^{3+}O_4^{2-}$ structure with alternating layers of $A^{2+}B^{3+}O_3^{2-}$ and $A^{2+}O^{2-}$(Bhattar, Abedin, Kanitkar, et al., 2020).

The activity and/or stability of this type of oxide is closely associated with the choice of the metal that will compose its structure (Mousavi et al., 2020), with site B being considered the most relevant because it is the active metal in DRM (Xianglei et al., 2021). In addition to the high dispersion of the active site, oxides structured as perovskites are attractive for DRM because they have abundant oxygen vacancies (Xianglei et al., 2021), excellent thermal stability (Kryuchkova et al., 2020; Xianglei et al., 2021), and high catalytic activity (Kryuchkova et al., 2020). Kryuchkova et al. and Aramouni et al. still highlight perovskites as economically more effective compared to conventional supports and their respective catalytic performances (Aramouni et al., 2018; Kryuchkova et al., 2020).

Something interesting about perovskites is the fact that their catalytic characteristics can be adjusted through partial replacement of sites A and B (Kryuchkova et al., 2020) and also through doping with other metals (Xianglei et al., 2021). When metal A is doped, with a low valence cation, oxygen vacancies are generated and some metals B change the valence to maintain the electronic neutrality of the structure (Xianglei et al., 2021). In contrast, the substitution of metal B influences the chemical properties of perovskite as activity and redox capacity (Kryuchkova et al., 2020; Xianglei et al., 2021).

Looking more specifically at the DRM context, Kim et al. assert that the modification of site A influences the improvement of oxygen mobility and the ability of the support to adsorb CO_2, while modifications at site B, with transition metals, are responsible for potentiating the adsorption and activation of methane (Kim et al., 2019). An example reported by Mousavi et al. illustrates this peculiarity of the perovskite, when reporting the catalytic improvement of the perovskite $LaNiO_3$, through the induction of defects and oxygen mobility in the network, caused by substitutions in positions A and B with cations of different valences (Mousavi et al., 2020). $LaNiO_3$, which by the way, is also pointed out by Kim et al. as a promising catalyst in DRM due to its high initial activity, but despite this, it presents very low stability due to the carbon deposition during the reaction (Kim et al., 2019). As a way to solve this critical disadvantage, Kim et al. performed the doping of the catalyst with Mn, concluding that such a promoter improved the stability of the catalyst by mediating the interaction between active site and support, where, Co was also added as an active component to increase the reaction rates (Kim et al., 2019). Endorsing this report, Kryuchkova et al. points out that the introduction of cobalt at the B site of a perovskite crystalline structure leads to increased catalytic stability in DRM, and the partial replacement of Fe, Zr, Mn, and Cu by Ni, leads to changes in catalytic activity and suppression of coke deposit where the deposition of transition metal oxides on the perovskite surface is used as a substrate, serving to increase catalytic activity (Kryuchkova et al., 2020).

The layered double hydroxides (LDH), commonly called hydrotalcites, make up a class of anionic clays that have a 2D lamellar structure similar to brucite $[Mg(OH)_2]$. Thus, the structure of a hydrotalcite can be described as layers positively charged and compensated with anions such as CO_3^{2-}, NO_3^-, Cl^-, or OH^- in the interlayered spaces so that the structure of the hydrotalcite can be summarized using the formula $[A^{2+}_{1-x}B^{3+}_x(OH)_2]^{x+}[Anion^{n-}_{x/n}]^{x-}\cdot mH_2O$ where x is the molar ratio $B^{3+}/(A^{2+}+B^{3+})$ (Abdelsadek et al., 2021; Świrk et al., 2020). Within the context of the DRM reaction, hydrotalcites serve as precursors, which when calcined give rise to mixed oxides (Świrk et al., 2020; Tibra, Deepa, Selvakannan, Sadasivuni, & Bhargava, 2021).

The thermal decomposition of hydrotalcites is, in fact, crucial for catalytic activity and begins in the temperature range between 523.15 and 673.15 K, through dehydration, dehydroxylation, and loss of interlamellar anions (Abdelsadek et al., 2021). Abdelsadek et al., even report that when calcining a hydrotalcite Ni-Mg-Al around 723.15 K, with dehydration, dehydroxylation, and decarbonation, it was noted that this type of heat treatment facilitated the formation of mesoporous metal oxides, with high surface areas and hindered the formation of spinels, which normally have smaller surface areas due to high calcination temperatures. Abdelsadek et al. also emphasize that the thermal stability of hydrotalcites depends on the synthesis method, A^{2+}/B^{3+} ratio, and the nature of metals A^{2+} and B^{3+}, parameters that can lead to rehydration and reconstitution of the hydrotalcite structure (Abdelsadek et al., 2021).

In general, oxides derived from hydrotalcite have small particle sizes, high surface areas (Abdelsadek et al., 2021), mesoporosity, homogeneous dispersion of metals (Abdelsadek et al., 2021; Tibra et al., 2021), redox properties

(Świrk et al., 2020), thermal resistance (Bhattar, Abedin, Shekhawat, et al., 2020), resistance to sintering (Abdelsadek et al., 2021; Bhattar, Abedin, Kanitkar, et al., 2020) and resistance to carbon deposition (Bhattar, Abedin, Kanitkar, et al., 2020), compared to usual supported catalysts. In addition, mixed oxides derived from hydrotalcite have highly basic properties, which leads to a strong affinity for acidic molecules such as CO_2 (Abdelsadek et al., 2021; Cunha et al., 2020; Świrk et al., 2020; Xingyuan et al., 2020), which means, as explained in previous topics, a favorable aspect for the DRM reaction.

The importance of oxides derived from hydrotalcite is evidenced by the example reported by Xu et al., where $2Ni/Al_2O_3$ catalysts were tested, under the same reaction conditions, one of which was prepared by the conventional impregnation method and the other obtained by thermal decomposition of a Ni-Al hydrotalcite. It was observed, with this comparison, that when using the hydrotalcite-derived catalyst, there was a 15% increase in CH_4 conversion and a 75% decrease in carbon deposition (Xu et al., 2021), which highlights the superiority of hydrotalcite-derived supports in DRM.

Structured oxides such as SiO_2, CaO, La_2O_3, MgO, CeO_2, and ZrO_2, are also highly reported in DRM (Al-Doghachi et al., 2020; Al-Najar et al., 2020; Araiza, Arcos, et al., 2021; Azancot et al., 2020; Bekheet et al., 2021; Bian, Zhong, et al., 2021; Hambali et al., 2021; Hoyos et al., 2019; Hu et al., 2020; Hu & Ruckenstein, 2020; Kasim et al., 2020; Kaydouh et al., 2021; Khan et al., 2020; Li, Wang, et al., 2020; Liu et al., 2020; Lyu, Han, et al., 2020; Marin et al., 2020; de Araujo Moreira et al., 2020; Mozammel et al., 2020; Myltykbayeva et al., 2020; Nambo et al., 2021; Schiaroli et al., 2020; Shin et al., 2020; Sun et al., 2021; Wan et al., 2020; Wang et al., 2019, 2020) because in some cases they provide high surface area and porosity for dispersion of the active metals, emphasizing that the physical-chemical properties of such oxides can affect the catalytic performance (Shi et al., 2021). An oxide with distinct characteristics in the context of DRM is CeO_2 (Araiza, Arcos, et al., 2021). Araiza et al. point out that the use of this oxide as support is accompanied by the advantage of improving catalytic stability due to the high oxygen storage capacity caused by the Ce^{4+}/Ce^{3+} pair, and it is a great challenge in the preparation of such material to obtain a high surface area since this is not a specific characteristic of this oxide, it can have a negative impact on the dispersion of the metal, making it a tendency to use noble metals as the active site of this support (Araiza, Arcos, et al., 2021).

A relatively recent but promising support in DRM (Li, Yuan, et al., 2020), hydroxyapatite $Ca_{10}(PO_4)_6(OH)_2$ (Boukha, Yeste, Cauqui, & González-Velasco, 2019; Rego de Vasconcelos et al., 2020a, 2020b) has attractive properties such as high porosity, high surface area (Tran, Pham Minh, Phan, Pham, & Nguyen Xuan, 2020), low water solubility (Li, Yuan, et al., 2020.), high thermal stability (Li, Yuan, et al., 2020), low sensitivity to water formed during DRM, and mainly, adjustable basicity, which facilitates the adsorption of CO_2 (Li, Yuan, et al., 2020). Some characteristics of this material are considered adjustable due to the fact that the variation of the molar Ca/P ratio at the moment of synthesis can lead to the formation of additional phases and cause changes in textural aspects and acid-base properties (Boukha et al., 2019; Rego de Vasconcelos et al., 2020a, 2020b; Tran et al., 2020), causing this to be yet another distinct hydoxyapatite peculiarity.

The pore size distribution of hydroxyapatite varies between 3 and 100 nm, the control of this distribution is one of the keys to the catalytic efficiency of this material in DRM (Li, Yuan, et al., 2020). Hydroxyapatite provides strong oxidative metal-support interactions at high temperatures due to the loss of OH groups on the surface (Boukha et al., 2020), being so thermally stable that its decomposition only occurs above 1273.15 K, and below 973.15 K no sintering is observed (Rego de Vasconcelos et al., 2020a,

2020b). Another notable feature of hydroxyapatite is the ability to exchange cations and anions, which highlights its surface properties.

Rego de Vasconcelos et al. tested hydroxyapatite impregnated with 5wt% Ni and noticed that calcination at 1473.15K for 5h made the material less active due to reduced surface area and basicity. However, this heat treatment allowed the prevention of high rates of deactivation by sintering and provided excellent catalytic stability. The rate of deactivation for the reaction time of 50h was 0.17% h^{-1} and for 300h it was 0.07% h^{-1} with CH_4 and CO_2 conversions around 70% and 75%, respectively (Li, Yuan, et al., 2020; Rego de Vasconcelos et al., 2020b; Tran et al., 2020).

Deactivation and regeneration

Within the context of heterogeneous catalysis, of reactions involving gas-solid interactions, as in the case of dry methane reform, inevitably, the catalyst will suffer a reduction in catalytic activity and consequently deactivation. Such a phenomenon is categorized by Zhou et al. as six main forms of occurrence (Zhou et al., 2020):

- Poisoning by chemisorption of certain molecules in the active sites;
- Encrustation by coke deposition;
- Thermal degradation;
- Leaching or vapor formation that results in transport components of the catalytic surface;
- Modification of the catalyst by solid-vapor or solid-solid reactions;
- Friction and/or crushing.

Zhang et al. point out that the main form of catalytic deactivation in dry methane reform is the encapsulation of the catalyst by coke, emphasizing that the rate of catalytic deactivation in methane reform is strongly influenced by the amount of accumulated coke, that is, how much the more coke is deposited, the faster the catalyst deactivates (Zhang, Wang, Song, & Zhang, 2020). Hambali et al. explain that the usual severe conditions in dry methane reform raise the molecular energy for cleavage of C—H bonds, which makes coke deposition inevitable in this reaction (Hambali et al., 2021). Abdelsadek et al. and Hambali et al. bring in their work concepts of how active site blocking occurs in nickel catalysts in dry methane reform (Abdelsadek et al., 2021; Hambali et al., 2021). According to Hambali et al., the reaction first occurs with the decomposition of CH_4 at the active site, producing H_2 and CH_x species, while $O_{adsorbed}$ species released by the activation of CO_2 oxidize the CH_x species to form CO and H_2. This leads to encapsulation by coke, according to Hambali et al., is the adsorption of carbonaceous species, which over time become polymerized carbons, which diffuse at the C/Ni interface, causing carbon filaments that in turn cause the mechanical blocking of the active site (Hambali et al., 2021). Abdelsadek et al. bases the encapsulation in three stages, where first the nucleation of the carbon molecules occurs in the Ni nanoparticle, then the growth of a carbon nanotube occurs and finally, the encapsulation is concluded with the agglomeration of these nanotubes, blocking the active site and removing its adsorption capacity from the reagents (Abdelsadek et al., 2021).

As previously mentioned in this chapter, coke deposition can be controlled through the catalyst design, taking into account factors such as surface acidity-basicity, reducibility (Hambali et al., 2021), particle diameter (Abdelsadek et al., 2021; Hambali et al., 2021), dispersion of the active site (Abdelsadek et al., 2021) and metal-support interaction (Hambali et al., 2021). However, despite the alternatives provided within the catalyst design, regeneration techniques are ceaselessly used in dry methane reform with the aim of reversing catalytic deactivation and prolonging the life of the catalyst, therefore, the regeneration capacity of a catalyst has enormous importance in dry methane reform (Chong, Setiabudi, et al., 2020).

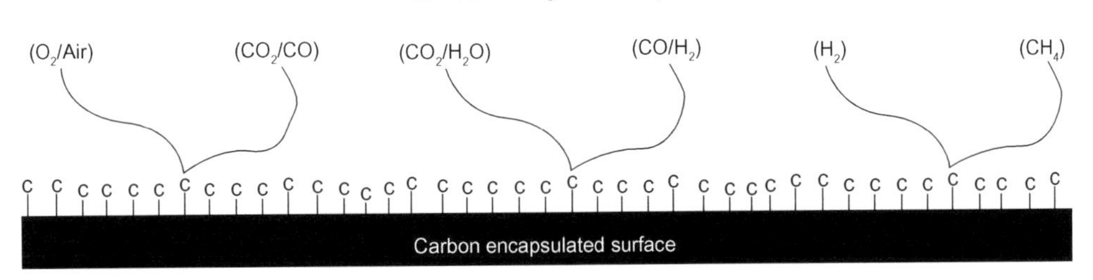

FIG. 7.5 The main regeneration methods applied in DRM: oxidation, gasification, and hydrogenation.

The regenerative processes aim to eliminate the coke formed in the pores and on the surface of the catalyst, preserving its structure. The three main regenerative processes in DRM are shown in Fig. 7.5. These processes receive great attention due to the economic advantages that exist in regenerating a catalyst instead of purchasing a new one. In addition to these advantages, regenerating a catalyst instead of disposing of it as waste with components that are toxic to the environment, sounds like a more environmentally friendly option (Chong, Cheng, et al., 2020; Rego de Vasconcelos, Pham Minh, Sharrock, & Nzihou, 2018). Among the main methods of regeneration can include oxidation, gasification, and hydrogenation. Each of these methods has its advantages and disadvantages, depending on the type of catalyst, deactivation mechanisms, and conditions for regeneration (Zhou et al., 2020).

Oxidation regeneration is the most used in the industrial context, where coke reacts with an oxidizing agent such as O_2, air, and NO_x, regenerating the spent catalyst (Chong, Cheng, et al., 2020; Zhou et al., 2020), and forming products such as CO, CO_2, H_2O, and NO_x (Zhou et al., 2020). As it is less flammable and more available than other oxidizing agents, the air is usually used more in this process (Chong, Cheng, et al., 2020; Rego de Vasconcelos et al., 2018). On the other hand, oxidation regeneration also faces challenges, of which we can highlight the high exothermicity of combustion of coke, which can cause hot spots and/or local gradients of high temperature, and with this, thermally degrade the catalyst, and the possibility of changing the characteristic of the residual coke from aliphatic to aromatic, which can make regeneration more complicated (Zhou et al., 2020). Some methods can be used to optimize oxidation regeneration, such as using O_3 as an oxidizing agent, something that allows the process to be carried out at low temperatures, and extractions with solvents or supercritical fluids to remove the coke from the catalyst pores (Zhou et al., 2020).

Although coke oxidation is more industrially applied, coke can also be eliminated by gasification (Rego de Vasconcelos et al., 2018), a method where coke reacts with steam or CO_2 (Zhou et al., 2020), with CO_2 being considered a more interesting agent since it promotes the highly endothermic reverse Boudouard reaction, where the coke is carbonated giving way to CO (Chong, Cheng, et al., 2020). Compared with air oxidation regeneration processes, which emit about 40%–45% of the CO_2 emitted in refineries, in gasification regeneration processes, whether with H_2O or CO_2, the main product is synthesis gas, which is more environmentally advantageous (Zhou et al., 2020). Although CO_2 can react with coke at high temperatures, for the sake of thermal stability, gasification at low temperatures is desirable, with the CO_2 gasification rate being elevated by the addition of promoters (Zhou et al., 2020), which reinforces, within the context of the design of catalyst, the need to match catalysts with metals that facilitate catalytic regenerability.

Chong et al. studied regeneration cycles of the Ni/dendritic fibrous catalyst SBA-15 by oxidation with air and by gasification with CO_2, which occurred after 30 h of dry reform reaction, with an activation step with H_2 between the cycles. When regenerating with CO_2, Chong et al. observed that the initial conversions of CO_2 and CH_4 fell from 87.64% and 89.61% to 77.16% and 76.30%, in the first cycle of regeneration, and in the second cycle, they fell from 83.64% and 82.69% to 68.11% and 72.28%. When the regeneration was oxidative, there was a drop in initial conversions in the first cycle of 90.42% and 93.55% to 85.12% and 87.01%, and in the second cycle, they fell from 90.42% and 88.37% to 84.88% and 82.63%. This result allowed Chong et al. to conclude that air regeneration was more effective in removing coke, implying that after the gasification cycles, relatively less efficient, more carbon molecules remained blocking the contact between active sites and reagents, reducing catalytic activity and directly affecting conversion percentages (Chong, Cheng, et al., 2020).

An example that shows how catalytic regeneration should be rationally chosen according to the catalytic structure is reported by Zhang et al. where after performing the regeneration of the Ni/ZrO_2 catalyst with CO_2, the initial CH_4 conversion was noted to be higher after the regeneration than before the regeneration, while the initial CO_2 conversion followed the opposite script. The explanation given by Zhang et al. is that gasification with CO_2 in Ni/ZrO_2 removed the deposited coke but promoted the formation of portions of oxygen in the form of carbonate species. Therefore, when the mixture of reagents was introduced, the carbonate intermediate first reacted with CH_x species derived from CH_4 forming CO, implying that the initial CH_4 conversion was promoted by these oxygen portions, while the initial CO_2 conversion was suppressed due to the covering of the active sites by carbonate (Zhang, Wang, et al., 2020).

Hydrogenation is a regenerative, nonoxidative treatment, where carbon reacts with H_2 to form methane. This form of catalytic regeneration is considered less efficient in removing coke than the other processes already mentioned. Zhou et al. define regeneration with H_2 as a time and energy-consuming process, classifying the relative rates of coke removal from regenerative processes at 1073.15 K in the order $O_2 > H_2O > CO_2 > H_2$ (Zhou et al., 2020).

In addition to the rational choice of the regeneration method, methods of synthesis and addition of promoters are crucial for catalytic regenerability and for increasing resistance against deactivation. Because of this, numerous experimental and theoretical works are reported in the literature, with the purpose of investigating the effects of the modification of catalysts in prolonging the catalytic activity in the dry reform of methane and making them more productive. Świrk et al. tested in the dry reform reaction of methane, Ni-based oxides, HDL derivatives of Mg/Al, adding Zr (5 wt%) and Y (0.4 wt%) by co-precipitation, and, for effect comparison, Y (0.2, 0.4, 0.6 wt%) by impregnation. The catalyst that showed the greatest methane conversion was Ni-Zr- (Mg/Al) synthesized by co-precipitation, and a reduction in CH_4 conversion was observed as more Y was added to the catalyst, which was justified by the partial obstruction of available Ni sites. The reactivity of the carbon formed during DRM was determined using H_2-TPSR, where it was noted that the addition of yttrium contributed significantly to the hydrogenation of the coke, and, consequently, to the regeneration of the catalyst. It was also shown that there was no correlation between the amount of methane produced and the amount of coke formed, which indicates that a certain amount of coke is not reactive (Świrk et al., 2020).

Lyu et al. synthesized catalysts produced by different methods, such as dry impregnation, strong electrostatic adsorption, co-precipitation, and combustion. In this work, all catalysts suffered from the nanoparticles of Ni nanoparticles while they were in reactive conditions of dry methane reform, which shows, in this case, that the efforts to promote the dispersion of nickel were not enough. However, it was possible to notice that the formation of coke is irreversibly related to the concentration of active oxygen on the surface of the $Ce_xZr_{1-x}O_2$ support. Lyu et al. also concluded that the dominant deactivation mechanism for catalysts synthesized by the combustion method is the encapsulation of Ni particles by the support itself. Such a distinct deactivation mechanism was attributed to the peculiarly porous structure formed during the combustion synthesis and also to the annealing of the Ce-Zr crystallites, reinforcing once again that the catalytic deactivation is also related to the structure and consequently to the synthesis method of the catalyst (Lyu, Jocz, Xu, Stavitski, & Sievers, 2020).

Chen et al. point out that cobalt catalysts tend to be deactivated mainly by oxidation and carbon deposition, emphasizing that the deactivation mechanism is caused according to the metal concentration in the catalyst. That is, with greater loads of metal, deactivation is observed due to coke deposition, and with lower loads, deactivation is observed due to the oxidation of Co. In this theoretical work, Chen et al. also emphasize that to balance activity and stability in DRM catalysts, the addition of a dopant is productive if it increases the carbon adsorption energy, so that the CH_4 dissociation is the limiting step to improve the carbon resistance so that reducing the CH_4 and CO_2 dissociation barriers result in maintaining good activity. With regard to deactivation by oxidation, Chen et al. recommend forming an alloy with another metal or adding a dopant in order to regulate the adsorption of O on the metal surface (Chen, Zaffran, & Yang, 2020).

7.3 Final considerations

However, for economic and environmental reasons, it is extremely important to overcome these challenges and spread the benefits offered by this reaction, which converts greenhouse gases into synthetic gas. In this context, the design of a rational catalyst becomes crucial for DRM, considering thermal stability, catalytic activity, and resistance over deactivation. The main strategies are to combine different metals as assets or promoters with different supports in a strategic way. To achieve good viability for DRM industrially, it needs to put efforts into studies focused on its improvement, which largely depends on the design of the catalysts.

References

Abbasi, S., Abbasi, M., Tabkhi, F., & Akhlaghi, B. (2020). Syngas production plus reducing carbon dioxide emission using dry reforming of methane: Utilizing low-cost Ni-based catalysts. *Oil and Gas Science and Technology, 75*(10). https://doi.org/10.2516/ogst/2020016.

Abdelsadek, Z., Holgado, J. P., Halliche, D., Caballero, A., Cherifi, O., Gonzalez-Cortes, S., et al. (2021). Examination of the deactivation cycle of NiAl- and NiMgAl-hydrotalcite derived catalysts in the dry reforming of methane. *Catalysis Letters*. https://doi.org/10.1007/s10562-020-03513-4.

Abdullah, N., Ainirazali, N., & Ellapan, H. (2020). Structural effect of Ni/SBA-15 by Zr promoter for H2 production via methane dry reforming. *International Journal of Hydrogen Energy*. https://doi.org/10.1016/j.ijhydene.2020.07.060.

Abdulrasheed, A., Jalil, A. A., Gambo, Y., Ibrahim, M., Hambali, H. U., & Shahul Hamid, M. Y. (2019). A review on catalyst development for dry reforming of methane to syngas: Recent advances. *Renewable and Sustainable Energy Reviews, 108*, 175–193. https://doi.org/10.1016/j.rser.2019.03.054.

Aguiar, M., Cazula, B. B., SaragiottoColpini, L. M., Borba, C. E., Alves da Silva, F., Noronha, F. B., et al. (2019). Si-MCM-41 obtained from different sources of silica and its application as support for nickel catalysts used in dry reforming of methane. *International Journal of Hydrogen Energy, 44*(60), 32003–32018. https://doi.org/10.1016/j.ijhydene.2019.10.118.

Akiki, E., Akiki, D., Italiano, C., Vita, A., Abbas-Ghaleb, R., Chlala, D., et al. (2020). Production of hydrogen by methane dry reforming: A study on the effect of cerium and lanthanum on Ni/$MgAl_2O_4$ catalyst performance. *International Journal of Hydrogen Energy, 45*(41), 21392–21408. https://doi.org/10.1016/j.ijhydene.2020.05.221.

Alabi, W. O., Wang, H., Adesanmi, B. M., Shakouri, M., & Hu, Y. (2021). Support composition effect on the structures, metallic sites formation, and performance of Ni-

Co-Mg-Al-O composite for CO_2 reforming of CH_4. *Journal of CO_2 Utilization, 43*. https://doi.org/10.1016/j.jcou.2020.101355, 101355.

Al-Doghachi, F. A. J., Jassim, A. F. A., & Taufiq-Yap, Y. H. (2020). Enhancement of CO_2 reforming of CH_4 reaction using Ni,Pd,Pt/$Mg_{1-x}Ce_x^{4+}O$ and Ni/$Mg_{1-x}Ce_x^{4+}O$ catalysts. *Catalysts, 10*(11), 1–24. https://doi.org/10.3390/catal10111240.

Al-Fatesh, A. S., Arafat, Y., Kasim, S. O., Ibrahim, A. A., Abasaeed, A. E., & Fakeeha, A. H. (2021). In situ auto-gasification of coke deposits over a novel Ni-Ce/W-Zr catalyst by sequential generation of oxygen vacancies for remarkably stable syngas production via CO_2-reforming of methane. *Applied Catalysis B: Environmental, 280*. https://doi.org/10.1016/j.apcatb.2020.119445.

Al-Fatesh, A. S., Kumar, R., Fakeeha, A. H., Kasim, S. O., Khatri, J., Ibrahim, A. A., et al. (2020). Promotional effect of magnesium oxide for a stable nickel-based catalyst in dry reforming of methane. *Scientific Reports, 10*(1). https://doi.org/10.1038/s41598-020-70930-1.

Al-Najar, A. M. A., Al-Doghachi, F. A. J., Al-Riyahee, A. A. A., & Taufiq-Yap, Y. H. (2020). Effect of La_2O_3 as a promoter on the Pt,Pd,Ni/MgO catalyst in dry reforming of methane reaction. *Catalysts, 10*(7). https://doi.org/10.3390/catal10070750.

Angeli, A., Gossler, S., Lichtenberg, S., Agrawal, A., Valerius, M., & Kinzel, K. (2021). Reduction of CO_2 emission from off-gases of steel industry by dry reforming of methane. *Angewandte Chemie, 60*(21), 11852–11857.

Araiza, D. G., Arcos, D. G., Gómez-Cortés, A., & Díaz, G. (2021). Dry reforming of methane over Pt-Ni/CeO_2 catalysts: Effect of the metal composition on the stability. *Catalysis Today*, 46–54. https://doi.org/10.1016/j.cattod.2019.06.018.

Araiza, D. G., González-Vigi, F., Gómez-Cortés, A., Arenas-Alatorre, J., & Díaz, G. (2021). Pt-based catalysts in the dry reforming of methane: Effect of support and metal precursor on the catalytic stability. *Journal of the Mexican Chemical Society, 65*(1), 1–19. https://doi.org/10.29356/jmcs.v65i1.1262.

Aramouni, N. A. K., Touma, J. G., Tarboush, B. A., Zeaiter, J., & Ahmad, M. N. (2018). Catalyst design for dry reforming of methane: Analysis review. *Renewable and Sustainable Energy Reviews, 82*, 2570–2585. https://doi.org/10.1016/j.rser.2017.09.076.

Arbag, H., Yasyerli, S., Yasyerli, N., & Dogu, G. (2010). Activity and stability enhancement of Ni-MCM-41 catalysts by Rh incorporation for hydrogen from dry reforming of methane. *International Journal of Hydrogen Energy, 35*(6), 2296–2304. https://doi.org/10.1016/j.ijhydene.2009.12.109.

Azancot, L., Bobadilla, L., Centeno, M., & Odriozola, J. A. (2020). IR spectroscopic insights into the coking-resistance effect of potassium on nickel-based catalyst during dry reforming of methane. *Applied Catalysis B: Environmental, 285*, 119822. https://doi.org/10.1016/j.apcatb.2020.119822.

Azancot, L., et al. (2021). IR spectroscopic insights into the coking-resistance effect of potassium on nickel-based catalyst during dry reforming of methane. *Applied Catalysis B: Environmental, 285*(2020), 119822.

Aziz, M. A. A., Jalil, A. A., Wongsakulphasatch, S., & Vo, D. V. N. (2020). Understanding the role of surface basic sites of catalysts in CO_2 activation in dry reforming of methane: A short review. *Catalysis Science and Technology, 10*(1), 35–45. https://doi.org/10.1039/c9cy01519a.

Bahari, M. B., Setiabudi, H. D., Duy Nguyen, T., Phuong, P. T. T., Duc Truong, Q., Abdul Jalil, A., et al. (2020). Insight into the influence of rare-earth promoter (CeO_2, La_2O_3, Y_2O_3, and Sm_2O_3) addition toward methane dry reforming over Co/mesoporous alumina catalysts. *Chemical Engineering Science, 228*. https://doi.org/10.1016/j.ces.2020.115967.

Beheshti Askari, A., Al Samarai, M., Hiraoka, N., Ishii, H., Tillmann, L., Muhler, M., et al. (2020). In situ X-ray emission and high-resolution X-ray absorption spectroscopy applied to Ni-based bimetallic dry methane reforming catalysts. *Nanoscale, 12*(28), 15185–15192. https://doi.org/10.1039/d0nr01960g.

Bekheet, M. F., Delir Kheyrollahi Nezhad, P., Bonmassar, N., Schlicker, L., Gili, A., Praetz, S., et al. (2021). Steering the methane dry reforming reactivity of Ni/La_2O_3 catalysts by controlled in situ decomposition of doped La_2NiO_4 precursor structures. *ACS Catalysis, 11*(1), 43–59. https://doi.org/10.1021/acscatal.0c04290.

Bhattar, S., Abedin, M. A., Kanitkar, S., & Spivey, J. J. (2020). A review on dry reforming of methane over perovskite derived catalysts. *Catalysis Today*. https://doi.org/10.1016/j.cattod.2020.10.041.

Bhattar, S., Abedin, M. A., Shekhawat, D., Haynes, D. J., & Spivey, J. J. (2020). The effect of La substitution by Sr- and Ca- in Ni substituted Lanthanum Zirconate pyrochlore catalysts for dry reforming of methane. *Applied Catalysis A: General, 602*. https://doi.org/10.1016/j.apcata.2020.117721.

Bian, Z., Zhong, W., Yu, Y., Wang, Z., Jiang, B., & Kawi, S. (2021). Dry reforming of methane on Ni/mesoporous-Al_2O_3 catalysts: Effect of calcination temperature. *International Journal of Hydrogen Energy, 46*, 31041–31053. https://doi.org/10.1016/j.ijhydene.2020.12.064.

Boaro, M., Colussi, S., & Trovarelli, A. (2019). Ceria-based materials in hydrogenation and reforming reactions for CO_2 valorization. *Frontiers in Chemistry, 7*. https://doi.org/10.3389/fchem.2019.00028.

Boukha, Z., Choya, A., Cortés-Reyes, M., de Rivas, B., Alemany, L. J., González-Velasco, J. R., et al. (2020). Influence of the calcination temperature on the activity of hydroxyapatite-supported palladium catalyst in the methane oxidation reaction. *Applied Catalysis B: Environmental, 277*. https://doi.org/10.1016/j.apcatb.2020.119280.

Boukha, Z., Yeste, M. P., Cauqui, M.Á., & González-Velasco, J. R. (2019). Influence of Ca/P ratio on the catalytic performance of Ni/hydroxyapatite samples in dry reforming of methane. *Applied Catalysis A: General, 580*, 34–45. https://doi.org/10.1016/j.apcata.2019.04.034.

Cakiryilmaz, N., Arbag, H., Oktar, N., Dogu, G., & Dogu, T. (2019). Catalytic performances of Ni and Cu impregnated MCM-41 and Zr-MCM-41 for hydrogen production through steam reforming of acetic acid. *Catalysis Today, 323*, 191–199. https://doi.org/10.1016/j.cattod.2018.06.004.

Cao, P., Adegbite, S., Zhao, H., Lester, E., & Wu, T. (2018). Tuning dry reforming of methane for F-T syntheses: A thermodynamic approach. *Applied Energy, 227*, 190–197. https://doi.org/10.1016/j.apenergy.2017.08.007.

Chen, X., Yin, L., Long, K., Sun, H., Sun, M., Wang, H., et al. (2020). The reconstruction of Ni particles on SBA-15 by thermal activation for dry reforming of methane with excellent resistant to carbon deposition. *Journal of the Energy Institute, 93*(6), 2255–2263. https://doi.org/10.1016/j.joei.2020.06.008.

Chen, S., Zaffran, J., & Yang, B. (2020). Dry reforming of methane over the cobalt catalyst: Theoretical insights into the reaction kinetics and mechanism for catalyst deactivation. *Applied Catalysis B: Environmental, 270*. https://doi.org/10.1016/j.apcatb.2020.118859.

Chong, C. C., Cheng, Y. W., Setiabudi, H. D., Ainirazali, N., Vo, D. V. N., & Abdullah, B. (2020). Dry reforming of methane over Ni/dendritic fibrous SBA-15 (Ni/DFSBA-15): Optimization, mechanism, and regeneration studies. *International Journal of Hydrogen Energy, 45*(15), 8507–8525. https://doi.org/10.1016/j.ijhydene.2020.01.056.

Chong, C. C., Setiabudi, H. D., & Jalil, A. A. (2020). Dendritic fibrous SBA-15 supported nickel (Ni/DFSBA-15): A sustainable catalyst for hydrogen production. *International Journal of Hydrogen Energy, 45*(36), 18533–18548. https://doi.org/10.1016/j.ijhydene.2019.05.034.

Cunha, A. F., Morales-Torres, S., Pastrana-Martínez, L. M., Martins, A. A., Mata, T. M., Caetano, N. S., et al. (2020). Syngas production by bi-reforming methane on an Ni-K-promoted catalyst using hydrotalcites and filamentous carbon as a support material. *RSC Advances, 10*(36), 21158–21173. https://doi.org/10.1039/d0ra03264f.

Daoura, O., Fornasieri, G., Boutros, M., El Hassan, N., Beaunier, P., Thomas, C., et al. (2021). One-pot prepared mesoporous silica SBA-15-like monoliths with embedded Ni particles as selective and stable catalysts for methane dry reforming. *Applied Catalysis B: Environmental, 280*. https://doi.org/10.1016/j.apcatb.2020.119417.

de Araujo Moreira, T. G., de Carvalho Filho, J. F. S., Carvalho, Y., de Almeida, J. M. A. R., Romano, P. N., & Sousa-Aguiar, E. F. (2020). Highly stable low noble metal content rhodium-based catalyst for the dry reforming of methane. *Fuel, 287*, 119536. https://doi.org/10.1016/j.fuel.2020.119536.

de Dios García, I., Stankiewicz, A., & Nigar, H. (2021). Syngas production via microwave-assisted dry reforming of methane. *Catalysis Today, 362*, 72–80. https://doi.org/10.1016/j.cattod.2020.04.045.

de la Cruz-Flores, V. G., Martinez-Hernandez, A., & Gracia-Pinilla, M. A. (2020). Deactivation of Ni-SiO_2 catalysts that are synthetized via a modified direct synthesis method during the dry reforming of methane. *Applied Catalysis A: General, 594*. https://doi.org/10.1016/j.apcata.2020.117455.

Ergazieva, G. E., Telbayeva, M. M., Popova, A. N., Ismagilov, Z. R., Dossumov, K., Myltykbayeva, L. K., et al. (2021). Effect of preparation method on the activity of bimetallic Ni-Co/Al_2O_3 catalysts for dry reforming of methane. *Chemical Papers*. https://doi.org/10.1007/s11696-021-01516-y.

Fertout, R. I., Ghelamallah, M., Helamallah, M., Kacimi, S., López, P. N., & Corberán, V. C. (2020). Nickel supported on alkaline earth metal–doped γ-Al_2O_3–La_2O_3 as catalysts for dry reforming of methane. *Russian Journal of Applied Chemistry, 93*(2), 289–298. https://doi.org/10.1134/S1070427220020196.

Gao, X., Tan, Z., Hidajat, K., & Kawi, S. (2017). Highly reactive Ni-Co/SiO_2 bimetallic catalyst via complexation with oleylamine/oleic acid organic pair for dry reforming of methane. *Catalysis Today, 281*, 250–258. https://doi.org/10.1016/j.cattod.2016.07.013.

Gholami, Z., Tišler, Z., & Rubáš, V. (2020). Recent advances in Fischer-Tropsch synthesis using cobalt-based catalysts: A review on supports, promoters, and reactors. *Catalysis Reviews—Science and Engineering*. https://doi.org/10.1080/01614940.2020.1762367.

Gomez, R., Lopez-Martin, A., Ramirez, A., Gascon, J., & Caballero, A. (2020). Elucidating the promotional effect of cerium in the dry reforming of methane. *ChemCatChem*, 553–563. https://doi.org/10.1002/cctc.202001527.

Goula, M. A., Charisiou, N. D., Papageridis, K. N., Delimitis, A., Pachatouridou, E., & Iliopoulou, E. F. (2015). Nickel on alumina catalysts for the production of hydrogen rich mixtures via the biogas dry reforming reaction: Influence of the synthesis method. *International Journal of Hydrogen Energy, 40*(30), 9183–9200. https://doi.org/10.1016/j.ijhydene.2015.05.129.

Guo, D., Lu, Y., Ruan, Y., Zhao, Y., Zhao, Y., Wang, S., et al. (2020). Effects of extrinsic defects originating from the interfacial reaction of CeO_{2-x}-nickel silicate on catalytic performance in methane dry reforming. *Applied Catalysis B: Environmental, 277.* https://doi.org/10.1016/j.apcatb.2020.119278.

Hambali, H. U., Jalil, A. A., Abdulrasheed, A. A., Siang, T. J., Abdullah, T. A. T., Ahmad, A., et al. (2020). Fibrous spherical Ni-M/ZSM-5 (M: Mg, Ca, Ta, Ga) catalysts for methane dry reforming: The interplay between surface acidity-basicity and coking resistance. *International Journal of Energy Research, 44*(7), 5696–5712. https://doi.org/10.1002/er.5327.

Hambali, H. U., Jalil, A. A., Abdulrasheed, A. A., Siang, T. J., Owgi, A. H. K., & Aziz, F. F. A. (2021). CO_2 reforming of methane over Ta-promoted Ni/ZSM-5 fibre-like catalyst: Insights on deactivation behavior and optimization using response surface methodology (RSM). *Chemical Engineering Science, 231.* https://doi.org/10.1016/j.ces.2020.116320, 116320.

Hassani Rad, S. J., Haghighi, M., Alizadeh Eslami, A., Rahmani, F., & Rahemi, N. (2016). Sol-gel vs. impregnation preparation of MgO and CeO_2 doped Ni/Al_2O_3 nanocatalysts used in dry reforming of methane: Effect of process conditions, synthesis method and support composition. *International Journal of Hydrogen Energy, 41*(11), 5335–5350. https://doi.org/10.1016/j.ijhydene.2016.02.002.

Hongmanorom, P., Ashok, J., Das, S., Dewangan, N., Bian, Z., Mitchell, G., et al. (2020). Zr–Ce-incorporated Ni/SBA-15 catalyst for high-temperature water gas shift reaction: Methane suppression by incorporated Zr and Ce. *Journal of Catalysis, 387*, 47–61. https://doi.org/10.1016/j.jcat.2019.11.042.

Hoyos, L. A. S., Faroldi, B. M., & Cornaglia, L. M. (2019). A coke-resistant catalyst for the dry reforming of methane based on Ni nanoparticles confined within rice husk-derived mesoporous materials. *Catalysis Communications, 135*, 105898. https://doi.org/10.1016/j.catcom.2019.105898.

Hu, J., Hongmanorom, P., Galvita, V. V., Li, Z., & Kawi, S. (2020). Bifunctional Ni-Ca based material for integrated CO_2 capture and conversion via calcium-looping dry reforming. *Applied Catalysis B: Environmental, 284*, 119734. https://doi.org/10.1016/j.apcatb.2020.119734.

Hu, Y. H., & Ruckenstein, E. (2020). Comment on "Dry reforming of methane by stable Ni–Mo nanocatalysts on single-crystalline MgO". *Science, 368*(6492). https://doi.org/10.1126/science.abb5459.

Hu, J., et al. (2021). Bifunctional Ni-Ca based material for integrated CO_2 capture and conversion via calcium-looping dry reforming. *Applied Catalysis B: Environmental, 284*(2020), 119734.

Ibrahim, A. A., Al-Fatesh, A. S., Kumar, N. S., Abasaeed, A. E., Kasim, S. O., & Fakeeha, A. H. (2020). Dry reforming of methane using ce-modified ni supported on $8\%PO_4 + ZrO_2$ catalysts. *Catalysts, 10*(2). https://doi.org/10.3390/catal10020242.

Jang, W. J., Shim, J. O., Kim, H. M., Yoo, S. Y., & Roh, H. S. (2019). A review on dry reforming of methane in aspect of catalytic properties. *Catalysis Today*, 15–26. https://doi.org/10.1016/j.cattod.2018.07.032.

Joo, S., Seong, A., Kwon, O., Kim, K., Lee, J. H., Gorte, R. J., et al. (2020). Highly active dry methane reforming catalysts with boosted in situ grown Ni-Fe nanoparticles on perovskite via atomic layer deposition. *Science Advances, 6*(35). https://doi.org/10.1126/sciadv.abb1573.

Kasim, S. O., Al-Fatesh, A. S., Ibrahim, A. A., Kumar, R., Abasaeed, A. E., & Fakeeha, A. H. (2020). Impact of Ce-loading on Ni-catalyst supported over $La_2O_3+ZrO_2$ in methane reforming with CO_2. *International Journal of Hydrogen Energy, 45*(58), 33343–33351. https://doi.org/10.1016/j.ijhydene.2020.08.289.

Kaydouh, M. N., El Hassan, N., Davidson, A., & Massiani, P. (2021). Optimization of synthesis conditions of Ni/SbA-15 catalysts: Confined nanoparticles and improved stability in dry reforming of methane. *Catalysts, 11*(1), 1–17. https://doi.org/10.3390/catal11010044.

Khalighi, R., Bahadoran, F., Panjeshahi, M. H., Zamaniyan, A., & Tahouni, N. (2020). High catalytic activity and stability of $X/CoAl_2O_4$ (X = Ni, Co, Rh, Ru) catalysts with no observable coke formation applied in the autothermal dry reforming of methane lined on cordierite monolith reactors. *Microporous and Mesoporous Materials, 305.* https://doi.org/10.1016/j.micromeso.2020.110371.

Khan, I. S., Ramirez, A., Shterk, G., Garzón-Tovar, L., & Gascon, J. (2020). Bimetallic metal-organic framework mediated synthesis of Ni-Co catalysts for the dry reforming of methane. *Catalysts, 10*(5). https://doi.org/10.3390/catal10050592.

Kim, W. Y., Jang, J. S., Ra, E. C., Kim, K. Y., Kim, E. H., & Lee, J. S. (2019). Reduced perovskite $LaNiO_3$ catalysts modified with Co and Mn for low coke formation in dry reforming of methane. *Applied Catalysis A: General, 575*, 198–203. https://doi.org/10.1016/j.apcata.2019.02.029.

Kryuchkova, T. A., Kost'. V. V., Sheshko, T. F., Chislova, I. V., Yafarova, L. V., & Zvereva, I. A. (2020). Effect of cobalt in $GdFeO_3$ catalyst systems on their activity in the dry reforming of methane to synthesis gas. *Petroleum Chemistry, 60*(5), 609–615. https://doi.org/10.1134/S0965544120050059.

Li, X., Li, D., Tian, H., Zeng, L., Zhao, Z. J., & Gong, J. (2017). Dry reforming of methane over Ni/La_2O_3 nanorod catalysts with stabilized Ni nanoparticles. *Applied Catalysis B: Environmental, 202*, 683–694. https://doi.org/10.1016/j.apcatb.2016.09.071.

Li, K., Pei, C., Li, X., Chen, S., Zhang, X., Liu, R., et al. (2020). Dry reforming of methane over $La_2O_2CO_3$-modified Ni/ Al_2O_3 catalysts with moderate metal support interaction. *Applied Catalysis B: Environmental, 264*. https://doi.org/10.1016/j.apcatb.2019.118448.

Li, Y., Wang, Z., Zhang, B., Liu, Z., & Yang, T. (2020). Dry reforming of methane (Drm) by highly active and stable ni nanoparticles on renewable porous carbons. *Catalysts, 10*(5). https://doi.org/10.3390/catal10050501.

Li, L., Wang, Y., Zhao, Q., & Hu, C. (2021). The effect of Si on CO_2 methanation over Ni-xSi/ZrO_2 catalysts at low temperature. *Catalysts, 11*(1), 1–14. https://doi.org/10.3390/catal11010067.

Li, B., Yuan, X., Li, B., & Wang, X. (2020). Impact of pore structure on hydroxyapatite supported nickel catalysts (Ni/HAP) for dry reforming of methane. *Fuel Processing Technology, 202*, 106359. https://doi.org/10.1016/j.fuproc.2020.106359.

Liang, T. Y., Chen, H. H., & Tsai, D. H. (2020). Nickel hybrid nanoparticle decorating on alumina nanoparticle cluster for synergistic catalysis of methane dry reforming. *Fuel Processing Technology, 201*. https://doi.org/10.1016/j.fuproc.2020.106335.

Liqing, W., Xiangjuan, X., Hailian, R., & Xingyuan, G. (2020). A short review on nickel-based catalysts in dry reforming of methane: Influences of oxygen defects on anti-coking property. *Materials Today: Proceedings*. https://doi.org/10.1016/j.matpr.2020.10.697.

Liu, X., Yan, J., Mao, J., He, D., Yang, S., Mei, Y., et al. (2020). Inhibitor, co-catalyst, or intermetallic promoter? Probing the sulfur-tolerance of MoOx surface decoration on Ni/ SiO_2 during methane dry reforming. *Applied Surface Science, 548*, 149231. https://doi.org/10.1016/j.apsusc.2021.149231.

Liu, X., et al. (2021). Inhibitor, co-catalyst, or intermetallic promoter? Probing the sulfur-tolerance of MoOx surface decoration on Ni/SiO_2 during methane dry reforming. *Applied Surface Science, 548*(2020), 149231.

Luisetto, I., Tuti, S., Romano, C., Boaro, M., & Di Bartolomeo, E. (2019). Dry reforming of methane over Ni supported on doped CeO_2: New insight on the role of dopants for CO_2 activation. *Journal of CO_2 Utilization, 30*, 63–78. https://doi.org/10.1016/j.jcou.2019.01.006.

Lyu, L., Han, Y., Ma, Q., Makpal, S., Sun, J., Gao, X., et al. (2020). Fabrication of Ni-based bimodal porous catalyst for dry reforming of methane. *Catalysts, 10*(10), 1–15. https://doi.org/10.3390/catal10101220.

Lyu, Y., Jocz, J., Xu, R., Stavitski, E., & Sievers, C. (2020). Nickel speciation and methane dry reforming performance of Ni/$Ce_xZr_{1-x}O_2$ prepared by different synthesis methods. *ACS Catalysis, 10*(19), 11235–11252. https://doi.org/10.1021/acscatal.0c02426.

Marin, C., Popczun, E., Nguyen-Phan, T., Tafen, D., Alfonso, D., Waluyo, I., et al. (2020). Designing perovskite catalysts for controlled active-site exsolution in the microwave dry reforming of methane. *Applied Catalysis B: Environmental, 284*, 119711. https://doi.org/10.1016/j.apcatb.2020.119711.

Marin, C. M., et al. (2021). Designing perovskite catalysts for controlled active-site exsolution in the microwave dry reforming of methane. *Applied Catalysis B: Environmental, 284*(2020), 119711.

Marinho, A. L. A., Toniolo, F. S., Noronha, F. B., Epron, F., Duprez, D., & Bion, N. (2021). Highly active and stable Ni dispersed on mesoporous CeO_2-Al_2O_3 catalysts for production of syngas by dry reforming of methane. *Applied Catalysis B: Environmental, 281*. https://doi.org/10.1016/j.apcatb.2020.119459.

Mark, M. F., Mark, F., & Maier, W. F. (1997). Reaction kinetics of the CO_2 reforming of methane. *Chemical Engineering and Technology, 20*(6), 361–370. https://doi.org/10.1002/ceat.270200602.

Meynen, V., Cool, P., & Vansant, E. F. (2009). Verified syntheses of mesoporous materials. *Microporous and Mesoporous Materials, 125*(3), 170–223. https://doi.org/10.1016/j.micromeso.2009.03.046.

Mourhly, A., Kacimi, M., Halim, M., & Arsalane, S. (2020). New low cost mesoporous silica (MSN) as a promising support of Ni-catalysts for high-hydrogen generation via dry reforming of methane (DRM). *International Journal of Hydrogen Energy, 45*(20), 11449–11459. https://doi.org/10.1016/j.ijhydene.2018.05.093.

Mousavi, M., Nakhaei Pour, A., Gholizadeh, M., Mohammadi, A., & KamaliShahri, S. M. (2020). Dry reforming of methane by $La_{0.5}Sr_{0.5}NiO_3$ perovskite oxides: Influence of preparation method on performance and structural features of the catalysts. *Journal of Chemical Technology and Biotechnology, 95*(11), 2911–2920. https://doi.org/10.1002/jctb.6451.

Mozammel, T., Dumbre, D., Hubesch, R., Yadav, G. D., Selvakannan, P. R., & Bhargava, S. K. (2020). Carbon dioxide reforming of methane over mesoporous alumina supported Ni(Co), Ni(Rh) bimetallic, and Ni(CORh) trimetallic catalysts: Role of nanoalloying in improving the stability and nature of coking. *Energy and Fuels, 34*(12), 16433–16444. https://doi.org/10.1021/acs.energyfuels.0c03249.

Myltykbayeva, L. K., Ergazieva, G. E., Telbayeva, M. M., Ismagilov, Z. R., Dossumov, K., Popova, N., et al. (2020). Effect of cobalt oxide content on the activity of NiO-CO_2O_3/γ-Al_2O_3 catalyst in the reaction of dry reforming of methane to synthesis gas. *Eurasian Chemico-Technological Journal, 22*(3), 187–195. https://doi.org/10.18321/ectj978.

Nambo, A., Atla, V., Vasireddy, S., Kumar, V., Jasinski, J. B., Upadhyayula, S., et al. (2021). Nanowire-based materials as coke-resistant catalyst supports for dry methane reforming. *Catalysts, 11*(2), 1–13. https://doi.org/10.3390/catal11020175.

Okutan, C., Arbag, H., Yasyerli, N., & Yasyerli, S. (2020). Catalytic activity of SBA-15 supported Ni catalyst in CH_4 dry reforming: Effect of Al, Zr, and Ti co-impregnation and Al incorporation to SBA-15. *International Journal of Hydrogen Energy, 45*(27), 13911–13928. https://doi.org/10.1016/j.ijhydene.2020.03.052.

Ranjekar, A. M., & Yadav, G. D. (2021). Dry reforming of methane for syngas production: A review and assessment of catalyst development and efficacy. *Journal of the Indian Chemical Society, 98*(1). https://doi.org/10.1016/j.jics.2021.100002, 100002.

Ray, K., Bhardwaj, R., Singh, B., & Deo, G. (2018). Developing descriptors for CO_2 methanation and CO_2 reforming of CH_4 over Al_2O_3 supported Ni and low-cost Ni based alloy catalysts. *Physical Chemistry Chemical Physics, 20*(23), 15939–15950. https://doi.org/10.1039/c8cp01859f.

Rego de Vasconcelos, B., Pham Minh, D., Martins, E., Germeau, A., Sharrock, P., & Nzihou, A. (2020a). A comparative study of hydroxyapatite- and alumina-based catalysts in dry reforming of methane. *Chemical Engineering and Technology, 43*(4), 698–704. https://doi.org/10.1002/ceat.201900461.

Rego de Vasconcelos, B., Pham Minh, D., Martins, E., Germeau, A., Sharrock, P., & Nzihou, A. (2020b). Highly-efficient hydroxyapatite-supported nickel catalysts for dry reforming of methane. *International Journal of Hydrogen Energy, 45*(36), 18502–18518. https://doi.org/10.1016/j.ijhydene.2019.08.068.

Rego de Vasconcelos, B., Pham Minh, D., Sharrock, P., & Nzihou, A. (2018). Regeneration study of Ni/hydroxyapatite spent catalyst from dry reforming. *Catalysis Today, 310*, 107–115. https://doi.org/10.1016/j.cattod.2017.05.092.

Saelee, T., Lerdpongsiripaisarn, M., Rittiruam, M., Somdee, S., Liu, A., Praserthdam, S., et al. (2021). Experimental and computational investigation on underlying factors promoting high coke resistance in NiCo bimetallic catalysts during dry reforming of methane. *Scientific Reports, 11*(1). https://doi.org/10.1038/s41598-020-80287-0.

Salazar Hoyos, L. A., Faroldi, B. M., & Cornaglia, L. M. (2020). A coke-resistant catalyst for the dry reforming of methane based on Ni nanoparticles confined within rice husk-derived mesoporous materials. *Catalysis Communications, 135*(2019), 105898.

Scaccia, S., Della Seta, L., Mirabile Gattia, D., & Vanga, G. (2021). Catalytic performance of $Ni/CaO\text{-}Ca_{12}Al_{14}O_{33}$ catalyst in the green synthesis gas production via CO_2 reforming of CH_4. *Journal of CO_2 Utilization, 45*. https://doi.org/10.1016/j.jcou.2021.101447.

Schiaroli, N., Lucarelli, C., Iapalucci, M. C., Fornasari, G., Crimaldi, A., & Vaccari, A. (2020). Combined reforming of clean biogas over nanosized ni–rh bimetallic clusters. *Catalysts, 10*(11), 1–17. https://doi.org/10.3390/catal10111345.

Seo, J. C., Kim, H., Lee, Y. L., Nam, S., Roh, H. S., Lee, K., et al. (2021). One-pot synthesis of full-featured mesoporous Ni/Al_2O_3 catalysts via a spray pyrolysis-assisted evaporation-induced self-assembly method for dry reforming of methane. *ACS Sustainable Chemistry and Engineering, 9*(2), 894–904. https://doi.org/10.1021/acssuschemeng.0c07927.

Shah, C. M. (2021). Nickel impregnated silica catalyst for dry reforming of methane. *Doctoral dissertation*. Buffalo: State University of New York.

Shahnazi, A., & Firoozi, S. (2020). Improving the catalytic performance of $LaNiO_3$ perovskite by manganese substitution via ultrasonic spray pyrolysis for dry reforming of methane. *Journal of CO_2 Utilization, 45*, 101455. https://doi.org/10.1016/j.jcou.2021.101455.

Shanshan, X., Huanhao, C., Christopher, H., & Xiaolei, F. (2021). Non-thermal plasma catalysis for CO_2 conversion and catalyst design for the process. *Journal of Physics D: Applied Physics, 54*(23). https://doi.org/10.1088/1361-6463/abe9e1, 233001.

Shi, C., Wang, S., Ge, X., Deng, S., Chen, B., & Shen, J. (2021). A review of different catalytic systems for dry reforming of methane: Conventional catalysis-alone and plasma-catalytic system. *Journal of CO_2 Utilization, 46*. https://doi.org/10.1016/j.jcou.2021.101462.

Shin, S. A., Eslami, A. A., Noh, Y. S., Song, H. T., Kim, H. D., Saeidabad, N. G., et al. (2020). Preparation and characterization of $Ni/ZrTiAlO_x$ catalyst via sol-gel and impregnation methods for low temperature dry reforming of methane. *Catalysts, 10*(11), 1–22. https://doi.org/10.3390/catal10111335.

Simonov, M., Bespalko, Y., Smal, E., Valeev, K., Fedorova, V., Krieger, T., et al. (2020). Nickel-containing ceria-zirconia doped with Ti and Nb. Effect of support composition and preparation method on catalytic activity in methane dry reforming. *Nanomaterials, 10*(7), 1–19. https://doi.org/10.3390/nano10071281.

Singh, R., Dhir, A., Mohapatra, S. K., & Mahla, S. K. (2020). Dry reforming of methane using various catalysts in the process: Review. *Biomass Conversion and Biorefinery, 10*(2), 567–587. https://doi.org/10.1007/s13399-019-00417-1.

Singh, S., Kumar, R., Setiabudi, H. D., Nanda, S., & Vo, D. V. N. (2018). Advanced synthesis strategies of mesoporous SBA-15 supported catalysts for catalytic reforming applications: A state-of-the-art review. *Applied Catalysis A:*

General, *559*, 57–74. https://doi.org/10.1016/j.apcata.2018.04.015.

Singh, S., Nguyen, T. D., Siang, T. J., Phuong, P. T. T., HuyPhuc, N. H., Truong, Q. D., et al. (2020). Boron-doped Ni/SBA-15 catalysts with enhanced coke resistance and catalytic performance for dry reforming of methane. *Journal of the Energy Institute*, *93*(1), 31–42. https://doi.org/10.1016/j.joei.2019.04.011.

Song, Z., Wang, Q., Guo, C., Li, S., Yan, W., Jiao, W., et al. (2020). Improved effect of Fe on the stable NiFe/Al_2O_3 catalyst in low-temperature dry reforming of methane. *Industrial and Engineering Chemistry Research*, *59*(39), 17250–17258. https://doi.org/10.1021/acs.iecr.0c01204.

Sun, Y., Zhang, G., Cheng, H., Liu, J., & Li, G. (2021). Kinetics and mechanistic studies of methane dry reforming over Ca promoted 1Co–1Ce/AC-N catalyst. *International Journal of Hydrogen Energy*, *46*(1), 531–542. https://doi.org/10.1016/j.ijhydene.2020.09.192.

Sun, Y., Zhang, G., Liu, J., Xu, Y., & Lv, Y. (2020). Production of syngas via CO_2 methane reforming process: Effect of cerium and calcium promoters on the performance of Ni-MSC catalysts. *International Journal of Hydrogen Energy*, *45*(1), 640–649. https://doi.org/10.1016/j.ijhydene.2019.10.228.

Sun, Y., et al. (2020). Kinetics and mechanistic studies of methane dry reforming over Ca promoted 1Co–1Ce/AC-N catalyst. *International Journal of Hydrogen Energy*, *46*(1), 531–542. https://doi.org/10.1016/j.ijhydene.2020.09.192.

Świrk, K., Rønning, M., Motak, M., Grzybek, T., & Costa, D. (2020). Synthesis strategies of Zr- and Y-promoted mixed oxides derived from double-layered hydroxides for syngas production via dry reforming of methane. *International Journal of Hydrogen Energy*, *46*(22), 12128–12144. https://doi.org/10.1016/j.ijhydene.2020.04.239.

Tian, J., Ma, B., Bu, S., Yuan, Q., & Zhao, C. (2018). One-pot synthesis of highly sintering- and coking-resistant Ni nanoparticles encapsulated in dendritic mesoporous SiO_2 for methane dry reforming. *Chemical Communications*, *54*(99), 13993–13996. https://doi.org/10.1039/c8cc08284g.

Tibra, Mozammel, Deepa, Dumbre, Selvakannan, P. R., Sadasivuni, K. K., & Bhargava, S. K. (2021). Calcined hydrotalcites of varying Mg/Al ratios supported Rh catalysts: Highly active mesoporous and stable catalysts toward catalytic partial oxidation of methane. *Emergent Materials*, 469–481. https://doi.org/10.1007/s42247-020-00158-2.

Tran, T. Q., Pham Minh, D., Phan, T. S., Pham, Q. N., & Nguyen Xuan, H. (2020). Dry reforming of methane over calcium-deficient hydroxyapatite supported cobalt and nickel catalysts. *Chemical Engineering Science*, *228*. https://doi.org/10.1016/j.ces.2020.115975.

Wan, C., Song, K., Pan, J., Huang, M., Luo, R., Li, D., et al. (2020). Ni–Fe/Mg(Al)O alloy catalyst for carbon dioxide reforming of methane: Influence of reduction temperature and Ni–Fe alloying on coking. *International Journal of Hydrogen Energy*, *45*(58), 33574–33585. https://doi.org/10.1016/j.ijhydene.2020.09.129.

Wang, L., Hu, R., Liu, H., Wei, Q., Gong, D., Mo, L., et al. (2019). Encapsulated Ni@La_2O_3/SiO_2 catalyst with a one-pot method for the dry reforming of methane. *Catalysts*, *10*(1), 38. https://doi.org/10.3390/catal10010038.

Wang, H., Mo, W., He, X., Fan, X., Ma, F., Liu, S., et al. (2020). Effect of Ca promoter on the structure, performance, and carbon deposition of Ni-Al_2O_3 catalyst for CO_2-CH_4 reforming. *ACS Omega*, *5*(45), 28955–28964. https://doi.org/10.1021/acsomega.0c02558.

Wang, N., Yu, X., Wang, Y., Chu, W., & Liu, M. (2013). A comparison study on methane dry reforming with carbon dioxide over $LaNiO_3$ perovskite catalysts supported on mesoporous SBA-15, MCM-41 and silica carrier. *Catalysis Today*, *212*, 98–107. https://doi.org/10.1016/j.cattod.2012.07.022.

Xianglei, Y., Shen, W., Baoyi, W., & Laihong, S. (2021). Perovskite-type $LaMn_{1-x}B_xO_{3+\delta}$ (B = Fe, Co and Ni) as oxygen carriers for chemical looping steam methane reforming. *Chemical Engineering Journal*, 128751. https://doi.org/10.1016/j.cej.2021.128751.

Xingyuan, G., Jangam, A., & Sibudjing, K. (2020). Smart designs of anti-coking and anti-sintering Ni-based catalysts for dry reforming of methane: A recent review. *Reactions*, 162–194. https://doi.org/10.3390/reactions1020013.

Xu, Y., Du, X., Shi, L., Chen, T., Wan, H., Wang, P., et al. (2021). Improved performance of Ni/Al_2O_3 catalyst deriving from the hydrotalcite precursor synthesized on Al_2O_3 support for dry reforming of methane. *International Journal of Hydrogen Energy*. https://doi.org/10.1016/j.ijhydene.2021.01.189.

Yasyerli, S., Filizgok, S., Arbag, H., Yasyerli, N., & Dogu, G. (2011). Ru incorporated Ni-MCM-41 mesoporous catalysts for dry reforming of methane: Effects of Mg addition, feed composition and temperature. *International Journal of Hydrogen Energy*, *36*(8), 4863–4874. https://doi.org/10.1016/j.ijhydene.2011.01.120.

Yentekakis, I. V., Goula, G., Hatzisymeon, M., Betsi-Argyropoulou, I., Botzolaki, G., Kousi, K., et al. (2019). Effect of support oxygen storage capacity on the catalytic performance of Rh nanoparticles for CO_2 reforming of methane. *Applied Catalysis B: Environmental*, *243*, 490–501. https://doi.org/10.1016/j.apcatb.2018.10.048.

Yu, S., Hu, Y., Cui, H., Cheng, Z., & Zhou, Z. (2021). Ni-based catalysts supported on $MgAl_2O_4$ with different properties for combined steam and CO_2 reforming of methane. *Chemical Engineering Science*, *232*. https://doi.org/10.1016/j.ces.2020.116379.

Yusuf, M., Farooqi, A. S., Keong, L. K., Hellgardt, K., & Abdullah, B. (2021). Contemporary trends in composite Ni-based catalysts for CO_2 reforming of methane.

Chemical Engineering Science, 229. https://doi.org/10.1016/j.ces.2020.116072.

Zhang, G., Liu, J., Xu, Y., & Sun, Y. (2018). A review of CH_4–CO_2 reforming to synthesis gas over Ni-based catalysts in recent years (2010–2017). *International Journal of Hydrogen Energy, 43*(32), 15030–15054. https://doi.org/10.1016/j.ijhydene.2018.06.091.

Zhang, X., Wang, F., Song, Z., & Zhang, S. (2020). Comparison of carbon deposition features between Ni/ZrO_2 and Ni/SBA-15 for the dry reforming of methane. *Reaction Kinetics, Mechanisms and Catalysis, 129*(1), 457–470. https://doi.org/10.1007/s11144-019-01707-5.

Zhang, M., Zhang, J., Zhou, Z., Chen, S., Zhang, T., Song, F., et al. (2020). Effects of the surface adsorbed oxygen species tuned by rare-earth metal doping on dry reforming of methane over Ni/ZrO_2 catalyst. *Applied Catalysis B: Environmental, 264*. https://doi.org/10.1016/j.apcatb.2019.118522.

Zhenghong, B., & Fei, Y. (2018). Catalytic conversion of biogas to syngas via dry reforming process. *3. Advances in bioenergy* (pp. 43–76). Elsevier.

Zhou, J., Zhao, J., Zhang, J., Zhang, T., Ye, M., & Liu, Z. (2020). Regeneration of catalysts deactivated by coke deposition: A review. *Chinese Journal of Catalysis, 41*(7), 1048–1061. https://doi.org/10.1016/S1872-2067(20)63552-5.

Further reading

Dębek, R., Galvez, M. E., Launay, F., Motak, M., Grzybek, T., & Da Costa, P. (2016). Low temperature dry methane reforming over Ce, Zr and CeZr promoted Ni–Mg–Al hydrotalcite-derived catalysts. *International Journal of Hydrogen Energy, 41*(27), 11616–11623. https://doi.org/10.1016/j.ijhydene.2016.02.074.

Kamaruzaman, M. F., Taufiq-Yap, Y. H., & Derawi, D. (2020). Green diesel production from palm fatty acid distillate over SBA-15-supported nickel, cobalt, and nickel/cobalt catalysts. *Biomass and Bioenergy, 134*. https://doi.org/10.1016/j.biombioe.2020.105476.

Kong, Wenbo, Fu, Yu, Shi, Lei, Li, S., Vovk, E., Zhou, X., et al. (2020). Nickel nanoparticles with interfacial confinement mimic noble metal catalyst in methane dry reforming. *Applied Catalysis B: Environmental, 285*, 119837. https://doi.org/10.1016/j.apcatb.2020.119837.

CHAPTER

8

Dry reforming of methane for catalytic valorization of biogas

Muriel Chaghouri, Sara Hany, Haingomalala Lucette Tidahy, Fabrice Cazier, Cédric Gennequin, and Edmond Abi-Aad

Environmental Chemistry and Life Interactions Unit (UCEIV), University of the Littoral Opal Coast, UR 4492, SFR Condorcet—FR CNRS 3417, Dunkerque, France

8.1 Introduction

Since the second industrial revolution, the temperature of the planet increases continuously. Nowadays, this rise is becoming quite significant. This phenomenon is referred to as global warming. Many researches have been done to identify the source of this increase. Currently, it is understood by the scientific community that the temperature increase is directly linked to the emission of greenhouses gases (GHGs). These types of gases contribute to the warming process by absorbing the infrared radiation and trapping it on the surface of the earth (Borduas & Donahue, 2018; Charlson et al., 1992; Snyder, Bruulsema, Jensen, & Fixen, 2009). The main GHGs that contribute to global warming are carbon dioxide, methane, and nitrous oxide (Maucieri, Barbera, Vymazal, & Borin, 2017). Even though these gases are produced through natural events such as forest fires, volcanoes, and permafrost, the increase in GHG emissions is highly due to human activity. The main reason for the overproduction of these GHGs remains the use of fossil fuels for industrial and residential usage (Streletskiy, Anisimov, & Vasiliev, 2015; Yue & Gao, 2018).

Fossil energies such as petroleum, coal, and natural gas represent 34%, 25%, and 21%, respectively, of global energy consumption. They are called fossil fuels because they are produced over millions of years ago by the transformation of biomass in oxygen-poor environments. This means that we are currently consuming the stock of fossil fuels that have been produced for millions of years. With the current increase in the world population, the consumption of fossil fuels greatly exceeds their speed of production. Recent studies estimate that if the current rate of consumption is maintained, the fossil fuel reservoir could be completely spent within 50 years (Höök & Tang, 2013; Lieberei & Gheewala, 2017).

In addition to this, the increase in waste is closely linked with the increase in the worldwide population. Getting rid of this waste also becomes a polluting process leading to the production of

Heterogeneous Catalysis
https://doi.org/10.1016/B978-0-323-85612-6.00008-5

GHGs. Several solutions are in place to reduce the pollution produced by waste such as recycling, composting, waste recovery, etc.

In summary, the three major problems that we just mentioned are all part of the same cycle directly linked with the increase of the worldwide population (Fig. 8.1).

To break this cycle, the most interesting solution seems to be the substitution of fossil fuels with clean and renewable sources of energy. In this regard, several processes are being developed to use sources that are unlimited and that cannot be completely consumed. Wind and solar are examples of clean and unlimited sources of energy that can be converted into electricity and heat. However, wind and solar radiations are not equal everywhere on earth because different regions have different weather conditions. Moreover, the usage of these intermittent sources of energy becomes more complicated with the obstacles related to storage. Finally, the structures built for solar and wind processes take up a large surface of land that could instead be used for agriculture purposes (Asadi & Pourhossein, 2019; Jamshidi, Pourhossein, & Asadi, 2021; Santos, Barros, & Fiho, 2020). In this chapter, we will focus our interest on biomass as an unlimited source of energy. The biological degradation of biomass produces a biogas that can be converted and used as a clean and renewable substitute for fossil fuels.

First, we will discuss the reactions that take place during the degradation of organic waste. The industrial processes and the conditions influencing the composition of biogas are described. We will also explore different methods to use biogas and focus on the catalytic reforming reaction.

Second, we will present the dry reforming of methane reaction, its advantages and disadvantages. The focus of this part will be the different methods of deactivation such as coking, sintering, poisoning, etc.

Finally, we will discuss the different reaction conditions that can be used for the dry reforming of methane, the catalytic materials that are mostly used for this reaction, and the purification techniques developed in order to eliminate the impurities that can render the biogas unusable.

In the end, a small conclusion will summarize the discussed topics.

FIG. 8.1 Global warming and fossil fuel depletion cycle. Effect of population increase on the fossil fuel reservoir and on global warming.

8.2 Biogas production, composition, and valorization

The management of this waste from industry, agriculture, or residential is done by different methods. For example, incineration is often used to destroy materials. However, this method produces gaseous compounds that are harmful to the environment and to human health (Jay & Stieglitz, 1995). There are also other more environmentally friendly techniques for treating waste (Di Maria & Micale, 2015). Recycling is used in order to reduce all waste. This material reconversion process is only applicable to certain metals, plastic, glass, or cardboard. For organic waste, biodegradation techniques such as composting and methanation can be applied. Composting is the decomposition of organic waste in an aerobic environment. This process permits the usage of organic matter as a fertilizer for crops. On the other hand, methanation occurs in an anaerobic environment and can also be used as a method of waste disposal and valorization (Vaverková, 2019; Zakir Hossain, Hasna Hossain, Uddin Monir, & Ahmed, 2014).

Biogas production by anaerobic digestion

Methanation is the biodegradation of organic matter in the absence of oxygen. It takes place in the presence of microorganisms that can decompose complex organic matter into simpler and smaller molecules. This process is called anaerobic digestion and is made of a series of biochemical reactions: hydrolysis, acidogenesis, acetogenesis, and methanation (Fig. 8.2) (Cioabla, Ionel, Dumitrel, & Popescu, 2012; Rashed, Mamun, & Torii, 2015; Ray, Mohanty, & Mohanty, 2013).

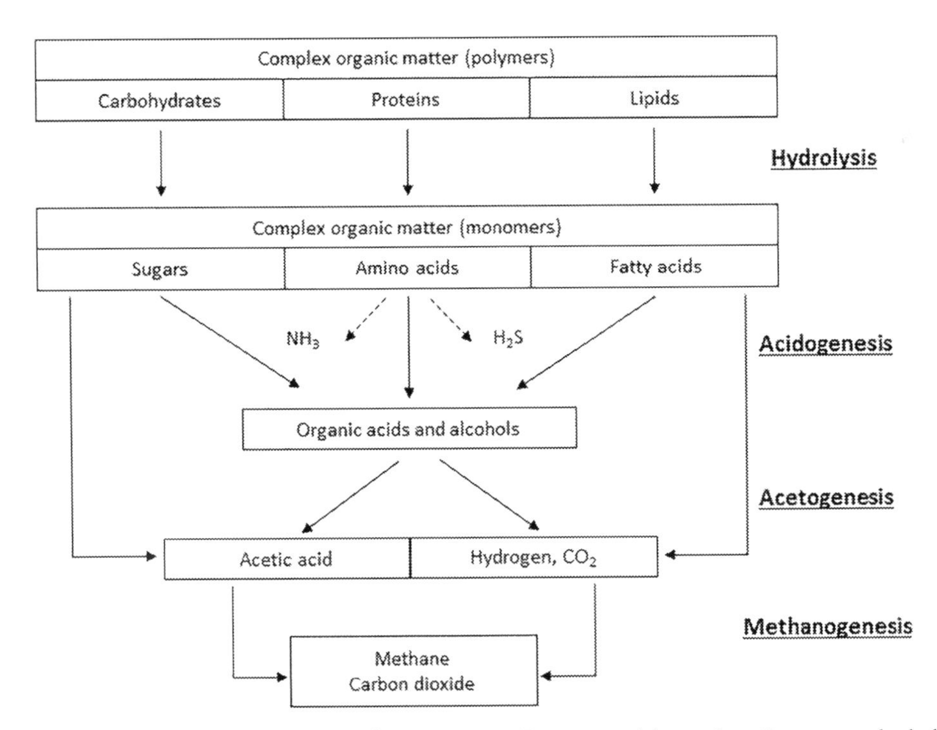

FIG. 8.2 Process of anaerobic decomposition of organic waste. Four steps of the methanation process: hydrolysis, acidogenesis, acetogenesis, and methanogenesis.

Biochemical reactions

Hydrolysis

After deposition of the organic waste in the storage center, the first step is their decomposition in the presence of oxygen (hydrolysis). Microorganisms break down complex organic matter (carbohydrates, lipids, and proteins) into simple organic matter (sugar, fatty acids, amino acids). Eq. (8.1) is an example of the hydrolysis of carbohydrates that produce monomers of sugar.

$$(C_6H_{10}O_5)n + nH_2O \leftrightarrow nC_6H_{12}O_6 + CO_2 \tag{8.1}$$

These reactions produce a large amount of CO_2.

Acidogenesis

After total consumption of the oxygen initially present in the medium, anaerobic conditions are established. During this reaction, the monomers (sugars, amino acids, and fatty acids) obtained from the degradation of organic matter are transformed into acetic (Eq. 8.2), propanoic (Eq. 8.3), and butanoic (Eq. 8.4) acids and ethanol (Eq. 8.5).

$$C_6H_{12}O_6 \leftrightarrow 3CH_3COOH + H_2 \tag{8.2}$$

$$C_6H_{12}O_6 \leftrightarrow 2CH_3CH_2COOH + 2H_2O \tag{8.3}$$

$$C_6H_{12}O_6 \leftrightarrow CH_3CH_2CH_2COOH + 2CO_2 + 2H_2O \tag{8.4}$$

$$C_6H_{12}O_6 \leftrightarrow 2CH_3CH_2OH + 2H_2 \tag{8.5}$$

Acetogenesis

Then, these compounds are transformed into acetic acid, CO_2, and H_2. This step is called acetogenesis because it leads to the formation of acetic acid (Eqs. 8.6, 8.7, and 8.8).

$$CH_3CH_2COOH + 2H_2O \leftrightarrow CH_3COOH + CO_2 + 3H_2 \tag{8.6}$$

$$CH_3CH_2CH_2COOH + 2H_2O \leftrightarrow 2CH_3COOH + 2H_2 \tag{8.7}$$

$$CH_3CH_2OH + H_2O \leftrightarrow CH_3COOH + 2H_2 \tag{8.8}$$

Methanogenesis

Acetic acid, CO_2, and H_2 produced during the previous reactions are transformed into methane, carbon dioxide, and H_2O. This step can be summarized into two reactions: acetotrophic methanogenesis (Eq. 8.9) and hydrogenotrophic methanogenesis (Eq. 8.10) (Demirel & Scherer, 2008).

$$CH_3COOH \leftrightarrow CH_4 + CO_2 \tag{8.9}$$

$$CO_2 + 4H_2 \leftrightarrow CH_4 + 2H_2O \tag{8.10}$$

During each reaction, several types of microorganisms are responsible for transforming the products of the previous reaction into reactants for the next reaction.

Microorganisms

Studies show that fungus can degrade organic matter, however bacteria and archaea are mostly known for taking part in the anaerobic digestion (Kazda, Langer, & Bengelsdorf, 2014; Langer et al., 2019). Both of these species are unicellular organisms belonging to the prokaryote kingdom (He, Zhao, Zhou, & Huang, 2009). Table 8.1 lists the bacteria and archaea that have been identified as involved in the degradation of organic matter.

The hydrolysis step is carried out by hydrolytic bacteria such as *Firmicutes*, *Chloroflexi*, *Bacteroids*, and *Spirochaetes* (Guo et al., 2014; Nguyen et al., 2019; Wang et al., 2018; Ziganshin et al., 2011). These microorganisms are responsible for the degradation of macro-elements (proteins, lipids, and carbohydrates) into monomers. These phyla include Bacilli, Bacteroidia, Flavobacteriia, and they are usually observed in mesophilic conditions, at temperatures around 40°C. Acidogenesis begins in the presence of acidogenic bacteria, belonging to the proteobacteria, firmicutes, and

TABLE 8.1 List of microorganisms involved in the anaerobic digestion of organic waste.

Kingdom	Phylum	Reaction	Reference
Bacteria	*Actinobacteria*	Hydrolysis, Acidogenesis	Ziganshin et al. (2011), Guo, Wang, Sun, Zhu, and Wu (2014), Wang, Wang, Qiu, Ren, and Jiang (2018), Nguyen, Nguyen, and Nghiem (2019)
	Bacteroidetes	Hydrolysis and nitrification	Díaz, Oulego, Laca, González, and Díaz (2019), Guo et al. (2014)
	Chloroflexi	Hydrolysis, Acidogenesis	Guo et al. (2014), Nguyen et al. (2019), Wang et al. (2018)
	Firmicutes	Hydrolysis, acidogenesis, acetogenesis, and sulfur reduction	Ziyang, Luochun, Nanwen, and Youcai (2015), Díaz et al. (2019), Ziganshin et al. (2011)
	Proteobacterium	Acidogenesis, nitrification, and sulfur reduction	Díaz et al. (2019), Guo et al. (2014), Ziyang et al. (2015)
	Spirochabacter	Hydrolysis	Guo et al. (2014)
	Spirochaetes	Hydrolysis, Acidogenesis	Guo et al. (2014), Wang et al. (2018), Nguyen et al. (2019)
	Synergistetes	Acidogenesis	Guo et al., 2014, Ziganshin et al. (2011)
	Tenericutes	Acidogenesis	Díaz et al. (2019), Guo et al. (2014)
	Thermotogae	Hydrolysis	Guo et al. (2014), Deublein and Steinhauser (2011)
Archaea	*Crenarchaeote*	Methanogenesis	Demirel and Scherer (2008), Ziyang et al. (2015).
	Euryarchaeota	Methanogenesis	Demirel and Scherer (2008), Ziyang et al. (2015).

Different bacteria and archaea are involved in the major and minor reactions taking place during methanation.

actinobacteria phylum. These microorganisms decompose the monomers produced from the previous reactions into volatile fatty acids such as propanoic acid and butanoic acid and alcohols such as ethanol. Following the acidogenesis, the acetogenic step can begin in the presence of acetogenic bacteria (thermatogex and firmicutes). Acetic acid, carbon dioxide, and hydrogen are the main products of the reactions occurring during this step. Finally, the methanogenesis step is carried out by the H_2 methanogenic or acetate methanogenic archaea. These prokaryotes belong to the phylum of euryarchaeota and are known to survive in extreme conditions. In this case, they are responsible for the production of methane through two different pathways: H_2 based or acetic acid based methanogenesis. The mentioned bacteria and archaea are responsible for the production of the major part of biogas (CO_2 and CH_4). However, several microorganisms are involved in the production of by-products or impurities such as sulfur and nitrogen compounds. Hydrogen sulfide is one of the most abundant impurities in biogas. This molecule derives from the degradation of sulfur-containing amino acids such as methionine, cysteine, homocysteine, and taurine (Brosnan, 2006). H_2S can also be produced through the reduction of sulfate. These reactions are possible in the presence of sulfur-reducing bacteria such as *Peptococcaceae*, *Desulfobacteraceae*, *Desulfobulbaceae* belonging to the phylum *Firmicutes* and *Proteobacterium* (Houari, 2020; St-Pierre & Wright, 2017). Moreover, nitrogen compounds such as NO_x, NH_x, and N_2 can also be found in biogas (Duan,

Scheutz, & Kjeldsen, 2021; Lou, Wang, Zhao, & Huang, 2015). These compounds are produced through the degradation of organic waste with high nitrogen content such as poultry, fish, etc. The denitrification of nitrogen-containing amino-acids takes place in the presence of *Proteobacterium* and *Bacteroidetes* (Mackie, Stroot, & Varel, 1998).

Phases of anaerobic digestion

In general, anaerobic digestion takes place in four phases (Fig. 8.3) (Asadi & Pourhossein, 2019; Jamshidi et al., 2021; Santos, Jong, Costa, & Torres, 2020).

Phase I: aerobic decomposition (O_2 consumption and CO_2 production): This phase takes place right after the organic waste is introduced into the container. During this phase, oxygen is still present which is why it is considered as an aerobic phase. Hydrolysis takes place and polymers are decomposed into monomers. In parallel, carbon dioxide is produced in a high concentration. This phase is known to be the limiting (slowest) stage of the methanation process (Guo et al., 2014; Nguyen et al., 2019; Wang et al., 2018; Ziganshin et al., 2011). It usually lasts between one and four years.

Phase II: Anaerobic non-methanogenic (production of CO_2 and H_2): After phase I, oxygen content is very low which is why all the next steps take place in an anaerobic environment. Phase II is mainly made of acidogenesis and acetogenesis. During this phase, we observe a peak of CO_2 and H_2 while methane is not yet being produced. This phase lasts between one and three years.

Phase III: Unstable anaerobic methanogenesis: The compounds produced during this phase are part of the final biogas composition. Methane and carbon dioxide are both being produced but in an unstable manner. The stabilization of the production of CH_4 and CO_2 requires between one and seven years.

Phase IV: Stable anaerobic methanogenesis: During the final phase, the methanogenic bacteria remain active and produce a constant amount of CH_4 and CO_2. In general, the lifespan of a landfill is estimated to be between 20 and 50 years. Methanation will therefore persist until the essential nutrients for the survival of the microorganisms are completely exhausted, or until extreme variations in pH occur. This will cause the shutdown of microbial metabolism (Duan et al., 2021; Ziyang et al., 2015; SCDHEC, 2021).

At the end of the methanation cycle, three units are obtained (EPA, 2021; United States Environmental Protection Agency, 2021):

- a solid phase called digestat and made up of undigested organic matter. This digestate can then be used as a natural fertilizer
- a liquid phase called leachate. This solution can be filtered and released into nature after quality analysis
- a gas phase called biogas.

Biogas composition

Overall, the biogas obtained from methanation is mostly made of methane, CO_2, N_2, oxygen, and water. In addition to these main components, some minor compounds (also called impurities or by-products) can also be found. They can be classified into 8 groups according to their chemical structure and functional group (Duan et al., 2021; Rasi, 2009; Schuetz et al., 2003; Tan, Zhao, Ling, Wang, & Wang, 2017; Wu et al., 2017)

- halogens
- aromatic compounds
- terpenes
- aliphatic compounds
- sulfur compounds
- nitrogen compounds
- oxygenated compounds
- siloxanes

These molecules are found in biogas in very small quantities (usually less than 1%). The concentration of the main components and the by-products varies according to the nature of the initial substrate, the microbial consortium, the process type, and conditions (temperature, pH, humidity, etc.). Methanation of organic

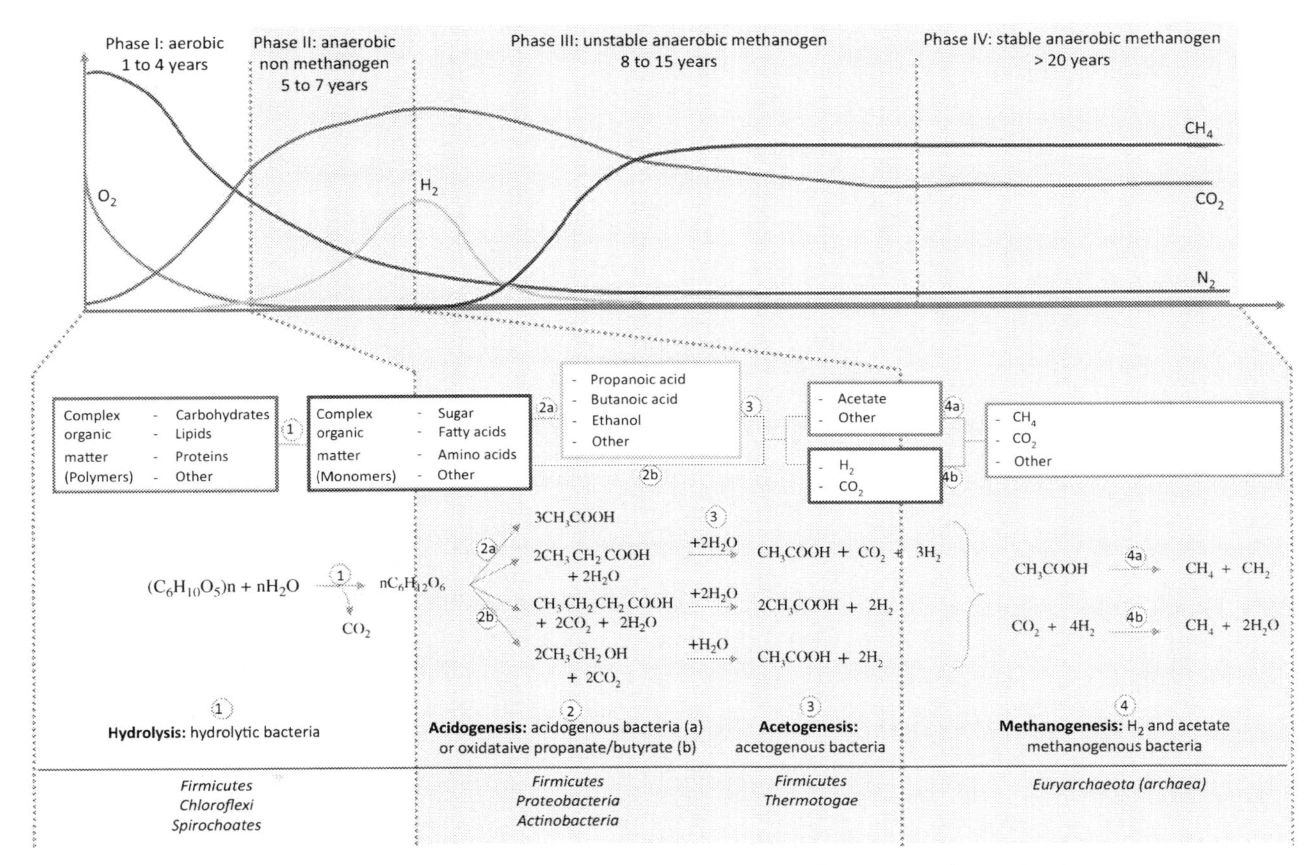

FIG. 8.3 Methanation, from organic waste to biogas. Microorganisms involved in the different steps of anaerobic digestion.

TABLE 8.2 Composition of biogas generated from different plants.

	Landfill	Biogas plant	Wastewater treatment plant
CH_4 (%)	30–60	60–70	55–65
CO_2 (%)	30–55	30–40	35–45
N_2 (%)	5–30	<1	<1
O_2 (%)	0–5	0–1	<1
H_2O (%)	1–5	5–10	n.a.
NH_3 (ppm)	0–5	0–500	n.a.
H_2S (ppm)	0–10,000	10–2000	10–40
VOC (ppm)	100–1000	<100	n.a.
Siloxanes (ppm)	<36	<1	<150
Halogens (ppm)	<225	<1	<1
References	Allen, Braithwaite, and Hills (1997), Anneli and Athur (2006), Arnold and Kajolinna (2010), Dumont (2015), Jaffrin, Bentounes, Joan, and Makhlouf (2003), Johansson (2012), Rasi, Läntelä, and Rintala (2011), Schweigkofler and Niessner (1999), Themelis and Ulloa (2007)	Dumont (2015), Rasi et al. (2011), Rasi, Veijanen, and Rintala (2007), Bharathiraja et al. (2018).	Dewil, Appels, and Baeyens (2006), Dumont (2015), Rasi et al. (2007, 2011), Spiegel and Preston (2003)

Content of major and minor compounds found in biogas depends on the source.

waste can take place in different structures such as a wastewater treatment plant, a biogas plant, a landfill, etc. The composition of the biogas is directly related to the type of plant it's produced in (Table 8.2).

- The methanation process in a biogas plant is a relatively clean and controlled operation. The initial organic substrate is sorted beforehand and is introduced in a digester where temperature, pH, agitation, and oxygen content can be monitored and controlled. This allows the production of biogas that has relatively low impurities content because the parameters that may influence the biogas composition were stabilized (Holm-Nielsen & Oleskowicz-Popiel, 2013).
- In a landfill, the waste is usually collected from residential or industrial sources. It undergoes first a "preselection" process during which the nonorganic fraction is separated. Then, the organic fraction is deposited in an underground unit where methanation takes place. This unit is very well isolated so that the gaseous and liquid compounds produced do not migrate into the soil or nearby water source. In a landfill, the anaerobic digestion process is less controlled which allows the possibility of having more by-products. This type of biogas usually has a high content of sulfur and halogen compounds (Allen et al., 1997; Bouzonville, Peng, & Atkins, 2013).

- The wastewater treatment plant is a recovery facility installed for the removal of pollutants from a polluted water source, residential drainage, etc. In this case, biogas can be produced through the anaerobic treatment of the water. This type of biogas also contains a large quantity of by-products (Spiegel & Preston, 2003).

In biogas, by-products usually come from the biodegradation of organic matter, from the volatilization of compounds, and from reactions between major or minor products. They are produced at different levels of the anaerobic digestion process. In general, a first peak of secondary compounds is observed during the aerobic phase. It is caused by the evaporation of volatile compounds like aerosols. A second peak is observed during anaerobic methanogenesis. This peak is caused by the production of by-products and intermediates of anaerobic degradation (Bierer, Kress, Nägele, Lemmer, & Palzer, 2019; Bouzonville et al., 2013; Cioabla et al., 2012; Parra-Orobio, Donoso-Bravo, Ruiz-Sánchez, Valencia-Molina, & Torres-Lozada, 2018).

In addition to the impurities found in biogas, the methane/carbon dioxide content also changes depending on the methanation process, the microbial consortium, the nutrient available for the microorganisms, etc.

As it was mentioned before, biogas is mostly made of methane and carbon dioxide but also contains minor products such as water, nitrogen, hydrogen, hydrogen sulfide, siloxanes, etc. Biogas can be valorized directly from the digester or it undergoes a cleaning and upgrading process and is used for other applications such as power production.

Biogas upgrading and purification

In some cases, biogas can be used to generate electricity and heat directly after production. However, in the case of indirect utilization, biogas undergoes a two-step valorization.

Biogas cleaning methods

The first step is the cleaning of biogas which is defined by the elimination of minor compounds that can be dangerous and cause damage to the equipment. This purification step can take place in situ (inside the digester) ex situ (in a different compartment following the digester) or as a hybrid process. Chemical, physical and biological techniques can be used to eliminate different types of compounds. Currently, chemical and physical processes are already industrialized, whereas biological methods are not fully commercial yet (Guieysse et al., 2008; Rasi, 2009; Shah, Ahmad, Pant, & Vijay, 2021).

Hydrogen sulfide is the most common impurity found in biogas. It is a highly toxic product because it can oxidize into SO_2 and cause the corrosion of metal conducts in industrial structures. Some industrial processes such as proton exchange membranes and solid oxides fuel cells have a very low tolerance for H_2S (<1 ppm). This molecule is usually eliminated by adsorption on activated carbon (physical technique) or through a chemical reaction with ferric iron that produces a FeS precipitate that can be easily eliminated (chemical techniques). Even though both of these techniques are highly efficient, they demand a very high implementation and maintenance cost. However, biological methods are less costly in terms of maintenance but are still under study for the development of more practical models (Barbusiński & Kalemba, 2016; Khoshnevisan et al., 2017).

Siloxanes are larger molecules containing silica. Nowadays, siloxanes are detected more frequently in biogas because of the usage of material containing silica in the equipment of the industry. This type of molecule can be eliminated via adsorption on activated carbon or by absorption in which case the biogas passes through a solvent that can absorb siloxane. It can also be eliminated by deep chilling or by biological processes using a biofilter containing microorganisms that can degrade siloxanes.

Halogenated compounds (chlorine-, bromine-, and iodine-containing compounds) are frequently found in biogas produced in a landfill. They originate from certain solvents, refrigerants, and plastic products. The most common method of elimination used for halogenated compounds is by adsorption on specific activated carbon. After the adsorption of halogens, the activated carbon is regenerated by heating until 200°C which causes the desorption of the compounds (Allegue & Hinge, 2012).

Terpenes, aromatic, aliphatic, and oxygenated compounds are frequently found in biogas. They derive from various sources: terpenes are usually found in plants (wood, herbs fruits, etc.) and are used in medicine and detergents. Aromatics come from petrol-based products as well as solvents and food additives. Aliphatic compounds originate from detergents, solvents, and oils whereas oxygenated compounds derive from plastic packaging (Bierer et al., 2019; Bouzonville et al., 2013; Parra-Orobio et al., 2018). Several methods can be used for the removal of these VOCs. Adsorption on activated carbon and water scrubbing are used most frequently (Santos-Clotas et al., 2019). However, other methods such as membrane separation and biological removal through biofilters and biotrickling filters can also be applied (Deng & Hägg, 2010; Peishi, Xianwan, Ruohua, Bing, & Ping, 2004; Rasi et al., 2011).

Biogas upgrading methods

After cleaning, biogas can also be upgraded, in which case the CO_2 is eliminated through several methods. One of the most used techniques is the water scrubbing system which can be done in water or organic scrubbing. Scrubbing is a purification method based on the solubility of molecules in a certain solvent. For example, CO_2 is 25 times more soluble than CH_4 in water (Awe, Zhao, Nzihou, Minh, & Lyczko, 2017; Towler & Sinnott, 2013). Methanol or dimethyl ether can also be used as solvent to eliminate polar compounds (CO_2 and H_2S). CO_2 can also be eliminated by chemical absorption and pressure swing adsorption. This method allows the elimination of CO_2 through its adsorption on activated carbon or on molecular sieves such as zeolites (Lasocki, Kołodziejczyk, & Matuszewska, 2015). Its efficiency is directly related to the molecule size. The membrane separation, based on permeability of CO_2, can also be used as an upgrade method. Finally, cryogenic separation process allows the separation of liquefied CH_4 from CO_2 by decreasing the temperature and increasing the pressure (Awe et al., 2017).

Biogas valorization

The direct utilization of biogas can be done without prior purification. This type of process is usually performed in a household which allows the usage of biogas coming directly from a local digester. It's used mostly for small-scale applications such as domestic cooking (Kadam & Panwar, 2017; Sahota et al., 2018). However, the most interesting approach for biogas valorization remains after its purification. After cleaning and upgrading, biogas can be separated into two fractions: a methane-rich portion and a carbon dioxide-rich portion. The clean and/or upgraded biogas can then be used in different pathways (Kapoor et al., 2020).

Methane utilization

The methane-rich portion contains a very high content of CH_4. Biomethane can be used as transport fuel in light and heavy-duty vehicles. Currently, it is being used as fuel for cars, buses, and trucks in several countries including Sweden, Switzerland, Brazil, and Germany (International Renewable Energy Agency, 2019; Kapoor et al., 2020). Upgraded biogas can also be used in dual-fuel engines that can function with methane and diesel fuel to generate power and heat (Balat & Balat, 2009; Rafiee, Khalilpour, Prest, & Skryabin, 2021). Moreover, upgraded biogas can be added into distribution networks to produce green electricity and heat. Grid-connected systems for natural gas are already implemented in many parts of the world. This provides a safe and convenient system for the distribution of biomethane (Murphy, McKeogh, & Kiely, 2004; Yang, Ge, Wan, Yu, & Li, 2014). Finally, some modern fuel cells use methane as their fuel. In some cases, the upgrade of biogas is not mandatory and the fuel cells can operate with biogas containing CO_2 (< 35%). However, purification of H_2S and siloxane is necessary for fuel cell applications (Benito, García, Ferreira-Aparicio, Serrano, & Daza, 2007; Sun et al., 2015). Biomethane can be used for the same applications as natural gas (heating, transport), or can be transformed through other processes into hydrogen or syngas.

Carbon dioxide utilization

Currently, CO_2 is known for being one of the main greenhouse gases causing global warming. However, CO_2 has also various industrial applications. This molecule is used as a nutrient for algae cultivation. It enhances photosynthesis which increases plant growth (Amosa, Mohammed, & Yaro, 2010). CO_2 can also be used as a neutralizing agent in the process of aluminum extraction from bauxite. In the food industry, highly purified carbon dioxide can be used to produce carbonated drinks. The production of a high-quality CO_2 (99.99%) is very costly and demands high energy which is why bioCO_2 is usually used as a refrigerant instead (dry ice) (Prussi, Padella, Conton, Postma, & Lonza, 2019; Sun et al., 2015). In the chemical industry, CO_2 is used as a feedstock for the production of chemicals and synthetic fuels. Urea, used in agriculture, fertilizers, and in the medical industry, is produced through the reaction between carbon dioxide and ammonia (Alper & Yuksel Orhan, 2017). By CO_2 reduction or catalytic hydrogenation, methanol can be produced. This molecule is then used in the production of solvents, plastics, vitamins, fuel, etc. (Pérez-Fortes, Bocin-Dumitriu, & Tzimas, 2014). Finally, CO_2 can also be used in the methanation reaction to produce CH_4 (Eq. 8.11).

$$CO_2 + 3H_2 \leftrightarrow CH_4 + H_2O \tag{8.11}$$

This power to gas technology allows the production of biomethane that can subsequently be added to the existing infrastructure for natural gas. This technique however is usually implemented before the upgrading step by adding H_2 to the biogas mixture (Götz et al., 2016; Prussi et al., 2019).

Clean biogas utilization

Seeing as CO_2 is one of the major compounds of biogas, its elimination is a high-cost process. For this reason, processes for the valorization of both CO_2 and CH_4 are being developed.

Even though both methane and carbon dioxide fractions can be used individually, in some cases, biogas can be utilized without an upgrade step. In this case, biogas can be used in several industrial processes. Biogas boilers produce thermal energy that can be applied for heat production. A cogeneration or a combined heat and power system can also be used for the production of heat and electricity (Allegue & Hinge, 2012). Reciprocating internal combustion engines converts fuel energy, in this case biogas, into electric power and/or heat. This category includes spark ignition gas engines, gas turbines, and micro-turbines stirring engines and fuel cells. These engines can tolerate the presence of CO_2 in the fuel gas which is why the biogas upgrading step is not mandatory. However, siloxane, H_2S, and water need to be eliminated before the biogas enters the pipeline because the formation of acid can deteriorate the engine (Allegue & Hinge, 2012; Kapoor et al., 2020). A different strategy for biogas utilization is through its conversion into chemicals with a higher added value (second-generation biogas). This method involves chemical reactions such as steam reforming (Eq. 8.12), dry reforming (Eq. 8.13), or oxidation of methane (Eq. 8.14) to produce hydrogen or synthetic gas (also called syngas) (Baena-Moreno, Sebastia-Saez, Pastor-Pérez, & Reina, 2021; Guerrero et al., 2020).

$$CH_4 + H_2O \leftrightarrow 3H_2 + CO\ \Delta H_{298K} = 206 kJ/mol \tag{8.12}$$

$$CH_4 + CO_2 \leftrightarrow 2H_2 + 2CO\ \Delta H_{298K} = 247 kJ/mol \tag{8.13}$$

$$CH_4 + 1/2O_2 \leftrightarrow CO + 2H_2\ \Delta H_{298K} = -36 kJ/mol \tag{8.14}$$

The produced mixture can be used as fuel, mainly as hydrogen, or as raw material for the production of chemicals such as gasoline, methanol, and Fischer-Tropsch oil. Steam reforming of methane takes place simultaneously with the water gas shift reaction (Eq. 8.15).

$$CO + H_2O \leftrightarrow CO_2 + H_2\ \Delta H_{298K} = -41 kJ/mol \tag{8.15}$$

The final outcome of both of these reactions is a mixture of 4 molecules of H_2 and one molecule of CO_2. This process is mainly used for the production of hydrogen. Nowadays, 48% of the hydrogen produced globally is generated by this reaction (da Silva Veras, Mozer, da Costa Rubim Messeder dos Santos, & da Silva César, 2017; Ewan & Allen, 2005; Figoli, Cassano, & Basile, 2016). Partial oxidation of methane leads to the formation of H_2 and CO in a ratio of 2. This ratio is highly in demand as a feedstock for Fischer-Tropsch processes that allow the production of liquid hydrocarbons with a higher added value. However, the mentioned reactions are usually reduced due to the presence of a high concentration of CO_2 in the initial biogas composition (Baena-Moreno et al., 2021; Minh et al., 2020). This is where the dry reforming of methane reaction becomes interesting.

8.3 Dry reforming of methane reaction: Advantages and disadvantages

This reaction takes place between methane and carbon dioxide, the two main greenhouse gases and the main products of biogas. This reaction makes it possible to transform biogas into H_2 and CO with a ratio of 1 and to reduce the emissions of CH_4 and CO_2.

An H_2/CO ratio between 1 and 3 is required for the formation of methanol, a molecule highly sought after in the chemical industry. From this compound, it is possible to synthesize different types of products such as formaldehyde, dimethyl ether, acetic acid, dimethylformaldehyde, etc. (Gandhi & Patel, 2015; Naeem, Al-Fatesh, Abasaeed, & Fakeeha, 2014). These compounds can be used in the textile and pharmaceutical industry. In addition, this ratio can be used for Fischer-Tropsch processes. The synthesis gas is then converted by a series of reactions into liquid hydrocarbons. These compounds are formed by carbon chains of different lengths. This system can be summarized by the following reaction (Eq. 8.16):

$$nCO + 2nH_2 \leftrightarrow (CH_2-)n + nH_2O \tag{8.16}$$

This process takes place in the presence of a catalyst at temperatures between 200°C and 350°C. The fuel synthesized by this process is easy to store and transport (Mahmoudi et al., 2017; Tristantini, Lögdberg, Gevert, Borg, & Holmen, 2007). Depending on the H_2 content of the syngas and the H_2/CO ratio, several compounds of different sizes (carbon number) can be produced. In general, long carbon chain hydrocarbons (C_{10+}) are obtained with an H_2/CO ratio = 1.5. A synthesis gas with 3 times more hydrogen than CO is more selective for the production of short compounds (C_1–C_4). The valorization of these compounds is less interesting than that of long-chain compounds (Loic, 2005). Some studies show that a ratio close to 2 would be ideal because it allows the production of kerosenes (alkanes) and olefins (alkenes) (Kumar, Shojaee, & Spivey, 2015). After a secondary treatment of hydrogenation and hydrocracking, the compounds obtained can be used as fuel. Furthermore, syngas with a ratio of 1 obtained by the DRM reaction is ideal for the synthesis of oxo-alcohols (Milanov, Schunk, Schwab, & Wasserschaff, 2013). A CO-rich syngas reacts with olefin to obtain alcohol (Gobina, 2000). They are used as solvents and as a plasticizer in the plastic industry (Grandviewresearch, 2021).

Context

The dry reforming reaction is currently not yet industrialized. The experience on the commercialization of this reaction is so far very limited (Nahar, Mote, & Dupont, 2017). The dry reforming reaction is highly endothermic and is only possible at temperatures above 650°C. From an industrial point of view, such a high temperature translates into a considerable energy input and a high financial cost. This first drawback prevents the dry reforming reaction from being profitable on an industrial scale. The use of an active catalyst is therefore required in order to increase the reaction yield at lower temperatures. In addition, several side reactions can take place in the temperature range favorable to DRM. These reactions can lead to the production of undesirable by-products. They can also cause the consumption of the products of the dry reforming reaction and subsequently alter the H_2/CO ratio. The presence of impurities in the real biogas can also strongly influence resistance of the catalysts toward different types of deactivation.

In order to understand the deactivation mechanism of the catalyst, it would be interesting to focus on the specific mechanism of the dry methane reforming reaction.

Mechanism of dry reforming of methane

Several mechanisms are proposed in the literature for this reaction depending on the temperature and the nature of the active phase (Charisiou et al., 2019; Lobo, 2017; Rostrup-Nielsen & Bak Hansen, 1993; Rostrup-Nielsen, Sehested, & Nørskov, 2002). In general, three steps are used for all mechanisms. These steps are summarized in Fig. 8.4:

First step: Methane activation and decomposition

The dry reforming reaction starts with the adsorption of methane on an active site. The methane is then dissociated to form two hydrogen molecules that are desorbed and released into the gas phase, and carbon which remains adsorbed on the active site. This step, considered the slowest, does not take place spontaneously. It is a successive dissociation of hydrocarbon species (CH_4, $-CH_3$, $-CH_2$, $-CH$) which forms at the end of the carbon alone $-C$ (Eq. 8.17).

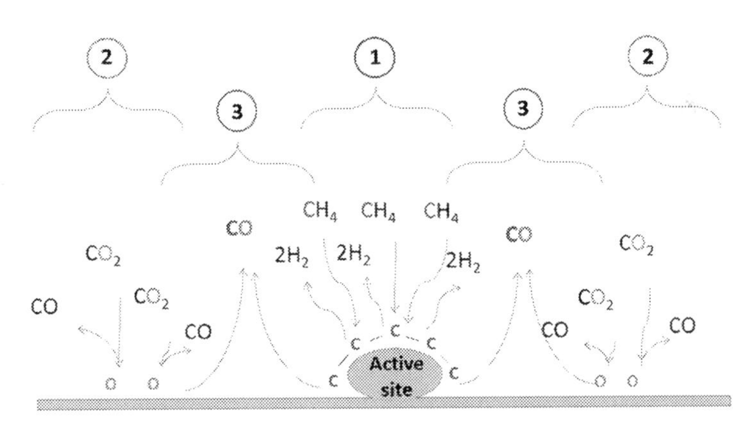

FIG. 8.4 Mechanism of dry reforming of methane. 3 step mechanism of DRM.

$$CH_4 + * \leftrightarrow *CH_x + (4 - x)H = 2H_2 + *C \quad (8.17)$$

With (*): active site.

Second step: Carbon dioxide activation and decomposition

The second step taking place during the DRM is the adsorption and dissociation of the CO_2 molecule (Eq. 8.18). In general, this step is considered as a fast step. It allows the release of a CO molecule and the formation of oxygen adsorbed on the active site.

$$CO_2 + * \leftrightarrow CO + *O \quad (8.18)$$

Third step: oxidation of adsorbed carbon

The third reaction takes place between the adsorbed species. The oxygen formed by the CO_2 activation on support oxygen vacancies reacts with the adsorbed carbon following the dissociation of CH_4. Following this step, a CO molecule is formed and desorbed and both active sites are thus released (Eq. 8.19).

$$*O + *C \leftrightarrow CO + * + * \quad (8.19)$$

In the literature, another mechanism close to this one is also used (Chen, Zaffran, & Yang, 2020; Zhu, Chen, Zhou, & Yuan, 2009; Zuo et al., 2018). It differs from this one by the formation of —OH groups and —CH_xO groups. The hydroxyl group would be formed by the reaction of the gas with water, while the CH_xOs are formed by the oxidation of dissociated methane (CH_x) by oxygen from the dissociation of CO_2. These intermediate species could be formed during the dry reforming reaction.

Deactivation of the catalytic material

When all three steps of the DRM mechanism are carried out simultaneously, the carbon formed on the catalytic surface is directly reoxidized by the adsorbed oxygen from the CO_2. On the other hand, if there is an imbalance between the first two reactions, it could lead to a deactivation of the catalysts.

Catalyst deactivation by reoxydation

First case scenario, if the dissociation of carbon dioxide is faster than the dissociation of methane, the complete reoxidation of carbon species formed on the catalytic surface is favored. On the other hand, this phenomenon can also lead to the deactivation of the catalysts by reoxidation of the active sites.

In general, before performing the dry reforming of methane reaction, most catalysts undergo an activation step. The reduction of catalytic material produces active sites and oxygen vacancies that contribute to the DRM.

If the catalyst undergoes a reoxidation during the dry reforming of methane reaction, the quantity of reduced active sites will decrease and therefore the catalytic activity will be lower. If all metal sites are reoxidized, we speak of deactivation by reoxidation. This phenomenon was observed by Luisetto et al. on Ni-based catalyst during the dry reforming of methane (Luisetto et al., 2017). In the presence of a large quantity of CO_2, the number of Ni^0 active sites was reduced after reoxidation into Ni^{2+} which caused a rapid deactivation of the catalyst (Mutz, Carvalho, Mangold, Kleist, & Grunwaldt, 2015). This phenomenon is sometimes observed in dry reforming of methane but especially takes place in the presence of H_2O and O_2 because of the stronger oxidizing potential of these elements (Dissanayake, 1991; Zhou et al., 2008). In case of complete reoxidation, the catalyst can be regenerated by reduction at high temperature.

Catalyst deactivation by coking

Second case scenario, methane dissociation is faster than carbon dioxide dissociation. In this case, the carbon formed during the methane dissociation reaction remained adsorbed on the catalytic surface. This compound can be considered as a reaction intermediate when it is directly oxidized by the oxygen formed by the dissociation of CO_2. If it is not oxidized, it can accumulate on the catalytic surface and cause the progressive or complete deactivation of the catalyst. This type of carbon is less reactive and can damage the solids in different ways. In addition to the DRM, several side reactions such as methane cracking reaction (Eq. 8.20), Boudouard reaction (Eq. 8.21), reverse carbon gasification reaction (Eq. 8.22), and water gas shift reaction (Eq. 8.23) can occur.

$$CH_4 \leftrightarrow C + 2H_2 \, \Delta H_{298K} = 75 \text{kJ/mol} \quad (8.20)$$

$$2CO \leftrightarrow C + CO \, \Delta H_{298K} = -171 \text{kJ/mol} \quad (8.21)$$

$$CO + H_2 \leftrightarrow C + H_2O \, \Delta H_{298K} = -131 \text{kJ/mol} \quad (8.22)$$

$$CO_2 + H_2 \leftrightarrow CO + H_2O \, \Delta H_{298K} = 41 \text{kJ/mol} \quad (8.23)$$

These reactions may compete with the dry reforming of methane reaction. They can reduce the yield and the selectivity of the catalysts as well as cause the production of a carbon deposit that can cover the surface of the catalyst. These reactions are thermodynamically possible at temperatures between 500°C and 900°C, which is the same temperature range as the DRM (Fig. 8.5) (Aouad, 2018; Nikoo & Amin, 2011; Zhang, Wang, & Dalai, 2007).

These reactions produce solid carbon that can deposit on the surface of the catalyst. This phenomenon can cause a physical blockage of the active sites that will strongly affect the stability and the activity of the catalyst.

Different types of carbon deposits can form. Some are known to be able to cause deactivation of the catalyst while others can behave as intermediates in the dry reforming reaction.

At low temperatures, α-carbon is formed (Gurav, Dama, Samuel, & Chilukuri, 2017; Serrano-Lotina & Daza, 2013). This type of carbon is unstable and can easily be re-oxidized. It can react with water, oxygen, or CO_2 to produce CO. If this type of carbon accumulates on the surface, a more stable carbon can be formed. The α-carbon can then transform into β-carbon (amorphous filamentous) or C-carbon (crystalline graphitic) (Gurav et al., 2017; Serrano-Lotina & Daza, 2013). These forms of carbon can deactivate the catalyst by encapsulating or distancing the catalytic sites (Gohier, Ewels, Minea, & Djouadi, 2008; Wittich, Krämer, Bottke, & Schunk, 2020). These types of carbon are not involved in the dry reforming reaction and can only be removed at high temperatures and in the presence of a strong oxidant (Argyle &

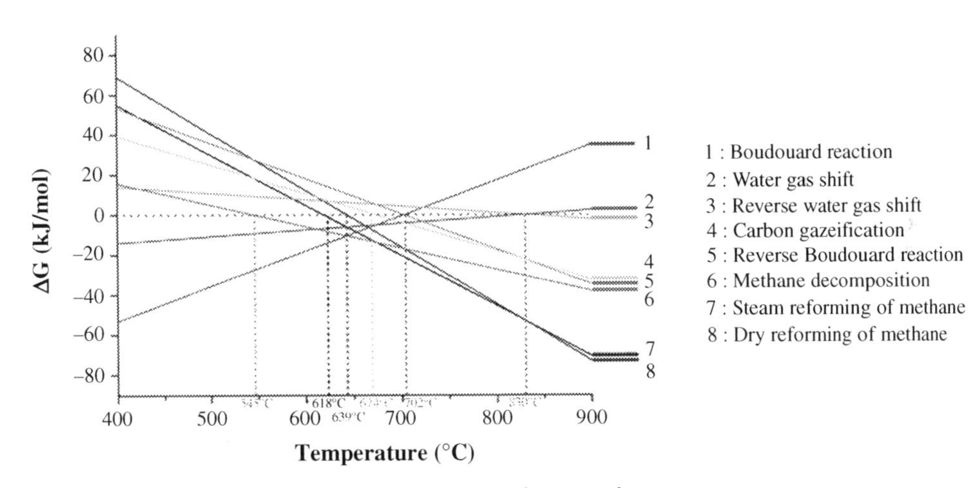

FIG. 8.5 Gibbs energy of main and secondary reactions as a function of temperature.

Bartholomew, 2015; Ginsburg, Piña, El Solh, & De Lasa, 2005).

There are three main carbon types that can cause the deactivation of the catalyst (Fig. 8.6).

First, the carbon deposit can form one or more layers over the catalytic surface and encapsulate the metal particles (Fig. 8.6 (1)). This blocks the active sites and prevents their interaction with the reactants. This phenomenon subsequently prevented the interaction between the reactants and the active sites of the catalyst and led to the deactivation of the catalysts (He et al., 2009).

Carbon adsorbed on the catalytic surface can also produce filaments of carbon. After the activation and dissociation on the catalytic site, if the carbon is not oxidized, it can be subject to a polymerization reaction on top of certain catalytic sites. In that case, the carbon filament extends upward and forms graphitic cylinders (Kumar et al., 2015).

In some cases, ramifications can also be observed which makes the lattice a lot more complex (Fig. 8.6 (2)). This mechanism is known as the base-growth model and can cause complete deactivation of the catalyst. On the other hand, carbon may migrate to the bottom of the metal particle and subsequently lift it to cause it to move away from the support or other particles (Fig. 8.6 (3)). This will make it difficult for the oxygen adsorbed on the CO_2 dissociation site to interact with the carbon on the methane activation site (Aouad, 2018; Fakeeha, Alfatish, Soliman, & Ibrahim, 2006). The carbon will then accumulate until it causes the reactor to block (Duan et al., 2021; Farquhar & Rovers, 1973; Santos, Barros, & Fiho, 2020; US EPA, 2016; Vaverková, 2019; Xu, Tsukuda, Sato, Gu, & Katayama, 2018; Ziyang et al., 2015). This type of mechanism is referred to as the tip-growth model and can occur when the interaction between the support and the active site is not too strong, or when particle size is small (<5 nm). In both cases, deactivation can be observed following the reactor plugging (Kumar, 2011). Carbon formed through carbon gasification of the Boudouard reaction follows similar mechanisms but with different active sites involved in the process.

It is important to underline that the deactivation of the catalyst can take place depending on the type of as well as the quantity of the carbon. Some researchers (Ahmed, Aitani, Rahman, Al-Dawood, & Al-Muhaish, 2009; Suelves, Lázaro, Moliner, Corbella, & Palacios, 2005) noticed that despite the large amount of carbon formed at low temperature (14 times the mass of catalyst), the catalysts were not deactivated whereas at high temperature, a much smaller amount of carbon was sufficient to deactivate them. In this case, the type and not the amount of carbon caused catalyst deactivation. In other cases, the excessive amount of carbon formed causes the plugging of the reactor (Dehimi, Benguerba, Virginie, & Hijazi, 2017; Ha et al., 2017; Newson, 1975). In addition to temperature, the type of catalyst and particle size also influences the type of carbon deposit.

Regeneration of the deactivated catalyst by the formation of carbon deposit is an expensive process. This makes it difficult to use this reaction for industrial purposes, hence the interest in developing an active and selective catalyst for the dry reforming reaction that is stable and resistant to sintering and carbon deposition.

Catalyst deactivation by sintering

The sintering of a catalyst is a degradation that can induce a loss in active surface, in particular by causing the increase of the crystalline phase or the decrease in the size of the pores caused by a high temperature. This will lead to a decrease in catalytic activity in an irreversible way (Gandhi & Patel, 2015; Naeem et al., 2014a). In general, this phenomenon is caused by the migration of crystallites or metal atoms which will agglomerate with other crystallites generally of greater size. They will regroup and form larger particles and thus result in a smaller active surface. The main parameters that influence sintering are the temperature, the

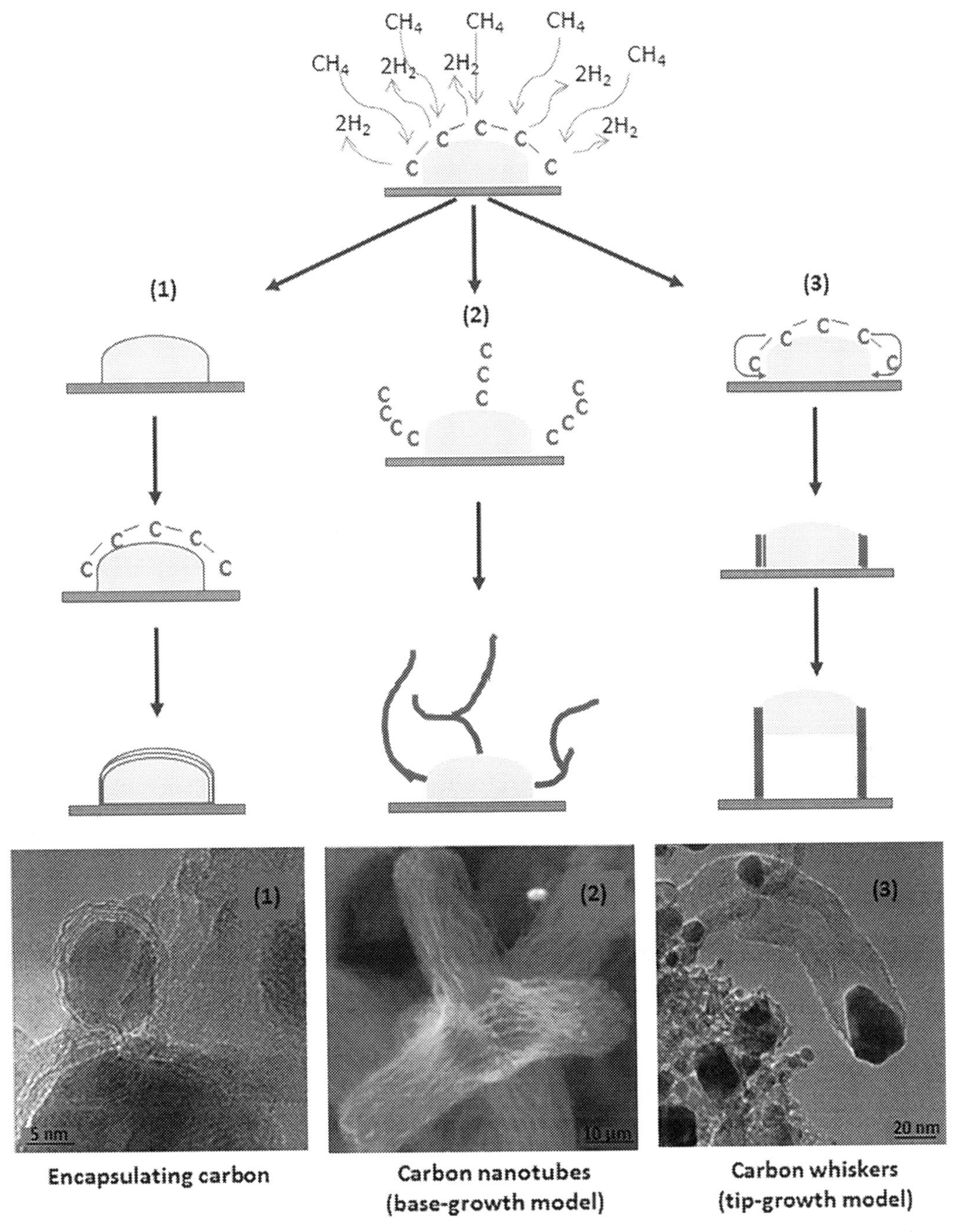

FIG. 8.6 Types of carbon formation causing the deactivation of the catalysts, (1): encapsulating carbon, (2): carbon nanotubes following a base-growth model, (3): carbon whiskers following a tip-growth model. Formation of filamentous and encapsulating carbon through three different mechanisms. *From Sehested, J. (2006). Four challenges for nickel steam-reforming catalysts.* Catalysis Today, 111*(1–2), 103–110. https://doi.org/10.1016/j.cattod.2005.10.002.*

nature of the metal, the support and the promoters, the dispersion of the metals, and the porosity of the material (Argyle & Bartholomew, 2015).

Catalytic deactivation by poisoning

By definition, it is the strong chemisorption of an impurity on the catalytic surface. These molecules may bind to the active sites specific to the reactants, or bind to an adjunct site and subsequently prevent the interaction between the reactants and the site of interest. This will lead to a reduction in the activity of the catalyst. In some cases, the presence of an impurity can lead to the rapid and total deactivation of the catalytic system. This phenomenon is associated with the presence of siloxanes, sulfurous, nitrogenous, or halogenated molecules. In the case of H_2S, a complete deactivation can even take place at very low concentration of this gas in the reaction mixture. Mancino, Cimino, and Lisi (2016) studied the effect of the presence of 30 ppm of H_2S on the activity of an Rh/Al_2O_3 catalyst. The effect of H_2S is apparent on the activity of this catalyst even at concentrations close to 1 ppm (Mancino et al., 2016). The sulfur present in the H_2S molecule adsorbs rapidly onto the metal sites through the following reaction (Eq. 8.24):

$$Me + H_2S \leftrightarrow Me - S + H_2 \tag{8.24}$$

The sulfur then forms a layer on the surface of the catalyst. This compound thus causes the deactivation of the catalytic system (Duan et al., 2021; Farquhar & Rovers, 1973; Santos, Barros, & Fiho, 2020; US EPA, 2016; Vaverková, 2019; Xu et al., 2018; Ziyang et al., 2015). It can be regenerated by thermal treatment at 300°C in the presence of water by eliminating the H_2S in the form of SO_2 or SO_3 (Eqs. 8.25–8.26): (Bogild Hansen & Rostrup-Nielsen, 2010)

$$Me - S + H_2O = Me - O + H_2S \tag{8.25}$$

$$H_2S + 2H_2O \leftrightarrow SO_2 + 3H_2 \tag{8.26}$$

To avoid this type of deactivation, the biogas must undergo purification before being valorized.

8.4 Catalytic reforming of biogas

It is possible to avoid deactivation of the catalysts as well as increase the catalytic activity by modifying the chemical formulation, the preparation method of the catalyst, and by changing the conditions of the reaction. In order to compare the efficiency of the different catalysts in all the conditions, many parameters such as the temperature and duration of the test, GHSV, metal loading, calcination, reduction temperature, etc. need to be taken into consideration. While it is very interesting to observe the specific effect of each of the mentioned parameters on the performance of the catalysts, it remains very hard and unreliable to compare the efficiency of catalysts that have been synthesized or tested in different conditions. For this reason, each parameter will be developed individually while examples and references are included in Table 8.3.

Catalytic material

Certain parameters are known in the literature as enhancing the resistance of the catalyst to sintering and carbon formation (Aouad, 2018; Nikoo & Amin, 2011; Zhang et al., 2007). A synthesis of several articles on different materials used in dry reforming showed that certain properties strongly influence the performance of the catalysts:

- the basicity of the materials increases the affinity to CO_2 and consequently facilitates its adsorption and activation by dissociation (Aziz, Jalil, Wongsakulphasatch, & Vo, 2020; Fan, Abdullah, & Bhatia, 2010; García, Fernández, Ruíz, Mondragón, & Moreno, 2009; Tungkamani et al., 2013)

TABLE 8.3 Catalysts used in dry reforming of methane reaction

Synthesis				Test conditions				Test results			
Metal	Support	Method	Calcination temperature (°C)	Temperature (°C)	Time (h)	Feed composition ($CH_4/CO_2/Ar$)	GHSW (mL/gcat/h)	CH_4 conversion (%)	CO_2 conversion (%)	Carbon deposition (%)	Reference
Co	CeO_2	Coprecipitation	550	750	20	20/20/60	30,000	82–83	–	6%	Luisetto, Tuti, and Di Bartolomeo (2012)
Ni								83–79	–	25%	
Co-Ni								89–89	–	6%	
Co	CeO_2	Incipient wet impregnation	900	700	6	20/20/60		08–06	20–10	0%	Ay and Üner (2015)
Ni								77–55	77–58	20%	
Co-Ni								80–55	80–60	13%	
Co	TiO_2	Incipient wet impregnation	400	750	25		60,000	59–48	–	0%	Takanabe, Nagaoka, Nariai, and Aika (2005)
Ni	ZrO_2	Coprecipitation	450	600	25	10/10/80	15,000	70–67	63–56	67.1 mmol	Wolfbeisser et al. (2016)
	CeO_2							38–20	40–26	33.4 mmol	
	$CeZrO_2$							55–30	70–55	34.2 mmol	
Co	MgO	Impregnation	500	700	12	30/30/40	60,000	68–63	74–71	57%	Li, Li, and Zhu (2018)
Ni	$Al2O_3$	Wet impregnation	800	800	12	45/45/10	185,000	23–10	35–20		Luisetto et al. (2017)
Ni-Ru								42–38	56–50		
Ni + La_2O_3	ZrO_2	Doping	700	700	67	45/45/10	42,000	60–44	75–57	1%	Lanre et al. (2020)
Ni	ZrO_2	Incipient wet impregnation	700	650	12	20/20/60	32,000	50–5		49%	Franz et al. (2020)
Ni + Na								40–5		40%	
Ni + Mn								40–5		40%	
Ni + Cs								40–5		38%	
Ni + K								40–5		38%	
Ni + 2Cs								30–5		29%	
Ni + 2K								30–5		30%	

Continued

TABLE 8.3 Catalysts used in dry reforming of methane reaction—cont'd

	Synthesis			Test conditions				Test results			
Metal	Support	Method	Calcination temperature (°C)	Temperature (°C)	Time (h)	Feed composition ($CH_4/CO_2/Ar$)	GHSW (mL/gcat/h)	CH_4 conversion (%)	CO_2 conversion (%)	Carbon deposition (%)	Reference
Ru	SiO_2	Wet impregnation	400	800	20	45/45/10	20,000	25–08	0–39		Whang et al. (2017)
Zr	SiO_2							10–00	04–0		
Ru	ZrO_2-SiO_2							89–88	100–99		
Co-Ru	ZrO_2-SiO_2							89–88	100–99		
Rh	CaO-SiO_2	Wet impregnation	550	550°C	48	30/30/40	45,000	29%	41		Faroldi et al. (2017)
						30/60/10		32%	30		
Ni	MCM-41	One-pot synthesis method	550	550	22	33/33/33	45,000	38–35	41–30		Salazar Hoyos, Faroldi, and Cornaglia (2020)
				750	100			7200%	81%		
Ni	SBA-15	Solid state grinding method	550	700	100	50/50	22,500	78–76	87–86	6,3	Zhang et al. (2017)
	MCM-41							79–70	85–77	32,6	
	KIT-6							76–77	87–86	8,1	
	SBA-15	Wet impregnation						75–58	85–72	38,8	
Ni-Ru	KIT-6	Wet impregnation	550	800	12	20/20/60	60,000	65	76–74	1	Mahfouz et al. (2021)
	Ce-KIT-6							91–97	90–95	40	
Cu	Mg(Al)O	Coprecipitation	800	600°C	25	25/25/50	60,000	5–4	1	34	Song et al. (2018)
Ni-Cu (Cu/Ni=0.5)								46–45	55	1.1	
Ni-Cu (Cu/Ni=1)								45–41	50–43	72.9	
Ni-Mn	Al_2O_3	Wet impregnation	450	700°C	20	50/50	12,000	66–68	75–78		Ramezani, Meshkani, and Rezaei (2018)
10Ni	ZSM-5		560	700	24	33/33/33	3000	67–24	62–36	48	

5Mn10Ni		Wet impregnation						24–13	37–22	12	Najfach, Almquist, and Edelmann (2019)
10Mn10Ni								23–17	40–27	2	
10Ni	NH4Y							40–22	56–35	41	
5Mn10Ni								35–23	54–34	64	
10Mn10Ni								49–23	56–33	1.6	
10Ni	NaY							37–25	53–34		
10Mn10Ni								4.4–4	3–10		
$LaNiFeO_3$		Coprecipitation	850	800	35	50/50	13,700	70–59	91–89	5%	Komarala, Komissarov, and Rosen (2020)
$LaNiMnO_3$			750					49–81	70–91	3%	
$LaNiFeO_3$			850	650				10–15	22–29	5%	
$LaNiMnO_3$			750					30–42	44–59	4%	
CoMgAl		Coprecipitation hydrotalcite	800	750	12	20/20/60	60,000	81–77		25	Chaghouri et al. (2020)
CoNiMgAl								83–83		8	
NiMgAl								88–87		6	
2NiCo	ZSM-5	Wet impregnation	450	700	12	20/20/60	60,000	65–53	71–60	20.6	Estephane et al. (2015)
Ni2Co								60–55	68–62	5	
4Co	MCM-41	Hydrothermal	550°C	750°C	72	25/25/50	50,000	70–0			Liu et al. (2010)
4Ni								70–0			
0.8Pt		Wet impregnation						85–30			
0.8Pt/4Co								85–78			
0.8Pt/4Ni								90–91			

Summary of different materials and synthesis methods used for dry reforming of methane.

- the dispersion of the active phases on the catalytic surface could increase the activity of the catalyst and improve its resistance to sintering and agglomeration of particles. This gives the solid a larger specific surface area and induces a larger contact area between the catalyst and the gas mixture. The dispersion of the active phases thus influences the activity of the catalyst and also its resistance to sintering and carbon formation (Dissanayake, 1991; Zhou et al., 2008)
- the oxygen storage capacity influences the oxidation rate of the carbon deposited on the catalyst surface. The more active oxygen species there are, the more they will react with the carbon on the surface. This will increase the resistance of the catalyst to deactivation by carbon deposition (Bian, Das, Wai, Hongmanorom, & Kawi, 2017; Shamsi, 2004)
- the reducibility of the active sites is an important factor that influences the activity and efficiency of the catalysts for this type of reaction (Ay & Üner, 2015; Shamsi, 2004)
- the thermal stability is an important parameter because dry reforming reactions take place at high temperatures (>650°C) (Dębek, Motak, Grzybek, Galvez, & Da Costa, 2017).

All these parameters fall into two categories: the Red/Ox properties and the metal-support interactions. We can improve these parameters by modifying the nature and the physicochemical properties of the catalyst.

Catalysts composition

Active phase

Several materials have been studied for the dry methane reforming reaction (Table 8.3). Historically, group VIII metals (Rh, Pt, Pd, Ru, Ni, Co, Fe) are researched. More specifically, the first materials studied for this reaction are the noble metals. They have proved to be very active in the dry reforming reaction.

Ruthenium and rhodium are noble metals that have been extensively studied in DRM. They show high activity and a strong resistance to carbon formation (Takami, Yamamoto, & Yoshida, 2020; Zhang et al., 2020). Moreover, carbon formation in the presence of these catalysts is very law. The problem however is the price of these materials. In order to reduce the cost of the catalytic treatment, researchers considered studying transition metals instead of noble metals because they are more abundant and therefore cheaper. Nowadays, some studies still consider the usage of noble metal as a necessity to increase catalytic activity and stability. For this reason, they tend to use noble metals in smaller quantities as promoters (Gennequin et al., 2016; Jabbour, El Hassan, Casale, Estephane, & El Zakhem, 2014).

The transition metals most commonly used in dry methane reforming are nickel (Ni), cobalt (Co), copper (Cu), and iron (Fe). In general, they show a good activity in the DRM reaction but have a higher risk toward deactivation by carbon deposition (Table 8.3).

The performance of nickel-based catalysts is comparable to that of noble metals in terms of activity. This transition metal is cheaper and therefore more affordable than noble metals. The higher activity of nickel catalysts comes from their ability to enhance the CH_4 activation and the dissociation of the C-H bond. Because of this affinity to carbon, Ni-catalysts are prone to carbon formation (Li, Li, & Wang, 2011). In addition, Ni particles have a tendency to form agglomerates due to thermal sintering (Bian et al., 2017). In order to increase the thermal and chemical stability of nickel-based catalysts, transition and noble metals can be added to the catalytic formulation.

Cobalt is frequently used for dry reforming of methane reaction as an active phase (Aramouni, Touma, Tarboush, Zeaiter, & Ahmad, 2017; Bao & Yu, 2018; Liu, Li, & He, 2020; Usman, Wan Daud, & Abbas, 2015; Zhang & Verykios, 1994). Several studies reported that the presence of this transition metal enhanced the stability of the catalyst. There exists however a strong controversy around this metal's activity. In a monometallic Co-based catalyst, some researchers have observed a lack of activity compared to other transition metals. In other cases, cobalt seems to have slightly less or similar conversion rate as highly active transition metals. Cobalt and nickel are frequently combined to form a bimetallic catalyst with better activity and stability. The efficiency of this catalyst was attributed to the synergistic effect between nickel and cobalt as well as the improved interaction between the metal and the support (Guo et al., 2014; Nguyen et al., 2019; Wang et al., 2018; Ziganshin et al., 2011).

In addition to cobalt and nickel, copper and iron have also been studied in the dry reforming of methane reaction. Both of these transition metals can be used to enhance the stability of the catalyst. Fe increases the affinity of the catalytic material to CO_2 enhances the adsorption and dissociation of this molecule which increases the resistance to carbon formation (More, Bhavsar, & Veser, 2016; Tomishige, Li, Tamura, & Nakagawa, 2017). Cu also has a more beneficial effect on the stability than the activity of the catalysts. It has been used as a "dissociation barrier" to reduce the C—H decomposition on Ni catalysts to allow the equilibrium between the dissociation of CO_2 and CH_4 (Chen et al., 2020; Zhu et al., 2009; Zuo et al., 2018). Other metals such as manganese and chrome can also be used as an active phase for the dry reforming of methane reaction. However, unless they are coupled with other elements, these metals tend to show a low activity. For this reason, these metals are less studied as an active phase and are more promising as a promoter or as part of the support (Rouibah et al., 2017; Wang, Zhao, Li, Hu, & Da Costa, 2021).

Support

Several oxides such as MgO, Al_2O_3, SiO_2, and TiO_2 have been studied for the dry reforming of methane reaction.

In general, magnesium is used as a support because of its high basicity. It increases the activation and dissociation of CO_2 which leads to a decrease of carbon formation by reoxidation of the adsorbed carbon (Akbari, Alavi, & Rezaei, 2017). In addition, this metal strongly enhances the thermal stability of the overall structure of the catalyst. Moreover, it has been reported that magnesium can form a solid solution with nickel (NiO-MgO) increasing the resistance to coke formation. The presence of this metal has proved to be beneficial for the stability of the catalyst but has not shown any activity toward the DRM reaction.

Alumina (Al_2O_3) is frequently used as a support in dry reforming of methane reactions because it increases the thermal and mechanical stability of the material. Alumina also enhances the surface area of the catalyst (Adans, Ballarini, Martins, Coelho, & Carvalho, 2017; Melo & Morlanés, 2005). This metal does not play a role in the activity of the catalyst but reduces the risk of thermal sintering. The acidic properties of this material can represent a disadvantage because of the affinity of CO_2 (an acidic molecule) to basic sites (Bang, Hong, Baek, & Shin, 2018).

Silica (SiO_2) and titania (TiO_2) are also conventional supports for catalytic materials used in this reaction. Titania offers the advantage of being reducible which helps in managing the carbon deposition. However, the transformation at high temperature of amorphous titania phase into anatase phase limits the application of this material in DRM (Al-Fatesh, Naeem, Fakeeha, & Abasaeed, 2014; Budiman, Song, Chang, Shin, & Choi, 2012). Silica is a neutral support that increases the affinity of the catalyst to CO_2. Compared to alumina, silica-based catalysts tend to show a weak metal-support interaction. This causes the formation of larger particles which promotes carbon production leading to the deactivation of the catalyst (Xu et al., 2019).

In order to improve the dispersion of the metals and increase the surface area, ordered mesoporous silica-based materials such as

MCM-41, KIT-6, and SBA-15 were developed and studied as a support for catalysts in DRM (Zhang et al., 2017). These materials enhance the activity of catalysts by increasing the surface area. These materials also show a more narrow pore size distribution. Both of these features induce a better incorporation of active metals (Liu et al., 2010). In addition, the well-ordered hexagonal structure increases the catalytic stability of the catalyst and offers a uniform particle size (Baena-Moreno et al., 2021; Minh et al., 2020). SBA-15 also shows the advantage of being less acidic than other supports which helps to prevent coke formation in the dry reforming of methane (Erdogan, Arbag, & Yasyerli, 2018). In comparison to SBA-15 and MCM-41, KIT-6 possesses a larger pore size which increases its effectiveness while dealing with larger molecules. As for MCM-41, it is believed that the smaller particle size could prevent carbon deposition (Zhang, Li, et al., 2017).

Promoter

Cerium is usually used as a support or as a promoter for catalytic material. This metal is known for its basic properties, oxygen retention abilities, and for increasing the metal-support interactions. These parameters contribute to the resistance of the catalyst to carbon formation (Iglesias, Baronetti, & Mariño, 2017; Liu, Da Costa, Hadj Taief, Benzina, & Gálvez, 2017). In addition to ceria, lanthane, sodium, and potassium can also be used as promoters. The addition of these elements can enhance catalytic properties by increasing the basic character and/or the oxygen storage capacity of the material (Ibrahim et al., 2019). However, the incorporation of these materials in the catalytic formulation is not beneficial in terms of activity. Several authors have described the effect of these metals as unrelated to the activity of the catalyst but effective toward the resistance to carbon formation. Lanre et al. studied the effect of La addition to a Ni/ZrO_2 series of catalysts. The results show a decrease in carbon formation in the presence of an optimal amount of lanthanum ($10La_2O_3$) (Lanre et al., 2020).

Synthesis methods

The method of synthesis can also have an impact on the catalysts' stability and/or activity. Several methods of synthesis such as coprecipitation, surfactant-assisted coprecipitation, thermal hydrolysis, sol-gel, pseudo sol-gel, dry and wet impregnation have been developed. The usage of different methods induces an impact on the physicochemical properties which allows us to control the catalytic performances. Using the same elements, a better performance can be acquired if the method of synthesis is modified. This parameter mostly influences the metal–support interaction, the size and the dispersion of the particle as well as the surface area.

Coprecipitation method is frequently used for catalyst synthesis because of its simplicity and reproducibility (Wolfbeisser et al., 2016). In addition, Terribile et al. noticed that coprecipitation positively influences the surface area of the catalyst (Terribile, Trovarelli, Llorca, De Leitenburg, & Dolcetti, 1998). While studying the effect of different preparation methods on Ni-based catalysts, Gurav et al. compared coprecipitation to wet impregnation and citrate gel method (Gurav et al., 2017a; Serrano-Lotina & Daza, 2013). Results showed that catalysts synthesized by coprecipitation had a smaller crystallite size which increased the specific surface area. With better dispersion, the material had higher reducibility at lower temperature. This increased the activity of the sample. Both wet impregnation and citrate gel method produced samples with lower activity correlated to larger NiO crystallite, harder to reduce and block the pores of the support (Gurav et al., 2017b). Herrera et al. compared coprecipitation, nitrate deposition, and freeze-drying methods. Structural, textural, redox, and morphological properties were affected by the preparation method. A better catalytic activity was observed for the catalyst with higher porosity, greater support metal interaction, and better metallic distribution. These characteristics were provided by the coprecipitation method (Torrez-Herrera, Korili, & Gil, 2021). Another study showed that sol-gel method favored the stronger interaction and better dispersion of Ni particles on Al_2O_3, CeO_2, and MgO support (Hassani Rad, Haghighi, Alizadeh Eslami, Rahmani, & Rahemi, 2016).

Currently, new technologies such as hydrothermal (Xie et al., 2020; Zhang, Zhang, Liu, & Zhang, 2016), microemulsion (Dekkar et al., 2020), and combustion are being developed. These methods can enhance the dispersion of active species and reduce the particle size distribution. In addition, microwave-assisted and ultrasound wave-assisted syntheses are being used to develop upgraded materials. These techniques are considered as green technologies because they allow a more efficient synthesis process and less equipment requirement (Das, Sengupta, Bag, Shah, & Bordoloi, 2018). Several researchers have studied the effect of adding these upgrades into an already established and patented synthesis protocol. These studies have shown that by adding a microwave or ultrasound step, catalysts show lower crystallite size, higher concentration of oxygen vacancies, and a better catalytic activity (Hosseini et al., 2018; Medeiros et al., 2021). Compared to traditional methods of synthesis, catalysts prepared by new or assisted technologies are cheaper, easily repeatable, and show an overall better performance.

Effect of pretreatment

Thermal treatment under oxidative atmosphere: Effect of oxidation

The calcination step has an important influence on catalytic performance because it affects the thermal stability and the particle size of the materials. In some cases, authors have observed a decrease in catalytic activity with increasing temperature of calcination. This was linked to the agglomeration of particles and the sintering effect occurring at high temperatures (Kusuma & Chandrappa, 2019; Serrano-Lotina et al., 2011; Sifontes, Rosales, Méndez, Oviedo, & Zoltan, 2013). In other and less frequent cases, the increase in the temperature of calcination decreased crystallite size, enhanced metal dispersion, and positively influenced the resistance to sintering (Katheria, Gupta, Deo, & Kunzru, 2016).

A study made by Ay and Üner (2015) showed that an increase in calcination temperature from 700°C to 900°C did not directly influence the activity of the catalyst but had an effect on carbon formation properties. The catalyst calcined at 700°C had a higher quantity of carbon formation. However, differential thermal analysis on the used samples showed that the carbon formed on the samples calcined at 900°C was oxidized at higher temperature. This means that the carbon deposit was more stable with higher calcination temperature (Ay & Üner, 2015).

Thermal treatment under reductive atmosphere: Effect of reduction

Several studies have shown that without reduction, the solids are not active in the dry reforming reaction (Gennequin, Safariamin, Siffert, Aboukaïs, & Abi-Aad, 2011; Pernicone & Traina, 1978). A study performed on Ni-Mg-Al catalysts showed that the reduction temperature strongly influences the activity and selectivity of the solids. Well dispersed metal particles are obtained after treatment in reduced atmosphere. These metal sites are responsible for the adsorption and activation of reactants by dissociation (Dębek et al., 2015; Perez-Lopez, Senger, Marcilio, & Lansarin, 2006). Usman and Wan Daud (2016) also studied the influence of the reduction temperature on the activity and stability of Ni/MgO catalysts in DRM. The results obtained show a decrease in activity and stability at high reduction temperature. Catalysts reduced at 800°C are more at risk of sintering, and therefore a decrease in activity, than those reduced at 550°C. In addition, different gas mixtures can be used for the reduction of the catalytic sites. In general, hydrogen is used as the reducing molecule. A study by Chen et al. showed that syngas produced during the reforming reaction could be sufficient for solids reduction (Chen & Ren, 1994). Methane can also drive the reduction of mixed and simple oxides. A study carried out on reductions in the presence of different gas mixtures shows that hydrogen and methane are less effective than an H_2+CO mixture for the activation of catalysts by reduction (Dekkar et al., 2020; Shang, Li, Li, Liu, & Liang, 2017; Chen & Ren, 1994).

Reaction conditions

Reaction temperature

One of the most important parameters in dry reforming of methane reaction is the temperature. DRM is an endothermic reaction that is thermodynamically possible at temperatures higher than 650°C (Fig. 8.5). However, secondary reactions can also occur at higher temperatures. Boudouard reaction is favored at temperatures under 700°C and methane cracking is possible at temperatures higher than 550°C. these two reactions produce carbon as a by-product which means that in the temperature range between 550°C and 700°C, carbon formation is at its peak (Charisiou et al., 2019). For this reason, it is important to increase the temperature of the reaction above the mentioned range in order to reduce the occurrence of secondary reactions, limit carbon formation and favor the dry reforming of methane.

To study the efficiency of catalysts in DRM, two types of tests can be used. First, reactant conversion and product formation can be monitored while increasing the temperature. This type of test is used to study the activity of the catalysts as well as the selectivity toward H_2 and CO formation at certain temperatures. For these tests, the temperature range is between 400°C and 900°C. In general, the conversion of CH_4 and CO_2 increases with the temperature rise, which is due to the endothermic nature of the reaction. Second, stability tests can be done at a constant temperature. This type of test allows us to study the catalytic stability while monitoring its activity as a function of time. These tests are usually performed at temperatures around 650–850°C. If at any point the conversion of the reactants decreases, it means a deactivation is taking place. The used samples can be analyzed after catalytic test in order to identify the deactivation mechanism. Table 8.4 summarizes the parameters that are directly influenced by the variation of temperature.

Most catalysts seem to have similar behavior when it comes to the temperature ranges lower or higher than 700°C. The difference in reactant

TABLE 8.4 Parameters influenced by temperature variation.

	Temperatures < 700°C	**Temperatures > 700°C**
Reactant conversion	$CH_4 > CO_2$	$CO_2 > CH_4$
H_2/CO	> than 1	Close to 1
Carbon balance	Less than 100%	Close to 100%
Quantity of carbon formation	High quantity	Low quantity
Type of carbon	Tends to be unstable	Tends to be more stable

Influence of temperature on reactant conversion, H_2/CO ratio, and carbon formation.

conversion is related to the occurrence of the methane cracking reaction, the Boudouard reaction, and the reverse water gas shift reaction. The first two are carbon forming reactions favored at lower temperatures. While these reactions take place, H_2 is being produced and CO is consumed, which explains de H_2/CO ratio > 1. At temperatures higher than 700°C, the occurrence of Boudouard reaction and methane cracking is attenuated and the reverse water gas shift reaction is favored and dry reforming of methane is attenuated (Fig. 8.5). At this temperature, methane is mostly reaction in DRM but CO_2 is involved in DRM and RWGS, which is why CO_2 consumption is higher than that of CH_4 and $H_2/CO = 1$. Both carbon-forming reactions are repressed which leads to a decrease in carbon formation and a higher carbon balance. However, studies show that the carbon produced at higher temperature (usually graphitic or whiskers) is harder to oxidize than the lower temperature carbon (adsorbed or nanofibers). These results show the benefit of increasing the temperature for a seemingly better overall catalytic performance. However, thermal sintering has been reported to cause irreversible damage to catalytic activity at higher temperatures (Liu et al., 2010; Saha, Khan, Ibrahim, & Idem, 2014; Takanabe et al., 2005a). This observation was especially frequent in Ni-based catalysts (Donphai, Witoon, Faungnawakij, & Chareonpanich, 2016). Knowing that both carbon formation and sintering can influence the catalytic performance, it is important to select a temperature range enhancing the optimal catalytic behavior.

In summary, the increase of reaction temperature not only increases the activity of the catalyst but also favors the DRM reaction and consequently decreases the carbon formation. Considering the cost of temperature increase and the deactivation by sintering, it is important to choose the lowest temperature that allows an acceptable syngas yield.

Pressure

This parameter is usually not used in the methane reforming reactions because it does not favor the formation of syngas from an equilibrium perspective ($n = 2 < n = 4$). Higher pressure usually causes a decrease in catalytic activity and an increase in carbon formation (Jean-Michel, 2014; Usman et al., 2015). Nikoo et al. suggested that higher pressure could suppress the effect of higher temperature on catalytic activity. DRM shifts to the left (reactant side) following LeChatelier's principle (Aouad, 2018; Nikoo & Amin, 2011a; Zhang et al., 2007a). There also seems to be a link between the increase in pressure and in carbon formation. While studying the effect of pressure increase on carbon formation, Shamsi and Johnson noticed that the carbon produced by methane cracking reaction was decreased. However, pressure had the opposite effect on the Boudouard reaction which was favored at higher pressures (Shamsi & Johnson, 2003). For these reasons, DRM reaction is usually performed at atmospheric pressure (Alipour, Rezaei, & Meshkani, 2014; Khajeh Talkhoncheh & Haghighi, 2015; Whang et al., 2017).

Inlet gas composition and flow rate

In addition to temperature and pressure, the inlet gas composition and flow rate can have a high influence on the performance of catalysts.

Effect of flow rate

For this part, flow rate is considered per g of catalyst (gas hourly space velocity, GHSV). Generally, inlet gas mixture is introduced into a reactor where the catalytic bed is placed. At this point, the contact between the reactants and the catalyst occurs and the reaction takes place. The duration of the contact depends on the flow rate. When the contact time is increased, a larger quantity of CH_4 and CO_2 will react. Therefore, when the flow rate increases, the contact time becomes shorter and the conversion of reactants decreases. Akbari et al. studied the influence of the GHSV on catalytic activity at 700°C. Their results showed a decrease in the conversion of methane and carbon dioxide as the GHSV increased (Akbari et al., 2017). Several authors have studied the influence of this parameter on the catalytic performance. Their results have been similar to the one mentioned earlier (Gao, Jiang, Meng, Yan, & Aihemaiti, 2018; Serrano-Lotina & Daza, 2014). However, the link between the variation of flow rate and carbon formation is not clear yet because of the implication of several parameters such as surface area, temperature, active sites, etc.

Effect of gas composition

Methane and carbon dioxide ratio is another factor that could influence the carbon formation and the catalytic activity is the composition of the inlet feed. In general, inlet gas used for DRM contains a ratio of $CO_2/CH_4=1$. This mixture is used to follow the stoichiometric conditions of this reaction ($CH_4+CO_2=2H_2+2CO$). Some studies were done to determine the effect of the CO_2/CH_4 ratio on the catalytic activity. Serrano-Lotina et al. studied the performance of a Ni-based catalyst under different CO_2/CH_4 ratios between 0.4 and 1.5 (Serrano-Lotina & Daza, 2014). A higher CH_4 conversion was observed when the ratio was higher than 1. However, the selectivity to hydrogen decreased because the reverse water gas shift reaction was favored (converts H_2 to H_2O). At a ratio lower than 1, conversion of CH_4 and production of water decreased since these conditions did not favor the RWGS reaction (Serrano-Lotina & Daza, 2014). Another interesting study done by Tungkamani et al. correlated the limiting reagent (CO_2 or CH_4) to the amount of carbon formation (Tungkamani et al., 2013). This study showed that when CH_4 is the limiting reactant ($CO_2/CH_4>1$), methane conversion is at 100% and the H_2/CO ratio is equal to unity which means that 100% of methane reacted with CO_2 through the dry reforming reaction. However, when CO_2 is the limiting reagent, the conversion of CH_4 decreases and carbon formation increases because the excess of methane undergoes decomposition and produces carbon (Tungkamani et al., 2013). A study on thermodynamic equilibrium of dry reforming and a design on dynamic modeling also show similar results: a decrease in CO_2/CH_4 ratio favors the occurrence of secondary reactions (especially methane decomposition and water gas shift reaction) resulting in an increased carbon formation (Chein, Chen, Yu, & Chung, 2015; Luyben, 2014).

It is very important to take into consideration the effect of the inlet gas composition on the dry reforming of methane for biogas valorization. As we mentioned before, the CO_2/CH_4 ratio in biogas is usually lower than 1 which means carbon formation will be favored. This is why in some cases, CO_2 is added to the initial biogas mixture in order to increase the CO_2 concentration and the reoxidation of carbon.

Influence of impurities As previously mentioned, biogas is mainly composed of CH_4 and CO_2. It contains however a variety of impurities depending on several factors in the waste degradation process (nature and age of the waste, conditions of the process, microbial consortium, etc.).

H_2O and O_2 are considered impurities in biogas because they represent less than 5% of its standard composition. However, they have been heavily studied as additives in the reforming process. Several studies have compared the activity and stability of catalytic materials in the presence of these two oxidative molecules. By adding even a small quantity of H_2O or O_2, carbon formation can be reduced which increases the shelf life of the catalysts.

A study performed on dry reforming and combined reforming shows that carbon formation is attenuated in the presence of water. A small deactivation is observed in the presence of water compared to the DRM-only catalyst which deactivates completely (Kumar et al., 2015). A second study performed on combined reforming in the presence of cobalt-based catalysts shows similar results. In this case, the presence of water in the reaction mixture improves the hydrogen yield and suppresses carbon formation (Itkulova, Zakumbaeva, Nurmakanov, Mukazhanova, & Yermaganbetova, 2014). Furthermore, a study performed on catalysts also based on cobalt shows that the presence of water caused the decrease of the temperature at which CH_4 and CO_2 are fully converted.

Several studies have compared the amount of carbon formed in the presence and absence of oxygen during the dry methane reforming reaction (Li et al., 2011). They noted that oxyreforming influences not only the amount but also the type of carbon formed. In dry reforming, graphitic carbon is predominant. While in oxyreforming, the most abundant carbon is in amorphous or atomic form (Kumar, Wang, Kanitkar, & Spivey, 2016). This type of carbon is easier to remove by oxidation. In addition, several studies performed on oxyreforming have also highlighted the increase in CH_4 conversion as well as the decrease in CO_2 conversion in the presence of oxygen (Nematollahi, Rezaei, & Khajenoori, 2011).

The tri-reforming process can also be interesting. In this case, methane reacts with carbon dioxide, water, and oxygen molecules. Usually, the ratio between the oxidant ($CO_2+H_2O+O_2$) and methane used for these tests is close to 1 (Baena-Moreno et al., 2021; Minh et al., 2020). In this case, the steam reforming, the dry reforming, and the partial oxidation of methane reactions can take place. This process generates a syngas with a ratio H_2/CO between 1.5 and 2 (Song & Pan, 2004; Świrk, Grzybek, & Motak, 2017). This ratio can be modified by changing the initial feed composition. Zhang et al. studied the influence of the concentration of CO_2, H_2O, and O_2 on the carbon formation and H_2 yield. They observed an increase in carbon formation after decreasing the concentration of oxidative molecules (Zhang, Zhang, Gossage, Lou, & Benson, 2014). The disadvantage of having a high concentration of these molecules is the reoxidation of the reduced metallic sites active in DRM. In this case, the catalyst undergoes a progressive deactivation by reoxidation. This type of deactivation is reversible and can be managed by a reduction treatment.

There are, however, other molecules such as halogens, siloxanes, and sulfur compounds that are more dangerous to the catalyst and can cause its deactivation by poisoning.

Halogen compounds containing bromine, chlorine, and iodine have been known to cause the deactivation of catalysts used in many reactions (formaldehyde oxidation (Chen et al., 2020; Zhu et al., 2009; Zuo et al., 2018), hydrogenation (Mork, 1976), dry reforming (Argyle & Bartholomew, 2015), etc.). In some studies, the deactivation by chlorine poisoning is described as a reversible phenomenon (Mortensen et al., 2014). In others, the deactivation process seems to be dependent on the concentration, the temperature, and the size of the halogen molecule (Mork, 1976). In most cases, the regeneration of a deactivated catalyst after halogen poisoning leads to the recovery of less than 50% of the initial activity.

The presence of siloxane in the inlet gas is also a major issue because silicon is known to be a severe poison for catalysts (Gao et al., 2018b; Serrano-Lotina & Daza, 2014). Researches show that the deactivation is directly related to the nature of the siloxane. For example, dimethoxydimethylsilane induces a rapid deactivation whereas dimethylsiloxane has no effect on the activity of the catalysts (Dubreuil et al., 2017). Siloxane concentration as well as the operating conditions have an impact on the poisoning effect of these molecules (Elsayed, Elwell, Joseph, & Kuhn, 2017). The deactivation mechanism of these molecules has not yet been established (Finocchio et al., 2009; Gao et al., 2018a). It has, however, been recognized that deactivation by siloxane poisoning ultimately leads to catalyst replacement which represents a large financial loss for industries (Chainet, Lienemann,

Courtiade, Ponthus, & Xavier Donard, 2011; Dubreuil et al., 2017).

H_2S can cause the deactivation of catalysts at very low concentrations (<1 ppm) (Mancino et al., 2016b). First, this molecule is adsorbed on the surface of the catalyst and undergoes a transforms into sulfur. Then, sulfur forms a covalent bond with the metal and blocks the active site, which reduces the activity of the catalyst and causes its deactivation. Several studies described the toxic effect of H_2S as dependent on the nature of the metal and the temperature of the reaction (Hulteberg, 2012; Saha et al., 2014). Noble metals such as ruthenium, rhodium, palladium, and platine are more resistant to the adsorption of H_2S than transition metals (cobalt and nickel) especially at temperatures higher than 600°C (Hulteberg, 2012). Researchers and studying several parameters In order to address this issue. Some studies show that a temperature of 900°C can be used to increase the resistance of catalysts to poisoning (Mancino et al., 2016a). Moreover, the addition of different metals such as molybdenum to the catalyst composition can also decrease the vulnerability of catalysts to the presence of H_2S (Gaillard, Virginie, & Khodakov, 2017).

The most common way to overcome the presence of H_2S, siloxanes, and halogens in biogas remains their elimination beforehand (Chung, Ho, & Tseng, 2006; Coppola & Papurello, 2018).

8.5 Conclusions and perspectives

With the current economic and environmental problems, researches have been done to find an alternative source of energy to replace fossil fuels. Biomass is a renewable and unlimited source of energy. It can be transformed into biogas through the anaerobic digestion of organic waste. This series of reactions takes place in the presence of microorganisms such as bacteria and archaea and in an oxygen-free environment. This process allows the transformation of nonrecyclable and organic waste into a gas mixture called biogas that can be converted into chemicals of higher added value.

Biogas is mainly made of methane (CH_4) and carbon dioxide (CO_2) and can contain different amounts of oxygen (O_2), water (H_2O), and nitrogen (N_2). It contains however several minor compounds such as VOCs, sulfur compounds, halogens, oxygenated compounds, nitrogen compounds, etc. that can be considered as impurities for the valorization of biogas. The concentration of these compounds depends on several factors such as the type and conditions of the process, the microbial consortium, the nature and the age of waste, etc. For this reason, several technics of biogas purification and upgrading must be applied before using biogas for industrial purposes.

Many of the impurities found in biogas can cause damage to the internal industrial structure. Chemical and physical methods have already been put in place for the elimination of H_2S, siloxane, VOC, and halogen compounds. Biological methods of elimination are currently being developed. After the cleaning process, biogas is subjected to an upgrading step which consists of the elimination of CO_2. Biomethane can then be used as natural gas or can be transformed into other compounds with higher added value. Purified CO_2 can also be used as a primary feedstock in several chemical industries. An equally interesting approach for biogas valorization is its usage in chemical reactions such as dry, steam reforming and oxidation of methane. Synthesis gas made of H_2 and CO is produced through the mentioned reactions. This gas can be transformed through many industrial processes into synthetic fuel and chemicals with a higher added value such as ammonia, methanol, oxo-alcohols, etc.

Dry reforming of methane provides the advantage of utilizing both of the major compounds in biogas (CH_4 and CO_2). It also produces a syngas with a ratio H_2/CO of 1 that can be used for industrial applications such as oxo-alcohol synthesis. However, this reaction is thermodynamically possible at temperatures higher than 650°C because of its endothermic nature. The usage of a catalytic material is therefore necessary to increase the kinetics of the reaction at lower temperatures. At high temperature (>650°C) and in the presence of biogas, catalytic materials risk deactivation through several processes such as sintering, carbon deposition, poisoning, and reoxidation. In order to increase the resistance of catalysts to deactivation as well as a higher reactivity, several parameters can be modified.

The elemental composition of the catalysts and the method of preparation strongly influence catalytic properties. Some noble metals such as ruthenium and rhodium are active in the dry reforming of methane reaction but are too expensive to be used on a large scale. Transition metals have shown high activity in this reaction but are at high risk of deactivation, especially by sintering and carbon deposition. The performance of these metals has led to the synthesis of a large variety of materials using different metals such as nickel, copper, and cobalt and with different amounts. Furthermore, the synthesis method can be modified in order to alter the chemical properties of the materials. Sol-gel, coprecipitation, and wet impregnation are the main synthesis methods used for the preparation of catalysts. These methods are used to elaborate specific structures such as hydrotalcite, perovskite, mesoporous materials, etc. in order to increase the dispersion of the elements. The preparation method strongly influences the efficiency of the catalyst.

In general, a pretreatment takes place prior to the dry reforming of methane reaction in order to activate the catalysts. Calcination, a thermal treatment in an oxidative environment, allows the decomposition of the structured material into spinels, simple and mixed oxides. These solids have a higher surface area and a much better thermal and chemical stability. A reductive pretreatment can also be used to activate the catalytic sites of the material. This step takes place in the presence of a reductive gas, such as H_2, methane, and carbon monoxide. The temperatures and the gas mixtures used in both of these steps affect the performance of the catalysts.

The conditions of the test such as temperature and pressure have a big influence on the final yield of the reaction. Atmospheric pressure is usually used for DRM. At higher pressure, this reaction is repressed and carbon formation is favored, especially through the Boudouard reaction. To sustain a stable conversion of reactants, a temperature of at least 700°C must be used for the dry reforming of methane. By increasing the temperature, this reaction is favored as opposed to other secondary reactions such as methane cracking and the Boudouard reaction. For this reason, carbon formation can be minimized at higher temperatures. However, sintering is favored at higher temperatures.

Finally, inlet gas flow rate and composition influence the performance of the materials. At higher flow rate, the time of contact between the catalysts and the reactants is reduced which, in some cases, decreases the conversion. Theoretically, DRM is a stoichiometric reaction ($CH_4/CO_2 = 1$). If the initial gas inlet contains a ratio of methane and carbon dioxide higher than 1, carbon formation is favored. On the other hand, a ratio below 1 reduces the production of a carbon deposit. Water and/or oxygen can also be added to the inlet gas. In general, the presence of either of these molecules tends to increase the stability and the activity of the catalysts by increasing their resistance to carbon formation. Both of these molecules are found in biogas and can cause the reoxidation of carbon. Other molecules, however, can increase the risk of deactivation. A rapid loss of activity can be observed in the presence of siloxanes, halogens, and sulfur compounds. These molecules cause the deactivation of catalysts by poisoning.

Nowadays, there are a lot of materials that are being studied for the industrialization of the dry reforming of methane reaction. These catalysts are relatively active and stable but are not efficient enough to sustain a commercialization. None has stood out yet for this reaction which is why more and more research are being done to develop faster and more efficient ways to synthesize materials. While nickel has shown great promise in terms of activity, its downside remains the susceptibility to sintering and carbon formation. For this reason, different formulations and methods of preparation are being developed and studied in order to achieve and sustain a high activity for the application of dry reforminane reaction on biogas.

References

Adans, Y. F., Ballarini, A. D., Martins, A. R., Coelho, R. E., & Carvalho, L. S. (2017). Performance of nickel supported on Γ-alumina obtained by aluminum recycling for methane dry reforming. *Catalysis Letters, 147*(8), 2057–2066. https://doi.org/10.1007/s10562-017-2088-3.

Ahmed, S., Aitani, A., Rahman, F., Al-Dawood, A., & Al-Muhaish, F. (2009). Decomposition of hydrocarbons to hydrogen and carbon. *Applied Catalysis A: General, 359*(1–2), 1–24. https://doi.org/10.1016/j.apcata.2009.02.038.

Akbari, E., Alavi, S. M., & Rezaei, M. (2017). Synthesis gas production over highly active and stable nanostructured Ni-MgO-Al_2O_3 catalysts in dry reforming of methane: Effects of Ni contents. *Fuel, 194*, 171–179. https://doi.org/10.1016/j.fuel.2017.01.018.

Al-Fatesh, A., Naeem, M., Fakeeha, A., & Abasaeed, A. (2014). Role of La_2O_3 as promoter and support in Ni/γ-Al_2O_3 catalysts for dry reforming of methane. *Chinese Journal of Chemical Engineering, 22*(1), 28–37. https://doi.org/10.1016/S1004-9541(14)60029-X.

Alipour, Z., Rezaei, M., & Meshkani, F. (2014). Effect of alkaline earth promoters (MgO, CaO, and BaO) on the activity and coke formation of Ni catalysts supported on nanocrystalline Al_2O_3 in dry reforming of methane. *Journal of Industrial and Engineering Chemistry, 20*(5), 2858–2863. https://doi.org/10.1016/j.jiec.2013.11.018.

Allegue, L. B., & Hinge, J. (2012). *Biogas and bio-syngas upgrading* (pp. 1–97). Danish Technological Institute.

Allen, M. R., Braithwaite, A., & Hills, C. C. (1997). Trace organic compounds in landfill gas at seven U.K. waste disposal sites. *Environmental Science and Technology, 31*(4), 1054–1061. https://doi.org/10.1021/es9605634.

Alper, E., & Yuksel Orhan, O. (2017). CO_2 utilization: Developments in conversion processes. *Petroleum, 3*(1), 109–126. https://doi.org/10.1016/j.petlm.2016.11.003.

Amosa, M., Mohammed, I., & Yaro, S. (2010). Sulphide scavengers in oil and gas industry – A review. *Nafta, 61*, 85–98.

Anneli, & Athur, W. (2006). Biogas upgrading technologies – Developments and innovations. *IEA Bioenergy, 34*, 1–20.

Aouad, S. (2018). *A review on the dry reforming processes for hydrogen production.* Catalytic Materials and Technologies.

Aramouni, N. A. K., Touma, J. G., Tarboush, B. A., Zeaiter, J., & Ahmad, M. N. (2017). Catalyst design for dry reforming of methane: Analysis review. *Renewable and Sustainable Energy Reviews*, 1–16.

Argyle, M. D., & Bartholomew, C. H. (2015). Heterogeneous catalyst deactivation and regeneration: A review. *Catalysts, 5*(1), 145–269. https://doi.org/10.3390/catal5010145.

Arnold, M., & Kajolinna, T. (2010). Development of on-line measurement techniques for siloxanes and other trace compounds in biogas. *Waste Management, 30*(6), 1011–1017. https://doi.org/10.1016/j.wasman.2009.11.030.

Asadi, M., & Pourhossein, K. (2019). Wind and solar farms site selection using geographical information system (GIS), based on multi criteria decision making (MCDM) methods: A case-study for East-Azerbaijan. In *2019 Iranian conference on renewable energy and distributed generation, ICREDG 2019*Institute of Electrical and Electronics Engineers Inc. https://doi.org/10.1109/ICREDG47187.2019.190216.

Awe, O. W., Zhao, Y., Nzihou, A., Minh, D. P., & Lyczko, N. (2017). A review of biogas utilisation, purification and upgrading technologies. *Waste and Biomass Valorization, 8*(2), 267–283. https://doi.org/10.1007/s12649-016-9826-4.

Ay, H., & Üner, D. (2015). Dry reforming of methane over CeO_2 supported Ni, Co and Ni-Co catalysts. *Applied Catalysis B: Environmental, 179*, 128–138. https://doi.org/10.1016/j.apcatb.2015.05.013.

Aziz, M. A. A., Jalil, A. A., Wongsakulphasatch, S., & Vo, D. V. N. (2020). Understanding the role of surface basic sites of catalysts in CO_2 activation in dry reforming of methane: A short review. *Catalysis Science and Technology, 10*(1), 35–45. https://doi.org/10.1039/c9cy01519a.

Baena-Moreno, F. M., Sebastia-Saez, D., Pastor-Pérez, L., & Reina, T. R. (2021). Analysis of the potential for biogas upgrading to syngas via catalytic reforming in the United Kingdom. *Renewable and Sustainable Energy Reviews, 144*. https://doi.org/10.1016/j.rser.2021.110939.

Balat, M., & Balat, H. (2009). Biogas as a renewable energy sourcea review. *Energy Sources, Part A: Recovery, Utilization and Environmental Effects, 31*(14), 1280–1293. https://doi.org/10.1080/15567030802089565.

Bang, S., Hong, E., Baek, S. W., & Shin, C.-H. (2018). Effect of acidity on Ni catalysts supported on P-modified Al_2O_3 for dry reforming of methane. *Catalysis Today, 303*, 100–105. https://doi.org/10.1016/j.cattod.2017.08.013.

Bao, Z., & Yu, F. (2018). Catalytic conversion of biogas to syngas via dry reforming process. In *Vol. 3. Advances in Bioenergy* Elsevier.

Barbusiński, K., & Kalemba, K. (2016). Use of biological methods for removal of H_2S from biogas in wastewater treatment plants – A review. *Architecture, Civil Engineering, Environment*, 103–112. https://doi.org/10.21307/acee-2016-011.

Benito, M., García, S., Ferreira-Aparicio, P., Serrano, L. G., & Daza, L. (2007). Development of biogas reforming Ni-La-Al catalysts for fuel cells. *Journal of Power Sources, 169*(1), 177–183. https://doi.org/10.1016/j.jpowsour.2007.01.046.

Bharathiraja, B., Sudharsana, T., Jayamuthunagai, J., Praveenkumar, R., Chozhavendhan, S., & Iyyappan, J. (2018). Biogas production – A review on composition, fuel properties, feed stock and principles of anaerobic digestion. *Renewable and Sustainable Energy Reviews, 90*, 570–582. https://doi.org/10.1016/j.rser.2018.03.093.

Bian, Z., Das, S., Wai, M. H., Hongmanorom, P., & Kawi, S. (2017). A review on bimetallic nickel-based catalysts for CO_2 reforming of methane. *ChemPhysChem, 18*(22), 3117–3134. https://doi.org/10.1002/cphc.201700529.

Bierer, B., Kress, P., Nägele, H. J., Lemmer, A., & Palzer, S. (2019). Investigating flexible feeding effects on the biogas quality in full-scale anaerobic digestion by high resolution, photoacoustic-based NDIR sensing. *Engineering in Life Sciences, 19*(10), 700–710. https://doi.org/10.1002/elsc.201900046.

Bogild Hansen, J., & Rostrup-Nielsen, J. (2010). Sulfur poisoning on Ni catalyst and anodes. In *Vol. 6. Handbook of fuel cells* (pp. 1–13). https://doi.org/10.1002/9780470974001.f500064.

Borduas, N., & Donahue, N. M. (2018). *The natural atmosphere*.

Bouzonville, A., Peng, S., & Atkins, S. (2013). Review of long term landfill gas monitoring data and potential for use to predict emissions influenced by climate change. In *Clean air society of Australia and New Zealand Conference Sydney, Australia* (p. 2).

Brosnan, J., & Brosnan, M. E. (2006). Amino acid assessment workshop. *The Journal of Nutrition, 136*, 16365–16405.

Budiman, A. W., Song, S. H., Chang, T. S., Shin, C. H., & Choi, M. J. (2012). Dry reforming of methane over cobalt catalysts: A literature review of catalyst development. *Catalysis Surveys from Asia, 16*(4), 183–197. https://doi.org/10.1007/s10563-012-9143-2.

Chaghouri, M., Hany, S., Cazier, F., Tidahy, L., Gennequin, C., & Aad, E. A. (2020). Catalytic reforming of biogas for syngas production: A study of stability and carbon deposition. In *11th international renewable energy congress, IREC 2020*Institute of Electrical and Electronics Engineers Inc. https://doi.org/10.1109/IREC48820.2020.9310412.

Chainet, F., Lienemann, C. P., Courtiade, M., Ponthus, J., & Xavier Donard, O. F. (2011). Silicon speciation by hyphenated techniques for environmental, biological and industrial issues: A review. *Journal of Analytical Atomic Spectrometry, 26*(1), 30–51. https://doi.org/10.1039/c0ja00152j.

Charisiou, N. D., Douvartzides, S. L., Siakavelas, G. I., Tzounis, L., Sebastian, V., Stolojan, V., et al. (2019). The relationship between reaction temperature and carbon deposition on nickel catalysts based on Al_2O_3, ZrO_2 or SiO_2 supports during the biogas dry reforming reaction. *Catalysts, 9*(8). https://doi.org/10.3390/catal9080676.

Charlson, R. J., Schwartz, S. E., Hales, J. M., Cess, R. D., Coakley, J. A., Hansen, J. E., et al. (1992). Climate forcing by anthropogenic aerosols. *Science, 255*(5043), 423–430. https://doi.org/10.1126/science.255.5043.423.

Chein, R. Y., Chen, Y. C., Yu, C. T., & Chung, J. N. (2015). Thermodynamic analysis of dry reforming of CH4 with CO2 at high pressures. *Journal of Natural Gas Science and Engineering, 26*, 617–629. https://doi.org/10.1016/j.jngse.2015.07.001.

Chen, S., Zaffran, J., & Yang, B. (2020). Dry reforming of methane over the cobalt catalyst: Theoretical insights into the reaction kinetics and mechanism for catalyst deactivation. *Applied Catalysis B: Environmental, 270*. https://doi.org/10.1016/j.apcatb.2020.118859.

Chen, Y. G., & Ren, J. (1994). Conversion of methane and carbon dioxide into synthesis gas over alumina-supported nickel catalysts. Effect of Ni-Al_2O_3 interactions. *Catalysis Letters, 29*(1–2), 39–48. https://doi.org/10.1007/BF00814250.

Chung, Y. C., Ho, K. L., & Tseng, C. P. (2006). Treatment of high H_2S concentrations by chemical absorption and biological oxidation process. *Environmental Engineering Science, 23*(6), 942–953. https://doi.org/10.1089/ees.2006.23.942.

Cioabla, A. E., Ionel, I., Dumitrel, G. A., & Popescu, F. (2012). Comparative study on factors affecting anaerobic digestion of agricultural vegetal residues. *Biotechnology for Biofuels, 5*. https://doi.org/10.1186/1754-6834-5-39.

Coppola, G., & Papurello, D. (2018). Biogas cleaning: Activated carbon regeneration for H_2S removal. *Clean Technologies*, 40–57. https://doi.org/10.3390/cleantechnol1010004.

da Silva Veras, T., Mozer, T. S., da Costa Rubim Messeder dos Santos, D., & da Silva César, A. (2017). Hydrogen: Trends, production and characterization of the main process worldwide. *International Journal of Hydrogen Energy, 42*(4), 2018–2033. https://doi.org/10.1016/j.ijhydene.2016.08.219.

Das, S., Sengupta, M., Bag, A., Shah, M., & Bordoloi, A. (2018). Facile synthesis of highly disperse Ni-Co nanoparticles over mesoporous silica for enhanced methane dry reforming. *Nanoscale, 10*(14), 6409–6425. https://doi.org/10.1039/c7nr09625a.

Dębek, R., Motak, M., Grzybek, T., Galvez, M. E., & Da Costa, P. (2017). A short review on the catalytic activity of hydrotalcite-derived materials for dry reforming of methane. *Catalysts, 7*(1). https://doi.org/10.3390/catal7010032.

Dębek, R., Zubek, K., Motak, M., Galvez, M. E., Da Costa, P., & Grzybek, T. (2015). Ni-Al hydrotalcite-like material as the catalyst precursors for the dry reforming of methane at low temperature. *Comptes Rendus Chimie, 18*(11), 1205–1210. https://doi.org/10.1016/j.crci.2015.04.005.

Dehimi, L., Benguerba, Y., Virginie, M., & Hijazi, H. (2017). Microkinetic modelling of methane dry reforming over Ni/Al_2O_3 catalyst. *International Journal of Hydrogen Energy, 42*(30), 18930–18940. https://doi.org/10.1016/j.ijhydene.2017.05.231.

Dekkar, S., Tezkratt, S., Sellam, D., Ikkour, K., Parkhomenko, K., Martinez-Martin, A., et al. (2020). Dry reforming of methane over $Ni–Al_2O_3$ and $Ni–SiO_2$ catalysts: Role of preparation methods. *Catalysis Letters, 150*(8), 2180–2199. https://doi.org/10.1007/s10562-020-03120-3.

Demirel, B., & Scherer, P. (2008). The roles of acetotrophic and hydrogenotrophic methanogens during anaerobic conversion of biomass to methane: A review. *Reviews in Environmental Science and Biotechnology, 7*(2), 173–190. https://doi.org/10.1007/s11157-008-9131-1.

Deng, L., & Hägg, M. B. (2010). Techno-economic evaluation of biogas upgrading process using CO_2 facilitated transport membrane. *International Journal of Greenhouse Gas Control, 4*(4), 638–646. https://doi.org/10.1016/j.ijggc.2009.12.013.

Deublein, D., & Steinhauser, A. (2011). *Biogas from waste and renewables energy*. Weinheim: Wiley-VCH.

Dewil, R., Appels, L., & Baeyens, J. (2006). Energy use of biogas hampered by the presence of siloxanes. *Energy Conversion and Management, 47*(13–14), 1711–1722. https://doi.org/10.1016/j.enconman.2005.10.016.

Di Maria, F., & Micale, C. (2015). Life cycle analysis of incineration compared to anaerobic digestion followed by composting for managing organic waste: The influence of system components for an Italian district. *International Journal of Life Cycle Assessment, 20*(3), 377–388. https://doi.org/10.1007/s11367-014-0833-z.

Díaz, A. I., Oulego, P., Laca, A., González, J. M., & Díaz, M. (2019). Metagenomic analysis of bacterial communities from a nitrification–denitrification treatment of landfill leachates. *Clean: Soil, Air, Water, 47*(11). https://doi.org/10.1002/clen.201900156.

Dissanayake, D. (1991). Partial oxidation of methane to carbon monoxide and hydrogen over a Ni/Al_2O_3 catalyst. *Journal of Catalysis*, 117–127. https://doi.org/10.1016/0021-9517(91)90252-y.

Donphai, W., Witoon, T., Faungnawakij, K., & Chareonpanich, M. (2016). Carbon-structure affecting catalytic carbon dioxide reforming of methane reaction over Ni-carbon composites. *Journal of CO_2 Utilization, 16*, 245–256. https://doi.org/10.1016/j.jcou.2016.07.011.

Duan, Z., Scheutz, C., & Kjeldsen, P. (2021). Trace gas emissions from municipal solid waste landfills: A review. *Waste Management, 119*, 39–62. https://doi.org/10.1016/j.wasman.2020.09.015.

Dubreuil, A. C., Chainet, F., de Sousa Bartolomeu, R. M., Marques Mota, F. M., Janvier, J., & Lienemann, C. P. (2017). Compréhension de l'impact de composés silicés sur des catalyseurs métalliques par une méthodologie couplant expérimentation et analyses multi-techniques. *Comptes Rendus Chimie, 20*(1), 55–66. https://doi.org/10.1016/j.crci.2016.05.020.

Dumont, E. (2015). H2S removal from biogas using bioreactors: A review. *International Journal of Energy and Environment, 6*, 479–498.

Elsayed, N. H., Elwell, A., Joseph, B., & Kuhn, J. N. (2017). Effect of silicon poisoning on catalytic dry reforming of simulated biogas. *Applied Catalysis A: General, 538*, 157–164. https://doi.org/10.1016/j.apcata.2017.03.024.

EPA. (2021). *Basic information about landfill gas*. Retrieved from https://www.epa.gov/lmop/basic-information-about-landfill-gas. (Accessed 23 October 2020).

Erdogan, B., Arbag, H., & Yasyerli, N. (2018). SBA-15 supported mesoporous Ni and Co catalysts with high coke resistance for dry reforming of methane. *International Journal of Hydrogen Energy, 43*(3), 1396–1405. https://doi.org/10.1016/j.ijhydene.2017.11.127.

Estephane, J., Aouad, S., Hany, S., El Khoury, B., Gennequin, C., El Zakhem, H., et al. (2015). CO_2 reforming of methane over Ni-CO/ZSM5 catalysts. Aging and carbon deposition study. *International Journal of Hydrogen Energy, 40*(30), 9201–9208. https://doi.org/10.1016/j.ijhydene.2015.05.147.

Ewan, B. C. R., & Allen, R. W. K. (2005). A figure of merit assessment of the routes to hydrogen. *International Journal of Hydrogen Energy, 30*(8), 809–819. https://doi.org/10.1016/j.ijhydene.2005.02.003.

Fakeeha, A. H., Alfatish, A. S., Soliman, M. A., & Ibrahim, A. A. (2006). Effect of changing CH_4/CO_2 ratio on hydrogen production by dry reforming reaction. In *Vol. 1. 16th World hydrogen energy conference 2006, WHEC 2006* (pp. 245–256).

Fan, M. S., Abdullah, A. Z., & Bhatia, S. (2010). Utilization of greenhouse gases through carbon dioxide reforming of methane over $Ni-Co/MgO-ZrO_2$: Preparation, characterization and activity studies. *Applied Catalysis B: Environmental, 100*(1–2), 365–377. https://doi.org/10.1016/j.apcatb.2010.08.013.

Faroldi, B., Múnera, J., Falivene, J. M., Ramos, I. R., García, Á. G., Fernández, L. T., et al. (2017). Well-dispersed Rh nanoparticles with high activity for the dry reforming of methane. *International Journal of Hydrogen Energy, 42*(25), 16127–16138. https://doi.org/10.1016/j.ijhydene.2017.04.070.

Farquhar, G. J., & Rovers, F. A. (1973). Gas production during refuse decomposition. *Water, Air, & Soil Pollution, 2*(4), 483–495. https://doi.org/10.1007/BF00585092.

Figoli, A., Cassano, A., & Basile, A. (2016). Membrane Technologies for Biorefining. In *Membrane technologies for biorefining* (pp. 1–500). Elsevier Inc. https://doi.org/10.1016/C2014-0-03660-X.

Finocchio, E., Montanari, T., Garuti, G., Pistarino, C., Federici, F., Cugino, M., et al. (2009). Purification of biogases from siloxanes by adsorption: On the regenerability of activated carbon sorbents. *Energy and Fuels, 23*(8), 4156–4159. https://doi.org/10.1021/ef900356n.

Franz, R., Franz, R., Kühlewind, T., Shterk, G., Abou-Hamad, E., Parastaev, A., et al. (2020). Impact of small promoter amounts on coke structure in dry reforming of methane over Ni/ZrO2. *Catalysis Science and Technology, 10*(12), 3965–3974. https://doi.org/10.1039/d0cy00817f.

Gaillard, M., Virginie, M., & Khodakov, A. Y. (2017). New molybdenum-based catalysts for dry reforming of methane in presence of sulfur: A promising way for biogas valorization. *Catalysis Today, 289*, 143–150. https://doi.org/10.1016/j.cattod.2016.10.005.

Gandhi, S., & Patel, S. (2015). Syngas production by dry reforming of methane over co-precipitated catalysts. *International Journal of Advanced Research in Engineering and Technology, 6*(11), 1–17.

Gao, Y., Jiang, J., Meng, Y., Yan, F., & Aihemaiti, A. (2018). A review of recent developments in hydrogen production via biogas dry reforming. *Energy Conversion and Management, 171*, 133–155. https://doi.org/10.1016/j.enconman.2018.05.083.

García, V., Fernández, J. J., Ruíz, W., Mondragón, F., & Moreno, A. (2009). Effect of MgO addition on the basicity of Ni/ZrO_2 and on its catalytic activity in carbon dioxide reforming of methane. *Catalysis Communications, 11*(4), 240–246. https://doi.org/10.1016/j.catcom.2009.10.003.

Gennequin, C., Hany, S., Tidahy, H. L., Aouad, S., Estephane, J., Aboukaïs, A., et al. (2016). Influence of the presence of ruthenium on the activity and stability of Co–Mg–Al-based catalysts in CO_2 reforming of methane for syngas production. *Environmental Science and Pollution Research, 23*(22), 22744–22760. https://doi.org/10.1007/s11356-016-7453-z.

Gennequin, C., Safariamin, M., Siffert, S., Aboukaïs, A., & Abi-Aad, E. (2011). CO_2 reforming of CH4 over CO-Mg-Al mixed oxides prepared via hydrotalcite like precursors. *Catalysis Today, 176*(1), 139–143. https://doi.org/10.1016/j.cattod.2011.01.029.

Ginsburg, J. M., Piña, J., El Solh, T., & De Lasa, H. I. (2005). Coke formation over a nickel catalyst under methane dry reforming conditions: Thermodynamic and kinetic models. *Industrial and Engineering Chemistry Research, 44*(14), 4846–4854. https://doi.org/10.1021/ie0496333.

Gobina, E. (2000). Inorganic membranes tackle oxo-alcohol synthesis gas production in a fluidised-bed reactor. *Membrane Technology, 2000*(125), 4–8. https://doi.org/10.1016/S0958-2118(00)80210-4.

Gohier, A., Ewels, C. P., Minea, T. M., & Djouadi, M. A. (2008). Carbon nanotube growth mechanism switches from tip- to base-growth with decreasing catalyst particle size. *Carbon, 46*(10), 1331–1338. https://doi.org/10.1016/j.carbon.2008.05.016.

Götz, M., Lefebvre, J., Mörs, F., McDaniel Koch, A., Graf, F., Bajohr, S., et al. (2016). Renewable power-to-gas: A technological and economic review. *Renewable Energy, 85*, 1371–1390. https://doi.org/10.1016/j.renene.2015.07.066.

Grandviewresearch. (2021). *Oxo alcohol market size, share & trends analysis report by application, regional outlook, competitive strategies, and segment forecasts, 2019 to 2025.* Retrieved from https://www.grandviewresearch.com/industry-analysis/oxo-alcohol-market#:~:text=. (Accessed 30 May 2021).

Guerrero, F., Espinoza, L., Ripoll, N., Lisbona, P., Arauzo, I., & Toledo, M. (2020). Syngas production from the reforming of typical biogas compositions in an inert porous media reactor. *Frontiers in Chemistry, 8*. https://doi.org/10.3389/fchem.2020.00145.

Guieysse, B., Hort, C., Platel, V., Munoz, R., Ondarts, M., & Revah, S. (2008). Biological treatment of indoor air for VOC removal: Potential and challenges. *Biotechnology Advances, 26*(5), 398–410. https://doi.org/10.1016/j.biotechadv.2008.03.005.

Guo, X., Wang, C., Sun, F., Zhu, W., & Wu, W. (2014). A comparison of microbial characteristics between the thermophilic and mesophilic anaerobic digesters exposed to elevated food waste loadings. *Bioresource Technology, 152*, 420–428. https://doi.org/10.1016/j.biortech.2013.11.012.

Gurav, H. R., Dama, S., Samuel, V., & Chilukuri, S. (2017). Influence of preparation method on activity and stability of Ni catalysts supported on Gd doped ceria in dry reforming of methane. *Journal of CO_2 Utilization, 20*, 357–367. https://doi.org/10.1016/j.jcou.2017.06.014.

Ha, Q., Armbruster, U., Atia, H., Schneider, M., Lund, H., Agostini, G., et al. (2017). Development of active and stable low nickel content catalysts for dry reforming of methane. *Catalysts, 7*(5), 157. https://doi.org/10.3390/catal7050157.

Hassani Rad, S. J., Haghighi, M., Alizadeh Eslami, A., Rahmani, F., & Rahemi, N. (2016). Sol-gel vs. impregnation preparation of MgO and CeO_2 doped Ni/Al_2O_3 nanocatalysts used in dry reforming of methane: Effect of process conditions, synthesis method and support composition. *International Journal of Hydrogen Energy, 41*(11), 5335–5350. https://doi.org/10.1016/j.ijhydene.2016.02.002.

He, Y., Zhao, Y., Zhou, G., & Huang, M. (2009). Evaluation of extraction and purification methods for obtaining PCR-amplifiable DNA from aged refuse for microbial community analysis. *World Journal of Microbiology and*

Biotechnology, *25*(11), 2043–2051. https://doi.org/10.1007/s11274-009-0106-3.

Holm-Nielsen, J. B., & Oleskowicz-Popiel, P. (2013). Process control in biogas plants. In *The biogas handbook: Science, production and applications* (pp. 228–247). Elsevier Inc. https://doi.org/10.1533/9780857097415.2.228.

Höök, M., & Tang, X. (2013). Depletion of fossil fuels and anthropogenic climate change—A review. *Energy Policy*, *52*, 797–809. https://doi.org/10.1016/j.enpol.2012.10.046.

Hosseini, S. A., Mehri, B., Niaei, A., Izadkhah, B., Alvarez-Galvan, C., & Fierro, J. G. L. (2018). Selective catalytic reduction of NOxby CO over $LaMnO_3$ nano perovskites prepared by microwave and ultrasound assisted sol–gel method. *Journal of Sol-Gel Science and Technology*, *85*(3), 647–656. https://doi.org/10.1007/s10971-017-4568-8.

Houari, A. E. (2020). Microorganisms abundance and diversity in municipal anaerobic sewage sludge digesters from. *PRO*, *8*, 1–22.

Hulteberg, C. (2012). Sulphur-tolerant catalysts in small-scale hydrogen production, a review. *International Journal of Hydrogen Energy*, *37*(5), 3978–3992. https://doi.org/10.1016/j.ijhydene.2011.12.001.

Ibrahim, A. A., Al-Fatesh, A. S., Khan, W. U., Kasim, S. O., Abasaeed, A. E., & Fakeeha, A. H. (2019). Kaolin-supported Ni catalysts for dry methane reforming: Effect of Cs and mixed K-Na promoters. *Journal of Chemical Engineering of Japan*, *52*(2), 232–238. https://doi.org/10.1252/jcej.18we125.

Iglesias, I., Baronetti, G., & Mariño, F. (2017). Ni/Ce0.95M0.05O2 − d (M = Zr, Pr, La) for methane steam reforming at mild conditions. *International Journal of Hydrogen Energy*, *42*(50), 29735–29744. https://doi.org/10.1016/j.ijhydene.2017.09.176.

International Renewable Energy Agency. (2019). *Global energy transformation: The REmap transition pathway.*, ISBN:978-92-9260-120-1.

Itkulova, S. S., Zakumbaeva, G. D., Nurmakanov, Y. Y., Mukazhanova, A. A., & Yermaganbetova, A. K. (2014). Syngas production by bireforming of methane over Co-based alumina-supported catalysts. *Catalysis Today*, *228*, 194–198. https://doi.org/10.1016/j.cattod.2014.01.013.

Jabbour, K., El Hassan, N., Casale, S., Estephane, J., & El Zakhem, H. (2014). Promotional effect of Ru on the activity and stability of Co/SBA-15 catalysts in dry reforming of methane. *International Journal of Hydrogen Energy*, *39*(15), 7780–7787. https://doi.org/10.1016/j.ijhydene.2014.03.040.

Jaffrin, A., Bentounes, N., Joan, A. M., & Makhlouf, S. (2003). Landfill biogas for heating greenhouses and providing carbon dioxide supplement for plant growth. *Biosystems Engineering*, *86*(1), 113–123. https://doi.org/10.1016/S1537-5110(03)00110-7.

Jamshidi, S., Pourhossein, K., & Asadi, M. (2021). Size estimation of wind/solar hybrid renewable energy systems without detailed wind and irradiation data: A feasibility study. *Energy Conversion and Management*, *234*. https://doi.org/10.1016/j.enconman.2021.113905, 113905.

Jay, K., & Stieglitz, L. (1995). Identification and quantification of volatile organic components in emissions of waste incineration plants. *Chemosphere*, *30*(7), 1249–1260. https://doi.org/10.1016/0045-6535(95)00021-Y.

Jean-Michel, L. (2014). Review on dry reforming of methane, a potentially more environmentally-friendly approach to the increasing natural gas exploitation. *Frontiers in Chemistry*, *2*, 81. https://doi.org/10.3389/fchem.2014.00081.

Johansson, K. A. U. (2012). *Characterisation of contaminants in biogas before and after upgrading to vehicle gas. Vol. 246.*

Kadam, R., & Panwar, N. L. (2017). Recent advancement in biogas enrichment and its applications. *Renewable and Sustainable Energy Reviews*, *73*, 892–903. https://doi.org/10.1016/j.rser.2017.01.167.

Kapoor, R., Ghosh, P., Tyagi, B., Vijay, V. K., Vijay, V., Thakur, I. S., et al. (2020). Advances in biogas valorization and utilization systems: A comprehensive review. *Journal of Cleaner Production*, *273*. https://doi.org/10.1016/j.jclepro.2020.123052.

Katheria, S., Gupta, A., Deo, G., & Kunzru, D. (2016). Effect of calcination temperature on stability and activity of Ni/$MgAl_2O_4$ catalyst for steam reforming of methane at high pressure condition. *International Journal of Hydrogen Energy*, *41*(32), 14123–14132. https://doi.org/10.1016/j.ijhydene.2016.05.109.

Kazda, M., Langer, S., & Bengelsdorf, F. R. (2014). Fungi open new possibilities for anaerobic fermentation of organic residues. *Energy, Sustainability and Society*, *4*(1), 1–9. https://doi.org/10.1186/2192-0567-4-6.

Khajeh Talkhoncheh, S., & Haghighi, M. (2015). Syngas production via dry reforming of methane over Ni-based nanocatalyst over various supports of clinoptilolite, ceria and alumina. *Journal of Natural Gas Science and Engineering*, *23*, 16–25. https://doi.org/10.1016/j.jngse.2015.01.020.

Khoshnevisan, B., Tsapekos, P., Alfaro, N., Díaz, I., Fdz-Polanco, M., Rafiee, S., et al. (2017). A review on prospects and challenges of biological H2S removal from biogas with focus on biotrickling filtration and microaerobic desulfurization. *Biofuel Research Journal*, *4*(4), 741–750. https://doi.org/10.18331/BRJ2017.4.4.6.

Komarala, E. P., Komissarov, I., & Rosen, B. A. (2020). Effect of fe and mn substitution in $LaNiO_3$ on exsolution, activity, and stability for methane dry reforming. *Catalysts*, *10*(1). https://doi.org/10.3390/catal10010027.

Kumar, M. (2011). Carbon nanotube synthesis and growth mechanism. In *Carbon nanotubes – Synthesis, characterization, applications* IntechOpen.

Kumar, N., Shojaee, M., & Spivey, J. J. (2015). Catalytic bi-reforming of methane: From greenhouse gases to syngas. *Current Opinion in Chemical Engineering*, *9*, 8–15. https://doi.org/10.1016/j.coche.2015.07.003.

Kumar, N., Wang, Z., Kanitkar, S., & Spivey, J. J. (2016). Methane reforming over Ni-based pyrochlore catalyst: Deactivation studies for different reactions. *Applied Petrochemical Research*, 201–207. https://doi.org/10.1007/s13203-016-0166-x.

Kusuma, M., & Chandrappa, G. T. (2019). Effect of calcination temperature on characteristic properties of $CaMoO_4$ nanoparticles. *Journal of Science: Advanced Materials and Devices*, *4*(1), 150–157. https://doi.org/10.1016/j.jsamd.2019.02.003.

Langer, S. G., Gabris, C., Einfalt, D., Wemheuer, B., Kazda, M., & Bengelsdorf, F. R. (2019). Different response of bacteria, archaea and fungi to process parameters in nine full-scale anaerobic digesters. *Microbial Biotechnology*, *12*(6), 1210–1225. https://doi.org/10.1111/1751-7915.13409.

Lanre, M. S., Al-Fatesh, A. S., Fakeeha, A. H., Kasim, S. O., Ibrahim, A. A., Al-Awadi, A. S., et al. (2020). Catalytic performance of lanthanum promoted Ni/ZrO_2 for carbon dioxide reforming of methane. *PRO*, *8*(11), 1–15. https://doi.org/10.3390/pr8111502.

Lasocki, J., Kołodziejczyk, K., & Matuszewska, A. (2015). Laboratory-scale investigation of biogas treatment by removal of hydrogen sulfide and carbon dioxide. *Polish Journal of Environmental Studies*, *24*(3), 1427–1434. https://doi.org/10.15244/pjoes/35283.

Li, J., Li, J., & Zhu, Q. (2018). Carbon deposition and catalytic deactivation during CO_2 reforming of CH_4 over Co/MgO catalyst. *Chinese Journal of Chemical Engineering*, 2344–2350. https://doi.org/10.1016/j.cjche.2018.05.025.

Li, Y., Li, D., & Wang, G. (2011). Methane decomposition to COx-free hydrogen and nano-carbon material on group 8-10 base metal catalysts: A review. *Catalysis Today*, *162*(1), 1–48. https://doi.org/10.1016/j.cattod.2010.12.042.

Lieberei, J., & Gheewala, S. H. (2017). Resource depletion assessment of renewable electricity generation technologies—Comparison of life cycle impact assessment methods with focus on mineral resources. *International Journal of Life Cycle Assessment*, *22*(2), 185–198. https://doi.org/10.1007/s11367-016-1152-3.

Liu, D., Cheo, W. N. E., Lim, Y. W. Y., Borgna, A., Lau, R., & Yang, Y. (2010). A comparative study on catalyst deactivation of nickel and cobalt incorporated MCM-41 catalysts modified by platinum in methane reforming with carbon dioxide. *Catalysis Today*, *154*(3–4), 229–236. https://doi.org/10.1016/j.cattod.2010.03.054.

Liu, H., Da Costa, P., Hadj Taief, H. B., Benzina, M., & Gálvez, M. E. (2017). Ceria and zirconia modified natural clay based nickel catalysts for dry reforming of methane. *International Journal of Hydrogen Energy*, *42*(37), 23508–23516. https://doi.org/10.1016/j.ijhydene.2017.01.075.

Liu, H., Li, Y., & He, D. (2020). Recent progress of catalyst design for carbon dioxide reforming of methane to syngas. *Energy Technology*, *8*(8). https://doi.org/10.1002/ente.201900493.

Lobo, L. S. (2017). Nucleation and growth of carbon nanotubes and nanofibers: Mechanism and catalytic geometry control. *Carbon*, *114*, 411–417. https://doi.org/10.1016/j.carbon.2016.12.005.

Loic, G. (2005). *Synthese de Fischer-Tropsch en réacteurs structures a catalyse supportée en paroi*.

Lou, Z., Wang, M., Zhao, Y., & Huang, R. (2015). The contribution of biowaste disposal to odor emission from landfills. *Journal of the Air & Waste Management Association*, 479–484. https://doi.org/10.1080/10962247.2014.1002870.

Luisetto, I., Sarno, C., De Felicis, D., Basoli, F., Battocchio, C., Tuti, S., et al. (2017). Ni supported on γ-Al_2O_3 promoted by Ru for the dry reforming of methane in packed and monolithic reactors. *Fuel Processing Technology*, *158*, 130–140. https://doi.org/10.1016/j.fuproc.2016.12.015.

Luisetto, I., Tuti, S., & Di Bartolomeo, E. (2012). Co and Ni supported on CeO_2 as selective bimetallic catalyst for dry reforming of methane. *International Journal of Hydrogen Energy*, *37*(21), 15992–15999. https://doi.org/10.1016/j.ijhydene.2012.08.006.

Luyben, W. L. (2014). Design and control of the dry methane reforming process. *Industrial and Engineering Chemistry Research*, *53*(37), 14423–14439. https://doi.org/10.1021/ie5023942.

Mackie, R. I., Stroot, P. G., & Varel, V. H. (1998). Biochemical identification and biological origin of key odor components in livestock waste. *Journal of Animal Science*, *76*(5), 1331–1342. https://doi.org/10.2527/1998.7651331x.

Mahfouz, R., Estephane, J., Gennequin, C., Tidahy, L., Aouad, S., & Abi-Aad, E. (2021). CO_2 reforming of methane over Ni and/or Ru catalysts supported on mesoporous KIT-6: Effect of promotion with Ce. *Journal of Environmental Chemical Engineering*, *9*(1). https://doi.org/10.1016/j.jece.2020.104662.

Mahmoudi, H., Mahmoudi, M., Doustar, O., Jajangiri, H., Tsolakis, A., & Mech Wyszynski, M. (2017). A review of Fischer Tropsch synthesis process, mechanism, surface chemistry and catalyst formulation. *Journal of Biofuels Energy*, *2*, 11–31. https://doi.org/10.1515/bfuel-2017-0002.

Mancino, G., Cimino, S., & Lisi, L. (2016). Sulphur poisoning of alumina supported Rh catalyst during dry reforming of methane. *Catalysis Today*, *277*, 126–132. https://doi.org/10.1016/j.cattod.2015.10.035.

Maucieri, C., Barbera, A. C., Vymazal, J., & Borin, M. (2017). A review on the main affecting factors of greenhouse gases emission in constructed wetlands. *Agricultural and Forest Meteorology*, *236*, 175–193. https://doi.org/10.1016/j.agrformet.2017.01.006.

Medeiros, R. L. B. A., Figueredo, G. P., Macedo, H. P., Oliveira, A. S., Rabelo-Neto, R. C., Melo, D. M. A., et al. (2021). One-pot microwave-assisted combustion synthesis of Ni-Al_2O_3 nanocatalysts for hydrogen production via dry reforming of methane. *Fuel, 287*. https://doi.org/10.1016/j.fuel.2020.119511, 119511.

Melo, F., & Morlanés, N. (2005). Naphtha steam reforming for hydrogen production. *Catalysis Today, 107–108*, 458–466. https://doi.org/10.1016/j.cattod.2005.07.028.

Milanov, A., Schunk, S., Schwab, E., & Wasserschaff, G. (2013). Dry reforming of methane dry reforming of methane with CO_2 at elevated pressures. In *vol. 2013. DGMK Tagungsbericht* (pp. 31–35). Deutsche Wissens. Gesell. fur Erdoel, Erdgas und Kohle EV.

Minh, D. P., Hernandez, A., Torres, A. H., Rego de Vasconcelos, B., Siang, T. J., & Vo, D. V. N. (2020). Conversion of biogas to syngas via catalytic carbon dioxide reforming reactions: An overview of thermodynamic aspects, catalytic design, and reaction kinetics. In *Biorefinery of alternative resources: Targeting green fuels and platform chemicals*. Singapore: Springer.

More, A., Bhavsar, S., & Veser, G. (2016). Iron–nickel alloys for carbon dioxide activation by chemical looping dry reforming of methane. *Energy Technology, 4*(10), 1147–1157. https://doi.org/10.1002/ente.201500539.

Mork, P. C. (1976). *Nickel-catalyzed hydrogenation: A study of the poisoning effect of halogen-containing compounds* (pp. 506–510). Springer.

Mortensen, P. M., Gardini, D., de Carvalho, H. W. P., Damsgaard, C. D., Grunwaldt, J.-D., Jensen, P. A., et al. (2014). Stability and resistance of nickel catalysts for hydrodeoxygenation: Carbon deposition and effects of sulfur, potassium, and chlorine in the feed. *Catalysis Science & Technology, 4*(10), 3672–3686. https://doi.org/10.1039/C4CY00522H.

Murphy, J. D., McKeogh, E., & Kiely, G. (2004). Technical/economic/environmental analysis of biogas utilisation. *Applied Energy, 77*(4), 407–427. https://doi.org/10.1016/j.apenergy.2003.07.005.

Mutz, B., Carvalho, H. W. P., Mangold, S., Kleist, W., & Grunwaldt, J. D. (2015). Methanation of CO2: Structural response of a Ni-based catalyst under fluctuating reaction conditions unraveled by operando spectroscopy. *Journal of Catalysis, 327*, 48–53. https://doi.org/10.1016/j.jcat.2015.04.006.

Naeem, M. A., Al-Fatesh, A. S., Abasaeed, A. E., & Fakeeha, A. H. (2014). Activities of Ni-based nano catalysts for CO_2-CH_4 reforming prepared by polyol process. *Fuel Processing Technology, 122*, 141–152. https://doi.org/10.1016/j.fuproc.2014.01.035.

Nahar, G., Mote, D., & Dupont, V. (2017). Hydrogen production from reforming of biogas: Review of technological advances and an Indian perspective. *Renewable and Sustainable Energy Reviews, 76*, 1032–1052. https://doi.org/10.1016/j.rser.2017.02.031.

Najfach, A. J., Almquist, C. B., & Edelmann, R. E. (2019). Effect of manganese and zeolite composition on zeolite-supported Ni-catalysts for dry reforming of methane. *Catalysis Today, 369*, 31–47.

Nematollahi, B., Rezaei, M., & Khajenoori, M. (2011). Combined dry reforming and partial oxidation of methane to synthesis gas on noble metal catalysts. *International Journal of Hydrogen Energy, 36*(4), 2969–2978. https://doi.org/10.1016/j.ijhydene.2010.12.007.

Newson, E. (1975). Catalyst deactivation due to pore-plugging by reaction products. *Industrial and Engineering Chemistry Process Design and Development, 14*(1), 27–33. https://doi.org/10.1021/i260053a005.

Nguyen, L. N., Nguyen, A. Q., & Nghiem, L. D. (2019). Microbial community in anaerobic digestion system: Progression in microbial ecology. In *Energy, environment, and sustainability* (pp. 331–355). Springer Nature. https://doi.org/10.1007/978-981-13-3259-3_15.

Nikoo, M. K., & Amin, N. A. S. (2011). Thermodynamic analysis of carbon dioxide reforming of methane in view of solid carbon formation. *Fuel Processing Technology, 92*(3), 678–691. https://doi.org/10.1016/j.fuproc.2010.11.027.

Parra-Orobio, B. A., Donoso-Bravo, A., Ruiz-Sánchez, J. C., Valencia-Molina, K. J., & Torres-Lozada, P. (2018). Effect of inoculum on the anaerobic digestion of food waste accounting for the concentration of trace elements. *Waste Management, 71*, 342–349. https://doi.org/10.1016/j.wasman.2017.09.040.

Peishi, S., Xianwan, Y., Ruohua, H., Bing, H., & Ping, Y. (2004). A new approach to kinetics of purifying waste gases containing volatile organic compounds (VOC) in low concentration by using the biological method. *Journal of Cleaner Production, 12*(1), 95–100. https://doi.org/10.1016/S0959-6526(02)00195-6.

Pérez-Fortes, M., Bocin-Dumitriu, A., & Tzimas, E. (2014). CO_2 utilization pathways: Techno-economic assessment and market opportunities. *Energy Procedia, 63*, 7968–7975. https://doi.org/10.1016/j.egypro.2014.11.834.

Perez-Lopez, O. W., Senger, A., Marcilio, N. R., & Lansarin, M. A. (2006). Effect of composition and thermal pretreatment on properties of Ni-Mg-Al catalysts for CO_2 reforming of methane. *Applied Catalysis A: General, 303*(2), 234–244. https://doi.org/10.1016/j.apcata.2006.02.024.

Pernicone, N., & Traina, F. (1978). Catalyst activation by reduction. *Pure and Applied Chemistry, 50*(9–10), 1169–1191. https://doi.org/10.1351/pac197850091169.

Prussi, M., Padella, M., Conton, M., Postma, E. D., & Lonza, L. (2019). Review of technologies for biomethane production and assessment of Eu transport share in 2030. *Journal of Cleaner Production, 222*, 565–572. https://doi.org/10.1016/j.jclepro.2019.02.271.

Rafiee, A., Khalilpour, K. R., Prest, J., & Skryabin, I. (2021). Biogas as an energy vector. *Biomass and Bioenergy, 144*. https://doi.org/10.1016/j.biombioe.2020.105935.

Ramezani, Y., Meshkani, F., & Rezaei, M. (2018). Preparation and evaluation of mesoporous nickel and manganese bimetallic nanocatalysts in methane dry reforming process for syngas production. *Journal of Chemical Sciences, 130*(1). https://doi.org/10.1007/s12039-017-1410-3.

Rashed, M., Mamun, A. L., & Torii, S. (2015). Possibility of anaerobic co-digestion of cafeteria, vegetable and fruit wastes for biogas production without inoculum source. In *4th world conference on applied sciences, engineering & technology, Kumamoto University, Kumamoto, Japan*.

Rasi, S. (2009). *Biogas composition and upgrading to biomethane*.

Rasi, S., Läntelä, J., & Rintala, J. (2011). Trace compounds affecting biogas energy utilisation – A review. *Energy Conversion and Management, 52*(12), 3369–3375. https://doi.org/10.1016/j.enconman.2011.07.005.

Rasi, S., Veijanen, A., & Rintala, J. (2007). Trace compounds of biogas from different biogas production plants. *Energy, 32*(8), 1375–1380. https://doi.org/10.1016/j.energy.2006.10.018.

Ray, N. H. S., Mohanty, M. K., & Mohanty, R. C. (2013). Biogas production and pretreatment of wastes, a review. *International Journal of Scientific and Research Publications, 3*, 2250–3153.

Rostrup-Nielsen, J. R., & Bak Hansen, J. H. (1993). CO_2-reforming of methane over transition metals. *Journal of Catalysis, 144*(1), 38–49. https://doi.org/10.1006/jcat.1993.1312.

Rostrup-Nielsen, J. R., Sehested, J., & Nørskov, J. K. (2002). Hydrogen and synthesis gas by steam- and CO_2 reforming. *Advances in Catalysis, 47*, 65–139. https://doi.org/10.1016/s0360-0564(02)47006-x.

Rouibah, K., Barama, A., Benrabaa, R., Guerrero-Caballero, J., Kane, T., Vannier, R. N., et al. (2017). Dry reforming of methane on nickel-chrome, nickel-cobalt and nickel-manganese catalysts. *International Journal of Hydrogen Energy, 42*(50), 29725–29734. https://doi.org/10.1016/j.ijhydene.2017.10.049.

Saha, B., Khan, A., Ibrahim, H., & Idem, R. (2014). Evaluating the performance of non-precious metal based catalysts for sulfur-tolerance during the dry reforming of biogas. *Fuel, 120*, 202–217. https://doi.org/10.1016/j.fuel.2013.12.016.

Sahota, S., Shah, G., Ghosh, P., Kapoor, R., Sengupta, S., Singh, P., et al. (2018). Review of trends in biogas upgradation technologies and future perspectives. *Bioresource Technology Reports, 1*, 79–88. https://doi.org/10.1016/j.biteb.2018.01.002.

Salazar Hoyos, L. A., Faroldi, B. M., & Cornaglia, L. M. (2020). A coke-resistant catalyst for the dry reforming of methane based on Ni nanoparticles confined within rice husk-derived mesoporous materials. *Catalysis Communications, 135*. https://doi.org/10.1016/j.catcom.2019.105898.

Santos, I., Barros, R., & Fiho, G. (2020). Biogas production from solid waste landfill. In *2010. Encyclopedia of renewable and sustainable materials* (pp. 11–19). Elsevier. https://doi.org/10.1016/b978-0-12-803581-8.10585-5.

Santos, J. A. F. D. A., Jong, P., Costa, C. A. D., & Torres, E. A. (2020). Combining wind and solar energy sources: Potential for hybrid power generation in Brazil. *Utilities Policy, 67*, 101084.

Santos-Clotas, E., Cabrera-Codony, A., Boada, E., Gich, F., Muñoz, R., & Martín, M. J. (2019). Efficient removal of siloxanes and volatile organic compounds from sewage biogas by an anoxic biotrickling filter supplemented with activated carbon. *Bioresource Technology, 294*. https://doi.org/10.1016/j.biortech.2019.122136.

SCDHEC. (2021). *How landfills work*. Retrieved from https://scdhec.gov/environment/land-and-waste-landfills/how-landfills-work. (Accessed 29 October 2020).

Schuetz, C., Bogner, J., Chanton, J., Blake, D., Morcet, M., & Kjeldsen, P. (2003). Comparative oxidation and net emissions of methane and selected non-methane organic compounds in landfill cover soils. *Environmental Science and Technology, 37*(22), 5150–5158. https://doi.org/10.1021/es034016b.

Schweigkofler, M., & Niessner, R. (1999). Determination of siloxanes and VOC in landfill gas and sewage gas by canister sampling and GC-MS/AES analysis. *Environmental Science and Technology, 33*(20), 3680–3685. https://doi.org/10.1021/es9902569.

Serrano-Lotina, A., & Daza, L. (2013). Highly stable and active catalyst for hydrogen production from biogas. *Journal of Power Sources, 238*, 81–86. https://doi.org/10.1016/j.jpowsour.2013.03.067.

Serrano-Lotina, A., & Daza, L. (2014). Influence of the operating parameters over dry reforming of methane to syngas. *International Journal of Hydrogen Energy, 39*(8), 4089–4094. https://doi.org/10.1016/j.ijhydene.2013.05.135.

Serrano-Lotina, A., Rodríguez, L., Muñoz, G., Martin, A. J., Folgado, M. A., & Daza, L. (2011). Biogas reforming over La-NiMgAl catalysts derived from hydrotalcite-like structure: Influence of calcination temperature. *Catalysis Communications, 12*(11), 961–967. https://doi.org/10.1016/j.catcom.2011.02.014.

Shah, G., Ahmad, E., Pant, K. K., & Vijay, V. K. (2021). Comprehending the contemporary state of art in biogas enrichment and CO2 capture technologies via swing adsorption. *International Journal of Hydrogen Energy, 46*(9), 6588–6612. https://doi.org/10.1016/j.ijhydene.2020.11.116.

Shamsi, A. (2004). Carbon formation on Ni-MgO catalyst during reaction of methane in the presence of CO_2 and CO. *Applied Catalysis A: General, 277*(1–2), 23–30. https://doi.org/10.1016/j.apcata.2004.08.015.

Shamsi, A., & Johnson, C. D. (2003). Effect of pressure on the carbon deposition route in CO_2 reforming of $13CH_4$. *Catalysis Today, 84*(1–2), 17–25. https://doi.org/10.1016/S0920-5861(03)00296-7.

Shang, Z., Li, S., Li, L., Liu, G., & Liang, X. (2017). Highly active and stable alumina supported nickel nanoparticle catalysts for dry reforming of methane. *Applied Catalysis B: Environmental, 201*, 302–309. https://doi.org/10.1016/j.apcatb.2016.08.019.

Sifontes, A. B., Rosales, M., Méndez, F. J., Oviedo, O., & Zoltan, T. (2013). Effect of calcination temperature on structural properties and photocatalytic activity of ceria nanoparticles synthesized employing chitosan as template. *Journal of Nanomaterials, 2013*. https://doi.org/10.1155/2013/265797.

Snyder, C. S., Bruulsema, T. W., Jensen, T. L., & Fixen, P. E. (2009). Review of greenhouse gas emissions from crop production systems and fertilizer management effects. *Agriculture, Ecosystems and Environment, 133*(3–4), 247–266. https://doi.org/10.1016/j.agee.2009.04.021.

Song, C., & Pan, W. (2004). Tri-reforming of methane: A novel concept for catalytic production of industrially useful synthesis gas with desired H2/CO ratios. *Catalysis Today, 98*(4), 463–484. https://doi.org/10.1016/j.cattod.2004.09.054.

Song, K., Lu, M., Xu, S., Chen, C., Zhan, Y., Li, D., et al. (2018). Effect of alloy composition on catalytic performance and coke-resistance property of Ni-Cu/Mg(Al)O catalysts for dry reforming of methane. *Applied Catalysis B: Environmental, 239*, 324–333. https://doi.org/10.1016/j.apcatb.2018.08.023.

Spiegel, R. J., & Preston, J. L. (2003). Technical assessment of fuel cell operation on anaerobic digester gas at the Yonkers, NY, wastewater treatment plant. *Waste Management, 23*(8), 709–717. https://doi.org/10.1016/S0956-053X(02)00165-4.

St-Pierre, B., & Wright, A. D. G. (2017). Implications from distinct sulfate-reducing bacteria populations between cattle manure and digestate in the elucidation of H2S production during anaerobic digestion of animal slurry. *Applied Microbiology and Biotechnology, 101*(13), 5543–5556. https://doi.org/10.1007/s00253-017-8261-1.

Streletskiy, D., Anisimov, O., & Vasiliev, A. (2015). Permafrost degradation. In *Snow and ice-related hazards, risks, and disasters* (pp. 303–344). Elsevier Inc. https://doi.org/10.1016/B978-0-12-394849-6.00010-X.

Suelves, I., Lázaro, M. J., Moliner, R., Corbella, B. M., & Palacios, J. M. (2005). Hydrogen production by thermo catalytic decomposition of methane on Ni-based catalysts: Influence of operating conditions on catalyst deactivation and carbon characteristics. *International Journal of Hydrogen Energy, 30*(15), 1555–1567. https://doi.org/10.1016/j.ijhydene.2004.10.006.

Sun, Q., Li, H., Yan, J., Liu, L., Yu, Z., & Yu, X. (2015). Selection of appropriate biogas upgrading technology-a review of biogas cleaning, upgrading and utilisation. *Renewable and Sustainable Energy Reviews, 51*, 521–532. https://doi.org/10.1016/j.rser.2015.06.029.

Świrk, K., Grzybek, T., & Motak, M. (2017). Tri-reforming as a process of CO_2 utilization and a novel concept of energy storage in chemical products. In *Vol. 14. E3S web of conferences*EDP Sciences. https://doi.org/10.1051/e3sconf/20171402038.

Takami, D., Yamamoto, A., & Yoshida, H. (2020). Dry reforming of methane over alumina-supported rhodium catalysts at low temperatures under visible and near-infrared light. *Catalysis Science and Technology, 10*(17), 5811–5814. https://doi.org/10.1039/d0cy00858c.

Takanabe, K., Nagaoka, K., Nariai, K., & Aika, K. I. (2005). Titania-supported cobalt and nickel bimetallic catalysts for carbon dioxide reforming of methane. *Journal of Catalysis, 232*(2), 268–275. https://doi.org/10.1016/j.jcat.2005.03.011.

Tan, H., Zhao, Y., Ling, Y., Wang, Y., & Wang, X. (2017). Emission characteristics and variation of volatile odorous compounds in the initial decomposition stage of municipal solid waste. *Waste Management, 68*, 677–687. https://doi.org/10.1016/j.wasman.2017.07.015.

Terribile, D., Trovarelli, A., Llorca, J., De Leitenburg, C., & Dolcetti, G. (1998). The preparation of high surface area CeO_2-ZrO_2 mixed oxides by a surfactant-assisted approach. *Catalysis Today, 43*(1–2), 79–88. https://doi.org/10.1016/S0920-5861(98)00136-9.

Themelis, N. J., & Ulloa, P. A. (2007). Methane generation in landfills. *Renewable Energy, 32*(7), 1243–1257. https://doi.org/10.1016/j.renene.2006.04.020.

Tomishige, K., Li, D., Tamura, M., & Nakagawa, Y. (2017). Nickel-iron alloy catalysts for reforming of hydrocarbons: Preparation, structure, and catalytic properties. *Catalysis Science and Technology, 7*(18), 3952–3979. https://doi.org/10.1039/c7cy01300k.

Torrez-Herrera, J. J., Korili, S. A., & Gil, A. (2021). Effect of the synthesis method on the morphology, textural properties and catalytic performance of La-hexaaluminates in the dry reforming of methane. *Journal of Environmental Chemical Engineering, 9*(4). https://doi.org/10.1016/j.jece.2021.105298.

Towler, G., & Sinnott, R. (2013). *Chemical engineering design: Principle, practice and economics of plant and process design. Vol. 53*. Elsevier.

Tristantini, D., Lögdberg, S., Gevert, B., Borg, Ø., & Holmen, A. (2007). The effect of synthesis gas composition on the Fischer-Tropsch synthesis over Co/γ-Al2O3 and Co-Re/γ-Al2O3 catalysts. *Fuel Processing Technology, 88*(7), 643–649. https://doi.org/10.1016/j.fuproc.2007.01.012.

Tungkamani, S., Phongaksorn, M., Narataruksa, P., Sornchamni, T., Kanjanabat, N., & Siri-Nguan, N. (2013). Developing carbon tolerance catalyst for dry methane reforming. *Chemical Engineering Transactions, 32*, 745–750. https://doi.org/10.3303/CET1332125.

United States Environmental Protection Agency. (2021). *Basic information about landfill gas*. Retrieved from https://www.epa.gov/lmop/basic-information-about-landfill-gas. (Accessed 23 October 2020).

US EPA. (2016). *LFG energy project development handbook*.

Usman, M., & Wan Daud, W. M. A. (2016). An investigation on the influence of catalyst composition, calcination and reduction temperatures on Ni/MgO catalyst for dry reforming of methane. *RSC Advances*, *6*(94), 91603–91616. https://doi.org/10.1039/c6ra15256b.

Usman, M., Wan Daud, W. M. A., & Abbas, H. F. (2015). Dry reforming of methane: Influence of process parameters – A review. *Renewable and Sustainable Energy Reviews*, *45*, 710–744. https://doi.org/10.1016/j.rser.2015.02.026.

Vaverková, M. D. (2019). Landfill impacts on the environment—Review. *Geoscience*, *9*, 1–16.

Wang, P., Wang, H., Qiu, Y., Ren, L., & Jiang, B. (2018). Microbial characteristics in anaerobic digestion process of food waste for methane production—A review. *Bioresource Technology*, *248*, 29–36. https://doi.org/10.1016/j.biortech.2017.06.152.

Wang, Y., Zhao, Q., Li, L., Hu, C., & Da Costa, P. (2021). Dry reforming of methane over Ni–ZrOx catalysts doped by manganese: On the effect of the stability of the structure during time on stream. *Applied Catalysis A: General*, *617*-(November 2020). https://doi.org/10.1016/j.apcata.2021.118120.

Whang, H. S., Choi, M. S., Lim, J., Kim, C., Heo, I., Chang, T. S., et al. (2017). Enhanced activity and durability of Ru catalyst dispersed on zirconia for dry reforming of methane. *Catalysis Today*, *293–294*, 122–128. https://doi.org/10.1016/j.cattod.2016.12.034.

Wittich, K., Krämer, M., Bottke, N., & Schunk, S. A. (2020). Catalytic dry reforming of methane: Insights from model systems. *ChemCatChem*, *12*(8), 2130–2147. https://doi.org/10.1002/cctc.201902142.

Wolfbeisser, A., Sophiphun, O., Bernardi, J., Wittayakun, J., Föttinger, K., & Rupprechter, G. (2016). Methane dry reforming over ceria-zirconia supported Ni catalysts. *Catalysis Today*, *277*, 234–245. https://doi.org/10.1016/j.cattod.2016.04.025.

Wu, C., Liu, J., Zhao, P., Li, W., Yan, L., Piringer, M., et al. (2017). Evaluation of the chemical composition and correlation between the calculated and measured odour concentration of odorous gases from a landfill in Beijing, China. *Atmospheric Environment*, *164*, 337–347. https://doi.org/10.1016/j.atmosenv.2017.06.010.

Xie, Y., Xie, F., Wang, L., Peng, Y., Ma, D., Zhu, L., et al. (2020). Efficient dry reforming of methane with carbon dioxide reaction on Ni@Y_2O_3 nanofibers anti-carbon deposition catalyst prepared by electrospinning-hydrothermal method. *International Journal of Hydrogen Energy*, *45*(56), 31494–31506. https://doi.org/10.1016/j.ijhydene.2020.08.202.

Xu, H.-B., Tsukuda, M., Sato, T., Gu, J.-D., & Katayama, Y. (2018). Lithoautotrophical oxidation of elemental sulfur by fungi including fusarium solani isolated from sandstone Angkor temples. *International Biodeterioration and Biodegradation*, *126*(October 2017), 95–102. https://doi.org/10.1016/j.ibiod.2017.10.005.

Xu, Y., Du, X., Li, J., Wang, P., Zhu, J., Ge, F., et al. (2019). A comparison of Al_2O_3 and SiO_2 supported Ni-based catalysts in their performance for the dry reforming of methane. *Journal of Fuel Chemistry and Technology*, *47*(2), 199–208. https://doi.org/10.1016/s1872-5813(19)30010-6.

Yang, L., Ge, X., Wan, C., Yu, F., & Li, Y. (2014). Progress and perspectives in converting biogas to transportation fuels. *Renewable and Sustainable Energy Reviews*, *40*, 1133–1152. https://doi.org/10.1016/j.rser.2014.08.008.

Yue, X. L., & Gao, Q. X. (2018). Contributions of natural systems and human activity to greenhouse gas emissions. *Advances in Climate Change Research*, *9*(4), 243–252. https://doi.org/10.1016/j.accre.2018.12.003.

Zakir Hossain, H. M., Hasna Hossain, Q., Uddin Monir, M. M., & Ahmed, M. T. (2014). Municipal solid waste (MSW) as a source of renewable energy in Bangladesh: Revisited. *Renewable and Sustainable Energy Reviews*, *39*, 35–41. https://doi.org/10.1016/j.rser.2014.07.007.

Zhang, J., Li, X., Chen, H., Qi, M., Zhang, G., Hu, H., et al. (2017). Hydrogen production by catalytic methane decomposition: Carbon materials as catalysts or catalyst supports. *International Journal of Hydrogen Energy*, *42*(31), 19755–19775. https://doi.org/10.1016/j.ijhydene.2017.06.197.

Zhang, J., Wang, H., & Dalai, A. K. (2007). Development of stable bimetallic catalysts for carbon dioxide reforming of methane. *Journal of Catalysis*, *249*(2), 300–310. https://doi.org/10.1016/j.jcat.2007.05.004.

Zhang, J. C., Ge, B. H., Liu, T. F., Yang, Y. Z., Li, B., & Li, W. Z. (2020). Robust ruthenium-saving catalyst for high-temperature carbon dioxide reforming of methane. *ACS Catalysis*, *10*(1), 783–791. https://doi.org/10.1021/acscatal.9b03709.

Zhang, L., Zhang, Q., Liu, Y., & Zhang, Y. (2016). Dry reforming of methane over Ni/MgO-Al_2O_3 catalysts prepared by two-step hydrothermal method. *Applied Surface Science*, *389*, 25–33. https://doi.org/10.1016/j.apsusc.2016.07.063.

Zhang, Q., Zhang, T., Shi, Y., Zhao, B., Wang, M., Liu, Q., et al. (2017). A sintering and carbon-resistant Ni-SBA-15 catalyst prepared by solid-state grinding method for dry reforming of methane. *Journal of CO_2 Utilization*, *17*, 10–19. https://doi.org/10.1016/j.jcou.2016.11.002.

Zhang, Y., Zhang, S., Gossage, J. L., Lou, H. H., & Benson, T. J. (2014). Thermodynamic analyses of tri-reforming reactions to produce syngas. *Energy and Fuels*, *28*(4), 2717–2726. https://doi.org/10.1021/ef500084m.

Zhang, Z. L., & Verykios, X. E. (1994). Carbon dioxide reforming of methane to synthesis gas over supported Ni catalysts. *Catalysis Today*, *21*(2–3), 589–595. https://doi.org/10.1016/0920-5861(94)80183-5.

Zhou, L., Guo, Y., Zhang, Q., Yagi, M., Hatakeyama, J., Li, H., et al. (2008). A novel catalyst with plate-type anodic alumina supports, Ni/NiAl2O4/γ-Al_2O_3/alloy, for steam reforming of methane. *Applied Catalysis A: General*, *347*(2), 200–207. https://doi.org/10.1016/j.apcata.2008.06.007.

Zhu, Y. A., Chen, D., Zhou, X. G., & Yuan, W. K. (2009). DFT studies of dry reforming of methane on Ni catalyst. *Catalysis Today*, *148*(3–4), 260–267. https://doi.org/10.1016/j.cattod.2009.08.022.

Ziganshin, A. M., Schmidt, T., Scholwin, F., Il'Inskaya, O. N., Harms, H., & Kleinsteuber, S. (2011). Bacteria and archaea involved in anaerobic digestion of distillers grains with solubles. *Applied Microbiology and Biotechnology*, *89*(6), 2039–2052. https://doi.org/10.1007/s00253-010-2981-9.

Ziyang, L., Luochun, W., Nanwen, Z., & Youcai, Z. (2015). Martial recycling from renewable landfill and associated risks: A review. *Chemosphere*, *131*, 91–103. https://doi.org/10.1016/j.chemosphere.2015.02.036.

Zuo, Z., Liu, S., Wang, Z., Liu, C., Huang, W., Huang, J., et al. (2018). Dry reforming of methane on single-site Ni/MgO catalysts: Importance of site confinement. *ACS Catalysis*, *8*(10), 9821–9835. https://doi.org/10.1021/acscatal.8b02277.

CHAPTER

9

Catalysts for steam reforming of biomass tar and their effects on the products

Mira Abou Rjeily[a], *Cédric Gennequin*[b], *Hervé Pron*[a], *Edmond Abi-Aad*[b], *and Jaona Harifidy Randrianalisoa*[a]

[a]Université de Reims Champagne-Ardenne, Institut de Thermique, Mécanique, Matériaux—ITheMM, EA 7548, SFR Condorcet - FR CNRS 3417, Reims Cedex 2, France

[b]Environmental Chemistry and Life Interactions Unit (UCEIV), University of the Littoral Opal Coast, UR 4492, SFR Condorcet—FR CNRS 3417, Dunkerque, France

9.1 Introduction

Aware of the rising demand on fuels for the energy supply and the depletion of the fossil resources with the population increase, intensive researches have been elaborated around the world to innovate the production of fuels deriving from renewable resources. As an alternative for fossil fuels, biomass has been emerging as a sustainable renewable resource for the production of biofuels having similar energy power to that of fossil fuels (Guan et al., 2016). Four major routes have been adopted for the conversion of biomass into multiple energy forms (such as gaseous, liquid, solid fuels, and heat), namely physical conversion, thermochemical and biochemical conversions, and direct combustion. The biomass has been converted in the thermochemical process by one of the four followings ways: direct liquefaction, torrefaction, gasification, and pyrolysis (Arregi et al., 2018; Guan et al., 2016).

The pyrolysis of biomass has been widely exploited for the production of useful products by thermochemically converting lignocellulosic biomass (Dickerson & Soria, 2013). Along with the production of biochar and numerous gases (methane, hydrogen, carbon monoxide and dioxide, etc.), the biomass pyrolysis results in the formation of a dark, viscous, complex liquid called "tar." Tar is commonly referred to as bio-oil or pyrolytic oil but can be also termed liquid wood, bio-crude oil, bio-fuel oil, pyrolysis oil (Isahak et al., 2012), wood distillate, wood oil, liquid smoke, pyroligneous tar, and pyroligneous acid (Balat et al., 2009). It is made of more than 300 compounds which can cause major issues during the process operation. Being a sticky material, it usually condenses in the low-temperature parts of the downstream process plugging narrow pipelines, corroding the metals, clogging the filters, polymerizing into more complex molecules, and deposing coke on the catalyst (Artetxe et al., 2017). Moreover,

Heterogeneous Catalysis
https://doi.org/10.1016/B978-0-323-85612-6.00009-7

elevated tar concentrations lead to intolerable need for engines' and turbines' maintenance. Additionally, tar is especially harmful to human health given its carcinogenic character (Li & Suzuki, 2009).

Several techniques are envisaged as possible solutions for the tar removal, classified into: physical methods using wet scrubbers or ceramic candle filters, thermochemical conversion technologies applying elevated temperatures, and catalytic reforming with the use of catalyst in order to transform the tar into syngas (Li & Suzuki, 2009). During the physical processes, the tar is removed from the produced gas through liquid/gas or solid/gas interactions. This is an efficient method with relatively easy maintenance. However, the problem is persistent since there is no destruction of the tar. In addition, it is difficult to dispose the filter loaded with tar in an environmentally responsible way. Thermochemical process relies on the considerable increase of the produced gas temperature where heavy aromatic compounds forming the tar are cracked into lighter species such as H_2, CO, and CH_4 causing fewer problems than tar. In order to effectively remove the tar, extremely elevated temperatures surpassing 1000°C are demanded for thermal cracking which might be challenging to attain in biomass thermochemical processes (Milne et al., 1998).

Catalytic reforming is regarded as a viable route to valorize the tar into syngas. It was introduced initially in 1940 by Universal Oil Products and since then, multiple reforming process types were developed (Babaqi et al., 2017). It can be operated at considerably lower temperatures of around 600°C to 800°C, when compared to the thermochemical processes, which eliminates the requirement of expensive alloys as the reactor materials Moreover, in contrast to the physical processes, catalytic reforming helps in destroying the tar completely and thereby removes the waste disposal issue. Two types of catalytic reforming exist: steam reforming and dry reforming. Water vapor and carbon dioxide are employed as oxidizing agents in steam and dry reforming, respectively. In both cases, the reforming reactions serve the aim of transforming the hydrocarbons into valuable gases mainly syngas rich in hydrogen and carbon monoxide. It is realized by applying heat in the presence of an appropriate catalyst that lowers the reforming temperature and allows the production of selective products (Guan et al., 2013).

Catalytic steam reforming has been widely applied for the generation of hydrogen H_2 and syngas which are used to produce methanol, ammonia or to directly generate energy. Hydrogen is abundant, nontoxic, nonpolluting, and exhibits a high energy density (Gao et al., 2018). Hydrogen is very reactive and largely valuable. It interferes in numerous and important industrial processes such as the synthesis of ammonia, which is the first consumer of hydrogen at the global level. This gas is also used during the petroleum reforming in the steps of hydrotreatment and hydrocracking, and for the hydrocarbon synthesis by the Fisher–Tropsh process (Rostrup-Nielsen, 2000). Nowadays, hydrogen is perceived as an alternative fuel with a special interest in fuel cells, which are able to provide a clean energy source for transport applications, replacing diesel and gasoline engines (Fajardo & Probst, 2006). The fuel cell plays the role of a direct converter of the chemical energy of the hydrogen into electrical energy and it is usually referred to as continuously operating batteries (Vozniuk et al., 2019).

The production of syngas and hydrogen is commonly ensured via the catalytic steam reforming of methane (SRM). The SRM is a strongly endothermic process that is generally operated at above 800°C (Park et al., 2019). It is an industrial cost-effective process that provides around 50% of the hydrogen global demand (Basile et al., 2015). Methane is the most abundant natural gas. It is a very stable gas requiring a considerable amount of energy input to use it as a renewable energy source (Park et al., 2019). The steam reforming of methane was first used in the United States given the abundant availability of the natural gas as feedstock. In 1930, Standard Oil of New Jersey was the first to implement the SRM for the industrial application (Nielsen & Jens, 1984).

Even though the SRM is a well-established process, it suffers from some drawbacks on which multiple researches are still being performed. These studies are dedicated to improve the performance of the catalyst to produce more hydrogen, reduce the carbon deposition, and resist to sintering (Basile et al., 2015).

In this chapter, the catalytic steam reforming of the biomass tar is the main topic discussed where first the theoretical approach of this process is investigated. The steam and dry reforming of methane are first introduced followed by the catalytic tar reforming. The issue of carbon formation during the process is also broached. Next, the reactor characteristics are defined in terms of the tar reforming parameters, the classification of the process, and the reactor design. Furthermore, the catalysts used for the tar reforming as well as the catalyst supports are detailed. Following the theoretical section, the experimental results obtained from the catalytic steam reforming of methane and different tar products are presented. The effects of the temperature, the catalyst loading, and the steam-to-carbon S/C ratio are investigated on methane, toluene, benzene, and some tar model compounds as well as on the biomass tar and the biomass fuel gas.

9.2 Tar classification and properties

The pyrolytic oil comprises several oxygenated organic compounds such as alcohols, acids, furans, ketones, aldehydes, and phenols in addition to complex oxygenates resulting from biomass lignin and carbohydrates (Wang et al., 1997). Tars with diverse compositions and reactivity are formed, depending on the thermal severity of the conversion of the vapor phase in the reactors, i.e., on the residence time and gas temperature (Dufour et al., 2011).

The major tar compounds produced are mainly oxygenated hydrocarbons. As the reaction temperature increases, the oxygenated hydrocarbons are converted into light hydrocarbons, olefins, and aromatics, which are next transformed into higher hydrocarbons and larger PAH (polycyclic aromatic hydrocarbons). The organic compounds formed become more stable as the temperature increases (Guan et al., 2016). Hence, tar can be considered as a mixture of the condensable organic compounds produced during the gasification process and consists of aromatic hydrocarbons such as toluene, benzene, phenol, naphthalene, and xylene (Gao et al., 2016). Additionally, water accounts for around 15 wt% of the bio-oil and is originated from dehydration reactions. The water present in the pyrolytic oil fraction is well dispersed. Therefore, the bio-oil is characterized by the hydrophilicity of its compounds derived from carbohydrates, representing the major fraction of the crude oil (Wang et al., 1997).

Bio-oil is miscible with water to an extent of 35%–40%, which leads to the gradual aging of the liquid due to the polymerization of the polyphenols resulting in a variable viscosity going from 10 to 10,000 cp (Dickerson & Soria, 2013). Tars represent the leading contaminant in the gas produced with content varying from 5 to $100\,g/Nm^3$. Nevertheless, their maximum acceptable amount in the gas turbines is $5\,mg/Nm^3$ whereas in internal combustion engines it is $100\,g/Nm^3$ (Artetxe et al., 2017). The undesirable properties labeling the bio-oil include acidity, incomplete volatility, and low heating value (French & Czernik, 2010). These unfavorable characteristics urge the need to investigate alternative technologies for the use or upgrade of the bio-oil to yield high added-value species (Dickerson & Soria, 2013).

One of the ways of classification of the tar is based on its appearance and the products can be categorized in three major classes as detailed in Table 9.1. Fig. 9.1 shows the typical composition of the biomass tar.

Dufour and colleagues (Dufour et al., 2011) performed the pyrolysis of spruce wood chips in a tubular reactor. They studied the evolution of the compositions of the tar and gas produced as a function of the reactor's wall temperature from 700°C to 1000°C. Solid-phase adsorbent (SPA) and impingers filled with methanol were used to sample the tar. The tar components were analyzed by gas chromatography/mass spectrometry (GC/MS) and some compounds were chosen to be quantified. The quantified tar components are listed in Table 9.2 with their corresponding elementary composition and molecular weight. The list of tar components includes benzene, toluene, m- and o-xylene, phenols, indene, m-, p-, and o-cresol, naphthalene, 1- and 2-methylnaphthalene, acenaphthylene, and phenanthrene. Being one of the most stable and abundant aromatic products derived from the pyrolysis or gasification, benzene is adopted as an aromatic tar. In addition, benzene can cause problems when used for advanced applications as in the conversions of catalytic gas.

TABLE 9.1 Tar classification based on its appearance (Devi et al., 2005; Dufour et al., 2011).

Tar class	Property
Primary	Low molecular weight oxygenates generated at low thermal severity (400–700°C) (i.e., hydroxyacetaldehyde, furfural, and levoglucosan)
Secondary	Olefins and phenolics, formed in a temperature range between 700°C and 850°C (i.e., xylene, cresol, and phenol)
Tertiary	Complex polycyclic aromatic hydrocarbons (PAHs) formed between 850°C and 1000°C (i.e., benzene, toluene, pyrene, and naphthalene)

9.3 Theoretical approach of biomass tar reforming

Catalytic steam reforming of methane

During the methane steam reforming process, the methane reacts with the water vapor (as a principal oxidizing agent), with a contact time of several seconds, over a catalyst following the reaction (Eq. 9.1) in order to produce hydrogen and carbon monoxide. When the reaction is driven in the stoichiometric conditions, the ratio H_2/CO is equal to 3. If an excess of steam is

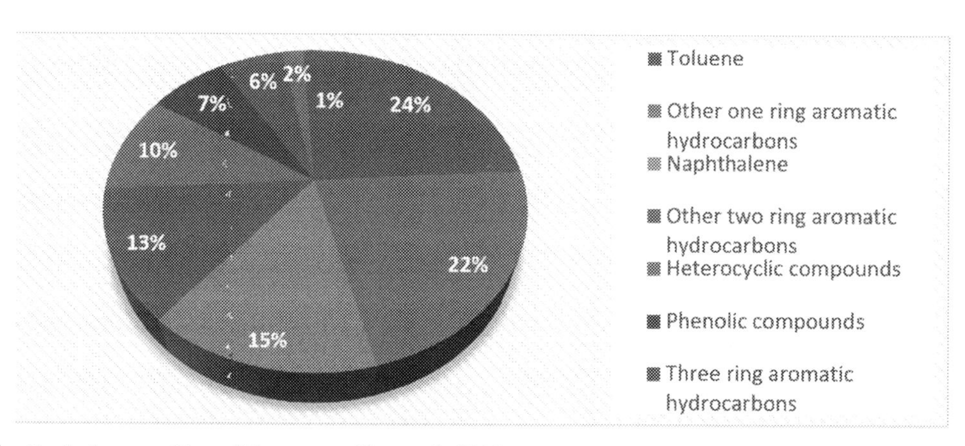

FIG. 9.1 Typical composition of biomass tar (Gao et al., 2016).

TABLE 9.2 List of quantified compounds of the tar derived from the pyrolysis of wood chips (Dufour et al., 2011).

Peak number	Name	Elementary composition	Molecular weight
1	Benzene	C_6H_6	78
2	Toluene-d_8	C_7D_8	100
3	Toluene	C_7D_8	92
4	m-Xylene	C_8H_{10}	106
5	o-Xylene		106
6	Phenol-d_6	C_6D_6O	100
7	Phenol	C_6H_6O	94
8	Indene	C_9H_8	116
9	o-Cresol	C_7H_8O	108
10	m- and p-Cresol		
11	Naphtalene-d_8	$C_{10}D_8$	136
12	Naphtalene	$C_{10}H_8$	128
13	2-Methylnaphtalene	$C_{11}H_{10}$	142
14	1-Methylnaphtalene		
15	Acenaphthylene	$C_{12}H_8$	152
16	Phenanthrene	$C_{14}H_{10}$	178

supplied, the water-gas shift reaction (WGSR) (Eq. 9.3) can take place despite its moderately exothermic nature. This reaction increases the hydrogen production resulting from the oxidation of CO into CO_2 (Park et al., 2019) while simultaneously reducing the CO formation which can be regarded as an advantage given its hazardous impact on human health and environment (Miller, 2011). The reactions involved in the catalytic steam reforming process are stated as follows (Eqs. 9.1–9.5) (Park et al., 2019):

Steam reforming of methane SRM1:

$$CH_4 + H_2O \leftrightarrow 3H_2 + CO \quad \Delta H_{298K} = +205.9\,kJ/mol \tag{9.1}$$

Steam reforming of methane SRM2:

$$CH_4 + 2H_2O \leftrightarrow 4H_2 + CO_2 \quad \Delta H_{298K} = +164.7\,kJ/mol \tag{9.2}$$

Water – gas shift reaction WGSR:

$$CO + H_2O \leftrightarrow H_2 + CO_2 \quad \Delta H_{298K} = -41.1\,kJ/mol \tag{9.3}$$

Reverse WGSR:

$$H_2 + CO_2 \leftrightarrow CO + H_2O \quad \Delta H_{298K} = +41.1\,kJ/mol \tag{9.4}$$

Boudouard reaction (BR):

$$CO_2 + C \leftrightarrow 2CO \quad \Delta H_{298K} = +172\,kJ/mol \tag{9.5}$$

As shown from Eqs. 9.1 and 9.2, SRM reactions represent volumetric expansion reactions and hence the steam reforming is usually operated at low pressure, making it more thermodynamically favorable. Nevertheless, in order to facilitate the overall operation and reduce the reactor size, an operating pressure above 0.5MPa is adopted in the reactor. Hence, the optimal operating conditions are to be chosen depending on the process scale, the quantity of the desired product, and the composition. Additionally, due to the low quantity of carbon dioxide generated in the process, the dry reforming of methane, described hereafter, might also take place (Park et al., 2019).

Catalytic dry reforming of methane

In this process, carbon dioxide is used as an oxidizing gas of methane instead of the steam according to the following reaction (Eq. 9.6) (Gao et al., 2018):

Dry reforming reaction:

$$CH_4 + CO_2 \leftrightarrow 2H_2 + 2CO \quad \Delta H_{298K} = +206.3\,kJ/mol \tag{9.6}$$

This process is not deprived of interest since, on one hand, it gives a ratio of H_2/CO close to one, useful in processes such as hydroformylation and the reactions of carbonylation, and on the other hand, it consumes two greenhouse gases (CO_2 and CH_4). However, this reaction is more endothermic than the steam reforming and requires higher reaction temperatures resulting in the deactivation of the catalyst by sintering of the active phase and by coke deposition. In addition, simultaneously to the principal water-gas shift reaction (WGSR) (Eq. 9.3), the reverse WGSR (Eq. 9.4), takes place, which translates in the conversion of CO_2 being always more important than the conversion of CH_4 but with an H_2/CO ratio inferior to 1 (Guo et al., 2004).

Catalytic tar reforming

The biomass pyrolysis produces, along with char and non-condensable gases, a complex liquid formed by more than 300 compounds. In order to represent the tar as a single product and to integrate it into the chemical reactions involved during the reforming process, the tar chemical formula was simplified by the scientific community of the catalytic reforming. The tar was therefore considered a single hydrocarbon compound having C_nH_x as a general chemical formula. The steam reforming of tar entails the reaction of tar with steam generating several non-condensable gases such as H_2, CO, CO_2... along with other hydrocarbons. The latter can also interact with steam producing additional syngas and other non-condensable gases (Abou Rjeily et al., 2021).

During this process, several reactions occur simultaneously and the competition between them leads to the distribution of the products. The reactions are listed as follows (Eqs. 9.7–9.11) (Abou Rjeily et al., 2021; Gao et al., 2015; Guan et al., 2016):

Thermal tar cracking:

$$pC_nH_x(\text{tar}) \rightarrow qC_mH_y(\text{smaller tar}) + rH_2 \qquad \Delta H > 0 \tag{9.7}$$

Steam tar reforming:

$$C_nH_x(\text{tar}) + nH_2O \rightarrow \left(n + \frac{x}{2}\right)H_2 + nCO \qquad \Delta H > 0 \tag{9.8}$$

Dry tar reforming:

$$C_nH_x(\text{tar}) + nCO_2 \rightarrow \left(\frac{x}{2}\right)H_2 + 2nCO \qquad \Delta H > 0 \tag{9.9}$$

Carbon formation:

$$C_nH_x(\text{tar}) \rightarrow nC + \left(\frac{x}{2}\right)H_2 \qquad \Delta H > 0 \tag{9.10}$$

Hydrocarbons (HC) reforming:

$$HC + H_2O \rightarrow H_2 + CO + CO_2 \qquad \Delta H > 0 \tag{9.11}$$

During the catalytic reforming process, these reactions are expected to coexist. Nonetheless, the reaction conditions as well as the type of catalyst used determine the dominant reaction based on the promoting or inhibiting interactions among the different compounds. According to the catalytic tar-reforming described by Eqs. (9.7)–(9.11), it is considered that all tar is expected to be transformed into lighter and simpler molecules like CO and H_2 by steam reforming (Guan et al., 2016). Therefore, catalytic steam reforming presents itself as a promising economical and technical alternative process where gas with a considerably high purity can be obtained in addition to the increase in the heating value of the gas product (Artetxe et al., 2017).

Carbon formation

One of the major issues faced during the catalytic dry reforming is the catalyst deactivation, which could be triggered by multiple mechanisms and numerous causes. In reforming reactions, the deposition of carbon or hydrocarbons (also referred to as coking), oxidation, poisoning, and sintering are the main causes behind the deactivation of the catalyst. In addition, the carbon formation can lead to technical issues such as increase in pressure drop and blocking of the reactor (Aouad et al., 2018).

It is important to make a distinction between the terms "carbon" and "coke." Although both terms are sometimes used interchangeably, by default their definition is related to their origin. Carbon is considered the product of CO disproportionation as in the Boudouard reaction (Eq. 9.5), whereas the term coke refers to the material that originates from the decomposition or condensation of hydrocarbons. Coke usually consists of polymerized heavy hydrocarbons. However, the composition of coke could vary from high molecular weight hydrocarbons to primary carbons such as graphite depending on reaction conditions (Abou Rjeily et al., 2021).

Carbon formation results from unfavorable reactions. It appears mainly during the reaction of tar steam reforming and dry reforming according to Eq. (9.10) (Guan et al., 2016). In the case of the methane steam reforming, the carbon deposition produced by the methane decomposition, Eq. (9.12), and the Boudouard reaction, Eq. (9.13), (Zhang et al., 2007) is to be avoided.

Methane decomposition (MD):

$$CH_4 \leftrightarrow C + 2H_2 \qquad \Delta H_{298K} = +75\,\text{kJ/mol} \tag{9.12}$$

Boudouard reaction (BR):

$$2CO \leftrightarrow C + CO_2 \qquad \Delta H_{298K} = -172\,\text{kJ/mol} \tag{9.13}$$

The formation of carbon during methane reforming can depend on numerous factors and the most important are the following (Swaan et al., 1994):

- The reaction conditions:
  - Temperature
  - Steam-to-carbon ratio
- The nature of catalysts:
  - Nature and morphological structure of the catalyst
  - Nature of support
  - Metal–support interaction

To prevent carbon deposition, two approaches have been adopted. The first approach consists in enhancing the steam adsorption on the catalyst surface by using carbon precursors (Swierczynski et al., 2008). The steam plays a favorable role via the reactions (Eqs. 9.14–9.15) that express the gasification potential of the carbon deposited on the catalyst by the steam (Zhang et al., 2007).

$$C + H_2O \leftrightarrow CO + H_2 \qquad \Delta H_{298K} = +131\,\text{kJ/mol} \tag{9.14}$$

$$CO + H_2O \leftrightarrow H_2 + CO_2 \qquad \Delta H_{298K} = -41.1\,\text{kJ/mol} \tag{9.15}$$

The second method is by modifying the catalyst by the addition of other metal additives (Swierczynski et al., 2008) such as CaO, MgO, K_2O, and La_2O_3. The use of these basic promoters helps to adsorb and to dissociate CO_2, which is acid, and thus results in reducing the carbon formation rate on the catalyst surface. Moreover, carbon can be oxidized by adding oxygen carriers such as La_2O_3, CeO_2, and ZrO_2 thereby decreasing its formation (Aouad et al., 2018). The use of bimetallic (or trimetallic) active phase systems has been suggested by some researchers in place of monometallic ones. They investigated the use of several metals such as Cu, Co, Fe, V, Cr, Rh, Mo, Mn, and Sn (Estephane et al., 2015; Zhang et al., 2007). It has been also proposed to add some quantities of noble metals such as Ru and Rh to the transition metals like Co or Ni (Gennequin et al., 2016). Moreover, it has been demonstrated that less carbon can be formed when the particle size is smaller (Perez-Lopez et al., 2006).

9.4 Reactor characteristics

Tar reforming parameters

Multiple parameters affect the yields of the catalytic reforming products and the model compound conversion including the reaction temperature, the steam-to-carbon ratio, the pressure, the nature of catalysts such as the metal loading, the time of the reaction, and the amount of steam feed.

Effect of temperature on the steam reforming of methane

The reaction temperature greatly affects the composition of the methane or tar reforming gaseous products. Knowing that higher temperatures are in favor of endothermic reactions, the temperature augmentation profoundly affects the production of CO, CO_2, and H_2 and strengthens the tar cracking. It is proposed that tars decompose more easily into small molecules such as CO,

CO_2, and H_2 with the temperature increase. Regarding the WGSR, given its exothermic nature, the temperature elevation becomes adverse for the production of H_2. Due to Le Chatelier's principle, the tendency is directed toward the reactants with the temperature increase. To improve the hydrogen yield and efficiency, the temperature must be balanced between the tar conversion and the energy consumption (Gao et al., 2015).

Several side reactions occur with the dry reforming of methane (DRM), Eq. (9.6), such as the SRM, Eq. (9.1), Boudouard reaction (BR), Eq. (9.5), reverse water-gas shift reaction (RWGSR), Eq. (9.4), carbon dioxide methanation (CDM), Eq. (9.16), carbon monoxide hydrogenation (CMH), Eq. (9.17), and methane decomposition (MD), Eq. (9.12) (Aouad et al., 2018).

Carbon dioxide methanation (CDM):

$$CO_2 + 4H_2 \leftrightarrow CH_4 + 2H_2O \qquad \Delta H^\circ = -164.7\,kJ/mol \tag{9.16}$$

Carbon monoxide hydrogenation (CMH):

$$CO + H_2 \leftrightarrow C + H_2O \qquad \Delta H^\circ = -131.3\,kJ/mol \tag{9.17}$$

The variation of the equilibrium constants with temperature is given in Fig. 9.2 for the main reactions. According to thermodynamic calculations at atmospheric pressure, the dry reforming reaction is found to be only realizable at temperatures above 633°C ($\Delta G > 0$). Indeed, for a good methane conversion, a high temperature is needed given the stability of the C–H bond in the methane molecule ($435\,kJ.mol^{-1}$). However,

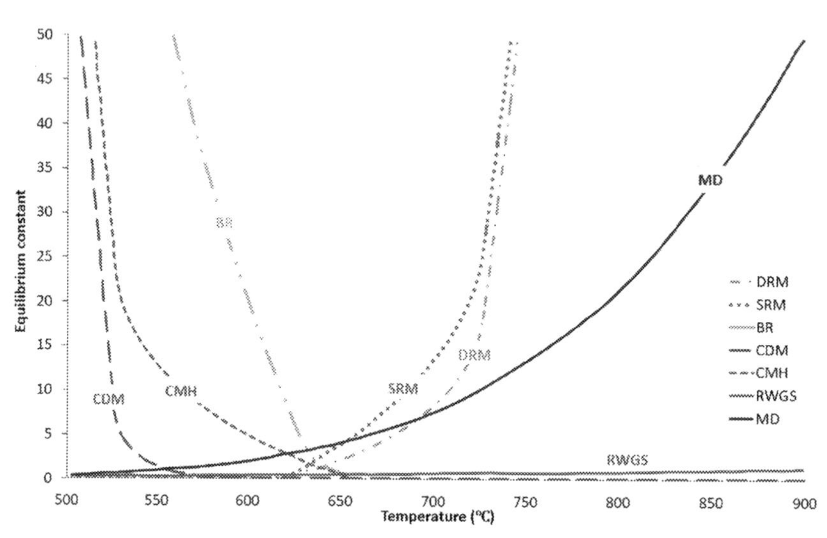

FIG. 9.2 Variation of the equilibrium constant as function of the temperature for the steam reforming of methane (SRM), dry reforming of methane (DRM), methane decomposition (MD), Boudouard reaction (BR), reverse-water shift reaction (RWSR), carbon dioxide methanation (CDM) and carbon monoxide hydrogenation (CMH). *Reproduced from Aouad, S., Labaki, M., Ojala, S., Seelam, P. K., Turpeinen, E., Gennequin, C., et al. (2018). A review on the dry reforming processes for hydrogen production.* Catalytic Materials and Technologies, *60–128. https://doi.org/10.2174/9781681087580118020007.*

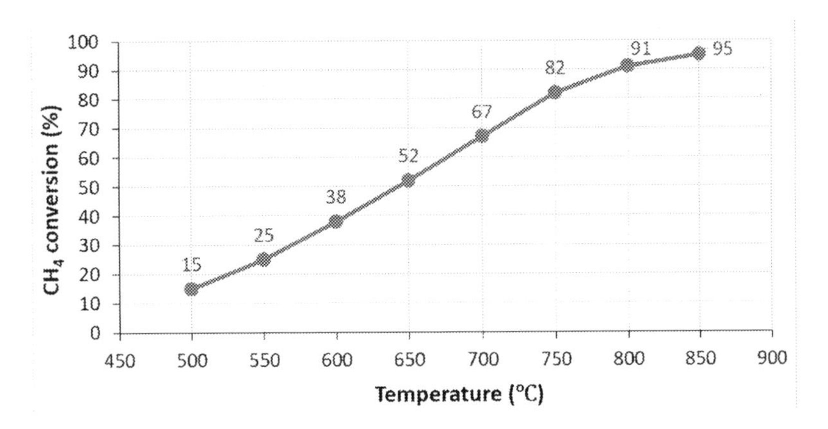

FIG. 9.3 Effect of temperature on the methane steam reforming (Park et al., 2019).

other reactions such as SRM and MD along with RWGSR at a lesser extent can occur when the temperature surpasses 633°C. These reactions have a significant impact on the dry reforming reaction. The equilibrium constant of the methane dry reforming reaction rises sharply with the temperature given its strong endothermic nature. Whereas the equilibrium constant of the MD and RWGS reactions increases moderately with the temperature given their moderate endothermic nature. Inversely, temperatures equal to or higher than 750°C are thermodynamically disfavoring for exothermic reactions (Zhang et al., 2007).

Practically, the residence time in the reactor is smaller than the interval of time required to attain the theoretical equilibrium. Hence, a catalyst is required to reduce the activation energy and to favor the DRM reaction over the other reactions. As a consequence, the methane dry reforming process is made more economical with the use of a catalyst (Aouad et al., 2018).

Park and colleagues (Park et al., 2019) worked on the steam reforming of methane at the lab scale and on the bench scale using powder- and pellet-type commercial Ni-based catalysts, respectively. The effects of the reaction temperature, the reaction pressure, and the steam/methane ratio on the methane steam reforming performance were evaluated.

The effect of the temperature on the SRM was evaluated at a pressure of 1 MPa, a gas hourly space velocity (GHSV) fixed at 4.8 L CH_4/(h.gcat), and a steam/methane ratio equaling to 3, as shown in Fig. 9.3. It indicates that the methane conversion increases with the temperature, the expected behavior from the highly endothermic methane steam reforming reaction, Eq. (9.1) (Park et al., 2019).

Effect of the reaction pressure on the steam reforming of methane

The effect of the reaction pressure on the SRM was studied at a reactor temperature of 830°C, a GHSV of 4.8 L CH_4/(h.gcat), and a steam/methane ratio of 3, and the results are reported in Fig. 9.4. It is observed that the methane conversion decreases as the pressure increases. It can be concluded that under the test conditions, the reaction rate becomes thermodynamically limited as the pressure increases (Park et al., 2019).

Effect of steam-to-carbon S/C ratio on the steam reforming of methane

The steam-to-carbon S/C ratio is defined as the ratio between the feed rate of steam for the tar reforming over the feed rate of tar containing carbon. It plays a crucial role in the tar reforming process and its value usually ranges between 0

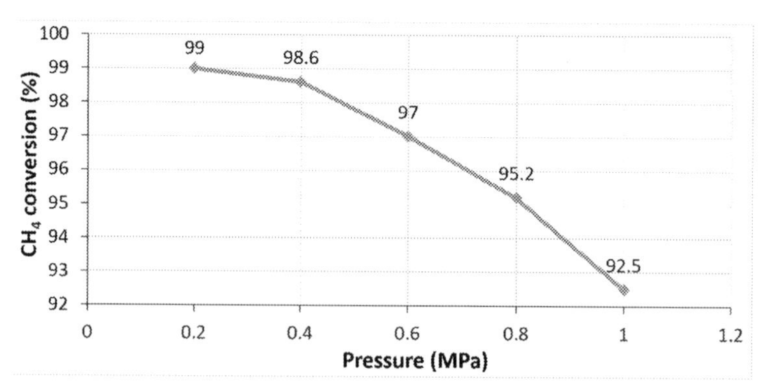

FIG. 9.4 Effect of pressure on the methane steam reforming (Park et al., 2019).

and 4. It is noteworthy to mention that the increase of the steam supply tends to favor the equilibrium of the tar reforming reaction and WGSR (Gao et al., 2015).

With S/C ratio equal to 1, the steam quantity would be insufficient to ensure the reforming of methane SRM2 and the WGSR. The increase of the S/C helps to improve the intensity of these reactions since they are greatly affected by the S/C ratio. An elevated steam partial pressure ameliorates the steam reforming reaction and the WGSR into increasing the formation of hydrogen. This can explain the fact that the increase of the S/C ratio to 2 results generally in a remarkable increase in H_2 and CO_2 yields in opposition to a dramatic decrease in CO and CH_4 yields. By further increasing the S/C ratio, a huge amount of heat for the gasification is absorbed by the water evaporation into steam, which negatively impacts the hydrogen production in the reforming reactions. Moreover, the active sites present on the surface of the catalyst would be saturated by the excess of water molecules increasing the reactants' partial pressure in the gas stream thus reducing the efficiency of the catalysis (Gao et al., 2018; Quan et al., 2018).

Practically, two issues are usually encountered in the real reforming reactions. On one hand, insufficient steam (i.e., low S/C) can lead to a low hydrogen yield and concentration, resulting from incomplete reactions occurring during the reforming. Consequently, the reforming reactions and the WGSR become unable to reach the complete reaction state. On the other hand, when high S/C ratio is applied, sufficient steam can be utilized to drive reforming and cracking of tar with consequently more elevated hydrogen and gas yields. Moreover, the equilibrium of the WGSR is shifted toward the hydrogen production with higher water partial pressure, and the gasification of carbonous intermediates is promoted, reducing the coke formation and deposition on the catalyst. Nonetheless, excessive steam has several undesirable effects. First, supplementary energy consumption results from excessive water, due to the separation of steam and the dryness of formed gas and in the condensation process. Secondly, the excess of steam leads to the absorption of the heat, thus decreasing the reforming temperature resulting in a drop of the tar decomposition. Therefore, an optimal steam-to-carbon ratio might exist at the different operating conditions (Gao et al., 2015).

The effect of the S/C ratio on the SRM was determined by Park et al. (2019) at the reaction temperature of 830°C, a GHSV of 4.8 L CH_4/(h.gcat) and a pressure of 1 MPa, as demonstrated in Fig. 9.5. If the SRM was the involved reaction, the excess of steam would not necessarily affect the methane conversion, as shown in Eq. (9.1). Nevertheless, since the increase of S/C ratio increases the methane conversion as seen in Fig. 9.5, then an additional reaction could

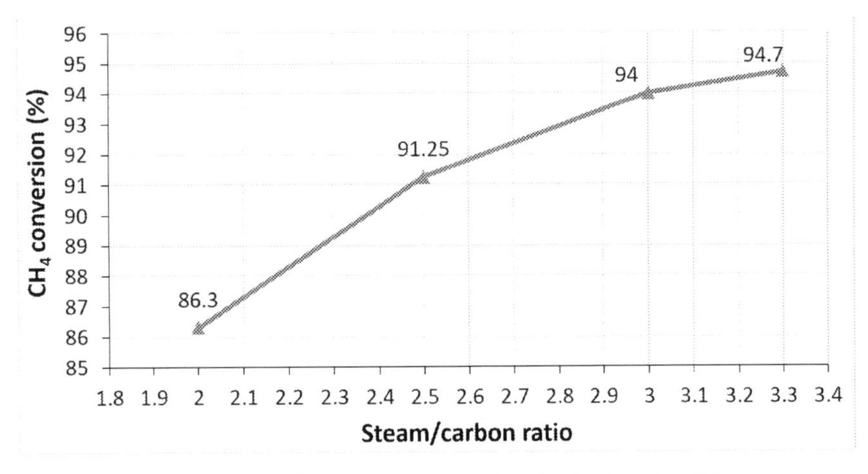

FIG. 9.5 Effect of steam/methane ratio on the methane steam reforming (Park et al., 2019).

have occurred which is the WGSR, Eq. (9.3). It shifts the equilibrium of the SRM in a way that the methane consumption accelerates as the S/C ratio is elevated (Park et al., 2019).

Process classification

The classification of the catalytic reforming processes is based on the system of the regeneration of the catalyst. It commonly divides the catalytic reforming processes into three types (Babaqi et al., 2017):

1. Semi-regenerative catalytic reformer process (SRCRP)
2. Cyclic regenerative catalytic reformer process (CRCRP)
3. Continuous catalytic regeneration reformer process (CCRRP).

These types are briefly described next.

Semi-regenerative catalytic reformer process (SRCRP)

SRCRP, the oldest reforming process, is used with a fixed-bed catalyst system and has three or four reactors in series. It is employed for producing rich aromatic compounds and gasoline. The in-situ regeneration of the catalyst is realized by shutting down the process once every 6 months to 2 years. The coke formation results in the gradual decrease of the catalyst activity which has a negative impact on the aromatics yield and the hydrogen by-product. With the drop of the catalyst activity, the reactor temperature must be raised to maintain a relatively constant reaction conversion. High pressures (13.8–20.7 bar) were applied in these units in order to reach a maximum time interval between two regenerations and to minimize the catalyst deactivation. The catalyst usually used in the SRCRP is the Pt—Re since it offers a large tolerance-to-coke level and enables operating at lower pressures. However, the activation of the catalyst requires the stop of the process operation. Moreover, the frequent interruption of the process for regeneration makes the energy conservation process difficult (Babaqi et al., 2017; Antos & Aitani, 2004; Rahimpour, Jafari, & Iranshahi, 2013).

Cyclic regenerative catalytic reformer process (CRCRP)

CRCRP consists of four reactors in series operating with fixed-bed catalyst system. It has a swing configuration allowing an in-situ regeneration to be applied in one of the reactors whereas normal operation is undergone in the other reactors. Therefore, this configuration removes only one reactor out of operation at a certain time for its regeneration while the rest of the reactors keep on operating. Few weeks to few months separate two regenerations for every reactor. The continuous operation of the reforming process is ensured by the CRCRP configuration. More severe operating conditions are applied during the CRCRP such as a lower operating pressure (13.8 bar), lower hydrogen-to-feed ratio, and a large range of boiling feed. These conditions increase the catalyst deactivation rate. Several advantages characterize the CRCRP including low operating pressure, consistent hydrogen purity, good conversion, and overall catalyst activity. However, the reactor switching control has a complex nature and thus stronger safety precautions are required (Babaqi et al., 2017; Rahimpour et al., 2013).

Continuous catalytic regeneration reformer process (CCRRP)

CCRRP operates with a moving-bed catalyst system and is formed by four reactors either placed in series or stacked one above another. A special regenerator is employed for the catalyst regeneration which is operated continuously, and the regenerated catalyst is recycled back to the operating reactors. In fact, the catalyst travels the stacked reactors from top to bottom by gravity. A continuous withdrawal of the consumed catalyst from the last reactor is realized followed by its transfer to the head of the regenerator where the coke is burned off. Then the regenerated catalyst returns to the top of the stacked reactors. In this process, the catalyst is transmitted between the reactors and the regenerator which is known as the gas lift method. The CCRRP design was a step change in the reforming process, in contrast to those of the CRCRP and the SRCRP. One of the major advantages of the CCRRP is the continuous production of hydrogen gas at higher yield and catalyst activity, lower operating pressure (3.4 bar) and inferior recycle ratio demanded. Furthermore, process integration can be applied to this process since it works continuously and therefore the operation does not need to be stopped. Moreover, the source of energy is continuous thereby applying heat integration becomes possible. The main catalyst used in the CCRRP is the platinum/tin alumina. Adding the tin to the platinum/alumina regeneration ability ameliorates the stability and the aromatics' selectivity of the catalyst (Babaqi et al., 2017; González-Marcos et al., 2005; Meyers, 2004).

Reactor design

Various designs of reactors have been adopted for the steam reforming. Nevertheless, the reactors are mainly classified according to the catalyst bed (Adeniyi et al., 2019). The reactor configurations for the steam reforming can be grouped under two main categories: (i) one-step processes including fixed bed, fluidized bed, and spouted bed reactors and (ii) two-step processes including two-stage fixed bed reactor and two-stage fixed-fluidized bed reactor (Ochoa et al., 2020). Fluidized bed and fixed bed reactors have been used to conduct several studies with both resulting in good yields and high hydrogen selectivity (Adeniyi et al., 2019). The five steam reforming reactor configurations are presented in Fig. 9.6.

One-step processes

Fixed bed

Methane steam reforming has been conventionally realized in a fixed bed reactor (Fig. 9.6a). Several tubes are used into which the catalyst is loaded and are introduced in a furnace whose operating temperature can go from 527°C up to 807°C. An efficient heat transfer and distribution is required for the extremely

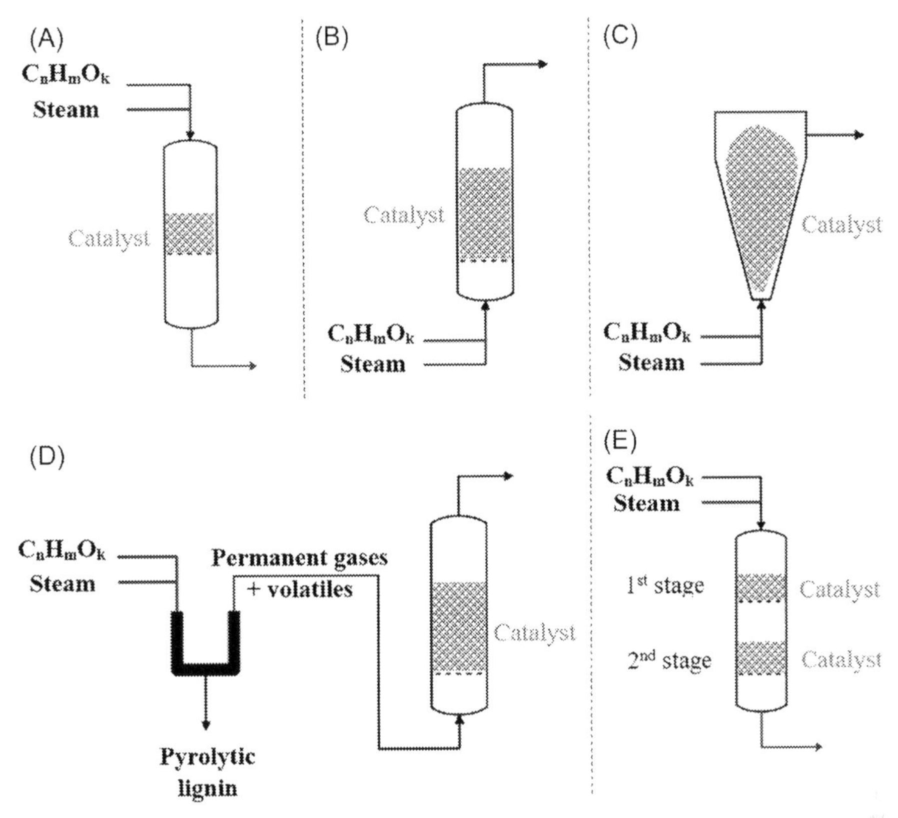

FIG. 9.6 Different configurations for the reactors of the steam reforming of biomass-derived oxygenates: (A) fixed bed reactor, (B) fluidized bed reactor, (C) spouted bed reactor, (D) two-stage fixed-fluidized bed reactor, and (E) two-stage fixed bed reactor. *Reproduced from Arregi, A., Amutio, M., Lopez, G., Bilbao, J., & Olazar, M. (2018). Evaluation of thermochemical routes for hydrogen production from biomass: A review.* Energy Conversion and Management, 165, *696–719. https://doi.org/10.1016/j.enconman.2018.03.089.*

endothermic reversible reaction of the steam reforming and hence the process is very challenging and expensive. Furthermore, the effectiveness factor is very low being in the order of 10^{-3} to 10^{-2} hence the limitations of diffusion are very severe (Adris et al., 1991; De Deken et al., 1982; Elnashaie & Abashar, 1993). Additionally, several drawbacks exist for the fixed bed reactor including huge temperature gradients, small heat transfer rates, limited mass transfer, and strong resistance to diffusion in the pores of the catalyst. It should be noted that the heat input efficiency strongly affects the highly endothermic steam methane reforming process (Roy et al., 1999).

Fluidized bed

Fluidized bed reactor for primary conversion The fluidized bed reactor (Fig. 9.6b) can operate in gasification, pyrolysis, and combustion operating modes. These modes of operation are determined by the ratio of the fuel to the air. In pyrolysis and gasification, the reactor is capable of processing up to 90-270 kg/h of biomass. While in combustion mode, the reactor processes around 20–60 kg/h. The fluidized bed represents a bubbling design operating at atmospheric pressure. Pressure taps and thermocouples are inserted along the axis of the fluidized bed reactor to ensure the measurement of the pressures and the temperatures, respectively (Brown, 2007).

Fluidization gas system The bed is fluidized by air which also ensures the oxygen supply for the gasification and combustion modes. The air is supplied to the distributor of the fluidized bed reactor by a positive displacement blower. During the pyrolytic gasification of the steam reforming, a co-injection of the steam with air is possible to fluidize the bed. Steam is generated by a boiler at flow rates that can reach 180 kg/h (Brown, 2007).

Fluidized bed reactor for catalytic reforming Fluidizing the reforming catalyst is possible if the particle size falls into the proper range and if the catalyst is strong enough to endure the mechanical environment inside the fluidized bed reactor. Successful use of the catalyst particles was achieved for mean diameters of around 170 to 200 μm and size ranges between 90 and 355 μm. The fluidized bed thermal uniformity was reached at a velocity five times higher than the minimum fluidization velocity (Roy et al., 1999).

Fluidized beds are characterized by their outstanding mass and heat transfer properties, a large variety of operating conditions, excellent gas/solid contact, uniformity of the catalyst bed, capability of producing elevated efficiencies as well as negligible emissions to multiple processes (Roy et al., 1999; Winter & Schratzer, 2013). In addition, it has the ability of shifting the thermodynamic equilibrium of the steam reforming reactions as well as realizing the reaction and the product separation simultaneously. The fluidized bed reactor has a more compact design in comparison to the fixed bed reactor and it can virtually eliminate the limitations of diffusion along with enhancing the selectivity and conversion (Abashar et al., 2003). Furthermore, fluidized bed processes have a strong ability to mitigate the deactivation of the catalyst which is especially motivating for feeds heavier than methanol, ethanol, butanol, and bio-oil. These products are more susceptible to the thermal degradation and lead to high amounts of coke deposition on the catalyst (Trane et al., 2012). Thereby, fluidized bed reactors are acquiring increasing attention and their field of application is extending specially in the industrial, chemical, and environmental processes (Winter & Schratzer, 2013).

Despite the fluidized bed ability to produce high H_2 yields (more than 80% of the maximum stoichiometric) and to perform a full conversion in the steam reforming of oxygenates (including biomass), the coke deposition and thus the catalyst deactivation remains a severe drawback of the fluidized bed (Trane et al., 2012).

Spouted bed reactor

The spouted bed reactor (Fig. 9.6c) was developed to mitigate the problem of coke deposition encountered with the use of fluidized bed. No coke was deposited when this reactor was evaluated for the steam reforming of the aqueous fraction of bio-oil (Kechagiopoulos et al., 2009). It was implemented after modifying the reactor by using an injection nozzle specially designed (Kechagiopoulos et al., 2007). This reactor was demonstrated to be very successful leading to a competent processing of ethylene glycol, chosen as a bio-oil model component. It was noticed that the coke formation is drastically low for three types of materials tested (olivine, Ni/olivine catalyst, and sand). The coke deposition is minimized in a spouted bed reactor thanks to advantageous hydrodynamics particularly in the catalyst continuous recirculation where the catalyst is exposed to changing reaction atmospheres (Kechagiopoulos et al., 2009).

Two-step processes

For the purpose of avoiding the issue of carbon formation and coke deposition in the bio-oil steam reforming processes, thereby leading to the deactivation of the catalyst, some works included the use of two-stage reactor configurations.

Two-stage fixed-fluidized bed reactor

A two-stage thermal-catalytic system (Fig. 9.6d) was used to study the steam reforming of raw bio-oil (Remiro et al., 2013c; Valle et al., 2018a; Valle et al., 2018b), bio-oil/ethanol mixtures (Remiro et al., 2014; Valle et al., 2014) and bio-oil aqueous fraction (Remiro et al., 2013a; Remiro et al., 2013b; Valle et al., 2013). In this system, some bio-oil compounds were repolymerized in the first stage leading to the deposition of "pyrolytic lignin" (also known as thermal pyrolytic lignin or thermal lignin). The second stage consisting of a fluidized bed was dedicated to the in-line reforming of gases produced in the first reactor along with the treated bio-oil.

Two-stage fixed bed reactor

Several authors have performed the steam reforming in a two-stage fixed bed reactor (Fig. 9.6e). Yao et al. (2014) first volatilized the bio-oil aqueous fraction at 400°C in a first fixed bed reactor and then performed the reforming in a second fixed bed reactor over modified Ni/Al catalysts. Wu et al. (2008) aimed at avoiding the direct contact between the catalyst and the bio-oil and therefore they relied on a two-stage fixed bed reactor system. In the first stage, dolomite was employed as catalyst for the reforming of bio-oil and in the second stage, the purification of the volatiles was realized using Ni/MgO catalyst.

9.5 Catalysts for catalytic steam reforming

Catalysts for steam reforming

The key to the success of the catalytic steam reforming is the choice of an effective catalyst. For this purpose, a wide selection of catalysts has been adopted going from natural minerals and alkaline earth to metal catalysts (Gao et al., 2015). For an efficient steam reforming, various metals can be employed counting noble metals (rhodium Rh, palladium Pd, platinum Pt), base metals (cobalt Co, iron Fe, nickel Ni), or their combination forming the so-called transition metals. For instance, although Rh is highly active and resistant to carbon deposition, it is rarely used at the industrial scale given its high cost, which is the case of all noble metals (Gao et al., 2018). On the other hand, a particular attention has been attributed to nickel-based systems among the transition metals. Compared to noble metals, nickel-based catalysts are less expensive and more available which attract their application on the industrial scale. However, their major drawback resides in their deactivation resulting from the coke deposition and the metallic phase sintering under certain operating conditions (Abou Rjeily et al., 2021).

In fact, during the steam reforming, the hydrocarbons are adsorbed over the nickel sites after which reactions of dehydrogenation and cracking occur. Consequently, smaller hydrocarbon fragments are formed and adsorbed. These smaller fragments can undergo polymerization to form undesirable intermediates and produce coke. There is a competition between these reactions and those with the adsorbed water species. Therefore, adsorbed water species must be sufficiently supplied by the catalyst to surpass the reactions of forming coke (Balagurumurthy et al., 2015). Moreover, to improve the coke resistance of Ni-based catalysts, multiple parameters can be adjusted including the catalyst preparation method, the insertion of promoters into the catalyst structure, the support characteristics, and its metal concentration. In fact, adding some metals such as magnesium in the support structure creates basic oxides which enable Ni catalysts to chemisorb oxidants like CO_2 and H_2O, thereby decreasing the coking (Gennequin et al., 2016).

Modifiers such as Fe, Pt, Cu, CaO, ZrO_2, CeO_2, and MnO_x are also adopted to enhance the Ni catalysts (Gao et al., 2018).

Based on the catalyst properties, the most convenient catalyst can be chosen according to several criteria (Abou Rjeily et al., 2021; Guan et al., 2016):

- The catalyst stability and reusability
- Moderate pressure drops during operation
- The catalyst efficacy of tar reforming
- The catalyst mechanical and thermal strengths
- The economical cost and availability for industrial deployment
- The capability of the catalyst to deliver suitable syngas ratio for a specified application
- Efficiency in reforming of aromatic compounds and heavy hydrocarbons
- Resistance to deactivation caused by impurity coking, sintering, and fouling

Bimetallic catalysts for steam reforming

Improving the metal dispersion is an essential feature in order to limit the fouling of the catalyst taking place generally at high temperatures. Interestingly, the use of bimetallic catalyst entails higher activities than monometallic catalysts due to the coexistence and good dispersion of the metals (Saw et al., 2015). The alloys of Ni, Co, and Fe have been commonly adopted and proven to be very promising for the catalyst steam reforming of tar. The important catalytic activity and the resistance to coking characterize these alloys (Ashok et al., 2020).

Nickel-iron bimetallic catalysts

Among the numerous bimetallic catalysts used for the steam reforming of biomass tar and its model compounds, the nickel-iron Ni—Fe bimetallic catalyst has been largely explored. The synergistic effect between Ni and Fe during tar reforming strongly impacts the catalyst performance. Water molecules are activated by the Fe sites which produces adsorbed oxygen while C—C and C—H linkages are activated in the hydrocarbons by the Ni sites. Incorporating iron into nickel increases the coverage of oxygen compounds due to the stronger iron-oxygen's affinity than nickel-oxygen's. Therefore, the formation of coke can be restrained by the rapid reaction of the carbonaceous products existing on the neighboring Ni sites and the oxygen atoms available. The uniformity of the Ni—Fe alloy structure is a key factor needed to improve the catalyst selectivity to produce the desired products as well as to minimize the carbon formation (Abou Rjeily et al., 2021; Ashok et al., 2020).

Nickel-cobalt bimetallic catalysts

The alloy of cobalt with nickel presents itself as an alternative effective catalyst for the tar removal. The similarity in the Ni and Co atomic radius favors the creation of Ni—Co alloy nanoparticles generating a synergistic effect between these two metals. The Co sites can efficiently restrict the reactions of tar decomposition and CO disproportionation causing the coke formation. Unfortunately, the oxidation of cobalt species produced Co oxides which leads to deactivation of the Ni—Co alloy catalyst. Subsequently, the exact amount of cobalt doping over nickel is crucial to reach a steady catalytic activity along with suppressed coke deposition. The synthesis method also impacts the uniformity of the nickel-cobalt bimetallic catalyst similar to the nickel-iron alloy (Abou Rjeily et al., 2021; Ashok et al., 2020).

Cobalt-iron bimetallic catalysts

The cobalt-iron Co—Fe bimetallic catalysts were also used to conduct the biomass tar steam reforming as well as that of the tar model compounds. Even though iron itself has a mediocre tar reforming performance, the alloy with cobalt helps significantly in ameliorating this activity given the synergetic effect existing between iron and cobalt. In addition, the Co—Fe alloy reduces greatly the coke formation. In fact, the tar molecules can be activated by the cobalt sites whereas the oxygen atom needed by the carbonaceous intermediates is provided by the iron sites (Ashok et al., 2020).

For instance, the Co/Al_2O_3 was co-impregnated with Fe by Wang et al. (2012) for the steam reforming of tar derived from cedar wood pyrolysis. The optimum Fe/Co ratio was found to be 0.25. The activity of the tar steam reforming with the bimetallic Co-Fe/Al_2O_3 was superior to that of monometallic catalysts Co and Fe, thanks to the synergy formed between the cobalt and iron at the optimal composition. Although the stability of Co/Al_2O_3 activity was constant with time, that of the bimetallic Co—Fe decreased due to the oxidation of the Co—Fe alloy particles, leading to the deactivation of the bimetallic catalyst. Adding H_2 to the reactant gas during the toluene steam reforming markedly enhanced the Co-Fe/Al_2O_3 activity and stability. The H_2 strongly interacted on the Co—Fe alloy surface resulting in the amelioration of the stability as well as restraining the Fe oxidation in the bimetallic alloy.

Catalyst supports for catalytic reforming

In the tar steam reforming, both of the reactants, the tar and the steam must be activated by the catalyst used. The balance between the reactants strongly controls the activity and stability of the catalyst. In general, a support is employed for the catalysts used for the tar removal where the cost-effective material such as silica, carbon, or alumina is adopted to disperse the metal catalyst. The catalyst support provides a strong linkage with the catalyst inhibiting the leaching out of the active metal. It also delivers acidic and basic centers, oxygen vacancy and enhances the performance of the catalyst (Ashok et al., 2020).

The nature of the catalyst and its support significantly affect the performance of the steam reforming process, along with reaction conditions. It also has a strong impact on the conversion of the raw materials and the influence on the selectivity of the products (Fajardo et al., 2010). In addition, the mechanisms of the steam reforming are markedly determined by the surface properties of the catalyst support (S. Li & Gong, 2014; Z. Li et al., 2016) In some cases, the support itself such as the zeolites might act as a catalyst (Guan et al., 2016).

If elevated metal loading is applied, the catalyst dispersion can be attenuated and the active sites' aggregation amplified which drops the catalyst efficiency. Therefore, for the tar steam reforming, low catalyst loading is usually more convenient (Quan et al., 2018). To be used at high temperatures, the dispersed metal particles should resist to sintering, melting, and coking which remain the major challenges encountered during the steam reforming of tar (Dickerson & Soria, 2013). The following parts describe the different supports and identify the impact of their properties on overcoming these drawbacks.

Particle bed supports

Small particles have been employed as catalyst support given their ability to augment the active sites' density (Quan et al., 2018). The good activity and stability of the supported nickel catalysts make them one of the most widely used catalyst support for the steam reforming of not only methane but also tar and aromatic products (benzene, toluene, naphtha). The most common Ni-based catalyst supports are particles of minerals (MgO, ZrO_2, and olivine), alumina (Al_2O_3), and alumina modifiers (α-Al_2O_3 and γ-Al_2O_3) (Kong et al., 2011).

Al_2O_3 based support

Alumina Al_2O_3 is extensively used for tar reforming as Ni-based catalyst support, being available, mechanically strong, chemically/physically stable, with a considerable ability to disperse the active metal phase. Alumina modifiers α- and γ-Al_2O_3 are often exploited for toluene steam reforming (He et al., 2018). The use of alumina alone as support leads to coke deposition over its surface sintering the metal catalyst. The surface properties of inorganic substances such as zeolites, silica, and alumina can be adjusted by applying silylation on the metal oxide surface which improves the hydrothermal stability of the support during the steam reforming process (Fidalgo & Ilharco, 2012).

Another alternative solution to improve the thermal stability is to dope Al_2O_3 in SiO_2 or Fe_2O_3 whom presence newly generate strong acids. These latter promote the tar cracking and the conversion reactivity of the different compounds reformed (toluene ...) (Adnan et al., 2017). Furthermore, redox metal oxides including ZrO_2 or CeO_2 can also be used to dope alumina. Ce and Zr presence covers the acidic sites and lower their densities (Castro et al., 2016). A substitute method to the modification of the alumina support can be the use of Ca, La, and La/Ce hexaaluminates based supports (Quitete et al., 2015). They are distinguished by their large surface area, good metal dispersion and favor the synergistic effect between the metal catalyst and its oxide support (Nichele et al., 2014).

Silica-based catalyst support

For high reforming temperatures, Silica SiO_2 is one of the most commonly employed supports given its large availability, high thermal stability, large surface area, and good resistance to metal sintering. Unfortunately, in environments of high steam reaction produced at elevated steam reforming temperatures, silica supports can leak. Thus, for commercial applications, the use of silica is limited to minimal amounts (Ashok et al., 2020). SBA-15 mesoporous silica has been also largely employed due to its large surface area and hexagonal, uniform, and ordered pores. SBA-15 can be modified as well with La and Ce which improves the support's stability and the oxygen mobility in addition to inhibiting cracking of big molecules. Moreover, La-doped Ni/SBA-15 produces oxy-carbonates which minimize the coke formation and extend the stability of the catalyst (Oemar et al., 2015). Additionally, MCM-41, a different type of mesoporous support, can be utilized to impregnate Nickel thus increasing the metal dispersion and the interaction between the metal and the support (Ashok et al., 2020). Despite these ameliorations, Silica-based catalyst supports are still less implied in the biomass tar reforming than alumina supports.

Biochar-based support

Biochar is one of the products formed during biomass gasification and pyrolysis. Thanks to its thermal stability, large surface area, large pore volume, and the availability of its surface active sites, biochar has been investigated as catalyst for tar reforming (Shen & Fu, 2018). Biochar is therefore the cheapest product employed as catalyst support. Char can be activated into carbon and used as catalyst support given the stability of the activated carbon formed under basic and acid environments. The carbon–metal interaction and the metal catalyst dispersion impact the carbon-based catalytic activity (Cao et al., 2018; Guo et al., 2018). Lately, biochar is being assessed by numerous researchers mostly for biomass pyrolysis and gasification for commercial applications (Gunarathne et al., 2019) and for producing biodiesel (Balajii & Niju, 2019).

Other supports

Numerous catalyst supports are applied for the tar steam reforming including mixed oxide, natural minerals, perovskite, and core–shell among others (Ashok et al., 2020). The $NiTiO_3$ mixed oxide permits the coexistence of smaller particles since it is very active and more resistant to sintering and catalyst deactivation during the ethanol steam reforming (Yaakob et al., 2013). Olivine, $(Mg_xFe_{1-x})SiO_4$, a mineral present in nature, is a support largely used in the biomass steam reforming (Pfeifer & Hofbauer, 2008). Oxides derived from perovskite have lattice oxygen and many oxygen vacancies useful for the steam activation thus improving the hydrocarbons' oxidation, the thermal stability, and the redox property (Oemar et al., 2016). Core–shell catalysts such as $Ni@ZrO_2$ help to increase the number of surface-active oxygen, reduce the metal sintering, and extend the interfacial perimeter of the metal–support. This facilitates the carbon deposit removal during the steam reforming of ethanol (Li et al., 2018).

Monolithic catalyst support

The Ni-based monolithic catalyst is an innovative catalyst characterized by its excellent resistance to coke deposition. It is prepared by impregnation of MgO and NiO over cordierite catalyst support. Its advantageous features include large surface area, smaller pressure drops for pellets catalyst used in fixed bed reactors (Lónyi et al., 2013) as well as being mechanically strong and highly efficient for catalyst recycling (Wang et al., 2010). Moreover, monoliths can withstand elevated temperatures, be effortlessly oriented in a reactor with numerous design options.

Ceramic foam as catalyst support

Ceramic foams as catalyst carriers are a three-dimensional cellular lightweight solid with a network of ceramic interconnected ligaments. They are known for their high porosity, robust thermal stability at high temperatures, moderate pressure drop, high specific heat capacity, strong ability for thermal isolation, and dual-scale specific surface area (Gao et al., 2016). Thanks to all these favorable physical properties, ceramic foam can be used as monolithic catalyst support for tar reforming. The tar removal rate can be potentially increased and the production of hydrogen largely enhanced (Gao et al., 2015).

Additional descriptions of the numerous catalyst supports introduced previously are present in the review papers of Abou Rjeily et al. (2021) and Ashok et al. (2020). An overview of the main catalysts and their supports applied for the catalytic steam reforming of methane and tar compounds as well as their advantages and drawbacks, can be found in the form of a summary table in (Abou Rjeily et al., 2021).

9.6 Catalytic steam reforming of methane and light hydrocarbons

Methane steam reforming

The methane steam reforming has been widely explored specially to produce hydrogen and syngas. Numerous works have been performed in this area with the study of the methane conversion and the effect of the operating parameters. In this section, the methane steam reforming is investigated at low temperatures (500–700°C) with the effect of two operating parameters: the temperature and the S/C ratio, based on the work of Khzouz and Gkanas (2017).

They performed both experimental and numerical analysis, but in this chapter, only the experimental results will be presented, to be consistent with the other sections. For the experimental tests, a fixed bed reactor was used with 10 wt% Ni/Al_2O_3 as catalyst. Two sets of trials were performed under atmospheric pressure where the temperature was varied from 500°C to 700°C with 50°C increment at S/C ratio equals to 2 and 3. The gas hourly space velocity (GHSV) was $1067.4\,h^{-1}$ and $1388.9\,h^{-1}$ for the steam reforming at S/C ratios of 2 and 3, respectively.

The conversion of methane is calculated as follows (Eq. 9.18):

$$X_{CH_4} = \frac{\dot{n}_{CH_4,in} - \dot{n}_{CH_4,out}}{\dot{n}_{CH_4,in}} \quad (9.18)$$

where $\dot{n}_{CH_4,in}$ is the molar flow rate of methane at the inlet in mol/min and $\dot{n}_{CH_4,out}$ is the molar flow rate of methane at the outlet in mol/min.

Similarly, the water conversion is determined as follows (Eq. 9.19):

$$X_{H_2O} = \frac{\dot{n}_{H_2O,in} \quad \dot{n}_{H_2O,out}}{\dot{n}_{H_2O,in}} \quad (9.19)$$

where $\dot{n}_{H_2O,in}$ is the molar flow rate of water at the inlet in mol/min and $\dot{n}_{H_2O,out}$ is the molar flow rate of water at the outlet in mol/min.

The yields of CO, H_2, and CO_2 are determined with respect to the methane inlet according to the following Eqs. (9.20)–(9.22):

$$yield_{CO} = \frac{\dot{n}_{CO,out}}{\dot{n}_{CH_4,in}} \quad (9.20)$$

$$yield_{H_2} = \frac{\dot{n}_{H_2,out}}{\dot{n}_{CH_4,in}} \quad (9.21)$$

$$yield_{CO_2} = \frac{\dot{n}_{CO_2,out}}{\dot{n}_{CH_4,in}} \quad (9.22)$$

Given that the number of moles of methane in the inlet is considered to be 1 and that the yields were published in the work of Khzouz and Gkanas (2017), we calculated the number of moles of the different gases at the outlet and derived the molar concentration of the gas products.

Effect of temperature on methane steam reforming

The effects of the temperature were established on the methane and water conversions as well as on the distribution of the different gas products as shown in Figs. 9.7 and 9.8. The increase of the temperature entailed an increase in the conversions of both methane and water as seen in Fig. 9.7. The methane conversion increased from 34% to 76% while that of water was elevated from 32% to 58% when the temperature was raised from 500°C to 700°C. This confirms the endothermic nature of the methane steam reforming which is favored by higher temperatures where the C—H bond is effectively activated.

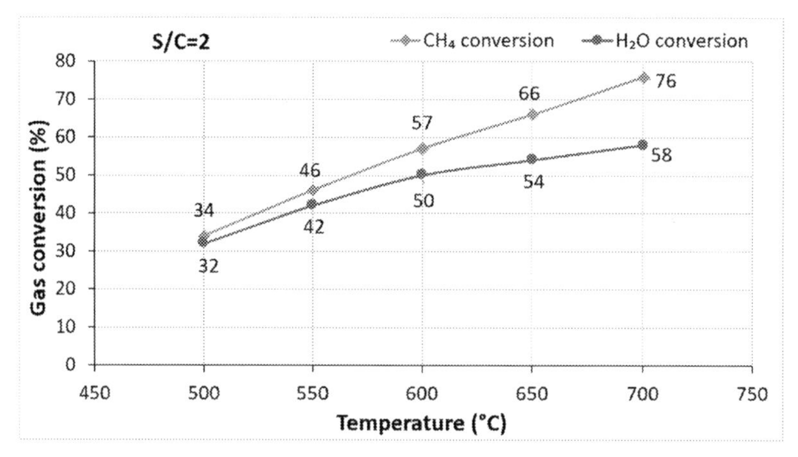

FIG. 9.7 Effect of temperature on methane and water conversion during methane steam reforming over 10% Ni/Al_2O_3 at S/C ratio of 2 and GHSV = 1067.4 h^{-1} (Khzouz & Gkanas, 2017).

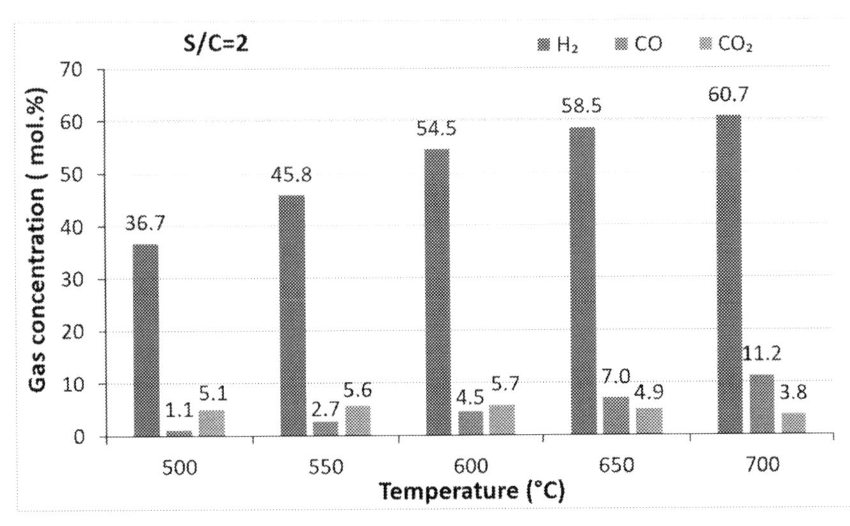

FIG. 9.8 Effect of temperature on gas products yields during methane steam reforming over 10% Ni/Al_2O_3 at S/C ratio of 2 and GHSV = 1067.4 h^{-1} (Khzouz & Gkanas, 2017).

Regarding the gases' composition, the impact of the temperature increase is clearly noticed over the evolution of the hydrogen production where it was improved from 36.7 mol% to 60.7 mol% with the temperature elevation from 500°C to 700°C. The increase of the carbon monoxide concentration is monotonic with the temperature increase being almost 10 times higher at 700°C (11.2 mol%) than at 500°C (1.1 mol%). In fact, the CO is produced by the methane steam reforming main reaction (Eq. 9.1) as well as by the reverse water-gas shift reaction (Eq. 9.4) and the Boudouard reaction (Eq. 9.5) which are enhanced by the temperature increase. The carbon dioxide concentration increases from 5.1 mol% at 500°C, it peaks at 600°C with 5.7 mol% and then drops to 3.8 mol% at 700°C, as a result of the reverse WGSR (Eq. 9.4).

Effect of steam-to-carbon ratio on methane steam reforming

In order to study the effect of the steam-to-carbon S/C ratio, the previous set of experimental conditions was applied to a new set of experiments while increasing the S/C ratio to 3 with the results shown in Figs. 9.9 and 9.10. It can be noticed that the same trend of variations are obtained as for S/C equals to 2, where the temperature increase leads to an upsurge in the conversions of methane and water. Comparing the experimental results for S/C equal 2 (Fig. 9.7) and S/C equal 3 (Fig. 9.9), it can be noticed that increasing the S/C ratio from 2 to 3 leads to a higher increase in the methane conversion which was augmented from 32% to 92%. On the other hand, the increase of the water inlet with S/C equals 3 leads to a lower water conversion which was 12% at 500°C and increased only to 30% at 700°C. This implies that a smaller amount of water was consumed when the S/C ratio was increased. In fact, the residence time for S/C ratio of 3 was determined to be 2.59 s, shorter than that of S/C ratio of 2 (3.37 s). Therefore, the contact time between the reactants was lowered and thus the consumption of water was decreased.

Although the gases' distribution follows a similar trend in the case of S/C equals 2 and 3, however, the hydrogen concentrations reached at S/C equals 3 are much lower than at S/C equals 2. At S/C of 3, the H_2 concentration increased from 21.6 mol% to 46.3 mol% with the temperature rising from 500°C to 700°C.

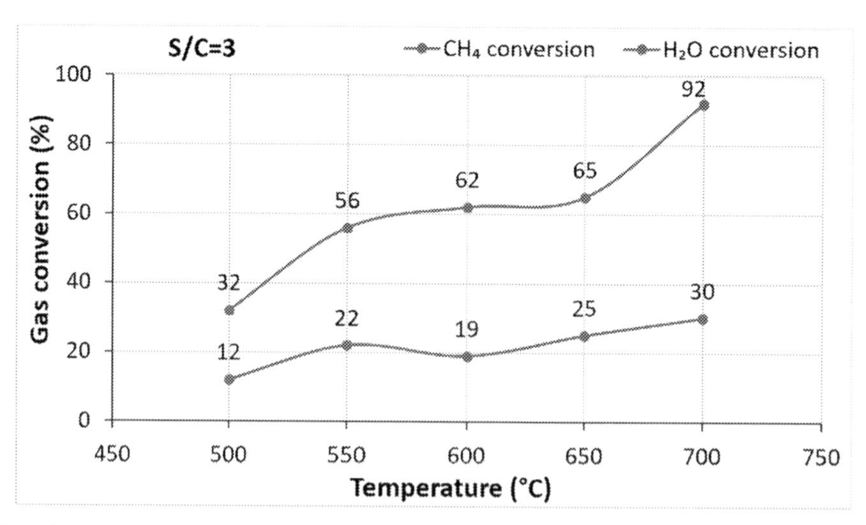

FIG. 9.9 Effect of temperature on methane and water conversion during methane steam reforming over 10% Ni/Al_2O_3 at S/C ratio of 3 and GHSV = 1388.9 h^{-1} (Khzouz & Gkanas, 2017).

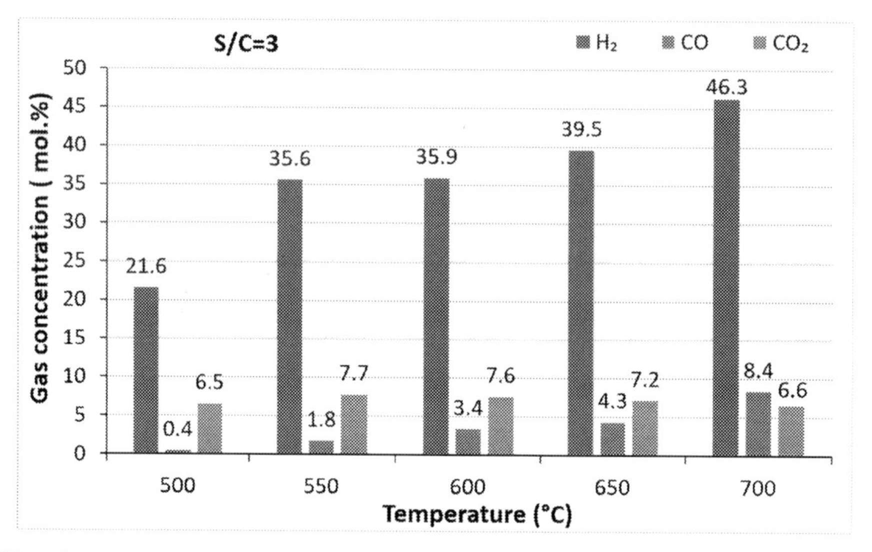

FIG. 9.10 Effect of temperature on gas products yields during methane steam reforming over 10% Ni/Al_2O_3 at S/C ratio of 3 and GHSV = 1388.9 h^{-1} (Khzouz & Gkanas, 2017).

Whereas, at S/C of 2, the H_2 concentration reached a maximum of 60.6 mol% at 700°C. Despite its increase with the temperature, the CO production at S/C of 3 is lower than that at S/C of 2 for each of the temperatures tested from 500°C to 700°C.

When the S/C ratio is increased from 2 to 3, the CO_2 concentration becomes higher than at S/C of 2 at the same temperatures tested. Between 500°C and 600°C, the CO_2 increases due to the equilibrium of the WGSR (Eq. 9.3) shifted toward the products with the rise of the S/C ratio, thereby ameliorating the CO_2 production. With the further increase in the temperature to 700°C, the CO_2 drops since the WGSR becomes less favorable at high temperature.

Methane, ethane, and propane steam reforming

Angeli and colleagues (Angeli et al., 2015) studied the effect of the presence of C_2 and C_3 alkanes in the feedstock on the coke formation over the catalyst during the methane steam reforming operated at low temperature (400–550°C). The conversion is improved by using a hydrogen-selective membrane. The catalysts used were nickel (10wt% Ni) and rhodium (1wt% Rh) while supported on lanthanum (5% La_2O_3) doped ceria-zirconia (17% CeO_2 -78% ZrO_2) mixed oxide. This combination minimizes the coke deposition hence enhancing the catalyst stability over time on steam. Operating at low temperature has several advantages such as reducing the cost for engineering, materials, and energy demands. Moreover, a quick start-up is ensured by a low-temperature process when compared to the conventional high-temperature steam reforming, and CO shift is not required thanks to the favorable equilibrium of water-gas shift reaction.

The success of this process relies on developing catalysts capable of activating methane at low temperature and driving its conversion up to equilibrium at small contact times. Furthermore, the process must resist deactivation phenomena like carbon deposition and coke formation (Halabi et al., 2010). For the series of experiments, a fixed bed reactor was employed. GHSV was 70,000 h^{-1}, 61,000 h^{-1} and 58,000 h^{-1} for methane, ethane and propane, respectively. The product stream composition for the steam reforming of methane, ethane, and propane at S/C ratio of 3 and under atmospheric pressure was elaborated using online gas chromatograph equipped with a thermal conductivity detector (TCD) as displayed in Figs. 9.11–9.13.

As seen in Fig. 9.11, the increase of the temperature leads to an almost linear increase in the hydrogen content in the outlet gas stream of the methane steam reforming from 45% at 400°C to 69.3% at 550°C. Controversially, the methane content decreases from 45% at 400°C to reach its lowest value of 12.3% at 550°C corresponding to 60% methane conversion. The CO

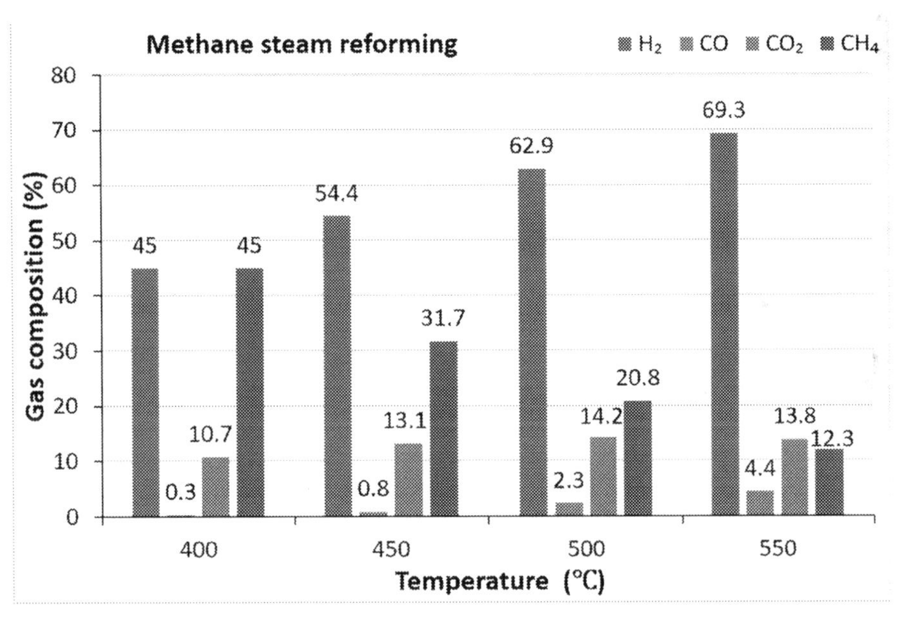

FIG. 9.11 Effect of temperature on outlet gas product of steam reforming of methane (GHSV = 70,000 h-1, P = 1 bar and S/C ratio = 3) (Angeli et al., 2015).

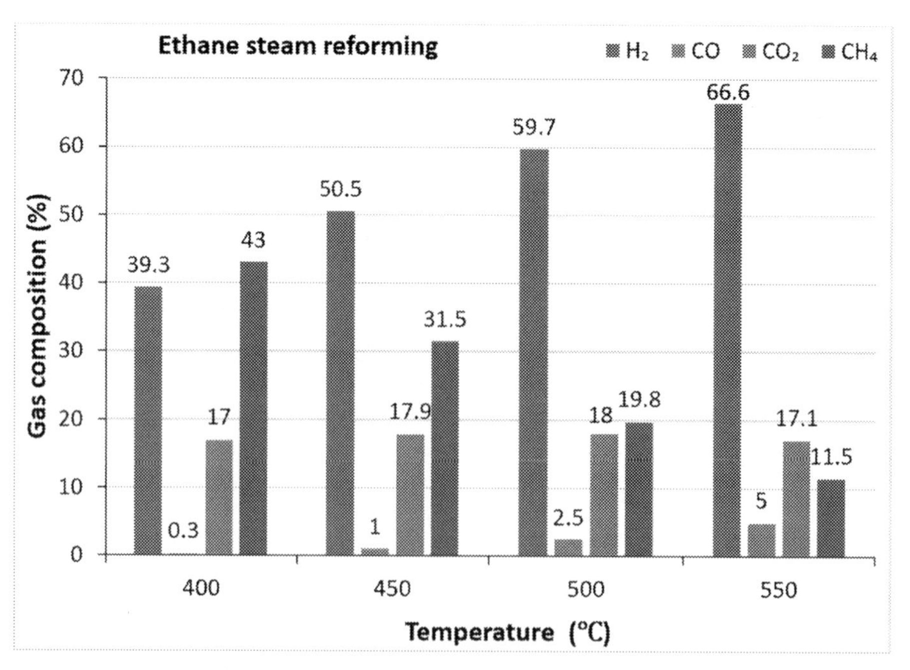

FIG. 9.12 Effect of temperature on outlet gas product of steam reforming of ethane (GHSV = 61,000 h^{-1}, P = 1 bar and S/C ratio = 3) (Angeli et al., 2015).

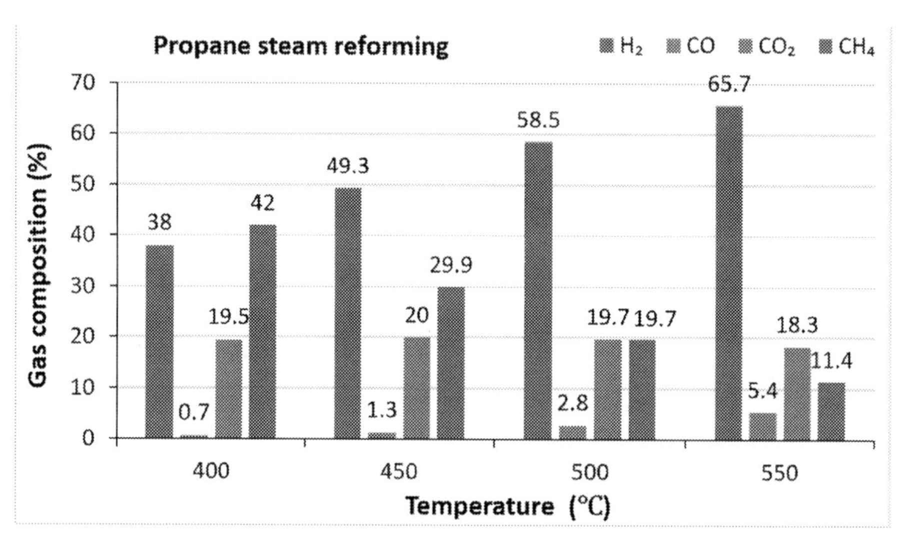

FIG. 9.13 Effect of temperature on outlet gas product of steam reforming of propane (GHSV = 58,000 h^{-1}, P = 1 bar and S/C ratio = 3) (Angeli et al., 2015).

content increases with the temperature increase but remains below 5% thanks to favorable equilibrium of the WGSR.

Regarding the ethane and propane steam reforming shown in Figs. 9.12 and 9.13, respectively, at the temperature range of 400–550°C, a complete conversion of ethane/propane is expected. The gas distribution is analogous to that obtained in the methane steam reforming except for CO_2 whose content augment as the C-number of the feed is increased.

Regarding the carbon deposition, temperature programmed oxidation (TPO) was used to assess and to quantify the coke deposited over the catalysts' surface. For this set of experiments, the catalytic performance of each catalyst among Ni(10)CeZrLa and Rh(1)CeZrLa was tested separately for the steam reforming of the three hydrocarbons. The same operating parameters values for the temperature (400–550°C), S/C ratio (3), and GHSV were applied. It was found that ethane and propane are highly reactive over both of the catalysts. The TPO analysis revealed the near absence of carbon accumulation over Rh(1)CeZrLa for the C_1-C_3 hydrocarbons feed. Whereas when the Ni(10)CeZrLa was employed, the carbon content increased with the increase of the carbon number in the hydrocarbons. Furthermore, both of the catalysts were not deactivated during the experimental tests where a mixture of the three hydrocarbons was reformed for 10h at 500°C. Interestingly, these catalysts' combination (Ni(10)CeZrLa and Rh(1)CeZrLa) was proven to be extremely encouraging for the natural gas steam reforming even at low operating temperatures.

9.7 Catalytic steam reforming of bio-oil compounds

Toluene reforming over Ni-CeO_2/SBA-15 packed bed catalysts

Tao and colleagues (Tao et al., 2013) studied the effect of temperature, steam-to-carbon ratio, and metal loading on the catalytic steam reforming of toluene as a model tar compound. Among the multiple catalysts used, the nickel-based catalysts were proved to be very efficient for tar reduction due to the promising cracking potential of the NiO/Ni catalyst. Nevertheless, this catalyst type is exposed to rapid deactivation resulting essentially from coke deposition. Therefore, it is crucial to ameliorate anti-coke deposition abilities of the Ni-based catalysts. As catalyst promoter, CeO_2 is selected since it is proven to be efficient in preventing coke formation and stabilizing the catalysts. Furthermore, SBA-15 was chosen as the catalyst support being a highly ordered mesoporous material, having a strong thermal and hydrothermal stability, which makes it a promising support for several catalysts. Toluene was selected as the model tar compound since it is a stable and major tar compound produced during the gasification process at high temperature.

The toluene steam reforming can be expressed as follows (Eq. 9.23):

Steam reforming of toluene:

$$C_7H_8 + 14H_2O \leftrightarrow 18H_2 + 7CO_2 \quad (9.23)$$

The water and the toluene molecules are adsorbed on the catalyst surface during the stream reforming process and react to produce H_2 and CO or CO_2. At elevated S/C ratios, most of the catalyst surface active sites could be occupied by the water molecules leading to the decrease of the adsorption and toluene conversion. However, the elevated water concentration can enhance the WGSR toward the formation of CO_2 and H_2. For this purpose, a fixed-bed lab-scale reactor set was employed to estimate the catalytic performances of the Nickel catalyst supported over SBA-15 doped with CeO_2 (Ni-CeO_2/SBA-15).

Effects of catalytic temperature and CeO_2 loading content on the steam reforming of toluene

The effect of catalytic temperature on the toluene conversion is evaluated at different CeO_2

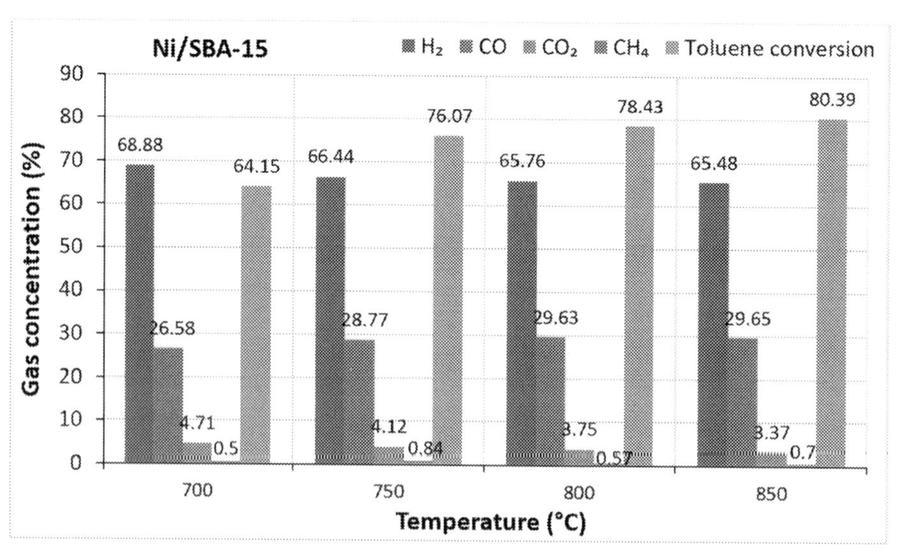

FIG. 9.14 Effect of temperature on toluene steam reforming over Ni/SBA-15 (S/C ratio = 3) (Tao et al., 2013).

loading rates (0, 1, and 3 wt%) with an S/C ratio of 3 and in the temperature range of 700–850°C. The results are shown in Figs. 9.14–9.16. It can be noticed that for the different CeO_2 loadings, the toluene conversion is enhanced with the temperature increase where the highest toluene conversion reached at 850°C was 98.9% when Ni-CeO_2 (3 wt%)/SBA-15 catalyst was used. It is clearly demonstrated that the Ni-CeO_2/SBA-15 catalyst is a very promising and effective catalyst for the toluene conversion and the increase of the CeO_2 loading promotes the

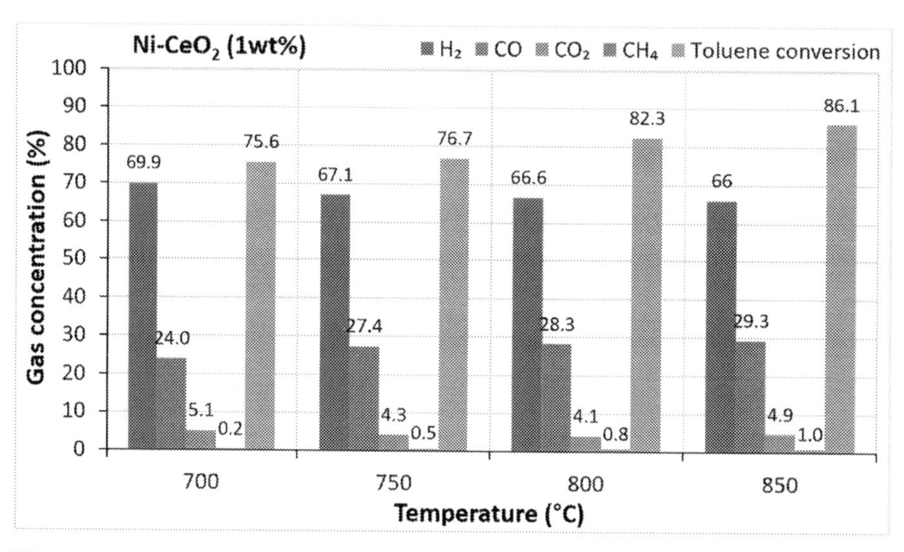

FIG. 9.15 Effect of temperature on toluene steam reforming over Ni-CeO_2 (1 wt%)/SBA-15 (S/C ratio = 3) (Tao et al., 2013).

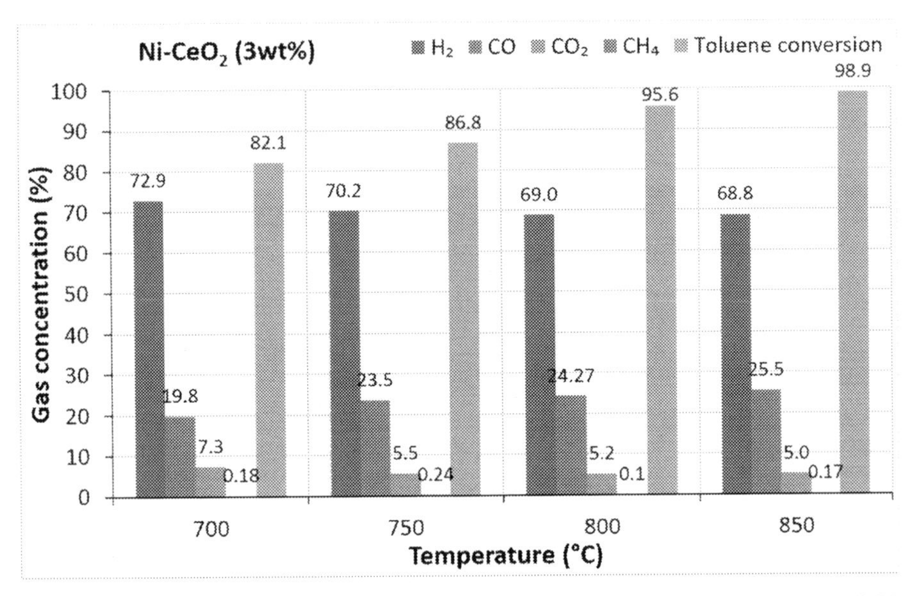

FIG. 9.16 Effect of temperature on toluene steam reforming over Ni-CeO_2 (3 wt%)/SBA-15 (S/C ratio = 3) (Tao et al., 2013).

catalytic activity. The major products of the toluene reforming are H_2 and CO along with CO_2 and CH_4 as minor products. The catalytic temperature remarkably affects the composition of these gaseous products while the CeO_2 loading had a smaller influence (Tao et al., 2013).

Effect of the S/C ratio on the steam reforming of toluene

The effect of the S/C ratios on the toluene steam reforming at 850 °C using the three catalysts was also evaluated and the results are summarized in Figs. 9.17–9.19. It can be deduced that the S/C ratio has a significant impact on the toluene conversion and the gas composition. Regarding the gas products, as the S/C ratio increases, the H_2 and CO_2 contents increase whereas CH_4 and CO decrease. The reasons explaining these trends can be attributed to the water-gas shift and toluene steam reforming reactions, Eq. (9.23). Referring to Figs. 9.17–9.19, an optimal S/C ratio of 3 was found for the toluene conversion and the product gas composition (Tao et al., 2013).

Benzene reforming over NiO/ceramic foam catalysts

Gao et al. (2016) selected benzene as tar model compound for the catalytic steam reforming because it is one of the major constituents of high-temperature gasification tar. In addition, benzene has a stable aromatic structure and is more reactive than naphthalene or toluene over a wide range of operating temperatures. The catalyst used was NiO/ceramic foam prepared by dissolving $Ni(NO_3)_2{\cdot}6H_2O$ into deionized water. The average amount of NiO loaded on the ceramic foam was around 3.5 wt%. The benzene complete reforming reaction is given by Eq. (9.24) (Gao et al., 2016):

Benzene steam reforming:

$$\frac{1}{6}C_6H_6 + 2H_2O \leftrightarrow CO_2 + \frac{5}{2}H_2 \quad \Delta H^\circ = +544.5\,\text{kJ/mol} \tag{9.24}$$

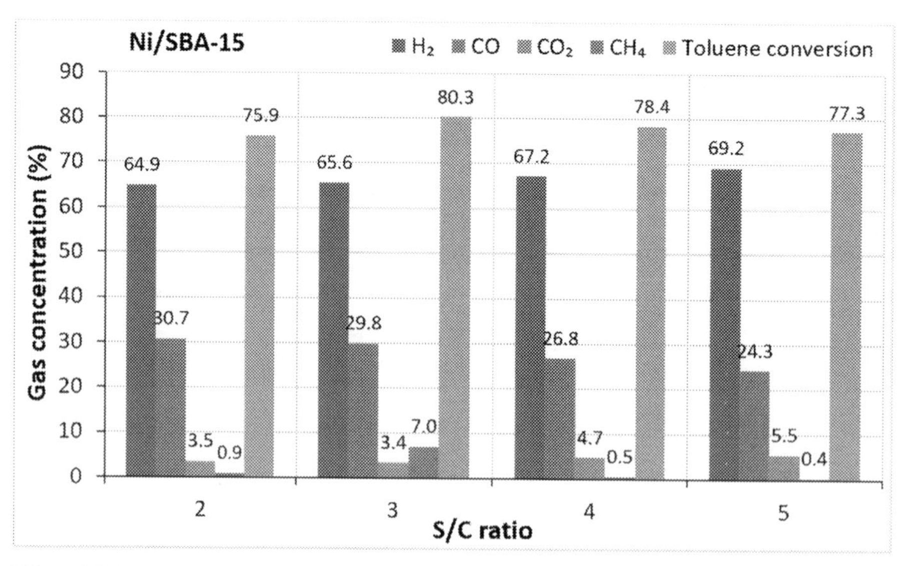

FIG. 9.17 Effect of S/C ratio on toluene steam reforming over Ni/SBA-15 ($T = 850°C$) (Tao et al., 2013).

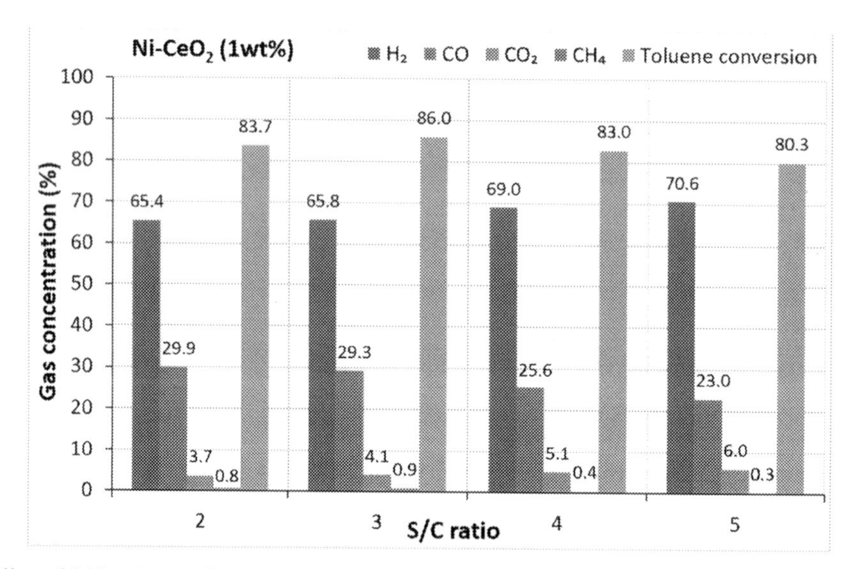

FIG. 9.18 Effect of S/C ratio on toluene steam reforming over Ni-CeO_2 (1 wt%)/SBA-15 ($T = 850°C$) (Tao et al., 2013).

Comparison to control experiment without NiO loading

Some metal oxide impurities such as Na_2O and Fe_2O_3 can be found in the ceramic foam. To eliminate their effect, ceramic foam alone was used in a control experiment without any NiO loading for benzene steam reforming at 750 °C, with an ER ratio of 0.1, S/C molar ratio equal to 1, and WHSV of 5.6 h^{-1}. The major gaseous products formed in both cases and their concentrations

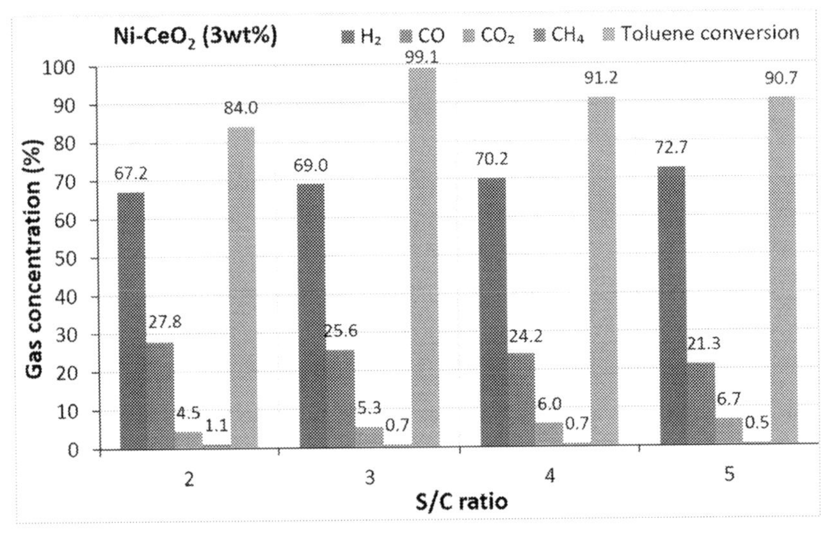

FIG. 9.19 Effect of S/C ratio on toluene steam reforming over Ni-CeO_2 (3wt%)/SBA-15 (T = 850°C) (Tao et al., 2013).

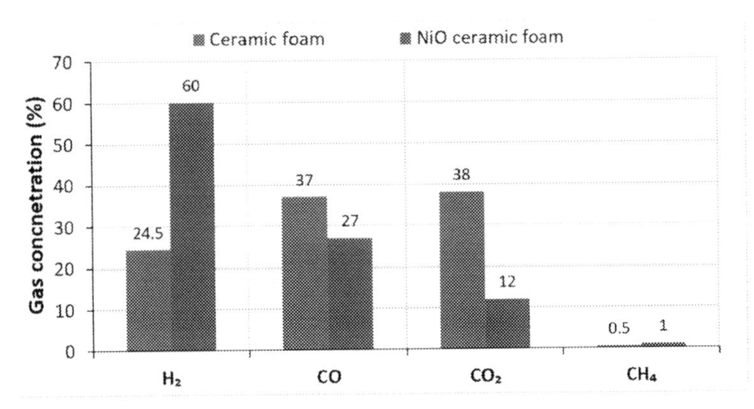

FIG. 9.20 Comparison between the steam reforming of benzene as tar model compound with/without NiO on ceramic foam at 750°C, S/C = 1, ER = 0.1 (Gao et al., 2016).

are shown in Fig. 9.20 (Gao et al., 2016). The H_2 yield is enhanced with the presence of the NiO/ceramic foam compared to the control experiment. The reason behind this is the dissociated adsorption of the steam and benzene on the surface of the NiO/ceramic foam, which promotes catalytic reforming reactions. This high activity can be accredited to the catalyst's excellent ability for the oxygen transfer (Gao et al., 2016).

Effect of the temperature on steam reforming of benzene

The distribution of the major gaseous products resulting from the steam reforming of benzene at an S/C ratio of 1 and equivalent ratio ER of 0.1 when the reaction temperature was increased from 700 °C to 900 °C is shown in Fig. 9.21. H_2 and CO are the dominant products revealing an excellent catalytic activity. As the

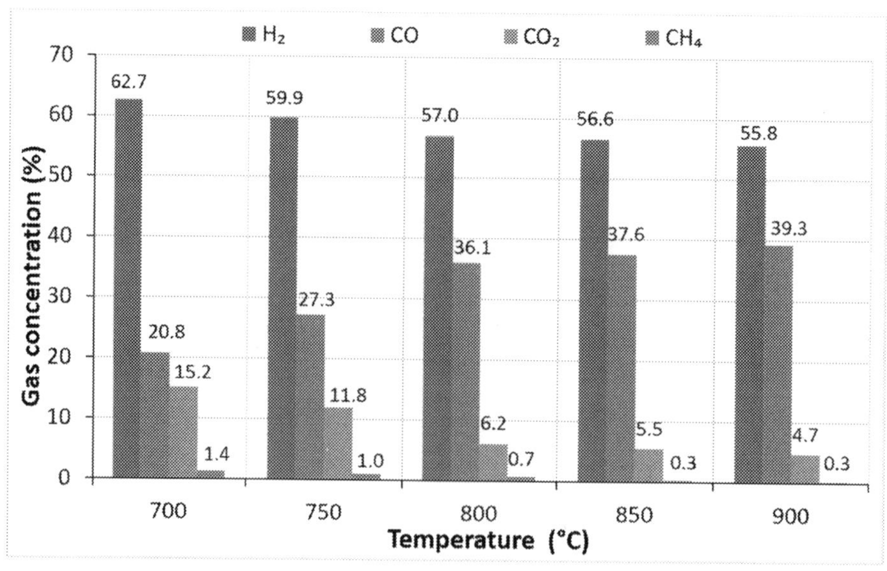

FIG. 9.21 Effect of temperature on the gaseous composition of benzene steam reforming at S/C=1, ER=0.1 (Gao et al., 2016).

temperature increases from 700°C to 900°C, H_2 concentration decreases from 62.7% to 55.75% and that of CO_2 dropped from 15.15% to 4.72%. On the other side, the CO demonstrated an opposite trend where its concentration increased from 20.81% to 39.28%. It can be deduced that benzene decomposes favorably at high temperature. It was reported that the steam reforming reaction and WGSR were promoted when the reaction temperature surpasses 700°C, which is beneficial for tar reduction using Ni-based catalyst. Nevertheless, at too elevated temperatures, carbidic carbon, precursor of graphitic carbon, can be easily formed leading to the catalyst deactivation. Moreover, the increase of the temperature inhibits the WGSR and the partial oxidation of benzene since both reactions are exothermic, thus H_2 concentration was slightly decreased (Gao et al., 2016).

The carbon conversion and H_2 yield variations with the temperature are demonstrated in Fig. 9.22. The carbon conversion increases significantly with the temperature from 54.43% at 700°C to 93.92% at 900°C. The hydrogen yield experienced some fluctuations but the general trend was an increase (Gao et al., 2016).

Effect of S/C ratio on the steam reforming of benzene

The catalytic effect of S/C molar ratio on the gaseous product distribution was evaluated for the benzene steam reforming at 750°C and with an ER ratio of 1 and the results are demonstrated in Fig. 9.23. The S/C ratio clearly influences the catalytic activity. For instance, CO concentration dropped from 31.03% to 19.83% whereas CO_2 was raised from 3.27% to 20.09% when the S/C ratio was increased from 0 to 3. At the same time, H_2 concentration stabilized at about 59% to 60% for the whole S/C ratio range (Gao et al., 2016).

The carbon conversion and H_2 yield variations with the S/C ratio are demonstrated in Fig. 9.24. H_2 yield increased from 102.38 to 192.36 (g $H_2.kg^{-1}$ benzene) when the S/C molar ratio increased from 0.0 to 1.5 but it remained constant with the further increase of S/C molar ratio to 3. The same trend was observed for the

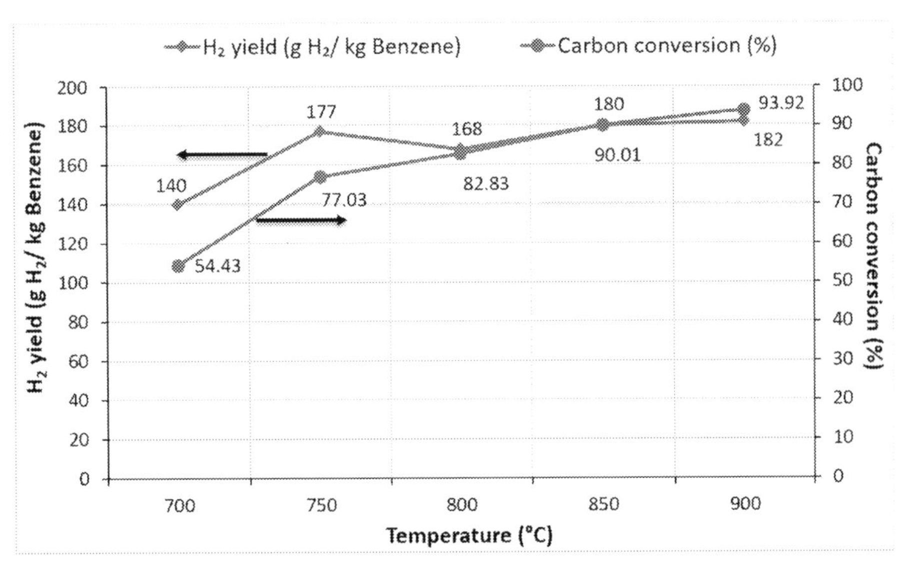

FIG. 9.22 Effect of temperature on the carbon conversion and H_2 yield of benzene steam reforming at S/C=1, ER=0.1 (Gao et al., 2016).

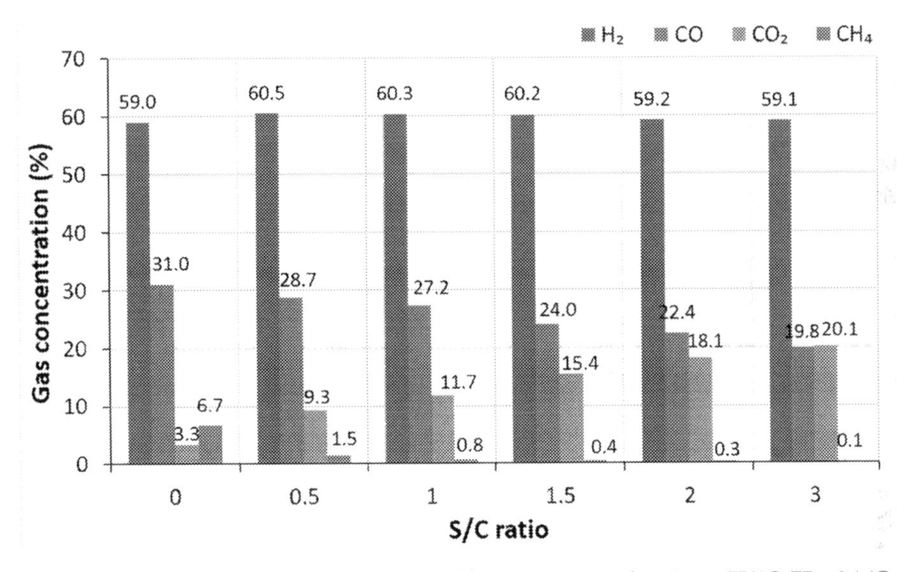

FIG. 9.23 Effect of S/C ratio on the gaseous composition of benzene steam reforming at 750°C, ER=0.1 (Gao et al., 2016).

carbon conversion variation. It is proposed that the increasing S/C molar ratio promotes the WGSR and the benzene complete reforming reaction enhancing the production of H_2. However, too elevated S/C molar ratio is not favored at industrial scale, due to the high energy inputs for steam generation, the associated cost of liquid separation, and the possible

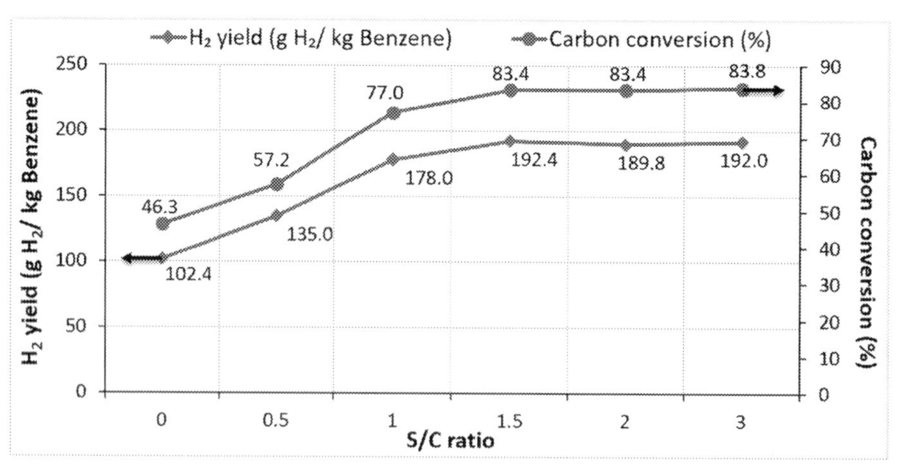

FIG. 9.24 Effect of S/C ratio on the carbon conversion and H_2 yield of benzene steam reforming at 750 °C, ER = 0.1 (Gao et al., 2016).

sintering of the active metal sites on the catalyst (Gao et al., 2016).

Tar model compounds reforming over Ni/Al_2O_3 packed bed catalyst

Artetxe and co-workers (Artetxe et al., 2017) conducted a study to better understand the steam reforming behavior of the most representative tar compounds over a catalyst having a high activity and a low-cost, the Ni/Al_2O_3. This was realized by taking into consideration the impact of each one of these compounds in the catalyst deactivation by carbon formation and coke deposition. This will enable the identification of the compounds responsible for the coke formation and the different reactivity. The model compounds investigated are toluene, furfural, anisole, indene, phenol, and methylnaphthalene. These species include a large variety of one- and two-ring aromatic hydrocarbons and oxygen-containing components found in the tars produced during the biomass gasification. Moreover, since some model compounds are solid at ambient room temperature, methanol was used as solvent for all the experiments, and it was previously subjected to steam reforming to evaluate its contribution to the final product stream. The steam reforming of these compounds and their mixture with the Ni/Al_2O_3 catalysts were realized in two-stage stainless steel tube reactor. In the first reactor, the mixture of the model compound with methanol and water was fed together. The steam reforming of the model compounds and their blend was performed in the second reactor at 750 °C and for 60 min. An S/C ratio of 3 was held constant during all of the experiments, which is an adequate value for tar conversion and product gas composition (Artetxe et al., 2017).

The carbon conversion of the different model compounds is given in Fig. 9.25. It is noticed that the conversion of all the model compounds is between 62% and 75% except that of methanol, being the highest at around 94%. Anisole and furfural are the most reactive compounds followed by toluene and indene. Methylnaphthalene is the most refractory compound. These results imply that significant differences exist in the mechanisms of the steam reforming of oxygenates and hydrocarbons. Therefore, the higher molecular weight cyclic hydrocarbons are less reactive while those having carbon–

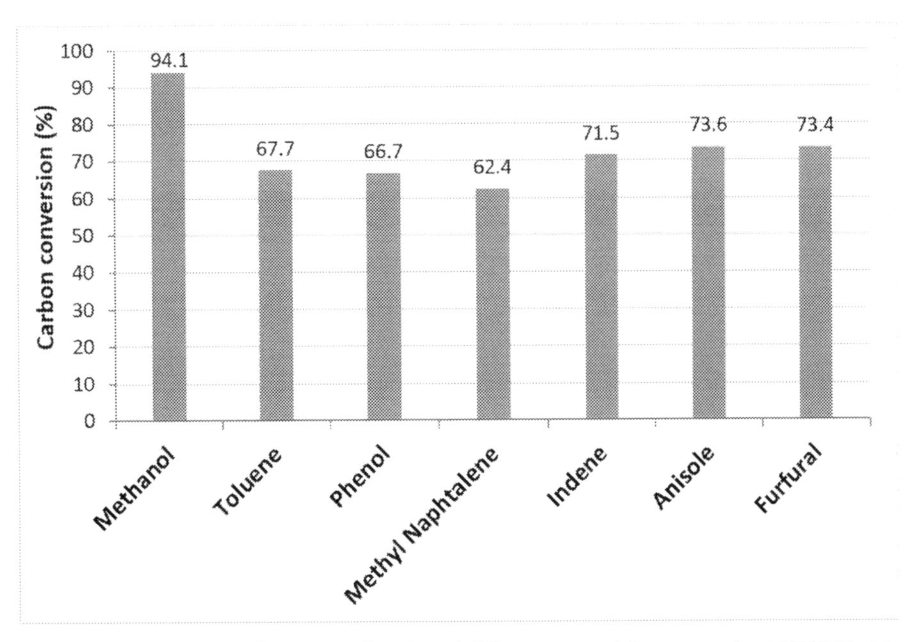

FIG. 9.25 The carbon conversion in the steam reforming of different tar model compounds at 750°C (Artetxe et al., 2017).

oxygen bonds can be easily formed (Artetxe et al., 2017).

The distribution of the gaseous compounds produced during the steam reforming of the tar model compounds is given in Fig. 9.26. The main products obtained are CO and CO_2 followed by H_2 confirming that the chief reaction occurring on the Ni/Al_2O_3 catalyst is the steam reforming followed by the WGSR. The highest CO_2 yields were favored for aromatic hydrocarbons such as toluene (62wt%), indene (61wt%), and methylnaphthalene (53wt%), while the highest CO yields were obtained from oxygenated hydrocarbons like anisole (52wt%) and phenol (47wt%). The CH_4 amounts were very low and those of C_2–C_4 were nearly negligible for all the compounds.

It must be mentioned that the anisole and toluene reforming leads to higher CH_4 concentrations than other compounds caused by the dealkylation of the methyl group in their structure. The coke formed leads to the blocking of the pores and poisoning of the active sites resulting in the loss of the catalyst activity, which will consequently reduce the hydrogen production. Nevertheless, coke deposition is reduced by the steam addition since the carbon produced can further react with the steam present in the medium. Hence, it is noteworthy to mention that the coke amount formed on all the cases was very low (<2.8wt%). This is an evidence that the operating conditions chosen ($S/C=3$, $T=750$°C and 1.5g of catalyst) limit the coke formation on the catalyst (Artetxe et al., 2017).

Effect of Ni loading on the carbon conversion of different tar model compounds

The impact of the Ni loading on the reforming efficiency for a mixture of all the model compounds previously stated and separately studied was evaluated at 750°C for 60min and is displayed in Fig. 9.27. It is observed that when the Ni loading is increased from 5 to 20wt%, the carbon conversion of the mixture rises from 65% to 90%. This increase in the Ni content

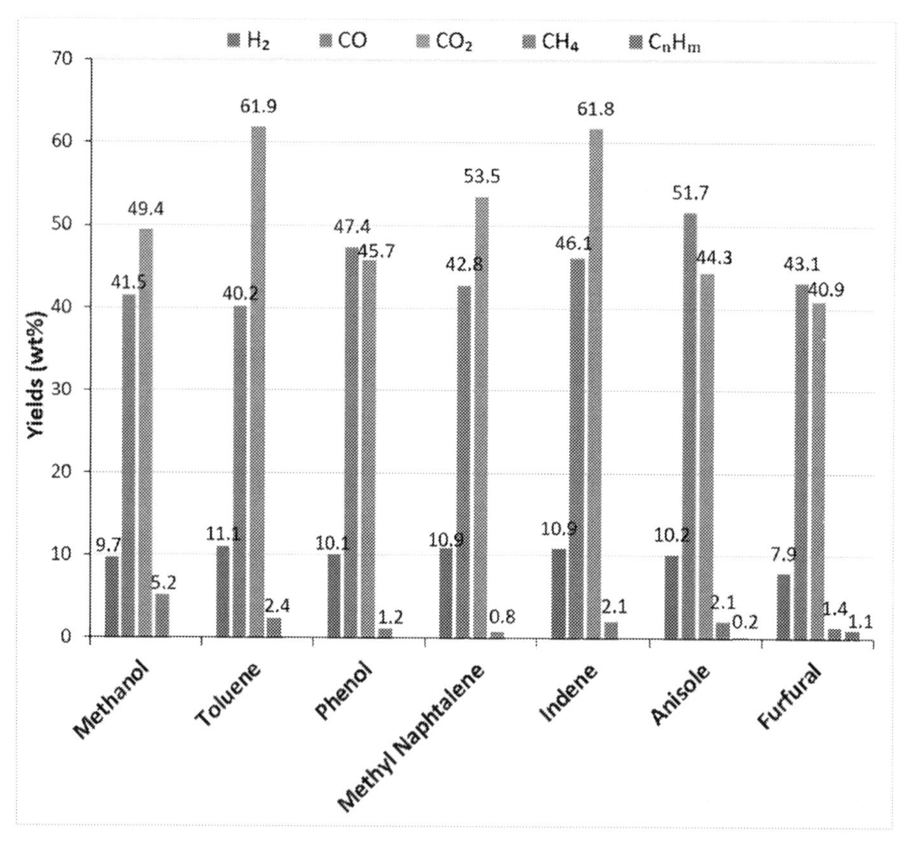

FIG. 9.26 Yields of the gaseous compounds in the steam reforming of different tar model compounds over Ni/Al_2O_3 at 750°C (Artetxe et al., 2017).

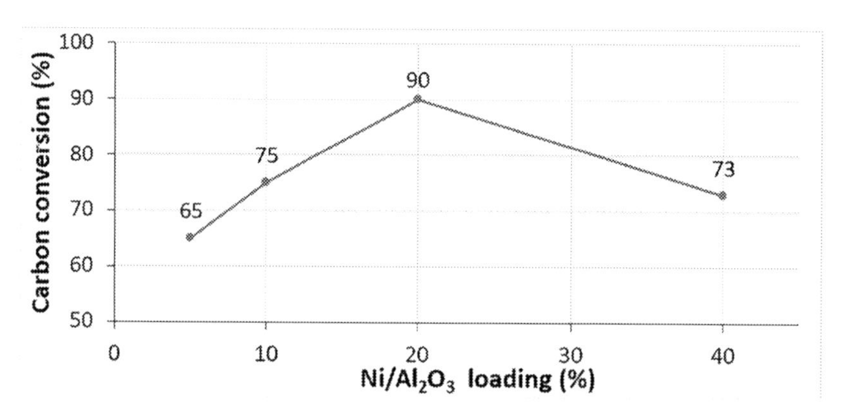

FIG. 9.27 Effect of Ni loading on carbon conversion of a mixture of all model compounds over Ni/Al_2O_3 at 750°C (Artetxe et al., 2017).

releases more active sites from the steam reforming and the WGSR. However, when the Ni loading is further increased, such as at 40 wt% of Ni, the carbon conversion decreases to 73%, since at higher Ni content the coke formation is accentuated, which deactivates the catalyst faster and consequently, lowers the tar conversion. The reason behind this would probably be the aggregation of the Ni species during the catalyst preparation step hence leading to the formation of large particles (Artetxe et al., 2017).

Effect of Ni loading on the yields of different tar model compounds

The distribution of the gaseous products as function of the Ni loading during the steam reforming of the mixture of all the compounds is demonstrated in Fig. 9.28. It can be observed that as the Ni loading increases up to 20%, the H_2, CO, and CO_2 mass yields also increase thanks to the improvement of the steam reforming, WGSR, and decomposition reactions. However, when the Ni loading is increased above 20%, the yields are decreased since the total conversion of the feed has also dropped (Artetxe et al., 2017).

Biomass tar reforming over NiO/ceramic foam catalyst

Gao and colleagues (Gao et al., 2015) investigated the influence of the temperature, equivalent ratio, and steam/carbon ratio on gas product distribution for the steam reforming of biomass tar over NiO/ceramic foam catalyst. The biomass tar was originally retrieved from rice husk gasification at 550°C. An ultimate analysis was realized on the biomass tar and it showed the following composition: 34.12% daf (dry and ash-free basis) carbon, 8.28% daf hydrogen, 2.00% daf nitrogen, and 55.60% daf oxygen (calculated by difference to 100%). The ceramic foam was dipped in excess of Ni $(NO_3)_2$,$6H_2O$. The amount of NiO loading was adjusted to be

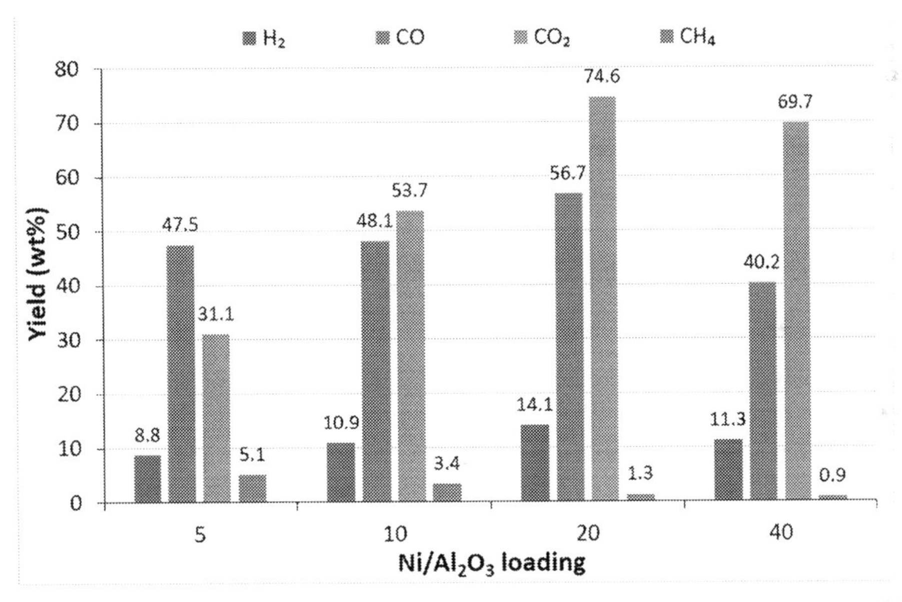

FIG. 9.28 Effect of Ni loading on gaseous compounds yields of a mixture of all model compounds over Ni/Al_2O_3 at 750°C (Artetxe et al., 2017).

3.5%. The experiments were realized in a tubular fixed-bed reactor under atmospheric pressure (Gao et al., 2015).

Effect of temperature on the steam reforming of biomass tar

The reaction temperature holds an important influence on tar reforming with NiO/ceramic foam catalyst. The temperature was varied from 500°C to 900°C with 100°C increment. The other experimental conditions are: tar feed rate of 6.78 g/h, equivalent ratio equal to zero, S/C of 1, and WHSV controlled at 4.68 h^{-1}. The main products resulting from the tar reforming are H_2, CH_4, CO, and CO_2 along with some small organic gases such as C_2H_4, C_2H_6, C_3H_6, and C_3H_8 (Gao et al., 2015). Their distribution and their yields as function of temperature are shown in Fig. 9.29. Note that H_2 concentration first increased from 50.04% at 500°C to reach a peak of 64.92% at 600°C. Then, it experiences a slight decrease to 57.97% at 900°C. As the reaction temperatures increase, the CO concentration increased significantly from 2.67% to 25.51% whereas the CO_2 concentration decreased from 44.45% to 14.37%. The amounts of alkane were low when compared to the other inorganic molecules. The CH_4 concentration was lower than 3%, and that of gaseous C_2-C_3 nearly remained less than 0.5% for the whole range of temperature. The results designated that the cracking of oxygenated organic compounds lead to the formation of the hydrocarbons, which were next easily catalyzed by NiO/ceramic foam catalyst.

The distribution of the yields and gas concentration are first accredited to the tar fast pyrolysis at the initiation of reforming. Next, the produced gases are reformed over the ceramic foam catalyst according to the tar reforming

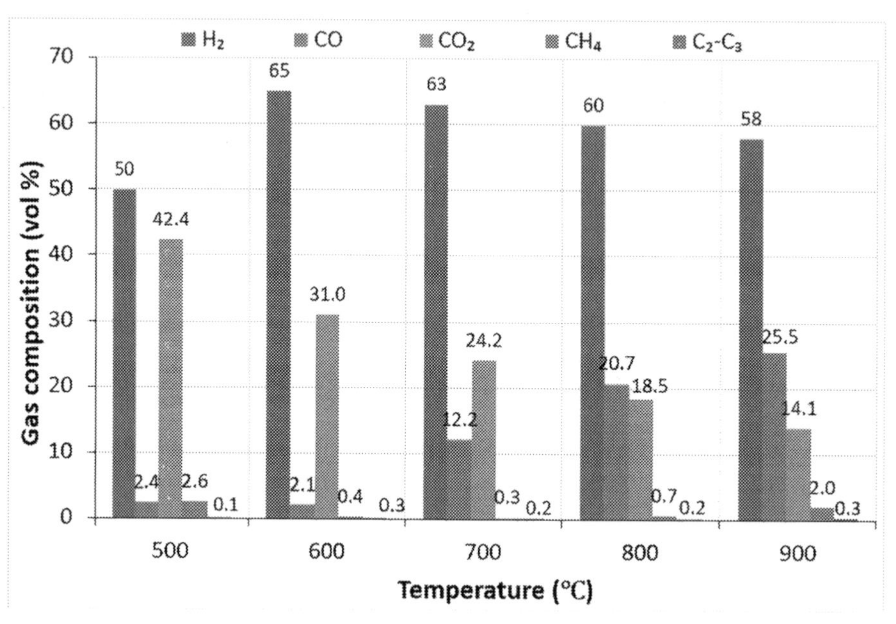

FIG. 9.29 Effect of temperature on gas yield by steam reforming of biomass tar over NiO/ceramic foam catalyst (Gao et al., 2015).

and Boudouard reactions. Practically, the reforming process happened in a reducing atmosphere, and the catalyst of NiO was reduced to nickel metal, the main active ingredient, in a short time. Since higher temperature favors endothermic reactions, the temperature increase possesses a considerable impact on the production of CO, CO_2, and H_2 as well as on the strengthening of the tar cracking. Furthermore, the ceramic foam has large heat capacity hence it is capable of stabilizing the reaction temperature, reducing the negative effects that can be caused by the fluctuating of the reaction temperature (Gao et al., 2015).

Effect of the S/C ratio on the steam reforming of biomass tar

The steam-to-carbon ratio effect is evaluated by testing a range of S/C ratios from 0 to 4 at 700°C. The tar feeding rate was held constant and the steam feed rate was modified to regulate S/C ratios. The equivalent ratio was zero; the tar feed rate is 6.78 g/h; and the WHSV is regulated at 4.68 h^{-1}. The results of the variation of the gas composition with the S/C ratio are displayed in Fig. 9.30.

It is noticed that the variation of each concentration presented monotonous trend with the increase of S/C ratio from 0 to 4. The reason behind this is that the augmented steam supply favors the equilibrium of tar reforming reaction and the WGSR. Consequently, H_2 concentration increased and that of CO decreased when S/C ratio was increased from 0 to 2. As the ratio of S/C is increased above 2, gas concentration remains almost constant due to the steam adsorption saturation on catalyst surface, decreasing the catalyst activity (Gao et al., 2015).

Biomass fuel gas reforming over NiO/monolithic porous ceramic catalysts

Quan et al. (2018) studied the catalytic steam reforming over monolithic NiO/porous ceramic

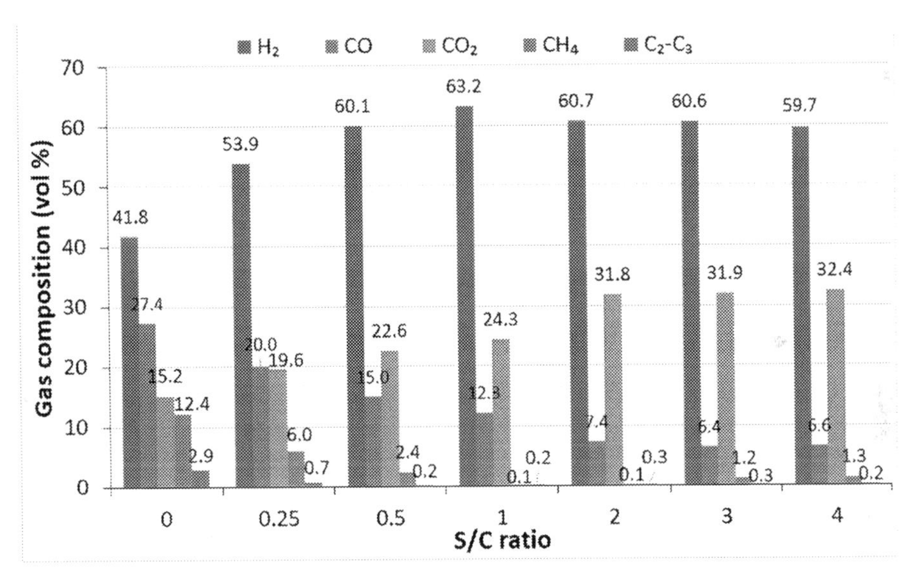

FIG. 9.30 Effect of S/C ratio on gas yield by steam reforming of biomass tar over NiO/ceramic foam catalyst (Gao et al., 2015).

catalyst for the production of high-quality syngas from biomass fuel gas in a fixed-bed reactor. They investigated the influence of the temperature, the S/C ratio, and the nickel loading on the catalyst performance. Usually, nickel is loaded on powder- or pellet-type catalyst support. Recently, monolithic catalyst has emerged as an interesting catalytic type given its strong resistance to coke deposition, high mechanical strength, and elevated effectiveness for catalyst recycling. Furthermore, the monolithic catalysts, in comparison to catalyst pellets loaded on fixed-bed reactors, are more suitable in heterogeneous catalysis applications due to their larger external surface area per unit volume and lower pressure drop. Porous ceramic, one of the monolithic catalyst supports, is loaded with NiO and employed for the reforming and upgrading of mixed model compounds of syngas. The composition of the model biomass fuel used for their study is as follows: 40 vol% CO_2, 30 vol% CO, 20 vol% H_2, and 10 vol% CH_4 (Quan et al., 2018).

Non-catalytic steam reforming of biomass fuel gas

No Nickel was loaded on the porous ceramic for this set of experiment. The temperature of the reaction was varied from 600°C to 900°C. It was demonstrated that no significant change occurs in the yield of the produced gases for the whole range of temperature as seen in Fig. 9.31. It can be deduced that with the absence of the catalyst, the reforming reaction barely occurs since the composition of the gases remained close to the initial composition of the model biomass fuel used for this study.

Catalytic steam reforming of biomass fuel gas

Effect of reaction temperature on the steam reforming of biomass fuel gas

The effect of temperature on gas composition using the NiO/monolithic porous ceramic catalyst was evaluated by varying the temperature from 550°C to 900°C while fixing the S/C ratio

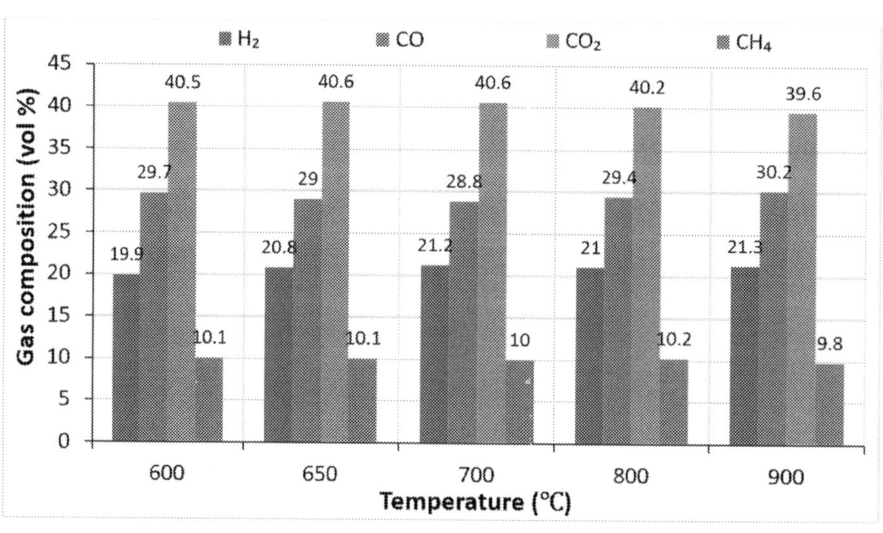

FIG. 9.31 Effect of reaction temperature on gas yields in non-catalytic steam reforming of biofuel gas (Quan et al., 2018).

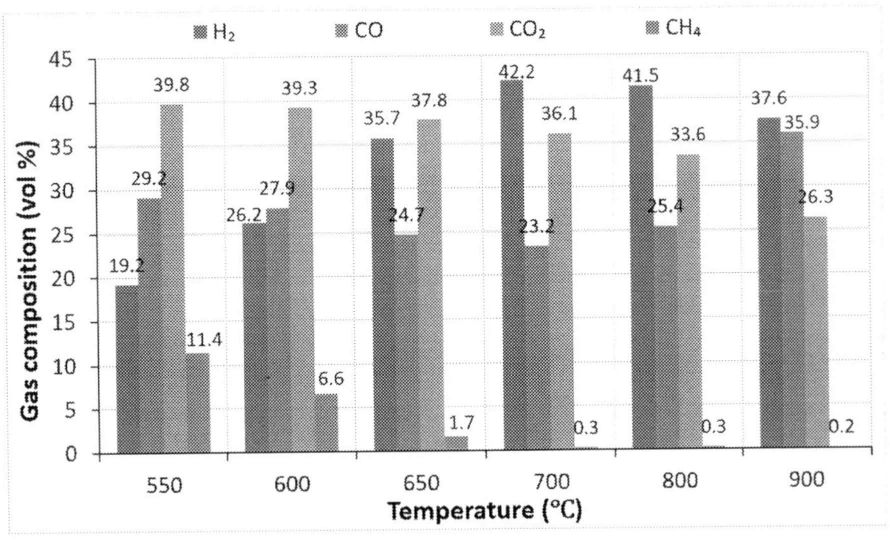

FIG. 9.32 Effect of reaction temperature on the gas composition from the steam reforming of bio-fuel gas at S/C = 2 and 2.5% metal loading (Quan et al., 2018).

at 2 and with 2.5% metal loading in the presence of steam. The results are delivered in Fig. 9.32. It is remarked that the temperature modifies the gas compositions. For instance, H_2 yield increased when the temperature was raised from 550°C to 700°C where it reached a peak of 42.2% after which it decreased with the further increase of the temperature up to 900°C. Controversially, CO content displays an opposite trend to that of H_2 where it reached a minimum at 700°C. On the other hand, the CO_2 content decreased as the temperature increased from 550°C to 900°C. The CH_4 yield decreased drastically from 11.5% to 0.3% from 550°C to 700°C indicating that the CH_4 was remarkably converted by the NiO/monolithic porous ceramic catalyst through the reforming of the model biomass fuel. The yield of CH_4 was stabilized at 0.3% with the further temperature increase up to 900°C.

The main reactions involved in the catalytic steam reforming of the mixed bio-fuel gas are those previously stated in Eqs. (9.2)–(9.4) and Eq. (9.6). The fact that the H_2 yield increased while that of CO and CH_4 decreased when the temperature was raised from 550°C to 700°C, reveals that the steam reforming reaction and WGSR are dominant for this temperature range when the reforming is done over NiO/monolithic porous ceramic catalyst. The CO_2 decrease can be rooted to the fact that the amount of CO_2 generated during the steam reforming reaction and the WGSR is inferior to its consumption by the dry reforming reaction. Regarding the results obtained between 700°C and 900°C, it can be deduced that the RWGSR was promoted which is thermodynamically favorable at high temperatures (Quan et al., 2018).

Effect of S/C ratio on the steam reforming of biomass fuel gas

The effect of the S/C ratio was studied at 700°C using 2.5% NiO/monolithic porous ceramic catalyst. The S/C ratio was regulated by modifying the steam amounts. The variation of the gas composition in the function of the S/C ratio is demonstrated in Fig. 9.33. H_2 yield increases from 28.1% to 40.2% with the increase of S/C ratio from 1 to 2, but it remains almost constant

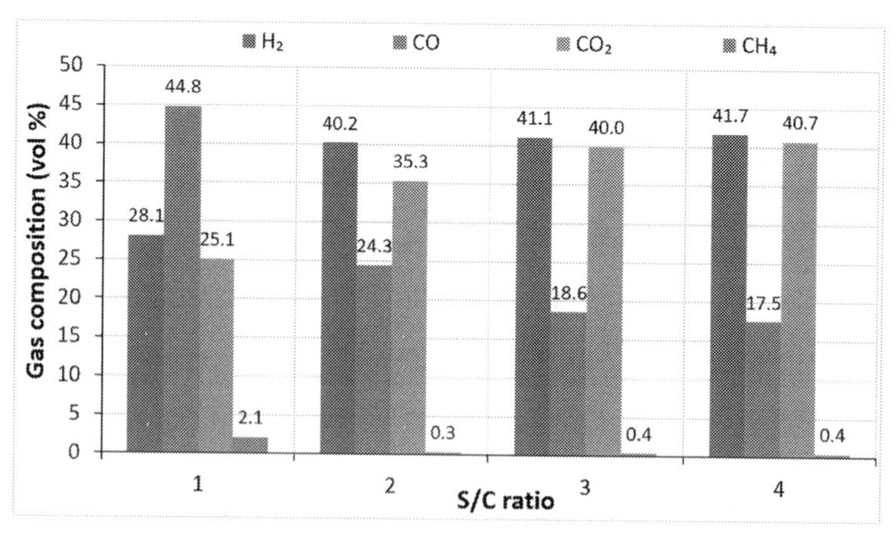

FIG. 9.33 Effect of S/C ratio on the gas composition from the steam reforming of bio-fuel gas at 700°C and 2.5% metal loading (Quan et al., 2018).

with the further increase of the S/C ratio. The CO_2 yield increases from 25.1% to 40% when the S/C ratio was raised from 0 to 3, while the CO yield drops down from 44.8% to 18.6%. CH_4 yield slightly decreased from 2.1 to 0.3% when the S/C ratio increased from 1 to 2, after which it remained almost stable.

At S/C ratio of 1, the water steam amount is inadequate for the steam reforming and water-gas shift reactions. The intensity of these latter reactions can be enhanced with the increase of the S/C ratio. Consequently, the increase of the S/C ratio from 1 to 2 leads to a remarkable elevation of the yields of CO_2 and H_2 on one hand and a dramatic decrease of the content of CH_4 and CO on the other hand. In fact, a very high S/C ratio results in the absorption of a large amount of heat for the evaporation of water into steam for the gasification, badly affecting the hydrogen production through the reforming reactions. Moreover, the active sites present on the catalyst surface would be occupied by the excess of H_2O molecules reducing the catalysis efficiency (Quan et al., 2018).

Effect of metal loading on the steam reforming of biomass fuel gas

The influence of the metal loading on the gas yields from the steam reforming of model biomass fuel gas was investigated at 700°C and S/C ratio of 2 as it is demonstrated in Fig. 9.34. Significant changes in gas composition can be noticed after the addition of NiO. Generally, an increasing trend was demonstrated by the CO content whereas a decreasing tendency was remarked for the CH_4 and CO_2. The H_2 content went through an increase and peaked at an S/C ratio of 2.5 followed by a decrease as the S/C ratio was further increased. When a high loading of NiO is applied, the dispersion of metal can be attenuated and active sites would be more aggregated reducing the efficiency. Hence, for the steam reforming of bio-oil or bio-fuel, the more suitable is to apply a lower metal loading amount. The major reason behind the decrease of the CH_4 and CO_2 yields can be speculated to be the dry reforming reaction (Quan et al., 2018).

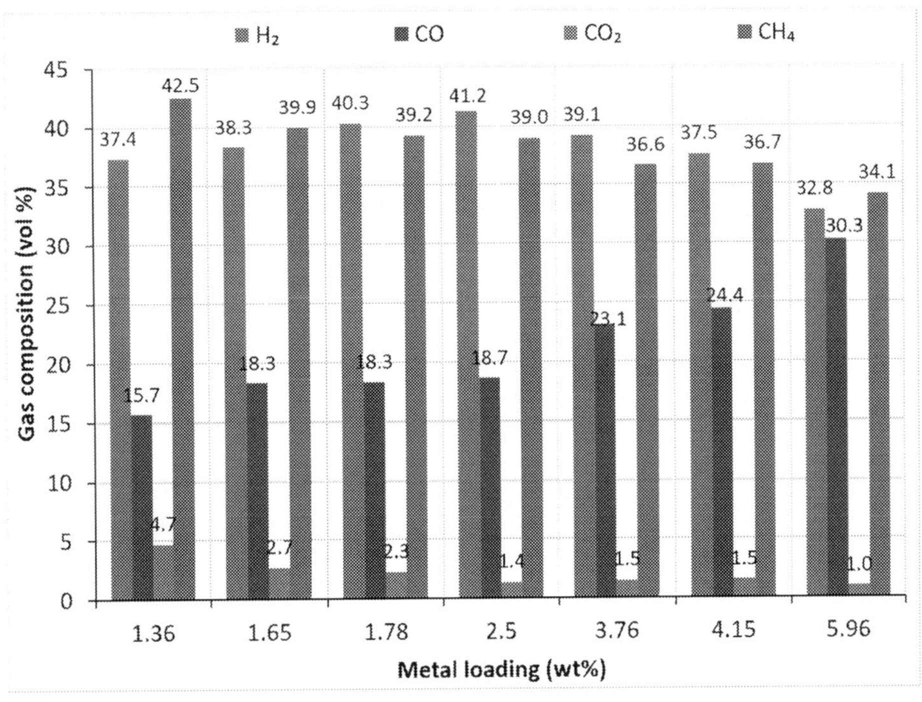

FIG. 9.34 Effect of metal loading on the gas composition from the steam reforming of bio-fuel gas at 700°C and S/C=2 (Quan et al., 2018).

9.8 Conclusion

This chapter discussed the catalytic steam reforming process applied to the removal of the tar produced during the biomass pyrolysis. A description of the different catalytic reforming types including steam and dry reforming was detailed along with a classification of the reforming process and the design of the reactors involved. In addition, a list of the main catalyst utilized for these processes with their advantages and drawbacks was elaborated. Coke formation and deposition on the catalyst surface is one of the main downsides encountered while operating a catalytic steam reforming process. Several reactions take place simultaneously during the catalytic reforming including the steam reforming, water-gas shift reaction, dry reforming, and Boudouard reaction. However, the dominant reaction might change with different catalysts and different reaction conditions due to the promoting or inhibiting interactions among different compositions.

It should be noted that the tar compositions vary depending on the biomass pyrolyzed. The tar is a complex mixture of hydrocarbons and it would be unreasonable to represent the tar by a single compound. The study of the evolution of all the tar compounds formed is technically very challenging. Therefore, in most studies, a model compound is chosen to characterize the tar and to study its catalytic reforming.

Essentially, it was found that nickel-based catalysts are the most widely investigated catalysts for tar removal given their low cost, large availability, and effectiveness in catalytic cracking of aromatic hydrocarbons. Alumina (Al_2O_3) has been largely used as Ni-based catalyst support. However, Ni/Al_2O_3 catalyst is known to be rapidly deactivated by carbon deposition on the surface of the catalyst, which can block the active sites, and result in the deactivation of catalyst. Natural minerals have been used as supports such as dolomites and olivine for Ni-based catalysts. Nevertheless, due to their low specific surface area and easy abrasion at high temperature, tar reduction using catalysts with dolomite or olivine support is not efficient enough for the downstream applications of the produced gas. On the other side, ceramic foams exhibit high porosity, large specific surface area, and moderate pressure drop making it an attractive catalyst support.

Following the theoretical background, the experimental results of the catalytic reforming of methane, ethane, propane were first presented with the impact of temperature and steam-to-carbon ratio on the conversion of reactants and the products' distribution. For the methane steam reforming, it was concluded that the increase of the temperature from 500°C to 700°C enhances the production of H_2 and CO. The CO_2 concentration increases with the temperature and reaches a maximum around 600°C above which it starts decreasing. These variations are the results of the advantageous effect of the temperature increase on the steam reforming reaction and the water-gas shift reaction leading to the production of H_2 and CO. Whereas the CO_2 production becomes limited with further temperature increase due to the less favorable WGSR. The increase of the S/C ratio reduces the concentrations of H_2 and CO but enhances the production of CO_2 due to the larger presence of water. As for the ethane and propane steam reforming, the gases products followed similar trends as for the methane with the temperature increase.

Next, the steam reforming of multiple tar compounds was displayed. The effects of temperature, steam-to-carbon ratio, and metal loadings on the gas yields produced from the steam reforming were largely investigated. Concerning the temperature's effect, for the different species studied including benzene, toluene, biomass tar, and bio-fuel gas, it was demonstrated that for a temperature falling between 500°C and 700°C, the H_2 yield increases and CO yield decreases, while for temperature exceeding 700°C and going up to 900°C, the opposite trends were obtained for both of the gases. On the other hand, for the temperature range going from 500°C to 900°C, the trend of variation of CO_2 and CH_4 was almost always decreasing. The toluene conversion was found to be enhanced by the increase in the temperature.

While investigating the effect of the steam-to-carbon (S/C) ratio on the steam reforming, it was found that the toluene conversion increases when the S/C ratio goes from 2 to 3 and decreases when the S/C ratio is further raised up to 5. For the different species tested, regarding the H_2 yield variation, when the S/C ratio goes from 0 to 1, H_2 yield increases, and when the S/C ratio is augmented from 1 to 4, the H_2 yield decreases. The trend of variation of the other gases was monotonous where the increase of the S/C ratio from 0 to 4 entailed the increase of the CO_2 yield and the decrease of that of CO and CH_4.

Analyzing the effect of the metal loading on the steam reforming, some contradictory results were obtained but it could be reasoned due to the difference in the catalyst employed as well as the amounts of loading. For instance, when the Ni/CeO_2 catalyst was used for the toluene catalytic steam reforming, raising the loading from 1 to 3 wt% leads to an increase in the H_2, CO_2, and CH_4 percentages and a decrease of that of CO. When NiO was used for the catalytic steam reforming of bio-fuel gas, the opposite trends were remarked. Increasing the loading from 1.36% to 6% results in an increase in the CO concentration and a decrease in that of CO_2 and CH_4. The H_2 concentration was increased when the metal loading was raised from 1.36% to 2.5% but it dropped when the metal loading was further elevated to 6%. When the Ni/Al_2O_3 was utilized in the catalytic steam reforming of a mixture of tar model compounds, it was noticed that when the loading was increased from 5% to 20%, the yields of H_2, CO, and CO_2 were increased while that of CH_4 was decreased. However, further increasing the metal loading up to 40% led to a drop in the yields of all the gases.

Although the values varied from a study to another, it can be deduced that the optimal operating temperature is between 700°C and 750°C. The most convenient steam-to-carbon ratio can be chosen to be around 2. The best metal loading was considered to be 20% Ni/Al_2O_3 when tested for a mixture of several tar model compounds. Operating above these values will lead to a decrease in the steam reforming efficiency mainly caused by coke deposition on the catalyst surface. Interestingly, the use of a specific catalyst's combination such as nickel (Ni) supported on lanthanum (La) doped ceria-zirconia (CeO_2-ZrO_2) mixed oxide can favor the operation at lower reaction temperatures (between 400°C and 550°C).

In this chapter, it was demonstrated that the catalytic reforming is an efficient method for converting hydrocarbons into syngas, and thereby it can be applied for the removal of the tar produced during the biomass pyrolysis. Therefore, it would be of great interest and significant efficiency to couple the process of the biomass pyrolysis with the catalytic reforming in order to valorize the gas produced and crack the tar or bio-oil formed into gases of high added value and wide application such as hydrogen and carbon monoxide.

Acknowledgments

We are grateful for the Reims Metropole, the University of Reims Champagne-Ardenne, and the SFR Condorcet—FR CNRS 3417 through the "Allocation doctorale" and "Appel à Projets 2021" Grants which supported this study, respectively.

References

Abashar, M. E. E., Alhumaizi, K. I., & Adris, A. M. (2003). Investigation of methane-steam reforming in fluidized bed membrane reactors. *Chemical Engineering Research and Design*, *81*(2), 251–258. https://doi.org/10.1205/026387603762878719.

Abou Rjeily, M., Gennequin, C., Pron, H., Abi-Aad, E., & Randrianalisoa, J. H. (2021). Pyrolysis-catalytic upgrading of bio-oil and pyrolysis-catalytic steam reforming of biogas: A review. *Environmental Chemistry Letters*. https://doi.org/10.1007/s10311-021-01190-2.

Adeniyi, A. G., Otoikhian, K. S., & Ighalo, J. O. (2019). Steam reforming of biomass pyrolysis oil: A review. *International Journal of Chemical Reactor Engineering*, *17*(4). https://doi.org/10.1515/ijcre-2018-0328.

Adnan, M. A., Muraza, O., Razzak, S. A., Hossain, M. M., & De Lasa, H. I. (2017). Iron oxide over silica-doped alumina catalyst for catalytic steam reforming of toluene as a surrogate tar biomass species. *Energy and Fuels*, *31*(7), 7471–7481. https://doi.org/10.1021/acs.energyfuels.7b01301.

Adris, A. M., Elnashaie, S. S. E. H., & Hughes, R. (1991). A fluidized bed membrane reactor for the steam reforming of methane. *The Canadian Journal of Chemical Engineering*, *69*(5), 1061–1070. https://doi.org/10.1002/cjce.5450690504.

Angeli, S. D., Pilitsis, F. G., & Lemonidou, A. A. (2015). Methane steam reforming at low temperature: Effect of light alkanes' presence on coke formation. *Catalysis Today*, *242*, 119–128. https://doi.org/10.1016/j.cattod.2014.05.043.

Antos, G. J., & Aitani, A. M. (2004). *Catalytic naphtha reforming, revised and expanded* (p. 571). CRC Press.

Aouad, S., Labaki, M., Ojala, S., Seelam, P. K., Turpeinen, E., Gennequin, C., et al. (2018). A review on the dry reforming processes for hydrogen production. *Catalytic Materials and Technologies*, 60–128. https://doi.org/10.2174/9781681087580118020007.

Arregi, A., Amutio, M., Lopez, G., Bilbao, J., & Olazar, M. (2018). Evaluation of thermochemical routes for hydrogen production from biomass: A review. *Energy Conversion and Management*, *165*, 696–719. https://doi.org/10.1016/j.enconman.2018.03.089.

Artetxe, M., Alvarez, J., Nahil, M. A., Olazar, M., & Williams, P. T. (2017). Steam reforming of different biomass tar model compounds over Ni/Al2O3 catalysts. *Energy Conversion and Management*, *136*, 119–126. https://doi.org/10.1016/j.enconman.2016.12.092.

Ashok, J., Dewangan, N., Das, S., Hongmanorom, P., Wai, M. H., Tomishige, K., et al. (2020). Recent progress in the development of catalysts for steam reforming of biomass tar model reaction. *Fuel Processing Technology*, *199*. https://doi.org/10.1016/j.fuproc.2019.106252.

Babaqi, B. S., Takriff, M. S., Kamarudin, S. K., Othman, N. T. B. A., & Ba-Abbad, M. M. (2017). Comparison of catalytic reforming processes for process integration opportunities: Brief review. *Advanced Engineering Research and Applications*, 1–11. https://www.researchgate.net/publication/309411348.

Balagurumurthy, B., Singh, R., & Bhaskar, T. (2015). Catalysts for thermochemical conversion of biomass. In *Recent advances in thermochemical conversion of biomass* (pp. 109–132). Elsevier Inc. https://doi.org/10.1016/B978-0-444-63289-0.00004-1.

Balajii, M., & Niju, S. (2019). Biochar-derived heterogeneous catalysts for biodiesel production. *Environmental Chemistry Letters*, *17*(4), 1447–1469. https://doi.org/10.1007/s10311-019-00885-x.

Balat, M., Balat, M., Kirtay, E., & Balat, H. (2009). Main routes for the thermo-conversion of biomass into fuels and chemicals. Part 2: Gasification systems. *Energy Conversion and Management*, *50*(12), 3158–3168. https://doi.org/10.1016/j.enconman.2009.08.013.

Basile, A., Liguori, S., & Iulianelli, A. (2015). Membrane reactors for methane steam reforming (MSR). In *Membrane reactors for energy applications and basic chemical production* (pp. 31–59). Elsevier Inc. https://doi.org/10.1016/B978-1-78242-223-5.00002-9.

Brown, R. C. (2007). *Biomass-Derived Hydrogen from a Thermally Ballasted Gasifier*. https://doi.org/10.2172/901792.

Cao, J. P., Ren, J., Zhao, X. Y., Wei, X. Y., & Takarada, T. (2018). Effect of atmosphere on carbon deposition of Ni/Al2O3 and Ni-loaded on lignite char during reforming of toluene as a biomass tar model compound. *Fuel*, *217*, 515–521. https://doi.org/10.1016/j.fuel.2017.12.121.

Castro, T., Robson, P., Raimundo, R. N., Borges, L., & Noronha, F. (2016). Steam reforming of toluene over Pt/Ce x Zr1 − x O2/Al2O3 catalysts. *Topics in Catalysis*, *59*, 1–292. https://doi.org/10.1007/s11244-015-0443-4.

De Deken, J. C., Devos, E. F., & Froment, G. F. (1982). *Steam reforming of natural gas: Intrinsic kinetics, diffusional influences, and reactor design* (pp. 181–197). American Chemical Society (ACS). https://doi.org/10.1021/bk-1982-0196.ch016.

Devi, L., Ptasinski, K. J., Janssen, F. J. J. G., Van Paasen, S. V. B., Bergman, P. C. A., & Kiel, J. H. A. (2005). Catalytic decomposition of biomass tars: Use of dolomite and untreated olivine. *Renewable Energy*, *30*(4), 565–587. https://doi.org/10.1016/j.renene.2004.07.014.

Dickerson, T., & Soria, J. (2013). Catalytic fast pyrolysis: A review. *Energies*, *6*(1), 514–538. https://doi.org/10.3390/en6010514.

Dufour, A., Masson, E., Girods, P., Rogaume, Y., & Zoulalian, A. (2011). Evolution of aromatic tar composition in relation to methane and ethylene from biomass pyrolysis-gasification. *Energy and Fuels*, *25*(9), 4182–4189. https://doi.org/10.1021/ef200846g.

Elnashaie, S. S. E. H., & Abashar, M. E. E. (1993). Steam reforming and methanation effectiveness factors using the dusty gas model under industrial conditions. *Chemical Engineering and Processing*, *32*(3), 177–189. https://doi.org/10.1016/0255-2701(93)80014-8.

Estephane, J., Aouad, S., Hany, S., El Khoury, B., Gennequin, C., El Zakhem, H., et al. (2015). CO2 reforming of methane over Ni-CO/ZSM5 catalysts. Aging and carbon deposition study. *International Journal of Hydrogen Energy*, *40*(30), 9201–9208. https://doi.org/10.1016/j.ijhydene.2015.05.147.

Fajardo, H. V., & Probst, L. F. D. (2006). Production of hydrogen by steam reforming of ethanol over Ni/Al2O3 spherical catalysts. *Applied Catalysis A: General*, *306*, 134–141. https://doi.org/10.1016/j.apcata.2006.03.043.

Fajardo, H. V., Longo, E., Mezalira, D. Z., Nuernberg, G. B., Almerindo, G. I., Collasiol, A., et al. (2010). Influence of support on catalytic behavior of nickel catalysts in the steam reforming of ethanol for hydrogen production. *Environmental Chemistry Letters*, *8*(1), 79–85. https://doi.org/10.1007/s10311-008-0195-5.

Fidalgo, A. M., & Ilharco, L. M. (2012). Tailoring the structure and hydrophobic properties of amorphous silica by silylation. *Microporous and Mesoporous Materials*, *158*, 39–46. https://doi.org/10.1016/j.micromeso.2012.03.009.

French, R., & Czernik, S. (2010). Catalytic pyrolysis of biomass for biofuels production. *Fuel Processing Technology*, *91*(1), 25–32. https://doi.org/10.1016/j.fuproc.2009.08.011.

Gao, N., Han, Y., & Quan, C. (2018). Study on steam reforming of coal tar over Ni–co/ceramic foam catalyst for hydrogen production: Effect of Ni/co ratio. *International Journal of Hydrogen Energy*, *43*(49), 22170–22186. https://doi.org/10.1016/j.ijhydene.2018.10.119.

Gao, N., Liu, S., Han, Y., Xing, C., & Li, A. (2015). Steam reforming of biomass tar for hydrogen production over NiO/ceramic foam catalyst. *International Journal of Hydrogen Energy*, *40*(25), 7983–7990. https://doi.org/10.1016/j.ijhydene.2015.04.050.

Gao, N., Wang, X., Li, A., Wu, C., & Yin, Z. (2016). Hydrogen production from catalytic steam reforming of benzene as tar model compound of biomass gasification. *Fuel Processing Technology*, *148*, 380–387. https://doi.org/10.1016/j.fuproc.2016.03.019.

Gennequin, C., Hany, S., Tidahy, H. L., Aouad, S., Estephane, J., Aboukaïs, A., et al. (2016). Influence of the presence of ruthenium on the activity and stability of co–mg–Al-based catalysts in CO2 reforming of methane for syngas production. *Environmental Science and Pollution Research*, *23*(22), 22744–22760. https://doi.org/10.1007/s11356-016-7453-z.

González-Marcos, M. P., Iñarra, B., Guil, J. M., & Gutiérrez-Ortiz, M. A. (2005). Development of an industrial characterisation method for naphtha reforming bimetallic Pt-Sn/Al2O3 catalysts through n-heptane reforming test reactions. In. *Catalysis Today*, *107–108*, 685–692. https://doi.org/10.1016/j.cattod.2005.07.052.

Guan, G., Kaewpanha, M., Hao, X., Zhu, A. M., Kasai, Y., Kakuta, S., et al. (2013). Steam reforming of tar derived from lignin over pompom-like potassium-promoted iron-based catalysts formed on calcined scallop shell. *Bioresource Technology*, *139*, 280–284. https://doi.org/10.1016/j.biortech.2013.04.007.

Guan, G., Kaewpanha, M., Hao, X., & Abudula, A. (2016). Catalytic steam reforming of biomass tar: Prospects and challenges. *Renewable and Sustainable Energy Reviews*, *58*, 450–461. https://doi.org/10.1016/j.rser.2015.12.316.

Gunarathne, V., Ashiq, A., Ramanayaka, S., Wijekoon, P., & Vithanage, M. (2019). Biochar from municipal solid waste for resource recovery and pollution remediation. *Environmental Chemistry Letters*, *17*(3), 1225–1235. https://doi.org/10.1007/s10311-019-00866-0.

Guo, F., Li, X., Liu, Y., Peng, K., Guo, C., & Rao, Z. (2018). Catalytic cracking of biomass pyrolysis tar over char-supported catalysts. *Energy Conversion and Management*, *167*, 81–90. https://doi.org/10.1016/j.enconman.2018.04.094.

Guo, J., Lou, H., Zhao, H., Chai, D., & Zheng, X. (2004). Dry reforming of methane over nickel catalysts supported on magnesium aluminate spinels. *Applied Catalysis A: General*, *273*(1–2), 75–82. https://doi.org/10.1016/j.apcata.2004.06.014.

Halabi, M. H., De Croon, M. H. J. M., Van Der Schaaf, J., Cobden, P. D., & Schouten, J. C. (2010). Intrinsic kinetics of low temperature catalytic methane-steam reforming and water-gas shift over Rh/CeαZr1-αO2 catalyst. *Applied Catalysis A: General*, *389*(1–2), 80–91. https://doi.org/10.1016/j.apcata.2010.09.005.

He, L., Hu, S., Jiang, L., Liao, G., Chen, X., Han, H., et al. (2018). Carbon nanotubes formation and its influence on steam reforming of toluene over Ni/Al2O3 catalysts: Roles of catalyst supports. *Fuel Processing Technology*, *176*, 7–14. https://doi.org/10.1016/j.fuproc.2018.03.007.

Isahak, W. N. R. W., Hisham, M. W. M., Yarmo, M. A., & Yun Hin, T. Y. (2012). A review on bio-oil production from biomass by using pyrolysis method. *Renewable and Sustainable Energy Reviews*, *16*(8), 5910–5923. https://doi.org/10.1016/j.rser.2012.05.039.

Kechagiopoulos, P. N., Voutetakis, S. S., Lemonidou, A. A., & Vasalos, I. A. (2007). Sustainable hydrogen production via reforming of ethylene glycol using a novel spouted bed reactor. *Catalysis Today*, *127*(1–4), 246–255. https://doi.org/10.1016/j.cattod.2007.05.018.

Kechagiopoulos, P. N., Voutetakis, S. S., Lemonidou, A. A., & Vasalos, I. A. (2009). Hydrogen production via reforming of the aqueous phase of bio-oil over Ni/olivine catalysts in a spouted bed reactor. *Industrial and Engineering Chemistry Research*, *48*(3), 1400–1408. https://doi.org/10.1021/ie8013378.

Khzouz, M., & Gkanas, E. I. (2017). Experimental and numerical study of low temperature methane steam reforming for hydrogen production. *Catalysts*, *8*(1). https://doi.org/10.3390/catal8010005.

Kong, M., Fei, J., Wang, S., Lu, W., & Zheng, X. (2011). Influence of supports on catalytic behavior of nickel catalysts in carbon dioxide reforming of toluene as a model compound of tar from biomass gasification. *Bioresource Technology*, *102*(2), 2004–2008. https://doi.org/10.1016/j.biortech.2010.09.054.

Li, S., & Gong, J. (2014). Strategies for improving the performance and stability of Ni-based catalysts for reforming reactions. *Chemical Society Reviews*, *43*(21), 7245–7256. https://doi.org/10.1039/c4cs00223g.

Li, C., & Suzuki, K. (2009). Tar property, analysis, reforming mechanism and model for biomass gasification-an overview. *Renewable and Sustainable Energy Reviews*, *13*(3), 594–604. https://doi.org/10.1016/j.rser.2008.01.009.

Li, Z., Jiang, B., Wang, Z., & Kawi, S. (2018). High carbon resistant Ni@Ni phyllosilicate@SiO2 core shell hollow sphere catalysts for low temperature CH4 dry reforming. *Journal of CO_2 Utilization*, *27*, 238–246. https://doi.org/10.1016/j.jcou.2018.07.017.

Li, Z., Li, M., Bian, Z., Kathiraser, Y., & Kawi, S. (2016). Design of highly stable and selective core/yolk-shell nanocatalysts-review. *Applied Catalysis B: Environmental*, *188*, 324–341. https://doi.org/10.1016/j.apcatb.2016.01.067.

Lónyi, F., Valyon, J., Someus, E., & Hancsók, J. (2013). Steam reforming of bio-oil from pyrolysis of MBM over particulate and monolith supported Ni/γ-Al2O3 catalysts. *Fuel*, *112*, 23–30. https://doi.org/10.1016/j.fuel.2013.05.010.

Meyers, R. A. (2004). *Handbook of petroleum refining processes* (p. 548). New York: McGraw-Hill Education. https://www.osti.gov/etdeweb/biblio/20686136.

Miller, B. G. (2011). The effect of coal usage on human health and the environment. In *Clean Coal Engineering Technology* (pp. 85–132). https://doi.org/10.1016/B978-1-85617-710-8.00004-2.

Milne, T. A., Evans, R. J., & Abatzaglou, N. (1998). *Biomass gasifier "'tars'": Their nature, formation, and conversion*. https://doi.org/10.2172/3726.

Nichele, V., Signoretto, M., Pinna, F., Menegazzo, F., Rossetti, I., Cruciani, G., et al. (2014). Ni/ZrO2 catalysts in ethanol steam reforming: Inhibition of coke formation by CaO-doping. *Applied Catalysis B: Environmental*,

150–151, 12–20. https://doi.org/10.1016/j.apcatb.2013.11.037.

Nielsen, R., & Jens, R. (1984). Catalytic steam reforming. In R. Gilles (Ed.), *Vol. 5. Advances in comparative and environmental physiology* (pp. 1–130). Berlin Heidelberg: Springer. https://doi.org/10.1007/978-3-642-93247-2_1.

Ochoa, A., Bilbao, J., Gayubo, A. G., & Castaño, P. (2020). Coke formation and deactivation during catalytic reforming of biomass and waste pyrolysis products: A review. *Renewable and Sustainable Energy Reviews, 119*. https://doi.org/10.1016/j.rser.2019.109600.

Oemar, U., Bian, Z., Hidajat, K., & Kawi, S. (2016). Sulfur resistant LaXCe1- xNi0.5Cu0.5O3 catalysts for an ultra-high temperature water gas shift reaction. *Catalysis Science and Technology, 6*(17), 6569–6580. https://doi.org/10.1039/c6cy00635c.

Oemar, U., Kathiraser, Y., Ang, M. L., Hidajat, K., & Kawi, S. (2015). Catalytic biomass gasification to syngas over highly dispersed lanthanum-doped nickel on SBA-15. *ChemCatChem, 7*(20), 3376–3385. https://doi.org/10.1002/cctc.201500482.

Park, H. G., Han, S. Y., Jun, K. W., Woo, Y., Park, M. J., & Kim, S. K. (2019). Bench-scale steam reforming of methane for hydrogen production. *Catalysts, 9*(7). https://doi.org/10.3390/catal9070615.

Perez-Lopez, O. W., Senger, A., Marcilio, N. R., & Lansarin, M. A. (2006). Effect of composition and thermal pretreatment on properties of Ni-mg-Al catalysts for CO2 reforming of methane. *Applied Catalysis A: General, 303*(2), 234–244. https://doi.org/10.1016/j.apcata.2006.02.024.

Pfeifer, C., & Hofbauer, H. (2008). Development of catalytic tar decomposition downstream from a dual fluidized bed biomass steam gasifier. *Powder Technology, 180*(1–2), 9–16. https://doi.org/10.1016/j.powtec.2007.03.008.

Quan, C., Gao, N., & Wu, C. (2018). Utilization of NiO/porous ceramic monolithic catalyst for upgrading biomass fuel gas. *Journal of the Energy Institute, 91*(3), 331–338. https://doi.org/10.1016/j.joei.2017.02.008.

Quitete, C. P. B., Bittencourt, R. C. P., & Souza, M. M. V. M. (2015). Steam reforming of tar model compounds over nickel catalysts supported on barium Hexaaluminate. *Catalysis Letters, 145*(2), 541–548. https://doi.org/10.1007/s10562-014-1405-3.

Rahimpour, M. R., Jafari, M., & Iranshahi, D. (2013). Progress in catalytic naphtha reforming process: A review. *Applied Energy, 109*, 79–93. https://doi.org/10.1016/j.apenergy.2013.03.080.

Remiro, A., Valle, B., Aramburu, B., Aguayo, A. T., Bilbao, J., & Gayubo, A. G. (2013c). Steam reforming of the bio-oil aqueous fraction in a fluidized bed reactor with in situ CO2 capture. *Industrial and Engineering Chemistry Research, 52*(48), 17087–17098. https://doi.org/10.1021/ie4021705.

Remiro, A., Valle, B., Oar-Arteta, L., Aguayo, A. T., Bilbao, J., & Gayubo, A. G. (2014). Hydrogen production by steam reforming of bio-oil/bio-ethanol mixtures in a continuous thermal-catalytic process. *International Journal of Hydrogen Energy, 39*(13), 6889–6898. https://doi.org/10.1016/j.ijhydene.2014.02.137.

Remiro, A., Valle, B., Aguayo, A. T., Bilbao, J., & Gayubo, A. G. (2013a). Operating conditions for attenuating Ni/La2O3- αAl2O3 catalyst deactivation in the steam reforming of bio-oil aqueous fraction. *Fuel Processing Technology, 115*, 222–232. https://doi.org/10.1016/j.fuproc.2013.06.003.

Remiro, A., Valle, B., Aguayo, A. T., Bilbao, J., & Gayubo, A. G. (2013b). Steam reforming of raw bio-oil in a fluidized bed reactor with prior separation of pyrolytic lignin. *Energy and Fuels, 27*(12), 7549–7559. https://doi.org/10.1021/ef401835s.

Rostrup-Nielsen, J. R. (2000). New aspects of syngas production and use. *Catalysis Today, 63*, 455–457. https://doi.org/10.1016/S0920-5861(00.

Roy, S., Pruden, B. B., Adris, A. M., Grace, J. R., & Lim, C. J. (1999). Fluidized-bed steam methane reforming with oxygen input. *Chemical Engineering Science, 54*(13–14), 2095–2102. https://doi.org/10.1016/S0009-2509(98)00300-5.

Saw, E. T., Oemar, U., Ang, M. L., Hidajat, K., & Kawi, S. (2015). Highly active and stable bimetallic nickel-copper Core-ceria Shell catalyst for Higherature water-gas shift reaction. *ChemCatChem, 7*(20), 3358–3367. https://doi.org/10.1002/cctc.201500481.

Shen, Y., & Fu, Y. (2018). Advances in: In situ and ex situ tar reforming with biochar catalysts for clean energy production. *Sustainable Energy & Fuels, 2*(2), 326–344. https://doi.org/10.1039/c7se00553a.

Swaan, H. M., Kroll, V. C. H., Martin, G. A., & Mirodatos, C. (1994). Deactivation of supported nickel catalysts during the reforming of methane by carbon dioxide. *Catalysis Today, 21*(2–3), 571–578. https://doi.org/10.1016/0920-5861(94)80181-9.

Swierczynski, D., Courson, C., & Kiennemann, A. (2008). Study of steam reforming of toluene used as model compound of tar produced by biomass gasification. *Chemical Engineering and Processing: Process Intensification, 47*(3), 508–513. https://doi.org/10.1016/j.cep.2007.01.012.

Tao, J., Zhao, L., Dong, C., Lu, Q., Du, X., & Dahlquist, E. (2013). Catalytic steam reforming of toluene as a model compound of biomass gasification tar using Ni-CeO2/SBA-15 catalysts. *Energies, 6*(7), 3284–3296. https://doi.org/10.3390/en6073284.

Trane, R., Dahl, S., Skjøth-Rasmussen, M. S., & Jensen, A. D. (2012). Catalytic steam reforming of bio-oil. *International Journal of Hydrogen Energy, 37*(8), 6447–6472. https://doi.org/10.1016/j.ijhydene.2012.01.023.

Valle, B., Aramburu, B., Benito, P. L., Bilbao, J., & Gayubo, A. G. (2018a). Biomass to hydrogen-rich gas via steam reforming of raw bio-oil over Ni/La2O3-A Al2O3 catalyst: Effect of space-time and steam-to-carbon ratio. *Fuel*, *216*, 445–455. https://doi.org/10.1016/j.fuel.2017.11.151.

Valle, B., Aramburu, B., Olazar, M., Bilbao, J., & Gayubo, A. G. (2018b). Steam reforming of raw bio-oil over Ni/La2O3-AAl2O3: Influence of temperature on product yields and catalyst deactivation. *Fuel*, *216*, 463–474. https://doi.org/10.1016/j.fuel.2017.11.149.

Valle, B., Aramburu, B., Remiro, A., Bilbao, J., & Gayubo, A. G. (2014). Effect of calcination/reduction conditions of Ni/La2O3-αAl2O3 catalyst on its activity and stability for hydrogen production by steam reforming of raw bio-oil/ethanol. *Applied Catalysis B: Environmental*, *147*, 402–410. https://doi.org/10.1016/j.apcatb.2013.09.022.

Valle, B., Remiro, A., Aguayo, A. T., Bilbao, J., & Gayubo, A. G. (2013). Catalysts of Ni/α-Al2O3 and Ni/La 2O3-αAl2O3 for hydrogen production by steam reforming of bio-oil aqueous fraction with pyrolytic lignin retention. *International Journal of Hydrogen Energy*, *38*(3), 1307–1318. https://doi.org/10.1016/j.ijhydene.2012.11.014.

Vozniuk, O., Tanchoux, N., Millet, J. M., Albonetti, S., Di Renzo, F., & Cavani, F. (2019). Spinel mixed oxides for chemical-loop reforming: From solid state to potential application. In *Vol. 178. Studies in surface science and catalysis* (pp. 281–302). Elsevier Inc. https://doi.org/10.1016/B978-0-444-64127-4.00014-8.

Wang, L., Hisada, Y., Koike, M., Li, D., Watanabe, H., Nakagawa, Y., et al. (2012). Catalyst property of co-Fe alloy particles in the steam reforming of biomass tar and toluene. *Applied Catalysis B: Environmental*, *121–122*, 95–104. https://doi.org/10.1016/j.apcatb.2012.03.025.

Wang, D., Czernik, S., Montané, D., Mann, M., & Chornet, E. (1997). Biomass to hydrogen via fast pyrolysis and catalytic steam reforming of the pyrolysis oil or its fractions. *Industrial and Engineering Chemistry Research*, *36*(5), 1507–1518. https://doi.org/10.1021/ie960396g.

Wang, C., Wang, T., Ma, L., Gao, Y., & Wu, C. (2010). Steam reforming of biomass raw fuel gas over NiO-MgO solid solution cordierite monolith catalyst. *Energy Conversion and Management*, *51*(3), 446–451. https://doi.org/10.1016/j.enconman.2009.10.006.

Winter, F., & Schratzer, B. (2013). Applications of fluidized bed technology in processes other than combustion and gasification. In *Fluidized bed technologies for near-zero emission combustion and gasification* (pp. 1005–1033). Elsevier Ltd. https://doi.org/10.1533/9780857098801.5.1005.

Wu, C., Huang, Q., Sui, M., Yan, Y., & Wang, F. (2008). Hydrogen production via catalytic steam reforming of fast pyrolysis bio-oil in a two-stage fixed bed reactor system. *Fuel Processing Technology*, *89*(12), 1306–1316. https://doi.org/10.1016/j.fuproc.2008.05.018.

Yaakob, Z., Bshish, A., Ebshish, A., Tasirin, S. M., & Alhasan, F. H. (2013). Hydrogen production by steam reforming of ethanol over nickel catalysts supported on sol gel made alumina: Influence of calcination temperature on supports. *Materials*, *6*(6), 2229–2239. https://doi.org/10.3390/ma6062229.

Yao, D., Wu, C., Yang, H., Hu, Q., Nahil, M. A., Chen, H., et al. (2014). Hydrogen production from catalytic reforming of the aqueous fraction of pyrolysis bio-oil with modified Ni-Al catalysts. *International Journal of Hydrogen Energy*, *39*(27), 14642–14652. https://doi.org/10.1016/j.ijhydene.2014.07.077.

Zhang, J., Wang, H., & Dalai, A. K. (2007). Development of stable bimetallic catalysts for carbon dioxide reforming of methane. *Journal of Catalysis*, *249*(2), 300–310. https://doi.org/10.1016/j.jcat.2007.05.004.

CHAPTER

10

Heterogeneous catalysts for biomass-derived alcohols and acid conversion

Gheorghita Mitran[a], *Octavian Dumitru Pavel*[a], *and Dong-Kyun Seo*[b]

[a]Laboratory of Chemical Technology and Catalysis, Department of Organic Chemistry, Biochemistry & Catalysis, University of Bucharest, Bucharest, Romania

[b]School of Molecular Sciences, Arizona State University, Tempe, AZ, United States

10.1 Introduction

The petrochemical industry is dominated by acid catalysts, whereas the fine chemical industry is dominated by oxidation catalysts (Mallat & Baiker, 2004); oxidation processes represent 30% of the chemical industry.

Biomass-derived fuel and fuel additive production also requires the presence of acid catalysts. An efficient and suitable catalyst is able to catalyze selective oxidation reactions at atmospheric pressure and in the presence of oxygen from air (Sarmah, Satpati, & Srivastava, 2018).

The production of alcohols and acids from lignocellulosic biomass is a promising method due to the regeneration capabilities of biomass. Alcohols obtained from biomass are denoted as bioalcohols like biomethanol, bioethanol, and biobutanol (Demirbas & Demirbas, 2010), and they are obtained from cellulosic residues, waste materials, and energy crops.

Biofuels, such as ethanol, from polysaccharide fermentation, belong to the first generation of biofuels, whereas biomethanol is a biofuel of the second generation (Adelabu, Kareem, Oluwafemi, & Abideen Adeogun, 2019), and, recently, third and fourth generations of biofuels like biobutanol have been developed.

These liquid biofuels obtained from lignocellulosic biomass have both advantages and disadvantages. For example, compared with bioethanol, biobutanol has the advantage that it is less corrosive, has a low vapor pressure, a higher flash point, and a higher energy content (Festel, 2008; Skevis, 2010), but its main disadvantage is its lower productivity.

Biomass-derived carboxylic acids could be transformed by hydrogenation of alcohols and by deoxygenation of ketones, with both types

Heterogeneous Catalysis
https://doi.org/10.1016/B978-0-323-85612-6.00010-3

of products having applications in the pharmaceutical, fragrance, and detergent industries. Alcohols, obtained from carboxylic acid hydrogenation, could be converted into intermediates for the fine chemical industry by dehydrogenation, selective oxidation, and dehydrogenative coupling. The most difficult reaction of carboxylic acids is hydrogenation because of their kinetic and thermodynamic stability.

10.2 Alcohol conversion

Methanol

Methanol can be converted into high value-added molecules such as dimethyl ether, olefins, and gasoline. Some of the products obtained from methanol are presented in Fig. 10.1.

Dimethyl ether (DME) is a clean diesel fuel and a precursor for olefin, methyl acetate, and dimethyl sulfate (Mondal & Yadav, 2019; Unglert et al., 2020) production. Traditionally, it is directly manufactured from syngas, but because of high energy requirements, alternative methods for its production are being studied. Methanol conversion to hydrocarbons involves a series of reactions, the first of which is its dehydration to DME and water.

For methanol conversion to dimethyl ether, a series of catalysts has been investigated. SAPO-11 (Chen, Li, et al., 2018; Chen, Yu, et al., 2018) is the most efficient catalyst for dimethyl ether production from methanol dehydration reaction. The incorporation of Lewis acid sites and zeolite-based catalysts plays an important role in methanol conversion as they are more efficient in decreasing crystal size. To improve DME yield, zeolite catalysts were modified by the introduction of heteroatoms such as ferrierite zeolites (Catizzone et al., 2019; Migliori et al., 2018; Sai Prasad, Bae, Kang, Lee, & Jun, 2008). Other catalysts such as H-mordenite (Bandiera & Naccache, 1991) and γ-Al_2O_3 that were composited with ZSM-5 molecular sieves (Zeng et al., 2020) have been studied. In the latter case, ZSM-5 was added to further improve the reactivity and stability of alumina in a severe hydrothermal environment, given that the ZSM-5 zeolite exhibits high hydrothermal stability. By interaction between the two phases, the thermal stability of γ-Al_2O_3 is improved, and, furthermore, the surface acidity and porosity of ZSM-5 are modulated to attenuate coke formation.

Methanol conversion to *light olefins* is an alternative for production of platform chemicals such as ethylene and propylene. Industrial catalysts for these transformations are based on SAPO-34 zeolites, given that selectivity in the desired product is high. They have a hierarchical nanostructure and high acidity, but the disadvantage of these zeolites includes coke agglomeration on the surface with decrease in catalyst

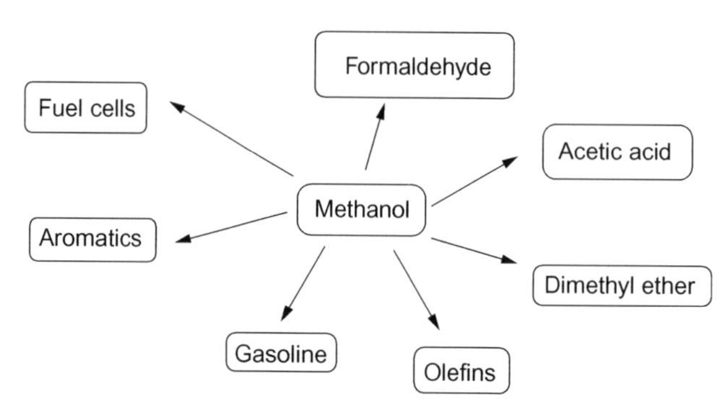

FIG. 10.1 Products from methanol conversion.

activity. Catalyst acidity can be modified by different methods in order to improve selectivity to olefins. Among them, desilication or dealumination of structure, metal incorporation, and ion exchange in the solid or liquid phase have been used.

A multitude of metals were used for incorporation in the structure of zeolites including Ca, Ti, Cu, Fe, Mg, Mn, Ce, Zn, Co, and Zr. Most of these metals decreased catalyst acidity while at the same time reducing the deposition of coke and improving the selectivity to olefins.

Other efficient catalysts are proved to be those based on ZSM zeolites, i.e., MFI materials (Koempel & Liebner, 2007); however, they have a disadvantage, which is the olefin formation is inhibited by the presence of a nanocrystalline structure.

Heteropoly acids represent the best acidic materials used as catalysts for redox and acid reactions, but their main drawback is their small surface area. Therefore, in order to resolve this inconvenience, they were supported on zeolites such as SBA-15 (Sheng et al., 2014) and SAPO-34 (Hashemi, Taghizadeh, & Rami, 2020). The results obtained using these materials are summarized in Fig. 10.2.

Incorporation of a moderate amount of heteropoly acids with W into the SAPO-34 framework results in an increase in surface area and Brønsted acidity, thus improving both catalyst lifetime and olefin selectivity.

Propylene is the most important petrochemical, mainly produced in petroleum refineries. Propylene demands are higher than its production, which is the reason why the requirement of new processes for propylene production with a high yield has increased. Methanol conversion to propylene represents a promising alternative process for propylene production from nonpetroleum resources. H-ZSM-5 zeolites with aluminum phosphate, alumina, or aluminum phosphate silica (single binder or with binary binder) were used as catalysts for this reaction with the observation that the selectivity to propylene was increased from 24% to 40% (Lee, Lee, & Ihm, 2010). This behavior is explained by the decrease in the strength and number of strong acid sites due to the binder addition.

The catalytic performances of $AlPO_4$/H-ZSM-5 extrudates were compared with those of conventional alumina and silica extrudates and the zeolite showed superior stability, activity, and propylene selectivity. However, the binder presence might

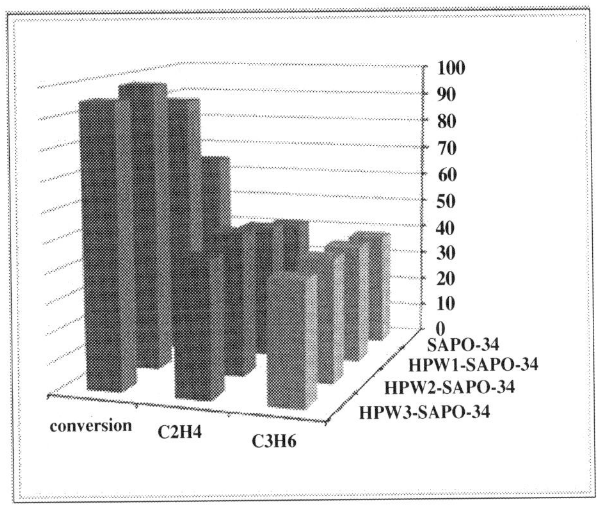

FIG. 10.2 Methanol conversion to ethylene and propylene (HPW: heteropoly acids with W; HPW1, HPW2, HPW3; HPW/Al_2O_3 ratio: 0.007, 0.01, 0.016, respectively).

partially block the micropore mouths of the zeolite, leading to a decrease in reactant accessibility to the active sites. On the other hand, methanol conversion is accompanied by the presence of water as a byproduct. Its presence at high temperatures leads to zeolite dealumination, a suitable method for enhancing the accessibility to active sites by mesopore creation.

Gasoline is a crucial fuel used in transport industries; therefore, methanol transformation in gasoline represents a good choice. The most appropriate options for catalysts are zeolites and zeolite-based catalysts due to their stability, environmental protection, product selectivity, and, additionally, their noncorrosive nature. It is more favorable to convert methanol into gasoline than to use direct methanol as fuel because of its affinity toward water, toxicity, volatility, corrosivity, and low calorific energy value.

Among the first catalysts used for methanol conversion into gasoline were ZSM-5 zeolites due to their unique channel structure, which allows shape-selective formation of gasoline.

Methanol conversion to *aromatics* (benzene, toluene, and xylene) has been of interest due to an increase in the demand for aromatics. Compared to the H-ZSM-5 zeolites, methanol conversion follows a dual cycle mechanism consisting of both aromatic and intermediate olefin production. Aromatic hydrocarbons can further be converted into polycyclic aromatics and can determine a rapid deactivation of the catalyst. To improve this type of catalyst, its structure was modified by metals like Ga (Freeman, Wells, & Hutchings, 2001), Cu, Ni (Dai et al., 2018), and Zn (Wang, Yu, Wang, Chu, & Liu, 2013). In Ga-modified zeolites, intermediate alkenes and cyclic hydrocarbons were converted into aromatics with high selectivity and hydrogen release. However, because of the dehydrogenation properties of metal, coke is deposited on the active sites, leading to catalyst deactivation. The encapsulation of metal nanoparticles has been an efficient method for their protection against sintering and poisoning (Gu et al., 2015).

Ethanol

Ethanol is one of the most used platform chemical due to its conversion to many products, namely, ethylene, propylene, and 1,3-butadiene. Ethanol transformation in a series of platform molecules is illustrated in Fig. 10.3.

Diethyl carbonate (DEC) is an important derivative that is obtained from ethanol and is used as a green solvent due to its low toxicity, its environment-friendly nature, low bioaccumulation, and also its use as a synthetic intermediate. Due to its higher oxygen content, it can successfully be used as a substitute for traditional fuel additives like methyl tert-butyl ether.

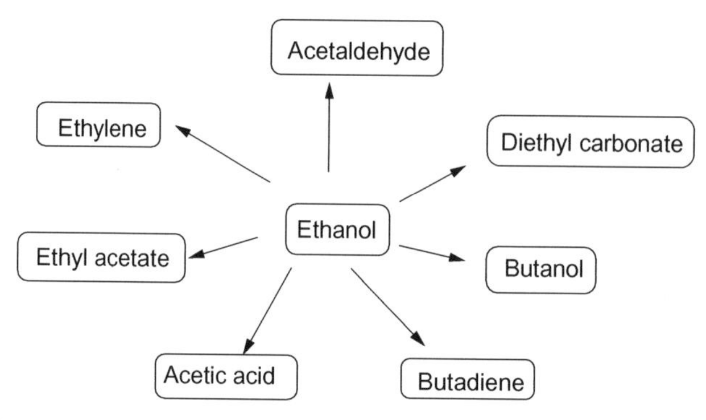

FIG. 10.3 Ethanol transformation in chemical platforms.

An efficient method for DEC production from ethanol is ethanolysis of urea, which consists of two steps: ethyl carbonate intermediate formation from ethanol and urea and the second reaction between ethyl carbonate and ethanol to DEC production.

Initially, as catalysts for this synthesis, metal oxides based on ZnO were used; but, these catalysts have the disadvantage of dissolving and forming a homogeneous complex during reaction. They were substituted with layered double hydroxides with versatile and adjustable chemical composition and structure (Pang et al., 2016).

Bioethanol obtained from renewable resources by fermentation is an excellent automobile fuel blended with gasoline. However, *n-butanol* is more suitable as an additive to gasoline than ethanol due to its less hydrophilicity and much closer energy density to gasoline.

Ethanol conversion to *n*-butanol includes three steps: dehydrogenation to acetaldehyde, condensation to crotonaldehyde, and hydrogenation to butanol. To achieve success in all three steps, multifunctional catalysts with activity in dehydrogenation, condensation, and hydrogenation need to be developed.

The Guerbet reaction for *n*-butanol production from ethanol also consists of three steps: ethanol dehydrogenation to acetaldehyde, acetaldehyde aldol condensation, and hydrogenation of the aldol adduct to form *n*-butanol.

A series of solid acid/base catalysts with active sites for both dehydrogenation and condensation has been investigated, and among them are Mg/Al oxides (Carvalho, De Avillez, Rodrigues, Borges, & Appel, 2012; Ramasamy, Gray, Job, Smith, & Wang, 2016) and transition-metal-functionalized catalysts (Apuzzo, Cimino, & Lisi, 2018; Sun et al., 2017). The presence of transition metals as active sites plays an important role, as they influence not only the dehydrogenation of ethanol but also the hydrogenation step, which is an extremely important reaction in this process.

Carbon materials that are used as a support for good dispersion of nanometals (Huang, Fung, Wu, & Jiang, 2020) have the advantages of high stability and good activity even at low temperatures and high selectivity to butanol.

In ethanol conversion to *1,3-butadiene*, the reaction involves two steps: the first is ethanol dehydrogenation to acetaldehyde and the second is acetaldehyde reaction with an ethanol molecule to form butadiene; this process is economically uncompetitive as butadiene obtained from petroleum sources using this method is approximately 95%. The recent availability of ethanol, with low cost, has led to an increase in interest in this type of butadiene production.

The catalytic systems used for this reaction included $MgO–SiO_2$, $ZrO_2–SiO_2$, and zeolite catalysts, with a high selectivity to butadiene (70%–90%) (Huang et al., 2020). The key role of all studied catalytic systems is represented by the balance between acid and basic sites. For example, in metal oxides supported on silica, the presence of weak basic sites is responsible for the rate-limiting stage, namely, ethanol dehydrogenation to acetaldehyde, whereas the presence of strong basic sites promotes aldol condensation of acetaldehyde. A series of metals (Zn, Cr, Cu, Mn, Ni) was added to modify the ratio between acid and basic sites. The conversion of acetaldehyde to 1,3-butadiene is affected especially by the weak Lewis acid sites; their concentration increases, leading to a decrease in 1,3-butadiene selectivity due to ethanol dehydration to ethylene and diethyl ether.

In addition, in zeolite materials, the presence of acid sites promotes ethanol dehydration to ethylene and subsequently oligomerization or cracking reaction to yield higher hydrocarbons.

Ethanol can be converted into other gaseous products such as *hydrogen*, with fuel cell application, by steam reforming, partial oxidation, and decomposition. The hydrogen yield is high in ethanol steam reforming, but the reaction requires a high operation temperature. Partial oxidation of ethanol is an exothermic reaction, but its control is difficult because of simultaneous combustion reactions, thus decreasing the yield of hydrogen.

The decomposition of ethanol is not a common reaction for hydrogen production due to undesirable byproduct formation.

Ethanol conversion leads to other gaseous products such as ethylene by dehydration, isobutylene by dimerization, and metathesis of ethylene and also to liquid products such as benzene, toluene, and xylene.

For ethanol conversion in *isobutylene* (Apuzzo et al., 2018; Sun et al., 2017) bifunctional Zn–Zr oxides were found to be highly selective. The presence of Lewis acid–base site pairs on the surface and the redox properties of the surface are responsible for catalyst selectivity.

Ethylene is used in many syntheses like polyethylene, polyethylene terephthalate, polystyrene, and ethylene glycol. Ethanol transformation to ethylene by dehydration has been evaluated using many types of catalysts such as H-ZSM-5, alumina, silica, clay, and phosphates.

In H-ZSM-5 catalysts, the yield of ethylene was up to 95% (Zhang, Wang, Yang, & Zhang, 2008), and silicoaluminophosphate (SAPO) catalysts doped with Mn and Zn had a yield of 98% ethylene (Chen, Li, et al., 2018). The strong acidity of zeolites and unequally distribution of active sites could determine the obtained C_4 byproducts. The ethanol conversion and ethylene selectivity on the doped SAPO catalysts decreased from SAPO-34 to SAPO-11 and from the samples doped with Mn to the samples doped with Zn.

Another utilization of ethanol includes its conversion into *propylene* using In_2O_3 catalysts (Huang, Cao, et al., 2020). The reactions (Eqs. 10.1–10.3) implied are:

$$CH_3-CH_2-OH \rightarrow CH_3-CH=O+H_2\,(\text{ethanol dehydrogenation}) \tag{10.1}$$

$$CH_3-CH=O+CH_3-CH=O \rightarrow CH_3COCH_3+CO_2+H_2O\,(\text{acetaldehyde coupling}) \tag{10.2}$$

and the final reaction is acetone transformation in propylene through an isopropoxy intermediate:

$$CH_3COCH_3 \rightarrow CH_2=CH-CH_3+O(s) \tag{10.3}$$

Ethanol coupling is other important reaction for reducing dependence on fossil resources. Derivatives of ethanol C–C coupling are propylene, *n*-butanol, and ethyl acetate.

Ethyl acetate is mainly synthesized from acetaldehyde disproportionation, whereas butanol is obtained from Guerbet condensation. These reactions require acetaldehyde as an intermediate, which in turn is obtained by ethanol dehydrogenation.

In ethanol coupling, a series of catalysts has been studied, especially multifunctional catalysts, which contain acidic and basic centers and also redox centers for dehydrogenation. The catalyst design must take into consideration the complexity of reactions implied in ethanol coupling such as dehydrogenation, dehydration, aldolization, and Tishchenko reactions.

The reactions implied in ethanol coupling are shown in Fig. 10.4. The distribution of products can be controlled by modification of surface acid–base composition.

Among the studied catalysts, perovskite-type materials such as $LaFeO_3$ and $LaMnO_3$ were found to be highly selective to C_4 and C_5 products like butanol, 2-pentanone, and ethyl butyrate (Chen, Yu, et al., 2018; Tesquet et al., 2016; Yu, Chang, Chung, & Lin, 2019). In perovskites, aldolization activity has been improved by introduction of silica as a support that induces acid–base pairs.

Many studies have attempted to improve the catalyst behavior in ethanol conversion. The most active catalysts for these applications were based on Co and Ni (Barakat, Motlak, Elzatahry, Khalil, & Abdelghani, 2014; Karim, Su, Engelhard, King, & Wang, 2011; Martono & Vohs, 2011). However, cobalt has the disadvantage of higher reactivity toward oxygen, resulting in an unstable Co–CoO_x couple, whereas

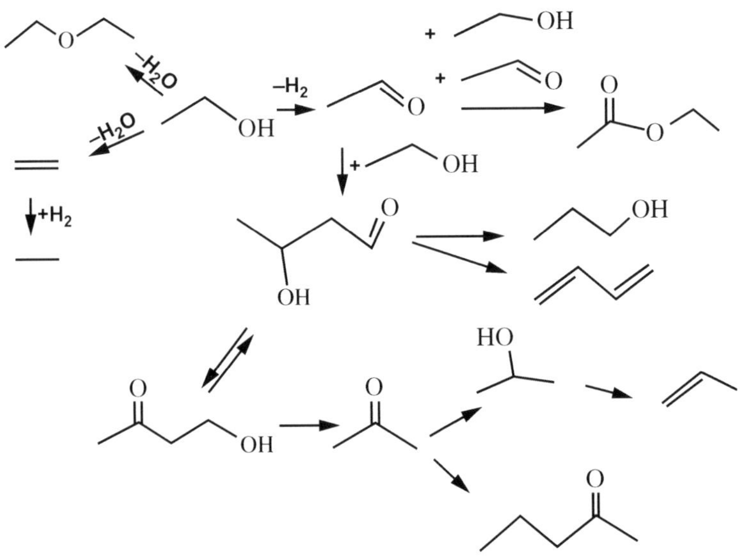

FIG. 10.4 Reactions implied in ethanol coupling.

catalysts based on Ni show lower reactivity toward oxygen, but they sinter and the quantity of coke that agglomerates on their surface is high, thus contributing to deactivation. The sinterization process could be reduced by utilization of a support, such as highly porous silica, which strongly interacts with the metal.

Propanol and isopropanol

The most important products obtained from isopropanol and propanol conversion are illustrated in Fig. 10.5.

Isopropanol can be dehydrated to *propene*, mainly used for propylene and diisopropyl ether synthesis, which is used as a fuel additive. The catalyst used at the industrial scale is γ-alumina due to its low cost. Lewis acid sites and basic sites on the surface have an important role in dehydration, as they are strong enough for dehydration (Kwak, Mei, Peden, Rousseau, & Szanyi, 2011; Larmier et al., 2015). Alkene was formed on the (100) facets of γ-Al_2O_3; however, the orientation of dehydration to alkene or ether depends on the relative stability of adsorbed intermediates on the surface as a function of temperature.

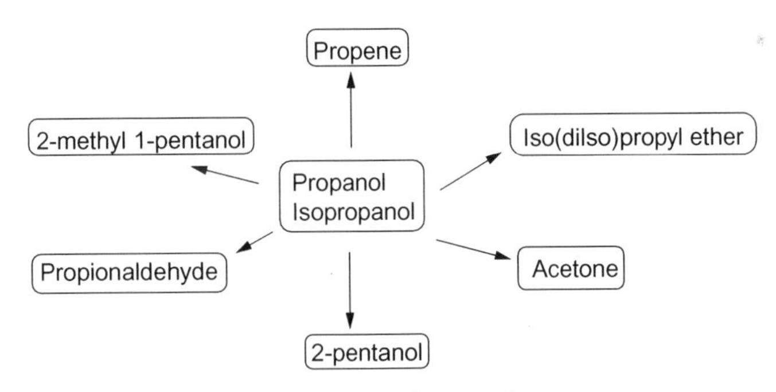

FIG. 10.5 Products obtained from propanol and isopropanol conversion.

Pentacoordinated Al presence on the surface, as a Lewis acid site, and the presence of basic species in its vicinity lead to direct formation of propylene or diisopropyl ether, with the catalyst activity depending on the amount of (100) facets exposed. Selectivity could also be improved by optimization of reaction conditions like temperature.

Heteropoly acids on η-Al_2O_3 were studied for isopropanol conversion to propene and diisopropyl ether (Guseinova, Adzhamov, & Yusubova, 2020). As heteropoly acids are thermally stable and easily regenerable, they could be used for multiple reactions during isopropanol transformation. The advantage of these catalysts is represented by their bifunctionality with oxidative and acidic properties at the same time. However, they have a small surface area; that is why they need to be deposited on a support like Al_2O_3, SiO_2, TiO_2, or ZrO_2. The polymorph η-Al_2O_3 has the largest surface area (η-Al_2O_3 > γ-Al_2O_3 > δ-Al_2O_3 > θ-Al_2O_3) and hence it is chosen as a support for heteropoly acids. The isopropanol conversion on phosphomolybdic heteropoly acids supported on η-Al_2O_3 is between 42% and 59%, propylene yield is 30%–40%, and diisopropyl ether yield is under 10%. For these catalysts, strong acid/base Lewis sites are responsible for propylene formation.

Other promising materials for isopropanol conversion are mixed metal oxides with perovskite-type structures due to their high thermal stability and good capacity to store oxygen, but, just like heteropoly acids, they suffer from a low surface area ($<10\,m^2/g$). In order to enhance their surface area, mesoporous materials like SBA-15 or MCM-41 have been used for perovskite dispersion (Phan, Nikoloski, Bahri, & Li, 2019; Wang et al., 2013).

However, zeolite utilization as a support leads to an increase in acidity; a too high acidity is a disadvantage for many reactions because coke deposition is increased. In isopropanol transformation, acidity is responsible for its dehydration to propene, whereas the presence of larger domains of perovskite on the surface is responsible for isopropanol dehydrogenation and oxidation to acetone (Khalil, Elhamdy, & Said, 2020).

Isopropanol can be directly alkylated with ethanol to *alcohols* and *ketones* with 5 or 7 carbons. The results obtained regarding conversion and selectivity, for individual alcohols and for mixture, on the Au-Cu/Al_2O_3 catalysts (Zharova et al., 2017) are shown in Table 10.1.

The condensation products in self-ethanol condensation are 1-butanol, 1-hexanol, and traces of 1-octanol. The self-condensation of isopropanol leads to acetone and propane as the primary

TABLE 10.1 Isopropanol and ethanol alkylation on Au-Cu/Al_2O_3 catalysts.

	EtOH	iPrOH	EtOH–iPrOH (mixture)
Conversion (%)	33.4	31.2	64.6–49.3
Selectivities (%)			
1-Butanol	74.4		12.3
1-Hexanol	17.8		1.4
Propane		12.7	1.5
Acetone		83.2	6.7
2-Pentanol			30.6
2-Pentanone			13.1
4-Heptanone			8.8

products. In the alkylation of isopropanol with ethanol, the total selectivity of alkylation products was between 64% and 70%, with the most important primary byproducts being diethyl ether (3%), 1-butanol (12%–14%), and acetone (7%–8%). Only traces of aldehyde were observed, concluding that it was rapidly converted into 1-butanol, while acetone that is less active was evidenced in the reaction products. The main product observed in the alkylation mixture of ethanol and isopropanol was 2-pentanol (30%–40%). Unsaturated alcohols and ketones were not evidenced in the reaction products.

Selective oxidation of isopropanol to *acetone* has been investigated on Au/α-Fe_2O_3 nanosheets (Zhang et al., 2008). There are three methods for acetone synthesis: grain fermentation, but the cost of this process is too high; as a byproduct in cumene oxidation (prepared from benzene and propene), with this being the main method; and selective oxidation of isopropanol as a supplementary process. However, on Au/α-Fe_2O_3 nanosheets, besides acetone, some traces of reaction products, including acetaldehyde, acetic acid, propylene, isopropyl ether, and 2,4-dimethyl furan, have been obtained. The yield in acetone is in the range of 31%–47%.

Propanol condensation to higher alcohols by Guerbet reaction has been studied using MgO as the catalyst (Ndou & Coville, 2004). With this catalyst, *propionaldehyde* and *2-methyl-1-pentanol* have been obtained with a selectivity of 35% and 45%, respectively.

The reactions (Eqs. 10.4–10.7) implied in propanol self-condensation are:

$$CH_3-CH_2-CH_2-OH \rightarrow CH_3-CH_2-CH=O+H_2 \tag{10.4}$$

$$2CH_3-CH_2-CH=O \rightarrow CH_3-CH_2-CH(OH)-CH(CH_3)-CH=O \tag{10.5}$$

$$CH_3-CH_2-CH(OH)-CH(CH_3)-CH=O \rightarrow CH_3-CH_2-CH=C(CH_3)-CH=O+H_2O \tag{10.6}$$

$$CH_3-CH_2-CH=C(CH_3)-CH=O+2H_2 \rightarrow CH_3-CH_2-CH_2-CH(CH_3)-CH_2-OH \tag{10.7}$$

In the first step, propanol is dehydrogenated to propionaldehyde and hydrogen, which is further required for the hydrogenation step. Propionaldehyde then undergoes an aldol condensation reaction in order to form a new C—C bond, which is rapidly converted to 2-methyl-1-pentanal. The presence of hydrogen in the reaction mixture determines the reaction product, 2-methyl-pentanal or 2-methyl-1-pentanol.

Butanol

Butanol is also a bioplatform for high value-added chemical synthesis since it could be obtained from biomass fermentation. It can be transformed into aldehyde, acid, and alkene by dehydrogenation, oxidation, or dehydration (Fig. 10.6).

Catalysts used for butanol conversion include zeolites (Phung, Proietti Hernández, Lagazzo, & Busca, 2015) heteropoly acids (Pylinina & Mikhalenko, 2011), or molybdenum/vanadium oxides (Bagheri & Muhd Julkapli, 2017). The difference in activity and product selectivity is related to the nature of active species on the surface. The possible reaction scheme for *n*-butanol conversion is shown in Fig. 10.7.

Dehydration of *n*-butanol takes place through a parallel consecutive manner that consists of *1-butene*, *2-butene*, and *dibutyl ether* formation. 2-Butene could be obtained from both 1-butanol dehydration and double-bond isomerization reaction. The isomerization of double bond, with isobutene formation, has not been observed at low temperatures.

Acid strength has a major influence on alcohol dehydration to alkenes. As is already known, for zeolites, the Si/Al ratio influences the acid strength, which is correlated with a number of Al atoms. An increase in Al content

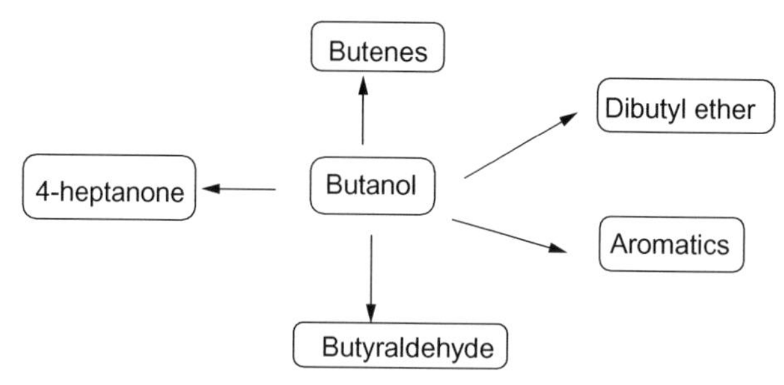

FIG. 10.6 Butanol as a bioplatform for high value-added products.

leads to a decrease in activity toward dehydration.

Dibutyl ether (DBE) is another product obtained from dehydration of *n*-butanol, which is used as a solvent and as a fuel additive since its oxygen content is reduced compared that of *n*-butanol. The ether can be obtained with high selectivity using heteropoly acids. The important petrochemicals obtained from *n*-butanol are paraffins (C_1–C_8), olefins (C_2–C_6), and aromatics (C_7–C_{10}) (Varvarin, Khomenko, & Brei, 2013).

The olefins, C_6–C_{12}, obtained in the first step, undergo further cyclization reaction with cyclic paraffins and are then transformed into aromatics by dehydrogenation (Palla, Shee, & Maity, 2016). Selectivity to xylenes and ethyl benzene depends on the reaction conditions, with observation that m- and o-xylenes are more readily isomerized than p-xylene.

Zirconia-supported Cu systems have proved to be promising catalysts for *n*-butanol conversion to *butyraldehyde* (Requies et al., 2012). Butyraldehyde is used as an accelerator in rubber vulcanization, as a solvent, and as an intermediate in fine chemical production. Its production from alcohol oxidation or dehydrogenation is a more attractive method compared with its production from alkene hydroformylation.

The catalyst with the highest Cu dispersion exhibits the best catalytic performance in *n*-butanol dehydrogenation. An increase in copper content leads to sinterization of the active phase due to less interaction between the active phase (copper) and the zirconia support. The catalyst was active for 25h and after that the activity showed a small decrease. The copper loading and its dispersion play a key role in the catalyst's stability, activity, and selectivity. The partial oxidation of *n*-butanol to butyraldehyde is influenced by the copper content; the catalyst with the highest content of copper exhibits the best catalytic behavior. It was observed that although the catalysts are

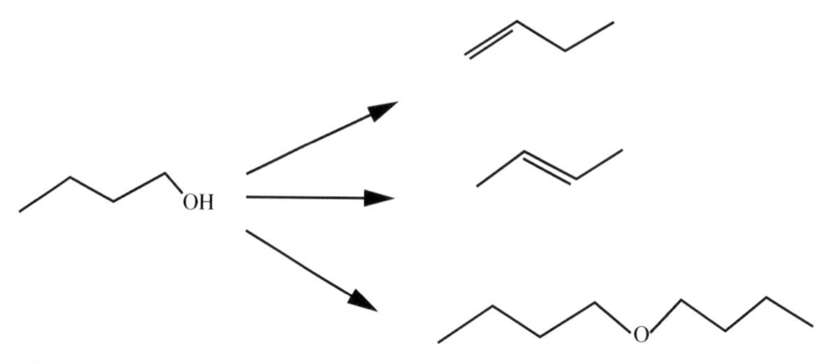

FIG. 10.7 Products from *n*-butanol dehydration.

active for both reactions, selectivity to butyraldehyde from *n*-butanol partial oxidation is higher than that from dehydrogenation reaction.

Zinc ferrite with a spinel structure is an active catalyst for both *n*-butanol dehydrogenation and bimolecular condensation to ketone (Klimkiewicz et al., 2009).

Primary alcohol ketonization in the gas phase using heterogeneous catalysts represents an interesting method to obtain ketone in addition to reaction in the liquid phase. The directions of conversion depend on the active centers' configuration.

One of the reaction mechanisms assumes the alcohol transformation to aldehyde, followed by aldehyde transformation in hemiacetal, ester, β-ketoester, and, finally, in ketone. The other mechanism assumes the aldol participation or the intermediate form of carboxylic acid participation.

Butanol ketonization to 4-heptanone has been successfully implemented in other catalytic systems such as molybdenum supported on alumina coated with active carbon (Mitran, Saab, Charisiou, Polychronopoulou, & Goula, 2020). Carbon-based catalysts are interesting materials especially as coverage for supports like alumina to prevent coke deposition.

Alcohol ketonization to 4-heptanone is influenced by molybdenum dispersion, with the catalyst with the highest dispersion being the most selective. The agglomeration of molybdenum on the surface leads to the production of butyraldehyde, but its further condensation to ketone is inhibited.

10.3 Diols

1,4-Butanediol

From 1,4-butanediol, a series of important chemicals such as γ-butyrolactone (GBL), tetrahydrofuran (THF), polyester polyols, polyurethane, etc. (Raju et al., 2019) can be obtained.

Catalysts based on Cu (Cu/MgO, Cu/SiO_2, Cu/CeO_2, $Cu/ZnO–ZrO_2–Al_2O_3$) (Bhanushali et al., 2020; Reddy, Anand, Prasad, Rao, & Raju, 2011) and Au (Au/TiO_2, Au/Fe_2O_3) (Huang, Dai, Li, & Fan, 2007; Huang, Dai, & Fan, 2008) have been used for 1,4-butanediol dehydrogenation.

In *γ-butyrolactone* synthesis from 1,4-butanediol dehydrogenation, an important role is also played by the support. Supports with acidic nature such as Al_2O_3 and $Al_2O_3–SiO_2$ lead to tetrahydrofuran production, whereas those with basic properties like MgO and ZnO favor the production of γ-butyrolactone.

Catalysts based on copper as the active metal are one of the best catalysts for dehydrogenation due to their properties like surface area variation, particle size modification, and good interaction with the support.

γ-Butyrolactone is an important precursor for pyrrolidone derivative production and is produced from both 1,4-butanediol dehydrogenation and maleic anhydride hydrogenation. Maleic anhydride hydrogenation is a costly process and at the same time unfriendly to the environment due to the release of carbon dioxide as a side product. In addition, for this process, three molecules of hydrogen are required, making it expensive. The process that uses 1,4-butanediol is a green route since it is synthesized from biomass resources like glucose. Two hydrogen molecules are generated from this method, which makes the process more economical.

An interesting process consists of combining two reactions: one of dehydrogenation with one of hydrogenation. An example is represented by 1,4-butanediol dehydrogenation combined with benzaldehyde hydrogenation. Dehydrogenation of 1,4-butanediol is an endothermic process, whereas hydrogenation of benzaldehyde is an exothermic process.

After combining the two processes, the reaction becomes exothermic, contributing to a reaction economy, simultaneously with the production of two important molecules, namely, γ-butyrolactone and benzyl alcohol. The selectivities to products from simultaneous reactions are higher than those from individual reactions. The difficulty of this method includes product separation because of their close boiling points. This reaction can be performed on basic catalysts (Reddy et al., 2011).

One of the most interesting reactions is copolymerization of 1,4-butanediol with carbon dioxide to aliphatic polycarbonates. Polycarbonates are interesting materials due to their biodegradability and biocompatibility. The conventional method for their production is the carbonylation of diols with phosgene ($COCl_2$), but that is a hazardous compound. Carbon dioxide utilization is attractive from two points of view: it is nontoxic and it is an abundant renewable resource especially due to its increased emission into the atmosphere.

The copolymerization process of α,ω-diols with carbon dioxide was enabled by catalysts such as CeO_2 (Tamura et al., 2016), using 2-cyanopyridine as the dehydration agent. Besides 1,4-butendiol and CO_2, poly(butylene carbonate) synthesis requires massive amounts of solvents (methanol and tetrahydrofuran) at the laboratory scale. To improve these process, scenarios with less solvents and their evaporation were discussed and investigated.

Another reaction includes 1,4-butanediol dehydration to 1,3-butadiene using rare earth oxides (Wang et al., 2013) as catalysts. The route of 1,4-butanediol dehydration consists of its transformation in the vapor phase to 3-buten-1-ol in the first step and to 1,3-butadiene in the second step. In this reaction, it has been observed that the presence of weak basic sites leads to a high selectivity to 3-buten-1-ol, whereas the presence of acid sites has an important role in 1,3-butadiene production. Among the rare oxide catalysts studied, Yb-based ones are the most effective.

10.4 Carboxylic acids

Carboxylic acids obtained from biomass have various applications in the chemical, pharmaceutical, and food industries. The chemical structures of many carboxylic acids obtained from anaerobic fermentation are shown in Fig. 10.8.

Acetic acid

Photocatalytic decomposition of acetic acid to hydrogen and biogas was performed using Cu/TiO_2 as the photocatalyst (Amorós-Pérez, Cano-Casanova, Lillo-Ródenas, & Román-Martínez, 2017). Photodegradation was initiated in the presence of ultraviolet, visible, or infrared radiation. TiO_2 is well known due to its semiconducting properties, high thermodynamic stability, low cost, resistance to corrosion, and, finally, its crystalline structures, which determine high activity.

Titania has a drawback that it is active only in the UV region. To improve its efficiency, transition metals are introduced into the titania

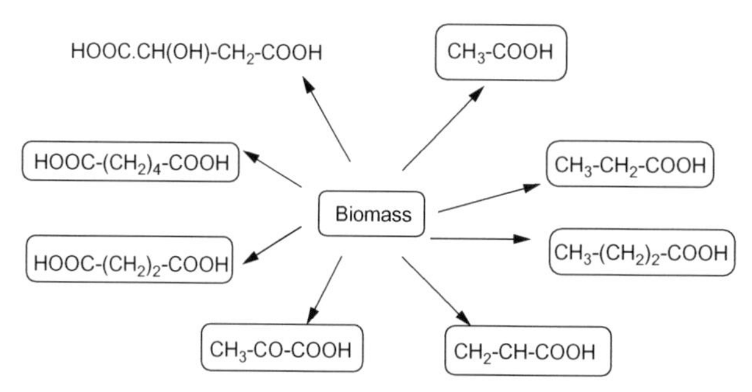

FIG. 10.8 Carboxylic acids obtained from anaerobic fermentation of biomass.

lattice, which acts as trapping sites by accepting electrons or holes from TiO_2, thus extending the catalyst's response to the visible light region.

Copper oxide incorporation into the titania lattice leads to an activity improvement in acetic acid conversion to hydrogen and hydrocarbons like methane. Hydrogen production from acetic acid can be achieved by steam reforming. Hydrogen is the best candidate for fossil fuel replacement and is produced by two routes from syngas or from bio-oil reforming.

Acetic acid represents 5%–10% of biomass, a significant amount of which is used in steam reforming reactions. The major inconveniences in the steam reforming process include side product formation and deposition of coke on the surface with deactivation of the catalyst.

Acetic acid steam reforming was carried out using a series of active metals such as Ni, Mg, Co, and Fe as catalysts over supports such as SiO_2, Al_2O_3, and La_2O_3 (Ozel, Meric, Arbag, Degirmenci, & Oktar, 2020).

Among them, Ni-based catalysts are one of the most efficient in steam reforming. However, these catalysts cannot prevent the formation of coke. Alkaline or alkaline earth metals (K, Ca, Mg) and rare earth metals (Ce, La) were added in order to prevent and suppress coke formation due to their oxygen storage capacity (Pu et al., 2018).

Catalysts with core–shell structures, synthesized by metal particle coating in porous silica materials, exhibit high catalytic performances in acetic acid steam reforming, providing a homogeneous distribution of metal in the silica structure.

Acetic acid steam reforming involves the following reactions (Eqs. 10.8–10.14):

$$CH_3-COOH+2H_2O \rightarrow 2CO_2+4H_2 \quad (10.8)$$

$$CH_3-COOH \leftrightarrow 2CO + 2H_2(\text{thermal decomposition}) \quad (10.9)$$

$$CO+H_2O \leftrightarrow H_2+CO_2(\text{water gas shift reaction}) \quad (10.10)$$

$$CH_3-COOH \leftrightarrow CH_4+CO_2(\text{decarboxylation}) \quad (10.11)$$

$$2CH_3-COOH \rightarrow (CH_3)_2CO+CO_2 + H_2O(\text{ketonic decarboxylation}) \quad (10.12)$$

$$2CO \leftrightarrow C+CO_2(\text{coke formation}) \quad (10.13)$$

$$CH_4 \leftrightarrow C+2H_2(\text{methane craking}) \quad (10.14)$$

The activity of 5% Ni/SiO_2 catalysts achieved 97% acetic acid conversion at 650 °C, and the selectivity to hydrogen was between 55% and 61%.

Acetic acid decomposition to hydrogen was evaluated using Co, Pt, and Pd as metals and Al_2O_3 and TiO_2 as supports (Brijaldo, Caytuero, Martínez, Rojas, & Passos, 2020). Acetic acid adsorption is different as a function of support nature. On the alumina support, a good dispersion of species on the surface of Al_2O_3 was evidenced and adsorption was realized as acetate species, whereas on the titania support, acetic acid was adsorbed as molecular acid species.

The performance of transition metals (Co, Ni) and noble metals (Pt, Pd) is presented in Table 10.2.

Reaction conditions: total flow = 100 mL/min (5% CH_3COOH, 95% He); 700 °C, 0.1 g catalyst.

This behavior leads to the conclusion that noble metals are more active in acetic acid decomposition than transition metals due to their capacity to adsorb dissociative acetic acid species favoring its decomposition.

An important utilization of acetic acid is as an acylation agent. It can participate in esterification reactions with different alcohols. The most important is its reaction with butanol with butyl acetate production, which has important applications such as its use as a solvent, an essence, and an intermediate in organic synthesis (Dash & Parida, 2007; Hu et al., 2020).

Among the solid acid catalysts studied in this reaction, zeolites and transition-metal oxides

TABLE 10.2 Acetic acid steam reforming on transition and noble metals.

Catalyst	Acetic acid conversion (%)	Hydrogen selectivity (%)
Pt/Al_2O_3	80.4	51.3
Pd/Al_2O_3	79.0	25.4
Co/Al_2O_3	61.7	19.9
Ni/Al_2O_3	35.4	11.0

(V, Mo, W) (Mitran, Yuzhakova, Popescu, & Marcu, 2015) have been reported. Materials with a metal–organic framework (MOF) represent other samples studied in acetic acid esterification. They have the advantage of high surface area, specific structure, and morphology but are not used at the industrial scale because of their poor chemical stability and low Brønsted acidity. Other disadvantages include a complex synthesis method that requires utilization of expensive functionalized ligands.

Acetic acid hydrogenation to ethanol represents another of its important applications. The conversion of syngas to methanol, methanol transformation to acetic acid, and subsequent transformation of acetic acid to ethanol by hydrogenation represent a promising method to valorize syngas in order to prevent carbon loss. The major inconvenience of this method is that the reaction takes place in several steps, but the selectivity has a high level compared to those of other methods, making it a more promising process.

The reactions implied in acetic acid hydrogenation (Eq. 10.15) are:

$$CH_3COOH + 2H_2 \rightarrow C_2H_5OH + H_2O \quad (10.15)$$

and the side reactions (Eqs. 10.16–10.18) are:

$$CH_3COOH + C_2H_5OH \leftrightarrow CH_3COOC_2H_5 + H_2O \quad (10.16)$$

$$CH_3COOH + H_2 \rightarrow CH_3CHO + H_2O \quad (10.17)$$

$$CH_3COOH + 4H_2 \rightarrow 2CH_4 + 2H_2O \quad (10.18)$$

Supported noble metals (Rachmady & Vannice, 2002; Rakshit, Voolapalli, & Upadhyayula, 2018) were used as catalysts for acetic acid hydrogenation. Acetic acid molecules are adsorbed on the support and hydrogenated with hydrogen dissociate on the surface. Acetic acid conversion strongly depends on a number of factors such as the ability of the metal to dissociate the hydrogen molecule, the capacity of the support to adsorb the acetic acid molecule, the acidity of support, and the metal dispersion on the support. On the other hand, the catalyst must have the capacity to suppress the side reactions.

Pt has a high capacity to dissociate hydrogen, while oxide supports with the capacity to donate protons are active for the esterification reaction of acetic acid with ethanol leading to ethyl acetate. Catalysts with Lewis acidity lead to a competition between esterification reaction and hydrogenation toward ethanol. To improve catalyst stability and inhibit the side reactions, a second metal like Co, Ni, Sn, or Cu has been added.

The favorable route to ethanol formation involves acetyl formation, its hydrogenation to acetaldehyde, and finally the hydrogenation of acetaldehyde to ethanol, with the first reaction being the rate-determining step.

Pt-Sn/Al_2O_3 catalysts modified with K have been studied in this reaction (Zhou, Zhang, Ma, & Ying, 2016). In general, bimetallic catalysts have different reaction characteristics due to the interaction between the two components. Sn can improve the dispersion of Pt and the interaction between Pt and the support or promoter. Introduction of alkali metals as a promoter has been frequently done, as it can promote the reduction of support acidity due to its basic properties. The ionic radius of alkali metals plays an important role in the decrease in support acidity (K > Na > Li) (Siri, Bertolini, Casella, & Ferretti, 2005).

Acetic acid conversion is low as it is influenced by the K content, whereas ethanol selectivity increases with the K amount. The blocking effect of K on acid centers inhibits the esterification reaction and also the subsequent dehydration of ethanol to ethylene.

Propanoic acid

Propanoic acid, an important component of bio-oils, could be transformed into a lot of products, as shown in Fig. 10.9.

Propanoic acid is a constituent of bio-oils, with its mass ratio between C, H, and O being highly similar to that of bio-oil, which makes it the subject of a steam reforming process.

The mechanism of oxygenated compound steam reforming is similar to that of hydrocarbon steam reforming (Czernik, Evans, & French, 2007). Organic molecules are dissociatively adsorbed on the active centers, whereas water molecules are adsorbed on the support. Organic molecules are dehydrogenated to obtain hydrogen and react with the hydroxyl group from the support surface with carbon oxide formation. Compounds with oxygen are more reactive than hydrocarbons, and they are reactive even at low temperatures.

Propanoic acid steam reforming implies the following reactions (Sahebdelfar, 2017) (Eqs. 10.19–10.22, and 10.10):

$$CH_3-CH_2-COOH+4H_2O \leftrightarrow 2CO_2+7H_2(\text{steam reforming}) \quad (10.19)$$

$$CH_3-CH_2-COOH+H_2O \leftrightarrow 3CO+4H_2(\text{steam reforming}) \quad (10.20)$$

$$CO+3H_2 \leftrightarrow CH_4+H_2O(\text{methanation reaction}) \quad (10.21)$$

$$CH_3-CH_2-COOH \leftrightarrow 2CO+3H_2+C(\text{acid decomposition}) \quad (10.22)$$

From thermodynamic equilibrium calculation, it has been evidenced that propanoic acid is completely reformed even at low temperatures, but high temperatures are required in order to reduce the deposition of coke on the catalyst surface.

The production of ketones from carboxylic acids is a clean method for carboxylic acid valorization.

Propanoic acid *ketonic decarboxylation* represents another method to transform it, and

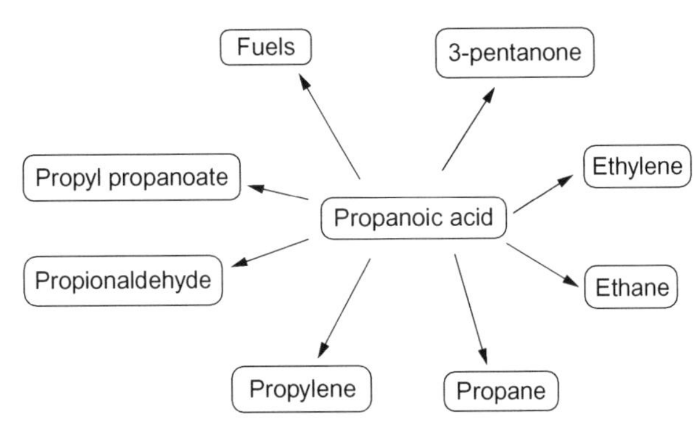

FIG. 10.9 Propanoic acid conversion to hydrocarbons and oxygenates.

beta-zeolites (Mitran, Chen, Dolge, Huang, & Seo, 2021) were used as catalysts in this case. With these catalysts, it has been observed that the ratio of Lewis acid sites to Brønsted acid sites plays an important role in *3-pentanone* production, with the presence of sites with excellent strength being favorable to propanol production, the proton from a Brønsted acid site being implied in this reaction, and deactivation of the catalyst being observed due to modification of the catalyst surface.

Propanoic acid ketonization has also been investigated using transition-metal oxides (Ce–Zr) (Ding, Wang, Han, Zhu, & Ge, 2018) as catalysts. Lewis acid–base pairs and metal cations that were coordinatively unsaturated were considered active centers for ketonization in these kinds of catalysts. Among the studied catalysts, Ce-based oxides are the most active, and addition of a second metal like Mn, Fe, Ni, or Cu enhances the activity of ceria catalysts. Ce–Zr oxides have been evidenced as the most selective to 3-pentanone, as they are more active than simple oxides such as CeO_2, ZrO_2, or TiO_2, which have high lattice energies.

The crystal structure also plays an important role in ketonization, especially due to the presence of both oxygen vacancies and coordinatively unsaturated metal cations.

The reaction rate strongly depends on the medium strength acid/base ratio and has a maximum value at a ratio equal to unity; the balance between the two types of centers is extremely important.

The presence of monodentate or bidentate carboxylates on the surface is crucial, as they are the precursors of enolate species formation and C–C coupling, and the reaction rate increases when monodentate concentration is high. Propanoic acid is adsorbed on Lewis acid sites through carbonyl and is subsequently dehydrogenated in the α-position by a vicinal base center with enolate formation. The enolate species are extremely active; they attack a new molecule of propanoic acid and thus a β-keto carboxylate is formed. Furthermore, the production of 3-pentanone is realized by decarboxylation and dehydration of these species.

Deoxygenation (ketonization) of propanoic acid has also been studied on bifunctional heteropoly acids (Alotaibi, Kozhevnikova, & Kozhevnikov, 2012). Dehydration of propanoic acid at about 500°C, with methyl ketene formation, has been observed using these catalysts.

Methyl ketene is formed from the following reaction (Eq. 10.23):

$$CH_3-CH_2-COOH \rightarrow CH_3-CH=C=O+H_2O \quad (10.23)$$

At the same time, propanoic acid could be simultaneously decarboxylated to CO_2 and ethane, according to the following reaction (Eq. 10.24):

$$CH_3-CH_2-COOH \rightarrow C_2H_6+CO_2 \quad (10.24)$$

Methyl ketene is further decomposed to ethane and CO as follows (Eq. 10.25):

$$CH_3-CH=C=O \rightarrow C_2H_4+CO \quad (10.25)$$

However, the principal product of propanoic acid decomposition is 3-pentanone from partial deoxygenation (ketonization), presented in the reaction (Eq. 10.26):

$$CH_3-CH_2-COOH \rightarrow (CH_3-CH_2)_2CO+CO_2+H_2O \quad (10.26)$$

Carboxylic acids can be partially or totally deoxygenated by hydrogenation, given that this is an environmentally friendly method compared to aldehyde production. The reaction is as follows (Eq. 10.27):

$$CH_3-CH_2-COOH+H_2 \rightarrow CH_3-CH_2-CHO+H_2O \quad (10.27)$$

Heteropoly acids have extremely strong Brønsted acidity, and they successfully catalyze C=C and C=O hydrogenation.

Molybdenum carbide doped with other metals is an active catalyst in propanoic acid hydrodeoxygenation (Lu et al., 2018). Three types of products, namely, hydrocarbons with three atoms of carbon (propane, propene), hydrocarbons with two atoms of carbon (ethane, ethylene), and propionaldehyde can be produced on undoped molybdenum carbide (Mo_2C).

The introduction of a second metal such as Ni contributes to the appearance of 3-pentanone among reaction products. Interestingly, the introduction of Ca as a second metal into the lattice of the molybdenum carbide catalyst results in a different behavior compared with both undoped and doped molybdenum carbide with Ni catalysts. The catalyst is almost inactive at low temperatures; only when the temperature is increased does the conversion start.

Propionaldehyde is produced at low temperatures, with 3-pentanone being the predominant product at high temperatures. Catalysts with molybdenum lose their hydrogenation function with time, but they maintain their ketonization function.

Gas-phase ketonization of propanoic acid over oxides of 32 chemical elements was also studied (Gliński, Zalewski, Burno, & Jerzak, 2014). Simple oxides of Ag, Bi, Cd, Cu, Pb, Re, Mg, Zn, Cr, Ga, and Sn have been studied in order to compare their activity.

In the first group of studied metals, the yield to 3-pentanone was as follows:

$$Cd > Bi > In > Pb > Cu > Ag > Re$$

In the second and the third studied groups, it was observed that the main group metals had a lower activity compared with transition elements, except Cr and Zn. The fourth group was composed of Mn, Zn, and Ce oxides, which were extremely active.

Oxides are supported on alumina and silica and a much stronger interaction between oxides and alumina has been evidenced compared with silica; this is reflected in the degree of oxide reduction, which is lower on alumina than on silica. Catalysts supported on titania show a lower 3-pentanone yield.

Carboxylic acid hydrogenation leads to alcohols, which can be used in the production of fragrances, detergents, and pharmaceuticals. The alcohols obtained can be derivatized by selective oxidation, dehydrogenation, and dehydrogenative coupling, leading to intermediates for fine chemical synthesis. However, carboxylic acid hydrogenation is a difficult reaction because of its thermodynamic and kinetic stability.

Propanoic acid can be reduced to *1-propanol* using bifunctional heterogeneous catalysts such as Ru-based materials promoted by molybdenum oxide (Chen, 2012). The Ru–Mo phases have a role in stabilizing the propanoyl intermediate. On the Ru sites, the C—C bond adjacent to carbonyl is cleaved, whereas on both Mo and Ru sites, propanoyl is formed as an intermediate and is further rapidly transformed into propanol.

Heteropoly acids doped with Cs were used as catalysts for gas-phase hydrogenation of propanoic acid (Benaiss, 2008). They catalyze a series of reactions due to the presence of both acid and redox properties. Although the substitution of Mo with V in heteropoly acids affects the redox properties and polyanion stability, it has no effect on the activity in propanoic acid hydrogenation. The strong acidity has an influence on catalyst deactivation by coke deposition.

Propane, propionaldehyde, and ethane are the main products, and no 3-pentanone was evidenced. The introduction of Cs enhances the catalytic activity and apparition of 3-pentanone has been observed.

Propanoic acid hydrogenation over Pt/TiO_2 and $Pt\text{-}Re/TiO_2$ (Lawal, Hart, Daly, Hardacre, & Wood, 2019) at 150 °C proceeds with a conversion of 47% and liquid products such as propanol (73% selectivity) and propyl propionate (27%) are obtained. An increase in conversion of about 90% has been observed with an increase of temperature at 20 °C but to the detriment of propanol selectivity that decreased to 62%. The introduction of Re as a promoter improved both propanoic acid conversion and propanol selectivity.

Butanoic acid

Butanoic acid has many applications such as its use as a solvent in the chemical and polymer industry, as a precursor to biofuels in the pharmaceutical industry, and has anticancer effects. Butanoic acid can be catalytically upgraded to aldehydes, alcohols, and alkanes by dehydration, hydrogenolysis, and hydrogenation. It can also be subjected to a series of C–C coupling reactions like ketonization and esterification.

In hydrogenation reactions using heterogeneous catalysts by adjusting the catalyst composition and operation conditions, the selectivity can be also adjusted to aldehyde, ester, or alcohol.

Butanoic acid transformation in a series of products is presented in Fig. 10.10.

Butanoic acid, in contrast to propanoic and acetic acid, is more resilient to degradation in the presence of microorganisms. Consequently, other methods for its degradation like photocatalytic processes need to be found. In the presence of $Cu_2O/BiWO_6$ (Zheng et al., 2017), butanoic acid can be decomposed in alkane and hydrogen. Bismuth-based materials are active photocatalysts for a series of reactions, given that they are n-type semiconductors with a unique layered structure that is favorable to charge transfer between electrons and holes. However, this quick recombination of electrons with holes is a disadvantage in photocatalytic reactions. In order to enhance its activity, another oxide such as a p-type semiconductor is used due to formation of n–p heteroconjunction and to generate electron–hole pairs.

An interesting example of a p-type semiconductor is cuprous oxide, due to its low cost production and high capacity to generate hydrogen.

Butanoic acid decarboxylation has been carried out both in the gas and liquid phases. Some catalysts like those based on Zn, W, Mo, Sn, and Pb are active in the gas phase but are rapidly deactivated in the presence of water. Similarly, many supports like alumina are stable in an organic medium and in the gas phase but are deactivated under hydrothermal conditions. The catalyst deactivation in decarboxylation reaction represents a disadvantage in the use of this route to valorize butanoic acid.

Pt-based catalysts are active and selective in acid decarboxylation under organic and hydrothermal conditions, especially when supported on carbon (Yeh, Linic, & Savage, 2014). A carbon support has the disadvantage of loss of its pore volume due to the collapse of pores in material and, in addition, Pt is deactivated by poisoning in the presence of water and by coke deposition.

Deoxygenation of butanoic acid to hydrocarbons takes place by decarbonylation, decarboxylation, or hydrogenation as a function of catalysts used or as a function of the reaction conditions.

Molybdenum carbide (Mo_2C) catalysts (Shi, Yang, Li, & Jiao, 2016) have shown excellent catalytic activity and selectivity in deoxygenation reactions and are preferred instead of Pt-based catalysts due to their resistance to poisoning and as they are not sintered as is the case with noble metals. In molybdenum carbide, the

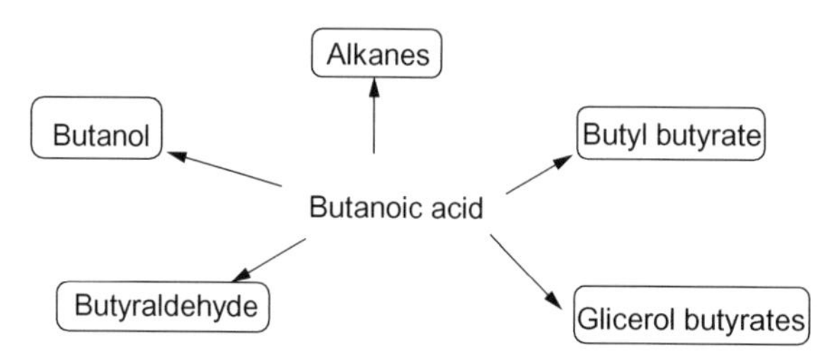

FIG. 10.10 Butanoic acid conversion to hydrocarbons and oxygenates.

carbon atoms are incorporated into the molybdenum interstitial sites, thus changing the electronic properties of hexagonal and orthorhombic Mo_2C phases. Each phase presents many facets containing different terminations as a function of preparation conditions, with the activity and selectivity of the catalyst being influenced by this structure.

On the basis of these facets, butanoic acid could be adsorbed as two stable configurations: in the first configuration, the carboxyl group is adsorbed on two molybdenum sites, with this adsorption being strong, and, in the second configuration, the carbonyl group is vertically adsorbed on the surface, with participation of the O atom from carbonyl that interacts with one molybdenum site. Butanoic acid can be transformed to butyraldehyde, butanol, or butane depending on the manner by which it binds with the surface.

Hydrogenation of butanoic acid can be achieved by four pathways:

(1) Dissociation of the OH group (Eq. 10.28):

$$CH_3-CH_2-CH_2-COOH+H \rightarrow CH_3-CH_2-CH_2-CO+H_2O \quad (10.28)$$

(2) Addition of hydrogen to the C=O group at the C atom (Eq. 10.29):

$$CH_3-CH_2-CH_2-COOH+H \rightarrow CH_3-CH_2-CH_2-CHO-OH \quad (10.29)$$

(3) Addition of H to the O atom from the C=O group (Eq. 10.30):

$$CH_3-CH_2-CH_2-COOH+H \rightarrow CH_3-CH_2-CH_2-C(OH)_2 \quad (10.30)$$

(4) Dissociation of the C=O group (Eq. 10.31):

$$CH_3-CH_2-CH_2-COOH+H \rightarrow CH_3-CH_2-CH_2-C-OH+OH \quad (10.31)$$

In the first step, the dissociation of carboxylic acid into R-CO + OH is favored over R-COO + H both thermodynamically and kinetically. In the second and third steps, for the hydrogenation of the C=O group bond, the preferred method is to form R-CHO(OH) instead of $RC(OH)_2$.

The first species adsorbed on the surface are R-CO and OH. Hydrogen is adsorbed onto R-CO with butyraldehyde formation. Butyraldehyde is further hydrogenated to butanol via R-CH_2-O as the intermediate. Butanol is then transformed to butane from R-CH_2 adsorbed on the surface.

Butanoic acid can be transformed to butanol via methyl butyrate. In the first step, esterification of butanoic acid to methyl butyrate occurs, and, in the second step, hydrogenolysis of methyl butyrate to butanol takes place. This method has a series of advantages such as low cost and availability of feedstock; thus, high-cost pretreatment steps are not required.

Pt–Co bimetallic catalysts have been evaluated as active catalysts for this reaction (Cho et al., 2019). Cobalt is a highly active metal for hydrogenolysis of the C—C bond and for hydrogenation of the C=O bond. Selectivity to the desired product could be improved by Pt addition.

The esterification of butanoic acid with glycerol has been studied using sulfated iron oxides as catalysts. Iron oxides are interesting catalysts due to their nontoxicity, availability, and low cost. The treatment of iron oxides with sulfuric acid has some inconveniences such as long preparation time and utilization of sulfuric acid for their sulfating. The advantage of the process is

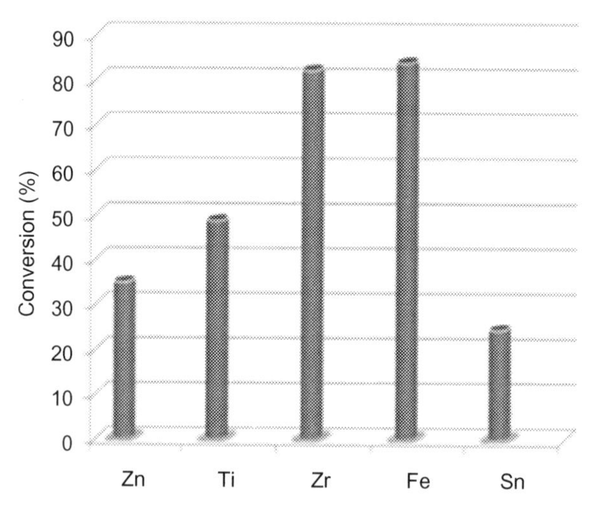

FIG. 10.11 Butanoic acid esterification with glycerol over sulfated oxides (time 6 h, $t = 95\,°C$, molar ratio of butanoic acid: glycerol is 3:5, catalyst loading 4 g/L).

an increase in the surface area compared initially with oxides. Other oxides, in addition to iron, have been sulfated (Zn, Ti, Sn, Zr), and the results obtained in butanoic acid esterification with glycerol are shown in Fig. 10.11 (Kaur, Wanchoo, & Toor, 2015). Sulfated iron oxides present the maximum value of conversion, followed by ZrO_2.

Succinic and adipic acids

Succinic acid is an ingredient in food that stimulates growth and is used in pharmaceuticals, bioplastics, and detergents. The active principles of drugs are succinate salts such as succinate tocopherol and hydrocortisone; monoethyl esters of succinic acid are used in diabetic treatment and monoesters with cellulose are characterized by their great capability to absorb water.

The chemicals generated from succinic acid are γ-butyrolactone, 1,4-butanediol, tetrahydrofuran, and fumaric acid (Fig. 10.12).

γ-Butyrolactone is used in the fine chemical industry as a solvent and in the pharmaceutical industry for *N*-methyl-2-pyrrolidone and 2-pyrrolidone synthesis. Conventionally, it is obtained from maleic anhydride hydrogenation, but its price is high and it is not environmental friendly. Succinic acid can successfully replace maleic anhydride as it is both inexpensive and biodegradable. Hydrogenation of succinic acid to γ-butyrolactone occurs in two consecutive reaction steps.

In the first step, reduction and dehydration of the C—O bond of succinic acid occurs, leading to succinic anhydride formation. In the second step, hydrogenation of the C=O bond of succinic anhydride leads to γ-butyrolactone formation. Succinic acid dehydration takes place on the acid sites, whereas C=O bond hydrogenation occurs on the metal sites. This is the reason why the reaction needs the presence of bifunctional catalysts, with both acidic and hydrogenation properties.

Therefore, for succinic acid hydrogenation, noble metals supported on alumina were chosen as catalysts due to the acid behavior of the support. For its hydrogenation, as for all other acids, noble metals (Pt, Pd, Ru) have been used successfully (Hong, Hwang, Seo, Lee, & Song, 2011).

1,4-Butanediol has been used for the synthesis of thermoplastic polymers such as polybutylene terephthalate or polybutylene succinate, whereas

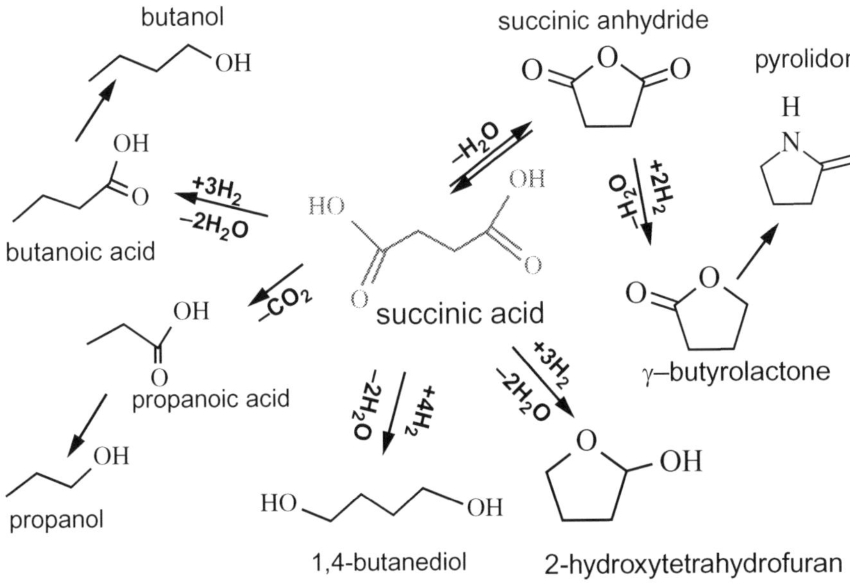

FIG. 10.12 Chemicals generated from succinic acid.

tetrahydrofuran has been used for polytetramethylene ether glycol production, as a solvent for PVC, or as a solvent in chemical reactions.

1,4-Butanediol is obtained from maleic anhydride hydrogenation, propylene oxide isomerization, and butadiene acetoxylation, but all these routes need petrochemicals derived from fossil fuels as raw materials. However, succinic acid hydrogenation to 1,4-butanediol represents an attractive route with respect to renewable resource utilization. The hydrogenation process takes place in two steps. Succinic acid is hydrogenated in the first step to γ-butyrolactone, and, by consecutive hydrogenation of γ-butyrolactone, 1,4-butanediol or tetrahydrofuran can be formed.

Hydrogenation of succinic acid on noble metals occurs under severe conditions, high reaction temperatures (>240 °C), and pressures (<80 bars). In hydrogenation processes, the metal plays an important role. As a function of catalytic nature, it has been evidenced that γ-butyrolactone has been mainly produced using monometallic catalysts, whereas 1,4-butanediol and tetrahydrofuran production has been facilitated using bimetallic catalysts, obtained by addition of a second metal as a promoter (Le & Nishimura, 2021).

For example, Re supported on mesoporous carbon and promoted with Ru as the second metal favors 1,4-butanediol formation, whereas Re/C promoted with Ir is the active catalyst in tetrahydrofuran production. The CuPd bimetallic catalyst supported on different supports such as γ-Al_2O_3, SiO_2, and TiO_2 exhibit a different catalytic behavior as a function of support nature. On the TiO_2 and SiO_2 supports, homogeneous phases of CuPd prevailed because of a strong or weak interaction between the metal and the support, whereas on the alumina support, the heterogeneous CuPd alloy or isolated Cu atoms prevailed because of a strong interaction between Cu and the support.

On the TiO_2 support, large CuPd nanoparticles have been evidenced because of sinterization, which leads to a low activity of this support for succinic acid hydrogenation and as an intermediate only γ-butyrolactone has been observed. CuPd/SiO_2 has a high activity and selectivity toward 1,4-butanediol due to the presence of smaller nanoparticles on this support. Furthermore, the presence of alumina as a support leads to the presence of strong Lewis acid sites on the surfaces, which favor tetrahydrofuran formation with high selectivity.

Another method to produce 1,4-butanediol from succinic acid is via dimethyl succinate using Re-Cu/C as the catalyst (Hong et al., 2014). Succinic acid could react with methanol in order to obtain dimethyl succinate (DMS), which is a reaction that can be achieved using Re-based catalysts that have metallic Re as an active site. For further demethylation of dimethyl succinate to 1,4-butanediol, a high activity of the catalyst is required; the catalyst must be suitable for both methylation and demethylation. Copper-based materials have proved to be active for demethylation; therefore, the couple Re–Cu is a potential candidate for succinic acid transformation to 1,4-butanediol via dimethyl succinate as the intermediate. Rh-Re-based catalysts are more expensive than catalysts based on Cu.

Mesoporous carbon as a support exhibits high porosity and has hydrophobic properties, and the dispersion of metal on this support is good.

The products obtained from succinic acid could be divided in four classes: (i) acyclic with O: 1,4-butanediol, esters; (ii) acyclic with O and N: 1,4-butanediamine, succinonitrile, and succinamide; (iii) cyclic with O: tetrahydrofuran, succinic acid anhydride, and γ-butyrolactone; and (iv) cyclic with O and N: 2-pyrrolidone and its derivatives. Diammonium succinate (DAS) represents a raw material for 2-pyrrolidone synthesis by hydrogenation using metallic catalysts deposited on a hydrothermally stable porous support because the reaction occurs in the aqueous phase. The presence of water leads to difficulties in final product separation.

Adipic acid can be transformed to 1,6-hexanediol, an important intermediate in the fine chemical industry, as it is an ingredient in polyester and polyurethane production. 1,6-Hexanediol is conventionally produced from adipic acid in two steps: (1) adipic acid esterification with methanol to dimethyl adipate (DMA) and (2) dimethyl adipate hydrogenation to 1,6-hexanediol; this step requires severe conditions (high hydrogen pressure, 10–20 MPa).

Noble metal-based catalysts are efficient in adipic acid hydrogenation in two steps. However, catalysts based on Cu/SiO_2 doped with Ni were found to be as effective in this reaction as noble metal-based catalysts (Tu et al., 2019). The presence of reduced Cu^{+} species was responsible for the higher activity.

Carboxylic acid hydrogenation to alcohols is a slower reaction compared to other reactions of hydrogenation. For example, acetic acid hydrogenation is approximately 10 times lower than the hydrogenation of furfural, 100 times smaller than acetaldehyde hydrogenation, and 1000 times lower than acetone hydrogenation because carboxylic acid is strongly adsorbed on the active species of heterogeneous catalysts, impeding hydrogen activation and further hydrogenation of the C=O group. At the same time, undesired side reactions can also occur.

Catalysts based on the precious metal Re (Li, Luo, & Liang, 2020) have proved to have a high affinity toward carbonyl group adsorption, given that they are active in the hydrogenation of aromatic carboxylic acids, dicarboxylic acids, and fatty acids. Their activity was increased by addition of a secondary precious metal (Ir, Pt, Pd).

The carboxylic acid group could be hydrogenated to alkane or alcohols as a function of reaction conditions. A good interaction between the two metals suppresses particle agglomeration and changes the size of the particles.

Re active species supported on Pt were enriched on the surface and increased the rate of C=O bond hydrogenation to alcohols. However, the presence of dicarboxylic acids in the reaction medium determines the oxidation of the Re species with decrease in the catalyst activity.

A monometallic Ir catalyst usually has low activity compared with those of other metals (Rh, Ru, Pt, Pd). Bimetallic Re–Ir catalysts have better activity and are selective in the hydrogenation of the C=O bond.

Malic and lactic acids

Malic acid is a platform for a series of chemicals, as illustrated in Fig. 10.13. It represents one of the promising sugar-derived products obtained from biomass conversion. Its behavior is similar to that of malonic acid with carbonyl and methylene groups.

Malic acid could be transformed to dimethyl malonate by selective oxidative decarbonylation. Malonic acid is quite unstable, so its aim is to transform into esters that are much more stable. As catalysts for this reaction, phosphovanadomolybdates were used, given that they are bifunctional catalysts (Liu et al., 2012).

In addition, malic acid may be subjected to a reaction of oxidative dehydrogenation to oxaloacetic acid, a precursor in amino acid production. Amino acids such as methionine, lysine, asparagine, and threonine could be synthesized from oxaloacetic acid. Pt-Bi/C (Drif et al., 2019) has been used as a catalyst in the transformation of malic acid to oxaloacetic acid with a conversion of 60% and a selectivity of 100%.

Lactic acid could be esterificated with ethanol to ethyl lactate, an environmentally friendly solvent, due to its biodegradability, nonvolatility, and noncarcinogenic properties.

Lactic acid is obtained by glucose fermentation; however, using this method, its concentration is low because the salts of lactic acid are produced instead of it. Esterification of lactic acid with ethanol has a low selectivity in ester because of low concentration of lactic acid and the presence of water. To achieve a high selectivity in ester, a higher concentration of lactic acid is required. However, a high concentration of lactic acid leads to its oligomerization, with this being additionally difficult.

The most studied catalysts for this reaction were Brønsted solid acids such as those based on sulfonated carbon (Nguyen et al., 2018) obtained from different carbon sources. The surface of sulfonated carbon possesses different functional groups with strong acidity like $-SO_3H$, phenolic –OH, and –COOH. The carbon skeleton has the advantage of being stable and insoluble in acidic or basic solutions.

Graphene oxides (GOs) can be obtained by graphite delamination and oxidation, as it is more active due to its numerous oxygenated functional groups. The acidity of graphene oxide is associated with the presence of free H^+ on the surface.

Lactic acid dehydrogenation to pyruvic acid has been carried out on Pb–Pt bimetallic catalysts supported on carbon (Zhang, Wang, & Ding, 2017). Conventionally, pyruvic acid is produced by dehydrative decarboxylation of tartaric acid. When the reaction takes place in the gas phase, the energy consumed is high because of the reaction temperature, 200 °C, and because of the side reactions. In the liquid phase, this inconvenience is eliminated.

Monometallic Pb/AC was inactive for reaction, and Pb served as an additive and not as an active phase. The oxidation state and molar ratio between Pb and Pt influence the catalyst activity, and, in addition, the metal particle size is influenced by the degree of graphitization.

Fatty acids

Saturated fatty acids can be deoxygenated, in the liquid or gas phase, to produce diesel-like hydrocarbons.

In the liquid phase, decarbonylation and decarboxylation reactions occur simultaneously, leading to the formation of alkane by carbon dioxide release and alkene production by carbon

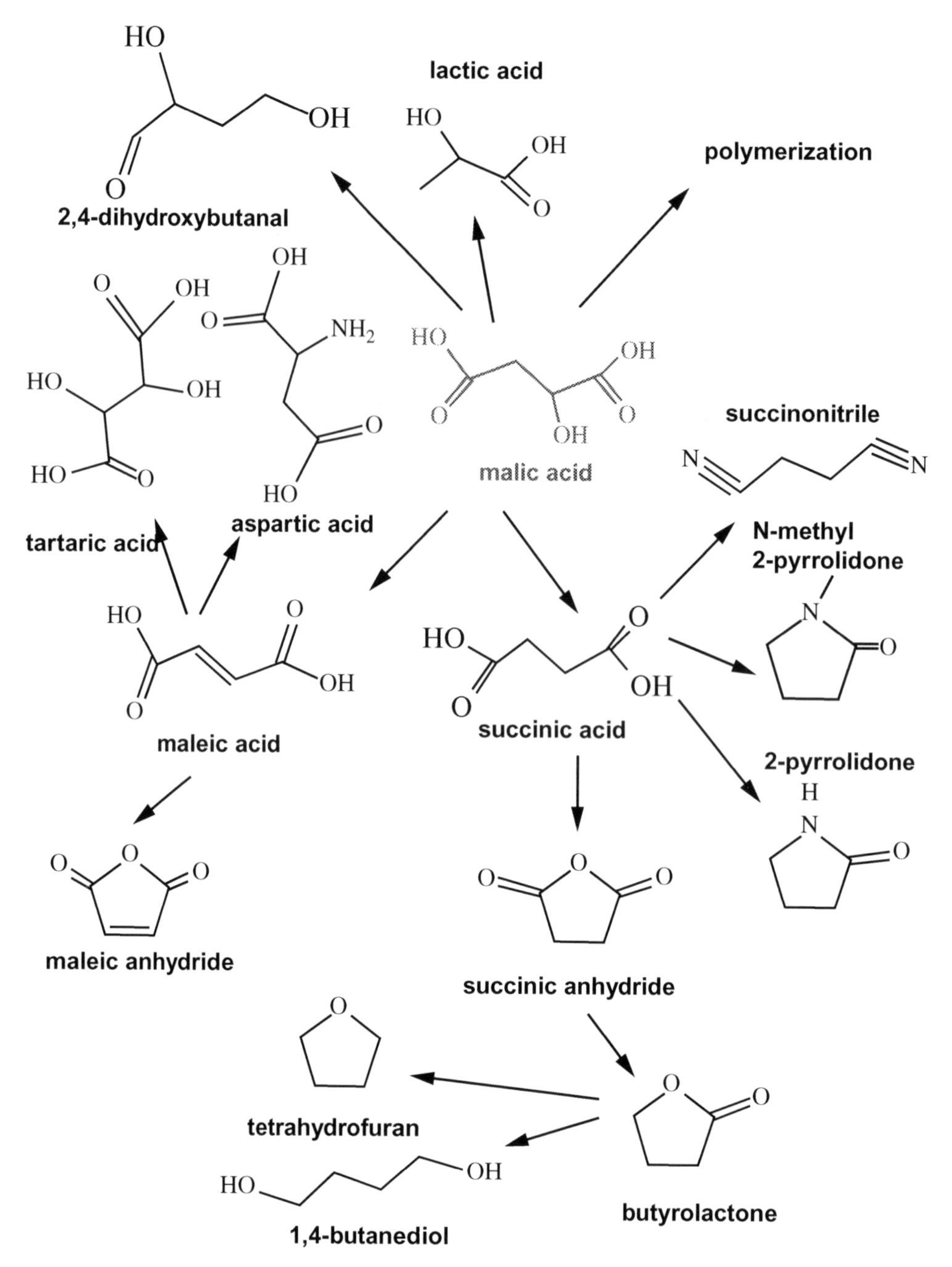

FIG. 10.13 Malic acid as a chemical platform for a series of products.

monoxide and water release, respectively. In the gas phase, methanation and water–gas shift reaction occur during the deoxygenation processes (Hermida, Abdullah, & Mohamed, 2015).

The reactions implied are as follow:

Liquid phase (Eqs. 10.32–10.35):

$$R-COOH \rightarrow R-H+CO_2(\text{decarboxylation}) \quad (10.32)$$

$$R-CH_2-CH_2-COOH \rightarrow R-CH=CH_2+CO+H_2O(\text{decarbonylation}) \quad (10.33)$$

$$R-COOH+H_2 \rightarrow R-H+CO+H_2O(\text{hydrogenation}) \quad (10.34)$$

$$R-COOH+3H_2 \rightarrow R-CH_3+2H_2O(\text{hydrogenation}) \quad (10.35)$$

Gas phase (Eqs. 10.10, 10.21 and 10.36):

$$CO_2+4H_2 \leftrightarrow CH_4+2H_2O(\text{methanation}) \quad (10.36)$$

Pt, Pd, and Ni on γ-Al_2O_3-based materials have been used as catalysts for these reactions. The Pd-based catalyst was more active than the Pt-based catalyst; the higher the content of the metal, the more active catalyst is evidenced, due to large particle size apparition (Theilgaard, 2011).

Solvents also play an important role in these types of reactions, contributing to the orientation of the reaction in a certain direction. For example, when the reaction is carried out in the absence of a solvent, no alkanes are obtained, whereas in the presence of dodecane as a solvent, alkanes are obtained with high selectivities.

Similarly, Pd supported on SBA-15 (Lestari et al., 2010) was studied as possible catalysts, for example, in stearic acid deoxygenation to *n*-heptadecane and in a mixture of palmitic and stearic acids with *n*-heptadecane and *n*-pentadecane production.

Unsaturated fatty acids can also be deoxygenated to diesel-like hydrocarbons by hydrogenation of their double bonds in the first step and by subsequent deoxygenation of products.

10.5 Conclusions

In the field of biomass-derived alcohols and acid conversion to a series of chemicals, great efforts have been made in order to develop heterogeneous catalysts with good yields and selectivities to desired products.

Monometallic and bimetallic catalysts based on noble and transition metals have been applied with success in this field, exhibiting excellent performances. Nanomaterials with controlled physicochemical properties were also efficient catalysts.

The catalyst's design represents an important factor to achieve high conversion and selectivity to the desired product. Catalysts with multifunctional properties have the advantage that they are active in multistep reactions, whereas for catalysts with acid/base properties, controlling high selectivity can be realized.

Other important factors to consider when designing a catalyst are that they must be both inexpensive and stable. Furthermore, an additional requirement is that they must be active at low temperatures in order to eliminate the side reactions.

Acknowledgments

This work was supported by a grant from the Romanian Ministry of Education and Research, CNCS—UEFISCDI, project number PN-III-P4-ID-PCE-2020-0580, within PNCDI III.

References

Adelabu, B. A., Kareem, S. O., Oluwafemi, F., & Abideen Adeogun, I. (2019). Bioconversion of corn straw to ethanol by cellulolytic yeasts immobilized in Mucuna urens matrix. *Journal of King Saud University—Science*, *31*(1), 136–141. https://doi.org/10.1016/j.jksus.2017.07.005.

Alotaibi, M. A., Kozhevnikova, E. F., & Kozhevnikov, I. V. (2012). Deoxygenation of propionic acid on heteropoly acid and bifunctional metal-loaded heteropoly acid catalysts: Reaction pathways and turnover rates. *Applied Catalysis A: General*, *447–448*, 32–40. https://doi.org/10.1016/j.apcata.2012.08.042.

Amorós-Pérez, A., Cano-Casanova, L., Lillo-Ródenas, M. Á., & Román-Martínez, M. C. (2017). Cu/TiO2 photocatalysts for the conversion of acetic acid into biogas and hydrogen. *Catalysis Today*, *287*, 78–84. https://doi.org/10.1016/j.cattod.2016.09.009.

Apuzzo, J., Cimino, S., & Lisi, L. (2018). Ni or Ru supported on MgO/γ-Al2O3 pellets for the catalytic conversion of ethanol into butanol. *RSC Advances*, *8*(45), 25846–25855. https://doi.org/10.1039/c8ra04310h.

Bagheri, S., & Muhd Julkapli, N. (2017). Mo3VOx catalyst in biomass conversion: A review in structural evolution and reaction pathways. *International Journal of Hydrogen Energy*, *42*(4), 2116–2126. https://doi.org/10.1016/j.ijhydene.2016.09.173.

Bandiera, J., & Naccache, C. (1991). Kinetics of methanol dehydration on dealuminated H-mordenite: Model with acid and basic active centres. *Applied Catalysis*, *69*(1), 139–148. https://doi.org/10.1016/S0166-9834(00)83297-2.

Barakat, N. A. M., Motlak, M., Elzatahry, A. A., Khalil, K. A., & Abdelghani, E. A. M. (2014). NixCo1-x alloy nanoparticle-doped carbon nanofibers as effective non-precious catalyst for ethanol oxidation. *International Journal of Hydrogen Energy*, *39*(1), 305–316. https://doi.org/10.1016/j.ijhydene.2013.10.061.

Benaiss, H. (2008). Heteropoly compounds as catalysts for hydrogenation of propanoic acid. *Journal of Catalysis*, *253*, 244.

Bhanushali, J. T., Prasad, D., Patil, K. N., Reddy, K. S., Kainthla, I., Rao, K. S. R., et al. (2020). Tailoring the catalytic activity of basic mesoporous Cu/CeO2 catalyst by Al2O3 for selective lactonization and dehydrogenation of 1,4-butanediol to γ-butyrolactone. *Catalysis Communications*, *143*. https://doi.org/10.1016/j.catcom.2020.106049.

Brijaldo, M. H., Caytuero, A. E., Martínez, J. J., Rojas, H., & Passos, F. B. (2020). Hydrogen production from acetic acid decomposition as bio-oil model molecule over supported metal catalysts. *International Journal of Hydrogen Energy*, *45*(53), 28732–28751. https://doi.org/10.1016/j.ijhydene.2020.07.205.

Carvalho, D. L., De Avillez, R. R., Rodrigues, M. T., Borges, L. E. P., & Appel, L. G. (2012). Mg and Al mixed oxides and the synthesis of n-butanol from ethanol. *Applied Catalysis A: General*, *415–416*, 96–100. https://doi.org/10.1016/j.apcata.2011.12.009.

Catizzone, E., Daele, S. V., Bianco, M., Di Michele, A., Aloise, A., Migliori, M., et al. (2019). Catalytic application of ferrierite nanocrystals in vapour-phase dehydration of methanol to dimethyl ether. *Applied Catalysis B: Environmental*, *243*, 273–282. https://doi.org/10.1016/j.apcatb.2018.10.060.

Chen, L. (2012). *Applied Catalysis A: General*, *411–412*, 95.

Chen, Z., Li, X., Xu, Y., Dong, Y., Lai, W., Fang, W., et al. (2018). Fabrication of nano-sized SAPO-11 crystals with enhanced dehydration of methanol to dimethyl ether. *Catalysis Communications*, *103*, 1–4. https://doi.org/10.1016/j.catcom.2017.09.002.

Chen, R. K., Yu, T. F., Wu, M. X., Tzeng, T. W., Chung, P. W., & Lin, Y. C. (2018). The aldolization nature of Mn4+-nonstoichiometric oxygen pair sites of perovskite-type LaMnO3 in the conversion of ethanol. *ACS Sustainable Chemistry & Engineering*, *6*(9), 11949–11958. https://doi.org/10.1021/acssuschemeng.8b02269.

Cho, S. H., Kim, J., Han, J., Lee, D., Kim, H. J., Kim, Y. T., et al. (2019). Bioalcohol production from acidogenic products via a two-step process: A case study of butyric acid to butanol. *Applied Energy*, *252*. https://doi.org/10.1016/j.apenergy.2019.113482.

Czernik, S., Evans, R., & French, R. (2007). Hydrogen from biomass-production by steam reforming of biomass pyrolysis oil. *Catalysis Today*, *129*(3–4), 265–268. https://doi.org/10.1016/j.cattod.2006.08.071.

Dai, W., Yang, L., Wang, C., Wang, X., Wu, G., Guan, N., et al. (2018). Effect of n-butanol cofeeding on the methanol to aromatics conversion over Ga-modified nano H-ZSM-5 and its mechanistic interpretation. *ACS Catalysis*, *8*(2), 1352–1362. https://doi.org/10.1021/acscatal.7b03457.

Dash, S. S., & Parida, K. M. (2007). Esterification of acetic acid with n-butanol over manganese nodule leached residue. *Journal of Molecular Catalysis A: Chemical*, *266*(1–2), 88–92. https://doi.org/10.1016/j.molcata.2006.10.031.

Demirbas, T., & Demirbas, A. H. (2010). Bioenergy, green energy. biomass and biofuels. *Energy Sources, Part A: Recovery, Utilization and Environmental Effects*, *32*(12), 1067–1075. https://doi.org/10.1080/15567030903058600.

Ding, S., Wang, H., Han, J., Zhu, X., & Ge, Q. (2018). Ketonization of propionic acid to 3-pentanone over $Ce_xZr_{1-x}O_2$ catalysts: The importance of acid-base balance. *Industrial and Engineering Chemistry Research*, *57*(50), 17086–17096. https://doi.org/10.1021/acs.iecr.8b04208.

Drif, A., Pineda, A., Morvan, D., Belliere-Baca, V., De Oliveira Vigier, K., & Jérôme, F. (2019). Catalytic oxidative dehydrogenation of malic acid to oxaloacetic acid. *Green Chemistry, 21*(17), 4604–4608. https://doi.org/10.1039/c9gc01768b.

Festel, G. W. (2008). Biofuels—Economic aspects. *Chemical Engineering and Technology, 31*(5), 715–720. https://doi.org/10.1002/ceat.200700335.

Freeman, D., Wells, R. P. K., & Hutchings, G. J. (2001). Methanol to hydrocarbons: Enhanced aromatic formation using a composite Ga2O3-H-ZSM-5 catalyst. *Chemical Communications, 1*(18), 1754–1755. https://doi.org/10.1039/b104844a.

Gliński, M., Zalewski, G., Burno, E., & Jerzak, A. (2014). Catalytic ketonization over metal oxide catalysts. XIII. Comparative measurements of activity of oxides of 32 chemical elements in ketonization of propanoic acid. *Applied Catalysis A: General, 470*, 278–284. https://doi.org/10.1016/j.apcata.2013.10.047.

Gu, J., Zhang, Z., Hu, P., Ding, L., Xue, N., Peng, L., et al. (2015). Platinum nanoparticles encapsulated in MFI zeolite crystals by a two-step dry gel conversion method as a highly selective hydrogenation catalyst. *ACS Catalysis, 5*(11), 6893–6901. https://doi.org/10.1021/acscatal.5b01823.

Guseinova, E. A., Adzhamov, K. Y., & Yusubova, S. E. (2020). Catalytic conversion of isopropanol on a heteropoly acid-η-aluminum oxide system. *Russian Journal of Physical Chemistry A, 94*(2), 301–309. https://doi.org/10.1134/S0036024420010082.

Hashemi, F., Taghizadeh, M., & Rami, M. D. (2020). Polyoxometalate modified SAPO-34: A highly stable and selective catalyst for methanol conversion to light olefins. *Microporous and Mesoporous Materials, 295*. https://doi.org/10.1016/j.micromeso.2019.109970.

Hermida, L., Abdullah, A. Z., & Mohamed, A. R. (2015). Deoxygenation of fatty acid to produce diesel-like hydrocarbons: A review of process conditions, reaction kinetics and mechanism. *Renewable and Sustainable Energy Reviews, 42*, 1223–1233. https://doi.org/10.1016/j.rser.2014.10.099.

Hong, U. G., Hwang, S., Seo, J. G., Lee, J., & Song, I. K. (2011). Hydrogenation of succinic acid to γ-butyrolactone (GBL) over palladium catalyst supported on alumina xerogel: Effect of acid density of the catalyst. *Journal of Industrial and Engineering Chemistry, 17*(2), 316–320. https://doi.org/10.1016/j.jiec.2011.02.030.

Hong, U. G., Kim, J. K., Lee, J., Lee, J. K., Yi, J., & Song, I. K. (2014). Conversion of succinic acid to 1,4-butanediol via dimethyl succinate over rhenium nano-catalyst supported on copper-containing mesoporous carbon. *Journal of Nanoscience and Nanotechnology, 14*(11), 8867–8872. https://doi.org/10.1166/jnn.2014.9943.

Hu, X., Ma, K., Sabbaghi, A., Chen, X., Chatterjee, A., & Lam, F. L. Y. (2020). Mild acid functionalization of metal-organic framework and its catalytic effect on esterification of acetic acid with n-butanol. *Molecular Catalysis, 482*. https://doi.org/10.1016/j.mcat.2019.110635.

Huang, R., Cao, C., Liu, J., Zheng, L., Zhang, Q., Gu, L., et al. (2020). Integration of metal single atoms on hierarchical porous nitrogen-doped carbon for highly efficient hydrogenation of large-sized molecules in the pharmaceutical industry. *ACS Applied Materials and Interfaces, 12*(15), 17651–17658. https://doi.org/10.1021/acsami.0c03452.

Huang, J., Dai, W. L., & Fan, K. (2008). Support effect of new Au/FeOx catalysts in the oxidative dehydrogenation of α,ω-diols to lactones. *Journal of Physical Chemistry C, 112*(41), 16110–16117. https://doi.org/10.1021/jp8043913.

Huang, J., Dai, W. L., Li, H., & Fan, K. (2007). Au/TiO2 as high efficient catalyst for the selective oxidative cyclization of 1,4-butanediol to γ-butyrolactone. *Journal of Catalysis, 252*(1), 69–76. https://doi.org/10.1016/j.jcat.2007.09.011.

Huang, R., Fung, V., Wu, Z., & Jiang, D. E. (2020). Understanding the conversion of ethanol to propene on In2O3 from first principles. *Catalysis Today, 350*, 19–24. https://doi.org/10.1016/j.cattod.2019.05.035.

Karim, A. M., Su, Y., Engelhard, M. H., King, D. L., & Wang, Y. (2011). Catalytic roles of Co0 and Co2+ during steam reforming of ethanol on Co/MgO catalysts. *ACS Catalysis, 1*(4), 279–286. https://doi.org/10.1021/cs200014j.

Kaur, K., Wanchoo, R. K., & Toor, A. P. (2015). Sulfated iron oxide: A proficient catalyst for esterification of butanoic acid with glycerol. *Industrial and Engineering Chemistry Research, 54*(13), 3285–3292. https://doi.org/10.1021/ie504916k.

Khalil, K. M. S., Elhamdy, W. A., & Said, A.-E.-A. A. (2020). Direct formation of LaFeO3/MCM-41 nanocomposite catalysts and their catalytic reactivity for conversion of isopropanol. *Materials Chemistry and Physics, 254*. https://doi.org/10.1016/j.matchemphys.2020.123412, 123412.

Klimkiewicz, R., Wolska, J., Przepiera, A., Przepiera, K., Jabłoński, M., & Lenart, S. (2009). The zinc ferrite obtained by oxidative precipitation method as a catalyst in n-butanol conversion. *Materials Research Bulletin, 44*(1), 15–20. https://doi.org/10.1016/j.materresbull.2008.08.004.

Koempel, H., & Liebner, W. (2007). Lurgi's Methanol To Propylene (MTP®) Report on a successful commercialisation. *Studies in Surface Science and Catalysis, 167*, 261–267. https://doi.org/10.1016/S0167-2991(07)80142-X.

Kwak, J. H., Mei, D., Peden, C. H. F., Rousseau, R., & Szanyi, J. (2011). (100) facets of γ-Al2O3: The active surfaces for alcohol dehydration reactions. *Catalysis Letters, 141*(5), 649–655. https://doi.org/10.1007/s10562-010-0496-8.

Larmier, K., Chizallet, C., Cadran, N., Maury, S., Abboud, J., Lamic-Humblot, A. F., et al. (2015). Mechanistic investigation of isopropanol conversion on alumina catalysts: Location of active sites for alkene/ether production. *ACS Catalysis*, *5*(7), 4423–4437. https://doi.org/10.1021/acscatal.5b00723.

Lawal, A. M., Hart, A., Daly, H., Hardacre, C., & Wood, J. (2019). Catalytic hydrogenation of short chain carboxylic acids typical of model compound found in bio-oils. *Industrial and Engineering Chemistry Research*, *58*(19), 7998–8008. https://doi.org/10.1021/acs.iecr.9b01093.

Le, S. D., & Nishimura, S. (2021). Effect of support on the formation of CuPd alloy nanoparticles for the hydrogenation of succinic acid. *Applied Catalysis B: Environmental*, *282*. https://doi.org/10.1016/j.apcatb.2020.119619, 119619.

Lee, K. Y., Lee, H. K., & Ihm, S. K. (2010). Influence of catalyst binders on the acidity and catalytic performance of HZSM-5 zeolites for methanol-to-propylene (MTP) process: Single and binary binder system. *Topics in Catalysis*, *53*(3–4), 247–253. https://doi.org/10.1007/s11244-009-9412-0.

Lestari, S., Mäki-Arvela, P., Eränen, K., Beltramini, J., Max Lu, G. Q., & Murzin, D. Y. (2010). Diesel-like hydrocarbons from catalytic deoxygenation of stearic acid over supported pd nanoparticles on SBA-15 catalysts. *Catalysis Letters*, *134*(3–4), 250–257. https://doi.org/10.1007/s10562-009-0248-9.

Li, X., Luo, J., & Liang, C. (2020). Hydrogenation of adipic acid to 1,6-hexanediol by supported bimetallic Ir-Re catalyst. *Molecular Catalysis*, *490*. https://doi.org/10.1016/j.mcat.2020.110976.

Liu, J., Du, Z., Yang, Y., Lu, T., Lu, F., & Xu, J. (2012). Catalytic oxidative decarboxylation of malic acid into dimethyl malonate in methanol with dioxygen. *ChemSusChem*, *5*(11), 2151–2154. https://doi.org/10.1002/cssc.201200489.

Lu, M., Lepore, A. W., Choi, J. S., Li, Z., Wu, Z., Polo-Garzon, F., et al. (2018). Acetic acid/propionic acid conversion on metal doped molybdenum carbide catalyst beads for catalytic hot gas filtration. *Catalysts*, *8*(12). https://doi.org/10.3390/catal8120643.

Mallat, T., & Baiker, A. (2004). Oxidation of alcohols with molecular oxygen on solid catalysts. *Chemical Reviews*, *104*(6), 3037–3058. https://doi.org/10.1021/cr0200116.

Martono, E., & Vohs, J. M. (2011). Active sites for the reaction of ethanol to acetaldehyde on Co/YSZ(100) model steam reforming catalysts. *ACS Catalysis*, *1*(10), 1414–1420. https://doi.org/10.1021/cs200404h.

Migliori, M., Catizzone, E., Aloise, A., Bonura, G., Gómez-Hortigüela, L., Frusteri, L., et al. (2018). New insights about coke deposition in methanol-to-DME reaction over MOR-, MFI- and FER-type zeolites. *Journal of Industrial and Engineering Chemistry*, *68*, 196–208. https://doi.org/10.1016/j.jiec.2018.07.046.

Mitran, G., Chen, S., Dolge, K. L., Huang, W., & Seo, D. K. (2021). Ketonic decarboxylation and esterification of propionic acid over beta zeolites. *Microporous and Mesoporous Materials*, *310*. https://doi.org/10.1016/j.micromeso.2020.110628.

Mitran, G., Saab, R., Charisiou, N., Polychronopoulou, K., & Goula, M. (2020). Molybdenum supported on carbon covered alumina: Active sites for n-butanol dehydrogenation and ketonization. *Molecular Catalysis*, *495*. https://doi.org/10.1016/j.mcat.2020.111159.

Mitran, G., Yuzhakova, T., Popescu, I., & Marcu, I. C. (2015). Study of the esterification reaction of acetic acid with n-butanol over supported WO3 catalysts. *Journal of Molecular Catalysis A: Chemical*, *396*, 275–281. https://doi.org/10.1016/j.molcata.2014.10.014.

Mondal, U., & Yadav, G. D. (2019). Perspective of dimethyl ether as fuel: Part I. Catalysis. *Journal of CO_2 Utilization*, *32*, 299–320. https://doi.org/10.1016/j.jcou.2019.02.003.

Ndou, A. S., & Coville, N. J. (2004). Self-condensation of propanol over solid-base catalysts. *Applied Catalysis A: General*, *275*(1–2), 103–110. https://doi.org/10.1016/j.apcata.2004.07.025.

Nguyen, V. C., Bui, N. Q., Mascunan, P., Vu, T. T. H., Fongarland, P., & Essayem, N. (2018). Esterification of aqueous lactic acid solutions with ethanol using carbon solid acid catalysts: Amberlyst 15, sulfonated pyrolyzed wood and graphene oxide. *Applied Catalysis A: General*, *552*, 184–191. https://doi.org/10.1016/j.apcata.2017.12.024.

Ozel, S., Meric, G. G., Arbag, H., Degirmenci, L., & Oktar, N. (2020). Steam reforming of acetic acid in the presence of Ni coated with SiO2 microsphere catalysts. *International Journal of Hydrogen Energy*, *45*(41), 21252–21261. https://doi.org/10.1016/j.ijhydene.2020.05.146.

Palla, V. C. S., Shee, D., & Maity, S. K. (2016). Conversion of n-butanol to gasoline range hydrocarbons, butylenes and aromatics. *Applied Catalysis A: General*, *526*, 28–36. https://doi.org/10.1016/j.apcata.2016.07.026.

Pang, J., Zheng, M., He, L., Li, L., Pan, X., Wang, A., et al. (2016). Upgrading ethanol to n-butanol over highly dispersed Ni–MgAlO catalysts. *Journal of Catalysis*, *344*, 184–193. https://doi.org/10.1016/j.jcat.2016.08.024.

Phan, T. T. N., Nikoloski, A. N., Bahri, P. A., & Li, D. (2019). Facile fabrication of perovskite-incorporated hierarchically mesoporous/macroporous silica for efficient photoassisted-Fenton degradation of dye. *Applied Surface Science*, *491*, 488–496. https://doi.org/10.1016/j.apsusc.2019.06.133.

Phung, T. K., Proietti Hernández, L., Lagazzo, A., & Busca, G. (2015). Dehydration of ethanol over zeolites, silica alumina and alumina: Lewis acidity, Brønsted acidity and confinement effects. *Applied Catalysis A: General*, *493*, 77–89. https://doi.org/10.1016/j.apcata.2014.12.047.

Pu, J., Luo, Y., Wang, N., Bao, H., Wang, X., & Qian, E. W. (2018). Ceria-promoted Ni@Al2O3 core-shell catalyst for steam reforming of acetic acid with enhanced activity and coke resistance. *International Journal of Hydrogen Energy*, *43*(6), 3142–3153. https://doi.org/10.1016/j.ijhydene.2017.12.136.

Pylinina, A. I., & Mikhalenko, I. I. (2011). Dehydrogenation of butyl alcohols on NASICON-type solid electrolytes of Na1-2x Cu x Zr2(PO4)3 composition. *Russian Journal of Physical Chemistry A*, *85*(12), 2109–2114. https://doi.org/10.1134/S0036024411110252.

Rachmady, W., & Vannice, M. A. (2002). Acetic acid reduction by H2 on bimetallic Pt-Fe catalysts. *Journal of Catalysis*, *209*(1), 87–98. https://doi.org/10.1006/jcat.2002.3623.

Raju, M. A., Gidyonu, P., Nagaiah, P., Rao, M. V., Raju, B. D., & Rao, K. S. R. (2019). Mesoporous silica–supported copper catalysts for dehydrogenation of biomass-derived 1,4-butanediol to gamma butyrolactone in a continuous process at atmospheric pressure. *Biomass Conversion and Biorefinery*, *9*(4), 719–726. https://doi.org/10.1007/s13399-019-00406-4.

Rakshit, P. K., Voolapalli, R. K., & Upadhyayula, S. (2018). Acetic acid hydrogenation to ethanol over supported Pt-Sn catalyst: Effect of Bronsted acidity on product selectivity. *Molecular Catalysis*, *448*, 78–90. https://doi.org/10.1016/j.mcat.2018.01.030.

Ramasamy, K. K., Gray, M., Job, H., Smith, C., & Wang, Y. (2016). Tunable catalytic properties of bi-functional mixed oxides in ethanol conversion to high value compounds. *Catalysis Today*, *269*, 82–87. https://doi.org/10.1016/j.cattod.2015.11.045.

Reddy, K. H. P., Anand, N., Prasad, P. S. S., Rao, K. S. R., & Raju, B. D. (2011). Influence of method of preparation of Co-Cu/MgO catalyst on dehydrogenation/dehydration reaction pathway of 1,4-butanediol. *Catalysis Communications*, *12*(10), 866–869. https://doi.org/10.1016/j.catcom.2011.02.013.

Requies, J., Güemez, M. B., Maireles, P., Iriondo, A., Barrio, V. L., Cambra, J. F., et al. (2012). Zirconia supported Cu systems as catalysts for n-butanol conversion to butyraldehyde. *Applied Catalysis A: General*, *423–424*, 185–191. https://doi.org/10.1016/j.apcata.2012.02.039.

Sahebdelfar, S. (2017). Steam reforming of propionic acid: Thermodynamic analysis of a model compound for hydrogen production from bio-oil. *International Journal of Hydrogen Energy*, *42*(26), 16386–16395. https://doi.org/10.1016/j.ijhydene.2017.05.108.

Sai Prasad, P. S., Bae, J. W., Kang, S. H., Lee, Y. J., & Jun, K. W. (2008). Single-step synthesis of DME from syngas on Cu-ZnO-Al2O3/zeolite bifunctional catalysts: The superiority of ferrierite over the other zeolites. *Fuel Processing Technology*, *89*(12), 1281–1286. https://doi.org/10.1016/j.fuproc.2008.07.014.

Sarmah, B., Satpati, B., & Srivastava, R. (2018). Selective oxidation of biomass-derived alcohols and aromatic and aliphatic alcohols to aldehydes with O2/air using a RuO2-supported Mn3O4 catalyst. *ACS Omega*, *3*(7), 7944–7954. https://doi.org/10.1021/acsomega.8b01009.

Sheng, X., Kong, J., Zhou, Y., Zhang, Y., Zhang, Z., & Zhou, S. (2014). Direct synthesis, characterization and catalytic application of SBA-15 mesoporous silica with heteropolyacid incorporated into their framework. *Microporous and Mesoporous Materials*, *187*, 7–13. https://doi.org/10.1016/j.micromeso.2013.12.007.

Shi, Y., Yang, Y., Li, Y. W., & Jiao, H. (2016). Theoretical study about Mo2C(101)-catalyzed hydrodeoxygenation of butyric acid to butane for biomass conversion. *Catalysis Science and Technology*, *6*(13), 4923–4936. https://doi.org/10.1039/c5cy02008e.

Siri, G. J., Bertolini, G. R., Casella, M. L., & Ferretti, O. A. (2005). PtSn/γ-Al2O3 isobutane dehydrogenation catalysts: The effect of alkaline metals addition. *Materials Letters*, *59*(18), 2319–2324. https://doi.org/10.1016/j.matlet.2005.03.013.

Skevis, G. (2010). Liquid biofuels: Biodiesel and bioalcohols, part 3. Gaseous and liquid fuels. In *Handbook of combustion* (pp. 359–388). Wiley Online Library. https://doi.org/10.1002/9783527628148.hoc051.pub2.

Sun, Z., Vasconcelos, A. C., Bottari, G., Stuart, M. C. A., Bonura, G., Cannilla, C., et al. (2017). Efficient catalytic conversion of ethanol to 1-butanol via the guerbet reaction over copper- and nickel-doped porous. *ACS Sustainable Chemistry & Engineering*, *5*(2), 1738–1746. https://doi.org/10.1021/acssuschemeng.6b02494.

Tamura, M., Ito, K., Honda, M., Nakagawa, Y., Sugimoto, H., & Tomishige, K. (2016). Direct copolymerization of CO2 and diols. *Scientific Reports*, *6*. https://doi.org/10.1038/srep24038.

Tesquet, G., Faye, J., Hosoglu, F., Mamede, A. S., Dumeignil, F., & Capron, M. (2016). Ethanol reactivity over La1+x FeO3+δ perovskites. *Applied Catalysis A: General*, *511*, 141–148. https://doi.org/10.1016/j.apcata.2015.12.005.

Theilgaard, M. (2011). Step changes and deactivation behavior in the continuous decarboxylation of stearic acid. *Industrial and Engineering Chemistry Research*, *50*.

Tu, C. C., Tsou, Y. J., To, T. D., Chen, C. H., Lee, J. F., Huber, G. W., et al. (2019). Phyllosilicate-derived CuNi/SiO2 catalysts in the selective hydrogenation of adipic acid to 1,6-hexanediol. *ACS Sustainable Chemistry & Engineering*, *7*(21), 17872–17881. https://doi.org/10.1021/acssuschemeng.9b04418.

Unglert, M., Bockey, D., Bofinger, C., Buchholz, B., Fisch, G., Luther, R., et al. (2020). Action areas and the need for research in biofuels. *Fuel, 268*. https://doi.org/10.1016/j.fuel.2020.117227.

Varvarin, A. M., Khomenko, K. M., & Brei, V. V. (2013). Conversion of n-butanol to hydrocarbons over H-ZSM-5, H-ZSM-11, H-L and H-Y zeolites. *Fuel, 106*, 617–620. https://doi.org/10.1016/j.fuel.2012.10.032.

Wang, N., Yu, X., Wang, Y., Chu, W., & Liu, M. (2013). A comparison study on methane dry reforming with carbon dioxide over LaNiO3 perovskite catalysts supported on mesoporous SBA-15, MCM-41 and silica carrier. *Catalysis Today, 212*, 98–107. https://doi.org/10.1016/j.cattod.2012.07.022.

Yeh, T., Linic, S., & Savage, P. E. (2014). Deactivation of Pt catalysts during hydrothermal decarboxylation of butyric acid. *ACS Sustainable Chemistry & Engineering, 2*(10), 2399–2406. https://doi.org/10.1021/sc500423b.

Yu, T. F., Chang, C. W., Chung, P. W., & Lin, Y. C. (2019). Unsupported and silica-supported perovskite-type lanthanum manganite and lanthanum ferrite in the conversion of ethanol. *Fuel Processing Technology, 194*. https://doi.org/10.1016/j.fuproc.2019.06.001.

Zeng, L., Wang, Y., Mou, J., Liu, F., Yang, C., Zhao, T., et al. (2020). Promoted catalytic behavior over γ-Al2O3 composited with ZSM-5 for crude methanol conversion to dimethyl ether. *International Journal of Hydrogen Energy, 45*(33), 16500–16508. https://doi.org/10.1016/j.ijhydene.2020.04.115.

Zhang, C., Wang, T., & Ding, Y. (2017). Oxidative dehydrogenation of lactic acid to pyruvic acid over Pb-Pt bimetallic supported on carbon materials. *Applied Catalysis A: General, 533*, 59–65. https://doi.org/10.1016/j.apcata.2017.01.010.

Zheng, X. J., Li, C. L., Zhao, M., Zheng, Z., Wei, L. F., Chen, F. H., et al. (2017). Photocatalytic degradation of butyric acid over Cu2O/Bi2WO6 composites for simultaneous production of alkanes and hydrogen gas under UV irradiation. *International Journal of Hydrogen Energy, 42*(12), 7917–7929. https://doi.org/10.1016/j.ijhydene.2016.12.131.

Zharova, P. A., Chistyakov, A. V., Tsodikov, M. V., Nikolaev, S. A., Rossi, F., & Manenti, F. (2017). Supercritical ethanol and isopropanol conversion into chemicals over Au-M catalysts. *Chemical Engineering Transactions, 57*, 31–36. https://doi.org/10.3303/CET1757006.

Zhang, X., Wang, R., Yang, X., & Zhang, F. (2008). Comparison of four catalysts in the catalytic dehydration of ethanol to ethylene. *Microporous and Mesoporous Materials, 116*(1–3), 210–215. https://doi.org/10.1016/j.micromeso.2008.04.004.

Zhou, M., Zhang, H., Ma, H., & Ying, W. (2016). The catalytic properties of K modified PtSn/Al_2O_3 catalyst for acetic acid hydrogenation to ethanol. *Fuel Processing Technology, 144*, 115–123.

CHAPTER

11

Zinc oxide or molybdenum oxide deposited on bentonite by the microwave-assisted hydrothermal method: New catalysts for obtaining biodiesel

Ana Flávia Felix Farias[a,b], *Marcos Antonio Gomes Pequeno*[a], *Suelen Alves Silva Lucena de Medeiros*[a], *Thiago Marinho Duarte*[a], *Herbet Bezerra Sales*[b], *and Ieda Maria Garcia dos Santos*[a]

[a]NPE-LACOM, Federal University of Paraiba, João Pessoa, PB, Brazil

[b]UAEMa, Federal University of Campina Grande, Campina Grande, PB, Brazil

11.1 Introduction

Biodiesel is a fuel obtained from the transesterification of oils and fats (Fig. 11.1) using strong acids (HCl, H_2SO_4) or bases (NaOH, KOH) as homogeneous catalysts (Chithambararaj & Bose, 2011; Lee, Bennett, Manayil, & Wilson, 2014; Mendieta et al., 2011; Rathod, Lande, Arbad, & Gambhire, 2014). Basic catalysts are associated with formation of emulsion when raw materials do not present high purity, whereas acid catalysts are associated with corrosion and present a smaller catalytic activity than basic catalysts (Berrios & Skelton, 2008; Tesser, Di Serio, Guida, Nastasi, & Santacesaria, 2005). For all of the homogeneous catalysts, a decrease in the activity is observed as the size of the alcohol chain increases, with extremely low activity or even no activity for alcohols with more than three carbon atoms (Stern, Hillion, Rouxel, & Leporq, 1999). For these reasons, a technological challenge for the development of the biodiesel industry is the search for alternative catalysts, which do not require pure raw materials, are not associated with corrosion, and present a high activity for bigger molecules, just like ethanol, leading to a biofuel obtained without any petroleum derivatives.

Traditional catalysts may be replaced by heterogeneous or enzymatic systems (Colombo & Barros, 2009; Lee et al., 2014; Suwannakarn, Lotero, Ngaosuwan, & Goodwin, 2009). Different aluminosilicates, oxides, hydroxides, and carbonates containing representative metals, transition metals, or organic polymers have been

Heterogeneous Catalysis
https://doi.org/10.1016/B978-0-323-85612-6.00011-5

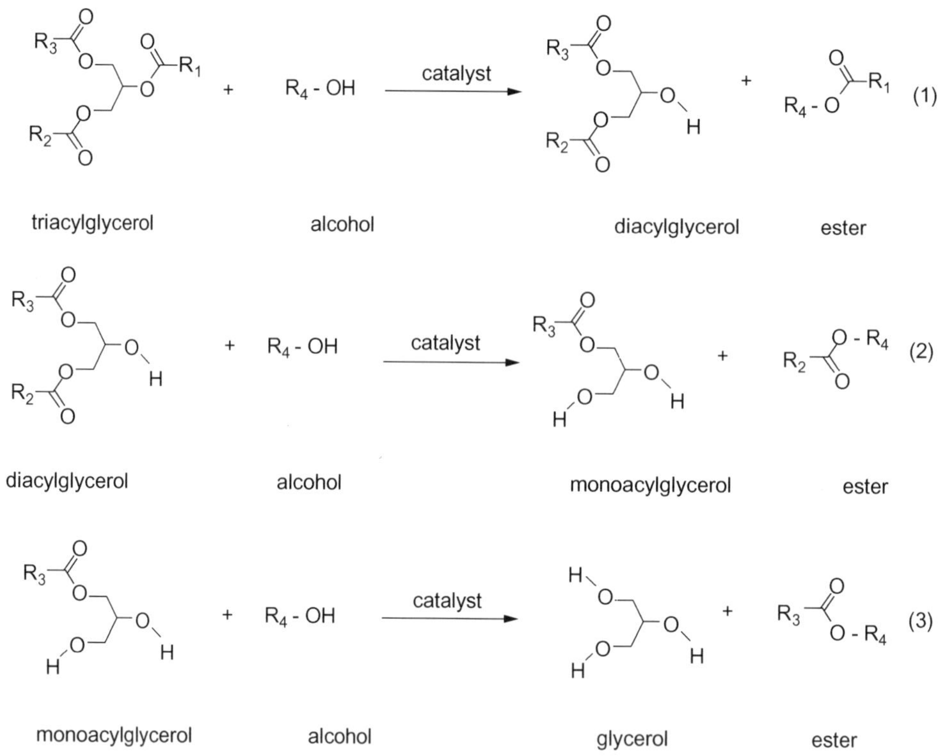

triacylglycerol + alcohol —catalyst→ diacylglycerol + ester (1)

diacylglycerol + alcohol —catalyst→ monoacylglycerol + ester (2)

monoacylglycerol + alcohol —catalyst→ glycerol + ester (3)

FIG. 11.1 Steps of the transesterification reaction.

studied for transesterification reactions (Kim et al., 2004; Mendieta et al., 2011; Pinto et al., 2005; Somidi, Das, & Dalai, 2016; Yigezu & Muthukumar, 2014). The catalytic performance of these materials is directly related to the nature of acid or basic sites found in the crystalline structures. Some of these heterogeneous catalysts, especially the acid ones, are active for high-molecular-weight alcohols attaining conversions above 95% for systems where traditional catalysts have small conversions, without promoting corrosion or emulsion formation, thus making the separation process easier; however, a smaller activity is usually attained (Stern et al., 1999). These solids may be used as single-phase materials, doped materials, mixed with other oxides, or may be supported or impregnated in a matrix (Antunes, Veloso, & Henriques, 2008; Farias et al., 2015; Tariq, Ali, & Khalid, 2012; Xie & Yang, 2007; Xie & Zhao, 2014). In spite of the low activity of clays in the synthesis of biodiesel, they can be used as a support due to their low cost associated with the possibility of use in composites.

In the last few years, metal oxides have received attention, especially transition-metal oxides, which have been highly investigated in theoretical and experimental research studies, due to their interesting physical properties, high versatility, and wide variety of applications in different areas including the technological area (Ha & Tuyen, 2014; Samadi et al., 2018). In this context, in this work, zinc oxide and molybdenum oxide were studied. The literature reports different papers about the use of ZnO as a catalyst for biodiesel, whereas fewer papers are reported for MoO_3. In this chapter, these two oxides are supported on bentonites by the microwave-assisted hydrothermal method and tested in the synthesis of biodiesel.

Considering the need of biodiesel production using low-cost, environmentally friendly catalysts, this research aims at developing materials based on ZnO and MoO_3 in their massic forms or deposited on bentonites by a one-step synthesis process based on the microwave-assisted hydrothermal method, which has been widely used to obtain nanometric materials with different compositions associated with the stoichiometric and morphological control of the particles (Fernandes, Capelli, Vaz, & Nunes, 2015; Ren, Sun, Cui, & Li, 2018).

Smectites applied as a catalytic support or a nanocomposite

Among the different clays described in the literature, this work focuses on bentonites, an interesting clay widely available in Brazil and mainly composed of montmorillonite (MMT), which belongs to the smectite group (Fig. 11.2). MMT has the general formula: $(M^{n+}_{y/n} \times mH_2O)(Al^{3+}_{4-y}Mg^{2+}_{y})(Si^{4+}_{8})O_{20}(OH)_4$, where "M" is the interlamellar cation (usually Na^+, K^+, Mg^{2+}, or Ca^{2+}), "*n*" is the valence, "*y*" indicates the lamella charge due to isomorphic replacements in the octahedral sites, and "*m*" is the number of water molecules coordinated to the M cation (Bergaya & Lagaly, 2013; Fonseca, Vaiss, Wypych, Diniz, & Leitão, 2017).

Due to their interesting properties such as a moderate negative surface charge, high cation exchange capacity, high surface area, formation of stable suspensions, possibility of intercalation of molecules in the interlamellar space, and high resistance to temperature and solvents (Alves, Ferraz, & Gamelas, 2020; Bergaya & Lagaly, 2013; Brigatti, Galán, & Theng, 2013; Cavalcanti, Fonseca, da Silva Filho, & Jaber, 2019), smectites have been widely used in industries and in

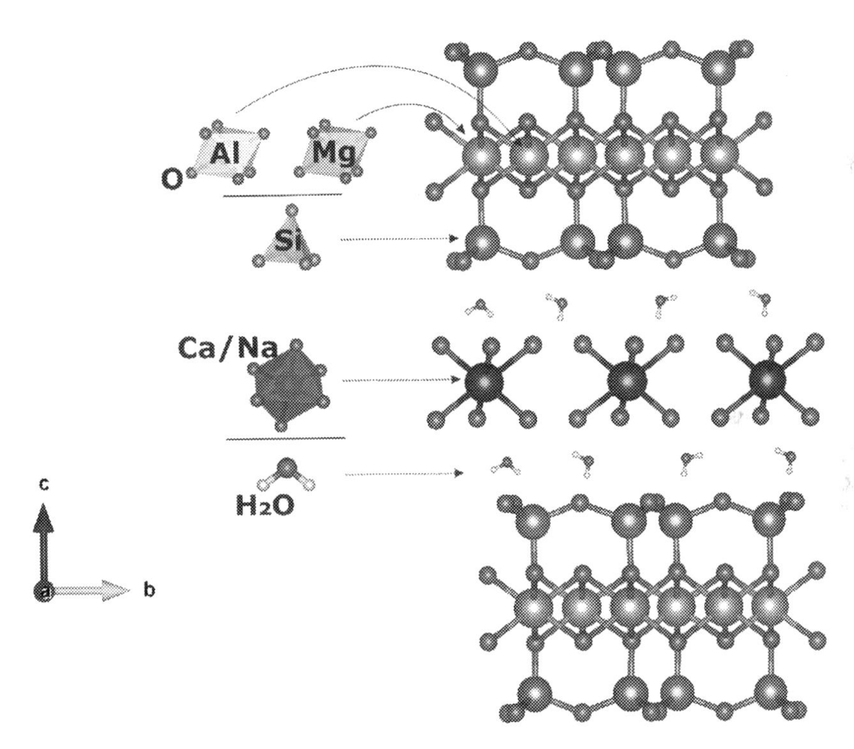

FIG. 11.2 Structure of smectite clay, obtained from the CIF file 34812.

scientific research studies, including drilling fluids, antisedimentation additives, detergents, drugs, cosmetics, catalysts, catalytic supports, elimination of radioactive residues, and agrochemicals (Milodowski, Norris, & Russell Alexander, 2016; Shen et al., 2020).

Some clay minerals have been widely used in nanocomposites, such as montmorillonites, hectorites, saponites, and palygorskites, with montmorillonites being the most important one (Galimberti, Cipolletti, & Coombs, 2013; Mccabe & Adams, 2013). These nanocomposites have been widely used in the last few years and may be classified as inorganic/inorganic or inorganic/organic (Deka and Maji, 2012; Lei, Hoa, & Ton-That, 2006; Tawade et al., 2021). With regard to inorganic/inorganic nanocomposites or hybrid materials, different researches have been described in the literature (Deka & Maji, 2012; Farias et al., 2015; Hu, Gu, Luan, Song, & Zhu, 2012; Meshram et al., 2011; Ye, Li, Hong, Chen, & Fan, 2015), such as TiO_2 (Jagtap & Ramaswamy, 2006), NiO (Park, Kim, Hwuang, & Choy, 2006), CuO (Farias et al., 2015; Khaorapapong, Khumchoo, & Ogawa, 2015), CeO_2 (Farias et al., 2015; Kamada, Kang, Paek, & Choy, 2012), ZnO (Farias et al., 2015; Hur et al., 2006), and MoO_3 (Salerno, Mendioroz, & Agudo, 2003), which were intercalated on aluminosilicates cas pillars, hybrid materials, or supported as the active phase on material surfaces. In most cases, intercalation of impregnation is attained by an ionic exchange reaction or even using the conventional hydrothermal method (Cecilia et al., 2015; Kumar, Ramacharyulu, Prasad, & Singh, 2015; Motshekga, Ray, Onyango, & Momba, 2015).

Chemical modifications have been usually performed on clay minerals applied in catalysis in order to improve their chemical and physical properties and to obtain active materials for adsorption and catalysis (Chen et al., 2016). These methods include acid activation (Dill et al., 2021), pillarization (De León, Castiglioni, Bussi, & Sergio, 2008), impregnation of metals (Velasco, Pérez-Mayoral, Mata, Rojas-Cervantes, & Vicente-Rodríguez, 2011), and pillarization associated with metal impregnation (Li et al., 2012). These modifications lead to solids with higher thermal stability and higher specific surface area.

Improvement of the adsorptive/catalytic activity is usually attained by better distribution of the active phase, besides protection against deactivation of the material (Farias et al., 2015; Mccabe & Adams, 2013). This aim may be attained by the use of a support, which increases the specific surface area of the active phase. Catalytic supports may also have high porosity and mechanical resistance and they may be active or inactive from the catalytic point of view (Li, Hu, Zuo, & Wang, 2018; Verga et al., 2016; Xu & Wei, 2018).

In this context, clay minerals are an interesting alternative for use as a catalytic support, and they have been attracting attention in the last few years as they are environmentally friendly and have low cost, high availability, and high specific surface area (Mccabe & Adams, 2013). These materials have also been used for the synthesis of biodiesel using transesterification or esterification reactions.

Acid activation has been used to obtain clays with catalytic activity in the esterification of lauric acid with methanol. Santos-Beltrán et al. (2017) used phosphoric acid in the acid activation of montmorillonite and obtained 90% conversion at 180°C for 90 min. Reinoso, Angeletti, Cervellini, and Boldrini (2020) used sulfuric acid in the activation and observed a meaningful increase in the amount and in the force of the acid sites when H_2SO_4 concentrations of 1 and $3\,mol\,L^{-1}$, respectively, were used.

Clay minerals have been used as a support for potassium compounds. For instance, Boz, Degirmenbasi, and Kalyon (2013) performed the functionalization of bentonite with KF, KOH, and K_2CO_3 for the methanolic transesterification of canola oil and obtained 98.2% conversion for KF. Li and Jiang (2018) reported the use of palygorskite as a catalytic support for KF/CaO by impregnation with 97.9% conversion in the methanolic transesterification of soybean oil using $KCaF_3$ as the catalyst.

Our research group has described the impregnation of bentonite with different metal oxides such as SnO_2 (Neris et al., 2015), ZnO, CuO, and CeO_2 (Farias et al., 2015), with a meaningful increase in the catalytic activity depending on the material.

Zinc oxide applied as a catalyst for the synthesis of biodiesel

Zinc oxide (ZnO) crystallizes with different structures such as wurtzite, rock salt, and zinc blend. Under ambient conditions, wurtzite is a thermodynamically stable phase (Fig. 11.3), with a hexagonal unit cell and a $P6_3mc$ space group (Özgür et al., 2005; Wojnarowicz, Chudoba, & Lojkowski, 2020). Due to its structure, the ZnO surface is polar with zinc-positive charges or oxygen-negative charges, resulting in a normal dipole moment and spontaneous polarization along the *c*-axis (Borysiewicz, 2019; Özgür et al., 2005; Wang, 2004). Moreover, zinc oxide has a low cost, is environmentally friendly, and highly available all around the world.

ZnO has been synthesized using different methods such as thermal oxidation (Dikici & Demirci, 2019; Liang, Pan, & Liu, 2008), combustion (Farias et al., 2020), microwave-assisted hydrothermal or solvothermal methods (Huang, Xia, Cao, & Zeng, 2008; Shaporev, Ivanov, Baranchikov, & Tret'yakov, 2007; Wojnarowicz et al., 2020), coprecipitation (Shweta & Thapa, 2019), polymeric precursors, and sol–gel (Lima, Cremona, Davolos, Legnani, & Quirino, 2007).

ZnO is described as an excellent catalyst applied in different types of reactions (Franco-Urquiza, May-Crespo, Escalante Velázquez, Pérez Mora, & González García, 2020; Guckan et al., 2020; Salem et al., 2017), including the synthesis of biodiesel (Alba-Rubio et al., 2010; Yan, Salley, & Simon Ng, 2009), with more than 140 works published in the last 5 years. For instance, ester conversions of around 96% were already attained by different authors using the methanolic route (Abdala, Nur, & Mustafa, 2020; Al-Saadi, Mathan, & He, 2020; Antunes et al., 2008; Gavhane, Kate, Pawar, Soudagar, & Fayaz, 2020; Soltani et al., 2020; Vasić, Podrepšek, Knez, & Leitgeb, 2020). In these works, ZnO is used in its massic form (Farias et al., 2020; Jadhav & Tandale, 2018; Mguni, Mukenga, Muzenda, Jalama, & Meijboom, 2013), mixed with other oxides (Kim et al., 2004; Mendieta et al., 2011; Pinto et al., 2005; Somidi et al., 2016; Yigezu & Muthukumar, 2014), doped (Borah, Devi, Borah, & Deka, 2019), or supported (Farias et al., 2015). In spite of this, very few works have reported the use of silica, alumina, or aluminosilicate as catalytic supports in the last 5 years.

ZnO was deposited on γ-Al_2O_3 and compared to CaO and MgO for the transesterification of soybean oil by the methanolic and butanolic routes (Navas, Lick, Bolla, Casella, & Ruggera,

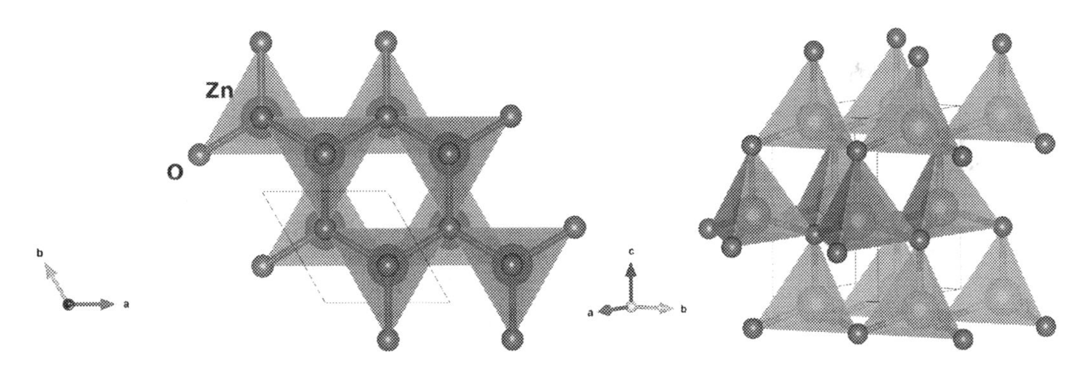

FIG. 11.3 Crystalline structure of ZnO, obtained from the CIF file 26170.

2018). Supported materials had a higher activity than massic ones, but only 41% of activity was obtained using $ZnO/\gamma\text{-}Al_2O_3$ after 6h of a catalytic test using the methanolic route. A better result was obtained when the cellulose–ZnO composite was supported on SiO_2 for the methanolic transesterification of coconut oil (Helmiyati & Suci, 2019). In this work, 95% of conversion was obtained after 270min at 60°C. Our research group has been evaluating the influence of ZnO deposition on bentonite for the ethylic transesterification of soybean oil (Farias et al., 2015). In this work, the influence of the pH of the solution in catalytic activity is carefully discussed.

Molybdenum oxide applied as a catalyst for the synthesis of biodiesel

Molybdenum trioxide (MoO_3) is a transition-metal oxide with some peculiarities concerning its crystalline structures and multiple valence states. It is usually found in the stable orthorhombic phase ($\propto -_3$), the metastable monoclinic phase ($-_3$), or the hexagonal phase ($h -_3$). The $\propto -_3$ and $h -_3$ were synthesized in this work and are presented in Fig. 11.4.

MoO_3 is usually obtained with a low surface area and many research studies have been conducted in order to improve its catalytic activity, including different strategies, such as morphological control, phase transition, oxygen vacancy formation, doping, and hybridization with other materials. These materials have been applied as catalysts in different areas such as energy, environmental, fuel cell, photocatalytic degradation, and selective thermocatalysis (Zhu et al., 2020).

Few papers report the use of MoO_3 as a catalyst for the synthesis of biodiesel. Moreover, most papers report this synthesis by an esterification reaction (Bail et al., 2013; De Almeida et al., 2014), and almost all papers use methanol as a reaction route, which is derived from fossil fuels. For the synthesis of a real green fuel, ethanol obtained from vegetable sources should be used. For instance, Sankaranarayanan, Pandurangan, Banu, and Sivasanker (2011) obtained a conversion of 90% during the methanolic transesterification of sunflower oil at 373 K for 24h, using 16wt% of MoO_3 supported on $\gamma\text{-}Al_2O_3$. A conversion of 81% was attained using the ethanolic route under the same conditions. The highest activity was obtained for the sample calcined at 950K, which was composed of $Mo_{13}O_{33}$, Mo_9O_{26}, MoO_3, and $Al_2(MoO_4)_3$. Xie and Zhao (2014) used CaO–MoO_3 supported on SBA-15 in the methanolic transesterification of soybean oil at 338K for 50h, and 83.2% of conversion was attained when 40% of the active phase was used. The Ca/Mo molar ratio was 6:1, and the catalyst was

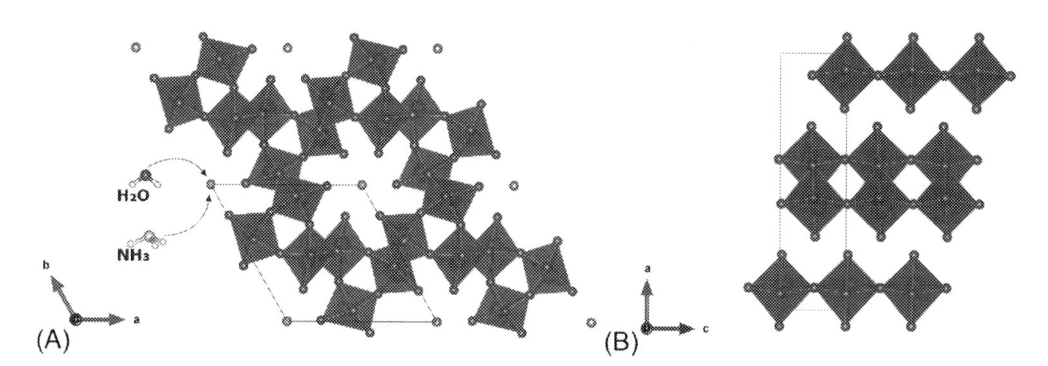

FIG. 11.4 Crystalline structures of MoO_3 representing orthorhombic (A) and hexagonal (B) phases, obtained from the CIF files 230017 and 38145, respectively.

composed of CaO, MoO_3, and $CaMoO_4$. MoO_3 was also doped with Fe and Mn and supported on ZrO_2 for the methanolic transesterification of waste cooking oil by Alhassan, Rashid, and Taufiq-Yap (2015). After an optimization process, the authors obtained 95.6% of conversion at 473 K for 5 h, with a meaningful increase in the activity after doping. MoO_3–Al_2O_3 was used in the simultaneous esterification and transesterification reaction of sunflower oil acidified with oleic acid by the methanolic route. Deposition of MoO_3 on Al_2O_3 favored the formation of stronger acid sites, in spite of the decrease in the surface area (Navajas et al., 2020). MoO_3 has also been used as a dopant to stabilize SO_4^{2-} in sulfated zirconia and has been applied in esterification and transesterification reactions (Li, Tong, & Hu, 2015).

11.2 Methods

Synthesis methods and impregnations of Zn and Mo oxides

The syntheses of oxides and hybrids were performed by the microwave-assisted hydrothermal method, as described below.

Synthesis of ZnO

For the synthesis of ZnO, dihydrated zinc acetate ($Zn_2(CH_3COO)_2 \cdot H_2O$; Synth P.A. > 98%) was completely dissolved in distilled water for 10 min in order to obtain a solution with a concentration of 0.04 mol L^{-1}. The solution was alkalinized with ammonium hydroxide (NH_4OH from Vetec P.A. 99%) to adjust the pH to 8.0 (Zn08) or 11.0 (Zn11) and to obtain the suspensions, which were transferred to a Teflon vessel and coupled with a microwave reactor (RMW-1, Inove Produtos and Tecnologia/Ltda) for hydrothermalization at 100 °C for 5 min. After cooling, the suspensions were centrifuged and the powders were washed with distilled water until complete neutralization (pH 7.0), dried in a conventional oven at 80 °C for 12 h, deagglomerated in mortar, and characterized.

Synthesis of the ZnO/bentonite hybrid

Synthesis of the hybrids was planned in order to obtain 2 g of the final material with 20% of ZnO and 80% of bentonite for each batch. The whole procedure was similar to the synthesis of ZnO, except for the addition of the bentonite (BENTONISA S.A., Brazil) into the final zinc acetate solution, followed by another 10 min of stirring, before alkalinization and adjustment of the pH to obtain the hybrids named CZ08 and CZ11, for suspensions with pH 8.0 and 11.0, respectively. The flowchart of the syntheses of ZnO and ZnO/bentonite hybrids by the microwave-assisted hydrothermal method is illustrated in Fig. 11.5.

Synthesis of h-MoO_3 and α-MoO_3

For the synthesis of MoO_3 with a hexagonal structure, 5.42 g of $(NH_4)_6Mo_7O_{24} \cdot 4H_2O$ was dissolved in 70 mL of distilled water with constant stirring for 30 min, followed by the addition of 1.0 mol L^{-1} HNO_3 (Dinâmica) solution until a pH of 1 was reached. The mixture was transferred to the Teflon reactor for hydrothermal heating under microwave radiation at 100 °C for 1 min. The suspension was centrifuged for separation of the precipitate, which was washed with distilled water and then with ethanol and dried in an oven at 100 °C for 8 h. This sample was named h-MoO_3 100 °C. The synthesis of the orthorhombic MoO_3 structure was performed using the same procedure but with microwave heating at 150 °C for 5 min (h-MoO_3 150 °C). After washing and drying, the precipitate was heat treated in a conventional furnace at 400 °C for 2 h (α-MoO_3 400 °C).

Synthesis of the MoO_3/bentonite hybrid

For the synthesis of the MoO_3/bentonite hybrid, a suspension containing 4.25 g of bentonite and 40 mL of distilled water was prepared by

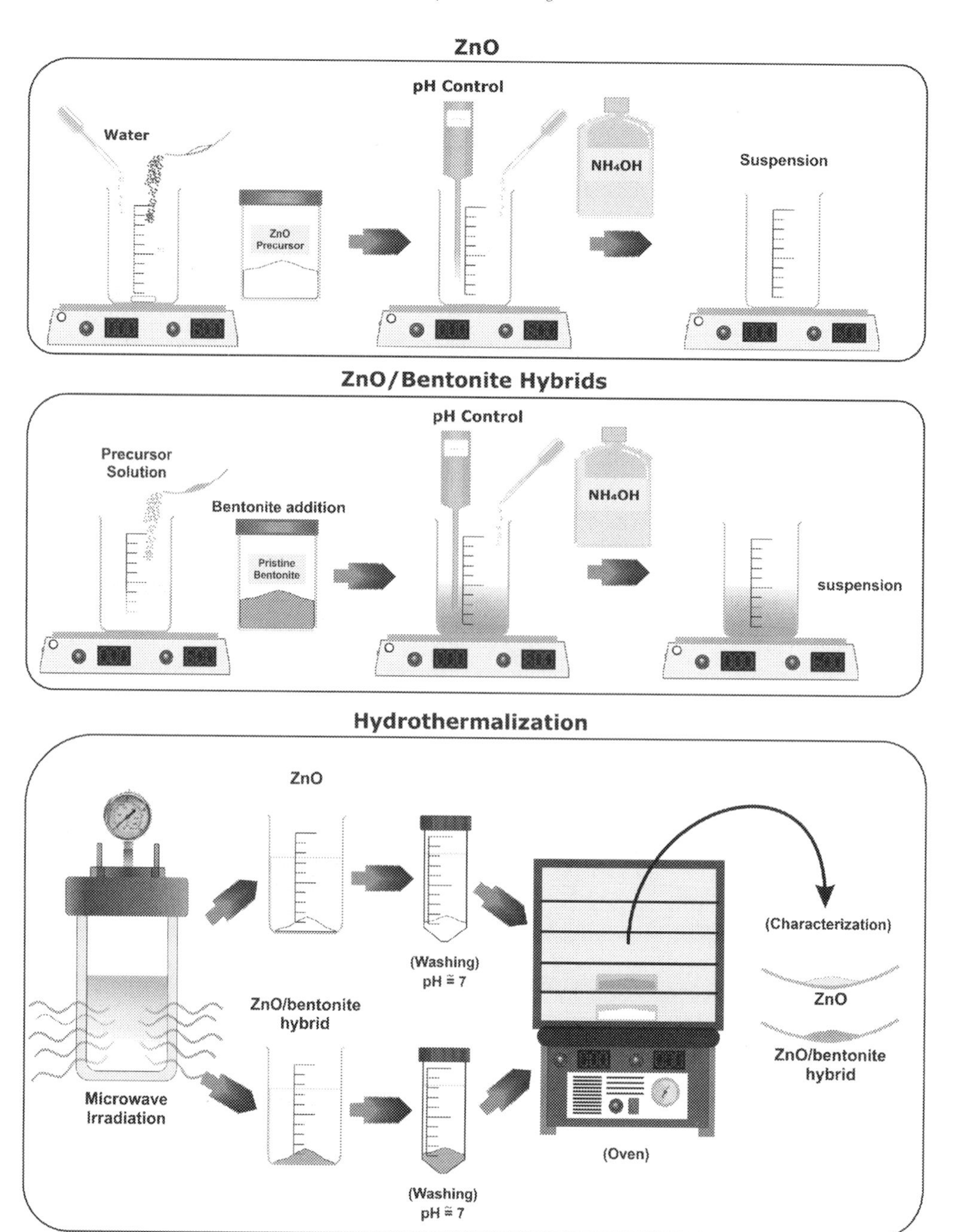

FIG. 11.5 Flowchart of synthesis of ZnO and ZnO/bentonite hybrids by the microwave-assisted hydrothermal method.

stirring with the pH adjusted to 5.0. The molybdenum solution was prepared as previously described, using 1.84 g of the precursor and 30 mL of water and added to the bentonite suspension, followed by another 30 min of stirring before acidification. The following procedure was similar to the synthesis of the pristine h-MoO_3 and α-MoO_3.

The flowchart of the syntheses of MoO_3 and MoO_3/bentonite hybrids by the microwave-assisted hydrothermal method is illustrated in Fig. 11.6.

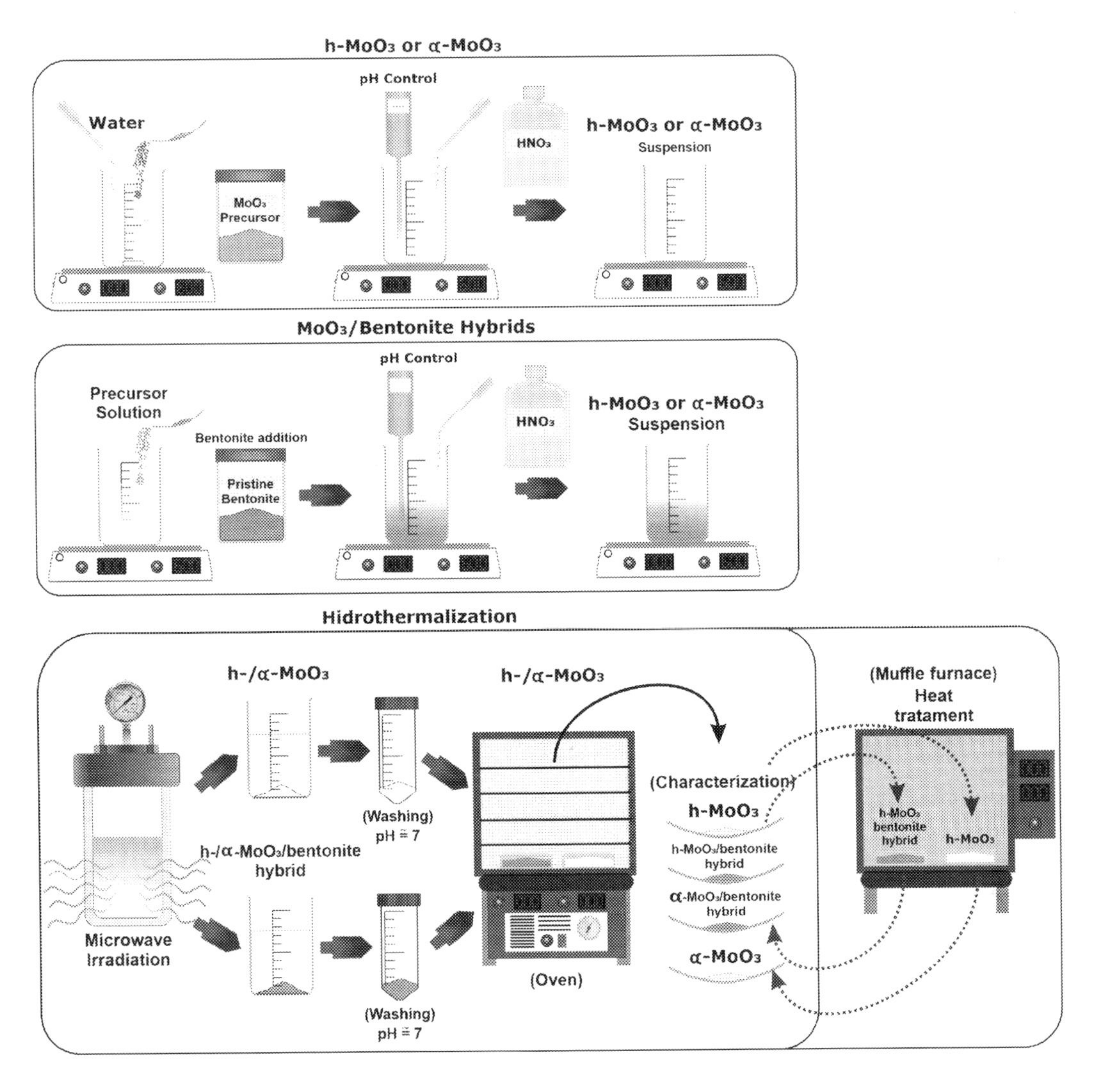

FIG. 11.6 Flowchart of synthesis of MoO_3 and MoO_3/bentonite hybrids by the microwave-assisted hydrothermal method.

Characterization of the catalysts and catalytic testing products

Powders were characterized by X-ray diffraction (XRD) using an XRD 6000 instrument from Shimadzu at 30 kV and 30 mA with nickel-filtered, Cu-$K_{\alpha 1}$ radiation with a 2θ range between 3 and 80 degrees, a step of 0.03 degree, and a rate of $0.02° s^{-1}$. X-ray fluorescence (XRF) was performed using a Shimadzu EDX-8000 X-ray fluorescence spectrometer to determine the percentage of the impregnated materials in the hybrids. Synthetic standards were prepared for the construction of an analytical curve for each material. The samples were prepared by homogenizing the bentonite with the oxide in a mortar, followed by pressing of the pellets.

The micrographs were obtained using a scanning electron microscope (SEM) (model LEO 1430, Zeiss), with gold metallization of the surface samples, and field-emission scanning electron microscopy (FE-SEM) analysis was performed using a FEG-VP Zeiss Supra 35 microscope for ZnO hybrids. Surface area measurements were performed by nitrogen (N_2) adsorption at a temperature of 77 K, using a BELSORP mini II device from Bel Japan. The values of the specific surface area were determined by the BET method, whereas pore volumes and the average pore size were obtained from the desorption curve using the BJH method. Raman spectroscopy measurements were performed using an InVia spectrophotometer from Renishaw, equipped with an argon laser source with a wavelength of 514 nm and an incident power of 100 mW.

The acid sites of the catalysts were determined by n-butylamine adsorption/desorption, according to the methodology described by da Silva et al. (2020), Guo et al. (2021), Kim, DiMaggio, Salley, and Simon Ng (2012), and Yan et al. (2009). The activation temperature used was at 120 °C/2 h for molybdenum trioxide, bentonites, and hybrids instead of 400 °C, due to the clay dehydroxylation reaction and the phase transformation of MoO_3. Quantification of *n*-butylamine adsorbed/chemisorbed on the acid sites was performed using the simultaneous equipment SDT 650 (TA Instruments). The analyses were carried out in alumina crucibles containing ca. 10 mg of the sample preadsorbed with n-butylamine. High-purity nitrogen was used as the carrier gas with a flow rate of $25 mL min^{-1}$. Samples were heated at a rate of $10 °C min^{-1}$ up to 700 °C. For this calculation, the mass loss assigned to dehydroxylation was subtracted from the total mass loss.

Catalytic tests

All of the catalytic tests were performed according to the ethylic route using commercial soybean oil (Campestre). For ZnO samples, ethanol from Química Moderna (P.A. 99.3%) was used, whereas ethanol from Dinâmica was used (95%) for MoO_3 samples. Reaction conditions for the different catalysts are displayed in Table 11.1. Before the catalytic tests, the materials were previously heated in an oven at 100 °C for 1.5 h for drying. Thereafter, they were mixed with ethanol and kept in a freezer for 12 h. The reagents were placed in a 4561 Parr reactor and heated up to the desired temperature at a heating rate of $2 °C min^{-1}$ under stirring at 600 rpm. The final pressure varied from 200 to 250 psi. Samples were removed from the reactor, centrifuged for separation of the catalysts, and transferred to a decantation balloon for washing and drying at 80 °C under vacuum. The whole procedure is summarized in Fig. 11.7.

For ZnO hybrids, catalysts were recovered, washed with ethyl alcohol, and dried in an oven at 100 °C for 2 h. Materials were tested in reuse reactions using the same procedure of activation and reaction conditions described above. Reused tests were named as R1 and R1.

The product samples of the catalytic tests were characterized by kinematic viscosity measurements using a V18 Julab viscometer, with a Cannon Fenske glass capillary immersed in a water bath at 40 °C. ^{1}H and ^{13}C NMR spectra

TABLE 11.1 Reaction conditions for the transesterification reaction using the different catalysts.

Material	Ethanol: oil ratio	Amount of catalyst[a]	Temperature (°C)	Time (h)
ZnO and bentonite	6:1	3%	200	4
ZnO/bentonite	6:1	3%	130 and 200	4
MoO_3	6:1	5%	150 and 200	0.5, 1, and 2
MoO_3/bentonite	6:1	5%	150 and 200	0.5, 1, and 2

[a] *Calculated in relation to the oil mass.*

were obtained using a GEMINI 300BB VARIAN spectrometer operating at a frequency of 200 MHz. Conversion of the oil into biodiesel was evaluated considering the peaks of ^1H NMR integration (Compton, Laszlo, Appell, Vermillion, & Evans, 2014; Ghesti, de Macedo, Resck, Dias, & Dias, 2007; Neto, Caro, Mazzuco, & Nascimento, 2004; Tariq et al., 2012), according to the methodology adapted by Ghesti et al. (2007) for ethyl esters, as displayed in Eq. (11.1).

$$C_{EE}\,(\%) = 100 \times \left(\frac{I_{TAG+EE} - I_{TAG}}{I_{\alpha CH_2}}\right) \quad (11.1)$$

where C_{EE} = conversion into ethyl ester; I_{TAG+EE} = integrated area of the superposed peaks between 4.1 and 4.2 ppm, assigned to the glycerol methyl hydrogen atoms and the hydrogen atoms of the CH_2 group of the formed ethoxy group; I_{TAG} = integrated area of the peaks between 4.25 and 4.35 ppm, assigned to the non-transesterified glycerol methyl hydrogen atoms; and $I_{\alpha\text{-}CH2}$ = integrated area of the peaks assigned to methyl hydrogen of the carbonyl group between 2.2 and 2.4 ppm.

The concentration of the ethylic esters was determined using a gas chromatograph with a flame ionizing detector (GC-FID) at a temperature of 380 °C, using a CGMS-2010 equipment from Shimadzu, with an automatic sampler, a split injector, and a Durabond DB-23 capillary column (Agilent Technologies). Helium was used as the carrier gas with a flow rate of 30.0 mL min^{-1} and an injection volume of 0.5 μL.

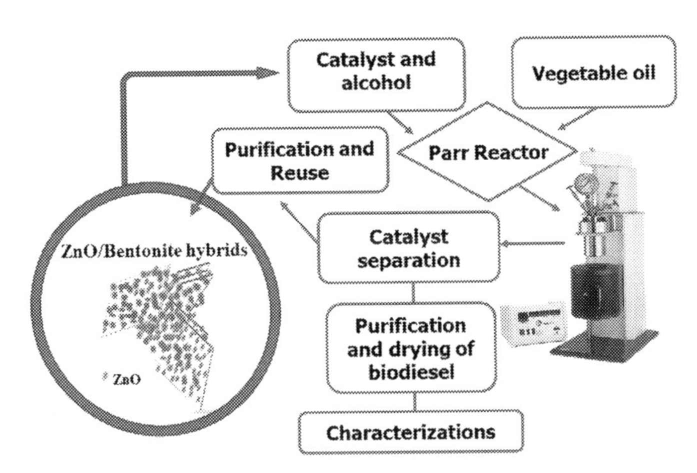

FIG. 11.7 Flowchart of the catalytic tests using transesterification reaction by the ethylic route.

11.3 Recent advances

ZnO and ZnO/bentonite hybrids applied as catalysts for obtaining biodiesel

XRD patterns of the bentonite before (BF) and after impregnation with ZnO (CZ08 and CZ11) are displayed in Fig. 11.8, and chemical compositions are presented in Table 11.2. XRD patterns of the pristine ZnO are also presented in Fig. 11.8.

The presence of clay minerals from the smectite group (montmorillonite and nontronite) was confirmed, as well as quartz, usually present as a secondary phase. Indexing was done considering the following ICDD crystallographic charts: 00–060-0318 ($(Ca,Na)_{0.3}Al_2(Si,Al)_4O_{10}(OH)_2 \cdot xH_2O$), 00-058-2038 ($(Ca,Na)_{0.3}Al_2(Si,Al)_4O_{10}(OH)_2 \cdot xH_2O$), 00-058-2007 ($(Ca_{0.2}(Al,Mg)_2Si_4O_{10}(OH)_2 \cdot xH_2O$), 00-029-1497 ($Na_{0.3}Fe_2Si_4O_{10}(OH)_2 \cdot 4H_2O$), and 00-046-1045 ($SiO_2$). This result is confirmed by a chemical analysis (Table 11.2), which showed a high amount of SiO_2, besides Al_2O_3 and Fe_2O_3, as well as MgO, which probably replaced Al in the octahedral site. XRD patterns also indicated a decrease of the (001) reflection intensity of the smectite after impregnation, besides its shift to larger angles, leading to d_{001} values of 1.19 and 1.34 nm for CZ08 and CZ11, respectively. This decrease in the basal distance indicated that a cation exchange may have occurred and Zn^{2+} may have intercalated in the interlamellar region, besides its deposition on the bentonite surface, for both synthesis procedures. In addition, the results indicated that the slightly alkaline pH favored ZnO crystallization. According to Svensson and Hansen (2013), the basal space of a hydrated Ca-montmorillonite was around ~1.9 nm, which corresponds to a hydration of about three layers of water. Considering the use of a natural bentonite, with a certain hydration degree, d_{001} values and the chemical compositions obtained in this work were similar to those of the literature data with regard to calcium and sodium montmorillonites (Fatimah, Wang, & Wulandari, 2011; Mishra & Rao, 2005).

XRD patterns of the pristine ZnO indicated that the wurtzite phase (ICDD crystallographic charts 00-036-1451) was obtained irrespective of the pH used during synthesis. This phase was also observed in the hybrids, especially in

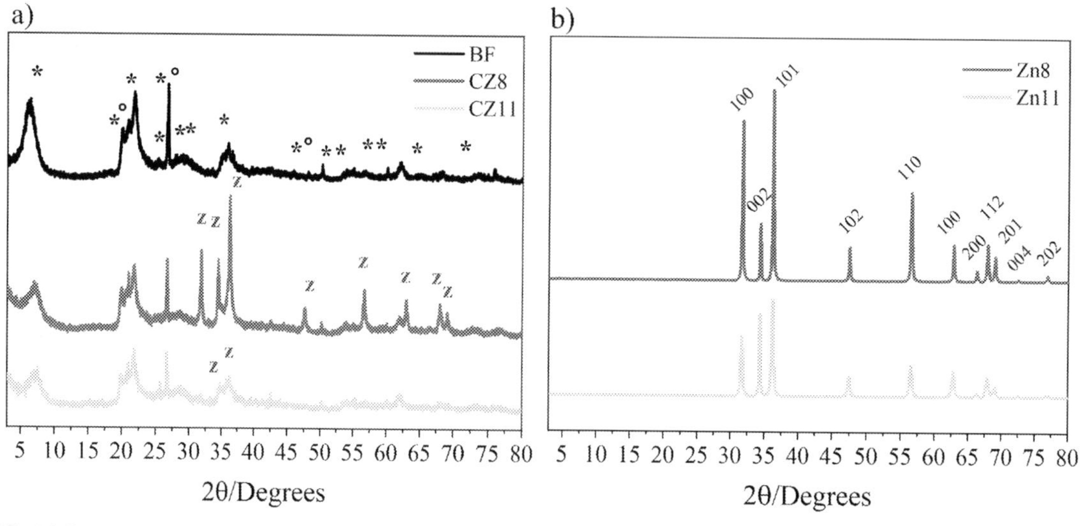

FIG. 11.8 XRD patterns of the pristine bentonite (BF), ZnO, and ZnO/bentonite hybrids. Both hybrids and ZnO were synthesized under the same conditions, using suspensions with a pH value of 8 or 11.

TABLE 11.2 Chemical composition (%) of the bentonite before and after impregnation with ZnO.

Catalyst	SiO_2	Al_2O_3	Fe_2O_3	MgO	TiO_2	CaO	Na_2O	K_2O
BF	77	13	6	2	<1	1	<1	<1
CZ08[a]	78	14	6	2	<1	<1	–	<1
CZ11[a]	78	14	6	2	<1	<1	–	<1

[a]*Percentage of the elements was calculated considering only the elements present in the pristine bentonite.*

the CZ08 hybrid, which indicated the success of the synthesis procedure. The increase of pH from 8 (Zn08) to 11 (Zn11) led to a smaller crystallinity of the ZnO particles, thus confirming the previous observation of the hybrid materials, which displayed higher-intensity ZnO peaks for the sample synthesized at pH 8 (CZ08) compared to those for the sample synthesized at pH 11 (CZ11).

According to the chemical composition after ZnO impregnation, the hydrothermalization process did not change the proportion among the elements present in the pristine bentonite, which indicated that dissolution of the clay components did not take place. Jozefaciuk and Bowanko (2002) have described the reaction of the Wyoming-type bentonite in solutions with pH 13.5 at 35 and 60 °C for 1–730 days and have reported the same stability.

The amount of ZnO was also quantified by XRF and values of 19% and 7% were obtained for the samples CZ08 and CZ11, respectively. These data were in agreement with the XRD patterns (Fig. 11.8), which showed ZnO peaks with higher intensity for the sample CZ08.

Micrographs of ZnO, bentonite, and the hybrids are illustrated in Figs. 11.9 and 11.10. The SEM images of zinc oxide (Fig. 11.9) obtained under the same conditions as the hybrid materials indicated different morphologies for the synthesis under different pH values. At pH 8, formation of rod-like particles and well-defined hexagonal rods was observed, whereas flower-like morphologies were formed at pH 11.

The FE-SEM images of the bentonite precursor (Fig. 11.10A and B) indicated a heterogeneous morphology, with some round aggregated particles probably assigned to the presence of quartz (Fig. 11.10A), and regions with lamellar stacking, which are characteristic of smectites. This behavior was expected due to the use of nonpurified clay.

For the hybrids, significant changes were observed in the morphology of ZnO for the different pH conditions used during synthesis, as observed in the FE-SEM images shown in Fig. 11.10C–F. The presence of bentonite in the reaction medium had a considerable interference in the morphology of the ZnO nanoparticles, as it acted as a seed for ZnO crystallization. Similar morphologies were observed for both hybrids, with two highly different regions, one with nanometric particles deposited on the material surface and another region with these nanometric particles distributed among the exfoliated lamellae of the smectite, which forms a plate-like morphology. Moreover, the pH changed the distribution of ZnO on the bentonite surface, and a higher amount of nanometric particles may be observed when synthesis was performed at pH 8.0.

The textural data displayed in Table 11.3 showed a decrease in the surface area, pore volume, and pore diameter of the bentonite after ZnO deposition. This behavior may be related to the formation of ZnO among the exfoliated lamellae and on the surface of the bentonite, as previously observed (Fig. 11.10). An average pore diameter of 3.75 nm was determined for the pristine bentonite and for the hybrids, confirming the

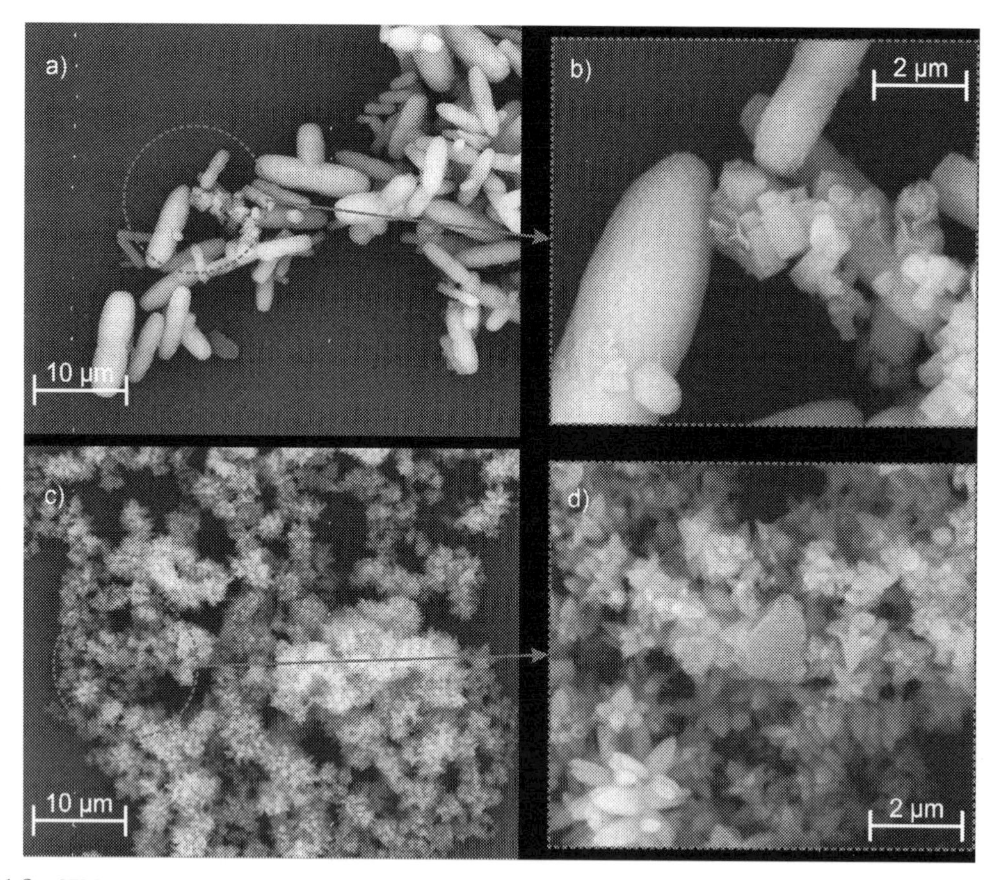

FIG. 11.9 SEM images of the ZnO samples: (A and B) Zn08 and (C and D) Zn11.

presence of mesopores (Hayati-Ashtiani, 2011; Noyan, Önal, & Sarikaya, 2006; Sing, 1982). In relation to the pristine ZnO, no meaningful difference was observed in the surface area of the two materials, in spite of the different morphologies, which indicates that rods observed in Zn08 are probably composed of agglomerated smaller particles, as indicated by the rugosity observed in the micrograph displayed in Fig. 11.9B.

Catalytic test using ZnO and ZnO/bentonite hybrids

Viscosity is not a suitable technique to measure conversion of triacylglycerol to ester, due to formation of intermediate compounds such as monoacylglycerol and diacylglycerol in addition to esters and free fatty acids during transesterification reactions (Eq. 11.1), which also change solution viscosity. In spite of this, it is a good indicator of catalytic activity.

According to the kinematic viscosity of the reaction products obtained with the different catalysts, no meaningful reduction occurred in relation to the viscosity of soybean oil ($34.9\,m^2\,s^{-1}$), when a reaction was performed at 130 °C, as displayed in Table 11.4. A greater decrease occurred for reactions at 200 °C, and viscosity reduction changed from 2.9% to 70.5% and from 0.6% to 71.9% for the CZ08 and CZ11 hybrids, respectively, during the first catalytic test (R0). These data indicated that temperature was an extremely important factor for

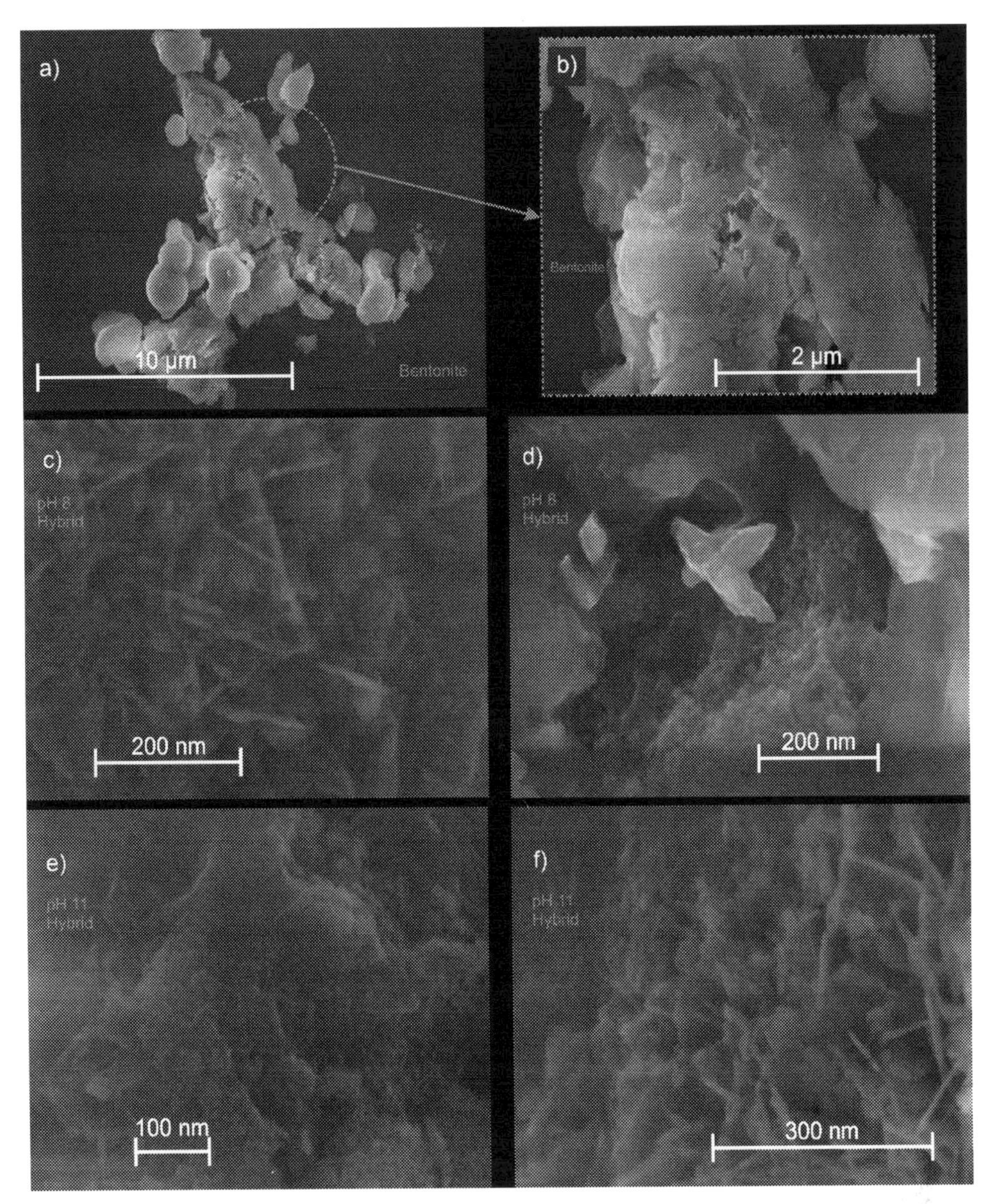

FIG. 11.10 FE-SEM images of the pristine bentonite (A and B) and ZnO/bentonite hybrids: CZ08 (C and D) and CZ11 (E and F).

the catalytic process, and similar viscosity reductions were obtained for both materials. Moreover, the pristine bentonite had a small activity and a viscosity reduction of 32.1% was obtained even for reactions at 200 °C.

According to the viscosity data, CZ11 had a smaller deactivation process during reuse (samples R2 and R3), associated with a variation of 5.7% in the viscosity reduction from the first to the third test, whereas CZ08 had a variation of

TABLE 11.3 Results of nitrogen adsorption measurements for the pristine bentonite (BF), ZnO, and ZnO/bentonite hybrids.

Catalyst	S_{BET} ($m^2\,g^{-1}$)[a]	V_p ($cm^3\,g^{-1}$)[b]	a_p ($m^2\,g^{-1}$)[c]	d_p (nm)[d]
Zn08	1.1	0.003	–	–
Zn11	2.1	0.004	–	–
BF	99	0.1016	60.2	3.75
CZ08	71	0.0945	50.6	3.75
CZ11	54	0.0961	47.4	3.75

[a] *BET-specific surface area.*
[b] *Pore volume.*
[c] *Pore area.*
[d] *Pore diameter.*

20.9% between the same tests. Although these results are usually associated with catalyst deactivation, they can also be assigned to a mass loss of the catalyst during the test itself, which was followed by catalyst recovery and washing.

When pristine ZnO was used as a catalyst, Zn08 was more effective in viscosity reduction (74.5%) compared to Zn11 (57.6%), whereas the hybrids had almost the same viscosity reduction. Considering the amount of the active phase in the hybrids and the low activity of the pristine bentonite, an important synergic effect was observed when the hybrids were used as catalysts, especially for CZ11, which had only 7% of ZnO and a smaller viscosity reduction when Zn11 was used as the catalyst.

A more accurate analysis of the ester conversion was performed by ^{1}H and ^{13}C NMR

TABLE 11.4 Data of viscosity reduction of the catalytic test products after transesterification reaction using different catalysts.

Temperature (°C)	Catalyst	Viscosity ($m^2\,s^{-1}$)	Reduction of viscosity (%)
130	CZ08R0	33.9	2.9
	CZ11R0	34.7	0.6
200	CZ08R0	10.3	70.5
	CZ08R1	11.8	66.2
	CZ08R2	17.6	49.6
	CZ11R0	9.8	71.9
	CZ11R1	10.6	69.6
	CZ11R2	10.8	66.2
	BF	23.7	32.1
	Zn08	8.9	74.5
	Zn11	14.8	57.6

spectroscopy, by which chemical group characteristics of ethyl esters, triacylglycerol, and intermediate compounds were identified in the reaction products. The ^{1}H NMR spectra of soybean oil (SO) before and after the catalytic test at 200 °C, using the pristine solids and hybrids as catalysts, are displayed in Fig. 11.11.

The spectrum range between 0 and 3.0 ppm did not change after the transesterification reaction because it was related to the peaks assigned to the fatty acid group characteristics of each triacylglycerol. The conversion of triacylglycerol (TAG) to ethyl esters (EE) was evaluated considering the doublet of doublets between 4.0 and 4.4 ppm, associated with the methylene hydrogen atoms of the glycerol group. This region was modified after conversion of triacylglycerol to ethyl ester, which has a quartet between 4.0 and 4.2, assigned to the hydrogen atoms of the ethoxy group (CH_3CH_2O–) (Ghesti et al., 2007; Guzatto, Defferrari, Reiznautt, Cadore, & Samios, 2012). Conversion to ester was calculated according to Eq. (11.1) and the results are displayed in Fig. 11.11.

Amplification of the ^{1}H NMR spectra, displayed in Fig. 11.11, indicated the formation of a quartet between 4.0 and 4.2 assigned to ethyl ester, for all the catalytic tests. In spite of this, high-intensity peaks due to the doublet of doublets between 4.0 and 4.4, assigned to the glycerol portion of TAG, indicated that a small conversion was observed when pristine bentonite was used as the catalyst. When the pristine ZnO (Zn08 and Zn11) and the hybrids (CZ08 and CZ11) were used as catalysts, small peaks were also observed between 4.25 and 4.35 ppm, which suggested that a partial conversion also occurred during these catalytic tests and diacylglycerol (DAG) and monoacylglycerol (MAG) may also be present in the reaction product along with triacylglycerol.

The conversion percentage obtained by ^{1}H NMR exhibited the same behavior as observed for the kinematic viscosity data and confirmed the low activity of the pristine bentonite (43.9%)

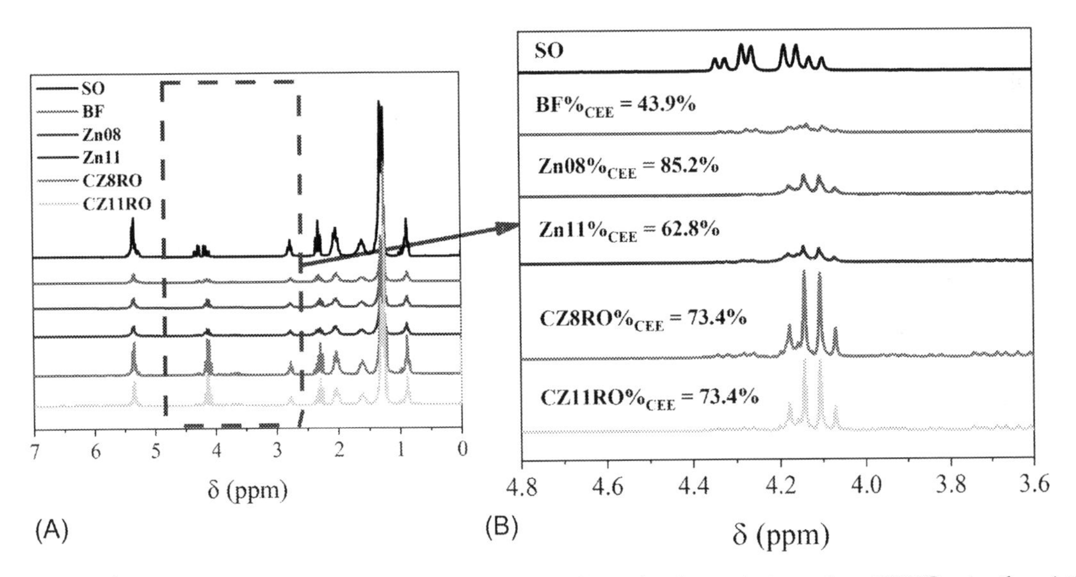

FIG. 11.11 ^{1}H NMR spectra of soybean oil and the reaction products after the catalytic reaction at 200 °C, using the pristine solids (BF, Zn08, and Zn11) and the ZnO/bentonite hybrids (CZ08 and CZ11) as catalysts. Amplification between 3.8 and 4.6 ppm and the ethylic conversion for each test (%C_{EE}) are presented.

and a 1.6-fold increase after ZnO deposition, irrespective of the hybrid used as the catalyst (73.4% of conversion for both hybrids). The synergic effect of the ZnO deposition on bentonite for both catalysts was also confirmed as previously discussed for the viscosity data. This effect was present even for the Zn08 and CZ08R0 catalysts, if the amount of ZnO deposited on the bentonite is considered. As the same catalyst mass was used during all of the catalytic tests and CZ08 had only 19% of ZnO, a smaller activity would be expected for CZ08, if this synergic effect was not present. For Zn11, this effect is even stronger, as this sample had a smaller catalytic activity and less ZnO was deposited on CZ11 (7%). This behavior may be assigned to the smaller particle size of ZnO after deposition.

^{13}C NMR was used to characterize the presence of TAG and EE, as well as the formation of intermediate compounds (DAG and MAG), during the catalytic tests. This evaluation was carried out in the range of 56–70 ppm, considering the following signals: methylene group of the alcohol portion of the ester group [$CH_3CH_2OC(=O)$-R] at 60 ppm; methylene groups of the glycerol portion of the TAG, DAG, and MAG compounds, which are observed at 62 and/or 69 ppm for TAG, at 65 and/or 68 for DAG, and at 63 ppm for MAG (Fernandes, De Souza, & De Vasconcellos Azeredo, 2012; Ren et al., 2018). This region of the spectra has been enlarged and it is shown in Fig. 11.12 for the reaction products obtained using the pristine catalysts (BF, Zn08, and Zn11) and the hybrids (CZ08R0 and CZ11R0).

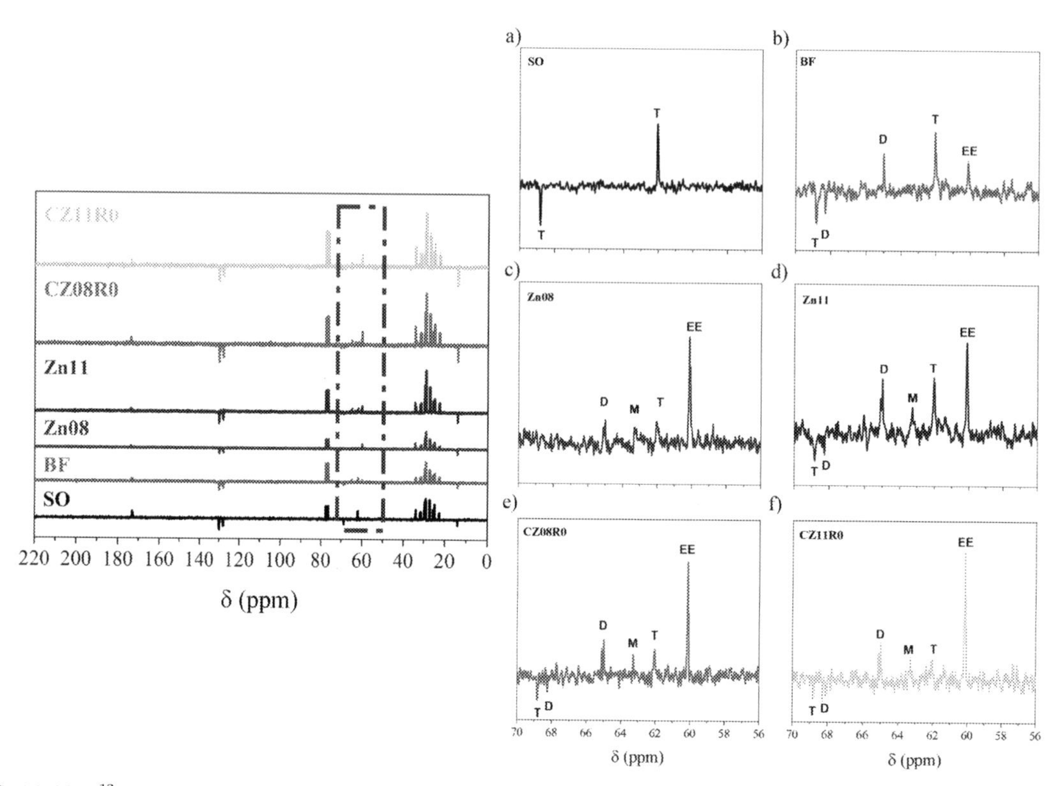

FIG. 11.12 ^{13}C NMR spectra of soybean oil and the reaction products after the catalytic reaction at 200 °C, using the pristine solids (BF, Zn08, and Zn11) and the ZnO/bentonite hybrids (CZ08 and CZ11) as catalysts. Amplification between 56 and 70 ppm is presented. Legend: T: triacylglycerol; D: diacylglycerol; M: monoacylglycerol; EE: ethyl ester.

The results obtained by ^{13}C NMR were in agreement with those obtained by ^{1}H NMR (Fig. 11.11) and confirmed the formation of the esters (peak identified as EE), intermediate compounds (M and D), and residual triacylglycerol (T), as shown in Fig. 11.12, thus confirming that the reaction was not complete. A small amount of ethyl esters was obtained when pristine bentonite was used as the catalyst, whereas a meaningful increase in the EE peak was obtained with the hybrids, which displayed spectra with a low intensity of the T, D, and M peaks compared to the EE peaks. The small decrease in viscosity of the oil during reaction with the pristine bentonite can be associated with the partial conversion of TAG into DAG, which is the first reaction during transesterification. The spectra confirmed the higher efficiency of Zn08, compared to Zn11, as indicated by the low intensity of the T, D, and M peaks compared to the EE peaks, similar to the hybrid materials.

da Silva et al. (2014) obtained biodiesel by methanolic and ethylic transesterification reactions using zinc-doped aluminum oxide as a catalyst. A higher activity was observed when the methanolic route was used compared to the ethylic one, which required a higher temperature (180 °C) to obtain a better conversion. Farias et al. (2020) also confirmed the viability of the ethylic route to obtain biodiesel by transesterification reaction, with a conversion of 83.9% at 180 °C using ZnO/bentonite as the catalyst. These results are in agreement with this work and confirm the viability of the ethylic transesterification route as long as a higher reaction temperature is used. The main aspect of the ethylic route is the use of renewable reagents in the whole catalytic process.

All of the results obtained in this work indicated the viability of the use of bentonite as a support for ZnO hybrids, with an important synergic factor that improves the catalytic activity toward transesterification reaction. Formation of the ZnO/bentonite hybrids by the microwave-assisted hydrothermal method at pH 8 or 11 favors formation of active sites, probably due to the deposition of nanometric particles on the bentonite, adding value to this natural material. Moreover, no previous treatment has been carried out and the bentonite can be used as received, which reduces the cost and time of synthesis.

MoO_3 and MoO_3/bentonite applied as catalysts for the synthesis of biodiesel

Evaluation of the massic form of MoO_3

XRD patterns and Raman spectra of the materials obtained using different conditions of synthesis are presented in Fig. 11.13.

The same hexagonal structure was obtained after hydrothermalization under different conditions (Fig. 11.4A), according to the ICDD index card 21-0569. A meaningful difference between the two patterns was observed in relation to the intensity of the (100) and (200) reflections, which was much higher for the sample h-MoO_3 100 °C, thus indicating that some orientation occurred. After heat treatment of the sample h-MoO_3 150 °C under conventional heating, a phase transition from the hexagonal to the orthorhombic structure (ICDD 05-0508) was observed, associated with the elimination of H_2O and NH_4^+ from the structure.

This hexagonal–orthorhombic phase transition was confirmed by Raman spectroscopy, as presented in Fig. 11.13B. For h-MoO_3, bands between 850 and 1000 cm^{-1} were assigned to the stretching modes of the bond Mo=O; bands between 200 and 700 cm^{-1} were assigned to the scissor, balance, or torsion mode of the O–Mo–O bond; and bands below 200 cm^{-1} were due to the deformation and torsion modes of the MoO_6 octahedron (Moura et al., 2018). For α-MoO_3, the stretching modes were observed between 450 and 1000 cm^{-1} (Atuchin, Gavrilova, Kostrovsky, Pokrovsky, & Troitskaia, 2008; Atuchin et al., 2011; Seguin, Figlarz, Cavagnat, & Lassègues, 1995; Zhang et al., 2011). Experimental data showed a good agreement with the literature data for both the phases (Chithambararaj, Yogamalar, & Bose, 2016;

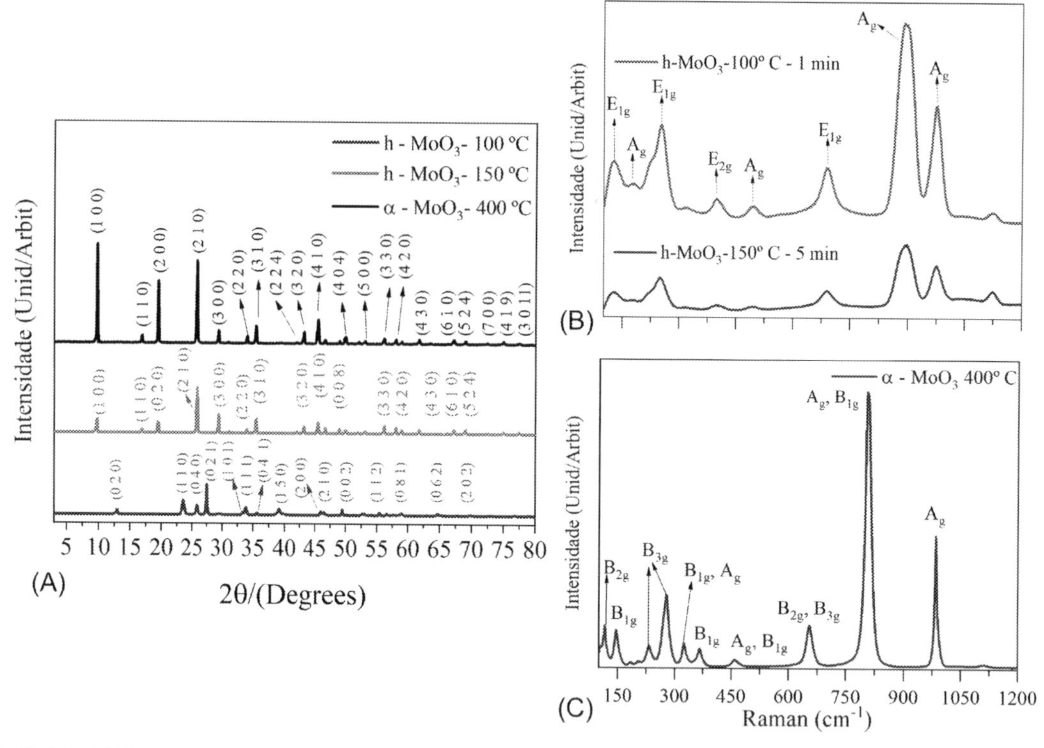

FIG. 11.13 XRD patterns (A), Raman spectra (B), and MoO_3 (C) obtained under different conditions. *unidentified phase.

Dieterle, Weinberg, & Mestl, 2002; Ramana et al., 2009).

According to Ramana et al. (2009), Chithambararaj et al. (2016), and Dieterle et al. (2002), during the hydrothermal synthesis of h-MoO_3, a self-assembly of MoO_6 took place assisted by NH_4^+ and H_2O molecules. When these molecules are not present inside the hexagonal tube, the hexagonal structure does not crystallize. At higher temperatures and longer times, the properties of the supercritical fluid changed depending on the solvent and the acidity of the medium and formation of α-MoO_3 was favored by a higher pressure. According to Yue-Ping, An-Wu, Ai-Miao, and Rui-Jin (2005) and Gong, Zeng, and Zhang (2015), the increase of the reaction temperature formed new growing nuclei due to the higher internal pressure inside the material, leading to the formation of new crystalline structures.

SEM micrographs of the samples obtained under different conditions are presented in Fig. 11.14. Samples with a hexagonal structure had similar morphologies with hexagonal rods of different diameters and lengths, but with bigger particles for the sample h-MoO_3 150 °C, which was obtained at higher temperatures and longer times. Similar results were obtained by Ramana et al. (2009) and Song, Ni, Gao, and Zheng (2007), who observed the same morphology using the hydrothermal method at higher temperatures. After conventional heat treatment, a meaningful change in the surface was observed with a more heterogeneous aspect. A cleavage seemed to occur in the longitudinal and diagonal directions of some rods. As the

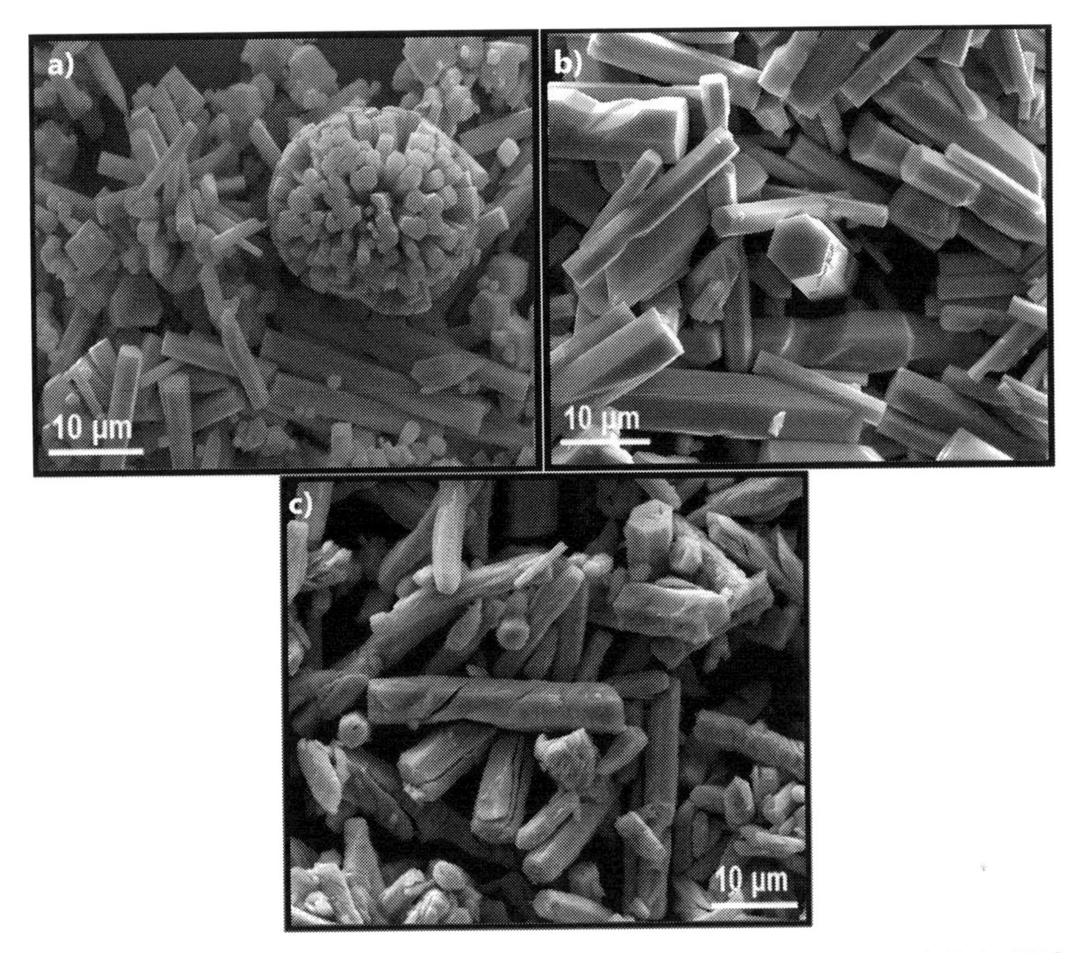

FIG. 11.14 SEM images of MoO_3 obtained under different conditions. (A) h-MoO_3 100 °C, (B) h-MoO_3 150 °C, and (C) α-MoO_3 400 °C.

hexagonal MoO_3 structure has been characterized by the presence of ammonium ions in the channels of the structure (Chithambararaj et al., 2016; Dieterle et al., 2002; Ramana et al., 2009), the cleavage process was assigned to the elimination of these ions during heat treatment.

The phase transition from hexagonal to orthorhombic led to an increase in the BET surface area, larger pore volume, and larger pore diameter as determined by N_2 adsorption measurements (Table 11.5). This behavior was directly related to the structural properties of the hexagonal and orthorhombic phases, as the last one has been characterized by a lamellar structure, with an increased amount of mesopores. This property may be interesting for catalytic purposes, as a higher surface area may enhance activity.

The samples h-MoO_3 100 °C and α-MoO_3 400 °C were used in the catalytic tests. In relation to the hexagonal structure, the sample obtained with shorter time (1 min) was chosen for

TABLE 11.5 Results of N_2 adsorption measurements of the MoO_3 catalysts.

Catalyst	$S_{BET}(m^2 g^{-1})^a$	V_p (cm^3/ g^{-1})^b	D_p(nm)^c
h-MoO_3 100 °C	0.4	0.0024	3.75
h-MoO_3 150 °C	0.6	0.0057	3.75
α-MoO_3 400 °C	3.5	0.0135	4.25

[a] *BET-specific surface area.*
[b] *Pore volume.*
[c] *Pore diameter.*

economical purposes. From now on, these samples will be simply named as h-MoO_3 and α-MoO_3. The acidity of these samples was measured by adsorption of *n*-butylamine under controlled conditions. Curves of mass loss after *n*-butylamine adsorption are displayed in Fig. 11.15.

Mass loss steps were determined considering the DTG curve for each sample. For both materials, the fourth step was not considered due to Mo^{6+} reduction. For calculation of the acid sites, the mass loss of the samples with *n*-butylamine was subtracted from the sample heat treated under the same conditions without *n*-butylamine. Moreover, it was not possible to distinguish between physisorption and weak chemisorption, both represented in the first mass loss step. Results are displayed in Table 11.6 and they indicated that α-MoO_3 had less acid sites than h-MoO_3, but all acid sites of h-MoO_3 were due to physisorption and weak and medium chemisorption, whereas most acid sites of sample α-MoO_3 were assigned to strong chemisorption. The behavior of h-MoO_3 may be assigned to the presence of NH_3 in the hexagonal structure, avoiding the adsorption of n-butylamine on these strong acid sites.

These two materials were applied in the synthesis of biodiesel. Results of viscosity reduction after catalytic tests are displayed in Table 11.7. Great values of viscosity reduction were obtained for MoO_3 with hexagonal and orthorhombic structures, indicating that these materials had a high activity for the transesterification process. In spite of this, it was not possible to ensure that the reaction was complete, as other compounds may still be present in the final product, as previously discussed. An extremely small variation in viscosity was obtained using the two different

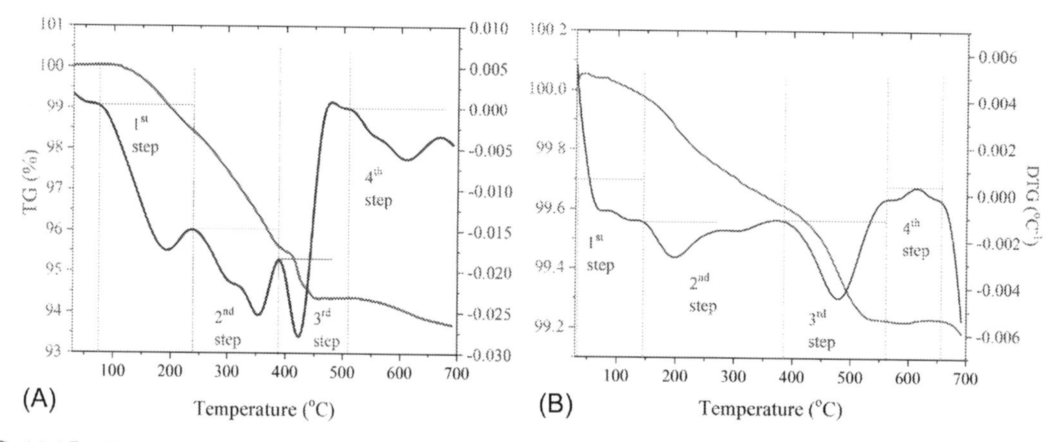

FIG. 11.15 Thermogravimetric analysis of samples h-MoO_3 (A) and α-MoO_3 (B) after *n*-butylamine adsorption.

TABLE 11.6 Acid sites of h-MoO_3 and α-MoO_3 according to the TG curves after *n*-butylamine adsorption.

Catalyst	Mass loss steps	Physisorption/weak acid sites	Medium acid sites	Strong acid sites	Total (mmol)
h-MoO_3	Temperature	75–236 °C		–	–
	Acid sites	0.0746 mmol		–	0.0746
α-MoO_3	Temperature	69–145 °C	145–377 °C	377–551 °C	–
	Acid sites	0.0096 mmol	0.0064 mmol	0.0316 mmol	0.0476

TABLE 11.7 Viscosity reduction values and ethyl ester conversion (C_{EE}) of the catalytic test products after transesterification reaction under different conditions.

		150 °C		200 °C	
Catalyst	Time (h)	Reduction (%)	C_{EE} (%)	Reduction (%)	C_{EE} (%)
h-MoO_3	0.5	76.8	76.2	83.2	73.0
	1	81.1	89.4	83.7	92.5
	2	81.8	88.2	83.5	97.1
α-MoO_3	0.5	72.4	91.4	83.5	70.8
	1	80.0	94.0	83.7	76.7
	2	80.6	94.0	83.8	97.1

MoO_3 structures, especially when the catalytic test was performed at 200 °C.

The transesterification reaction was confirmed by ^{1}H NMR spectroscopy as displayed in Fig. 11.16. As previously discussed, ^{1}H NMR spectra were analyzed considering the region between 4.0 and 4.4 ppm (Farias et al., 2015; Fatimah et al., 2011; Mendieta et al., 2011).

After the catalytic tests with both materials, a meaningful decrease in the amount of triacylglycerol occurred for all the reactions performed at 150 and 200 °C as the doublet of doublets between 4.2 and 4.4 ppm were only observed as extremely small peaks for reactions performed with both catalysts at 150 °C for 0.5 h. For the catalytic tests conducted at 150 °C, the quartet between 4.0 and 4.2 ppm was clearly observed even after 0.5 h of reaction for both catalysts, indicating that a meaningful amount of ethyl esters was formed, in agreement with the viscosity data (Table 11.7), which showed a high viscosity reduction after a short time of reaction. On the other hand, two different profiles were observed for the catalytic tests at 200 °C, but no peaks between 4.2 and 4.4 ppm were observed. This behavior may be assigned to the presence of intermediate compounds, whose peaks are superposed to the ethyl ester ones. Apparently, a high conversion was obtained at 200 °C for 2 h for both samples and also at 200 °C for 1 h for h-MoO_3.

^{1}H NMR spectra were used to quantify the conversion percentages of the transesterification reaction (Farias et al., 2015; Gelbard, Brès, Vargas, Vielfaure, & Schuchardt, 1995; Ghesti et al., 2007; Mendieta et al., 2011; Monteiro, Ambrozin, Lião, & Ferreira, 2009), as displayed in Table 11.7. A high conversion was attained

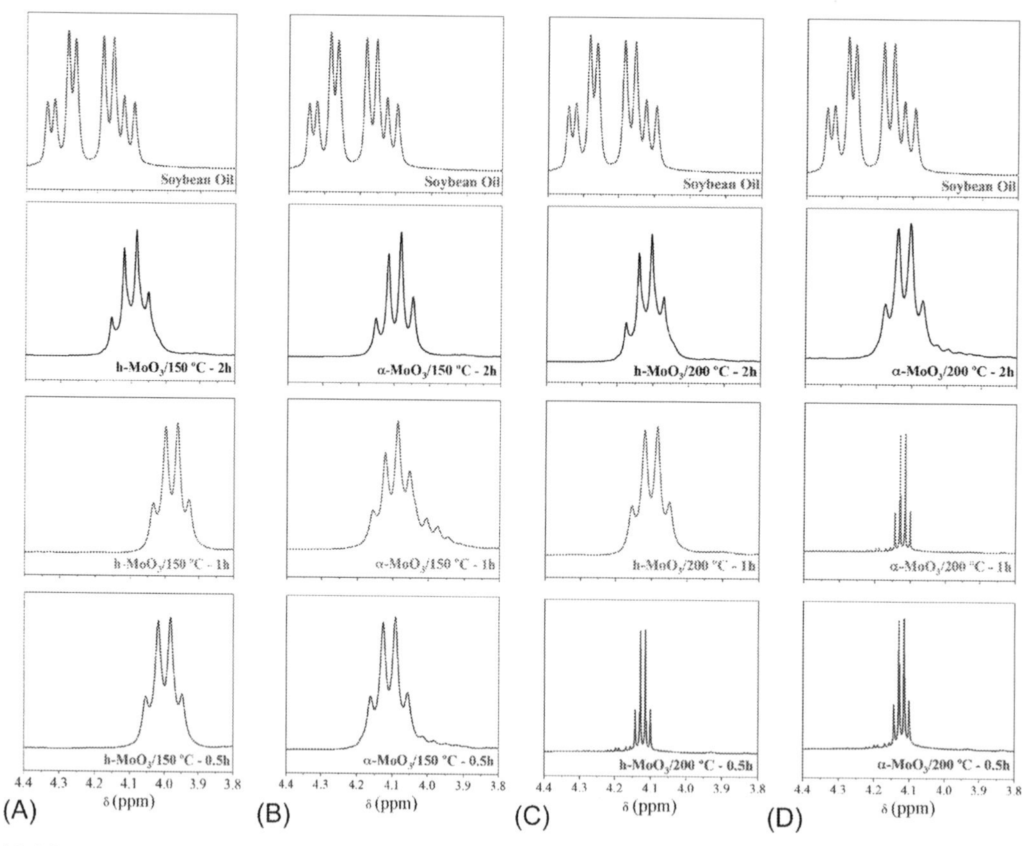

FIG. 11.16 Enlargement between 3.8 and 4.4 ppm of the ^{1}H NMR spectra of soybean oil and the products after the catalytic test at 150 °C with h-MoO_3 (A) and α-MoO_3 (B) and at 200 °C with h-MoO_3 (C), and α-MoO_3 (D).

after 2 h of reaction for both catalysts, confirming the efficiency of MoO_3 in the transesterification reaction. For h-MoO_3, a higher temperature was necessary to obtain a greater conversion, which may be assigned to the absence of strong acid sites available for adsorption. A better result was obtained when α-MoO_3 was used as the catalyst at 150 °C, which may be assigned to the higher surface area, in addition to a greater amount of strong acid sites. At 200 °C, the same conversion was attained for both catalysts after 2 h, but a greater activity was observed for h-MoO_3 after 1 h, probably related to the large amount of weak and medium acid sites, which were probably more active at this temperature.

The reaction intermediates, monoacylglycerols and diacylglycerides, usually formed during the transesterification reaction, were qualitatively evaluated by ^{13}C NMR spectroscopy (Fig. 11.17), as previously discussed (Casas et al., 2012; Farias et al., 2015; Fatimah et al., 2011; Fernandes et al., 2012; Moreira, Perez, Zanin, & de Castro, 2007; Siddiqui et al., 2003).

Evaluation of ^{13}C NMR between 56 and 70 ppm confirms the formation of ethyl esters (EE) even after 0.5 h of reaction for both catalysts at both temperatures, besides small residual peaks assigned to diacylglycerols (D) and monoacylglycerols (M). Triacylglycerol was completely consumed during all the catalytic tests. The best

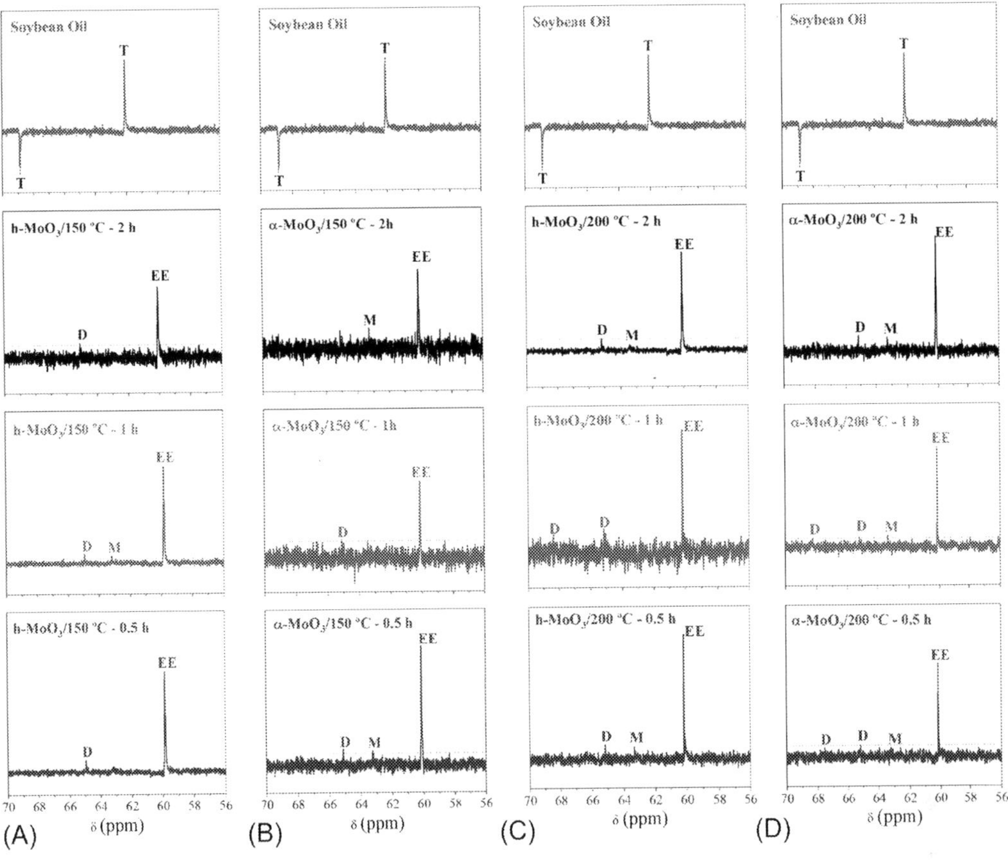

FIG. 11.17 Amplification between 56 and 70 ppm of the ^{13}C NMR spectra of soybean oil and the products after the catalytic test at 150 °C with h-MoO_3 (A) and α-MoO_3 (B) and at 200 °C with h-MoO_3 (C) and α-MoO_3 (D). Legend: M = monoacylglycerol, D = diacylglycerol, T = triacylglycerol and EE = ethyl ester.

result was obtained at 150 °C for 1 and 2 h with the α-MoO_3 catalyst, which did not display intermediate phases.

The transesterification reactions were also evaluated by gas chromatography to identify the fatty acid esters. The composition was similar to that of soybean oil (Farias et al., 2016; Likozar & Levec, 2014), presenting linoleic acid (C18:2) as a major compound, and smaller amounts of linoleic acid (C18:3), oleic acid (C18:1), stearic acid (C18:0), and palmitic acid (C16:0). According to the results obtained by GC–MS, it was possible to obtain biodiesel with an ester amount above 96.5%, in agreement with the European Standard (prEN 1403) and with the Resolution n. 07/2008, from the "Agência Nacional de Petróleo, Gás Natural e Biocombustível" (Resolução, 2014). The highest conversion was obtained using α-MoO_3 as the catalyst with a reaction period of 2 h at 150 °C, in agreement with the data obtained by ^{13}C NMR.

Use of MoO_3/bentonite hybrids as catalysts

XRD patterns of the pristine bentonite and MoO_3/bentonite hybrids are displayed in Fig. 11.18. The crystalline phases observed in the bentonite were the same as previously discussed in the section "ZnO and ZnO/bentonite hybrids applied as catalysts for obtaining

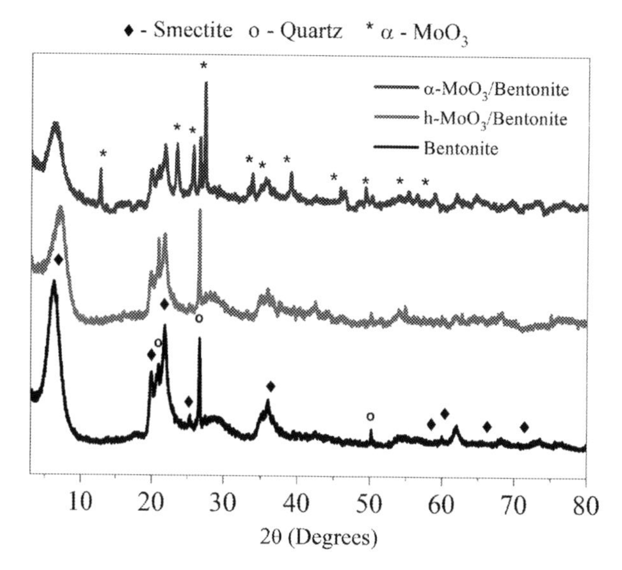

FIG. 11.18 XRD patterns of the bentonite before and after impregnation with h-MoO_3 and α-MoO_3.

biodiesel." The original composition and structure of the bentonite was maintained after hydrothermalization, as observed for ZnO/bentonite hybrids. The basal space of the smectite (1.41 nm) had no meaningful change after impregnation with α-MoO_3 (1.39 nm), whereas a small decrease was observed for h-MoO_3 (1.25 nm), which suggested that cation exchange was favored for this last case.

Analysis of the XRD pattern of the h-MoO_3/bentonite hybrid indicated that only peaks assigned to the bentonite were observed. This behavior may be related to the cation exchange with the NH_4^+ ions present in the solution, which did not become available for h-MoO_3 crystallization. On the other hand, peaks assigned to the MoO_3 orthorhombic phase were clearly observed in the XRD pattern of the α-MoO_3/bentonite hybrid, in agreement with the ICDD file 05-0508. As the α-MoO_3/bentonite hybrid was calcined at 400 °C after hydrothermalization, the crystallization of the orthorhombic phase was favored.

The amount of MoO_3 in the hybrids was quantified by XRF, which indicated similar amounts of MoO_3 in the two materials, 1.8% and 2.0% in the α-MoO_3/bentonite and h-MoO_3/bentonite hybrids, respectively. These results confirm that the crystallization process of h-MoO_3 was not favored, as discussed above.

Micrographs of the bentonite before and after impregnation are displayed in Fig. 11.19. The pristine bentonite (Fig. 11.19A and B) had a nonuniform morphology with plate-like particles, characteristic of smectites, in agreement with the literature data (Grim, 1968; Khobzaoui, Tennouga, Benabadji, Mansri, & Bouras, 2019; Pusch, Zwahr, Gerber, & Schomburg, 2003). No meaningful change in the morphology was observed after impregnation, as indicated in Fig. 11.19C and D, even when calcination was performed (Fig. 11.19E and F). The characteristic hexagonal or plate-like morphologies of the MoO_3 were not observed, which may be due to the deposition of the oxide on the bentonite surface.

The pristine bentonite and the hybrids were used as catalysts for the ethylic transesterification of soybean oil. Viscosity reduction data after the catalytic tests at 150 and 200 °C are presented in Table 11.8. When the pristine bentonite was used as a catalyst, an extremely small viscosity reduction was obtained, with percentages

FIG. 11.19 Micrographs of the pristine bentonite (A and B), h-MoO_3/bentonite (C and D), and α-MoO_3/bentonite (E and F) hybrids.

varying from 4.3% to 7.4%, depending on the temperature. When the MoO_3/bentonite hybrids were used, this reduction was much higher than the pristine bentonite one, but it was still extremely small and indicated that a partial conversion of the triacylglycerol took place, probably with formation of intermediate compounds. Similar to the catalytic tests with MoO_3, different behaviors were observed depending on the material. When the h-MoO_3/bentonite hybrid was used, the greatest viscosity reduction was obtained at 200 °C, whereas similar results were obtained for α-MoO_3/bentonite for both temperatures.

The catalytic behavior was confirmed by the conversion obtained from ^{1}H NMR results, which were calculated according to Eq. (11.1) and are displayed in Table 11.8. For the pristine bentonite, 10.4% of conversion was obtained. The improvement of the catalytic behavior after MoO_3 impregnation was extremely clear and the 1.9- to 7.8-fold increase in conversion was obtained depending on the reaction conditions. This result was obtained with only ~2% of

TABLE 11.8 Viscosity reduction data and ethyl ester conversion (C_{EE}) of the catalytic test products after transesterification reaction under different conditions.

		h-MoO_3/bentonite		α-MoO_3/bentonite	
Temperature (°C)	**Time (h)**	**Viscosity reduction (%)**	**C_{EE} (%)**	**Viscosity reduction (%)**	**C_{EE} (%)**
150	0.5	25.4	19.5	68.6	63.6
	1	30.6	31.3	79.4	71.5
	2	46.4	48.5	78.2	82.0
200	0.5	64.4	68.7	71.4	65.5
	1	66.0	89.0	69.6	64.2
	2	69.3	78.0	80.2	68.4

MoO_3 on the hybrid, which showed an important synergic effect of these hybrids.

The difference between the two hybrids was also confirmed by the C_{EE} conversion values, but numeric data were slightly different from viscosity reduction data, which may be related to the compounds present after each catalytic test, as previously discussed. For the h-MoO_3/bentonite hybrid, a greater catalytic activity was obtained at 200 °C, whereas the greatest conversion of the tests with α-MoO_3/bentonite was obtained at 150 °C. In relation to the hybrids, the different behaviors indicated that different acidities were obtained for different materials, as already observed for the pristine MoO_3 phases (Bail et al., 2013; De Almeida et al., 2014). Although the amount of MoO_3 impregnated on the bentonite in both hybrids was quite similar, the final structure of the active phase was extremely different, as it was amorphous in h-MoO_3/bentonite and crystalline in α-MoO_3/bentonite, with an orthorhombic structure. We believe that this behavior had a direct influence on the acid sites present on the material surface.

Ethylic transesterification reaction was evaluated by analysis of the region between 4.0 and 4.4 ppm (Ghesti et al., 2007; Guzatto et al., 2012), as discussed before and displayed in Fig. 11.20.

The decrease of the peaks between 4.2 and 4.4 ppm was extremely clear when the hybrids were used as catalysts and confirmed the synergic effect of MoO_3 impregnation on the TAG conversion, when compared to results presented by bentonite. The best result was obtained for the reaction performed with α-MoO_3/bentonite at 150 °C for 2 h, confirming the data displayed in Table 11.8.

Results displayed in Fig. 11.20 also indicated the superposition of doublet of doublets and the quartet between 4.0 and 4.2 ppm, as well as peaks probably assigned to diacylglycerol and monoacylglycerol, which indicated that the conversion was not complete (Ghesti et al., 2007; Ramana et al., 2009). This analysis was also performed by ^{13}C NMR spectroscopy, as displayed in Fig. 11.21, and allowed a better visualization of these intermediate compounds.

As discussed before, analysis of the reaction products was performed considering the region between 56 and 70 ppm. A meaningful decrease in the peaks assigned to triacylglycerol (62 and

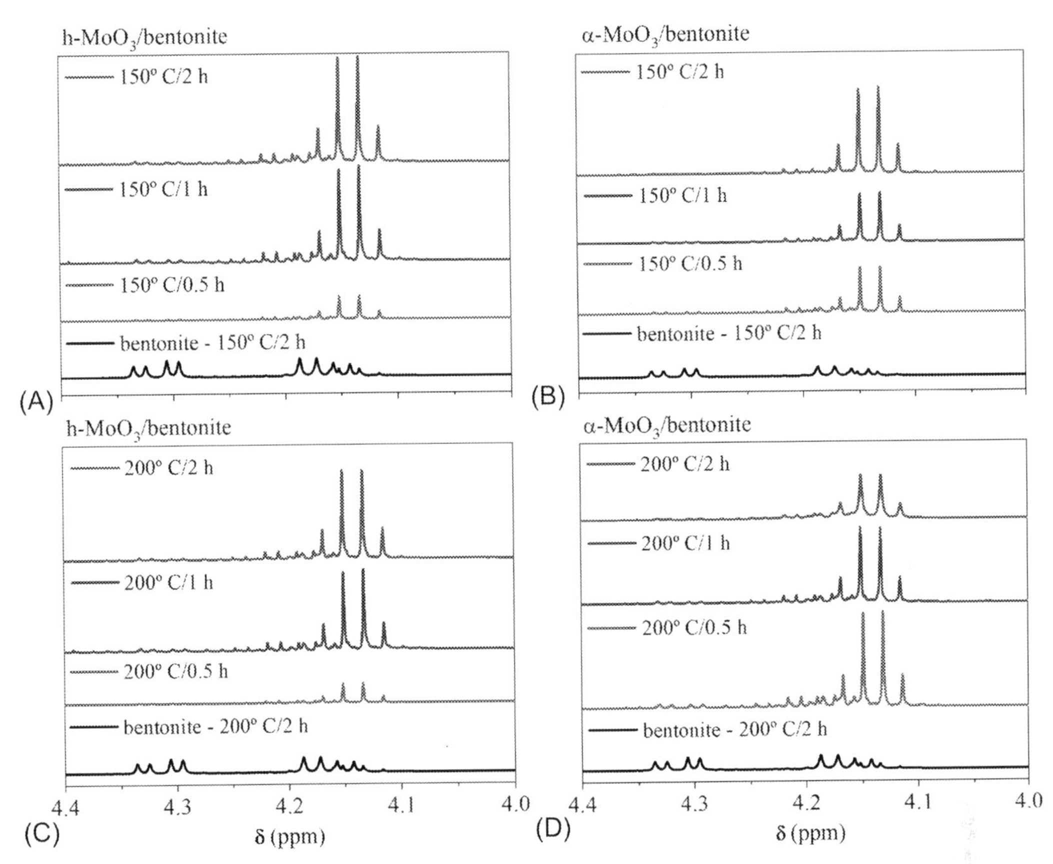

FIG. 11.20 ^1H NMR spectra of the reaction products obtained with different catalysts and reaction conditions: (A) h-MoO_3/bentonite at 150 °C, (B) α-MoO_3/bentonite at 150 °C, (C) h-MoO_3/bentonite at 200 °C, and (D) α-MoO_3/bentonite at 200 °C.

69 ppm) was observed for all the samples, in addition to the appearance of a peak assigned to the ester group at 60 ppm. The presence of the intermediate compounds, diacylglycerol and monoacylglycerol, between 63 and 69 ppm was also observed for all samples.

In spite of this, extremely meaningful differences were observed among the spectra. For the pristine bentonite, the high intensity of the TAG peak associated with the extremely low intensity of the EE peak confirmed that an extremely small conversion occurred. For the h-MoO_3/bentonite hybrid, conversion at 150 ° C was extremely small especially after 0.5 h of reaction, whereas a higher amount of EE seemed to be present when the reaction was performed at 200 °C for 1 and 2 h. The highest conversion was obtained when the α-MoO_3/bentonite hybrid was used as the catalyst, especially for the reaction at 150 °C for 2 h, when no peaks assigned to TAG were observed in the spectra and only small peaks assigned to DAG and MAG were obtained.

Results obtained by ^{13}C NMR spectroscopy were in close agreement with the data obtained by ^1H NMR and viscosity reduction and

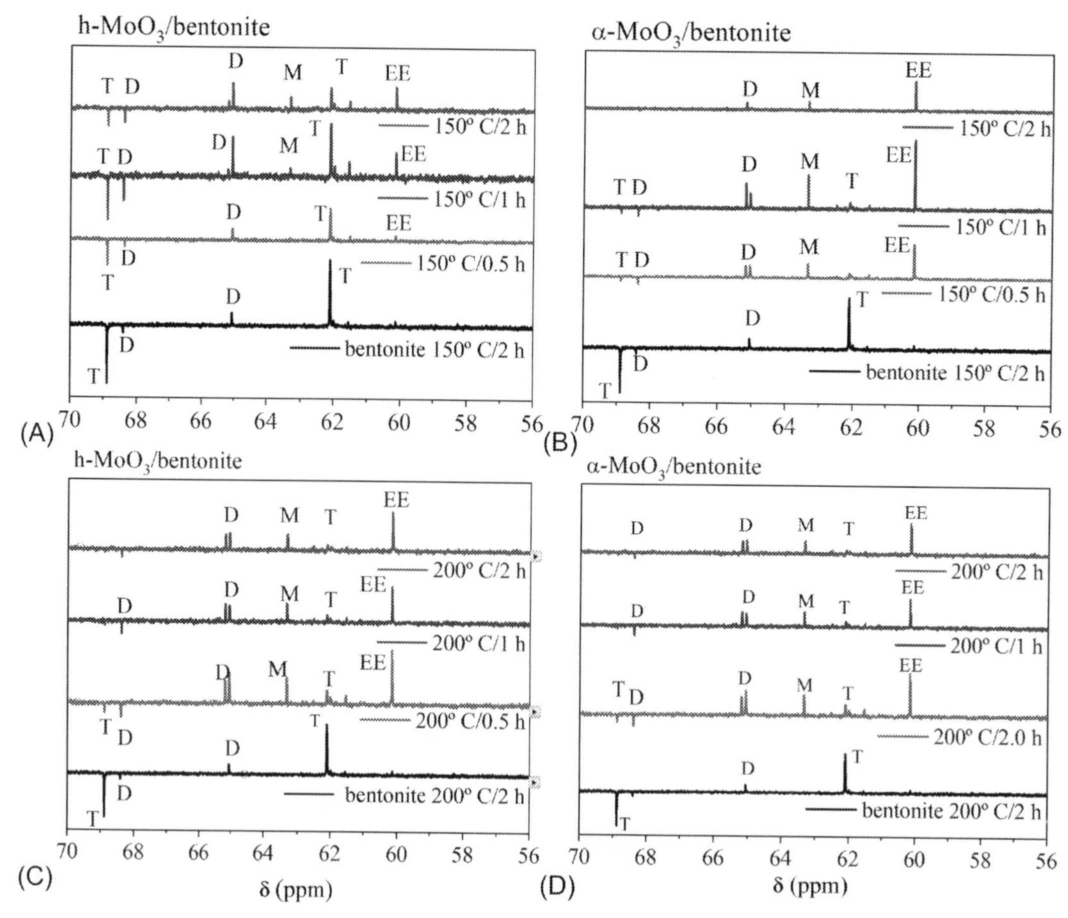

FIG. 11.21 ^{13}C NMR spectra of the reaction product obtained with different catalysts and reaction conditions: (A) h-MoO_3/bentonite at 150°C, (B) α-MoO_3/bentonite at 150°C, (C) h-MoO_3/bentonite at 200°C, and (D) α-MoO_3/bentonite at 200°C.

confirmed the formation of ethyl esters associated with the decrease in the amount of triacylglycerol. The best catalytic condition was obtained using the α-MoO_3/bentonite hybrid with reaction at 150°C for 2h. Similar to the ZnO/bentonite hybrids, a strong synergic effect was observed on MoO_3, with a meaningful increase in the catalytic activity of the bentonite after MoO_3 impregnation, even with only 2% of oxide in the hybrid. The crystallization of the orthorhombic phase in the hybrid had an important effect on the catalytic activity.

11.4 Conclusions

Impregnation of oxides on bentonite by the microwave-assisted hydrothermal method is an extremely interesting strategy to improve the catalytic activity of bentonite toward transesterification reaction, even using the ethylic route, which usually leads to smaller conversions. This behavior is associated with the formation of nanometric particles of the oxide on the bentonite surface, which causes a meaningful synergic effect in the catalytic activity. Comparing the two oxides tested in this work, an easier crystallization was observed for ZnO with impregnation of 19% and 7% at pH8 and 11, respectively. Despite the difference in the ZnO amount impregnated on bentonite, similar activities were obtained for both hybrids. Only 2% of MoO_3 was impregnated on bentonite, but a high conversion was obtained even with this small amount of active phase, which confirms the high catalytic activity of this oxide.

Conflict of interest

There is no conflict of interest in this work.

Acknowledgments

This work was supported by CT-INFRA/FINEP/MCTIC, CNPq/MCTIC (Grant Number 406857/2013-0-Edital 40/2013), Paraíba State Research Foundation (Grant Number 0012/2019-FAPESQ/CNPq), and CAPES/MEC (Coordenação de Aperfeiçoamento de Pessoal de Nível Superior-Finance Code: 001).

References

Abdala, E., Nur, O., & Mustafa, M. A. (2020). Efficient biodiesel production from algae oil using Ca-doped ZnO nanocatalyst. *Industrial and Engineering Chemistry Research, 59*(43), 19235–19243. https://doi.org/10.1021/acs.iecr.0c04118.

Al-Saadi, A., Mathan, B., & He, Y. (2020). Biodiesel production via simultaneous transesterification and esterification reactions over SrO–ZnO/Al2O3 as a bifunctional catalyst using high acidic waste cooking oil. *Chemical Engineering Research and Design, 162*, 238–248. https://doi.org/10.1016/j.cherd.2020.08.018.

Alba-Rubio, A. C., Santamaría-González, J., Mérida-Robles, J. M., Moreno-Tost, R., Martín-Alonso, D., Jiménez-López, A., et al. (2010). Heterogeneous transesterification processes by using CaO supported on zinc oxide as basic catalysts. *Catalysis Today, 149*(3–4), 281–287. https://doi.org/10.1016/j.cattod.2009.06.024.

Alhassan, F. H., Rashid, U., & Taufiq-Yap, Y. H. (2015). Synthesis of waste cooking oil based biodiesel via ferric-manganese promoted molybdenum oxide/zirconia nanoparticle solid acid catalyst: Influence of ferric and manganese dopants. *Journal of Oleo Science, 64*(5), 505–514. https://doi.org/10.5650/jos.ess14228.

Alves, L., Ferraz, E., & Gamelas, J. A. F. (2020). Composites of nanofibrillated cellulose with clay minerals: A review. *Advances in Colloid and Interface, 272*, 101994.

Antunes, W. M., Veloso, C. D. O., & Henriques, C. A. (2008). Transesterification of soybean oil with methanol catalyzed by basic solids. *Catalysis Today, 133–135*(1–4), 548–554. https://doi.org/10.1016/j.cattod.2007.12.055.

Atuchin, V. V., Gavrilova, T. A., Grigorieva, T. I., Kuratieva, N. V., Okotrub, K. A., Pervukhina, N. V., et al. (2011). Sublimation growth and vibrational microspectrometry of α-MoO3 single crystals. *Journal of Crystal Growth, 318*(1), 987–990. https://doi.org/10.1016/j.jcrysgro.2010.10.149.

Atuchin, V. V., Gavrilova, T. A., Kostrovsky, V. G., Pokrovsky, L. D., & Troitskaia, I. B. (2008). Morphology and structure of hexagonal MoO_3 nanorods. *Inorganic Materials, 44*, 622–627. https://doi.org/10.1134/s0020168508060149.

Bail, A., dos Santos, V. C., de Freitas, M. R., Ramos, L. P., Schreiner, W. H., Ricci, G. P., et al. (2013). Investigation of a molybdenum-containing silica catalyst synthesized by the sol-gel process in heterogeneous catalytic esterification reactions using methanol and ethanol. *Applied Catalysis B: Environmental, 130–131*, 314–324. https://doi.org/10.1016/j.apcatb.2012.11.009.

Bergaya, F., & Lagaly, G. (2013). Handbook of clay science. In *General introduction: Clays, clay minerals, and clay science* (5th ed.). Elsevier.

Berrios, M., & Skelton, R. L. (2008). Comparison of purification methods for biodiesel. *Chemical Engineering Journal, 144*(3), 459–465. https://doi.org/10.1016/j.cej.2008.07.019.

Borah, M. J., Devi, A., Borah, R., & Deka, D. (2019). Synthesis and application of Co doped ZnO as heterogeneous nanocatalyst for biodiesel production from non-edible oil. *Renewable Energy, 133*, 512–519. https://doi.org/10.1016/j.renene.2018.10.069.

Borysiewicz, M. A. (2019). ZnO as a functional material, a review. *Crystals, 9*(10). https://doi.org/10.3390/cryst9100505.

Boz, N., Degirmenbasi, N., & Kalyon, D. M. (2013). Transesterification of canola oil to biodiesel using calcium bentonite functionalized with K compounds. *Applied Catalysis B: Environmental, 138–139*, 236–242. https://doi.org/10.1016/j.apcatb.2013.02.043.

Brigatti, M. F., Galán, E., & Theng, B. K. G. (2013). Chapter 2—Structure and mineralogy of clay minerals. In *Handbook of clay science* (5th ed.). Elsevier.

Casas, A., Ramos, M. J., Pérez, A., Simón, A., Lucas-Torres, C., & Moreno, A. (2012). Rapid quantitative determination by ^{13}C NMR of the composition of acetylglycerol mixtures as byproduct in biodiesel synthesis. *Fuel,*

92(1), 180–186. https://doi.org/10.1016/j.fuel.2011.06.061.

Cavalcanti, G. R. S., Fonseca, M. G., da Silva Filho, E. C., & Jaber, M. (2019). Thiabendazole/bentonites hybrids as controlled release systems. *Colloids and Surfaces B: Biointerfaces, 176*, 249–255. https://doi.org/10.1016/j.colsurfb.2018.12.030.

Cecilia, J. A., Arango-Díaz, A., Franco, F., Jiménez-Jiménez, J., Storaro, L., Moretti, E., et al. (2015). CuO-CeO_2 supported on montmorillonite-derived porous clay heterostructures (PCH) for preferential CO oxidation in H_2-rich stream. *Catalysis Today, 253*, 126–136. https://doi.org/10.1016/j.cattod.2015.01.040.

Chen, L., Zhou, C. H., Fiore, S., Tong, D. S., Zhang, H., Li, C. S., et al. (2016). Functional magnetic nanoparticle/clay mineral nanocomposites: Preparation, magnetism and versatile applications. *Applied Clay Science, 127–128*, 143–163. https://doi.org/10.1016/j.clay.2016.04.009.

Chithambararaj, A., & Bose, A. C. (2011). Hydrothermal synthesis of hexagonal and orthorhombic MoO3 nanoparticles. *Journal of Alloys and Compounds, 509*, 8105–8110. https://doi.org/10.1016/j.jallcom.2011.05.067.

Chithambararaj, A., Yogamalar, N. R., & Bose, A. C. (2016). Hydrothermally synthesized h-MoO3 and α-MoO3 nanocrystals: New findings on crystal-structure-dependent charge transport. *Crystal Growth & Design, 16*(4), 1984–1995.

Colombo, K., & Barros, A. A. (2009). Utilização de catalisadores heterogêneos na produção de Biodiesel. *Evidência—Ciência E Biotecnologia, 9*(1–2), 7–16.

Compton, D. L., Laszlo, J. A., Appell, M., Vermillion, K. E., & Evans, K. O. (2014). Synthesis, purification, and acyl migration kinetics of 2-monoricinoleoylglycerol. *Journal of the American Oil Chemists' Society, 91*(2), 271–279. https://doi.org/10.1007/s11746-013-2373-2.

da Silva, F. M., Pinho, D. M. M., Houg, G. P., Reis, I. B. A., Kawamura, M., Quemel, M. S. R., et al. (2014). Continuous biodiesel production using a fixed-bed Lewis-based catalytic system. *Chemical Engineering Research and Design, 92*(8), 1463–1469. https://doi.org/10.1016/j.cherd.2014.04.024.

da Silva, A. L., Luna, C. B. B., de Farias, A. F. F., de Medeiros, S. A. S. L., Meneghetti, S. M. P., Rodrigues, A. M., et al. (2020). From disposal to reuse: Production of sustainable fatty acid alkyl esters derived from residual oil using a biphasic magnetic catalyst. *Sustainability (Switzerland), 12*(23), 1–18. https://doi.org/10.3390/su122310159.

De Almeida, R. M., Souza, F. T. C., Júnior, M. A. C., Albuquerque, N. J. A., Meneghetti, S. M. P., & Meneghetti, M. R. (2014). Improvements in acidity for TiO_2 and SnO_2 via impregnation with MoO_3 for the esterification of fatty acids. *Catalysis Communications, 46*, 179–182. https://doi.org/10.1016/j.catcom.2013.12.020.

De León, M. A., Castiglioni, J., Bussi, J., & Sergio, M. (2008). Catalytic activity of an iron-pillared montmorillonitic clay mineral in heterogeneous photo-Fenton process. *Catalysis Today, 133–135*(1–4), 600–605. https://doi.org/10.1016/j.cattod.2007.12.130.

Deka, B. K., & Maji, T. K. (2012). Effect of nanoclay and ZnO on the physical and chemical properties of wood polymer nanocomposite. *Journal of Applied Polymer Science, 124*(4), 2919–2929. https://doi.org/10.1002/app.35314.

Dieterle, M., Weinberg, G., & Mestl, G. (2002). Raman spectroscopy of molybdenum oxides—Part I. structural characterization of oxygen defects in MoO_{3-x} by DR UV/VIS, Raman spectroscopy and X-ray diffraction. *Physical Chemistry Chemical Physics, 4*(5), 812–821. https://doi.org/10.1039/b107012f.

Dikici, T., & Demirci, S. (2019). Influence of thermal oxidation temperature on the microstructure and photoelectrochemical properties of ZnO nanostructures fabricated on the zinc scraps. *Journal of Alloys and Compounds, 779*, 752–761. https://doi.org/10.1016/j.jallcom.2018.11.241.

Dill, L. P., Kochepka, D. M., Lima, L. L., Leitão, A. A., Wypych, F., & Cordeiro, C. S. (2021). Brazilian mineral clays: Classification, acid activation and application as catalysts for methyl esterification reactions. *Journal of the Brazilian Chemical Society, 32*(1), 145–157. 10.21577/0103-5053.20200164.

Farias, A. F. F., Moura, K. F., Souza, J. K. D., Lima, R. O., Nascimento, J. D. S. S., Cutrim, A. A., et al. (2015). Biodiesel obtained by ethylic transesterification using CuO, ZnO and CeO_2 supported on bentonite. *Fuel, 160*, 357–365. https://doi.org/10.1016/j.fuel.2015.07.102.

Farias, A. F. F., Da Conceição, M. M., Cavalcanti, E. H. S., Melo, M. A. R., Dos Santos, I. M. G., & De Souza, A. G. (2016). Analysis of soybean biodiesel additive with different formulations of oils and fats. *Journal of Thermal Analysis and Calorimetry, 123*(3), 2121–2127. https://doi.org/10.1007/s10973-015-4772-0.

Farias, A. F. F., de Araújo, D. T., da Silva, A. L., Leal, E., Pacheco, J. G. A., Silva, M. R., et al. (2020). Evaluation of the catalytic effect of ZnO as a secondary phase in the $Ni_{0.5}Zn_{0.5}Fe_2O_4$ system and of the stirring mechanism on biodiesel production reaction. *Arabian Journal of Chemistry, 13*(6), 5788–5799. https://doi.org/10.1016/j.arabjc.2020.04.016.

Fatimah, I., Wang, S., & Wulandari, D. (2011). ZnO/montmorillonite for photocatalytic and photochemical degradation of methylene blue. *Applied Clay Science, 53*(4), 553–560. https://doi.org/10.1016/j.clay.2011.05.001.

Fernandes, C. I., Capelli, S. C., Vaz, P. D., & Nunes, C. D. (2015). Highly selective and recyclable MoO3 nanoparticles in epoxidation catalysis. *Applied Catalysis A: General, 504*, 344–350. https://doi.org/10.1016/j.apcata.2015.02.027.

Fernandes, J. L. N., De Souza, R. O. M. A., & De Vasconcellos Azeredo, R. B. (2012). 13C NMR quantification of mono and diacylglycerols obtained through the solvent-free lipase-catalyzed esterification of saturated fatty acids. *Magnetic Resonance in Chemistry, 50*(6), 424–428. https://doi.org/10.1002/mrc.3814.

Fonseca, C. G., Vaiss, V. S., Wypych, F., Diniz, R., & Leitão, A. A. (2017). Structural and thermodynamic investigation of the hydration-dehydration process of Na +-montmorillonite using DFT calculations. *Applied Clay Science, 143*, 212–219. https://doi.org/10.1016/j.clay.2017.03.025.

Franco-Urquiza, E. A., May-Crespo, J. F., Escalante Velázquez, C. A., Pérez Mora, R., & González García, P. (2020). Thermal degradation kinetics of ZnO/polyester nanocomposites. *Polymers, 12*(8), 1753. https://doi.org/10.3390/POLYM12081753.

Galimberti, M., Cipolletti, V. R., & Coombs, M. (2013). Chapter 4.4 – Applications of clay–polymer nanocomposites. *Vol. 5B. Handbook of clay science*. Elsevier.

Gavhane, R. S., Kate, A. M., Pawar, A., Soudagar, M. E. M., & Fayaz, H. (2020). Effect of soybean biodiesel and copper coated zinc oxide nanoparticles on enhancement of diesel engine characteristics. *Energy Sources, Part A: Recovery, Utilization and Environmental Effects*. https://doi.org/10.1080/15567036.2020.1856237.

Gelbard, G., Brès, O., Vargas, R. M., Vielfaure, F., & Schuchardt, U. F. (1995). 1H nuclear magnetic resonance determination of the yield of the transesterification of rapeseed oil with methanol. *Journal of the American Oil Chemists' Society, 72*(10), 1239–1241. https://doi.org/10.1007/BF02540998.

Ghesti, G. F., de Macedo, J. L., Resck, I. S., Dias, J. A., & Dias, S. C. L. (2007). FT-Raman spectroscopy quantification of biodiesel in a progressive soybean oil transesterification reaction and its correlation with ^1H NMR spectroscopy methods. *Energy and Fuels, 21*(5), 2475–2480. https://doi.org/10.1021/ef060657r.

Gong, J., Zeng, W., & Zhang, H. (2015). Hydrothermal synthesis of controlled morphologies of MoO_3 nanobelts and hierarchical structures. *Materials Letters, 154*, 170–172. https://doi.org/10.1016/j.matlet.2015.04.092.

Grim, R. E. (1968). *Clay mineralogy (second)*. McGraw-Hill.

Guckan, V., Altunal, V., Ozdemir, A., Tsiumra, V., Zhydachevskyy, Y., & Yegingil, Z. (2020). Calcination effects on europium doped zinc oxide as a luminescent material synthesized via sol-gel and precipitation methods. *Journal of Alloys and Compounds, 823*, 153878. https://doi.org/10.1016/j.jallcom.2020.153878.

Guo, M., Jiang, W., Chen, C., Qu, S., Lu, J., Yi, W., et al. (2021). Process optimization of biodiesel production from waste cooking oil by esterification of free fatty acids using La^{3+}/ZnO-TiO_2 photocatalyst. *Energy Conversion and Management, 229*, 113745. https://doi.org/10.1016/j.enconman.2020.113745.

Guzatto, R., Defferrari, D., Reiznautt, Q. B., Cadore, I. R., & Samios, D. (2012). Transesterification double step process modification for ethyl ester biodiesel production from vegetable and waste oils. *Fuel, 92*(1), 197–203. https://doi.org/10.1016/j.fuel.2011.08.010.

Ha, T. T., & Tuyen, N. V. (2014). Copper oxide nanomaterials prepared by solution methods, some properties, and potential applications: A brief review. *International Scholarly Research Notices*, 856592. https://doi.org/10.1155/2014/856592.

Hayati-Ashtiani, M. (2011). Characterization of nano-porous bentonite (montmorillonite) particles using FTIR and BET-BJH analyses. *Particle and Particle Systems Characterization, 28*(3–4), 71–76. https://doi.org/10.1002/ppsc.201100030.

Helmiyati, H., & Suci, R. P. (2019). Nanocomposite of cellulose-ZnO/SiO_2 as catalyst biodiesel methyl ester from virgin coconut oil. *Vol. 2168*. American Institute of Physics Inc.

Hu, C. H., Gu, L. Y., Luan, Z. S., Song, J., & Zhu, K. (2012). Effects of montmorillonite-zinc oxide hybrid on performance, diarrhea, intestinal permeability and morphology of weanling pigs. *Animal Feed Science and Technology, 177*(1–2), 108–115. https://doi.org/10.1016/j.anifeedsci.2012.07.028.

Huang, J., Xia, C., Cao, L., & Zeng, X. (2008). Facile microwave hydrothermal synthesis of zinc oxide one-dimensional nanostructure with three-dimensional morphology. *Materials Science & Engineering, B: Solid-State Materials for Advanced Technology, 150*(3), 187–193. https://doi.org/10.1016/j.mseb.2008.05.014.

Hur, S. G., Kim, T. W., Hwang, S. J., Hwang, S. H., Yang, J. H., & Choy, J. H. (2006). Heterostructured nanohybrid of zinc oxide-montmorillonite clay. *Journal of Physical Chemistry B, 110*(4), 1599–1604. https://doi.org/10.1021/jp0543633.

Jadhav, S. D., & Tandale, M. S. (2018). Optimization of transesterification process using homogeneous and nanoheterogeneous catalysts for biodiesel production from Mangifera indica oil. *Environmental Progress & Sustainable Energy, 37*(1), 533–545. https://doi.org/10.1002/ep.12690.

Jagtap, N., & Ramaswamy, V. (2006). Oxidation of aniline over titania pillared montmorillonite clays. *Applied Clay Science, 33*(2), 89–98. https://doi.org/10.1016/j.clay.2006.04.001.

Jozefaciuk, G., & Bowanko, G. (2002). Effect of acid and alkali treatments on surface areas and adsorption energies of selected minerals. *Clays and Clay Minerals*, *50*(6), 771–783. https://doi.org/10.1346/000986002762090308.

Kamada, K., Kang, J. H., Paek, S. M., & Choy, J. H. (2012). CeO_2-layered aluminosilicate nanohybrids for UV screening. *Journal of Physics and Chemistry of Solids*, *73*(12), 1478–1482. https://doi.org/10.1016/j.jpcs.2011.11.042.

Khaorapapong, N., Khumchoo, N., & Ogawa, M. (2015). Preparation of copper oxide in smectites. *Applied Clay Science*, *104*, 238–244. https://doi.org/10.1016/j.clay.2014.11.038.

Khobzaoui, S., Tennouga, L., Benabadji, I. K., Mansri, A., & Bouras, B. (2019). Preparation of rigid bentonite/PAM nanocomposites by an adiabatic process: Influence of load content and nano-structure on mechanical properties and glass transition temperature. *Journal of Inorganic and Organometallic Polymers and Materials*, *29*(4), 1111–1118. https://doi.org/10.1007/s10904-019-01073-8.

Kim, H. J., Kang, B. S., Kim, M. J., Park, Y. M., Kim, D. K., Lee, J. S., et al. (2004). Transesterification of vegetable oil to biodiesel using heterogeneous base catalyst. *Catalysis Today*, *93–95*, 315–320. https://doi.org/10.1016/j.cattod.2004.06.007.

Kim, M., DiMaggio, C., Salley, S. O., & Simon Ng, K. Y. (2012). A new generation of zirconia supported metal oxide catalysts for converting low grade renewable feedstocks to biodiesel. *Bioresource Technology*, *118*, 37–42. https://doi.org/10.1016/j.biortech.2012.04.035.

Kumar, J. P., Ramacharyulu, P. V. R. K., Prasad, G. K., & Singh, B. (2015). Montmorillonites supported with metal oxide nanoparticles for decontamination of sulfur mustard. *Applied Clay Science*, *116–117*, 263–272. https://doi.org/10.1016/j.clay.2015.04.007.

Lee, A. F., Bennett, J. A., Manayil, J. C., & Wilson, K. (2014). Heterogeneous catalysis for sustainable biodiesel production via esterification and transesterification. *Chemical Society Reviews*, *43*(22), 7887–7916. https://doi.org/10.1039/c4cs00189c.

Lei, S. G., Hoa, S. V., & Ton-That, M. T. (2006). Effect of clay types on the processing and properties of polypropylene nanocomposites. *Composites Science and Technology*, *66*(10), 1274–1279. https://doi.org/10.1016/j.compscitech.2005.09.012.

Li, Y., & Jiang, Y. (2018). Preparation of a palygorskite supported KF/CaO catalyst and its application for biodiesel production via transesterification. *RSC Advances*, *8*(29), 16013–16018. https://doi.org/10.1039/c8ra02713g.

Li, Y., Liu, J. R., Jia, S. Y., Guo, J. W., Zhuo, J., & Na, P. (2012). TiO_2 pillared montmorillonite as a photoactive adsorbent of arsenic under UV irradiation. *Chemical Engineering Journal*, *191*, 66–74. https://doi.org/10.1016/j.cej.2012.02.058.

Li, J., Hu, M., Zuo, S., & Wang, X. (2018). Catalytic combustion of volatile organic compounds on pillared interlayered clay (PILC)-based catalysts. *Current Opinion in Chemical Engineering*, *20*, 93–98. https://doi.org/10.1016/j.coche.2018.02.001.

Li, X., Tong, D., & Hu, C. (2015). Efficient production of biodiesel from both esterification and transesterification over supported SO_4^{2-}-MoO_3-ZrO_2-Nd_2O_3/SiO_2 catalysts. *Journal of Energy Chemistry*, *24*(4), 463–471. https://doi.org/10.1016/j.jechem.2015.06.010.

Liang, H.-Q., Pan, L.-Z., & Liu, Z. (2008). Synthesis and photoluminescence properties of ZnO nanowires and nanorods by thermal oxidation of Zn precursors. *Materials Letters*, *62*(12 – 13), 1797–1800. https://doi.org/10.1016/j.matlet.2007.10.010.

Likozar, B., & Levec, J. (2014). Transesterification of canola, palm, peanut, soybean and sunflower oil with methanol, ethanol, isopropanol, butanol and tert-butanol to biodiesel: Modelling of chemical equilibrium, reaction kinetics and mass transfer based on fatty acid composition. *Applied Energy*, *123*, 108–120. https://doi.org/10.1016/j.apenergy.2014.02.046.

Lima, S. A. M., Cremona, M., Davolos, M. R., Legnani, C., & Quirino, W. G. (2007). Electroluminescence of zinc oxide thin-films prepared via polymeric precursor and via sol-gel methods. *Thin Solid Films*, *516*(2–4), 165–169. https://doi.org/10.1016/j.tsf.2007.06.106.

Mccabe, R. W., & Adams, J. M. (2013). Chapter 4.3 – Clay minerals as catalysts. *5A. Handbook of clay science*. Elsevier.

Mendieta, L. J., del García, C. T. N. J., Galindo, M. E. E., Nájera, M. R., de la Cruz González, V. M., Alcántara, F. J. L., et al. (2011). Kinetic study by ^1H nuclear magnetic resonance spectroscopy for biodiesel production from castor oil. *Chemical Engineering Journal*, *178*, 391–397. https://doi.org/10.1016/j.cej.2011.10.038.

Meshram, S., Limaye, R., Ghodke, S., Nigam, S., Sonawane, S., & Chikate, R. (2011). Continuous flow photocatalytic reactor using ZnO-bentonite nanocomposite for degradation of phenol. *Chemical Engineering Journal*, *172*(2–3), 1008–1015. https://doi.org/10.1016/j.cej.2011.07.015.

Mguni, L. L., Mukenga, M., Muzenda, E., Jalama, K., & Meijboom, R. (2013). Expanding the synthesis of Stöber spheres: Towards the synthesis of nano-magnesium oxide and nano-zinc oxide. *Journal of Sol-Gel Science and Technology*, *66*(1), 91–99. https://doi.org/10.1007/s10971-013-2971-3.

Milodowski, A. E., Norris, S., & Russell Alexander, W. (2016). Minimal alteration of montmorillonite following long-term interaction with natural alkaline groundwater: Implications for geological disposal of radioactive waste. *Applied Geochemistry*, *66*, 184–197. https://doi.org/10.1016/j.apgeochem.2015.12.016.

Mishra, B. G., & Rao, G. R. (2005). Cerium containing Al- and Zr-pillared clays: Promoting effect of cerium (III) ions on structural and catalytic properties. *Journal of Porous Materials*, *12*(3), 171–181. https://doi.org/10.1007/s10934-005-1645-0.

Monteiro, M. R., Ambrozin, A. R. P., Lião, L. M., & Ferreira, A. G. (2009). Determination of biodiesel blend levels in different diesel samples by ^1H NMR. *Fuel*, *88*(4), 691–696. https://doi.org/10.1016/j.fuel.2008.10.010.

Moreira, A. B. R., Perez, V. H., Zanin, G. M., & de Castro, H. F. (2007). Biodiesel synthesis by enzymatic transesterification of palm oil with ethanol using lipases from several sources immobilized on silica-PVA composite. *Energy and Fuels*, *21*(6), 3689–3694. https://doi.org/10.1021/ef700399b.

Motshekga, S. C., Ray, S. S., Onyango, M. S., & Momba, M. N. B. (2015). Preparation and antibacterial activity of chitosan-based nanocomposites containing bentonite-supported silver and zinc oxide nanoparticles for water disinfection. *Applied Clay Science*, *114*, 330–339. https://doi.org/10.1016/j.clay.2015.06.010.

Moura, J. V. B., Silveira, J. V., da Silva Filho, J. G., Filho, A. G. S., Luz-Lima, C., & Freire, P. T. C. (2018). Temperature-induced phase transition in h-MoO_3: Stability loss mechanism uncovered by Raman spectroscopy and DFT calculations. *Vibrational Spectroscopy*, *98*, 98–104. https://doi.org/10.1016/j.vibspec.2018.07.008.

Navajas, A., Reyero, I., Jiménez-Barrera, E., Romero-Sarria, F., Llorca, J., & Gandía, L. M. (2020). Catalytic performance of bulk and Al_2O_3-supported molybdenum oxide for the production of biodiesel from oil with high free fatty acids content. *Catalysts*, *10*(2), 158. https://doi.org/10.3390/catal10020158.

Navas, M. B., Lick, I. D., Bolla, P. A., Casella, M. L., & Ruggera, J. F. (2018). Transesterification of soybean and castor oil with methanol and butanol using heterogeneous basic catalysts to obtain biodiesel. *Chemical Engineering Science*, *187*, 444–454. https://doi.org/10.1016/j.ces.2018.04.068.

Neris, A. M., Araújo, D., Cavalcante, Y., Farias, A. F. F., Moura, K. F., Cutrin, A. A., et al. (2015). Avaliação de argilas pura e impregnada com SnO_2 como catalisador para a produção de biodiesel. *Cerâmica*, *61*(359), 323–327. https://doi.org/10.1590/0366-69132015613591910.

Neto, P. R. C., Caro, M. S. B., Mazzuco, L. M., & Nascimento, M. D. G. (2004). Quantification of soybean oil ethanolysis with ^1H NMR. *Journal of the American Oil Chemists' Society*, *81*(12), 1111–1114. https://doi.org/10.1007/s11746-004-1026-0.

Noyan, H., Önal, M., & Sarikaya, Y. (2006). The effect of heating on the surface area, porosity and surface acidity of a bentonite. *Clays and Clay Minerals*, *54*(3), 375–381. https://doi.org/10.1346/CCMN.2006.0540308.

Özgür, U., Alivov, Y. I., Liu, C., Teke, A., Reshchikov, M. A., Doğan, S., et al. (2005). A comprehensive review of ZnO materials and devices. *Journal of Applied Physics*, *98*(4), 1–103. https://doi.org/10.1063/1.1992666.

Park, H. M., Kim, T. W., Hwuang, S.-J., & Choy, J.-H. (2006). Chemical bonding nature and mesoporous structure of nickel intercalated montmorillonite clay. *Bulletin of the Korean Chemical Society*, *27*(9), 1323–1328.

Pinto, A. C., Guarieiro, L. L. N., Rezende, M. J. C., Ribeiro, N. M., Torres, E. A., Lopes, W. A., et al. (2005). Biodiesel: An overview. *Journal of the Brazilian Chemical Society*, *16*(6 B), 1313–1330. https://doi.org/10.1590/S0103-50532005000800003.

Pusch, R., Zwahr, H., Gerber, R., & Schomburg, J. (2003). Interaction of cement and smectitic clay—Theory and practice. *Applied Clay Science*, *23*(1–4), 203–210. https://doi.org/10.1016/S0169-1317(03)00104-2.

Ramana, C. V., Atuchin, V. V., Troitskaia, I. B., Gromilov, S. A., Kostrovsky, V. G., & Saupe, G. B. (2009). Low-temperature synthesis of morphology controlled metastable hexagonal molybdenum trioxide (MoO_3). *Solid State Communications*, *149*, 6–9. https://doi.org/10.1016/j.ssc.2008.10.036.

Rathod, S. B., Lande, M. K., Arbad, B. R., & Gambhire, A. B. (2014). Preparation, characterization and catalytic activity of MoO_3/CeO_2-ZrO_2 solid heterogeneous catalyst for the synthesis of β-enaminones. *Arabian Journal of Chemistry*, *7*(3), 253–260. https://doi.org/10.1016/j.arabjc.2010.10.027.

Reinoso, D. M., Angeletti, S., Cervellini, P. M., & Boldrini, D. E. (2020). Study of structural properties of acid-treated natural sediment and its application as a sustainable catalyst. *Brazilian Journal of Chemical Engineering*, *37*(4), 679–690. https://doi.org/10.1007/s43153-020-00066-2.

Ren, H., Sun, S., Cui, J., & Li, X. (2018). Synthesis, functional modifications, and diversified applications of molybdenum oxides micro-/nanocrystals: A review. *Crystal Growth and Design*, *18*(10), 6326–6369. https://doi.org/10.1021/acs.cgd.8b00894.

Resolução, A. (2014). *Biodiesel (B100), especificado conforme Resolução ANP n° 45/2014. Vol. 17.*

Salem, A. M. S., El-Sheikh, S. M., Harraz, F. A., Ebrahim, S., Soliman, M., Hafez, H. S., et al. (2017). Inverted polymer solar cell based on MEH-PPV/PC 61 BM coupled with ZnO nanoparticles as electron transport layer. *Applied Surface Science*, *425*, 156–163. https://doi.org/10.1016/j.apsusc.2017.06.322.

Salerno, P., Mendioroz, S., & Agudo, A. L. (2003). Al-pillared montmorillonite-based Mo catalysts: Effect of the impregnation conditions on their structure and hydrotreating activity. *Applied Clay Science, 23*, 287–297. https://doi.org/10.1016/s0169-1317(03)00128-5.

Samadi, M., Sarikhani, N., Zirak, M., Zhang, H., Zhang, H. L., & Moshfegh, A. Z. (2018). Group 6 transition metal dichalcogenide nanomaterials: Synthesis, applications and future perspectives. *Nanoscale Horizons, 3*(2), 90–204. https://doi.org/10.1039/c7nh00137a.

Sankaranarayanan, T. M., Pandurangan, A., Banu, M., & Sivasanker, S. (2011). Transesterification of sunflower oil over MoO_3 supported on alumina. *Applied Catalysis A: General, 409–410*, 239–247. https://doi.org/10.1016/j.apcata.2011.10.013.

Santos-Beltrán, M., Paraguay-Delgado, F., García, R., Antúnez-Flores, W., Ornelas-Gutiérrez, C., & Santos-Beltrán, A. (2017). Fast methylene blue removal by MoO_3 nanoparticles. *Journal of Materials Science: Materials in Electronics, 28*(3), 2935–2948. https://doi.org/10.1007/s10854-016-5878-2.

Seguin, L., Figlarz, M., Cavagnat, R., & Lassègues, J. C. (1995). Infrared and Raman spectra of MoO_3 molybdenum trioxides and $MoO_3 \cdot xH_2O$ molybdenum trioxide hydrates. *Spectrochimica Acta Part A: Molecular and Biomolecular Spectroscopy, 51*(8), 1323–1344. https://doi.org/10.1016/0584-8539(94)00247-9.

Shaporev, A. S., Ivanov, V. K., Baranchikov, A. E., & Tret'yakov, Y. D. (2007). Microwave-assisted hydrothermal synthesis and photocatalytic activity of ZnO. *Inorganic Materials, 43*(1), 35–39. https://doi.org/10.1134/S0020168507010098.

Shen, C. C., Petit, S., Li, C. J., Li, C. S., Khatoon, N., & Zhou, C. H. (2020). Interactions between smectites and polyelectrolytes. *Applied Clay Science, 198*. https://doi.org/10.1016/j.clay.2020.105778.

Shweta, K. P., & Thapa, K. B. (2019). Synthesis and characterization of ZnO nano-particles for solar cell application by the cost effective co-precipitation method without any surfactants. In *Vol. 2142. AIP conference proceedings*American Institute of Physics Inc. https://doi.org/10.1063/1.5122336.

Siddiqui, N., Sim, J., Silwood, C. J. L., Toms, H., Iles, R. A., & Grootveld, M. (2003). Multicomponent analysis of encapsulated marine oil supplements using high-resolution ^1H and ^{13}C NMR techniques. *Journal of Lipid Research, 44*(12), 2406–2427. https://doi.org/10.1194/jlr.D300017-JLR200.

Sing, K. S. W. (1982). Reporting physisorption data for gas/solid systems with special reference to the determination of surface area and porosity (provisional). *Pure and Applied Chemistry, 54*(11), 2201–2218. https://doi.org/10.1351/pac198254112201.

Soltani, S., Khanian, N., Shean Yaw Choong, T., Rashid, U., Arbi Nehdi, I., & Mohamed Alobre, M. (2020). PEG-assisted microwave hydrothermal growth of spherical mesoporous Zn-based mixed metal oxide nanocrystalline: Ester production application. *Fuel, 279*, 118489. https://doi.org/10.1016/j.fuel.2020.118489.

Somidi, A. K. R., Das, U., & Dalai, A. K. (2016). One-pot synthesis of canola oil based biolubricants catalyzed by MoO_3/Al_2O_3 and process optimization study. *Chemical Engineering Journal, 293*, 259–272. https://doi.org/10.1016/j.cej.2016.02.076.

Song, J., Ni, X., Gao, L., & Zheng, H. (2007). Synthesis of metastable h-MoO_3 by simple chemical precipitation. *Materials Chemistry and Physics, 102*(2–3), 245–248. https://doi.org/10.1016/j.matchemphys.2006.12.011.

Stern, R., Hillion, G., Rouxel, J.-J., & Leporq, S. (1999). *Process for the production of esters from vegetable oils or animal oils alcohol*. (5,908,946). U.S. Patent.

Suwannakarn, K., Lotero, E., Ngaosuwan, K., & Goodwin, J. G. (2009). Simultaneous free fatty acid esterification and triglyceride transesterification using a solid acid catalyst with in situ removal of water and unreacted methanol. *Industrial and Engineering Chemistry Research, 48*(6), 2810–2818. https://doi.org/10.1021/ie800889w.

Svensson, P. D., & Hansen, S. (2013). Combined salt and temperature impact on montmorillonite hydration. *Clays and Clay Minerals, 61*(4), 328–341. https://doi.org/10.1346/CCMN.2013.0610412.

Tariq, M., Ali, S., & Khalid, N. (2012). Activity of homogeneous and heterogeneous catalysts, spectroscopic and chromatographic characterization of biodiesel: A review. *Renewable and Sustainable Energy Reviews, 16*(8), 6303–6316. https://doi.org/10.1016/j.rser.2012.07.005.

Tawade, B. V., Apata, I. E., Singh, M., Das, P., Pradhan, N., Al-Enizi, A. M., et al. (2021). Recent developments in the synthesis of chemically modified nanomaterials for use in dielectric and electronics applications. *Nanotechnology, 32*(14). https://doi.org/10.1088/1361-6528/abcf6c.

Tesser, R., Di Serio, M., Guida, M., Nastasi, M., & Santacesaria, E. (2005). Kinetics of oleic acid esterification with methanol in the presence of triglycerides. *Industrial and Engineering Chemistry Research, 44*(21), 7978–7982. https://doi.org/10.1021/ie050588o.

Vasić, K., Podrepšek, G. H., Knez, Ž., & Leitgeb, M. (2020). Biodiesel production using solid acid catalysts based on metal oxides. *Catalysts, 10*(2), 237. https://doi.org/10.3390/catal10020237.

Velasco, J., Pérez-Mayoral, E., Mata, G., Rojas-Cervantes, M. L., & Vicente-Rodríguez, M. A. (2011). Cesium-saponites as excellent environmental-friendly catalysts for the synthesis of N-alkyl pyrazoles. *Applied Clay Science, 54*(2), 125–131. https://doi.org/10.1016/j.clay.2011.06.007.

Verga, L. G., Aarons, J., Sarwar, M., Thompsett, D., Russell, A. E., & Skylaris, C. K. (2016). Effect of graphene support on large Pt nanoparticles. *Physical Chemistry Chemical Physics, 18*(48), 32713–32722. https://doi.org/10.1039/c6cp07334d.

Wang, Z. L. (2004). Zinc oxide nanostructures: Growth, properties and applications. *Journal of Physics. Condensed Matter*, *16*(25), R829–R858. https://doi.org/10.1088/0953-8984/16/25/R01.

Wojnarowicz, J., Chudoba, T., & Lojkowski, W. (2020). A review of microwave synthesis of zinc oxide nanomaterials: Reactants, process parameters and morphologies. *Nanomaterials*, *10*(6), 1086. https://doi.org/10.3390/nano10061086.

Xie, W., & Yang, Z. (2007). Ba-ZnO catalysts for soybean oil transesterification. *Catalysis Letters*, *117*(3–4), 159–165. https://doi.org/10.1007/s10562-007-9129-2.

Xie, W., & Zhao, L. (2014). Heterogeneous CaO-MoO_3-SBA-15 catalysts for biodiesel production from soybean oil. *Energy Conversion and Management*, *79*, 34–42. https://doi.org/10.1016/j.enconman.2013.11.041.

Xu, M., & Wei, M. (2018). Layered double hydroxide-based catalysts: Recent advances in preparation, structure, and applications. *Advanced Functional Materials*, *28*(47). https://doi.org/10.1002/adfm.201802943.

Yan, S., Salley, S. O., & Simon Ng, K. Y. (2009). Simultaneous transesterification and esterification of unrefined or waste oils over ZnO-La_2O_3 catalysts. *Applied Catalysis A: General*, *353*(2), 203–212. https://doi.org/10.1016/j.apcata.2008.10.053.

Ye, J., Li, X., Hong, J., Chen, J., & Fan, Q. (2015). Photocatalytic degradation of phenol over ZnO nanosheets immobilized on montmorillonite. *Materials Science in Semiconductor Processing*, *39*, 17–22. https://doi.org/10.1016/j.mssp.2015.04.039.

Yigezu, Z. D., & Muthukumar, K. (2014). Catalytic cracking of vegetable oil with metal oxides for biofuel production. *Energy Conversion and Management*, *84*, 326–333. https://doi.org/10.1016/j.enconman.2014.03.084.

Yue-Ping, F., An-Wu, X., Ai-Miao, Q., & Rui-Jin, Y. (2005). Selective synthesis of hexagonal and tetragonal dysprosium orthophosphate nanorods by a hydrothermal method. *Crystal Growth & Design*, *5*, 1221–1225. https://doi.org/10.1021/cg0495781.

Zhang, C. C., Zheng, L., Zhang, Z. M., Dai, R. C., Wang, Z. P., Zhang, J. W., et al. (2011). Raman studies of hexagonal MoO_3 at high pressure. *Physica Status Solidi B: Basic Research*, *248*(5), 1119–1122. https://doi.org/10.1002/pssb.201000633.

Zhu, Y., Yao, Y., Luo, Z., Pan, C., Yang, J., Fang, Y., et al. (2020). Review—Nanostructured MoO_3 for efficient energy and environmental catalysis. *Molecules*, *25*, 18.

CHAPTER

12

Assisted catalysis: An overview of alternative activation technologies for the conversion of biomass

C. Coutanceau, F. Jérôme, and K. De Oliveira Vigier

IC2MP, University of Poitiers, UMR CNRS 7285, Poitiers, France

12.1 Introduction

With the exponential increase of the world's population, and the awareness of our impact on the planet, the chemical industry is now facing two main challenges: (1) the search for cutting-edge technologies to lower the environmental impact of chemical processes, while improving the intrinsic performances of chemicals, and (2) the substitution of fossil carbon by renewable carbon (Marion et al., 2017). Lignocellulose is mainly composed of cellulose (38%–50%), hemicellulose (23%–32%), and lignin (15%–25%) and is a source of carbon for various applications (biofuels, polymers, and chemical building blocks). Sugars, polyols, organic acids (such as succinic, levulinic, itaconic, and lactic acids), and furans are some of the key building blocks that can be obtained from lignocellulosic biomass. The chemical conversion of this biomass is complex and therefore energy-efficient processes are required. The complexity of the raw material and its variability in structure hamper catalyst activity (Corma Canos, Iborra, & Velty, 2007). Thus, innovative approaches are needed to convert biomass with limited waste. Physical activation has gained interest in the last few decades. Thus, nonconventional technologies such as mechanical activation, ultrasound (US), and electroactivation (Fig. 12.1) are being studied. Mechanocatalysis is widely studied for biomass pretreatment to break the linkages between its components, leading to an increase in the reactivity of the lignocellulosic components. If mechanical milling is combined with the use of a catalyst, depolymerization of lignocellulosic biomass can be performed using an acid catalyst. US is a processing technique that can create severe physicochemical environments via acoustic cavitation, to improve the accessibility of recalcitrant lignocellulosic biomass for conversion into high-value chemicals (Tabasso, Carnaroglio, Calcio Gaudino, & Cravotto, 2015). The use of US in homogeneous and heterogeneous catalytic reaction systems was found to be beneficial for lignocellulosic biomass valorization (Chatel, Oliveira Vigier, & Jérôme, 2014; Kuna, Behling, Valange,

Heterogeneous Catalysis
https://doi.org/10.1016/B978-0-323-85612-6.00012-7

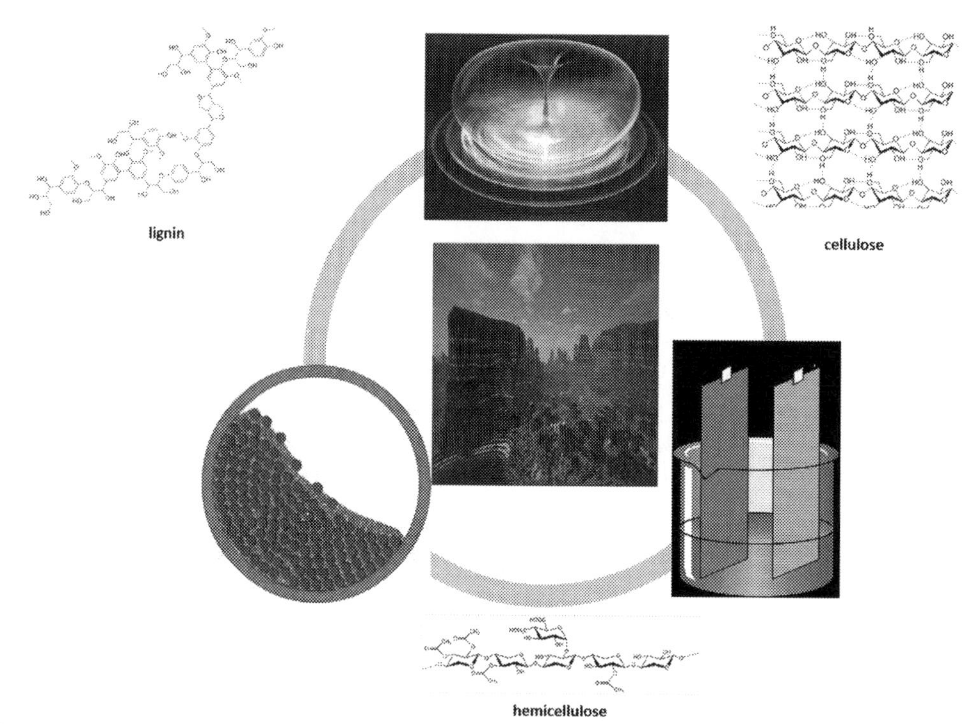

FIG. 12.1 Assisted catalysis for the conversion of biomass.

Chatel, & Colmenares, 2017). Ultrasonic cavitation leads to physical and chemical effects such as local heating (around 5000°C) and pressure (around 1000 atm). The asymmetric collapse of cavitation bubbles and shock waves creates microjet streams that are conducive to accelerating reaction rates and increasing both yields and selectivities by promoting better energy and mass transfer in the presence of homogeneous and heterogeneous catalysts. More and more research studies are dedicated to the US-assisted catalytic conversion of biomass (Chatel et al., 2014; Kuna et al., 2017).

For a long time, electrochemical methods were considered for converting organic products into value-added compounds. In 1834, Faraday himself described the production of ethane from electrolysis of an aqueous solution of acetate. Later, Kolbe generalized the concept to different carboxylate structures through anodic decarboxylation reactions to yield hydrocarbons (Vijh & Conway, 1967). Since then, numerous other reactions using electrochemistry have been studied, and electrosynthesis has been proposed as a powerful green tool for organic synthesis with many industrial applications (Cardoso, Šljukić, Santos, & Sequeira, 2017).

With regard to the conversion of lignocellulosic biomass, electrochemical processes have several advantages over biotechnology and heterogeneous catalysis processes. The former allows extremely high selectivity but suffers from slow conversion, multicomponent media, and complex separation processes (Beltrame, Comotti, Della Pina, & Rossi, 2004; Tathod, Kane, Sanil, & Dhepe, 2014), whereas the latter often necessitates the use of oxidants or reductants and temperatures higher than 40°C to activate the molecules (Della Pina & Falletta, 2011; Rafaïdeen, Baranton, & Coutanceau, 2019). Indeed, electrochemical processes avoid utilization of strong chemical oxidants/reductants, only using the molecule to be transformed, with water and electrons as reactants, and they can be carried out under mild conditions (room temperature and atmospheric pressure). In addition to these advantages, the development of renewable energy sources such as hydraulic, wind, solar, and tidal powers to produce electricity (International Agency of Energy, 2019) contributes to making electrocatalysis a sustainable nonthermal activation method (Li & Sun, 2018; Möhle et al., 2018).

Assisted catalysis using these three technologies will be highlighted in this chapter. The main idea is to show the benefits and the drawbacks of such technologies for the catalytic conversion of biomass.

12.2 Synergistic effect between catalysis and mechanical forces: Cellulose as a case study

Cellulosic biomass has become an area of growing interest as it represents a huge reservoir of (nonedible) renewable carbon, from which a myriad of downstream chemicals can be theoretically produced through catalytic, thermochemical, or enzymatic processes. In this field, the depolymerization of cellulosic biomass to glucose is a prerequisite step (Climent, Corma, & Iborra, 2011; Dhepe & Fukuoka, 2008; Yabushita, Kobayashi, & Fukuoka, 2014). A cocktail of enzymes can be used to selectively depolymerize cellulose to glucose. If this route is implemented on a large scale to produce ethanol, its deployment in the chemical industry will be difficult due to the production of highly diluted feed of glucose, low space-time yield, and complex purification process, which often raise the price of glucose to unacceptable levels for further processing to specialty chemicals. Besides, acid catalysis has also been proposed. Although acid catalysts are much cheaper than enzymes, the recalcitrance of cellulose to hydrolysis requires harsh conditions of temperatures and pressures, thus inducing the formation of tar-like materials (also called humins) stemming from the in situ degradation of glucose (Final report of E4tech, RE-CORD and WUR for the European Commission Directorate, 2021).

Cellulose: Hurdles faced by catalysis

Developing efficient and clean technologies capable of depolymerizing cellulose requires a better understanding of the cellulose structure. Recent advances in the field of DFT (density functional theory) calculations and analysis have revealed that nature has designed cellulose as a near perfect polymer, with defenses at the macro and molecular levels against hydrolysis (Box, 1990, 1991; Cramer, Truhlar, & French, 1997; Jeffrey, Pople, Binkley, & Visheveshwara, 1978). It is known that, at the macromolecular level, the packing of cellulosic chains is ensured by a dense hydrogen bond network, preventing the diffusion of the catalyst within the cellulosic crystal (Klemm et al., 2011; Klemm, Heublein, Fink, & Bohn, 2005; Moon, Martini, Nairn, Simonsen, & Youngblood, 2011; Siro & Plackett, 2010). In addition, the cellulosic chains are packed together through hydrophobic-hydrophilic interaction (Bergenstråhle, Wohlert, Himmel, & Brady, 2010; Biermann, Hädicke, Koltzenburg, & Müller-Plathe, 2001; Medronho, Romano, Miguel, Stigsson, & Lindman, 2012; Miyamoto et al., 2009; Yamane et al., 2006). Hence, it drastically limits the diffusion of water, an essential component for the depolymerization of cellulose, within the cellulosic crystal. One of the solutions to facilitate the diffusion of acid catalysts and water within the cellulosic backbone consists of breaking the cohesive H bond and van der Waals networks of cellulose, for instance, by dissolution/regeneration or by grinding. Although cellulose with

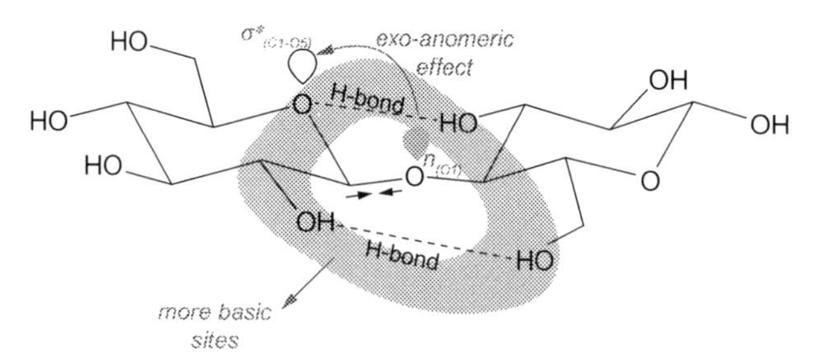

FIG. 12.2 Electronic effect protecting the glycosidic bond of cellobiose/cellulose.

a low crystallinity index has been obtained, the catalytic depolymerization of this pretreated cellulose remains extremely difficult. One of the reasons behind this result is the ability of "decrystallized" cellulose to quickly recrystallize in water to cellulose II, which is thermodynamically even more stable than native cellulose (Ago, Endo, & Hirotsu, 2004; Howsmon & Marchessault, 1959; Ouajai & Shanks, 2006). Unfortunately for chemists, this recrystallization is much faster than the hydrolysis of glycosidic bonds. However, all these macromolecular effects alone cannot explain the high recalcitrance of cellulose. More information at a molecular level (i.e., electronic nature and chemical environment of the β-(1→4) glycosidic bond) has been reported using DFT calculations and Car-Parrinello MD (molecular dynamics) simulations first on cellobiose, which is the repeating unit of cellulose (Loerbroks, Rinaldi, & Thiel, 2013).

When using an acid catalyst, the first step (at the molecular level) is the protonation of the glycosidic bond. Unfortunately, the oxygen atoms surrounding the glycosidic bond are more basic and thus scavenge the proton, thus protecting the glycosidic bond against protonation (Fig. 12.2). In particular,–CH_2OH is the most basic site and it lies at an ideal distance from the glycosidic bond to prevent its protonation (Loerbroks et al., 2013). Furthermore, the glycosidic bond is stabilized by electronic effects. For instance, there is an *exo*-anomeric effect stemming from an electronic donation from the oxygen in the glycosidic position to the antibonding orbital of the glucose ring. This *exo*-anomeric effect reduces the length of the glycosidic bond, resulting in a stabilization of 18.1 kcal/mol (in the cellobiose model) (Fig. 12.2). Furthermore, this bond length reduction allowed two hydrogen bonds to be established, thus locking the structure of cellobiose units and resulting in further stabilization of 8.5–14.5 kcal/mol.

These effects at the molecular level account for about 90% of the free energetic cost to protonate the glycosidic position in cellobiose (31 kcal/mol) (Loerbroks et al., 2013). However, this is not the end of the story. Even if one may succeed in protonating the glycosidic bond, which should result in a lowering of the *exo*-anomeric effect and in an elongation of the glycosidic bond, the anhydroglucose unit should undergo a conformational change from chair to nonchair, which accounts for about 10% of the apparent activation energy for the glycosidic bond cleavage.

Moving from cellobiose to cellulose, the protonation and cleavage of the glycosidic bond becomes even more challenging, due to the lack of freedom of cellulosic chains, thus making all these conformational changes even more difficult. Last but not least, it was shown that all these electronic effects are maximized in the presence of water. In contrast to what was previously done in recent years, one may question whether cellulose could advantageously depolymerize in its dry form, i.e., without any solvent. There is, in theory, enough water physisorbed on cellulose (~5%) to promote its complete hydrolysis.

"Physical" activation of cellulose

This concept was recently demonstrated by subjecting a powder of cellulose to a nonthermal atmospheric plasma (NTAP) treatment. Highly excited species (radicals, atoms, ions, etc.) are created in the plasma gas, which activate water present in cellulose, presumably through the formation of OH radicals, leading to the cleavage of the glycosidic bond and the formation of short-chain cellulosic fragments (Benoit et al., 2011; Delaux et al., 2016). As the NTAP proceeds, it concomitantly dries cellulose. Hence, when the amount of water in the powder of cellulose is not high enough, the short cellulosic chains repolymerize, mainly through 1,6 linkages, conferring to the final polysaccharide a much improved solubility in water. The same phenomenon was observed by applying a suspension of cellulose in water to an ultrasonic irradiation at a high frequency (550 kHz) (Haouache et al., 2020). Cavitation bubbles nucleate on the particles of cellulose. When these cavitation bubbles implode on the surface of cellulose, they provide energy and radicals (water dissociation to OH and H radicals), resulting in a cleavage of the glycosidic bond of cellulose. In contrast to NTAP, using high-frequency ultrasound, water is no longer a limitation as it is the solvent, which results in a much deeper depolymerization of cellulose, thus leading to the formation of glucose with 30% yield, without assistance of any catalyst. These two examples clearly demonstrate that depolymerization of cellulose could be achieved by looking at the problem from a radically different angle.

The mechanocatalytic process

To better control the selectivity of reactions involved in these "physical" activations, the coupling with catalysis is often investigated. In this context, the concept of mechanocatalysis has emerged as a promising tool for the selective saccharification of cellulose. It has been known for a long time that ball milling of cellulose disrupts H bonding and van der Waals interactions governing the cohesion of the cellulosic crystal (Barakat, de Vries, & Rouau, 2013). It is also used to reduce the particle size of cellulose fibers close to 15 μm. During ball milling of cellulose, the mechanical forces induce a torsion of the cellulosic chain, resulting in a lowering of the *exo*-anomeric effect, thus making the glycosidic bond more reactive. When the ball milling stops, the system relaxes and the cellulosic chains recover their initial conformation. However, if during the ball milling, an acid catalyst is introduced, then there is a synergistic effect between mechanical forces (torsion of the cellulosic chain) and the acid catalyst (protonation of the glycosidic bond), resulting in a deep depolymerization of cellulose. For instance, Rinaldi and Schuth reported that the ball milling of cellulose impregnated with a mineral acid such as HCl or H_2SO_4 led to a selective depolymerization of cellulose to water-soluble low-molecular-weight oligosaccharides (Fig. 12.3) (Meine, Rinaldi, & Schüth, 2012). Using mass spectrometry, a degree of polymerization of about 4–7 was claimed. As observed above with NTAP, during the mechanocatalytic depolymerization of cellulose, depolymerization and repolymerization occurred concomitantly, leading to the formation of branched oligosaccharides, in particular through 1,6 linkages (Shrotri, Lambert, Tanksale, & Beltramini, 2013). As the ball milling does not alter H_2SO_4, this process was used to further convert these oligosaccharides to valuable downstream products such as hexitol, glucose, or 5-hydroxymethylfurfural (Carrasquillo-Flores, Käldström, Schüth, Dumesic, & Rinaldi, 2013; Liao et al., 2014; Liu et al., 2015).

Solid acid catalysts such as kaolinite (clays) have been proposed to replace H_2SO_4 (Hick et al., 2010). During ball milling, kaolinite was delaminated, thus facilitating the contact between cellulose and kaolinite. Albeit not

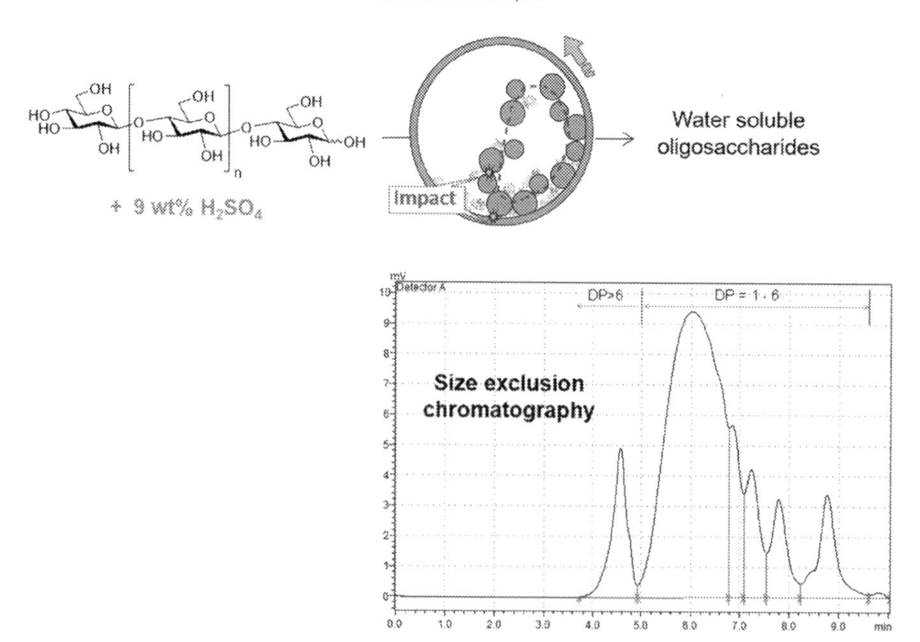

FIG. 12.3 Mechanocatalytic depolymerization of cellulose in the presence of H_2SO_4.

investigated in the original report, one may suspect that the recovery of dealuminated kaolinite at the end of the reaction is not an easy task. More recently, the mechanocatalytic depolymerization of cellulose has been investigated in the presence of Aquivion (Fig. 12.4) (Karam et al., 2018). Aquivion, a perfluorinated sulfonic acid membrane, is considered as a solid superacid with a Hammett acidity function of −12, which is like that of H_2SO_4 and higher than that of classical sulfonated polystyrenes such as Amberlyst 15, for instance (H0 = −2). As a result, Aquivion is active at lower temperatures than the usual solid acid catalysts. Furthermore, the mechanical and chemical integrity of Aquivion is preserved at temperatures within the 150–180°C range, making it an extremely attractive solid catalyst for the mechanocatalytic depolymerization of cellulose. Remarkably, Aquivion led to similar results as those obtained with H_2SO_4, with the formation of water-soluble oligosaccharides with a degree of polymerization in the 4–7 range. In contrast to H_2SO_4, which remained tightly bonded to oligosaccharides, Aquivion was easily separated from the as-formed oligosaccharides, by solubilization of the latter in water and by filtration of Aquivion. Owing to its high mechanical and chemical resistance, Aquivion was successfully recycled at least 10 times, without an obvious loss of its performances (Karam et al., 2018). As in the case of H_2SO_4, depolymerization and repolymerization reactions occurred concomitantly during the mechanocatalytic process with Aquivion. Further inspections by mass spectrometry, gas chromatography, and high-performance anion-exchange chromatography with pulse amperometric detection revealed the presence of different types of α/β glycosidic linkages, namely, (1→2)–, (1→3)–, (1→4)–, and (1→6), with the β-(1→4) linkage being dominant (79.5%) (Karam et al., 2018). Only a small proportion (3.8%) of the glucose unit is doubly glycosylated at positions O-4 and O-6. Remarkably,

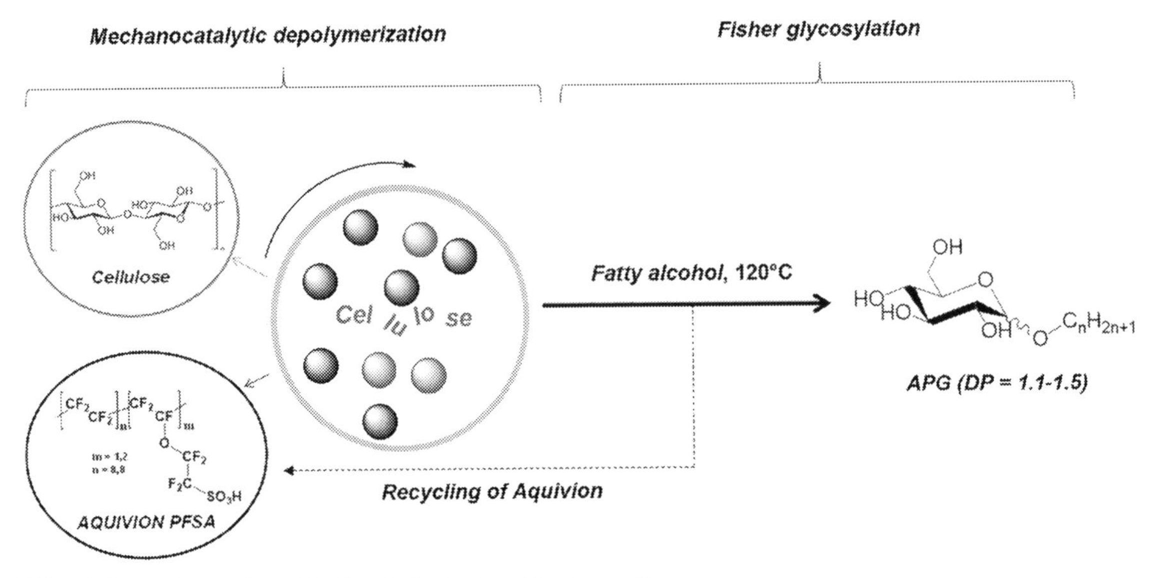

FIG. 12.4 Synthesis of amphiphilic alkyl polyglycosides from cellulose.

this mechanocatalytic process was highly selective, affording oligosaccharides as exclusive products, with no traces of furanic derivatives observed. The residual amount of water contained in Aquivion and cellulose plays an important role. Indeed, water buffers the mechanical forces and thus lowers the efficiency of the mechanocatalytic process. Hence, an excess of water is not preferred.

This mechanocatalytic process was successfully extended to the saccharification of lignocellulosic biomass waste (Käldström, Meine, Farès, Rinaldi, & Schüth, 2014; Käldström, Meine, Farès, Schüth, & Rinaldi, 2014; Schüth et al., 2014). In particular, it was employed for the synthesis of amphiphilic alkyl polyxylosides and alkyl polyglucosides, which are biobased nonionic surfactants used in the food, cosmetic, and detergent industries (Boissou et al., 2015). A life cycle assessment revealed that the burning of lignin, the coproduct of the reaction, provides 98% of the energy required for the glycosylation reaction and for the generation of steam (Brière et al., 2018). This work has led to the creation of the startup BIOSEDEV, which is now producing sugars from lignocellulosic biomass waste using this technology. It should be noted that the mechanocatalytic depolymerization of cellulose can be scaled up to 1 kg, leading to a drastic reduction in energy expenditure (Biosedev, n.d.).

12.3 Sonocatalysis for the conversion of biomass

In chemical reactions, the combination of sonochemistry and catalysis is known to enhance the conditions used in terms of sustainability. In the valorization of lignocellulose, for instance, the combination of a heterogeneous or a homogeneous catalyst with ultrasound (US) can be beneficial. Hence, sonochemistry is the use of ultrasonic waves in chemical synthesis aiming at reducing solvents, energy consumption, and increasing the selectivity. Many sonochemical reactors can be found and they are classified into two categories: (1) ultrasonic horn

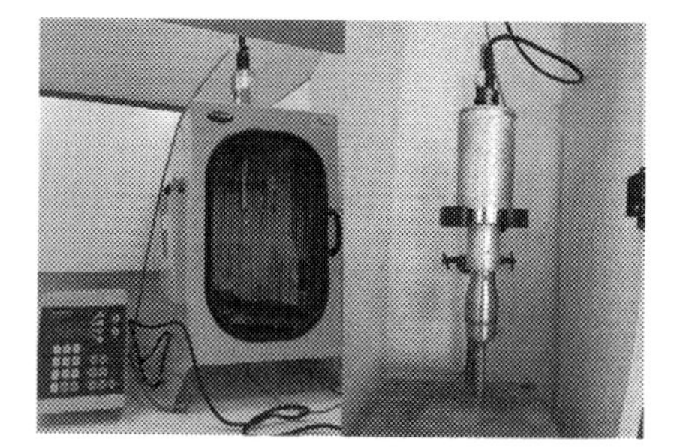

FIG. 12.5 High- and low-frequency US reactors.

devices, equipped with a single transducer for the direct irradiation of a solution (cup horn systems are less common) and (2) ultrasonic bath devices, equipped with multiple transducers at the bottom of a reactor for the indirect irradiation of a solution. We can distinguish between low-frequency US (20–80 kHz) and high-frequency US (>150 kHz) that can generate radicals (Fig. 12.5). High-frequency US produces less violent cavitation and leads to chemical effects, whereas low-frequency US is responsible for the mechano-physical effects, during which rapid cavitation leads to localized high temperatures and pressures.

Sonocatalytic conversion of lignocellulose

US-assisted hydrolysis of cellulose contained in lignocellulosic biomass to monomeric sugars was investigated in depth using both homogeneous and heterogeneous catalysts (Fig. 12.6). For example, Hu et al. showed that monosaccharides were produced from soybean straw and corn straw in a US bath for 120 min at 70°C in the presence of H-3-methylimidazolium chloride. US helps to increase the interaction between the ionic liquid and cellulose. Hence, ionic liquids are known to help dissolve cellulose. Thus, the ultrasonic cavitation enhances heat transfer and decreases the degree of polymerization of cellulose, leading to an increase of the reducing sugar yields than using an oil bath. In this study, there was no need to use a catalyst. Wood waste from municipalities was also recycled using acid hydrolysis assisted by US (Kunaver, Jasiukaityte, & Čuk, 2012). It was observed that condensation reactions were prevented due to cavitation effects and that the reaction time was decreased. Complete liquefaction was obtained with an energy of 0.44 $kW\,kg^{-1}$ after 10 min of sonification, whereas 1.35 $kW\,h^{-1}$ was acquired without US after 90 min of reaction. Following the same trend, US-assisted hydrolysis of hemicellulose to xylose was investigated (Montalbo-Lomboy, Johnson, Khanal, (Hans) van Leeuwen, & Grewell, 2010; Yunus, Salleh, Abdullah, & Biak, 2010). In all, 52% of xylose was observed starting from palm empty fruit bunch fibers at 100°C with 2% sulfuric acid as the catalyst within 45 min of reaction time. Starch was also depolymerized in the presence of sulfuric acid (5 wt%) under ultrasound at 100°C, resulting in a glucose yield of 97% after 2 h of reaction. The localized energy and pressure created during the irradiation of liquids positively affect the activation energy of a reaction.

Interesting molecules that can be obtained from sugars in the presence of an acid catalyst are furfural and 5-hydroxymethylfurfural (HMF). Thus, US-assisted conversion of cellulose or biomass to furan derivatives was studied. The solvents used were ionic liquids (ILs). Sarwono et al. used 1-butyl-

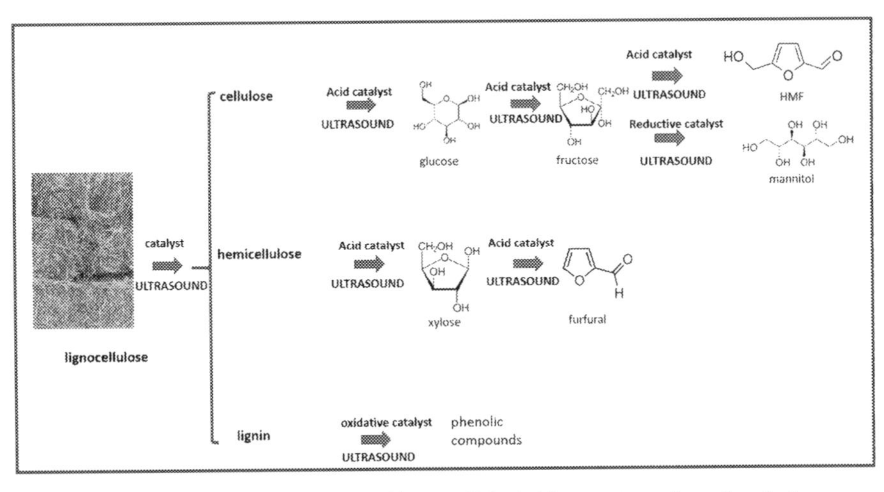

FIG. 12.6 US-assisted catalysis for the conversion of lignocellulosic biomass to various chemicals.

3-methylimidazolium and $CrCl_3$ to produce HMF from glucose, cellulose, and bamboo. To this end, they compared probe and batch sonication to conventional heating. They demonstrated that yields increase from bath to probe sonication (42% vs 1%) due to a greater localized effect. However, starting from glucose, similar yields to HMF were obtained using conventional heating than using a probe sonication (Sarwono et al., 2017). The only parameter that was affected was the reaction time that was lower using US (10 min) than using conventional heating (3 h). Upon prolonged reaction time under US, humins that limit the increase of the HMF yield were produced. One can mention that by tuning the acoustic power, the yield of HMF could be increased. Another study showed that 73% of HMF could be obtained from treatment of fructose at 40 Hz for 0.5 h in the presence of ILs and the HY zeolite catalyst. Energy consumption was thus reduced owing to the reduction of the reaction time compared to conventional heating (Bozell & Petersen, 2010). The direct production of HMF from cellulose or biomass can be performed using US, but low yields are achieved. However, by combining ball milling and US-assisted one-pot reaction in ionic liquids, higher yields were obtained than without ball milling due to size reduction and crystallinity decrease of cellulose. In this method, biomass was pretreated by ball milling to decrease the crystallinity. Another interesting study combining US and an acid (Amberlyst 15) was investigated to produce 5,5-(oxy-bis (methylene))bis-2-furfural from the etherification of HMF. US influences the kinetic and catalyst efficiency (Mliki & Trabelsi, 2015). A maximum yield of around 96%, in 5,5-(oxy-bis(methylene))bis-2-furfural was obtained using dichloromethane as a solvent at a temperature lower than 35°C after 120 min. This high yield was explained by the increase of the interface surface between the catalyst and the liquid phase. One can mention that the catalyst was recyclable.

Furfural was also produced by US-assisted acid hydrolysis of cellulose (Marullo, Rizzo, Meli, & D'Anna, 2019). Batch, cup horn, and probe sonication were investigated, and it was shown that the acoustic power affects the furfural yield. In a US bath and a US probe, no furfural was observed. This was explained by the lowest power (around 2 W) in the US bath and by the formation of an excess of radicals in the case of the US probe due to the highest power (14 W) [b]. When a cup horn was used (3 W), furfural was produced. After optimizing the reaction conditions, a conversion of 78% into furfural after 60 min of reaction at 30°C in the presence of nitric acid was achieved. To investigate the effect of frequency, a dual-frequency ultrasonic reactor was used (Hernoux-Villière, Lassi, & Lévêque, 2013; Santos et al., 2018). This rector allowed the use of low frequency (20 kHz) and high frequency (500 kHz). It was demonstrated that high frequency leads to the production of 35% of reducing sugars from starch due to a higher accessibility to the active sites than under low-frequency irradiation. A combination of low and high frequencies afforded 48% of reducing sugars from potato peels due to intensive cavitational collapse.

Other interesting reactions from lignocellulosic biomass are hydrogenation reactions. The process of combining US with catalysts such as Ni was studied in the hydroprocessing of lignin obtained from Miscanthus x giganteus (Finch et al., 2012). In the US reactor used, high hydrogen pressure could not be reached, leading to less depolymerization than in the batch reactor. Moreover, US can lead to the agglomeration of small particles, thus reducing the catalytic activity. Furthermore, US-assisted hydroprocessing was carried out at room temperature, and it is known that high temperatures are necessary to disrupt lignin structure in a batch reactor. The main advantage of US in these conditions is the selectivity. Hydrogenation of D-fructose into D-mannitol was performed combining US and heterogeneous catalysts (Raney-Ni, Cu/SiO_2, $Cu/ZnO/Al_2O_3$) in the presence of 50 bar of H_2 (Toukoniitty, Kuusisto, Mikkola, Salmi, & Murzin, 2005). The catalytic nature affects US influence, and the best catalyst was Raney-Ni. However, US improves the activity of the catalysts.

US-assisted biomass oxidation is the most studied reaction. Using high-frequency US, radicals that can be of interest in oxidation reactions are produced. However low-frequency US was used due to their availability. Mishra et al. studied the oxidation of primary hydroxyl groups on cellulose polymer chains using US irradiation at 68 and 170 kHz (Mishra, Thirree, Manent, Chabot, & Daneault, 2011). Nanocellulose with higher carboxyl content than silent conditions was obtained in the presence of TEMPO (2,2,6,6-tetramethylpiperidine-1-oxyl) combined with US. This was explained by an increase in the C6 hydroxyl accessibility due to the collapse of cavitation bubbles, leading to an increase in the reaction rate and carboxyl moiety content. One can mention that the yield was increased when high-frequency US was used (170 kHz). Another study demonstrated that using a semi-continuous mode in a flow-through sonoreactor (Paquin, Loranger, Hannaux, Chabot, & Daneault, 2013), the nominal input power was decreased by 87.5%, compared to the batch mode, with a 36% increase in the production rate of radicals. Owing to better homogenization and improved distribution of fibers and reagents than classical mixing, reproducibility was enhanced. Moreover, it was shown that US stabilized the heterogeneous catalyst in the presence of hydrogen peroxide by improving the stability of Au/SiO_2 suspensions. In another study, it was shown that mechanical effect leads to surface cleaning, resulting in excellent conversion and yield (Bujak, Bartczak, & Polanski, 2012). The Au/SiO_2 catalyst was recyclable up to the fourth cycle.

The implosion of cavitation bubbles on a solid surface generates high-speed jets of liquid directed at the surface. Recently, research groups have shown that this physical behavior is an effective means to concentrate radicals on the surfaces of catalysts, thus allowing increased control of reaction selectivity. In this context, Amaniampong et al. investigated the selective conversion of carbohydrates to fine chemicals by high-frequency US (HFUS) (Amaniampong et al., 2017, 2018, 2019). It was shown that the mechanistic pathway for the radical-driven conversion of carbohydrates by HFUS strongly depended on carbohydrate concentration. For example, at a high glucose concentration (>40 wt%), the reaction mostly occurred at the liquid-bubble interface, leading to a pyrolysis-like mechanism, with the in situ formation of levoglucosan. In contrast, at low concentrations (<10 wt%), glucose reacted with the H• and •OH radicals (stemming from water sonolysis) propelled into the bulk solution, leading to its selective oxidation. Guided by density functional theory (DFT) calculations, it was reported that, by adding a CuO catalyst in the HFUS reactor, the selectivity of glucose oxidation can be steered toward glucuronic acid. This result indicates a significant state-of-the-art advancement because conventional catalytic oxidation of glucose usually produces gluconic acid (oxidation on the anomeric position). Through a combined experimental computational approach, a synergy between the CuO catalyst and HFUS was highlighted (Fig. 12.7). In particular, it was shown that under ultrasound irradiation at 550 kHz, the surface lattice oxygen of CuO trapped H• radicals that stemmed from water sonolysis, thus inhibiting the ring opening of glucose. The annihilation of H• radicals also increased the coverage of •OH radicals on the CuO surface. These •OH radicals steered the highly selective oxidation of glucose toward glucuronic acid. This work also points toward the importance of optimizing CuO nanoparticle size for HFUS chemistry. Efficient transfer of radicals from cavitation bubbles to the catalyst surface required nanostructured CuO with specific size and morphology.

Oxidative depolymerization of Kraft lignin was also performed in the presence of H_2O_2 and metal salt catalysts (Napoly et al., 2015). Using low-frequency US (20 kHz) increases the catalytic activity due to an increase in mass transfer between the liquid and the catalyst surface. It also allows the decrease in particle size and the acceleration of suspended particle motion by shock waves and microstreaming. MCM-41-supported PTA was also used in the US-assisted depolymerization of stalk lignin. The depolymerization of lignin to phenolic monomers and low-molecular-weight lignin bio-oil was favored by US (Du et al., 2020).

Sonobiocatalysis

US combined with enzymes as catalysts for the conversion of biomass, owing to their ability to increase the activity and stability in enzymatic reactions, was also investigated. US helps to (i) homogenize and emulsify reactants and catalysts, (ii) activate chemical and biological catalysts, and (iii) improve dispersion owing to micro-convection and intense shock waves (Fig. 12.8).

Low-frequency US allows the esterification of vegetable oils under mild temperatures and short reaction times. The transesterification of glycerol carbonate from dimethyl carbonate (DMC) using commercial immobilized lipase

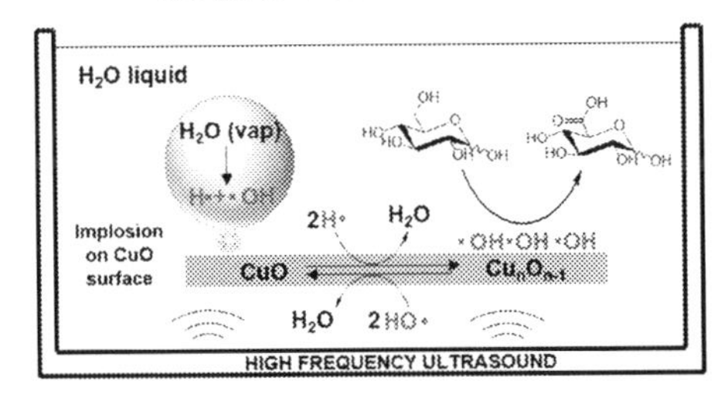

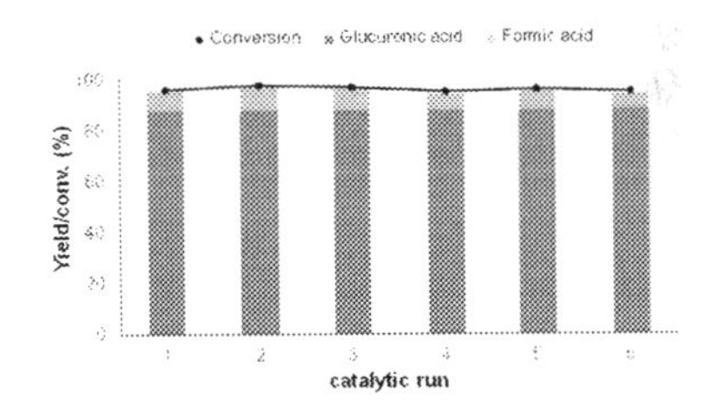

FIG. 12.7 Synergistic effect between CuO and HFUS during glucose oxidation.

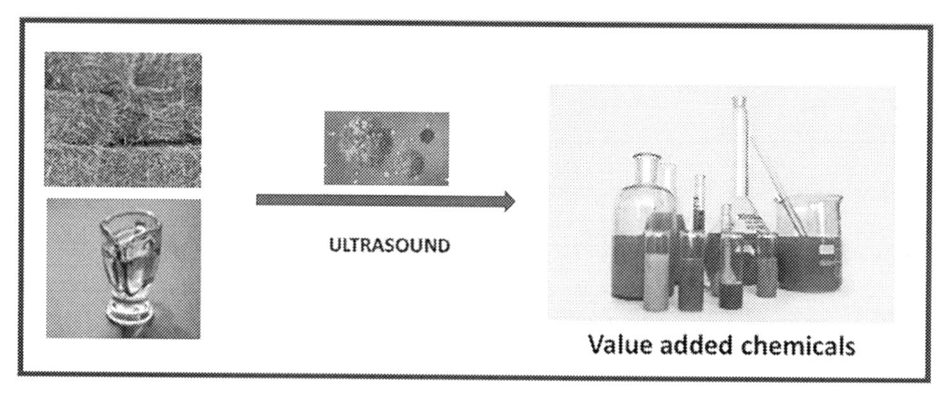

FIG. 12.8 US-assisted biocatalysis for the conversion of biomass.

(Novozym 435) with US frequency of 25 kHz, 50% amplitude, and 100 W power was studied (Gharat & Rathod, 2013). High conversion (~100%) was obtained at 60°C after 4 h compared to 14 h in the silent mode. The enzyme was recycled, but a loss in activity was observed, probably due to the breakage of the bonding between the enzyme and supporting media using US. Another example is the hydrolysis of sugarcane bagasse (Lunelli et al., 2014). US-assisted hydrolysis (40 kHz, 132 W) increases the amount of fermentable sugars (35%) compared to the process carried out in the silent mode (17.0%). Moreover, lower enzyme amounts, compared with silent conditions, are needed to obtain higher yields of fermentable sugars with US. It was recently demonstrated that the use of US under appropriate conditions generally increased the catalytic activity of cellulase in the hydrolysis of a lignocellulosic substrate. It was shown that, under optimal conditions, the relative total cellulase (FPase) activity of Celluclast 1.5 L increased by 42%, compared to silent conditions. In contrast, the FPase activity of Cellic CTec2 decreased by 4% under sonication. Celluclast 1.5 L in the absence or presence of a pure cellulosic substrate was investigated in depth (de Carvalho Silvello, Martínez, & Goldbeck, 2020). Enzyme activity that was ascribed to modifications of the enzyme structure increased under sonication (180.8 $W cm^{-2}$). Celluclast 1.5 L released 8.83 $g L^{-1}$ of reducing sugars with a hydrolysis yield of 42% due to an increase in mass transfer rates and in the enzyme-substrate affinity, which resulted in improved enzymatic hydrolysis kinetics.

Using US-assisted catalysis allows increase in the reaction rates owing to the cavitation phenomenon and the secondary effects of microjets and shock wave generations, which overall improves the efficiency and economics of the biomass conversion process.

12.4 Electroconversion of biomass

This section does not have the objective to be exhaustive but will exemplify the potency of the electrochemical conversion of biomass for obtaining selective value-added compounds from typical examples, among many others.

General principle of electrochemical conversion

Electrocatalysis consists of assisting reactions using a dedicated catalyst with the electrode potential. Therefore, the control of the catalyst composition/structure, electrode potential or current, electrolyte pH value, reactant concentrations, residence time in the reactor, etc., is expected to procure high tunability and controllability of the conversion process in terms of both conversion rate (activity) and selectivity toward a given reaction product (Simões, Baranton, & Coutanceau, 2012). Electro-oxidation or electroreduction of biomass-derived compounds is performed using specific catalysts dedicated to the considered reactant and desired product, at the anode or the cathode, respectively, of an electrosynthesis reactor according to the following reactions (Eqs. 12.1 and 12.2), assuming the formation of only one product:

Oxidation of biomass at the anode

$$C_xH_yO_z + aH_2O \rightarrow \text{product} + n(H^+ + e^-) \quad (12.1)$$

Reduction of biomass at the cathode

$$C_xH_yO_z + n(H^+ + e^-) \rightarrow \text{product} + bH_2O \quad (12.2)$$

Fig. 12.9A shows the schematic representation of an electrosynthesis reactor for the simultaneous production of a compound at the anode and hydrogen at the cathode. According to Eq. (12.1), the electro-oxidation of biomass also produces electrons and protons. Generally, those electrons and protons are consumed at the cathode of the electrosynthesis reactor to form hydrogen as the byproduct according to Eq. (12.3).

$$nH^+ + ne^- \rightarrow n/2H_2 \quad (12.3)$$

The electrical balance between reactions in Eqs. (12.1) and (12.3) leads to the global equation (Eq. 12.4):

$$C_xH_yO_z + aH_2O \rightarrow \text{product} + n/2H_2 \quad (12.4)$$

The reversible cell voltage of such an electroreforming reactor corresponds to the difference between the anode potential E_a and the cathode potential E_c. Using platinum nanoparticles disseminated over a high specific surface area and conductive carbon powder (Pt-NPs/C) as a cathode catalyst, which is an outstanding catalyst for hydrogen evolution reaction (HER), the HER at the cathode occurs with extremely high kinetics (low overpotential) and the cathode potential remains close to 0.0 V vs SHE (standard hydrogen electrode) (Lamy, Coutanceau, & Baranton, 2020; Rafaïdeen et al., 2019). Therefore, the reversible cell voltage is imposed by

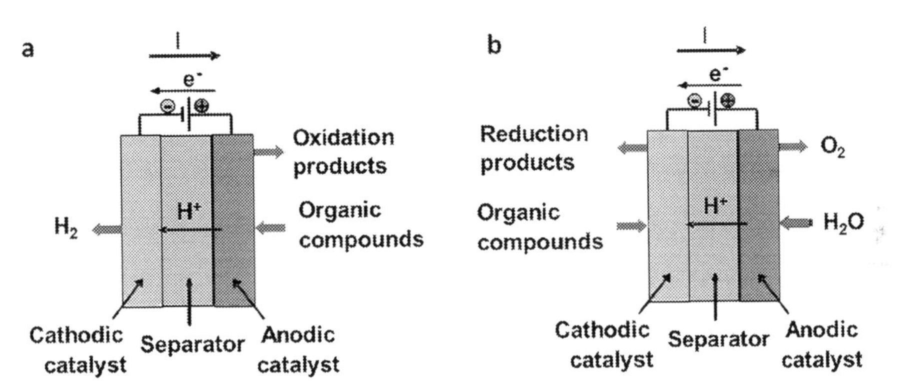

FIG. 12.9 Schematic representation of electrosynthesis reactors for the electroconversion of oxygenated organic compounds (A) by electro-oxidation with simultaneous production of hydrogen at the cathode and (B) by electroreduction with simultaneous production of oxygen at the anode.

the anode potential. Under the given experimental conditions, the reversible anode potential is directly related to the thermodynamic data of the reaction, i.e., the ΔG and ΔH of formation of the product ($\Delta G_{f,\ prod}$ and $\Delta H_{f,prod}$), water ($\Delta G_{f,\ H2O}$ and $\Delta H_{f,\ H2O}$), and the biomass compound ($\Delta G_{f,\ CxH_yO_z}$, $\Delta H_{f,\ CxH_yO_z}$). The thermodynamic data (ΔG_a and ΔH_a) associated with the reaction in Eq. (12.1) can then be calculated (Eqs. 12.5 and 12.6):

$$\Delta G_a = \Delta G_{f,\mathrm{prod}} - \Delta G_{f,\mathrm{C}_x\mathrm{H}_y\mathrm{O}_z} - a\Delta G_{f,\mathrm{H_2O}} \quad (12.5)$$

$$\Delta H_a = \Delta H_{f,\mathrm{prod}} - \Delta H_{f,\mathrm{C}_x\mathrm{H}_y\mathrm{O}_z} - a\Delta H_{f,\mathrm{H_2O}} \quad (12.6)$$

The reversible cell voltage corresponding to the minimum cell voltage that must be applied to the electrosynthesis reactor for initiating the oxidative electroconversion of biomass can then be expressed as follows (Eq. 12.7):

$$U_{\mathrm{cell}} = E_a - E_c \approx E_a = \frac{(\Delta G_a^0)}{nF} \quad (12.7)$$

For many oxygenated organic compounds (saccharides, oligosaccharides, polyols, alcohols, etc.), the reversible anode potential is lower than that of water (1.23 V vs SHE at 25°C) (Möhle et al., 2018). Indeed, considering the reaction of water oxidation (Eq. 12.8):

$$\mathrm{H_2O} \rightarrow 1/2\mathrm{O_2} + 2\mathrm{H^+} + 2\mathrm{e^-} \quad (12.8)$$

where $\Delta G_{f,\ H_2O}$ is 237.1 kJ mol L^{-1} at 25°C, two electrons are exchanged per oxidized water molecule ($n=2$), and F, the Faraday constant, is 96,485 C mol^{-1}, the anode potential is calculated as follows (Eq. 12.9):

$$E_{\mathrm{a}}^{\mathrm{O_2/H_2O}} = \frac{\Delta G_{\mathrm{H_2O}}}{nF} = \frac{237.1 \times 10^3}{2 \times 96485} \approx 1.23\ \mathrm{V\ vs\ SHE} \quad (12.9)$$

Therefore, the reversible cell voltage of the electrosynthesis cell will be lower than 1.23 V if hydrogen evolution is the counter reaction at the cathode. However, for certain biosourced compounds, fatty acids (Ziogas, Pennemann, & Kolb, 2020), and lignins (Shao, Liang, Cui, Xu, & Yan, 2014), for example, the minimum anode potential for their activation can be higher than that for water oxidation, which imposes some specific characteristics for the anode materials (high overpotential for the oxygen evolution reaction) or reaction media (nonaqueous media). However, clean hydrogen formed at the cathode is a valuable byproduct that can be used further for electrical energy delivery through a fuel cell (Coutanceau & Baranton, 2016) or for other chemical processes (Fraile, Lanoix, Maio, Rangel, & Torre, 2015), limiting the production cost involved in energy consumption.

The same reasoning as mentioned above is applicable to the electroreduction of biomass. In this case, the counter reaction at the anode is often the oxygen evolution reaction (Fig. 12.9B). Then, the anode potential will be higher than 1.23 V vs SHE, and the electrosynthesis cell voltage will be higher than 1.23 V to initiate the reductive electroconversion of biomass. This means that such a process will involve extremely high electrical energy consumption; in addition, in contrast to what happens in the case of biomass electro-oxidation process, oxygen formed at the anode is not a valuable byproduct and therefore cannot help to make the whole process more economically viable. To decrease the cell voltage, sacrificial anodes made of transition metals can be used, but, obviously, such a solution is questionable from industrial and sustainable points of view (Boissou, Baranton, Tarighi, De Oliveira Vigier, & Coutanceau, 2019). A highly interesting method for minimizing the energy requirement of the process consists of developing "paired" electrosynthesis reactors in which compounds from biomass can selectively be converted into value-added products on both the cathode and the anode of the reactor (Park, Pintauro, Baizer, & Nobe, 1985; Pintauro, Johnson, Park, Baizer, & Nobe, 1984).

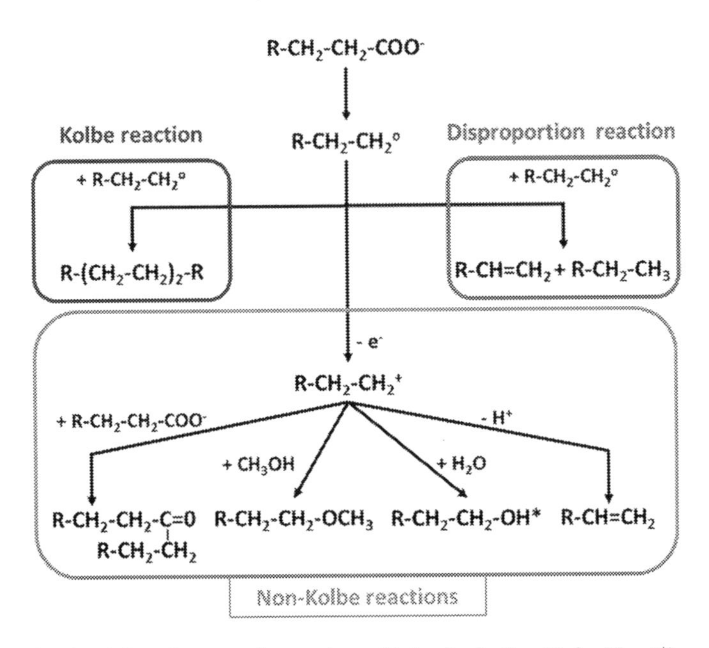

FIG. 12.10 Pathways for the Kolbe, disproportion, and non-Kolbe (including Hofer-Moest*) reactions.

Electroactivation of biosourced carboxylic acids

Carboxylic acids constitute a large accessible variety of compounds from biomass that can be considered as raw materials to produce value-added compounds or biofuel (Holzhäuser, Mensah, & Palkovits, 1849). According to Kolbe (1849), carboxylic acid can undergo decarboxylation reaction and dimerization through the formation of radical intermediates according to the following scheme (Eqs. 12.10 and 12.11):

$$2R-COO^- \rightarrow 2CO_2 + 2R^\circ + 2e^- \quad (12.10)$$

$$2R^\circ \rightarrow R-R \quad (12.11)$$

The mechanism of the Kolbe electrolysis reaction is expected to form two radicals, R°, which react together and dimerize. However, each radical formed can be further transformed into carbocation after loss of a second electron, which further allows non-Kolbe electrolysis reactions, including the Hofer-Moest reaction, as shown in Fig. 12.10. These non-Kolbe reactions can lead to the formation by solvolysis of esters, ethers, olefins, amides, alcohols, etc. (Schäfer, 1990). The versatility of this process allows the large accessible variety of carboxylic acids from biomass to serve as raw materials and to make the electrified value chains as part of the biorefinery concepts (Holzhäuser, Mensah, & Palkovits, 2020).

The control of the electrode material, pressure, medium, presence of additives, chain length, applied current density, etc., allows limiting the occurrence of (non)-Kolbe electrolysis and orienting the reaction toward Kolbe electrolysis (Levy, Sanderson, & Cheng, 1984). For example, platinum electrodes are known to favor the coupling of radicals (Holzhäuser et al., 2020; Schäfer, 1990) and organic or biphasic (aqueous/organic) solvents to allow solubilization of the reaction products (Zhang, Liu, & Wu, 2018). Fig. 12.11 shows the linear scan voltammetry curve obtained in the case of

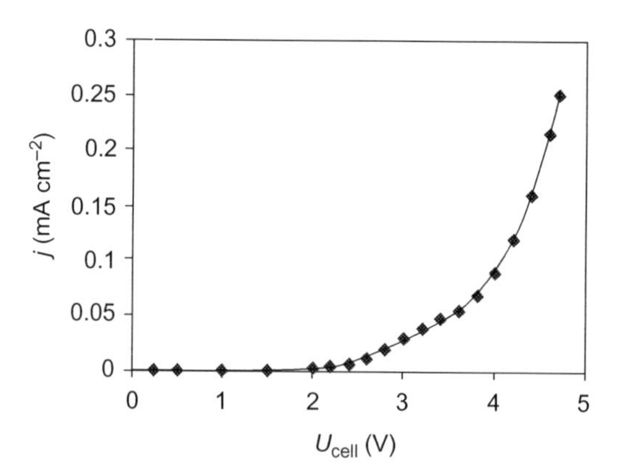

FIG. 12.11 Stationary polarization curve recorded in a filter-press-like electrolysis cell with two $25\,cm^2$ platinum grids as electrodes in an ethanoic acid/ethanol mixture in alkaline aqueous solution ($T = 20°C$).

Kolbe electrolysis of hexanoic acid in a water/ethanol mixture. It can be observed that the reaction occurs at an extremely high electrode potential, higher than 1.23 V, which corresponds to the oxygen evolution reaction. Indeed, it has been proposed that oxygen formation by water splitting at the anode is suppressed due to the high surface coverage of the electrode by the substrate (Palkovits & Palkovits, 2019). However, it has also been shown that Kolbe electrolysis could be carried out in aqueous media (Levy et al., 1984; Sanderson, Levy, Cheng, & Barnard, 1983; Ziogas et al., 2020).

One of the most important applications of Kolbe electrolysis for the future consists of the synthesis of biodiesel and biokerosene of extremely high grades from fatty acids (Dos Santos, Harnisch, Nilges, & Schröder, 2015; Ziogas et al., 2020). Mixtures of molecules with linear chain lengths containing between 8 and 20 carbon atoms and between 8 and 16 carbon atoms are preferred for biodiesel and jet fuel, respectively (Ziogas et al., 2020). Therefore, carboxylic acids with short-length chains issued from biomass fermentation are generally considered for such applications.

In the case where vegetable oils are used as starting materials, their transesterification using methanol (Eq. 12.12) leads to the production of glycerol and a mixture of fatty acid methyl esters that are used as biodiesel (Simões et al., 2012).

$$\begin{array}{l} CH_2O\text{-}\overset{O}{\overset{\|}{C}}\text{-}R_1 \\ | \\ CHO\text{-}\overset{O}{\overset{\|}{C}}\text{-}R_2 \\ | \\ CH_2O\text{-}\underset{O}{\underset{\|}{C}}\text{-}R_3 \end{array} + 3\,CH_3OH \longrightarrow \begin{array}{l} R_1\text{-}\overset{O}{\overset{\|}{C}}\text{-}OCH_3 \\ + R_2\text{-}\overset{O}{\overset{\|}{C}}OCH_3 + CH_2OH\text{-}CHOH\text{-}CH_2OH \\ + R_3\text{-}\underset{O}{\underset{\|}{C}}\text{-}OCH_3 \end{array} \tag{12.12}$$

Their saponification forms glycerol and fatty carboxylate salts according to reactions (12.13):

$$\begin{array}{l} CH_2O\text{-}\overset{O}{\overset{\|}{C}}\text{-}R_1 \\ | \\ CHO\text{-}\overset{O}{\overset{\|}{C}}\text{-}R_2 \\ | \\ CH_2O\text{-}\underset{O}{\underset{\|}{C}}\text{-}R_3 \end{array} + 3\,(K^+, OH^-) \longrightarrow \begin{array}{l} R_1\text{-}\overset{O}{\overset{\|}{C}}\text{-}O^-, K^+ \\ + R_2\text{-}\overset{O}{\overset{\|}{C}}O^-, K^+ + CH_2OH\text{-}CHOH\text{-}CH_2OH \\ + R_3\text{-}\underset{O}{\underset{\|}{C}}\text{-}O^-, K^+ \end{array} \tag{12.13}$$

The first step of the Kolbe reaction will lead, in the case of the example in Eq. (12.13), to the formation of $R_1°$, $R_2°$, and $R_3°$ radicals. However, it has been shown for a long time by Wurtz (1855) that different radicals can react together, leading to nonsymmetrical coupling. Therefore, the coupling of the different radicals will produce several linear hydrocarbon molecules, $R_1–R_1$, $R_1–R_2$, $R_1–R_3$, and $R_2–R_3$, in the case of the example in Eq. (12.13). Fatty acids from vegetable oils have generally long linear carbon chains (from C_{16} to C_{18} in the case of colza), and their coupling through Kolbe electrolysis will lead to paraffins and olefins with chain lengths from C_{30} to C_{34} of high viscosities and boiling points, which therefore cannot be used as fuels.

Two solutions are generally considered to avoid the formation of long chains. The first one consists of shortening the chain length by, for example, metathesis of the starting fatty acids (Mol & Buffon, 1998) and to proceed with Kolbe electrolysis on the shortened chains. The second one consists of performing Kolbe electrolysis in the presence of excess of acetates (Sumera & Sadain, 1990; Ziogas et al., 2020).

Oxidative electroconversion

A heterogeneous catalytic oxidation of biosourced molecules has been proposed as a highly promising method for producing valuable chemicals (Carrettin et al., 2004; Climent et al., 2011; Gallezot, 2012). For example, the first generation of biofuels from the transesterification of vegetable oils leads to 10 wt% of glycerol as the byproduct (Fig. 12.12). Glycerol is an interesting compound from a sustainable economy point of view. It is now considered as an industrial waste (Aarthy, Saravanan, Gowthaman, Rose, & Kamini, 2014), and, therefore, it is cheap and its use leads to an environmental factor (*E*-factor of 0 (Sheldon, 2017)). From a chemical point of view, all carbons in the molecule are activated due to the presence of an alcohol group and it is therefore possible to obtain valuable products from its electrooxidation (Simões et al., 2012).

Considering the second generation of biofuels, lignocellulosic starting materials may undergo hydrothermal liquefaction, leading to long-chain oxygenated molecules and biofuels after hydrodeoxygenation and to small-chain oxygenated molecules, mainly C5 and C6 sugars and alcohols that can be electroreformed to produce value-added compounds for fine chemistry, and hydrogen for either fuel cell or to help the hydrodeoxygenation reaction, as in a biorefinery (Fig. 12.13). However, in both cases, because reactions at both electrodes lead to valuable compounds, such an electrochemical system could help make biofuel industries more profitable.

It is generally admitted that oxygenated molecules such as alcohols, polyols, and sugars are more electroreactive in alkaline media than in acidic ones (Wang et al., 2003) since base-catalyzed reactions are favored in alkaline media (Kwon, Lai, Rodriguez, & Koper, 2011). In addition, alkaline media allows using nonnoble metal-based catalysts that are generally unstable in acidic media. Moreover, it has also been shown that the C—C bond cleavage was more difficult in alkaline than in acidic media (Gomes & Tremiliosi-Filho, 2011; Roquet, Belgsir, Léger, & Lamy, 1994), which is an excellent point in the perspective of producing selective value-added compounds together with hydrogen at the cathode. Several electrocatalysts were studied mainly based on Pt, Pd, and Au (Holade et al., 2016; Holade, Morais, Servat, Napporn, & Kokoh, 2013; Rafaïdeen et al., 2019;

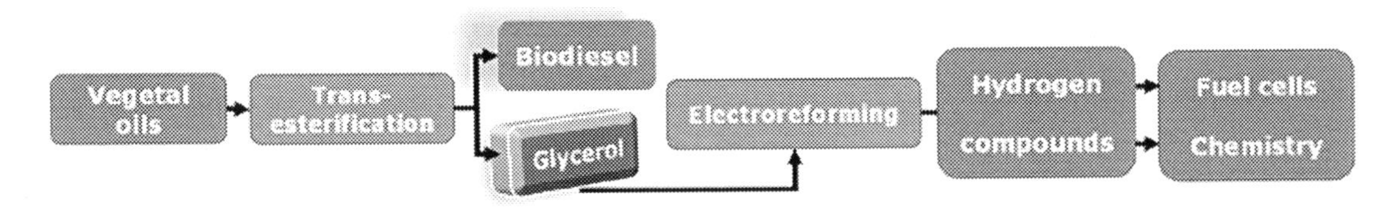

FIG. 12.12 Schematic of the possible value chain for first-generation biodiesel production.

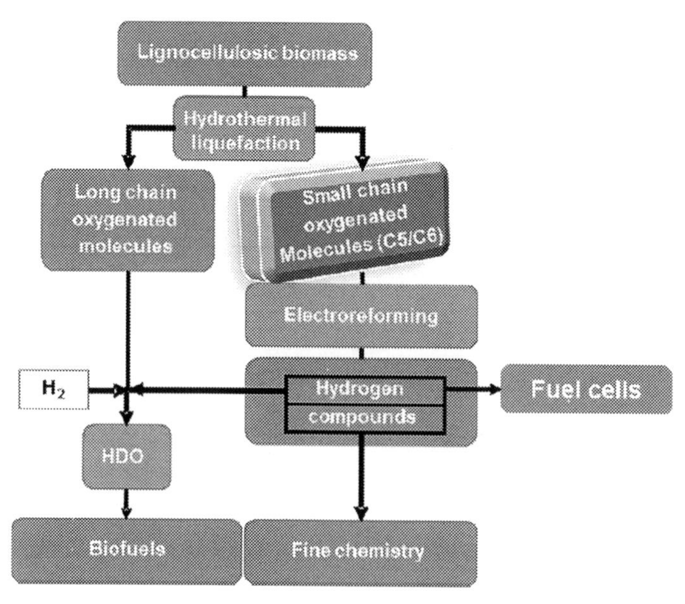

FIG. 12.13 Schematic of the possible value chain for second-generation biodiesel production.

Simões et al., 2012), but it has been known for a long time that platinum-bismuth catalysts are highly active for biomass conversion through heterogeneous catalysis (Gallezot, 1997; Mallat & Baiker, 1994), and it has been shown more recently that such catalysts are also the outstanding ones in terms of activity for electro-oxidation of both glycerol and sugars (Neha, Islam, Baranton, Coutanceau, & Pollet, 2020; Simões, Baranton, & Coutanceau, 2011), particularly Pt_9Bi_1/C in the atomic ratio (Fig. 12.14).

Besides the high electrocatalytic activity of platinum-bismuth nanomaterials, it is important to determine the selectivity of such catalysts. In situ Fourier transform infrared spectroscopy (FTIR), where infrared spectra are recorded as a function of the electrode potential that is linearly varied at a low scan rate of $1\,mV\,s^{-1}$, is an extremely convenient method to study the selectivity of a catalyst for a given electrochemical reaction.

In the case of the electro-oxidation of $0.10\,mol\,L^{-1}$ glycerol in $0.10\,mol\,L^{-1}$ NaOH electrolyte between 0.05 and 1.15 V vs RHE (reversible hydrogen electrode), no absorption band appears on the spectra from 0.05 to 0.60 V vs RHE (Fig. 12.15A). For higher potential, bands related to the formation of dihydroxyacetone at ca. $1335\,cm^{-1}$ and to carboxylates at ca. 1310, 1490, and $1580\,cm^{-1}$ are expanding (Simões et al., 2011, 2012). In addition, in the wavenumber range from 1700 to $2000\,cm^{-1}$, an absorption band that is moving with the electrode potential is typical of the formation of multibound adsorbed CO species at the surface of the catalyst (Couto, Rincón, Pérez, & Gutiérrez, 2001) (Fig. 12.15B), which means that platinum can break the C—C bond, and, therefore, the carboxylate formed from 0.60 V is a mixture of the C3, C2, and C1 species.

In the case of the Pt_9Bi_1/C catalyst, two potential regions can be separated (Fig. 12.15C). In the first one, from 0.20 to 0.60 V vs RHE, three bands are visible, those at ca. 1225 and $1300\,cm^{-1}$ corresponding to the formation of glyceraldehyde and those at $1335\,cm^{-1}$ corresponding to the formation of dihydoxyacetone. For higher

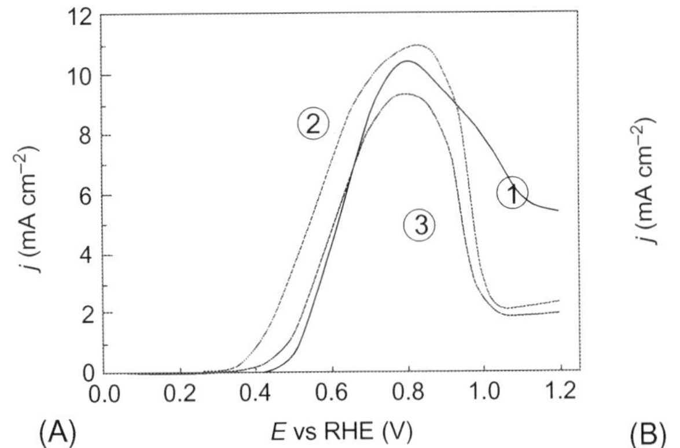

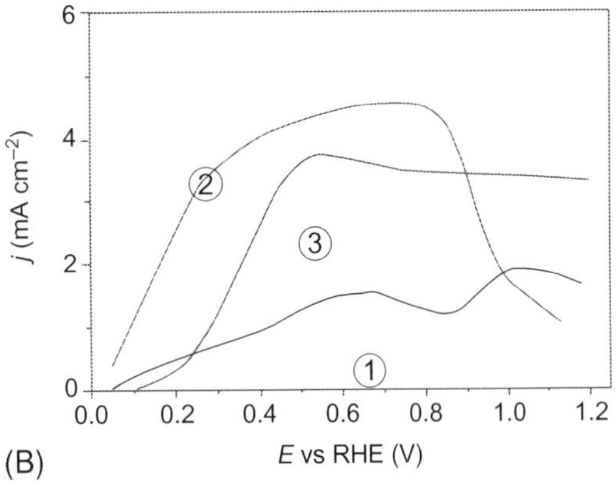

FIG. 12.14 Linear scan voltammetry curves recorded for the electroconversion of (A) $0.10\,mol\,L^{-1}$ glycerol and (B) $0.10\,mol\,L^{-1}$ glucose electro-oxidation (0.10M NaOH, $T = 20°C$, and scan rate $= 0.005\,V\,s^{-1}$) at (1) Pt/C, (2) Pt_9Bi_1/C, and (3) Pt_8Bi_2/C.

potentials, the same bands as those for platinum alone are present, indicating the formation of carboxylates. However, looking at the spectra (Fig. 12.15D), no band related to the adsorbed CO species is visible, indicating that the breaking of the C—C bond does not take place at this catalyst and therefore C3 carboxylates are formed (mainly glycerate and tartronate).

These are extremely important results as they show that such a platinum-bismuth catalyst is not only highly active but also highly selective toward the C3 species, at low and high electrode potentials.

With regard to glucose electro-oxidation, the spectra recorded as a function of the electrode potential on a Pt/C catalyst (Fig. 12.16A) show the typical band of linearly adsorbed CO at ca. $2050\,cm^{-1}$ (Couto et al., 2001), two small bands related to the formation of gluconate from 0.20V vs RHE at ca. 1410 and $1590\,cm^{-1}$, two small bands related to the formation of gluconolactone from 0.60V vs RHE at ca. 1360 and $1710\,cm^{-1}$, and a strong band at ca. $2345\,cm^{-1}$ related to the formation of CO_2 (Beden, Largeaud, Kokoh, & Lamy, 1996) from also ca. 0.60V vs RHE. In the case of the Pt_9Bi_1/C catalyst (Fig. 12.16B), the band related to the adsorbed CO is again missing, the band related to gluconate is stronger than on platinum, and bands related to CO_2 and gluconolactone from 0.60V vs RHE are smaller, with gluconolactone remaining as a shoulder. From these results, the following mechanism is proposed in the case of the electro-oxidation of glucose on platinum-bismuth catalysts.

For potentials lower than 0.60V, glucose molecules in solution diffuse toward the electrode and adsorb on platinum sites. Then, because at such low potential platinum is not expected to activate water, an Eley-Rideal mechanism is proposed to produce gluconate according to the following equation:

$$C_6H_{12}O_6 + OH^- \rightarrow C_6H_{11}O_{7-} + 2H_2O + 2e^- \tag{12.14}$$

For potentials higher than 0.60V, glucose adsorbs as a lactone. Lactone can be desorbed and can undergo hydrolysis reaction in the presence of hydroxyl ions in solution to produce gluconate or can remain adsorbed and undergo a bifunctional mechanism owing to the presence of adsorbed OH species at the platinum surface, thus leading to gluconate in the case of platinum-bismuth catalysts and to gluconate, degradation products, and CO_2 in the case of platinum catalysts. Indeed, the relative intensity

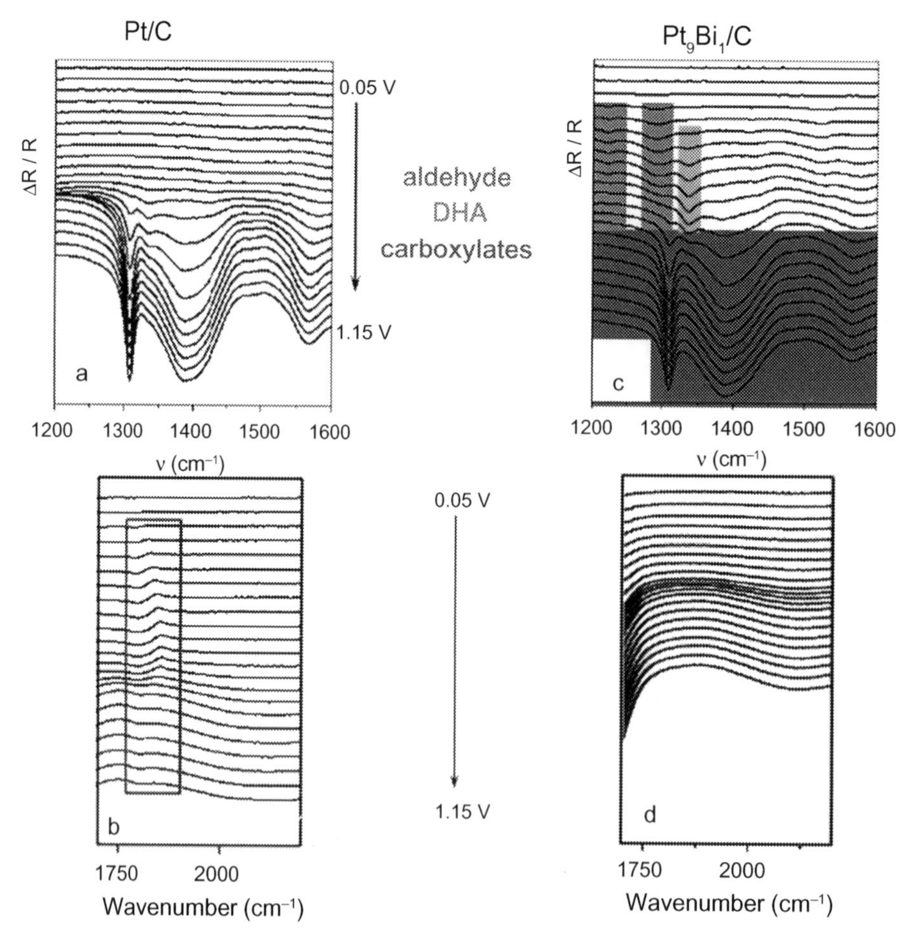

FIG. 12.15 In situ FTIR spectra recorded for the electroconversion of 0.10 mol L^{-1} glycerol in 0.10 mol L^{-1} NaOH ($T = 20°C$ and scan rate = 0.001 V s^{-1}) at (A, B) Pt/C and (C, D) Pt_9Bi_1/C.

of the CO_2 infrared absorption compared to those of the bands relative to the formation of gluconate indicates that the Pt_9Bi_1/C catalyst displays a much higher selectivity toward gluconate than the Pt/C catalyst.

The electroconversions of glycerol and glucose have also been studied in an electrolysis cell.

In the case of glycerol (2.0 mol L^{-1}), the anode was loaded with a Pt_9Bi_1/C catalyst and the cathode with a Pt/C catalyst at 1.6 mg_{metal} cm^{-2}. Two cell voltages were considered, i.e., 0.55 and 0.75 V. The current achieved is higher at 0.75 V than at 0.55 V, and, in both cases, it decreases monotonously over time. However, after the solution was renewed after 4 h of electrolysis, the initial current was recovered, which showed that the phenomenon responsible for the current decrease was reversible and not related to the continuous degradation of the electrodes. The reaction products after 4 h of electrolysis were analyzed by HPLC. At 0.55 V, the selectivity toward glyceraldehyde reached 80%, whereas that at 0.75 V was 58% and more C3 carboxylates were produced, in agreement with in situ infrared spectroscopy measurements.

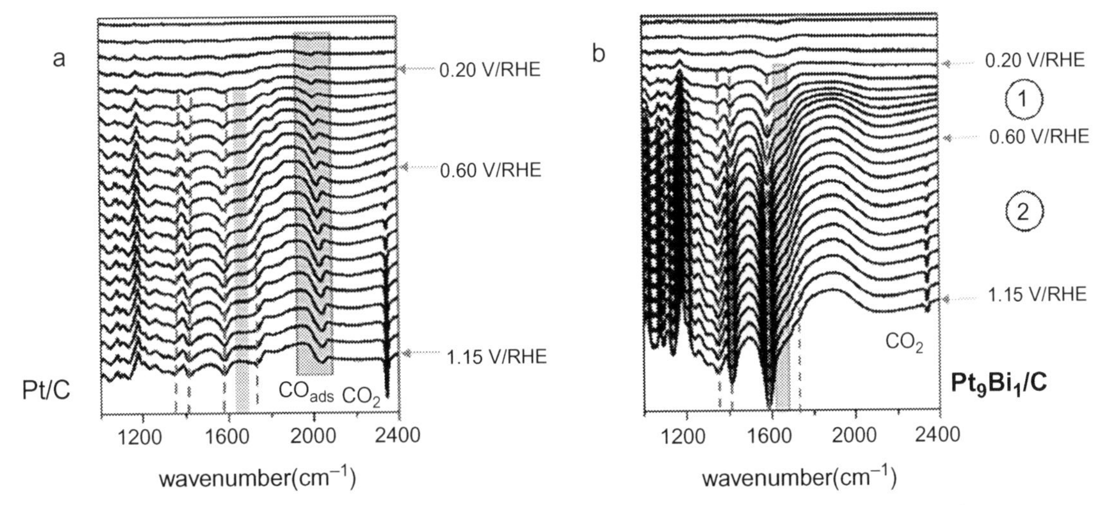

FIG. 12.16 In situ FTIR spectra recorded for the electroconversion of 0.10 mol L^{-1} glucose in 0.10 mol L^{-1} NaOH (T = 20°C and scan rate = 0.001 V s^{-1}) at (A) Pt/C and (B) Pt_9Bi_1/C.

The same kind of experiments was also performed for glucose electroconversion in an electrolysis cell at 0.3 V. The decrease of current with time was also reversible and has been shown to be due to the depletion of glucose in solution (Rafaïdeen et al., 2019). The reaction products were analyzed by HPLC every hour of electrolysis, and only one peak increasing with electrolysis time was observed at a retention time corresponding to gluconic acid. To confirm the nature of the reaction product, ^{13}C NMR and mass spectroscopy measurements were performed, and the results indicated without any doubt that the only reaction product detected by HPLC was gluconate. However, CO_2 and other volatile compounds could be formed. To discard this possibility, gluconate concentration as a function of time determined by HPLC was compared to that calculated from the electric charges involved in the electrolysis process, assuming the only formation of gluconate involving the exchange of two electrons according to Eq. (12.14).

Both curves are extremely close to each other. This clearly indicates that the selectivity is close to 100% as is the faradaic yield. At 0.30 V, the electrical energy consumed is 0.71 kWh per standard cubic meter of hydrogen produced. Considering the production of 1 ton of sodium gluconate, 9 kg of hydrogen will also be evolved, and the electrical energy consumed will correspond to circa 8.05 kWh/kg hydrogen. The average cost of electricity in Europe is circa 0.15 € per kWh, and the energy cost for 1 ton of sodium gluconate and 9 kg of hydrogen is 11 €, 13 USD, or 17 CAD, which is extremely low.

However, such approaches have several drawbacks. The first is that the concentration in sugars is limited. Rafaïdeen et al. (2019) studied the oxidative electroconversion of glucose and xylose on alloyed Pd_xAu_{1-x}/C catalysts and found that the Pd_3Au_7/C composition led to higher activity and selectivity toward gluconate or xylonate, in agreement with previous work on nonalloyed (Pt_x+Au_{1-x})/C catalysts by Yan et al. (2014), although the alloyed catalysts led to higher activity (lower-onset potentials) than the nonalloyed ones. They also changed the concentrations and cell voltages and concluded that in terms of selectivity and faradaic yield, the best conditions were low concentration and low cell voltages, indicating a low production rate of the desired compounds (Lamy, Jaubert, Baranton, & Coutanceau, 2014). Moreover, such approaches lean on catalytic materials from platinum group metals (PGMs) that have high cost, low availability, and are strategical and for which the achieved current densities, as well as the production rate of hydrogen and value-added compounds, are extremely low. Although the electric energy consumed in electroreforming systems is independent of time (Coutanceau & Baranton, 2016), time is money. An increase in the reaction rate will necessitate the increase of PGM loading in the electrode, indicating an increase in the capital expenditure to implement the system.

The coupling of electroactivation methods with other nonthermal activation processes can help enhance the conversion rate. For example, sonoelectrochemistry (application of power ultrasound in electrochemistry) may offer many advantages, such as: (a) enhanced electrochemical diffusion processes, (b) increase in electrochemical rates and yields, (c) increase in process efficiencies, (d) decrease in cell voltages and electrode overpotentials, and (e) improved electrode surface activation (Pollet, 2018; Pollet and Ashokkumar, 2019). Neha et al. (2020) performed electro-oxidation of glucose and methylglucoside in 0.10 mol L^{-1} NaOH on polycrystalline Pt in the potential range from 0.00 to 1.20 V vs RHE under silent and low-frequency ultrasonic (bath, 45 kHz, $P_{acous} = 11.20$ W) conditions. Electrochemical and chromatography studies indicated that ultrasound significantly enhanced the electro-oxidation rate of both glucose and methylglucoside but at the expense of the selectivity. However, studies on the effect of the ultrasound frequency and energy, dedicated catalysts, and operation conditions need to be carried out to ameliorate such kind of coupled activation processes.

The development of non-PGM catalysts, highly electroactive at potentials as low as possible represents another interesting solution. Such catalysts will certainly be based on nickel as nickel is known to be an active material for alcohol electro-oxidation (Houache et al., 2020).

The electro-oxidation of lignin was also studied. It has been shown that electrochemical oxidation was a green and inexpensive method for the depolymerization of lignins as this reaction can be achieved at room temperature and ambient pressure. For example, Stiefel, Schmitz, Peters, Di Marino, and Wessling (2016) used nickel as anode materials with different structures (plate, wire, fleece, foam, and foam stack) and found that 3D structures with subsequent membrane filtration of the products increased the reaction yield. Shao et al. (2014) used Ti/Sb-SnO_2 and Ti/PbO_2 for the degradation of lignin. In those studies, the process occurs at extremely high cell voltages, i.e., extremely high electrical energy consumption that makes it not viable economically, involving oxygen evolution as a competitive reaction, which also leads to extremely low selectivity, i.e., to the formation of numerous products. However, Di Marino, Stöckmann, Kriescher, Stiefel, and Wessling (2016) reported on the use of pure deep eutectic solvents (DESs) to dissolve lignin in combination with electrochemical oxidative depolymerization at the nickel electrode, thus also avoiding the oxygen evolution reaction. Under these conditions, lignin depolymerization was successfully achieved, leading to guaiacol and vanillin as the main products.

Reductive electroconversion

The electrochemical reduction of glucose, mannose, and invert sugar to their corresponding alcohols has been studied rather early. In 1939, Creighton studied the effect of current density, temperature, and electrolyte composition on the yield of these electrochemical conversions in alkaline media at mercury and amalgamated lead cathodes (Jermain, 1939). Such electrode materials are known to have extremely high hydrogen evolution overpotentials, which led to limit this competitive reaction and to increase the faradaic yield (Bin Kassim, Rice, & Kuhn, 1981). Creighton also discussed the possibility of commercial development of the electrochemical reduction of sugars, and the process was completely industrial since the end of the second war where it was replaced by the catalytic hydrogenation process using a Raney-Ni catalyst. Indeed, high electricity prices at the time, insufficient development of electrochemical technologies, and environmental issues due to the use of mercury or amalgamated lead cathodes led to the abandonment of the electrochemical route. Now, growth of renewable electricity production makes electrochemical methods for biomass conversion more profitable. However, developments of electrocatalytic materials to decrease cathode overpotential, their environmental footprint, e.g., using more eco-friendly materials than mercury or lead, and their cost, e.g., less strategical and costly metal than pgm, are mandatory.

For this purpose, nickel is again a good material that allows the electroreduction of glucose to sorbitol at lower overpotentials than other monometallic electrodes (Kwon & Koper, 2013). However, nickel belongs to the category of metals with low hydrogen evolution overpotential and therefore this later reaction will occur simultaneously, decreasing the target product yield. Therefore, important efforts must be taken to develop nickel-based catalysts that are active toward the glucose reduction reaction, selective toward the formation of sorbitol with a high faradaic yield, and stable.

As explained before, if the counter reaction at the anode is the oxygen evolution reaction, the anode potential will be higher than 1.23 V vs SHE, and the electrosynthesis cell voltage will be higher than 1.23 V to initiate the reductive electroconversion of biomass. This will then lead to extremely high electrical energy consumption, and further to high cost of production, since oxygen evolved at the anode is not a valuable compound. The use of a sacrificial anode can be a solution. For example, Boissou et al. (2019) performed electrocarboxylation of furfural by CO_2 in γ-valerolactone as the solvent at the Pt cathode of an electrosynthesis reactor and limited the cell voltage using a copper cathode that was oxidized into Cu^{2+} as the counter reaction. However, these authors concluded that the use of a Cu sacrificial anode was questionable from both industry and sustainability points of view and that pairing reduction and oxidation reactions to produce value-added chemicals at the anode and the cathode of a single cell should be a more efficient electrochemical process. For example, glucose electrolysis was performed in a paired electrochemical reactor, in which glucose was reduced to sorbitol using a Raney-Ni cathode and oxidized to gluconic acid by HOBr generated at a graphite anode (Park et al., 1985). More recently, Houache et al. paired glycerol electro-oxidation with formiate at a Ni-based anode with CO_2 reduction into CO at an Ag cathode (Houache et al., 2020).

12.5 Conclusions

Mechanocatalysis, sonocatalysis, and electrocatalysis are interesting approaches that can be applied to lignocellulosic biomass conversion. Their role is a promising research field due to the large number of possibilities provided by biomass-based reactions. Despite the great potential of bio-derived compounds, there is a lack of literature examples dealing with catalytic biomass conversion under physical activation. One can mention that for mechanocatalysis and sonocatalysis, their characterizations are highly in demand, in order to ensure that experiments and processes are reproducible. A focus should be placed on the applied frequency for sonocatalysis and rotation speed in ball-milling devices in order to correlate with catalytic activity. This will be helpful in designing a reactor to develop an efficient and economic process. Moreover, some predictive methods should be developed to understand the chemical reactions and mechanisms.

With regard to other nonconventional technologies, microwaves have been widely explored for biomass conversion. Some other techniques, involving the generation of reactive species (such as photochemical and plasma activation), have also been investigated but still need further investigations to avoid the formation of undesirable products.

In future experiments, the extended collection of data on equipment and parameters will therefore be of strategic importance. Studies related to energetics and the financial feasibility of these approaches, in comparison with the existing conventional processes, will allow the production of large volumes of chemicals to be scaled up in continuous processes and/or in the synthesis of high value-added products from biomass.

References

Aarthy, M., Saravanan, P., Gowthaman, M. K., Rose, C., & Kamini, N. R. (2014). Enzymatic transesterification for production of biodiesel using yeast lipases: An overview. *Chemical Engineering Research and Design*, *92*(8), 1591–1601. https://doi.org/10.1016/j.cherd.2014.04.008.

Ago, M., Endo, T., & Hirotsu, T. (2004). Crystalline transformation of native cellulose from cellulose I to cellulose II polymorph by a ball-milling method with a specific amount of water. *Cellulose*, *11*(2), 163–167. https://doi.org/10.1023/B:CELL.0000025423.32330.fa.

Amaniampong, P. N., Clément, J.-L., Gigmes, D., Ortiz Mellet, C., Garcia Fernandez, J. M., Blériot, Y., et al. (2018). Catalyst-free synthesis of alkylpolyglycosides induced by high-frequency ultrasound. *ChemSusChem*, *11*, 2673–2676. https://doi.org/10.1002/cssc.201801137.

Amaniampong, P. N., Karam, A., Trinh, Q. T., Xu, K., Hirao, H., Jérôme, F., et al. (2017). Selective and catalyst-free oxidation of D-glucose to D-glucuronic acid induced by high-frequency ultrasound. *Scientific Reports*, *7*. https://doi.org/10.1038/srep40650.

Amaniampong, P., Trinh, Q. T., Oliveira Vigier, K. D., Dao, D. Q., Tran, N. H., Wang, Y., et al. (2019). Synergistic effect of high frequency ultrasound with cupric-oxide catalyst resulting in a selectivity switch in glucose oxidation under argon. *Journal of the American Chemical Society*, *141*, 14772–14779.

Barakat, A., de Vries, H., & Rouau, X. (2013). Dry fractionation process as an important step in current and future lignocellulose biorefineries: A review. *Bioresource Technology*, *134*, 362–373. https://doi.org/10.1016/j.biortech.2013.01.169.

Beden, B., Largeaud, F., Kokoh, K. B., & Lamy, C. (1996). Fourier transform infrared reflectance spectroscopic investigation of the electrocatalytic oxidation of D-glucose: Identification of reactive intermediates and reaction products. *Electrochimica Acta*, *41*(5), 701–709. https://doi.org/10.1016/0013-4686(95)00359-2.

Beltrame, P., Comotti, M., Della Pina, C., & Rossi, M. (2004). Aerobic oxidation of glucose I. Enzymatic catalysis. *Journal of Catalysis*, *228*(2), 282–287. https://doi.org/10.1016/j.jcat.2004.09.010.

Benoit, M., Rodrigues, A., Zhang, Q., Fourré, E., De Oliveira Vigier, K., Tatibouët, J. M., et al. (2011). Depolymerization of cellulose assisted by a nonthermal atmospheric plasma. *Angewandte Chemie, International Edition*, *50*(38), 8964–8967. https://doi.org/10.1002/anie.201104123.

Bergenstråhle, M., Wohlert, J., Himmel, M. E., & Brady, J. W. (2010). Simulation studies of the insolubility of cellulose. *Carbohydrate Research*, *345*(14), 2060–2066. https://doi.org/10.1016/j.carres.2010.06.017.

Biermann, O., Hädicke, E., Koltzenburg, S., & Müller-Plathe, F. (2001). Hydrophilicity and lipophilicity of cellulose crystal surfaces. *Angewandte Chemie International Edition*, *40*(20), 3822–3825. https://doi.org/10.1002/1521-3773(20011015)40:20<3822::AID-ANIE3822>3.0.CO;2-V.

Bin Kassim, A., Rice, C. L., & Kuhn, A. T. (1981). Formation of sorbitol by cathodic reduction of glucose. *Journal of Applied Electrochemistry*, *11*(2), 261–267. https://doi.org/10.1007/BF00610988.

Biosedev. (n.d.). http://biosedev.com.

Boissou, F., Baranton, S., Tarighi, M., De Oliveira Vigier, K., & Coutanceau, C. (2019). The potency of γ-valerolactone as bio-sourced polar aprotic organic medium for the electrocarboxlation of furfural by CO_2. *Journal of Electroanalytical Chemistry*, *848*. https://doi.org/10.1016/j.jelechem.2019.113257.

Boissou, F., Sayoud, N., De Oliveira Vigier, K., Barakat, A., Marinkovic, S., Estrine, B., et al. (2015). Acid-assisted ball milling of cellulose as an efficient pretreatment process for the production of butyl glycosides. *ChemSusChem*, *8*(19), 3263–3269. https://doi.org/10.1002/cssc.201500700.

Box, V. G. S. (1990). The role of lone pair interactions in the chemistry of the monosaccharides. The anomeric effects. *Heterocycles*, *31*, 1157–1181.

Box, V. G. S. (1991). The role of lone pair interactions in the chemistry of the monosaccharides. Stereo-electronic effects in unsaturated monosaccharides. *Heterocycles*, *32*, 795–807.

Bozell, J. J., & Petersen, G. R. (2010). Technology development for the production of biobased products from biorefinery carbohydrates—The US Department of Energy's "Top 10" revisited. *Green Chemistry, 12,* 539–554. https://doi.org/10.1039/B922014C.

Brière, R., Loubet, P., Glogic, E., Estrine, B., Marinkovic, S., Jérôme, F., et al. (2018). Life cycle assessment of the production of surface-active alkyl polyglycosides from acid-assisted ball-milled wheat straw compared to the conventional production based on corn-starch. *Green Chemistry, 20*(9), 2135–2141. https://doi.org/10.1039/c7gc03189k.

Bujak, P., Bartczak, P., & Polanski, J. (2012). Highly efficient room-temperature oxidation of cyclohexene and d-glucose over nanogold Au/SiO_2 in water. *Journal of Catalysis, 295,* 15–21. https://doi.org/10.1016/j.jcat.2012.06.023.

Cardoso, D. S. P., Šljukić, B., Santos, D. M. F., & Sequeira, C. A. C. (2017). Organic electrosynthesis: From laboratorial practice to industrial applications. *Organic Process Research and Development, 21*(9), 1213–1226. https://doi.org/10.1021/acs.oprd.7b00004.

Carrasquillo-Flores, R., Käldström, M., Schüth, F., Dumesic, J. A., & Rinaldi, R. (2013). Mechanocatalytic depolymerisation of dry (ligno)cellulose as an entry point for high-yield production of furfurals. *ACS Catalysis, 3,* 993–997.

Carrettin, S., McMorn, P., Johnston, P., Griffin, K., Kiely, C. J., Attard, G. A., et al. (2004). Oxidation of glycerol using supported gold catalysts. *Topics in Catalysis, 27*(1–4), 131–136. https://doi.org/10.1023/B:TOCA.0000013547.35106.0d.

Chatel, G., Oliveira Vigier, K. D., & Jérôme, F. (2014). Sonochemistry: What potential for conversion of lignocellulosic biomass into platform chemicals. *ChemSusChem, 7,* 481–520.

Climent, M. J., Corma, A., & Iborra, S. (2011). Converting carbohydrates to bulk chemicals and fine chemicals over heterogeneous catalysts. *Green Chemistry, 13*(3), 520–540. https://doi.org/10.1039/c0gc00639d.

Corma Canos, A., Iborra, S., & Velty, A. (2007). Chemical routes for the transformation of biomass into chemicals. *Chemical Reviews, 107*(6), 2411–2502. https://doi.org/10.1021/cr050989d.

Coutanceau, C., & Baranton, S. (2016). Electrochemical conversion of alcohols for hydrogen production: A short overview. *Wiley Interdisciplinary Reviews: Energy and Environment, 5*(4), 388–400. https://doi.org/10.1002/wene.193.

Couto, A., Rincón, A., Pérez, M. C., & Gutiérrez, C. (2001). Adsorption and electrooxidation of carbon monoxide on polycrystalline platinum at pH 0.3–13. *Electrochimica Acta, 46*(9), 1285–1296. https://doi.org/10.1016/S0013-4686(00)00714-3.

Cramer, C. J., Truhlar, D. G., & French, A. D. (1997). Exo-anomeric effects on energies and geometries of different conformations of glucose and related systems in the gas phase and aqueous solution. *Carbohydrate Research, 298,* 1–14. https://doi.org/10.1016/S0008-6215(96)00297-2.

de Carvalho Silvello, M. A., Martínez, J., & Goldbeck, R. (2020). Low-frequency ultrasound with short application time improves cellulase activity and reducing sugars release. *Applied Biochemistry and Biotechnology, 191*(3), 1042–1055. https://doi.org/10.1007/s12010-019-03148-1.

Delaux, J., Ortiz Mellet, C., Canaff, C., Fourré, E., Gaillard, C., Barakat, A., et al. (2016). Impact of nonthermal atmospheric plasma on the structure of cellulose: Access to soluble branched glucans. *Chemistry – A European Journal, 22*(46), 16522–16530. https://doi.org/10.1002/chem.201603214.

Della Pina, C., & Falletta, E. (2011). Gold-catalyzed oxidation in organic synthesis: A promise kept. *Catalysis Science and Technology, 1*(9), 1564–1571. https://doi.org/10.1039/c1cy00283j.

Dhepe, P. L., & Fukuoka, A. (2008). Cellulose conversion under heterogeneous catalysis. *ChemSusChem, 1*(12), 969–975. https://doi.org/10.1002/cssc.200800129.

Di Marino, D., Stöckmann, D., Kriescher, S., Stiefel, S., & Wessling, M. (2016). Electrochemical depolymerisation of lignin in a deep eutectic solvent. *Green Chemistry, 18*(22), 6021–6028. https://doi.org/10.1039/c6gc01353h.

Dos Santos, T. R., Harnisch, F., Nilges, P., & Schröder, U. (2015). Electrochemistry for biofuel generation: Transformation of fatty acids and triglycerides to diesel-like olefin/ether mixtures and olefins. *ChemSusChem, 8*(5), 886–893. https://doi.org/10.1002/cssc.201403249.

Du, B., Chen, C., Sun, Y., Yu, M., Yang, M., Wang, X., et al. (2020). Catalytic conversion of lignin to bio-oil over PTA/MCM-41 catalyst assisted by ultrasound acoustic cavitation. *Fuel Processing Technology, 206.* https://doi.org/10.1016/j.fuproc.2020.106479.

Final report of E4tech, RE-CORD and WUR for the European Commission Directorate. (2021). Can Be Downloaded at https://Ec.Europa.Eu/Energy/Sites/Ener/Files/Documents/EC%20Sugar%20Platform%20final%20report.Pdf.

Finch, K. B. H., Richards, R. M., Richel, A., Medvedovici, A. V., Gheorghe, N. G., Verziu, M., et al. (2012). Catalytic hydroprocessing of lignin under thermal and ultrasound conditions. *Catalysis Today, 196*(1), 3–10. https://doi.org/10.1016/j.cattod.2012.02.051.

Fraile, D., Lanoix, J.-C., Maio, P., Rangel, A., & Torre, A. (2015). *Overview of the market segmentation for hydrogen across potential customer groups, based on key application areas.* CertifHy. In D1_2_Overview_of_the_market_segmentation_ Final_22_June_low-res.

Gallezot, P. (1997). Selective oxidation with air on metal catalysts. *Catalysis Today*, *37*(4), 405–418. https://doi.org/10.1016/S0920-5861(97)00024-2.

Gallezot, P. (2012). Conversion of biomass to selected chemical products. *Chemical Society Reviews*, *41*(4), 1538–1558. https://doi.org/10.1039/c1cs15147a.

Gharat, N., & Rathod, V. K. (2013). Ultrasound assisted enzyme catalyzed transesterification of waste cooking oil with dimethyl carbonate. *Ultrasonics Sonochemistry*, *20*(3), 900–905. https://doi.org/10.1016/j.ultsonch.2012.10.011.

Gomes, J. F., & Tremiliosi-Filho, G. (2011). Spectroscopic studies of the glycerol electro-oxidation on polycrystalline Au and Pt surfaces in acidic and alkaline media. *Electrocatalysis*, *2*(2), 96–105. https://doi.org/10.1007/s12678-011-0039-0.

Haouache, S., Karam, A., Chave, T., Clarhaut, J., Amaniampong, P. N., Garcia Fernandez, J. M., et al. (2020). Selective radical depolymerization of cellulose to glucose induced by high frequency ultrasound. *Chemical Science*, *11*(10), 2664–2669. https://doi.org/10.1039/d0sc00020e.

Hernoux-Villière, A., Lassi, U., & Lévêque, J. M. (2013). An original ultrasonic reaction with dual coaxial frequencies for biomass processing. *Ultrasonics Sonochemistry*, *20*(6), 1341–1344. https://doi.org/10.1016/j.ultsonch.2013.04.011.

Hick, S. M., Griebel, C., Restrepo, D. T., Truitt, J. H., Buker, E. J., Bylda, C., et al. (2010). Mechanocatalysis for biomass-derived chemicals and fuels. *Green Chemistry*, *12*(3), 468–474. https://doi.org/10.1039/b923079c.

Holade, Y., Morais, C., Servat, K., Napporn, T. W., & Kokoh, K. B. (2013). Toward the electrochemical valorization of glycerol: Fourier transform infrared spectroscopic and chromatographic studies. *ACS Catalysis*, *3*(10), 2403–2411. https://doi.org/10.1021/cs400559d.

Holade, Y., Servat, K., Napporn, T. W., Morais, C., Berjeaud, J. M., & Kokoh, K. B. (2016). Highly selective oxidation of carbohydrates in an efficient electrochemical energy converter: Cogenerating organic electrosynthesis. *ChemSusChem*, *9*(3), 252–263. https://doi.org/10.1002/cssc.201501593.

Holzhäuser, F. J., Mensah, J. B., & Palkovits, R. (1849). (Non-) Kolbe electrolysis in biomass valorization—A discussion of potential applications. *Annalen der Chemie und Pharmacie*, *22*, 257–294. https://doi.org/10.1039/c9gc03264a.

Holzhäuser, F. J., Mensah, J. B., & Palkovits, R. (2020). (Non-) Kolbe electrolysis in biomass valorization—A discussion of potential applications. *Green Chemistry*, *22*(2), 286–301. https://doi.org/10.1039/c9gc03264a.

Houache, M. S. E., Safari, R., Nwabara, U. O., Rafaïdeen, T., Botton, G. A., Kenis, P. J. A., et al. (2020). Selective electrooxidation of glycerol to formic acid over carbon supported $Ni_{1-x}M_x$(M = Bi, Pd, and Au) nanocatalysts and coelectrolysis of CO_2. *ACS Applied Energy Materials*, *3*(9), 8725–8738. https://doi.org/10.1021/acsaem.0c01282.

Howsmon, J. A., & Marchessault, R. H. (1959). The ball-milling of cellulose fibers and recrystallization effects. *Journal of Applied Polymer Science*, *1*, 163–167.

International Agency of Energy. (2019). *Renewables 2019 market analysis and forecast from 2019 to 2024*. Available at https://www.iea.org/reports/renewables-2019/power. Accessed on 11 May 2020.

Jeffrey, G. A., Pople, J. A., Binkley, J. S., & Visheveshwara, S. (1978). Application of ab initio molecular orbital calculations to the structural moieties of carbohydrates. *Journal of the American Chemical Society*, *100*(2), 373–379. https://doi.org/10.1021/ja00470a003.

Jermain, C. H. (1939). The electrochemical reduction of sugars. *Transactions of the Electrochemical Society*, *289*. https://doi.org/10.1149/1.3498378.

Käldström, M., Meine, N., Farès, C., Rinaldi, R., & Schüth, F. (2014). Fractionation of 'water-soluble lignocellulose' into C5/C6 sugars and sulfur-free lignins. *Green Chemistry*, *16*, 2454–2462. https://doi.org/10.1039/C4GC00168K.

Käldström, M., Meine, N., Farès, C., Schüth, F., & Rinaldi, R. (2014). Deciphering 'water-soluble lignocellulose' obtained by mechanocatalysis: New insights into the chemical processes leading to deep depolymerization. *Green Chemistry*, *16*, 3528–3538. https://doi.org/10.1039/C4GC00004H.

Karam, A., Amaniampong, P. N., Fernández, J. M. G., Oldani, C., Marinkovic, S., Estrine, B., et al. (2018). Mechanocatalytic depolymerization of cellulose with perfluorinated sulfonic acid ionomers. *Frontiers in Chemistry*, *6*. https://doi.org/10.3389/fchem.2018.00074.

Klemm, D., Heublein, B., Fink, H. P., & Bohn, A. (2005). Cellulose: Fascinating biopolymer and sustainable raw material. *Angewandte Chemie International Edition*, *44*(22), 3358–3393. https://doi.org/10.1002/anie.200460587.

Klemm, D., Kramer, F., Moritz, S., Lindström, T., Ankerfors, M., Gray, D., et al. (2011). Nanocelluloses: A new family of nature-based materials. *Angewandte Chemie International Edition*, *50*, 5438–5466. https://doi.org/10.1002/anie.201001273.

Kolbe, H. (1849). Untersuchungen uber die elektrolyse organisher verbindungen. *Annalen der Chemie und Pharmacie*, *69*, 257–294. https://doi.org/10.1002/jlac.18490690302.

Kuna, E., Behling, R., Valange, S., Chatel, G., & Colmenares, J. C. (2017). Sonocatalysis: A potential sustainable pathway for the valorization of lignocellulosic biomass and derivatives. *Topics in Current Chemistry*, *375*(2). https://doi.org/10.1007/s41061-017-0122-y.

Kunaver, M., Jasiukaityte, E., & Čuk, N. (2012). Ultrasonically assisted liquefaction of lignocellulosic materials.

Bioresource Technology, *103*(1), 360–366. https://doi.org/10.1016/j.biortech.2011.09.051.

Kwon, Y., & Koper, M. T. M. (2013). Electrocatalytic hydrogenation and deoxygenation of glucose on solid metal electrodes. *ChemSusChem*, *6*(3), 455–462. https://doi.org/10.1002/cssc.201200722.

Kwon, Y., Lai, S. C. S., Rodriguez, P., & Koper, M. T. M. (2011). Electrocatalytic oxidation of alcohols on gold in alkaline media: Base or gold catalysis? *Journal of the American Chemical Society*, *133*(18), 6914–6917. https://doi.org/10.1021/ja200976j.

Lamy, C., Coutanceau, C., & Baranton, S. (2020). Production of clean hydrogen by electrochemical reforming of oxygenated organic compounds. In B. Pollet (Ed.), *Hydrogen energy and fuel cells primers*. Amsterdam, The Netherland: Elsevier, ISBN:978-0-12-821500-5.

Lamy, C., Jaubert, T., Baranton, S., & Coutanceau, C. (2014). Clean hydrogen generation through the electrocatalytic oxidation of ethanol in a Proton Exchange Membrane Electrolysis Cell (PEMEC): Effect of the nature and structure of the catalytic anode. *Journal of Power Sources*, *245*, 927–936. https://doi.org/10.1016/j.jpowsour.2013.07.028.

Levy, P. F., Sanderson, J. E., & Cheng, L. K. (1984). Kolbe electrolysis of mixtures of aliphatic organic acids. *Journal of the Electrochemical Society*, *131*(4), 773–777. https://doi.org/10.1149/1.2115697.

Li, K., & Sun, Y. (2018). Electrocatalytic upgrading of biomass-derived intermediate compounds to value-added products. *Chemistry – A European Journal*, *24*(69), 18258–18270. https://doi.org/10.1002/chem.201803319.

Liao, Y., Liu, Q., Wang, T., Long, J., Zhang, Q., Ma, L., … Li, Y. (2014). Promoting hydrolytic hydrogenation of cellulose to sugar alcohols by mixed ball milling of cellulose and solid acid catalyst. *Energy Fuels*, *28*(9), 5778–5784.

Liu, S., Okuyama, Y., Tamura, M., Nakagawa, Y., Imai, A., & Tomishige, K. (2015). Production of renewable hexanols from mechanocatalytically depolymerized cellulose by using Ir-ReOx/SiO_2 catalyst. *ChemSusChem*, *8*, 628–635. https://doi.org/10.1002/cssc.201403010.

Loerbroks, C., Rinaldi, R., & Thiel, W. (2013). The electronic nature of the 1,4-β-glycosidic bond and its chemical environment: DFT insights into cellulose chemistry. *Chemistry – A European Journal*, *19*(48), 16282–16294. https://doi.org/10.1002/chem.201301366.

Lunelli, F. C., Sfalcin, P., Souza, M., Zimmermann, E., Dal Prá, V., Foletto, E. L., et al. (2014). Ultrasound-assisted enzymatic hydrolysis of sugarcane bagasse for the production of fermentable sugars. *Biosystems Engineering*, *124*, 24–28. https://doi.org/10.1016/j.biosystemseng.2014.06.004.

Mallat, T., & Baiker, A. (1994). Oxidation of alcohols with molecular oxygen on platinum metal catalysts in aqueous solutions. *Catalysis Today*, *19*(2), 247–283. https://doi.org/10.1016/0920-5861(94)80187-8.

Marion, P., Bernela, B., Piccirilli, A., Estrine, B., Patouillard, N., Guilbot, J., et al. (2017). Sustainable chemistry: How to produce better and more from less? *Green Chemistry*, 4973–4989. https://doi.org/10.1039/C7GC02006F.

Marullo, S., Rizzo, C., Meli, A., & D'Anna, F. (2019). Ionic liquid binary mixtures, zeolites, and ultrasound irradiation: A combination to promote carbohydrate conversion into 5-hydroxymethylfurfural. *ACS Sustainable Chemistry & Engineering*, *7*(6), 5818–5826. https://doi.org/10.1021/acssuschemeng.8b05584.

Medronho, B., Romano, A., Miguel, M. G., Stigsson, L., & Lindman, B. (2012). Rationalizing cellulose (in)solubility: Reviewing basic physicochemical aspects and role of hydrophobic interactions. *Cellulose*, *19*(3), 581–587. https://doi.org/10.1007/s10570-011-9644-6.

Meine, N., Rinaldi, R., & Schüth, F. (2012). Solvent-free catalytic depolymerization of cellulose to water-soluble oligosaccharides. *ChemSusChem*, *5*(8), 1449–1454. https://doi.org/10.1002/cssc.201100770.

Mishra, S. P., Thirree, J., Manent, A. S., Chabot, B., & Daneault, C. (2011). Ultrasound-catalyzed TEMPO-mediated oxidation of native cellulose for the production of nanocellulose: Effect of process variables. *BioResources*, *6*(1), 121–143. http://www.ncsu.edu/bioresources/BioRes_06/BioRes_06_1_0121_Mishra_TMCD_Sono_Cat_TEMPO_Ox_Cellu_Nano_Proc_Var_1191.pdf.

Miyamoto, H., Umemura, M., Aoyagi, T., Yamane, C., Ueda, K., & Takahashi, K. (2009). Structural reorganization of molecular sheets derived from cellulose II by molecular dynamics simulations. *Carbohydrate Research*, *344*(9), 1085–1094. https://doi.org/10.1016/j.carres.2009.03.014.

Mliki, K., & Trabelsi, M. (2015). Chemicals from biomass: Efficient and facile synthesis of 5,5′(oxy-bis(methylene)) bis-2-furfural from 5-hydroxymethylfurfural. *Industrial Crops and Products*, *78*, 91–94. https://doi.org/10.1016/j.indcrop.2015.10.026.

Möhle, S., Zirbes, M., Rodrigo, E., Gieshoff, T., Wiebe, A., & Waldvogel, S. R. (2018). Modern electrochemical aspects for the synthesis of value-added organic products. *Angewandte Chemie International Edition*, *57*(21), 6018–6041. https://doi.org/10.1002/anie.201712732.

Mol, J. C., & Buffon, R. (1998). Metathesis in oleochemistry. *Journal of the Brazilian Chemical Society*, *9*(1), 1–11. https://doi.org/10.1590/S0103-50531998000100002.

Montalbo-Lomboy, M., Johnson, L., Khanal, S. K., (Hans) van Leeuwen, J., & Grewell, D. (2010). Sonication of sugary-2 corn: A potential pretreatment to enhance sugar release. *Bioresource Technology*, *101*(1), 351–358. https://doi.org/10.1016/j.biortech.2009.07.075.

Moon, R. J., Martini, A., Nairn, J., Simonsen, J., & Youngblood, J. (2011). Cellulose nanomaterials review: Structure, properties and nanocomposites. *Chemical Society Reviews, 40*, 3941–3994. https://doi.org/10.1039/C0CS00108B.

Napoly, F., Kardos, N., Jean-Gérard, L., Goux-Henry, C., Andrioletti, B., & Draye, M. (2015). H_2O_2-mediated kraft lignin oxidation with readily available metal salts: What about the effect of ultrasound? *Industrial and Engineering Chemistry Research, 54*(22), 6046–6051. https://doi.org/10.1021/acs.iecr.5b00595.

Neha, N., Islam, M. H., Baranton, S., Coutanceau, C., & Pollet, B. G. (2020). Assessment of the beneficial combination of electrochemical and ultrasonic activation of compounds originating from biomass. *Ultrasonics Sonochemistry, 63*. https://doi.org/10.1016/j.ultsonch.2019.104934.

Ouajai, S., & Shanks, R. A. (2006). Solvent and enzyme induced recrystallization of mechanically degraded hemp cellulose. *Cellulose, 13*(1), 31–44. https://doi.org/10.1007/s10570-005-9020-5.

Palkovits, S., & Palkovits, R. (2019). The role of electrochemistry in future dynamic bio-refineries: A focus on (non-) Kolbe electrolysis. *Chemie-Ingenieur-Technik, 91*(6), 699–706. https://doi.org/10.1002/cite.201800205.

Paquin, M., Loranger, E., Hannaux, V., Chabot, B., & Daneault, C. (2013). The use of Weissler method for scale-up a Kraft pulp oxidation by TEMPO-mediated system from a batch mode to a continuous flow-through sonoreactor. *Ultrasonics Sonochemistry, 20*(1), 103–108. https://doi.org/10.1016/j.ultsonch.2012.08.007.

Park, K., Pintauro, P. N., Baizer, M. M., & Nobe, K. (1985). Flow reactor studies of the paired electro-oxidation and electroreduction of glucose. *Journal of the Electrochemical Society, 132*(8), 1850–1855. https://doi.org/10.1149/1.2114229.

Pintauro, P. N., Johnson, D. K., Park, K., Baizer, M. M., & Nobe, K. (1984). The paired electrochemical synthesis of sorbitol and gluconic acid in undivided flow cells. I. *Journal of Applied Electrochemistry, 14*(2), 209–220. https://doi.org/10.1007/BF00618739.

Pollet, B. G. (2018). A short introduction to sonoelectrochemistry. *Electrochemical Society Interface, 27*(3), 41–42. https://doi.org/10.1149/2.F03183if.

Pollet, B. G., & Ashokkumar, M. (2019). Introduction to ultrasound, sonochemistry and sonoelectrochemistry. In B. G. Pollet, & M. Ashokkumar (Eds.), *SpringerBriefs in molecular science*. Cham, Germany: Springer, ISBN:978-3-030-25862-7.

Rafaïdeen, T., Baranton, S., & Coutanceau, C. (2019). Highly efficient and selective electrooxidation of glucose and xylose in alkaline medium at carbon supported alloyed PdAu nanocatalysts. *Applied Catalysis B: Environmental, 243*, 641–656. https://doi.org/10.1016/j.apcatb.2018.11.006.

Roquet, L., Belgsir, E. M., Léger, J. M., & Lamy, C. (1994). Kinetics and mechanisms of the electrocatalytic oxidation of glycerol as investigated by chromatographic analysis of the reaction products: Potential and pH effects. *Electrochimica Acta, 39*(16), 2387–2394. https://doi.org/10.1016/0013-4686(94)E0190-Y.

Sanderson, J. E., Levy, P. F., Cheng, L. K., & Barnard, G. W. (1983). The effect of pressure on the product distribution in Kolbe electrolysis. *Journal of the Electrochemical Society, 130*(9), 1844–1848. https://doi.org/10.1149/1.2120109.

Santos, D., Silva, U. F., Duarte, F. A., Bizzi, C. A., Flores, E. M. M., & Mello, P. A. (2018). Ultrasound-assisted acid hydrolysis of cellulose to chemical building blocks: Application to furfural synthesis. *Ultrasonics Sonochemistry, 40*, 81–88. https://doi.org/10.1016/j.ultsonch.2017.04.034.

Sarwono, A., Man, Z., Muhammad, N., Khan, A. S., Hamzah, W. S. W., Rahim, A. H. A., et al. (2017). A new approach of probe sonication assisted ionic liquid conversion of glucose, cellulose and biomass into 5-hydroxymethylfurfural. *Ultrasonics Sonochemistry, 37*, 310–319. https://doi.org/10.1016/j.ultsonch.2017.01.028.

Schäfer, H.-J. (1990). Recent contributions of Kolbe electrolysis to organic synthesis. *Topics in Current Chemistry, 152*, 91–151.

Schüth, F., Rinaldi, R., Meine, N., Käldström, M., Hilgert, J., & Kaufman Rechulski, M. D. (2014). Mechanocatalytic depolymerization of cellulose and raw biomass and downstream processing of the products. *Catalysis Today, 234*, 24–30. https://doi.org/10.1016/j.cattod.2014.02.019.

Shao, D., Liang, J., Cui, X., Xu, H., & Yan, W. (2014). Electrochemical oxidation of lignin by two typical electrodes: $Ti/SbSnO_2$ and Ti/PbO_2. *Chemical Engineering Journal, 244*, 288–295. https://doi.org/10.1016/j.cej.2014.01.074.

Sheldon, R. A. (2017). The: E factor 25 years on: The rise of green chemistry and sustainability. *Green Chemistry, 19*(1), 18–43. https://doi.org/10.1039/c6gc02157c.

Shrotri, A., Lambert, L. K., Tanksale, A., & Beltramini, J. (2013). Mechanical depolymerisation of acidulated cellulose: Understanding the solubility of high molecular weight oligomers. *Green Chemistry, 15*(10), 2761–2768. https://doi.org/10.1039/c3gc40945g.

Simões, M., Baranton, S., & Coutanceau, C. (2011). Enhancement of catalytic properties for glycerol electrooxidation on Pt and Pd nanoparticles induced by Bi surface modification. *Applied Catalysis B: Environmental, 110*, 40–49. https://doi.org/10.1016/j.apcatb.2011.08.020.

Simões, M., Baranton, S., & Coutanceau, C. (2012). Electrochemical valorisation of glycerol. *ChemSusChem, 5*(11), 2106–2124. https://doi.org/10.1002/cssc.201200335.

Siro, I., & Plackett, D. (2010). Microfibrillated cellulose and new nanocomposite materials: A review. *Cellulose, 17*, 459–494.

Stiefel, S., Schmitz, A., Peters, J., Di Marino, D., & Wessling, M. (2016). An integrated electrochemical process to convert lignin to value-added products under mild conditions. *Green Chemistry, 18*(18), 4999–5007. https://doi.org/10.1039/c6gc00878j.

Sumera, F. C., & Sadain, S. (1990). Diesel fuel by Kolbe electrolysis of potassium salts of coconut fatty acids and acetic acid. *Philippine Journal of Science, 119*, 333–345.

Tabasso, S., Carnaroglio, D., Calcio Gaudino, E., & Cravotto, G. (2015). Microwave, ultrasound and ball mill procedures for bio-waste valorisation. *Green Chemistry, 17*(2), 684–693. https://doi.org/10.1039/c4gc01545b.

Tathod, A., Kane, T., Sanil, E. S., & Dhepe, P. L. (2014). Solid base supported metal catalysts for the oxidation and hydrogenation of sugars. *Journal of Molecular Catalysis A: Chemical, 388–389*, 90–99. https://doi.org/10.1016/j.molcata.2013.09.014.

Toukoniitty, B., Kuusisto, J., Mikkola, J. P., Salmi, T., & Murzin, D. Y. (2005). Effect of ultrasound on catalytic hydrogenation of D-fructose to D-mannitol. *Industrial and Engineering Chemistry Research, 44*(25), 9370–9375. https://doi.org/10.1021/ie050190s.

Vijh, A. K., & Conway, B. E. (1967). Electrode kinetic aspects of the Kolbe reaction. *Chemical Reviews, 67*(6), 623–664. https://doi.org/10.1021/cr60250a003.

Wang, Y., Li, L., Hu, L., Zhuang, L., Lu, J., & Xu, B. (2003). A feasibility analysis for alkaline membrane direct methanol fuel cell: Thermodynamic disadvantages versus kinetic advantages. *Electrochemistry Communications, 5*(8), 662–666. https://doi.org/10.1016/S1388-2481(03)00148-6.

Wurtz, M. A. (1855). Sur une nouvelle classe de radicaux organiques. *Annales de Chimie Physique, 44*, 275–313.

Yabushita, M., Kobayashi, H., & Fukuoka, A. (2014). Catalytic transformation of cellulose into platform chemicals. *Applied Catalysis B: Environmental, 145*, 1–9. https://doi.org/10.1016/j.apcatb.2013.01.052.

Yamane, C., Aoyagi, T., Ago, M., Sato, K., Okajima, K., & Takahashi, T. (2006). Two different surface properties of regenerated cellulose due to structural anisotropy. *Polymer Journal, 38*(8), 819–826. https://doi.org/10.1295/polymj.PJ2005187.

Yan, L., Brouzgou, A., Meng, Y., Xiao, M., Tsiakaras, P., & Song, S. (2014). Efficient and poison-tolerant PdxAuy/C binary electrocatalysts for glucose electrooxidation in alkaline medium. *Applied Catalysis B: Environmental, 150–151*, 268–274. https://doi.org/10.1016/j.apcatb.2013.12.026.

Yunus, R., Salleh, S. F., Abdullah, N., & Biak, D. R. A. (2010). Effect of ultrasonic pre-treatment on low temperature acid hydrolysis of oil palm empty fruit bunch. *Bioresource Technology, 101*(24), 9792–9796. https://doi.org/10.1016/j.biortech.2010.07.074.

Zhang, Y., Liu, G., & Wu, J. (2018). Electrochemical conversion of palmitic acid via Kolbe electrolysis for synthesis of n-triacontane. *Journal of Electroanalytical Chemistry, 822*, 73–80. https://doi.org/10.1016/j.jelechem.2018.05.018.

Ziogas, A., Pennemann, H., & Kolb, G. (2020). Electrochemical synthesis of tailor-made hydrocarbons from organic solvent free aqueous fatty acid mixtures in a micro flow reactor. *Electrocatalysis, 11*(4), 432–442. https://doi.org/10.1007/s12678-020-00600-3.

CHAPTER

13

Regenerable adsorbents for SOx removal, material efficiency, and regeneration methods: A focus on CuO-based adsorbents

Julie Schobing[a,b], *Moisés R. Cesário*[c], *Sophie Dorge*[a,b], *Habiba Nouali*[b,d], *David Habermacher*[a,b], *Joël Patarin*[b,d], *Bénédicte Lebeau*[b,d], *and Jean-François Brilhac*[a,b]

[a]University of Upper Alsace (UHA), LGRE UR 2334, Mulhouse, France [b]University of Strasbourg, Strasbourg, France [c]Materials Science and Engineering Postgraduate Program (PPCEM), Federal University of Paraíba (UFPB), João Pessoa, Brazil [d]University of Upper Alsace (UHA), CNRS, IS2M UMR 7361, Mulhouse, France

13.1 Introduction

Negative impacts of SO_2 emissions

Although gaseous sulfur species (sulfur dioxide (SO_2), sulfur trioxide (SO_3), etc.) can be naturally emitted by volcanoes, these pollutants also have numerous anthropogenic sources. For example, in the case of sulfur dioxide, nearly 70% of global emissions in 2019 are linked to the coal, oil, and gas industries (Dahiya et al., 2020). In fact, common fossil fuels usually contain 0.5%–5% of sulfur (Hanif, Ibrahim, & Abdul Jalil, 2020).

SO_2 is a colorless gas, heavier than air, nonflammable, and is an irritating gas with a pungent odor. In the atmosphere, this molecule can be converted into SO_3 that reacts with the humidity of the atmosphere to produce acidic compounds like sulfuric acid (H_2SO_4) (Lippmann, 2020; Miller & Miller, 2015). These sulfur species can then drop on the ground in the form of wet (rains, fog, etc.) or dry (particles, gas, etc.) acid depositions (Miller & Miller, 2015; Romero et al., 2019). These acid-generated species have a severe negative impact on both the health and the environment. Indeed, the acidification of water can

Heterogeneous Catalysis
https://doi.org/10.1016/B978-0-323-85612-6.00013-9

cause serious damages to vegetation (defoliation), ground (demineralization), aquatic fauna (loss of reproducibility), etc. (Miller & Miller, 2015, 2017; Likens & Bormann, 1974; Singh & Agrawal, 2008). Construction materials, and thus buildings and sculptures, are also mainly damaged by acid depositions. The latter cause an accelerated corrosion of metals and erosion of limestone and marble (Fan, Hu, & Luan, 2012; Livingston, 2016; Miller & Miller, 2015, 2017; Oesch & Faller, 1997).

Moreover, long-term exposure to a high concentration of SO_2 has serious impacts on human health by causing respiratory diseases and increasing the death rate (Amsalu et al., 2019; Berger, 2018; Hwang et al., 2020; Khaniabadi et al., 2017; Ku et al., 2016; Lippmann, 2020; Miller & Miller, 2015, 2017; Wu et al., 2020). A recent study has linked the increase in hospital admissions for cardiovascular diseases in China to the increase in SO_2 concentration in ambient air (Amsalu et al., 2019). Another work, based on air quality data from 2016, estimated that 884 and 27,854 cases per year of mortality and morbidity, respectively, are due to excessive SO_2 concentration in the ambient air in Beijing (Wu et al., 2020). Hwang et al. (2020) highlighted the significant relationship between the long-term exposure to SO_2 over a period of 19 years in South Korea and the high mortality rate in both rural and urban areas. SO_2 is also capable of reacting with fine particulate matter (PM) to produce inhalable secondary aerosols. The synergetic effect between these two pollutants is responsible for severe health issues such as chronic bronchitis or neurodegeneration (Berger, 2018; Ku et al., 2016). Using an air quality modeling system, Khaniabadi et al. (2017) showed that SO_2 and PM_{10} emissions of the growing industries and urban traffic of Khorramabad city (Iran) are responsible, between 2015 and 2016, for 3.9% and 1.7% of respiratory and cardiovascular deaths, respectively. Finally, SO_2 can provoke visual pollution, like smog, causing impaired visibility (Hanif et al., 2020; Miller & Miller, 2015).

Regulations of SO_2 emissions

Regarding the numerous negative impacts of SO_2, its emissions are regulated in the vast majority of developed nations. In Europe, the 2016/2284 directive (Directive (EU) 2016/2284, 2016) has imposed on each European nation (country-specific) a total SO_2 emission abatement rate (in %) compared to 2005 that needs to be adhered to and respected. An initial value has to be adhered to for the years between 2020 and 2029 and a stricter one beyond 2030. For example, France and Germany have to at least reduce their SO_2 emissions by 55% and 21%, respectively (by comparison with their emissions in 2005) until 2029. Beyond 2030, the targets are up to 77% for France and 58% for Germany. Other directives (Directive (EU) 2015/2193, 2015; Directive 2010/75/EU, 2010) have set the upper SO_2 emission limits for industrial plants according to their power and the type of fuel used. For instance, gas turbines are limited to 120 or 15 mg/Nm^3 based on whether they are burning liquid fuels or gaseous fuels, respectively. In the case of waste incineration plants, the daily average SO_2 emissions cannot exceed 50 mg/Nm^3 (Directive 2010/75/EU, 2010). Limits are not exactly the same for already existing facilities and future ones; generally, the latter have some more stringent limits.

In the United State of America, the United States Environmental Protection Agency (US EPA) has set, with the Clean Air Act, National Ambient Air Quality Standards (NAAQS) for several pollutants. In the case of SO_2, global concentrations cannot exceed 75 ppb (about 0.21 mg/Nm^3) in a 1-h period (US EPA, O, 2014). In 2010, only 26 areas, covering a population of about 213,000 (less than 1% of the national population), were considered as "nonattainment areas" in the whole country (EPA, U, 2010). SO_2 emission limits, according to the type of facilities, were also established by the New Source Performance Standards (NSPS) (US EPA, O, 2016). For example, in the case of stationary gas turbines with a heat input at a peak load equal to or greater than 10.7 GJ/h (about 3 MW), SO_2 emissions cannot exceed 0.015 vol% (at 15% of O_2 and on a dry basis). This value corresponds to about 430 mg of SO_2/Nm^3, which is more than 3.5 times higher than the European limits.

In 2014, China introduced the ultralow emissions (ULE) policy for regulating the pollutant emissions of the largest source of electricity supply of the country, i.e., coal-fired power plants. SO_2 emissions are at present limited to 35 mg/m^3 (Qin et al., 2021; Tang et al., 2019; Ye et al., 2020). It was evaluated that between 2014 and 2017, these standards induced a decrease by 65% in the annual power emissions of SO_2 in China (Tang et al., 2019). Compared to the values of SO_2 emission limits of European directives for solid fuels, and thus for coal (between 150 and 1100 mg/Nm^3 regarding the power of the installation (Directive (EU) 2015/2193, 2015; Directive 2010/75/EU, 2010), the Chinese directives are stricter for coal combustion plants. For these types of facilities, the American directives are limiting SO_2 emissions to 520 ng/J of heat input (US EPA, O, 2016).

Desulfurization methods

In order to reduce SO_2 emissions at the industrial level and to comply with the different worldwide standards, both primary and secondary techniques can be used. The Best available technique REFerence document (BREF) (Lecomte et al., 2017) presents the main desulfurization methods used nowadays and lists both their advantages and drawbacks. The primary techniques aim to directly reduce the sulfur content of the fuel used. Generally, sulfur is removed from fuel by hydrodesulfurization through high pressure and temperature treatment with H_2 and an alumina-supported catalyst containing cobalt and molybdenum (Toutov et al., 2017). Unfortunately, this method is not always industrially feasible. In that case, secondary techniques like postcombustion flue gas desulfurization (FGD) systems, trapping the sulfur species emitted in the exhaust gas, can be implemented.

Current FGD techniques

FGD systems have been commercialized since the early 1970s and are now the most used method to reduce SO_2 emissions worldwide (Hanif et al., 2020; Jacubowiez, 2000; Lecomte et al., 2017; Miller & Miller, 2015; Romero et al., 2019). FGD systems gather numerous techniques and can be classified into three categories according to the nature of the active phase involved: wet, semidry, or dry (Table 13.1) (Hanif et al., 2020; Miller & Miller, 2015; Romero et al., 2019; Srivastava & Jozewicz, 2001). Wet FGD techniques generally consist of wet scrubbers that spray slurry of sorbent and trap SO_2 in the droplets. Dry FGD processes are mainly based on the same mechanism, except that the sorbent is sprayed dry in the form of powders, pellets, etc. Semidry FGD techniques are intermediate and are used, for example, as a hydrated sorbent (Hanif et al., 2020; Miller & Miller, 2015; Romero et al., 2019).

In the case of once-through methods (principally employing alkali and alkali earth species, Table 13.1), the used sorbent is a throwback or sometimes used as a byproduct. Currently, the majority of FGD systems industrially applied are once-through methods and the most used are Ca-based ones. Regarding the coal industry, 95% of the installed capacities are equipped with a once-through wet FGD system mainly using limestone (Romero et al., 2019). For example, in 2000, the desulfurization of fumes by CaO was used for 85% of the total power installed worldwide, which corresponded to 60% of the facilities (Jacubowiez, 2000). Ca-based FGD systems are highly efficient and consist of the injection of lime, limestone, or other derivatives, trapping SO_2 to form hydrated $CaSO_4$ or $CaSO_3$ species (Jacubowiez, 2000; Miller & Miller, 2017; Vanderschuren & Thomas, 2010).

TABLE 13.1 Classification of FGD systems.

FGD type	Once-through methods	Regenerable methods
Wet FGD	• *Adsorption with chemical reaction* • *Ca-based scrubbing* • *Na-based scrubbing* • *Mg-based scrubbing* • *NH_3-based scrubbing* • *Seawater scrubbing* • *Wastewater scrubbing*	• Ionic liquid adsorption • Deep eutectic solvent adsorption • Ammonium halide adsorption • Bunsen reaction • Amine scrubbing • Alkyl aniline adsorption • Amino acid solution adsorption • Ca-based solution adsorption • Aluminum sulfate adsorption • *Magnesia scrubbing* • *Wellman-Lord process* • *Ammonia* • *Sodium carbonate*
Semidry FGD	• *$Ca(OH)_2$ sorption* • *Sorption by Ca-supported sorbent*	• Na-based sorption • Zn-based sorption
Dry FGD	• *Sorbent injection* • *Spray drying* • *Circulating fluidized bed (CFB)*	• *Activated carbon adsorption* • Mesoporous silica adsorption • Carbon-silica composite adsorption • Metal oxide adsorption • Zeolite adsorption • Metal-organic framework adsorption

Classification of FGD systems (techniques in italics are assumed to already have an industrial application) (Hanif et al., 2020; Miller & Miller, 2015; Romero et al., 2019; Srivastava & Jozewicz, 2001).

These byproducts have the drawbacks that they are poorly or not recoverable. Indeed, only gypsum ($CaSO_4{\cdot}2H_2O$), characterized by a high degree of purity, can be reused in the fabrication of plaster or cement (Jacubowiez, 2000). In addition to the formation of a large amount of nonrecoverable waste, wet FGD techniques are the most expensive once-through process due to the use of a large amount of water during the process and thus the treatment of wastewater. They are also highly energy-consuming (Hanif et al., 2020; Lecomte et al., 2017; Vanderschuren & Thomas, 2010).

Unlike once-through methods, regenerable processes present the advantages of reusing the spent adsorbent for SO_2 adsorption after a phase of regeneration. Sulfur species recovered in the regeneration phase can also be valorized. In this manner, even though they are costlier, the use of these regenerable techniques is preferable (Romero et al., 2019). Only few regenerable processes are already in use industrially (Table 13.1, indicated in italics), but, according to the literature, several processes are reported to be in the development stage (Hanif et al., 2020; Jacubowiez, 2000; Mathieu et al., 2013; Miller & Miller, 2015). These latter processes employ a large variety of active phases such as ionic liquids (Kumar, Singh, Shukla, & Singh, 2020), alkali-based sorbents (Egan & Felker, 1986), or supported adsorbents (Hanif et al., 2020). Regenerable processes using supported adsorbents will be discussed more intensively in the following sections. At present, the most noteworthy commercially accepted regenerable FGD processes are the Wellman-Lord process and regenerable magnesia scrubbing. The Wellman-Lord process is based on the trapping of SO_2 by Na_2SO_3 to form $NaHSO_3$. The latter is then regenerated in an evaporator-crystallizer by producing concentrated SO_2 gas. On the one hand, this method presents the advantages of producing little solid waste and consuming few alkali reagents. On the other hand, this voluminous system requires a large footprint and the energy consumption and maintenance are both high. Regenerable magnesia scrubbing is based on the reaction between $Mg(OH)_2$ and SO_2 to produce $MgSO_3$ or $MgSO_4$ (Miller & Miller,

2015). The literature reports that the efficiency of SO_2 removal mainly depends on the $Mg(OH)_2$ to SO_2 molar ratio (Egan & Felker, 1986). A ratio above ~1.4 (an excess of $Mg(OH)_2$) was required to remove more than 90% of SO_2 from gas streams containing 0.1%–1.0% of SO_2 in N_2 (for an inlet gas temperature between 107°C and 155°C). The products formed are $MgSO_3{\cdot}3H_2O$ and $MgSO_3{\cdot}6H_2O$, with the hexahydrate predominating at lower temperatures and at a higher rate of humidity. MgO is then recovered by heating treatment in the presence of coke or other reducing agents (some feasibility studies highlight H_2 and CO as the most promising reductive agents) (Lowell, Corbett, Brown, & Wilde, 1976; Miller & Miller, 2015). This technique presents really high SO_2 removal efficiency (up to 99%), produces low waste, is not significantly impacted by the SO_2 inlet levels, and is inexpensive in comparison to other regenerable processes. However, regenerable magnesia scrubbing suffers from its complexity, and, as the recovered SO_2 in the regeneration step is diluted, the production of sulfur will be expensive (Miller & Miller, 2015).

Dry regenerable FGD systems are particularly interesting in terms of preserving water resources and will be discussed in the following sections.

The potential of metal oxides as SOx adsorbents

Viable dry regenerable FGD processes are already in use (Table 13.1) (Hanif et al., 2020; Jacubowiez, 2000; Mathieu et al., 2013; Miller & Miller, 2015), and numerous novel ones are in the development stage (see Section 13.3 and following). A lot of them are based on the utilization of metal oxides (MO_x) because they appear to be highly promising regenerable active phases for SO_x adsorption processes based on the following general mechanism (Eqs. 13.1–13.3):

$$SO_2 + \frac{1}{2}O_2 \rightarrow SO_3 \qquad (13.1)$$

$$xSO_3 + MO_x \rightarrow M(SO_4)_x \qquad (13.2)$$

$$M(SO_4)_x \rightarrow xSO_3 + MO_x \qquad (13.3)$$

After being oxidized into SO_3 (Eq. 13.1), SO_2 is intended to be trapped under the sulfated form of the metal (Eq. 13.2). The latter is then decomposed in order to recover the original form of the metal oxide and it releases pure SOx species (Eq. 13.3), which can be recovered to produce, for example, sulfuric acid or sulfur (Centi et al., 1995; Lowell et al., 1976; Zhao, Liu, Jia, & Xing, 2007).

Lowell, Schwitzgebel, Parsons, and Sladek (1971) highlighted the potential for SO_2 removal of the oxides of several metals such as Ce, Co, Cu, Fe, V, and Zn, based on the literature and thermodynamics data. Among them, copper oxide (CuO) is widely used as an active phase in desulfurization systems because of its excellent oxidative and trapping capacities (Gaudin, Dorge, Nouali, Patarin, et al., 2015; Gavaskar & Abbasian, 2006). In fact, this metal presents a high catalytic activity for reaction described in Eq. 13.1 and an efficient chemisorption of SO_3 in the $CuSO_4$ form. Moreover, decomposition of such species is possible at a moderate temperature (under 600°C) (Centi, Passarini, Perathoner, & Riva, 1992; Gaudin, Dorge, Nouali, Patarin, et al., 2015; Pollack et al., 1988; Yoo, Kim, & Park, 1994). However, in the case of copper oxide, SO_2 adsorption mainly occurs on the surface of the particle and thus sulfation of the core of copper particles is poor because of diffusional limitations (Centi, Passarini, et al., 1992; Galtayries, Grimblot, & Bonnelle, 1996; Gaudin, Dorge, et al., 2016). The use of small particle size will be necessary to maximize the efficiency of CuO. This can be achieved using CuO-supported adsorbents on which CuO can be dispersed. However, a lot of parameters such as the type of support, preparation protocol, etc. have to be taken into account in order to obtain a relevant dispersion of the active phase. Indeed, the literature often highlights the agglomeration issue, leading to large CuO particles and thus poor DeSOx capacities (Berger et al., 2017; Centi et al., 1995; Gaudin et al., 2015, 2016; Gaudin, Dorge, et al., 2016).

State-of-the-art DeSOx activities of regenerable CuO-supported adsorbents described in the literature are discussed in the following sections. Particular attention is paid to the multiple parameters impacting the efficiency of these adsorbents for SOx trapping. First, the nature of the supports and then their textural properties are briefly discussed. The influence of the preparation protocol and treatment of the support as well as the loading of the active phase are then addressed. The modification of the SO_2 adsorption capacities through the addition of other metal oxides is also discussed. The impact of the conditions (temperature, gas composition, etc.) of both the adsorption and the regeneration phases is subsequently considered. Finally, the stability of the adsorbents over time is addressed.

13.2 Role of the CuO support

Several supports are used to synthesize CuO-supported adsorbents for SO_2 removal. The major ones presented in the following sections are:

- **Alumina** and particularly **γ-Al_2O_3** is widely used in the automotive and petroleum industries as a catalyst or a catalyst support. Alumina supports are generally competitive from an economic point of view (Centi et al., 1995; Trueba & Trasatti, 2005). Unfortunately, Al_2O_3 often suffers from a limited specific surface area (Gaudin, Michelin, et al., 2016). Moreover, in the case of an application in desulfurization, this support presents the disadvantage of not being inert with regard to SO_2 (Centi, Perathoner, Kartheuser, Rohan, & Hodnett, 1992; Pollack et al., 1988; Yu, Zhang, & Wang, 2007; Yu, Zhang, Wang, Zhang, & Lu, 2008). In fact, aluminum sulfate can be formed on the alumina surface. The decomposition of these sulfates is not easy and can lead to certain restructuring of the support, which can be detrimental to its textural properties (Centi et al., 1995). Even so, the desulfurization properties of CuO-alumina sorbents have been intensively studied in the literature (Bahrin, Subagjo, & Susanto, 2015; Buelna & Lin, 2003, 2004; Centi et al., 1995; Centi, Passarini, et al., 1992; Centi & Perathoner, 1997; Centi, Perathoner, et al., 1992; Deng & Lin, 1996; Gavaskar & Abbasian, 2006; Liu, Liu, & Huang, 2005; Liu, Liu, Huang, & Xie, 2004; Liu, Liu, & Wu, 2009; Liu, Liu, Zhu, Xie, & Wang, 2004; Macken & Hodnett, 1998; Macken, Hodnett, & Paparatto, 2000; Pollack et al., 1988; Waqif, Saur, Lavalley, Perathoner, & Centi, 1991; Xie, Liu, Zhu, Liu, & Ma, 2003; Yoo, Jeong, Kim, & Park, 1996; Yu, Zhang, & Wang, 2007; Yu, Zhang, Wang, Zhang, & Lu, 2008; Zhao et al., 2007). Moreover, these materials also exhibited promising behavior for simultaneous activities of DeSOx and DeNOx for flue gas (Buelna & Lin, 2003; Centi, Passarini, et al., 1992; Centi & Perathoner, 1997; Centi, Perathoner, et al., 1992; Liu et al., 2005, 2009; Liu, Liu, Huang, & Xie, 2004; Liu, Liu, Zhu, et al., 2004; Macken et al., 2000).
- **Titania (TiO_2)** exists under three main crystalline forms (anatase, rutile, and brookite, with brookite being the less used form) and is a well-known support for heterogeneous catalysis. This support has the advantage that it is stable at high temperatures and under acidic and oxidative atmospheres and also presents a high mechanical resistance. However, TiO_2 usually shows low specific surface area (Bagheri, Muhd Julkapli, & Bee Abd Hamid, 2014). In order to enhance this feature, alumina-titania-mixed supports are used

(Maity, Ancheyta, Rana, & Rayo, 2006). In the case of an application in desulfurization, as for alumina, titania also has the drawback that it is not inert toward SO_2 and suffers from surface sulfation (Centi, Passarini, et al., 1992). Only few studies (Centi, Passarini, et al., 1992; Centi, Perathoner, et al., 1992) dealing with CuO-based titania, which used supported adsorbents for desulfurization applications, have been found in the literature.

- **Activated carbons (ACs)** are commonly used as adsorbents or catalyst supports because they present the advantage of being relatively inexpensive and also being stable in both acidic and basic conditions. They are obtained by carbonization and activation of waste materials, biomass, coal, etc. Depending on the raw material sources and the preparation method used, these supports can show micropores, mesopores, and/or macropores and thus can be used in various applications. Gaseous adsorption on ACs generally goes through the physisorption mechanism but can also be driven by chemisorption in the case of modified activated carbons (acidic treatment, functionalization, impregnation, etc.) (Abdulrasheed et al., 2018; Daud & Houshamnd, 2010; Rodríguez-Reinoso, 1998; Tseng, Wey, & Fu, 2003). Unfortunately, ACs suffer from low durability at high temperatures in oxidative conditions because of decomposition and gasification issues (Rodríguez-Reinoso, 1998). As seen in the previous section, ACs already have an industrial application as a desulfurization adsorbent.
- **Silica (SiO_2)** is another suitable material for adsorbent and catalyst supports. SiO_2 shows high stability and mechanical strength (Ali, Rahman, Sarkar, & Hamid, 2014). In the case of desulfurization applications, this support presents the advantage of being inert toward SOx (Berger et al., 2017; Gaudin, Fioux, et al., 2016). The structure of silica can be tailored using, for example, self-assembly of amphiphilic molecules that act as porogen agents in order to synthesize organized mesoporous silica (OMS) like SBA-15, MCM-41, KIT-6, or COK-12 (Rahmat, Abdullah, & Mohamed, 2010; Zhao et al., 1998). OMS constitutes highly condensed amorphous silica wall covering by silanol groups and delimits pores with a regular diameter and spatial arrangement. OMS is characterized by a high specific surface area (between 800 and $1000\,m^2/g_{ads}$) and a large pore volume (higher than $1.0\,cm^3/g_{ads}$). Owing to pore diameters ranging between 2 and 15 nm, these supports allow the diffusion and adsorption of various molecules. Moreover, the literature shows that OMS allows an extremely high dispersion of CuO particles through formation of the Cu species in a strong interaction with a silica support (like the Si-O-Cu-OH and Si-O-Cu-O-Si species) (Berger et al., 2017; Gaudin, Fioux, et al., 2016; Gentry & Walsh, 1982; Huo, Ouyang, & Yang, 2014; Kong et al., 2004; Shao et al., 2012; Wang, Ying Wu, & Zhu, 2004). Unfortunately, OMS often suffers from a high cost of synthesis (Nada et al., 2019).

Owing to various conditions of the DeSOx processes used in the literature, it is not possible to reach a conclusion regarding the influence of the nature of the support on DeSOx efficiency.

13.3 Influence of the textural properties of the support

All the SO_2 adsorption capacities described in the following were measured, unless otherwise stated, using a fixed bed apparatus.

The initial textural properties of the support can significantly modify the final DeSOx efficiency of the adsorbent (Centi, Passarini, et al., 1992; Centi & Perathoner, 1997). For example, commercial silica, titania, and alumina supports with different specific surface areas (S_{area}) were loaded at 4.8 wt% with CuO by incipient wetness impregnation with a copper acetate precursor (Centi, Passarini, et al., 1992). In the case of CuO-SiO_2 adsorbents, the authors observed no significant modification of the DeSOx activity (tested under 0.8 vol% of SO_2 and 3 vol% of O_2 at 250°C with the particle size in the 0.25–0.42 mm range) when the S_{area} varied from approximately 100 to 500 m^2/g_{ads}. On the contrary, SO_2 adsorption capacities of CuO-TiO_2 adsorbents were multiplied by a factor of 4 when the S_{area} increased from approximately 50 to 200 m^2/g_{ads}. The authors also observed that the DeSOx efficiency of the CuO-γ-Al_2O_3 adsorbent increased with the S_{area} until 200 m^2/g_{ads} and then decreased. They attributed this increase in SO_2 adsorption capacity to the fact that, in the case of DeSOx reactions, the presence of "support sites" close to well-dispersed copper ions is necessary (higher S_{area} at the same CuO loading will increase the number of these types of sites) because it favors the tetragonal distorted octahedral form of copper ions and therefore DeSOx efficiency. Moreover, higher S_{area} can lead to a higher amount of TiO_2 and Al_2O_3 support sites that are able to react with SO_2 and be sulfated. The SO_2 adsorption capacities of four commercial γ-Al_2O_3 pellets loaded with 4.2 wt% of CuO and presenting different S_{area} and total pore volume (TPV) were also evaluated (Centi & Perathoner, 1997). A linear relation between the TPV and the adsorbent DeSOx efficiency was established. The greater the TPV, the greater was the catalyst DeSOx efficiency. The authors suggested that the presence of mesoporosity in the support has a large influence on the SO_2 sorbent efficiency. In fact, SO_2 breakthrough (under 0.2 vol% of SO_2, 3 vol% of O_2, 10 vol% of H_2O, 8 vol% of CO_2, 0.1 vol% of NO, and 0.1 vol % of NH_3 in He, at 350°C) of the CuO-Al_2O_3 pellets that had the highest S_{area} but the lowest fraction of mesoporosity (pellet A: S_{area} of 349 m^2/g_{ads} and TPV of 0.48 cm^3/g_{ads}) occurred five times faster than those of the two pellets showing the lowest S_{area} but the highest fractions of mesoporosity (pellet C: S_{area} of 120 m^2/g_{ads} and TPV of 1.08 cm^3/g_{ads} and pellet D: S_{area} of 116 m^2/g_{ads} and TPV of 1.21 cm^3/g_{ads}). SO_2 adsorption capacities measured at 95% of conversion for these three samples were equal to about 10, 27, and 30 mg_{SO2}/g_{ads} for pellets A, C, and D, respectively. They also showed that a larger diameter (in the 0.28–3.21 mm range) of the pellets led to a lower rate of SO_2 capture. This can be linked to the fact that the less apparent surface of the adsorbent (and thus the less actives sites) was in contact with the flue gas with larger pellets.

The impact of the textural properties on the DeSOx efficiencies of CuO-based AC-supported adsorbents has also been evaluated in the literature (Qin et al., 2021; Tseng, Wey, & Fu, 2003). Qin et al. (2021) studied the impact of carbonization temperature on the textural characteristics of AC adsorbents made from walnut shells. Irrespective of the carbonization temperature (between 500°C and 900°C), a SO_2 conversion of 100% is obtained for the undoped AC adsorbent under 200 ppmv of SO_2, 300 ppmv of NOx, 100 μm/cm^3 of Hg^0, 300 ppmv of NH_3, and 5 vol% O_2 in N_2 for temperatures between 100°C and 300°C. For higher adsorption temperatures (until 400°C), the AC adsorbent carbonized at 700°C showed the highest SO_2 conversion values. Moreover, this sample also presented the highest textural properties (S_{area} of 1426 m^2/g_{ads} and TPV of 0.62 cm^3/g_{ads}). The authors suggested that below 700°C, a complete carbonization of the AC sorbent was not achieved and that beyond this temperature the collapse of the microporosity of the carbon structure occurred. In the case of metal oxide-based AC-supported adsorbents, these higher textural properties also allowed the obtainment of a highly dispersed metal oxide active phase and thus better SO_2 adsorption capacities. In fact, for the same Fe loading of 5 wt%, the AC support carbonized at 500°C presented ferrous agglomerate and a SO_2 conversion of about 68% against about 88% for a carbonization temperature of 700°C. Finally, the authors showed that, for the same metal loading (5 wt% corresponding to about 6.3 wt% in the case of CuO), AC adsorbents functionalized with Al, Cu, Fe, or Mn showed extremely closed DeSOx efficiency. Slightly higher capacities were still observed for the Cu-AC adsorbent at 400°C (SO_2 conversion of about 80% for copper against about 70%–75% for the other metals). These adsorbents also showed promising efficiency for the simultaneous removal of SO_2, NOx, and Hg^0 of the flue gas.

Schobing et al. (2021) compared the DeSOx efficiency of two copper-based adsorbents loaded with about 15.5 wt% of CuO by wet impregnation with a copper nitrate precursor using two different OMS supports, namely, SBA-15 and COK-12. The samples were tested in the same conditions of adsorption (under 250 ppmv of SO_2 and 10 vol% of O_2 in N_2 at 400°C) and regeneration (under 0.5 vol% of H_2 in N_2 at 400°C). The study showed that using COK-12 as a support instead of SBA-15 allows an increase in the total SO_2 adsorption capacity by 6%. Although SBA-15 and COK-12 have the same 2D hexagonal mesostructures, the higher textural characteristics (mesoporous volume of $0.38\,cm^3/g_{SiO2}$ and pore diameter of 6.0 nm for CuO-SBA-15 against $0.47\,cm^3/g_{SiO2}$ and 6.9 nm for CuO-COK-12) of COK-12, which allow a better accessibility of SO_2 to the active copper sites, can explain its better reactivity. No significant agglomeration of the CuO particles was observed, and both adsorbents showed a good stability over 15 adsorption-regeneration cycles. These results are interesting from an industrial point of view because of the easier and inexpensive synthesis protocol of COK-12 compared to that of SBA-15.

Studies mainly showed that a higher S_{area} leads to a higher SO_2 adsorption capacity. Moreover, a more developed mesoporous structure through higher mesoporous volume and larger pore diameter also enhances the DeSOx efficiency. This can be attributed not only to a better access of the gaseous species to the active phase but also to a better dispersion of the CuO particles on the support surface.

13.4 Influence of the preparation protocol and support treatment

The literature has highlighted that the preparation protocol used to synthesize CuO-supported adsorbents and other treatments (such as acidic treatment of the support, thermal treatment before or after impregnation, etc.) influences their final DeSOx capacity (Buelna & Lin, 2004; Centi et al., 1995; Deng & Lin, 1996; Gavaskar & Abbasian, 2006; Liu et al., 2005; Liu, Liu, Huang, & Xie, 2004; Qin et al., 2021; Tseng, Wey, & Fu, 2003).

Several authors investigated the performance of sol-gel-derived alumina sorbents (Buelna & Lin, 2004; Deng & Lin, 1996; Gavaskar & Abbasian, 2006). This synthesis method generally leads to materials with high S_{area} and high mechanical strength (Gavaskar & Abbasian, 2006). Moreover, the incorporation of the copper active phases can be conducted during sol-gel synthesis (Buelna & Lin, 2004; Gavaskar & Abbasian, 2006). Buelna and Lin (2004) compared the reactivity of sol-gel-derived CuO-γ-Al_2O_3 granular adsorbents with two commercial CuO-γ-Al_2O_3 granular adsorbents (coated by the wet impregnation method). All the synthesized sol-gel-derived CuO-γ-Al_2O_3 samples had higher S_{area} than the commercial samples. For example, the S_{area} between 222 and $248\,m^2/g_{ads}$ was measured for sol-gel-derived CuO-γ-Al_2O_3 samples with a CuO loading of about 7.5 wt%, whereas the commercial support coated with 9.4 and 6.6 wt% of CuO had a S_{area} of 204 and $99\,m^2/g_{ads}$, respectively. All the samples showed close TPV (in the $0.46–0.51\,cm^3/g_{ads}$ range), with the exception of the commercial sample coated with 9.4 wt% ($0.24\,cm^3/g_{ads}$).

After three cycles of adsorption (under 2000 ppmv of SO_2 in air) and desorption (under 10% of CH_4 in nitrogen) at 400°C, the sol-gel sorbent containing 8.7 wt% presented an adsorption of about 113 mg_{SO2}/g_{ads}. In the same conditions, the two commercial sorbents containing 9.4 and 6.6 wt% of CuO had adsorption capacities of about 96 and 58 mg_{SO2}/g_{ads}, respectively. The higher efficiency of the sol-gel-derived CuO-γ-Al_2O_3 adsorbent was directly linked to a better CuO dispersion obtained with this sample. Therefore, the sol-gel method for support synthesis is recommended to obtain better DeSOx efficiency in the case of the γ-Al_2O_3 support. Moreover, no significant impact of the CuO precursor used ($CuCl_2$ or $Cu(NO_3)_2$) on the pore structure of the sorbent was observed at the same metal oxide loading (10 wt% of CuO) (Deng & Lin, 1996).

Liu, Liu, Huang, and Xie (2004) and Liu et al. (2005) studied the impact of the preparation protocol on the characteristics of monolithic cordierite-based CuO-γ-Al_2O_3 adsorbents. The authors showed that an acidic treatment prior to the γ-Al_2O_3 coating and CuO impregnation led to a significant enhancement of the SO_2 adsorption efficiency. For example, the amount of SO_2 adsorbed for a SO_2 conversion of 80% (under 1960 ppmv of SO_2, 500 ppmv of NO, 5.5 vol% of O_2, 2.5 vol% of H_2O, and 500 ppmv of NH_3 in Ar at 400°C) by a monolithic cordierite-based 8 wt% CuO-γ-Al_2O_3 adsorbent increased from 8 to 23 mg_{SO2}/g_{ads} owing to an oxalic acidic treatment (Liu, Liu, Huang, & Xie, 2004). The better DeSOx efficiencies obtained after acidic treatment were linked to the enhancement of the textural properties of the adsorbent. In fact, higher S_{area} and the creation of porosity allowed a better γ-Al_2O_3 coating and a better dispersion of the CuO active phase and potential additives (Liu et al., 2005; Liu, Liu, Huang, & Xie, 2004). No significant impact of the type of acid (oxalic acid, nitric acid, or hydrochloric acid) used was observed. Moreover, the higher the concentration of the acid, the higher was the DeSOx efficiency along with the creation of more pores. The quantity of SO_2 adsorbed at a SO_2 conversion of 80% by a monolithic cordierite-based 6 wt% CuO-γ-Al_2O_3 adsorbent doped with 1 wt % of Na_2O increased from 30 to 52 mg_{SO2}/g_{ads} when the oxalic acid concentration of the acidic treatment varied from 0% to 50%, respectively. A similar behavior was observed with a Na_2O loading, twice as higher (Liu et al., 2005). The authors also showed that in the absence of the γ-Al_2O_3 coating, monolithic cordierite-based CuO adsorbents were less efficient (with or without acidic treatment) (Liu, Liu, Huang, & Xie, 2004). Moreover, the amount of γ-Al_2O_3 coated on the monolithic cordierite impacted the DeSOx efficiency of the adsorbent. When the concentration of γ-Al_2O_3 coating increased from 5 to 21 wt%, the quantity of SO_2 adsorbed at a SO_2 conversion of 80% increased from 17 to 75 mg_{SO2}/g_{ads}, respectively. For higher γ-Al_2O_3 coating (up to 33 wt%), slightly lower SO_2 adsorption capacities were observed (Liu, Liu, Huang, & Xie, 2004). These types of adsorbents also showed good DeNOx capacities.

With regard to AC-supported metal oxide-based adsorbents, Tseng, Wey, and Fu (2003) studied the influence of the preparation protocol of several CuO-AC adsorbents on their DeSOx efficiencies. Cu-based adsorbents were prepared by the pore volume impregnation technique with an aqueous solution of $Cu(NO_3)_2{\cdot}5H_2O$. They first showed that an acidic pretreatment of the AC support (granulometry of 250–297 μm) with HNO_3 decreased the SO_2 adsorption capacities from approximately 14 to 5 mg_{SO2}/g_{ads} (under 200 ppmv of SO_2 in nitrogen at 300°C) for a CuO loading of about 6.3 wt%. On the other hand, the same treatment with HCl increased the SO_2 adsorption capacity up to around 30 mg_{SO2}/g_{ads} for the same CuO loading and support granulometry. The authors pointed out that this last acidic treatment tended to lower the hydrophobicity of the AC support, enhancing the accessibility of the aqueous precursor solution during the impregnation step and thus facilitating the dispersion of CuO (Rodríguez-Reinoso, 1998; Tseng, Wey, & Fu, 2003). Such a result seems to be in opposition with results of the literature reporting the enhancement of SO_2 capacity for AC-supported metal oxide-based adsorbents in general when HNO_3 is used because of the formation of more basic functional groups (pyronic and quinonic types that enhance SO_2 adsorption) and the prevention of metal oxide particle agglomeration (Abdulrasheed et al., 2018; Bandosz, Loureiro, & Kartel, 2006). The authors concluded that in the case of the AC support, the chemical characteristics were more important than the physical ones.

Since the last few years, the development of efficient CuO-SBA-15 adsorbents has been reported in the literature (Berger, 2018; Berger, Brillard, et al., 2020; Berger, Dorge, et al., 2018, 2020; Berger, Nouali, et al., 2018; Berger et al., 2017; Gaudin, Dorge, et al., 2016; Gaudin, Dorge, Nouali, Patarin, et al., 2015;Gaudin, Fioux, et al., 2016 ; Gaudin, Michelin, et al., 2016). In these studies, SBA-15 was synthesized using pluronic P123 as a surfactant and tetraethyl orthosilicate as the source of silica. Gaudin, Michelin, et al. (2016) showed that modifying the SBA-15 synthesis temperature from 90 to 130°C does not impact the DeSOx efficiency of the CuO-SBA-15 adsorbents. The team also investigated the impact of several preparation protocols of CuO-SBA-15 materials on their textural and structural properties and on their SO_2 adsorption capacities (Berger, 2018; Gaudin, Dorge, et al., 2016; Gaudin, Dorge, Nouali, Patarin, et al., 2015). The authors showed that irrespective of the solvent used, methanol or water, for the wet impregnation method, the textural properties of the SBA-15 support were not modified. On the other hand, because of the use of ammonia, the ion exchange method created a new microporosity inside the SBA-15 support via partial hydrolysis of the silica walls, therefore changing the textural properties. At the same CuO loading, different CuO impregnation methods led to different SO_2 adsorption capacities (Berger, 2018; Gaudin, Dorge, et al., 2016; Gaudin, Dorge, Nouali, Patarin, et al., 2015). For example, after three cycles of adsorption (under 250 ppmv of SO_2, 10 vol% of O_2 in N_2 at 400°C) and regeneration (under nitrogen at 600°C), the SO_2 adsorption capacity at 75 ppmv (here, adsorption capacities are calculated as the quantity of SO_2 adsorbed on the adsorbent when the SO_2 emissions reach 75 ppmv on the breakthrough curve) of an 8 wt% CuO-SBA-15 adsorbent was about 25 and 32 mg_{SO2}/g_{ads} for impregnation with solid-state grinding and wet impregnation (with a nitrate copper precursor in both cases), respectively. The authors linked the globally better SO_2 adsorption efficiency obtained with the wet impregnation method for three different CuO loadings (8, 15, and 31 wt%) to the achieved better CuO dispersion (Berger, 2018). Studies clearly pointed out that the key step of the preparation protocol of the CuO-SBA-15 adsorbents was calcination of the impregnated material (Gaudin, Dorge, et al., 2016; Gaudin, Dorge, Nouali, Patarin, et al., 2015). In fact, this phase had a signification impact on the dispersion of CuO particles. High temperature conditions led to the agglomeration of the CuO particles and thus to the formation of large CuO particles inside and outside the porosity. Calcination with a ramp, avoiding the thermal shocking of the sample, led to a homogeneous dispersion of the active phase without agglomeration of the CuO particles. For example, in the case of a CuO-SBA-15 adsorbent prepared by solid-state grinding, calcination with a ramp of 1°C/min until 500°C allowed a SO_2 uptake at 75 ppmv (under 250 ppmv of SO_2 and 10 vol% of O_2 in nitrogen at 400°C) of 29 mg_{SO2}/g_{ads} (CuO loading of 12.5 wt%) against only 8 mg_{SO2}/g_{ads} (CuO loading of 10.3 wt%) with a flash calcination at the same temperature (Gaudin, Dorge, et al., 2016). Gaudin, Dorge, et al. (2016) and Gaudin, Dorge, Nouali, Patarin, et al. (2015) also concluded that, in the case of the CuO-SBA-15 adsorbent loaded with 18 wt% of CuO, the preparation protocol producing the best SO_2 uptake at 75 ppmv (49 mg_{SO2}/g_{ads}) was wet impregnation in water using calcination with a ramp of 1°C/min until 500°C.

Besides a homogeneous CuO dispersion, the oxidation state of the copper species also has a significant impact on the DeSOx efficiency of CuO-SBA-15 adsorbents. After thermal pretreatment under N_2, better adsorption capacities were observed (Gaudin, Fioux, et al., 2016; Gaudin, Michelin, et al., 2016). The latter were attributed to the autoreduction of the Cu^{2+} species into the Cu^+ species that occurred under an inert atmosphere (Berger et al., 2017; Larsen, Aylor, Bell, & Reimer, 1994; Popova et al., 2014; Wang, Yang, & Heinzel, 2009; Yin, Tan, Liu, Zhu, & Sun, 2014). Moreover, higher temperatures favored this reaction (Gaudin, Fioux, et al., 2016; Liu & Robota, 1993; Wang, Yang, & Heinzel, 2009). Indeed, Gaudin, Fioux, et al. (2016) observed through XPS analyses higher concentration of the Cu^+ species after thermal treatment of CuO-SBA-15 under N_2 at 600°C rather than at 400°C and an increase in the SO_2 adsorption capacities by 50%. It is likely that the Cu^+ species is able to catalyze the oxidation of SO_2 into SO_3.

Compared to SBA-15 that has a two-dimensional pore network (hexagonal packing of cylindrical pores), KIT-6 has a three-dimensional pore network (cubic structure resulting in a bicontinuous interpenetrating network of cylindrical pores), suggesting a better diffusion of the gas phase. Unfortunately, the CuO-KIT-6 adsorbent (containing 6.2 wt% of CuO) showed an extremely low SO_x adsorption capacity of 1.7 mg_{SO2}/g_{ads} at 75 ppmv (under 250 ppmv of SO_2 and 10 vol% of O_2 in N_2 at 400°C) related to the relatively large size (>1 μm) of the CuO particles formed as well as the rapid formation of superficial sulfate species that subsequently have limited the diffusion of SO_2 in the core of the CuO particles (Gaudin, Dorge, Nouali, Kehrli, et al., 2015). This poor efficiency may be attributed to the unsuitable preparation conditions for the CuO phase rather than to the nature of the OMS support. The CuO-KIT-6 adsorbent was prepared by wet impregnation in methanol using a nitrate copper precursor with a drying step at 90°C and a calcination step at 550°C (with a ramp of 1°C/min). In another study (Gaudin, Dorge, Nouali, Patarin, et al., 2015), the same team showed that a CuO-SBA-15 adsorbent (about 15–18 wt% of CuO) prepared with the same protocol (except a calcination temperature of 500°C) had a similar DeSOx efficiency of 2 mg_{SO2}/g_{ads} at 75 ppmv (under the same adsorption conditions). Large particles of CuO (~2 μm) were also detected in this sample. Moreover, in the same conditions, the CeO_2-KIT-6 adsorbent (containing 11.7 wt% of CeO_2) showed a SO_2 uptake capacity (8.3 mg_{SO2}/g_{ads} at 75 ppmv) significantly higher than that of the CuO-KIT-6 adsorbent (containing 6.2 wt% of CuO) due to the smaller size of CeO_2 particles (~10 nm), its better dispersion, and the remarkable oxidative properties of cerium. In fact, theoretically, one cerium atom is able to adsorb 1.5 molecules of SO_2 against 1 for one copper atom (Gaudin, Dorge, Nouali, Kehrli, et al., 2015).

Studies showed that the preparation protocol of the adsorbents and pre- and postsynthesis treatments have a significant impact on the final DeSOx efficiency of CuO-supported adsorbents, irrespective of the nature of the support used. The results highlighted that, in general, preparation conditions have to be chosen in order to enhance the textural properties of the support and therefore the CuO dispersion, which plays a major role in the SO_2 adsorption capacity. Conditions of thermal treatments (such as calcination) have to be well controlled in order to avoid agglomeration of the CuO particles. Acidic treatments (modifying, for example, the surface of the support) also have to be considered in order to enhance the DeSOx efficiency. The literature has shown that, in the case of a silica support, the presence of the Cu^+ species, being more reactive, has to be privileged.

13.5 Influence of active-phase loading

As expected, the literature has shown that loading of the active phase, here CuO, has a significant impact on the DeSOx efficiency of CuO-based adsorbents (Berger, 2018; Buelna & Lin, 2004; Centi et al., 1995; Centi, Passarini, et al., 1992; Deng & Lin, 1996; Gaudin, Michelin, et al., 2016; Gavaskar & Abbasian, 2006; Tseng & Wey, 2004; Tseng, Wey, & Fu, 2003; Yu et al., 2007, 2008).

Some studies investigated the impact of the CuO loading on the DeSOx capacities of CuO-γ-Al_2O_3 adsorbent powders (Yu et al., 2007, 2008). Several CuO-γ-Al_2O_3 adsorbents (particle granulometry of 20–80 μm) were prepared by wet impregnation with a CuO concentration between 3 and 55 wt% of CuO from the $Cu(NO_3)_2 \cdot 3H_2O$ precursor. SO_2 adsorption tests in a thermogravimetric setup (at 350°C under 200 ppmv of SO_2, 5 vol% of O_2, and 3 vol% of H_2O in N_2) highlighted the existence of an optimal CuO loading of 12 wt% of CuO for CuO-γ-Al_2O_3 adsorbents, leading to the highest SO_2 adsorption efficiency (about 66 mg_{SO2}/g_{ads} after 5500 s of sulfation) (Yu et al., 2007). Beyond this loading, the decrease in activity with the increase in CuO concentration was linked to the formation of larger CuO particles. It is noteworthy that above 7 wt% of CuO, no sulfation of the alumina support was observed. This was attributed to the fact that, above this loading, the entire alumina surface was covered by copper oxide and could not react with gaseous SO_2 anymore. In a similar manner, above a CuO loading of 8 wt%, no more increase in DeSOx activities was observed in the case of γ-Al_2O_3 pellet supports (SO_2 adsorption capacity of 34 mg_{SO2}/g_{ads} after five cycles of adsorption—under 3200 ppmv of SO_2, 800 ppmv of NO, 800 ppmv of NH_3, and 3 vol% of O_2 at 350°C until a minimum of 95% of SO_2 removal is attained—and regeneration; conditions unspecified) (Centi et al., 1995).

In the case of sol-gel-derived CuO-γ-Al_2O_3 adsorbents, several studies also investigated the variation of DeSOx efficiencies with the CuO loading (Buelna & Lin, 2004; Deng & Lin, 1996; Gavaskar & Abbasian, 2006). By studying the impact of the CuO loading from 6 to 21 wt%, the authors highlighted an optimal loading of around 8 wt% for sol-gel-derived CuO-γ-Al_2O_3 samples (Buelna & Lin, 2004). Gavaskar and Abbasian (2006) observed the highest efficiency (48 mg_{SO2}/g_{ads}, measured in a fluidized bed reactor under 0.25 vol% of SO_2 and 3.7 vol% of O_2 in N_2 at 450°C) for a sol-gel-derived CuO-γ-Al_2O_3 sample containing 14.1 wt% of CuO. Contrary to the common behavior of an alumina support, no significant sulfation of the sol-gel-derived alumina support was observed. Deng and Lin (1996) were able to achieve an optimal SO_2 adsorption capacity (SO_3 uptake of about 20 wt% corresponding to 160 mg_{SO2}/g_{ads} for 1 h of sulfation under 1 vol% of SO_2 in air at 500°C) for the sol-gel-derived CuO-γ-Al_2O_3 adsorbent loaded with about 20 wt% of CuO using wet impregnation and $CuCl_2$ as the precursor.

The impact of the CuO loading was also investigated for AC-supported adsorbents (Tseng & Wey, 2004;Tseng, Wey, & Fu, 2003). The authors showed that irrespective of the performed acidic treatment, the increase in the CuO loading from 3 to 10 wt% led to an increase in DeSOx efficiencies (Tseng & Wey, 2004; Tseng, Wey, & Fu, 2003). For example, in the case of a HCl-treated AC support, a CuO loading of 10 wt% allowed a SO_2 adsorption (under 200 ppmv of SO_2 in N_2 at 250°C) of about 59 mg_{SO2}/g_{ads} against about only 14 mg_{SO2}/g_{ads} for 3 wt% of CuO (an increase by a factor of 4.2) (Tseng, Wey, & Fu, 2003). The notion of an optimal CuO loading beyond which the SO_2 adsorption capacity decreases was not approached in the case of AC-supported adsorbents in the CuO loading range studied.

Other works (Berger, 2018; Gaudin, Michelin, et al., 2016) highlighted the impact of the CuO loading on the characteristics of CuO-SBA-15 adsorbents. The adsorbents were prepared with CuO loadings of 8.8, 15.6 and 31.7 wt% by wet impregnation with a copper nitrate precursor (Gaudin, Michelin, et al., 2016). A decrease in the S_{area} and TPV was observed with an increase in CuO loading. These decreases were linked to a densification of the support. Irrespective of the concentration of copper, no modification of the SBA-15 structure was observed by XRD characterization. Moreover, no CuO peaks were detected on the X-ray diffractograms, indicating the high dispersion of copper even at high concentrations. The SO_2 adsorption capacities were measured over five cycles of adsorption (under 250 ppmv of SO_2 and 10 vol% of O_2 in N_2 at 400°C) and regeneration (under N_2 at 600°C). The results showed that a high copper concentration (31.7 wt%) led to a nonstable adsorbent. In fact, a severe agglomeration of the CuO particles was observed as soon as the second cycle started, causing a significant decrease in the total SO_2 adsorption capacity. The authors concluded that the optimal CuO loading providing a better compromise between high SO_2 adsorption capacity and adsorbent stability for a CuO-SBA-15 adsorbent was close to 15 wt% of CuO. The same value of the CuO optimal loading for CuO-SBA-15 adsorbents was found after impregnation with a solid-state grinding method (Berger, 2018). Until 15 wt%, an increase in the CuO loading led to an increase in the SO_2 adsorption capacities with no detrimental loss of the SBA-15 textural properties. For example, after three cycles of adsorption-desorption (in the same conditions (Gaudin, Michelin, et al., 2016)), the SO_2 adsorption capacity at 75 ppmv was about 25 and 60 mg_{SO2}/g_{ads} for a CuO loading of 8 and 15 wt%, respectively. In the case of a higher CuO loading (31 wt%), this impregnation method did not permit the preservation of the textural properties of the SBA-15 support. In fact, a pore blocking phenomenon was observed due to the formation of large CuO particles with a diameter higher than those of the mesopores of the support.

The literature has shown that DeSOx efficiency is directly linked to the amount of the active phase in the adsorbent. On the other hand, increasing the CuO loading does not systematically lead to an increase in SO_2 adsorption capacity. In fact, several studies highlighted the existence of an optimal CuO loading producing the highest efficiency. Beyond this loading, a decrease in SO_2 adsorption capacity was observed and was often linked to the formation of large particles of CuO. Moreover, the copper optimal loading appeared to vary according to the properties of the support and the conditions of the adsorption tests.

13.6 Doping CuO-based adsorbents with other metal oxides

Doping CuO-based adsorbents with other metal oxides can significantly enhance their DeSOx activities. First, the impact of vanadium was studied by several authors (Carabineiro et al., 2003; Centi, Perathoner, et al., 1992; Liu et al., 2009; Yang et al., 2020). For example, a previous work showed that doping a CuO-Al_2O_3-TiO_2 adsorbent with vanadium increased its SO_2 adsorption capacity (Centi, Perathoner, et al., 1992). Supports were doped by double impregnation, first with NH_4VO_3 and then with $CuSO_4$ salts. The CuO-Al_2O_3-TiO_2 adsorbent without vanadium and the ones with an addition of 1.5 and 4.0 wt% of V, showed a SO_2 adsorption capacity of 51, 54, and 55 mg_{SO2}/g_{ads} (at 300°C under 0.8 vol% of SO_2 and 2.5 vol% of O_2 in He), respectively. In this case, the impact of the vanadium loading on the SO_2 adsorption capacity was low.

The impact of the V_2O_5 loading on the SO_2 adsorption capacity of a monolithic cordierite-based CuO-γ-Al_2O_3 adsorbent (containing 7 wt % of CuO) was investigated by Liu et al. (2009). The results showed that an increase in the quantity of SO_2 adsorbed at a SO_2 conversion of 80% (under 1960 ppmv of SO_2, 500 ppmv of NO, 5.5 vol% of O_2, 2.5 vol% of H_2O, and 500 ppmv of NH_3 in Ar at 400°C) from 23 to

$55 mg_{SO2}/g_{ads}$ was observed when the V_2O_5 loading increased from 0 to 1.6 wt%, respectively. If this loading was further increased up to 2.1 wt%, the amount of SO_2 adsorbed at a SO_2 conversion of 80% dropped to $32 mg_{SO2}/g_{ads}$. The authors pointed out that the better DeSOx efficiency measured in the presence of vanadium was linked to an improvement in the dispersion of the CuO particles (Liu et al., 2009). After a second cycle (regeneration under 5 vol% of NH_3 in Ar at 400°C), all the V-doped samples showed closed SO_2 adsorption capacity (SO_2 adsorption capacity at a SO_2 conversion of 80% between 33 and $37 mg_{SO2}/g_{ads}$). These lower efficiencies were attributed to an incomplete regeneration of the sample. They concluded that, from an economic point of view and regarding the gain in SO_2 adsorption capacity, the optimal V_2O_5 loading was 0.5 wt% (SO_2 adsorption capacity at a SO_2 conversion of 80% of $38 mg_{SO2}/g_{ads}$).

The beneficial impact of vanadium was also observed by Carabineiro et al. (2003) through the study of a V-doped copper-AC adsorbent (containing 4 wt% of copper detected under both CuO and Cu forms). Initially, the copper-AC adsorbent showed a SO_2 adsorption capacity of about $58 mg_{SO2}/g_{ads}$ (under about 4660 ppmv of SO_2 in Ar at 20°C). After being doped by 4 wt % of V (detected under V_2O_5 and V_6O_{13} forms), the modified copper-AC adsorbent presented a SO_2 adsorption capacity of about $141 mg_{SO2}/g_{ads}$. This new DeSOx efficiency was even higher than the sum of the adsorption capacities of the two monometallic adsorbents (SO_2 adsorption capacity of about $58 mg_{SO2}/g_{ads}$ for the vanadium-AC adsorbent). The authors (Carabineiro et al., 2003) hypothesized that the synergetic effect between copper and vanadium was linked to a better dispersion of the active phase on the adsorbent surface. Moreover, with vanadium, they observed the presence of more oxygenated groups after sulfation that could enhance SO_2 adsorption. In the same study, Carabineiro et al. (2003) observed a synergetic effect between copper and iron and also between copper and nickel. The SO_2 adsorption capacity of the copper-AC adsorbent increased to about 134 and $90 mg_{SO2}/g_{ads}$ after doping by 4 wt% of Fe (detected under Fe_2O_3 and Fe_3O_4 forms) and Ni (detected under NiO and Ni forms), respectively.

Yang et al. (2020) compared the DeSOx efficiency of mono-, bi- and polymetallic-doped AC-supported adsorbents. They observed a synergetic effect between copper and iron, manganese, vanadium, and titanium. In fact, AC adsorbents containing 2.5 wt% of CuO and 2.5 wt% of one of these metal oxides presented higher SO_2 adsorption capacities (measured at 80°C under 3000 ppmv of SO_2, about 9 vol% of O_2 in humidified N_2 at 80% of relative humidity) than those of samples containing one single metal oxide at a 5 wt% loading. For example, the 5 wt% CuO-AC adsorbent showed a SO_2 adsorption capacity of $116 mg_{SO2}/g_{ads}$ and AC adsorbents containing 2.5 wt% of CuO and 2.5 wt% of Fe_2O_3, MnO_2, V_2O_5, or TiO_2 had SO_2 adsorption capacities of 148, 134, 123 or $123 mg_{SO2}/g_{ads}$, respectively. The better efficiency of the bimetallic adsorbent could be attributed to an enhancement of the textural properties of the adsorbent. In fact, the CuO-Fe_2O_3-AC adsorbent had a higher S_{area} ($429 m^2/g_{ads}$) and a higher mesoporous volume ($0.062 m^3/g_{ads}$) than both CuO-AC ($414 m^2/g_{ads}$ and $0.052 m^2/g_{ads}$) and Fe_2O_3-AC ($406 m^2/g_{ads}$ and $0.037 m^2/g_{ads}$) adsorbents. As already pointed out by Carabineiro et al. (2003), Yang et al. (2020) suggested that these metal oxides could also lead to a modification of the nature of the carbonaceous functional surface groups, providing more basic functional groups on the carbon surface that are known to have a beneficial impact on SO_2 adsorption (via strong physical adsorption) (Bandosz et al., 2006). Finally, the study showed that the presence of three metal oxides also led to a better efficiency than with one single metal oxide. AC adsorbents containing 2 wt% of CuO, 2 wt% of Fe_2O_3, and 1 wt% of Co_2O_3 or MnO_2

presented a SO_2 adsorption capacity of 137 or 161 mg_{SO2}/g_{ads}.

The impact of Na_2O additives on the DeSOx efficiency of a monolithic cordierite-based CuO-γ-Al_2O_3 adsorbent was also studied (Liu, Liu, Huang, & Xie, 2004; Liu, Liu, Zhu, et al., 2004). The authors investigated the impact of the addition of 0.5–3.0 wt% of Na_2O on the DeSOx activity of a monolithic cordierite-based CuO-γ-Al_2O_3 adsorbent containing 6 wt% of CuO (Liu, Liu, Zhu, et al., 2004). A significant enhancement of the SO_2 adsorption capacity was observed for Na_2O loadings higher than 1.0 wt%. Moreover, an optimal loading of Na_2O exhibiting the best DeSOx activity was determined at 2.0 wt%. At this loading, the quantity of SO_2 adsorbed at a SO_2 conversion of 80% (under 1960 ppmv of SO_2, 500 ppmv of NO, 5.5 vol% of O_2, 2.5 vol% of H_2O, and 500 ppmv of NH_3 in Ar at 400°C) was equal to 68 against 21 mg_{SO2}/g_{ads} for the Na-free sample. The authors attributed the beneficial impact of Na_2O to an improvement of the CuO dispersion on the adsorbent. Indeed, the smaller CuO particles obtained in the presence of Na_2O led to a higher S_{area} (less pore plugging) (Liu, Liu, Huang, & Xie, 2004; Liu, Liu, Zhu, et al., 2004). Moreover, in the presence of Na_2O, SO_2 was trapped to form $NaCu_2(SO_4)_2OH \cdot H_2O$ and Na_2SO_4 species in addition to the common $CuSO_4$ form (Liu, Liu, Zhu, et al., 2004).

The synergetic effect between copper and cerium was also investigated in the literature by several authors (Akyurtlu & Akyurtlu, 1999; Gaudin, Dorge, Nouali, Kehrli, et al., 2015; Wey, Lu, Tseng, & Fu, 2002; Zhang, Dong, & Cui, 2020). Zhang et al. (2020) studied the impact of the CeO_2 loading on the DeSOx efficiency of CuO-γ-Al_2O_3 pellets (with 8 wt% of CuO). When the CeO_2 loading varied between 2 and 8 wt%, an enhancement of the SO_2 adsorption capacities of the CuO-γ-Al_2O_3 adsorbent was observed. Moreover, the best DeSOx efficiency was measured for the sample containing 3 wt% of Ce_2O. A saturation of this adsorbent was obtained after 88 min of sulfation (under 1800 ppmv of SO_2, 5 vol% of O_2 in N_2 at 400°C) against 62 min for the undoped sample. The authors highlighted that cerium doping decreased the size of the CuO particles and led to a better dispersion of the latter on the adsorbent surface (Zhang et al., 2020).

Akyurtlu and Akyurtlu (1999) studied the SO_2 adsorption capacities of several alumina-supported adsorbents impregnated with different proportions of copper and cerium (total metal loading of 10 wt%). The results showed that between 450°C and 550°C, samples containing the two metal oxides had better DeSOx efficiency than those of the monometallic materials. For example, a specific sulfur capacity (normalized with respect to the theoretical maximum value of adsorption by CuO and CeO_2 sites) at 3000 s of sulfation (determined by thermogravimetry at 450°C under 3000 ppmv of SO_2 and 3 vol% of O_2 in N_2) of CuO-Al_2O_3 (with 13 wt% of CuO), CeO_2-Al_2O_3 (with 12.3 wt% of CeO_2), and CuO-CeO_2-Al_2O_3 (with 3.5 wt% of CuO and 8 wt% of CeO_2) was equal to 0.56, 0.70, and 1.14, respectively. Moreover, the best adsorption capacities were obtained with the sample containing a molar ratio of approximately 1:1 of copper and cerium. The authors suggested that the interaction between the copper atom and cerium atom was the key to reach high desulfurization capacities (Akyurtlu & Akyurtlu, 1999). They proposed the following mechanism: SO_2 was adsorbed on a CuO site and was oxidized and trapped as a sulfate by a close CeO_2 site, thus leaving the copper site free for further SO_2 adsorption.

A similar synergetic behavior between CuO and CeO_2 was observed in the case of γ-Al_2O_3-supported adsorbents (Wey et al., 2002). In fact, the CuO-CeO_2-Al_2O_3 material (containing 3.2 wt% of CuO and 6.8 wt% CeO_2) showed a better SO_2 adsorption capacity (about 80% of SO_2 removal) than those of the Al_2O_3-supported adsorbents containing 10 wt% of CuO or CeO_2 (SO_2 removal of about 72% or

75%, respectively). In this study, DeSOx efficiencies were measured in a pilot-scale bubbling fluidized bed incinerator burning simulated feed wastes containing a PE bag, sawdust, LDPE, PVC, and sulfur and equipped with a dry scrubber. It is noteworthy that the exact gas composition is not known (SO_2 concentration varied between 650 and 1000 ppmv and the temperature was between 500°C and 550°C). Surprisingly, in other study (Wey, Fu, Tseng and Chen, 2003); , AC-supported adsorbents, tested in the same pilot (Wey et al., 2002), showed lower DeSOx activities in the simultaneous presence of CuO and CeO_2 compared to those of the single metal oxide adsorbents. The authors (Wey, Fu, Tseng, & Chen, 2003) argued that this behavior was linked to the difference of the nature between the two types of support. Finally, Zuo, Yi, and Tang (2015) also observed the beneficial impact of cerium on a CuO-AC adsorbent (containing 7 wt% of CuO and doped by 8 wt% of $NaCO_3$). After the addition of 10 wt % of CeO_2, the SO_2 breakthrough time of the adsorbent increased from 55 to 70 min (under 1000 ppmv of SO_2, 600 ppmv of NO, and 15 vol % of O_2 in N_2 at 120°C).

Gaudin, Dorge, Nouali, Kehrli, et al. (2015) showed that a CuO-CeO_2-KIT-6 adsorbent presented a higher SO_2 adsorption capacity (16.6 mg_{SO2}/g_{ads} at 75 ppmv—as a reminder, here, adsorption capacities are calculated as the quantity of SO_2 adsorbed on the adsorbent when the SO_2 emissions reach 75 ppmv on the breakthrough curve—adsorption step under 250 ppmv of SO_2 and 10 vol% of O_2 in N_2 at 400°C) than that of the CuO-KIT-6 adsorbent (1.7 mg_{SO2}/g_{ads} at 75 ppmv). Moreover, the sum of the two SO_2 adsorption capacities obtained with the two KIT-6-supported single metal oxide adsorbents (with the same metal loading) prepared in the same conditions is smaller than the SO_2 adsorption capacity of the CuO-CeO_2-KIT-6 adsorbent (8.3 mg_{SO2}/g_{ads} at 75 ppmv for CeO_2-KIT-6). This was attributed to the synergetic effect between CuO and CeO_2 and could be explained by high dispersion of the CeO_2 particles, which was favored by the presence of copper and by some interactive electronic transfers between cerium and copper, as described in a previous work (Tsoncheva et al., 2013).

The literature has shown that the addition of another metal oxide, and in particular V_2O_5 and CeO_2, led to a significant enhancement of the DeSOx efficiency of CuO-supported adsorbents. This gain of activity can be linked to a synergetic effect between the metal oxides, generally leading to a better dispersion of the CuO particles. Additives can also catalyze the oxidation of SO_2 into SO_3 or have the capacity to trap SO_2 under a sulfated form. In the case of AC adsorbents, these metal oxides can create new carbon active sites on the carbon surface.

13.7 Influence of the operational conditions of the adsorption step

As seen previously, the nature of the support, the preparation protocol of the adsorbents, the loading of the active phase, and the presence of several metal oxides associated with CuO are important to obtain excellent DeSOx activities. Nevertheless, the conditions under which the SO_2 adsorption is going to take place are also essential. Parameters like temperature and gas composition (nature and concentration) can significantly impact the performance of the adsorbent. Thus, it is necessary to study the impact of these different parameters on the DeSOx adsorption capacity of the adsorbents with the perspective of an industrial application of this technology.

The literature points out that the temperature of the adsorption phase has an impact on the DeSOx efficiency of CuO-based adsorbents (Akyurtlu & Akyurtlu, 1999; Berger, Brillard, et al., 2020; Berger, Dorge, et al., 2018; Buelna & Lin, 2003, 2004; Gavaskar & Abbasian, 2006; Wang & Lin, 1998; Waqif et al., 1991; Yu et al., 2007, 2008; Zhang et al., 2020). An increase in the SO_2 adsorption capacities was generally observed with an increase in the temperature. For example, Yu et al. (2008) observed a constant increase in the quantity of SO_2 adsorbed on a CuO-γ-Al_2O_3 adsorbent (with 12wt% of CuO) by thermogravimetric analyses (under 2000ppmv of SO_2, 5vol% of O_2, and 3vol% of H_2O in N_2) when the temperature of the adsorption step increased from 300°C to 500°C. Berger, Brillard, et al. (2020) and Berger, Dorge, et al. (2018) investigated the impact of the temperature of the adsorption step on the DeSOx activity of a CuO-SBA-15 adsorbent (containing 15wt% of CuO). Under the same gaseous conditions (250ppmv of SO_2, 10vol% of O_2 in N_2 for the adsorption phase and 0.5vol% of H_2 in N_2 for the regeneration phase), an increase in temperature led to an increase in the SO_2 uptake at 75ppmv: 31 mg_{SO2}/g_{ads} at 350°C against 49 mg_{SO2}/gads at 450°C (after 10 adsorption-regeneration cycles) (Berger, Dorge, et al., 2018). The authors linked this gain of reactivity to an enhancement of the SO_2 oxidation rate with the temperature (Berger, Dorge, et al., 2018). Zhang et al. (2020) also showed that the DeSOx efficiency of a CuO-CeO_2-Al_2O_3 adsorbent (containing 8wt% of CuO and 3wt% of CeO_2) increased with the sulfation temperature. In fact, the saturation time of the adsorbent (measured under 1800ppmv of SO_2, 5vol% of O_2 in N_2 at 400°C) was nearly doubled when the temperature increased from 300°C (56min) to 500°C (100min).

However, in some cases, when the temperature of the adsorption phase exceeded a certain value, a decrease in the SO_2 adsorption capacity was observed (Ahmadian, Anbia, & Rezaie, 2020; Akyurtlu & Akyurtlu, 1999; Buelna & Lin, 2003, 2004; Gavaskar & Abbasian, 2006; Wang & Lin, 1998). For example, Gavaskar and Abbasian (2006) observed an enhancement of the DeSOx efficiency (evaluated under 2500pmv of SO_2 by thermogravimetry) of a CuO-AC adsorbent (14.4wt% of CuO) for a sulfation temperature ranging between 350°C and 450°C. If the latter was increased up to 550°C, lower DeSOx efficiencies were measured. The authors hypothesized that this loss of activity was linked to the agglomeration of the copper active phase above 450°C (Gavaskar & Abbasian, 2006). In the case of a sol-gel-derived CuO-γ-Al_2O_3 adsorbent (containing 8wt% of CuO), the quantity of SO_2 adsorbed at saturation (sulfation under 2000ppmv of SO_2 in air and regeneration under 10vol% of CH_4 in N_2) dropped from 115 to 93 mg_{SO2}/g_{ads} when the temperature increased from 350°C to 450°C, respectively (Buelna & Lin, 2004). In a similar manner, Wang and Lin (1998) observed a decrease in the SO_2 adsorption capacity from 93 to 17 mg_{SO2}/g_{ads} when the sulfation temperature increased from 300°C to 500°C, respectively, for a sol-gel-derived CuO-γ-Al_2O_3 adsorbent (containing 9wt% of CuO) and tested under 2000ppmv of SO_2 in dry air. The authors linked this loss of efficiency with the increase in the sulfation temperature to the higher conversion of SO_2 into SO_3. Since more CuO active sites are involved in the oxidation reaction, only few of them stay available for the adsorption step (Wang & Lin, 1998).

Akyurtlu and Akyurtlu (1999) investigated the SO_2 adsorption capacity of a CuO-Al_2O_3 adsorbent between 450°C and 550°C (by thermogravimetry under 3000ppmv of SO_2 and 3vol% of O_2 in N_2). They showed that above 500°C, the conversion at 3000s of the sample was affected by the fact that $CuSO_4$ was no more stable and thus it decomposed. For temperatures higher than 500°C, CuO only participated in the oxidation of SO_2 into SO_3, with the latter being mainly trapped by the alumina support, resulting in a lower sulfur capacity of copper.

Some authors studied the impact of the gas hourly space velocity (GHSV) on the DeSOx efficiency of CuO-based adsorbents (Ahmadian et al., 2020; Berger, Dorge, et al., 2018). For a GHSV ranging from 80,000 to 210,000 h^{-1}, Ahmadian et al. (2020) observed that, at a constant temperature (between 100°C and 400°C), the SO_2 adsorption capacity (measured under 500 ppmv of SO_2 and 5 vol% of O_2 in N_2) of a CuO-SBA-15 adsorbent (containing 8.7 wt% of CuO) decreased with the increase in the GHSV. At 400°C, for example, the SO_2 removal efficiency decreased from about 95% to 47% when the GHSV varied from 80,000 to 210,000 h^{-1}, respectively. A similar trend was observed by Berger, Dorge, et al. (2018) for a CuO-SBA-15 adsorbent (containing 15 wt% of CuO). The SO_2 adsorption at the breakthrough (under 250 ppmv of SO_2 and 10 vol% of O_2 in N_2 at 400°C) decreased from 54 to 14 mg_{SO2}/g_{ads} when the GHSV increased from 25,000 to 100,000 h^{-1}, respectively. The authors linked the higher SO_2 adsorption capacities at a low GHSV to the longer contact time between the adsorbent and the gaseous SO_2 (Berger, Dorge, et al., 2018). If the total quantity of SO_2 trapped is considered, the modification of the GHSV between 25,000 and 50,000 h^{-1} had no impact on the DeSOx efficiency (about 90 mg_{SO2}/g_{ads}) after 10 cycles of adsorption (under 250 ppmv of SO_2 and 10 vol% of O_2 in N_2 at 400°C during 10,800 s) and regeneration (under 0.5 vol% of H_2 in N_2 at 400°C) (Berger, Dorge, et al., 2018).

The gas composition of the adsorption phase can also highly influence the DeSOx efficiency of the CuO-based adsorbents. Several authors pointed out that the presence of O_2 in the flue gas during the adsorption phase is necessary to efficiently adsorb SO_2. In fact, as seen in Section 13.1, O_2 participates in the adsorption mechanism by oxidizing SO_2 into SO_3 (reaction (Eq. 13.1)). Waqif et al. (1991) showed that, in the temperature range of 250–350°C, an O_2 concentration below 1 vol% led to a decrease in the SO_2 adsorption capacity of a 4.88 wt% CuO-Al_2O_3 adsorbent. Above this value, no significant impact of the O_2 concentration was observed. For example, SO_2 adsorption capacities of about 5, 16, and 36 mg_{SO2}/g_{ads} were measured at 250°C in the presence of 10 ppmv, 420 ppmv, and 29 vol% of O_2, respectively (in the presence of 8000 ppmv of SO_2 in He) (Waqif et al., 1991). At the same SO_2 concentration in the flue gas (200 ppmv at 250°C), increasing the O_2 concentration from 0 to 5 vol% increased the SO_2 adsorption capacity of a CuO-AC adsorbent treated with HCl from 13 to 27 mg_{SO2}/g_{ads} (containing 3.76 wt% of CuO, adsorbent particle size between 250 and 297 μm), respectively. Increasing the O_2 concentration up to 10 vol% did not lead to a further increase in the SO_2 adsorption capacity (Tseng, Wey and Fu, 2003). For the same adsorbent (with a particle size between 125 and 177 μm), an increase in SO_2 adsorption capacity from 9 to 30 mg_{SO2}/g_{ads} was observed when the O_2 concentration increased from 0 to 7.68 vol% (under 400 ppmv of SO_2 in N_2 at 250°C), respectively. This increase by about a factor of 3 was observed in the whole temperature range of 100–250°C (Tseng & Wey, 2004). For a CuO-SBA-15 sorbent containing 15 wt% of CuO, the SO_2 adsorption capacities were not modified when the O_2 concentration varied from 1 to 50 vol% at 400°C (Berger, Dorge, et al., 2018). Studies showed that, globally, when O_2 was in excess regarding the oxidation of SO_2 into SO_3, there was no impact of its concentration on the DeSOx efficiency (Centi & Perathoner, 1997; Tseng & Wey, 2004; Tseng, Wey, & Fu, 2003; Waqif et al., 1991).

Gavaskar and Abbasian (2006) showed that the adsorption capacities of a sol-gel-derived 14.1 wt% CuO-Al_2O_3 adsorbent increased from 48 to 66 mg_{SO2}/g_{ads} when the SO_2 concentration in the flue gas increased from 1250 to 5000 ppmv (after 50 min of sulfation in a thermobalance at 450°C), respectively. The authors concluded that the order of the overall reaction concerning $[SO_2]$ was about 1. Similar results were found by Centi and Perathoner (1997) with 4.2 wt% CuO-Al_2O_3 pellets: after 120 min of sulfation under 3325 or 1498 ppmv of SO_2 (with 3 vol% of O_2 in He at 350°C), SO_2 adsorption capacities of 56 or 25 mg_{SO2}/g_{ads} were measured, respectively. At the same temperature (400°C), the total amount of SO_2 trapped, after 10 cycles of adsorption (variable SO_2 concentration with

10vol% of O_2 in N_2 during 10,800s) and regeneration (under 0.5vol% of H_2 in N_2), on a CuO-SBA-15 adsorbent (with 15wt% of CuO) also increased with an increase in the injected SO_2 concentration: about 90mg_{SO2}/g_{ads} with 250ppmv of SO_2 in the flue gas against almost 120mg_{SO2}/gads for 750ppmv of SO_2 (Berger, Dorge, et al., 2018). In the case of a CuO-AC treated with HCl and containing 3wt% of CuO, another behavior regarding the SO_2 concentration was observed (Tseng, Wey, & Fu, 2003). The total SO_2 adsorption capacity under 200, 400, or 600ppmv of SO_2 in N_2 at 250°C increased with the SO_2 concentration of the flue gas until 400ppmv of SO_2. Above this value, it tended to decrease because of the fast sulfation of the surface copper active sites, which blocked the access to the core copper active sites. This behavior may be linked to the microporous structure of the AC adsorbent studied.

During the utilization of the adsorbent in real-time conditions, numerous pollutants that may act as poison can be found in the flue gas to be depolluted. In the case of gas turbines, in addition to sulfur oxides, typical gas combustion contains CO_2, CO, NOx, water vapor, unburned hydrocarbons, and particles (Pavri & Moore, 2001). Heavy metal vapors (like Pb, Hg, or Cr) and HCl are also emitted by incineration and coal combustion gases (Li et al., 2020; Qin et al., 2021; Tseng, Wey, Liang, & Chen, 2003; Wey, Fu, Tseng, & Chen, 2003). Therefore, it is important to study the impact of the gaseous species on the SOx trapping performance of the adsorbents.

A study (Berger, Dorge, et al., 2020) showed that the presence of 180ppmv of CO or 4vol% of CO_2 (in a flue gas initially containing 250ppmv of SO_2 and 10vol% of O_2 in N_2 at 400°C) did not significantly impact the DeSOx efficiency of a CuO-SBA-15 adsorbent (15wt% of CuO). It is noteworthy that CO was totally oxidized into CO_2 by CuO. On the contrary, in the same conditions, the presence of NO_2 led to an enhancement of the DeSOx efficiency. With 200ppmv of NO_2 in the flue gas, an increase of 25% in the SO_2 uptake at 75ppmv was observed on the first cycle. This was attributed to a synergetic effect between NO_2 and SO_2, leading to a second SO_2 adsorption mechanism. A similar behavior was observed with the addition of 200ppmv of NO but to a lesser extent.

Using a pilot-scale incinerator burning simulated feed wastes (briefly described previously), a team studied the impact of the presence of HCl in the flue gas on the SO_2 adsorption capacity of CuO-doped ACs and alumina-supported adsorbents (Tseng, Wey, Liang, & Chen, 2003; Wey et al., 2002, 2003). It is noteworthy that the adsorption conditions were not exactly known. The authors (Wey et al., 2002) observed a decrease in SO_2 removal from about 75% to 59% for a CuO-Al_2O_3 adsorbent containing 10wt% of CuO (under a SO_2 concentration between 650 and 1000ppmv, a NO concentration between 40 and 60ppmv, and a temperature between 500°C and 550°C) in the presence of HCl (between 150 and 200ppmv). In the case of a CuO-AC adsorbent treated by HNO_3 and containing 12.5wt% of CuO, a similar decrease in SO_2 removal from about 70% to 55% (under a SO_2 concentration between 1750 and 2000ppmv and a temperature between 477°C and 547°C) was observed in the presence of HCl (between 425 and 530ppmv) (Tseng, Wey, Liang, & Chen, 2003). The team attributed this deactivation to the adsorption of HCl on copper sites (forming CuCl), thus competing with the SO_2 adsorption (Wey et al., 2002, 2003). Surprisingly, no significant impact of HCl (between 425 and 530ppmv) was observed in the case of the CuO-AC adsorbent treated by HCl and containing 12.5wt% of CuO (under a SO_2 concentration between 1500 and 2000ppmv, a NOx concentration between 150 and 200ppmv, and a temperature between 485°C and 547°C) (Tseng, Wey, Liang, & Chen, 2003). No hypothesis was proposed by the authors to explain this different behavior. In the same apparatus, the authors also showed that NH_3 had a beneficial impact on the SO_2 removal of the CuO-Al_2O_3 adsorbent (with 10wt% of CuO). In fact, the SO_2 removal increased from 75% to 90% (NH_3 concentration not reported) and formation of the $(NH_4)_2SO_4$ species was detected (Wey et al., 2002).

Rau, Tseng, Chiang, Wey, and Lin (2010) studied the impact of the presence of particles (here, fly ash) in the flue gas on the DeSOx efficiency of CuO-AC adsorbents. After acidic treatment with either HNO_3 or H_2SO_4, AC supports were impregnated with 3.75 wt% of CuO by wet impregnation. Simulated real flue gases were produced by the combustion of an artificial feedstock containing packing paper, sulfur powder, and wood. The authors showed that, in the presence of particles in the flue gas (diameter of 12 μm, concentration of about 240 mg/m^3), the DeSOx efficiency (measured in a fluidized bed reactor under about 1200 ppmv of SO_2 at 200°C during 60 min) of CuO-AC adsorbents decreased. In fact, without particles, the quantity of SO_2 adsorbed of the HNO_3- and H_2SO_4-treated CuO-AC adsorbents was equal to 19.2 and 16.7 mg_{SO2}/g_{ads} against 14.5 and 13.8 mg_{SO2}/g_{ads}, respectively, in the presence of particles. A decrease in the S_{area} from 835 to 712 m^2/g_{ads} for the HNO_3-treated CuO-AC sample and from 855 to 679 m^2/g_{ads} for the H_2SO_4-treated CuO-AC sample was observed after their exposition to the particles. The authors attributed this loss in efficiency to the blocking of the porosity of the adsorbent by the particles.

In conclusion of this part, the literature has shown that the temperature of the adsorption phase has an impact on the DeSOx activity of CuO-supported adsorbents. An increase in the SO_2 adsorption capacities was generally observed with an increase in the temperature because of the enhancement of the SO_2 oxidation rate. However, some studies highlighted a negative impact when the temperature exceeded a certain value (which depends on the experimental conditions and the type of adsorbent). The authors linked this behavior either to the agglomeration of the copper active phase or to the fact that CuO sites were no longer available for SO_2 adsorption because of the intensive oxidation of SO_2 into SO_3. The results also showed that the presence of O_2 was necessary for the SO_2 adsorption mechanism. However, when this oxidant was in excess (regarding the reaction of oxidation of SO_2 into SO_3), no more impact of its concentration on DeSOx efficiency was observed. In general, an increase in the SO_2 adsorption capacities was observed with an increase in the SO_2 concentration in the flue gas. Only few studies investigated the impact of other pollutants on the DeSOx efficiency of CuO-based adsorbents. Nevertheless, the results showed that CO and CO_2 had no significant impact on the adsorption of SO_2. On the contrary, the beneficial impact of the NOx species was observed by several authors and was attributed to a synergetic effect between NOx and SO_2. Studies also pointed out that HCl generally had a negative impact on SO_2 adsorption because of an adsorption competition on CuO active sites. Finally, by blocking the access to the porosity of the adsorbent, the presence of particles is deemed detrimental.

13.8 Influence of the regeneration conditions

The literature has highlighted that even if the adsorbent is well prepared, the conditions of the adsorption step (such as temperature, flue gas composition, etc.) can have a significant impact on its DeSOx performance. Nevertheless, the conditions for the regeneration step must also be chosen wisely. Generally, two types of regeneration methods are described in the literature: thermal regeneration under an inert atmosphere (N_2, He, etc.) and reactive regeneration under a reductive atmosphere (CH_4, H_2, etc.).

Centi et al. (1995) compared the efficiency, regarding the regeneration of a CuO-Al_2O_3 adsorbent (with 4.3 wt% of CuO), of thermal regeneration under He with reactive regeneration using 5 vol% of H_2 in He or 15 vol% of CH_4 in He. The results showed that temperatures higher than 600°C were necessary to decompose $CuSO_4$ under inert conditions, unlike the reactive regeneration where the $CuSO_4$ decomposition occurred at a lower temperature. Dihydrogen was a stronger reducing agent than CH_4: the highest SO_2 emissions were observed at around 330°C for H_2 against 510°C for CH_4. Moreover, contrary to CH_4, H_2 was able to decompose the $Al_2(SO_4)_3$ species (Centi et al., 1995; Macken & Hodnett, 1998). However, the authors reiterated that the decomposition of the $CuSO_4$ species on alumina by methane went by a single reaction forming Cu^0, SO_2, and CO_2, whereas this decomposition by H_2 suffers from several side reactions releasing H_2S and copper species in several oxidation states. A coating by TiO_2 of the alumina support (deposed by a reaction between the alumina surface hydroxyl groups and $TiCl_4$ in dichloromethane) could prevent some of these undesirable reactions (in particular, the ones between H_2 and the alumina support) (Centi et al., 1995). In isothermal conditions, increasing the temperature of the regeneration step, from 300°C to 500°C, allowed a significant decrease in the regeneration step duration under 5vol% of H_2 (Macken & Hodnett, 1998). In these conditions, less than 4min was necessary to regenerate a 4.3wt% $CuO-Al_2O_3$ (sulfated under 4800ppmv of SO_2 and 5150ppmv of O_2 at 350°C during 60min) at 500°C against more than 10min at 300°C. Moreover, below 500°C, some $CuSO_4$ was not decomposed (the quantity of undecomposed $CuSO_4$ at 400°C was estimated at 15%). However, temperatures higher than 500°C promoted the formation of side products like H_2S and CuS. Macken and Hodnett (1998) also observed the formation of some CuS during the regeneration of a 4.3wt% $CuO-Al_2O_3$ adsorbent under 5vol% of CH_4 (between 450°C and 550°C). The authors highlighted that the addition of water vapor (5 or 10vol%), during the regeneration step (under 5vol% of CH_4 at 500°C), inhibited the formation of these undesirable species but slowed down the regeneration (about 10min was necessary without water to go down a regeneration rate of 10 $\mu mol_{SO2}/g_{ads}/min$ against 15min with 10vol% of H_2O). They also showed that by performing TPR (temperature-programmed reduction) analyses between 200°C and 700°C, at the same concentration (5vol%), the reducing agents NH_3 and C_3H_8 exhibit a performance which is between those of H_2 and CH_4. The increase in the regeneration reaction rate with the increase in regeneration temperature has been described in the literature for other $CuO-Al_2O_3$ adsorbents under several different conditions of the regeneration step: 10vol% of CH_4 (Buelna & Lin, 2004), 5vol% of H_2 (Yoo et al., 1996; Yu et al., 2007), 5vol% of NH_3 (Xie et al., 2003), or air (Bahrin et al., 2015; ; ;). As for the adsorption step, extremely high temperatures for the regeneration step can lead to the agglomeration of the copper active phases (here, 700°C under air for a CuO-γ-Al_2O_3 adsorbent) (Bahrin et al., 2015; ; ;). It is noteworthy that in the case of alumina-supported sorbents, an additional phase of oxidation was often observed between two cycles of adsorption-regeneration (Buelna & Lin, 2004; Centi & Perathoner, 1997; Deng & Lin, 1996; Macken et al., 2000; Macken & Hodnett, 1998).

Tseng and Wey (2004) studied the impact of the conditions of the regeneration step on the DeSOx efficiency of a CuO-AC adsorbent. The authors observed no significant SO_2 desorption at 280°C or 310°C under a flue gas containing CH_4 from 400 to 1200ppmv. In the same conditions, the use of CO did not show better regeneration properties. Moreover, CO was converted into CO_2 by the copper sites of the adsorbent. The authors showed that a heat treatment up to 480°C under N_2 was sufficient to almost completely desorb all the SO_2 adsorbed, with a pretty stable efficiency over 10 cycles of adsorption and desorption. In these conditions of regeneration, the decomposition of the surface oxygen complexes of the AC into CO was observed. The latter then acted as a reducing agent for the regeneration of the copper phase.

Berger et al. (2017) showed that during the regeneration of CuO-SBA-15 adsorbents, a regeneration step under N_2 at 600°C allowed a better DeSOx efficiency of the adsorbent due to the reduction of the Cu^{2+} into the Cu^{+} active sites (as discussed in Section 13.4). Unfortunately, these new active sites tended to agglomerate at this temperature, resulting in a loss of SO_2 adsorption activity along cycles. They also showed that using a reductive atmosphere was a good compromise between high DeSOx efficiency, stability, and energy saving (Berger, Nouali, et al., 2018; Berger et al., 2017). In fact, under 0.5 vol% of H_2, a temperature of 280°C was sufficient to achieve the complete regeneration of the CuO-SBA-15 adsorbent. Increasing the temperature up to 400°C led to a faster regeneration step without any loss of activity. Moreover, performing the regeneration step at the same temperature as that of the adsorption step (400°C) allowed to avoid thermal shocking and was highly interesting in view of the fast and easy switch between the two steps. The possibility to regenerate the adsorbent in situ is extremely attractive for an industrial application (time and energy saving).

The literature has shown that, with regard to the regeneration conditions, globally, the use of a reductive atmosphere instead of an inert atmosphere is preferable. In fact, in this case, the softer conditions of temperature can be used for preventing thermal shocking of the adsorbent. Moreover, H_2 showed the highest regeneration power but suffered from side reactions (in particular, in the case of an alumina support).

13.9 Aging and stability of the SOx adsorbents over time

In order to be used in industrial applications, regenerable CuO-based adsorbents must have a good long-term stability. Deng and Lin (1996) studied the chemical and thermal stability via accelerated aging treatments of sol-gel-derived CuO-γ-Al_2O_3 adsorbents. After thermal treatment under air at 850°C for 168 h, all the studied sorbents showed a decrease in the S_{area} and in TPV and an increase in pore size because of the agglomeration of the CuO particles. Nevertheless, the authors considered that the S_{area} of these aged sorbents was still competitive with the those conventionally used in flue gas desulfurization processes. After chemical treatment under 22,000 ppmv of SO_2 in air at 850°C for 60 h, the textural properties were impacted in a similar manner. Besides the agglomeration of CuO particles, these textural modifications were also linked to the sulfation of the alumina support. The addition of four cycles of sulfation (under 22,000 ppmv of SO_2 in air) and regeneration (under 30 vol% of H_2 in nitrogen) at 850°C did not lead to further degradation of the textural properties of the sorbent. This behavior suggested that sol-gel-derived CuO-γ-Al_2O_3 adsorbents were stabilized during the initial chemical treatment. Finally, a second chemical treatment with the addition of 12 vol% of water vapor only led to a small modification of the pore structure of the sample. On the basis of these results, the authors evaluated, through a simple sintering model, that it would take about 3 and 13 years in mild sulfation (under 22,000 ppmv of SO_2 in air) and regeneration (under 30 vol% of H_2 in nitrogen) conditions (at 450°C) to lose 50% and 75%, respectively, of the S_{area} of a 20 wt% CuO-γ-Al_2O_3 adsorbent, suggesting the excellent stability of this material.

Gavaskar and Abbasian (2006) evaluated the DeSOx activity of a sol-gel-derived CuO-Al_2O_3 sorbent (with 14.1 wt% of CuO) over 25 cycles at 450°C of adsorption (under 2500 ppmv of SO_2 and 3.7 vol% of O_2 in N_2, the adsorption step was stopped when the exit SO_2 concentration was about 100 ppmv) and regeneration (under pure CH_4 until no more SO_2 is detected) in both a thermogravimetric analyzer (TGA) and a fluidized bed reactor. In both cases, the trend observed was the same: the DeSOx activity of the sorbent increased during the first three cycles and then decreased until reaching a plateau around the 20th cycle. A stabilized sulfur loading of about 40 mg_{SO2}/g_{ads} was attained in the fluidized bed reactor and about 46 mg_{SO2}/g_{ads} in the case of the TGA. After 25 cycles, a decrease in the specific surface area was also observed (150 m^2/g_{ads} for the fresh sample against 111 m^2/g_{ads} for the aged one). In another study, similar adsorption capacities were measured for 4.3 wt% CuO-γ-Al_2O_3 pellets after 150 cycles (24 mg_{SO2}/g_{ads}) or 5 cycles (26 mg_{SO2}/g_{ads}) at 350 °C of adsorption (under 3200 ppmv of SO_2, 800 ppmv of NO, 800 ppmv of NH_3, and 3 vol% of O_2) and regeneration (conditions nonprecise), confirming the good stability of the adsorbent (Centi et al., 1995).

Macken et al. (2000) tested the long-term stability of several CuO-γ-Al_2O_3 pellets under different conditions. The sorbent containing 4.3 wt% of CuO showed stable DeNOx and DeSOx activities over 750 cycles of adsorption (under 1 h in the standard conditions of 2000 ppmv of SO_2, 3 vol% of O_2, 0.1 vol% of NO, 0.1 vol% of NH_3, 10 vol% of CO_2, and 10 vol% of H_2O at 350°C for the first 370 cycles and then at 450°C), regeneration (at 500°C under 10 vol% of CH_4 for 20 min), and oxidation (10 min under air at 400°C for the first 370 cycles and then at 480°C). A SO_2 adsorption capacity of about 38 mg_{SO2}/g_{ads} was measured after 750 cycles. Moreover, no significant modification of the mechanical and textural properties was observed. The authors evaluated a lifetime of more than 1 year for this sorbent. The same adsorbent was tested over 36 cycles in a bench-scale apparatus with a commercial oil boiler and thiophene-doped fuel. Under real-time flue gas conditions, the sample showed a loss of about 3 mg_{SO2}/g_{ads} after the 36th cycle. Other CuO-γ-Al_2O_3 pellets containing 4.9 wt% of CuO was first tested under 117 accelerated aging cycles. For these types of cycles, compared to the standard conditions described previously, the temperature was increased to 450°C and the water vapor concentration doubled during the adsorption phase, and the temperature of the regeneration phase was increased to 550°C and the oxidation temperature was up to 480°C. Once again, no significant loss of DeSOx activities was observed over the 117 cycles. After 1525 cycles in the standard test conditions, this sample presented a loss of about 25% of its SO_2 adsorption capacities and S_{area} due to the agglomeration of the copper active phase (final SO_2 adsorption capacity of about 30 mg_{SO2}/g_{ads}). Surprisingly, 7.0 wt% CuO-γ-Al_2O_3 pellets showed an increase by 27% in SO_2 adsorption after 100 cycles in the standard conditions (final SO_2 adsorption capacity of about 59 mg_{SO2}/g_{ads}). The authors attributed this behavior to a better dispersion of the CuO particles through the cycles linked to the redispersion of the Cu^0 species (resulting from the decomposition of the $CuSO_4$ species during the regeneration under H_2) during the oxidation phase.

Tseng and Wey (2004) studied the stability of a CuO-AC adsorbent treated by HCl and containing 3.76 wt% of CuO over 10 cycles of adsorption (under 400 ppmv of SO_2, 7.68 vol% of O_2 in N_2 at 250°C) and regeneration (under N_2 in ramp from 250°C to 475°C). Only a small deactivation was observed after the 10th cycle. In fact, the SO_2 adsorption capacities were equal to 24 and 20 mg_{SO2}/g_{ads} after 1 and 10 cycles, respectively. Moreover, the S_{area} of the adsorbent was also slightly impacted (1054 m^2/g_{ads} for the fresh sample against 1011 m^2/g_{ads} after 5 cycles).

By regeneration just before the SO_2 breakthrough, Berger, Dorge, et al. (2018) stabilized over 25 cycles (under 250 ppmv of SO_2 with 10 vol% of O_2 in N_2 for the adsorption and under 0.5 vol% of H_2 in N_2 for the regeneration) the SOx trapping activity of a CuO-SBA-15 adsorbent around 35 mg_{SO2}/g_{ads} at 400°C and around 55 mg_{SO2}/g_{ads} at 450°C with no SO_2 emission during the adsorption step. Such performance is highly promising from an industrial point of view and is still compatible in the case of more stringent regulations. Moreover, a recent work

(Cesario et al., 2020) has highlighted the possibility to shape this CuO-SBA-15 adsorbent powder into beads, thus allowing an industrial application, for example, in a fluidized bed.

The literature has pointed out that several CuO-based adsorbents synthesized at the laboratory scale had relatively good stability over time, showing the possibility of further industrial applications.

13.10 Conclusions

FGD systems are now the most used method to reduce SO_2 emissions worldwide. The most current ones are CaO-based wet FGD techniques. Despite being highly efficient, these processes are expensive, water- and energy-consuming, and produce a large amount of nonrecoverable waste. This is why numerous dry regenerable FGD methods are now under development. The literature has highlighted that supported metal oxide adsorbents showed really promising DeSOx efficiencies. Nowadays, CuO-based adsorbents are the most used because of the excellent oxidative and trapping capacities of the copper active phase. The results showed that the better the CuO particle dispersion, the greater was the DeSOx efficiency. Data also highlighted that the Cu^+ species seemed to be more reactive for the catalysis of the reaction of the oxidation of SO_2 into SO_3.

The study of the impact of several parameters on the DeSOx efficiency of CuO-based adsorbents highlighted the following general conclusions. In the same conditions of preparation and adsorption/regeneration test, the nature of the support had an impact on the DeSOx efficiency because of the inherent properties of the support or different textural properties. In fact, better textural properties (higher specific surface area (S_{area}) and higher total pore volume (TPV)) usually led to higher SO_2 adsorption capacity because of not only the better access of the gaseous species to the active phase but also the better dispersion of the CuO particles achievable on the support surface. Irrespective of the nature of the support used, treatment and preparation protocols can have a significant impact on the final DeSOx efficiency of CuO-supported adsorbents. Mainly, preparation conditions have to be optimized in order to enhance the textural properties of the support in terms of S_{area} and pore size, and therefore the CuO dispersion, since the latter plays a major role in the SO_2 adsorption. Conditions of the different steps of the synthesis protocol, and in particular the calcination, must thus be well controlled and optimized. The addition of another metal oxide, and in particular V_2O_5 and CeO_2, can lead to significant enhancement of the DeSOx efficiency through a synergetic effect with CuO. Depending on the nature of the latter, a beneficial effect is not only linked to a better dispersion of the CuO particles but also to a catalytic effect of the additives on the oxidation of SO_2 into SO_3 and/or to the formation of supplementary active sites. Conditions of the adsorption phase must also be chosen wisely. An increase in the SO_2 adsorption capacities was generally observed with an increase in the temperature because of the enhancement of the SO_2 oxidation rate. However, some studies highlighted a negative impact when the temperature exceeded a certain value (which depends on the conditions and the adsorbent), attributed to either the agglomeration of the copper active phase or to the fact that CuO sites are no more available for SO_2 adsorption because of the intensive oxidation of SO_2 into SO_3. The presence of O_2 is necessary for the SO_2 adsorption mechanism. However, when this oxidant is in excess (regarding the reaction of oxidation of SO_2 into SO_3), no more impact of its concentration on the SO_2-trapping capacity was observed. In general, an increase in the SO_2 adsorption capacities was observed with an increase in the SO_2 concentration in the flue gas. The presence of CO and CO_2 had no significant impact on the adsorption of SO_2. On the contrary, an improvement of the latter was observed by several authors in the presence of NOx and was attributed to a synergetic effect between NOx and SO_2. Pollutants like HCl and particulate matter showed a detrimental impact on the efficiency of the adsorbent because of an adsorption competition on CuO active sites and blocking of the porosity, respectively. For the regeneration phase, the use of a reductive atmosphere instead of an inert atmosphere is generally preferable because of the softer conditions of temperature that can be used, thus preventing thermal shocking of the adsorbent. Furthermore, the possibility to carry out regeneration and adsorption in isothermal conditions is extremely interesting from an industrial point of view (no thermal shock and time- and energy-saving). Moreover, H_2 showed the highest regeneration power but suffered from side reactions (in particular, in the case of an alumina support). Finally, the long-term stability and efficiency at the laboratory scale of several regenerable CuO-based adsorbents presented in the literature is highly promising for the future industrial application of these processes.

References

Abdulrasheed, A. A., Jalil, A. A., Triwahyono, S., Zaini, M. A. A., Gambo, Y., & Ibrahim, M. (2018). Surface modification of activated carbon for adsorption of SO_2 and NOX: A review of existing and emerging technologies. *Renewable and Sustainable Energy Reviews, 94*, 1067–1085. https://doi.org/10.1016/j.rser.2018.07.011.

Ahmadian, M., Anbia, M., & Rezaie, M. (2020). Sulfur dioxide removal from flue gas by supported CuO nanoparticle adsorbents. *Industrial & Engineering Chemistry Research, 59*(50), 21642–21653. https://doi.org/10.1021/acs.iecr.0c05629.

Akyurtlu, J. F., & Akyurtlu, A. (1999). Behavior of ceria-copper oxide sorbents under sulfation conditions. *Chemical Engineering Science, 54*(15), 2991–2997. https://doi.org/10.1016/S0009-2509(98)00505-3.

Ali, M. E., Rahman, M. M., Sarkar, S. M., & Hamid, S. B. A. (2014). Heterogeneous metal catalysts for oxidation reactions. *Journal of Nanomaterials, 2014*. https://doi.org/10.1155/2014/192038, e192038.

Amsalu, E., Guo, Y., Li, H., Wang, T., Liu, Y., Wang, A., et al. (2019). Short-term effect of ambient sulfur dioxide (SO_2) on cause-specific cardiovascular hospital admission in Beijing, China: A time series study. *Atmospheric Environment, 208*, 74–81. https://doi.org/10.1016/j.atmosenv.2019.03.015.

Bagheri, S., Muhd Julkapli, N., & Bee Abd Hamid, S. (2014). Titanium dioxide as a catalyst support in heterogeneous catalysis. *The Scientific World Journal, 2014*. https://doi.org/10.1155/2014/727496.

Bahrin, D., Subagjo, S., & Susanto, H. (2015). Effect of regeneration temperature on particle characteristics and extent of regeneration of saturated SO_2-adsorption of CuO/γ-Al_2O_3 adsorbent. *Procedia Chemistry, 16*, 723–727. https://doi.org/10.1016/j.proche.2015.12.020.

Bandosz, T. J., Loureiro, J. M., & Kartel, M. T. (2006). *Carbonaceous materials as desulfurization media* (pp. 145–164). Netherlands: Springer. https://doi.org/10.1007/1-4020-5172-7_16.

Berger, M. (2018). *Matériaux adsorbants destinés à la désulfuration des fumées industrielles: Synthèse et étude des performances en milieux gazeux complexes*. http://www.theses.fr/s222237.

Berger, M., Brillard, A., Dorge, S., Habermacher, D., Nouali, H., Kerdoncuff, P., et al. (2020). Modeling SOx trapping on a copper-doped CuO/SBA-15 sorbent material. *Journal of Hazardous Materials, 385*. https://doi.org/10.1016/j.jhazmat.2019.121579, 121579.

Berger, M., Dorge, S., Nouali, H., Habermacher, D., Fiani, E., Vierling, M., et al. (2018). Role of the process conditions on the sulphation and stability of a CuO/SBA-15 type SO_x adsorbent in cycling operations. *Chemical Engineering Journal, 350*, 729–738. https://doi.org/10.1016/j.cej.2018.05.170.

Berger, M., Dorge, S., Nouali, H., Habermacher, D., Fiani, E., Vierling, M., et al. (2020). Desulfurization process: Understanding of the behaviour of the CuO/SBA-15 type SO_x adsorbent in the presence of NO/NO_2 and CO/CO_2 flue gas environmental pollutants. *Chemical Engineering Journal, 384*. https://doi.org/10.1016/j.cej.2019.123318, 123318.

Berger, M., Fioux, P., Dorge, S., Nouali, H., Habermacher, D., Fiani, E., et al. (2017). Structure-performance relationship in CuO/SBA-15-type SOx adsorbent: Evolution of copper-based species under different regenerative treatments. *Catalysis Science & Technology, 7*(18), 4115–4128. https://doi.org/10.1039/C7CY01010A.

Berger, M., Nouali, H., Dorge, S., Habermacher, D., Fiani, E., Vierling, M., et al. (2018). Long-term activity of a CuO/SBA-15 type SOx adsorbent: Impact of the regeneration step. *Chemical Engineering Journal, 347*, 202–213. https://doi.org/10.1016/j.cej.2018.04.066.

Buelna, G., & Lin, Y. S. (2003). Combined removal of SO_2 and no using sol-gel-derived copper oxide coated alumina sorbents/catalysts. *Environmental Technology, 24*(9), 1087–1095. https://doi.org/10.1080/09593330309385649.

Buelna, G., & Lin, Y. S. (2004). Characteristics and desulfurization-regeneration properties of sol–gel-derived copper oxide on alumina sorbents. *Separation and Purification Technology, 39*(3), 167–179. https://doi.org/10.1016/S1383-5866(03)00183-7.

Carabineiro, S. A. C., Ramos, A. M., Vital, J., Loureiro, J. M., Órfão, J. J. M., & Fonseca, I. M. (2003). Adsorption of SO_2 using vanadium and vanadium–copper supported on activated carbon. *Catalysis Today, 78*(1), 203–210. https://doi.org/10.1016/S0920-5861(02)00335-8.

Centi, G., Hodnett, B. K., Jaeger, P., Macken, C., Marella, M., Tomaselli, M., et al. (1995). Development of copper-on-alumina catalytic materials for the cleanup of flue gas and the disposal of diluted ammonium sulfate solutions. *Journal of Materials Research, 10*(3), 553–561. https://doi.org/10.1557/JMR.1995.0553.

Centi, G., Passarini, N., Perathoner, S., & Riva, A. (1992). Combined DeSOx/DeNOx reactions on a copper on alumina sorbent-catalyst. 1. Mechanism of sulfur dioxide oxidation-adsorption. *Industrial & Engineering Chemistry Research, 31*(8), 1947–1955. https://doi.org/10.1021/ie00008a016.

Centi, G., & Perathoner, S. (1997). Role of the size and texture properties of copper-on-alumina pellets during the simultaneous removal of SO_2 and NO_x from flue gas. *Industrial & Engineering Chemistry Research, 36*(8), 2945–2953. https://doi.org/10.1021/ie9604886.

Centi, G., Perathoner, S., Kartheuser, B., Rohan, D., & Hodnett, B. K. (1992). Assessment of copper-vanadium oxide on mixed alumina-titania supports as sulphur dioxide sorbents and as catalysts for the selective catalytic reduction of NOx by ammonia. *Applied Catalysis B: Environmental, 1*(2), 129–137. https://doi.org/10.1016/0926-3373(92)80038-2.

Cesario, M., Schobing, J., Bruder, F., Dorge, S., Nouali, H., Habermacher, D., et al. (2020). Impact of bentonite content on the structural, textural and mechanical properties of SBA-15 mesoporous silica beads. *Journal of Porous Materials, 27*(3), 905–910. https://doi.org/10.1007/s10934-020-00865-5.

Dahiya, S., Anhäuser, A., Farrow, A., Thierot, H., Kumar, A., & Myllyvirta, L. (2020). *Global SO_2 emission hot spot database* (p. 48). Center for Research on Energy and Clean Air & Green Peace India.

Daud, W. M. A. W., & Houshamnd, A. H. (2010). Textural characteristics, surface chemistry and oxidation of activated carbon. *Journal of Natural Gas Chemistry, 19*(3), 267–279. https://doi.org/10.1016/S1003-9953(09)60066-9.

Deng, S. G., & Lin, Y. S. (1996). Synthesis, stability, and sulfation properties of sol–gel-derived regenerative sorbents for flue gas desulfurization. *Industrial & Engineering Chemistry Research, 35*(4), 1429–1437. https://doi.org/10.1021/ie950399d.

Directive (EU) 2015/2193. (2015). *313: Vol. OJ L. European Parliament and Council of 25 November 2015 on the limitation of emissions of certain pollutants into the air from medium combustion plants (Text with EEA relevance).* http://data.europa.eu/eli/dir/2015/2193/oj/eng.

Directive (EU) 2016/2284. (2016). *344: Vol. OJ L. European Parliament and Council of 14 December 2016 on the reduction of national emissions of certain atmospheric pollutants, amending Directive 2003/35/EC and repealing Directive 2001/81/EC (Text with EEA relevance).* http://data.europa.eu/eli/dir/2016/2284/oj/eng.

Directive 2010/75/EU. (2010). *334: Vol. OJ L. European Parliament and Council of 24 November 2010 on industrial emissions (integrated pollution prevention and control) Text with EEA relevance.* http://data.europa.eu/eli/dir/2010/75/oj/eng.

Egan, B. Z., & Felker, L. K. (1986). Removal of sulfur dioxide from simulated flue gas by magnesia spray absorption: Parameters affecting removal efficiency and products. *Industrial & Engineering Chemistry Process Design and Development, 25*(2), 558–561. https://doi.org/10.1021/i200033a037.

EPA, U. (2010). *Green book.* US EPA. https://www3.epa.gov/airquality/greenbook/tbtc.html.

Fan, Y. F., Hu, Z. Q., & Luan, H. Y. (2012). Deterioration of tensile behavior of concrete exposed to artificial acid rain environment. *Interaction and Multiscale Mechanics, 5*(1), 41–56. https://doi.org/10.12989/IMM.2012.5.1.041.

Galtayries, A., Grimblot, J., & Bonnelle, J.-P. (1996). Interaction of SO_2 with different polycrystalline Cu, Cu_2O and CuO surfaces. *Surface and Interface Analysis, 24*(5), 345–354. https://doi.org/10.1002/(SICI)1096-9918(199605)24:5<345::AID-SIA126>3.0.CO;2-2.

Gaudin, P., Dorge, S., Nouali, H., Kehrli, D., Michelin, L., Josien, L., et al. (2015). Synthesis of Cu-Ce/KIT-6 materials for SOx removal. *Applied Catalysis A: General, 504*, 110–118. https://doi.org/10.1016/j.apcata.2014.11.024.

Gaudin, P., Dorge, S., Nouali, H., Patarin, J., Brilhac, J.-F., Fiani, E., et al. (2015). Synthesis of CuO/SBA-15 adsorbents for SOx removal applications, using different impregnation methods. *Comptes Rendus Chimie, 18*(10), 1013–1029. https://doi.org/10.1016/j.crci.2015.07.002.

Gaudin, P., Dorge, S., Nouali, H., Vierling, M., Fiani, E., Molière, M., et al. (2016). CuO/SBA-15 materials synthesized by solid state grinding: Influence of CuO dispersion and multicycle operation on DeSOX performances. *Applied Catalysis B: Environmental, 181*, 379–388. https://doi.org/10.1016/j.apcatb.2015.08.011.

Gaudin, P., Fioux, P., Dorge, S., Nouali, H., Vierling, M., Fiani, E., et al. (2016). Formation and role of Cu+ species on highly dispersed CuO/SBA-15 mesoporous materials for SOx removal: An XPS study. *Fuel Processing Technology, 153*, 129–136. https://doi.org/10.1016/j.fuproc.2016.07.015.

Gaudin, P., Michelin, L., Josien, L., Nouali, H., Dorge, S., Brilhac, J.-F., et al. (2016). Highly dispersed copper species supported on SBA-15 mesoporous materials for SOx removal: Influence of the CuO loading and of the support. *Fuel Processing Technology, 148*, 1–11. https://doi.org/10.1016/j.fuproc.2016.02.025.

Gavaskar, V. S., & Abbasian, J. (2006). Dry regenerable metal oxide sorbents for SO_2 removal from flue gases. 1. Development and evaluation of copper oxide sorbents. *Industrial & Engineering Chemistry Research, 45*(17), 5859–5869. https://doi.org/10.1021/ie060123d.

Gentry, S. J., & Walsh, P. T. (1982). Influence of silica and alumina supports on the temperature-programmed reduction of copper(II) oxide. *Journal of the Chemical Society, Faraday Transactions 1: Physical Chemistry in Condensed Phases, 78*(5), 1515–1523. https://doi.org/10.1039/F19827801515.

Hanif, M. A., Ibrahim, N., & Abdul Jalil, A. (2020). Sulfur dioxide removal: An overview of regenerative flue gas desulfurization and factors affecting desulfurization capacity and sorbent regeneration. *Environmental Science and Pollution Research, 27*(22), 27515–27540. https://doi.org/10.1007/s11356-020-09191-4.

Huo, C., Ouyang, J., & Yang, H. (2014). CuO nanoparticles encapsulated inside Al-MCM-41 mesoporous materials via direct synthetic route. *Scientific Reports, 4*(1), 1–9. https://doi.org/10.1038/srep03682.

Hwang, J., Kwon, J., Yi, H., Bae, H.-J., Jang, M., & Kim, N. (2020). Association between long-term exposure to air pollutants and cardiopulmonary mortality rates in South Korea. *BMC Public Health, 20*(1), 1402. https://doi.org/10.1186/s12889-020-09521-8.

Jacubowiez, I. (2000). Désulfuration des fumées. In *Techniques de l'Ingénieur*. J3924 V1.

Khaniabadi, Y. O., Polosa, R., Chuturkova, R. Z., Daryanoosh, M., Goudarzi, G., Borgini, A., et al. (2017). Human health risk assessment due to ambient PM10 and SO_2 by an air quality modeling technique. *Process Safety and Environmental Protection, 111*, 346–354. https://doi.org/10.1016/j.psep.2017.07.018.

Kong, Y., Zhu, H. Y., Yang, G., Guo, X. F., Hou, W. H., Yan, Q. J., et al. (2004). Investigation of the structure of MCM-41 samples with a high copper content. *Advanced Functional Materials, 14*(8), 816–820. https://doi.org/10.1002/adfm.200305111.

Ku, T., Chen, M., Li, B., Yun, Y., Li, G., & Sang, N. (2016). Synergistic effects of particulate matter (PM2.5) and sulfur dioxide (SO_2) on neurodegeneration via the microRNA-mediated regulation of tau phosphorylation†. *Toxicology Research, 6*(1), 7–16. https://doi.org/10.1039/c6tx00314a.

Kumar, A., Singh, R. M., Shukla, P., & Singh, P. (2020). A review on ionic liquids as novel absorbents for SO_2 removal. In *Environmental processes and management: Tools and practices* (pp. 285–307). Springer International Publishing. https://doi.org/10.1007/978-3-030-38152-3_15.

Lecomte, T., Ferreria de la Fuente, J. F., Neuwahl, F., Canova, M., Pinasseau, A., Jankov, I., et al. (2017). *Best available techniques (BAT) reference document for large combustion plants*. European Commission.

Lippmann, M. (2020). Sulfur oxides (SOx). In *Environmental toxicants* (pp. 927–971). John Wiley & Sons, Ltd. https://onlinelibrary.wiley.com/doi/abs/10.1002/9781119438922.ch25.

Liu, Q., Liu, Z., & Huang, Z. (2005). CuO supported on Al_2O_3-coated cordierite-honeycomb for SO_2 and NO removal from flue gas: Effect of acid treatment of the cordierite. *Industrial & Engineering Chemistry Research, 44*(10), 3497–3502. https://doi.org/10.1021/ie048881w.

Liu, Q., Liu, Z., Huang, Z., & Xie, G. (2004). A honeycomb catalyst for simultaneous NO and SO_2 removal from flue gas: Preparation and evaluation. *Catalysis Today, 93–95*, 833–837. https://doi.org/10.1016/j.cattod.2004.06.081.

Liu, Q., Liu, Z., & Wu, W. (2009). Effect of V_2O_5 additive on simultaneous SO_2 and NO removal from flue gas over a monolithic cordierite-based CuO/Al_2O_3 catalyst. *Catalysis Today, 147*, S285–S289. https://doi.org/10.1016/j.cattod.2009.07.013.

Liu, Q., Liu, Z., Zhu, Z., Xie, G., & Wang, Y. (2004). Al_2O_3-coated honeycomb cordierite-supported CuO for simultaneous SO_2 and NO removal from flue gas: Effect of Na_2O additive. *Industrial & Engineering Chemistry Research, 43*(15), 4031–4037. https://doi.org/10.1021/ie049942t.

Livingston, R. A. (2016). Acid rain attack on outdoor sculpture in perspective. *Atmospheric Environment, 146*, 332–345. https://doi.org/10.1016/j.atmosenv.2016.08.029.

Lowell, P. S., Corbett, W. E., Brown, G. D., & Wilde, K. A. (1976). *Feasibility of producing elemental sulfur from magnesium sulfite*. EPA-600/7-76-030.

Lowell, P. S., Schwitzgebel, K., Parsons, T. B., & Sladek, K. J. (1971). Selection of metal oxides for removing SO_2 from flue gas. *Industrial & Engineering Chemistry Process Design and Development, 10*(3), 384–390. https://doi.org/10.1021/i260039a018.

Macken, C., & Hodnett, B. K. (1998). Reductive regeneration of sulfated CuO/Al_2O_3 catalyst–sorbents in hydrogen, methane, and steam. *Industrial & Engineering Chemistry Research, 37*(7), 2611–2617. https://doi.org/10.1021/ie970870y.

Macken, C., Hodnett, B. K., & Paparatto, G. (2000). Testing of the CuO/Al_2O_3 catalyst–sorbent in extended operation for the simultaneous removal of NOx and SO_2 from flue gases. *Industrial & Engineering Chemistry Research, 39*(10), 3868–3874. https://doi.org/10.1021/ie000342d.

Maity, S. K., Ancheyta, J., Rana, M. S., & Rayo, P. (2006). Alumina–titania mixed oxide used as support for hydrotreating catalysts of Maya heavy crude effect of support preparation methods. *Energy & Fuels, 20*(2), 427–431. https://doi.org/10.1021/ef0502610.

Mathieu, Y., Tzanis, L., Soulard, M., Patarin, J., Vierling, M., & Molière, M. (2013). Adsorption of SOx by oxide materials: A review. *Fuel Processing Technology, 114*, 81–100. https://doi.org/10.1016/j.fuproc.2013.03.019.

Miller, B., & Miller, B. (2015). Sulfur oxides formation and control. In *Fossil fuel emissions control technologies* (pp. 197–242). Butterworth-Heinemann (Chapter 4) http://www.sciencedirect.com/science/article/pii/B978012801566700004X.

Miller, B. G., & Miller, B. G. (2017). The effect of coal usage on human health and the environment. In *Clean coal engineering technology* (2nd ed., pp. 105–144). Butterworth-Heinemann (Chapter 3) http://www.sciencedirect.com/science/article/pii/B978012811365300003X.

Nada, M., Jayalath, S., Gillan, E. G., Grassian, V. H., Larsen, S. C., Douhal, A., et al. (2019). Zeolites and mesoporous silica: From greener synthesis to surface chemistry of environmental and biological interactions. In *Vol. 2. Chemistry of silica and zeolite-based materials* (pp. 375–397). Elsevier (Chapter 20) https://www.sciencedirect.com/science/article/pii/B9780128178133000201.

Oesch, S., & Faller, M. (1997). Environmental effects on materials: The effect of the air pollutants SO_2, NO_2, NO and O_3 on the corrosion of copper, zinc and aluminium. A short literature survey and results of laboratory exposures. *Corrosion Science, 39*(9), 1505–1530. https://doi.org/10.1016/S0010-938X(97)00047-4.

Pavri, R., & Moore, G. D. (2001). *GER4211 – Gas turbines emissions and control*. GE Energy Services.

Pollack, S. S., Chisholm, W. P., Obermyer, R. T., Hedges, S. W., Ramanathan, M., & Montano, P. A. (1988). Properties of copper/alumina sorbents used for the removal of sulfur dioxide. *Industrial & Engineering Chemistry Research, 27*(12), 2276–2282. https://doi.org/10.1021/ie00084a013.

Qin, Y., Wang, C., Sun, X., Ma, Y., Song, X., Wang, F., et al. (2021). Defects on activated carbon determine the dispersion of active components and thus the simultaneous removal efficiency of SO_2, NO_x and Hg0. *Fuel, 293*. https://doi.org/10.1016/j.fuel.2021.120391, 120391.

Rau, J.-Y., Tseng, H.-H., Chiang, B.-C., Wey, M.-Y., & Lin, M.-D. (2010). Evaluation of SO_2 oxidation and fly ash filtration by an activated carbon fluidized-bed reactor: The effects of acid modification, copper addition and operating condition. *Fuel, 89*(3), 732–742. https://doi.org/10.1016/j.fuel.2009.10.017.

Rodríguez-Reinoso, F. (1998). The role of carbon materials in heterogeneous catalysis. *Carbon, 36*(3), 159–175. https://doi.org/10.1016/S0008-6223(97)00173-5.

Romero, C. E., Wang, X., Zhang, Y., Wang, T., Pan, W.-P., & Romero, C. E. (2019). Key technologies for ultra-low emissions from coal-fired power plants. In *Advances in ultra-low emission control technologies for coal-fired power plants* (pp. 39–79). Woodhead Publishing (Chapter 3) http://www.sciencedirect.com/science/article/pii/B9780081024188000036.

Schobing, J., Cesario, M., Dorge, S., Nouali, H., Patarin, J., Martens, J., et al. (2021). CuO supported on COK-12 and SBA-15 ordered mesoporous materials for temperature swing SOx adsorption. *Fuel Processing Technology, 211*. https://doi.org/10.1016/j.fuproc.2020.106586, 106586.

Shao, X.-C., Duan, L.-H., Wu, Y.-Y., Qin, Y.-C., Yu, W.-G., Wang, Y., et al. (2012). Effect of surface acidity of CuO-SBA-15 on adsorptive desulfurization of fuel oils. *Acta Physico-Chimica Sinica, 28*(06), 1467–1473. https://doi.org/10.3866/PKU.WHXB201203312.

Srivastava, R. K., & Jozewicz, W. (2001). Flue gas desulfurization: The state of the art. *Journal of the Air & Waste Management Association, 51*(12), 1676–1688. https://doi.org/10.1080/10473289.2001.10464387.

Tang, L., Qu, J., Mi, Z., Bo, X., Chang, X., Anadon, L. D., et al. (2019). Substantial emission reductions from Chinese power plants after the introduction of ultra-low emissions standards. *Nature Energy, 4*(11), 929–938. https://doi.org/10.1038/s41560-019-0468-1.

Toutov, A. A., Salata, M., Fedorov, A., Yang, Y.-F., Liang, Y., Cariou, R., et al. (2017). A potassium tert-butoxide and hydrosilane system for ultra-deep desulfurization of fuels. *Nature Energy, 2*(3), 1–7. https://doi.org/10.1038/nenergy.2017.8.

Trueba, M., & Trasatti, S. P. (2005). γ-Alumina as a support for catalysts: A review of fundamental aspects. *European Journal of Inorganic Chemistry, 2005*(17), 3393–3403. https://doi.org/10.1002/ejic.200500348.

Tseng, H.-H., & Wey, M.-Y. (2004). Study of SO_2 adsorption and thermal regeneration over activated carbon-supported copper oxide catalysts. *Carbon, 42*(11), 2269–2278. https://doi.org/10.1016/j.carbon.2004.05.004.

Tseng, H.-H., Wey, M.-Y., & Fu, C.-H. (2003). Carbon materials as catalyst supports for SO_2 oxidation: Catalytic activity of CuO–AC. *Carbon, 41*(1), 139–149. https://doi.org/10.1016/S0008-6223(02)00264-6.

Tseng, H.-H., Wey, M.-Y., Liang, Y.-S., & Chen, K.-H. (2003). Catalytic removal of SO_2, NO and HCl from incineration flue gas over activated carbon-supported metal oxides. *Carbon, 41*(5), 1079–1085. https://doi.org/10.1016/S0008-6223(03)00017-4.

Tsoncheva, T., Issa, G., Blasco, T., Concepcion, P., Dimitrov, M., Hernández, S., et al. (2013). Silica supported copper and cerium oxide catalysts for ethyl acetate oxidation. *Journal of Colloid and Interface Science, 404*, 155–160. https://doi.org/10.1016/j.jcis.2013.05.005.

US EPA, O. (2014). *NAAQS table*. US EPA. https://www.epa.gov/criteria-air-pollutants/naaqs-table.

US EPA, O. (2016). *New source performance standards*. US EPA. https://www.epa.gov/stationary-sources-air-pollution/new-source-performance-standards.

Vanderschuren, J., & Thomas, D. (2010). SO_2 (oxydes de soufre). In *Techniques de l'Ingénieur*. G1800 V1.

Wang, Z.-M., & Lin, Y. S. (1998). Sol–gel-derived alumina-supported copper oxide sorbent for flue gas desulfurization. *Industrial & Engineering Chemistry Research, 37*(12), 4675–4681. https://doi.org/10.1021/ie980343u.

Wang, Y. M., Ying Wu, Z., & Zhu, J. H. (2004). Surface functionalization of SBA-15 by the solvent-free method. *Journal of Solid State Chemistry, 177*(10), 3815–3823. https://doi.org/10.1016/j.jssc.2004.07.013.

Waqif, M., Saur, O., Lavalley, J. C., Perathoner, S., & Centi, G. (1991). Nature and mechanism of formation of sulfate species on copper/alumina sorbent-catalysts for sulfur dioxide removal. *Journal of Physical Chemistry, 95*(10), 4051–4058. https://doi.org/10.1021/j100163a031.

Wey, M. Y., Fu, C. H., Tseng, H. H., & Chen, K. H. (2003). Catalytic oxidization of SO_2 from incineration flue gas over bimetallic Cu–Ce catalysts supported on pre-oxidized activated carbon☆. *Fuel, 82*(18), 2285–2290. https://doi.org/10.1016/S0016-2361(03)00165-0.

Wey, M.-Y., Lu, C.-Y., Tseng, H.-H., & Fu, C.-H. (2002). The utilization of catalyst sorbent in scrubbing acid gases from incineration flue gas. *Journal of the Air & Waste Management Association, 52*(4), 449–458. https://doi.org/10.1080/10473289.2002.10470790.

Wu, Y., Li, R., Cui, L., Meng, Y., Cheng, H., & Fu, H. (2020). The high-resolution estimation of sulfur dioxide (SO_2) concentration, health effect and monetary costs in Beijing. *Chemosphere, 241*. https://doi.org/10.1016/j.chemosphere.2019.125031, 125031.

Xie, G., Liu, Z., Zhu, Z., Liu, Q., & Ma, J. (2003). Reductive regeneration of sulfated CuO/Al_2O_3 catalyst-sorbent in ammonia. *Applied Catalysis B: Environmental, 45*(3), 213–221. https://doi.org/10.1016/S0926-3373(03)00166-8.

Yang, L., Yao, L., Liu, Y., Zhao, X., Jiang, X., & Jiang, W. (2020). Bimetallic and polymetallic oxide modification of activated coke by a one-step blending method for highly efficient SO_2 removal. *Energy & Fuels, 34*(6), 7275–7283. https://doi.org/10.1021/acs.energyfuels.0c00704.

Ye, P., Xia, S., Xiong, Y., Liu, C., Li, F., Liang, J., et al. (2020). Did an ultra-low emissions policy on coal-fueled thermal power reduce the harmful emissions? Evidence from three typical air pollutants abatement in China. *International Journal of Environmental Research and Public Health, 17*(22), 8555. https://doi.org/10.3390/ijerph17228555.

Yoo, K. S., Jeong, S. M., Kim, S. D., & Park, S. B. (1996). Regeneration of sulfated alumina support in $CuO/\gamma\text{-}Al_2O_3$ sorbent by hydrogen. *Industrial & Engineering Chemistry Research, 35*(5), 1543–1549. https://doi.org/10.1021/ie950460e.

Yoo, K. S., Kim, S. D., & Park, S. B. (1994). Sulfation of Al_2O_3 in flue gas desulfurization by $CuO/Gamma\text{-}Al_2O_3$ sorbent. *Industrial & Engineering Chemistry Research, 33*(7), 1786–1791. https://doi.org/10.1021/ie00031a018.

Yu, Q., Zhang, S., & Wang, X. (2007). Thermogravimetric study of $CuO/\gamma\text{-}Al_2O_3$ sorbents for SO_2 in simulated flue gas. *Industrial & Engineering Chemistry Research, 46*(7), 1975–1980. https://doi.org/10.1021/ie0612358.

Yu, Q., Zhang, S., Wang, X., Zhang, J., & Lu, Z. (2008). Sulfation behavior of $CuO/\gamma\text{-}Al_2O_3$ sorbent for the removal of SO_2 from flue gas. *Journal of University of Science and Technology Beijing, Mineral, Metallurgy, Material, 15*(4), 500–504. https://doi.org/10.1016/S1005-8850(08)60094-8.

Zhang, Q.-L., Dong, Y., & Cui, L. (2020). Charcteristics and mechanism study of the $CuO/gamma\text{-}Al_2O_3$ modified by CeO_2 and NaCl. *IOP Conference Series: Materials Science and Engineering, 774*. https://doi.org/10.1088/1757-899X/774/1/012017, 012017.

Zhao, Y., Liu, Z., Jia, Z., & Xing, X. (2007). Elemental sulfur recovery through H_2 regeneration of a SO_2-adsorbed CuO/Al_2O_3. *Industrial & Engineering Chemistry Research, 46*(8), 2661–2664. https://doi.org/10.1021/ie0610041.

Zuo, Y., Yi, H., & Tang, X. (2015). Metal-modified active coke for simultaneous removal of SO_2 and NOx from sintering flue gas. *Energy & Fuels, 29*(1), 377–383. https://doi.org/10.1021/ef502103h.

Larsen, S. C., Aylor, A., Bell, A. T., & Reimer, J. A. (1994). Electron paramagnetic resonance studies of copper ion-exchanged ZSM-5. *Journal of Physical Chemistry, 98*(44), 11533–11540. https://doi.org/10.1021/j100095a039.

Li, G., Wu, Q., Wang, S., Li, J., You, X., Shao, S., et al. (2020). Promoting SO_2 resistance of a $CeO_2(5)\text{-}WO_3(9)/TiO_2$ catalyst for Hg0 oxidation via adjusting the basicity and acidity sites using a CuO doping method. *Environmental Science & Technology, 54*(3), 1889–1897. https://doi.org/10.1021/acs.est.9b04465.

Likens, G. E., & Bormann, F. H. (1974). Acid rain: A serious regional environmental problem. *Science, 184*(4142), 1176–1179. https://doi.org/10.1126/science.184.4142.1176.

Liu, D.-J., & Robota, H. J. (1993). In situ XANES characterization of the Cu oxidation state in Cu-ZSM-5 during NO decomposition catalysis. *Catalysis Letters, 21*(3), 291–301. https://doi.org/10.1007/BF00769481.

Popova, M., Ristić, A., Mazaj, M., Maučec, D., Dimitrov, M., & Tušar, N. N. (2014). Autoreduction of copper on silica and iron-functionalized silica nanoparticles with interparticle mesoporosity. *ChemCatChem, 6*(1), 271–277. https://doi.org/10.1002/cctc.201300463.

Rahmat, N., Abdullah, A. Z., & Mohamed, A. R. (2010). A review: Mesoporous Santa Barbara amorphous-15, types, synthesis and its applications towards biorefinery production. *American Journal of Applied Sciences,*

7(12), 1579–1586. https://doi.org/10.3844/ajassp.2010.1579.1586.

Singh, A., & Agrawal, M. (2008). Acid rain and its ecological consequences. *Journal of Environmental Biology*, *29*(1), 15–24. 18831326.

Wang, Y., Yang, R. T., & Heinzel, J. M. (2009). Desulfurization of jet fuel JP-5 light fraction by MCM-41 and SBA-15 supported cuprous oxide for fuel cell applications. *Industrial & Engineering Chemistry Research*, *48*(1), 142–147. https://doi.org/10.1021/ie800208g.

Yin, Y., Tan, P., Liu, X.-Q., Zhu, J., & Sun, L.-B. (2014). Constructing a confined space in silica nanopores: An ideal platform for the formation and dispersion of cuprous sites. *Journal of Materials Chemistry A*, 2(10), 3399–3406. https://doi.org/10.1039/C3TA14760F.

Zhao, D., Feng, J., Huo, Q., Melosh, N., Fredrickson, G. H., Chmelka, B. F., et al. (1998). Triblock copolymer syntheses of mesoporous silica with periodic 50 to 300 angstrom pores. *Science*, *279*(5350), 548–552. https://doi.org/10.1126/science.279.5350.548.

CHAPTER

14

Solid oxide cells (SOCs) in heterogeneous catalysis

Francisco J.A. Loureiro[a], *Allan J.M. Araújo*[a,b], *Daniel A. Macedo*[c], *Moisés R. Cesário*[c], *and Duncan P. Fagg*[a]

[a]Centre for Mechanical Technology and Automation (TEMA), Department of Mechanical Engineering, University of Aveiro, Aveiro, Portugal [b]Materials Science and Engineering Postgraduate Program—PPGCEM, Federal University of Rio Grande do Norte—UFRN, Natal, Brazil [c]Materials Science and Engineering Postgraduate Program (PPCEM), Federal University of Paraíba (UFPB), João Pessoa, Brazil

14.1 Introduction

The goal of modern scientific research has placed its focus on creating a more advanced and sustainable society, where the development of disruptive new technologies for energy conversion with reduced environmental impact is shown as an urgent need (EU 2030 Climate & Energy Framework, 2021). In this context, one of the main challenges is the reduction of CO_2 emissions by the development of new technologies for industries and processes (Loureiro, Souza, et al., 2019). An exciting and recent technology is that of solid oxide cells (SOCs) that can be used for energy conversion (solid oxide fuel cells, SOFCs), producing electrical energy, or, conversely, for chemical production upon the input of electrical energy (solid oxide electrolysis cells, SOECs) (Loureiro, Araújo, Paskocimas, Macedo, & Fagg, 2021). In this context, SOFCs represent a capable technology to efficiently convert chemical fuels to electricity, while the reverse, i.e., the electrolysis mode of these devices, SOECs, is especially fascinating due to their ability to produce hydrogen or other chemicals (e.g., syngas, etc.) by the input of renewable energy. Indeed, a fast growing investment in both public and private sectors in the electrolysis mode is related to the production of hydrogen using electricity to split water into hydrogen and oxygen. Such electrolyzers can range in size from small, appliance-sized equipment that is well-suited for small-scale distribution of hydrogen production to large-scale, central production facilities that could be tied directly to renewable or other nongreenhouse, gas-emitting forms of electricity production. Further applications of the electrolysis mode are in separation processes, e.g., for product purification, and/or in the enhancement of chemical reactions by selective manipulation of either reactants or product streams.

Heterogeneous Catalysis
https://doi.org/10.1016/B978-0-323-85612-6.00014-0

14.2 Background of separation processes in heterogeneous catalysis

Traditionally, separation processes have been a crucial step in chemical and related industries, such as in the production of pharmaceuticals, fertilizers, cosmetics, and foodstuffs, among others. Separation processes are mainly needed in the purification step of raw materials, in the separation of reaction products, and in the treatment of industrial effluents (Richardson, 1990). The most common processes include distillation, adsorption, and extraction. In this context, a large number of industrially important catalytic processes occur in the temperature range between 200°C and 600°C. For instance, the Haber-Bosch synthesis for ammonia (NH_3) production (450–550°C) involves the reaction of gaseous nitrogen and hydrogen using, typically, a Fe-based catalyst at high pressures (150–300bar) (Ouzounidou, Skodra, Kokkofitis, & Stoukides, 2007; Richardson, 1990). Another example is the synthesis of methanol (200–300°C), currently the second largest use of hydrogen, after that of ammonia synthesis (Struis, Stucki, & Wiedorn, 1996). Methanol is produced by the catalytic conversion of synthesis gas (a mixture of CO and H_2) by Fischer-Tropsch synthesis, an important and flexible synthesis route that can be used to produce a wide variety of fuels and valuable hydrocarbons up to C_5^+ (Ghareghashi, Ghader, & Hashemipour, 2013; Rahimpour, Jokar, & Jamshidnejad, 2012). In all these processes, the desired chemical product must be separated from the overall gas product to produce the desired pure chemical. In this context, direct membrane-based processes have recently become an attractive alternative to conventional separation processes in various chemical applications, due to providing simultaneous chemical separation combined with the potential for electrochemical enhancements (Athanassiou et al., 2007; de Lucas-Consuegra, Gutiérrez-Guerra, Endrino, Serrano-Ruiz, & Valverde, 2015; Kyriakou, Garagounis, Vasileiou, Vourros, & Stoukides, 2017; Rihko-Struckmann, Munder, Chalakov, & Sundmacher, 2010).

Electrochemical membrane reactors incorporate ion-conducting membranes that allow the selective permeation of single ionic species, e.g., H^+, O^{2-}, OH^-, N^{3-}, etc., whilst being impermeable to uncharged reaction species (Fig. 14.1). Therefore, these reactors are electrochemical cells, in which the oxidation and reduction reactions are carried out separately on catalyst/electrodes layers located on the opposite sides of the electrolyte. Thus, these devices are similar to fuel cells in the fact that they contain oxidation and reduction reactions at opposing electrodes. Nonetheless, these devices focus on producing increased yields of useful chemicals rather than electrical output (Athanassiou et al., 2007; Stoukides, 2000). The use of such kind of reactors presents several advantages such as enhanced catalytic activity and selectivity, better process integration, reduced feedstock, and easy reaction rate control (Athanassiou et al., 2007).

One interesting example is the one-step conversion of methane to C_2 compounds (e.g., ethane and ethylene), which has become one of the most challenging problems in catalysis research. In the presence of a variety of catalysts, and in the temperature range from 600°C to 900°C, methane and oxygen react to produce C_2 hydrocarbons, with CO and CO_2 being the major carbon-containing byproducts (Stoukides, 2000). In this context, Yentekakis, Jiang, Makri, and Vayenas (1995) obtained C_2 yields exceeding 85% either by the electrochemical supply of O^{2-} on a (80wt%) Ag—(19wt%) Sm_2O_3+(1wt%) CaO anode or using gaseous O_2 and a (80wt%) Ag—(20wt%) Sm_2O_3 catalyst (Fig. 14.2). These conversions were more than four times higher than those achieved with the best catalysts in the traditional configuration.

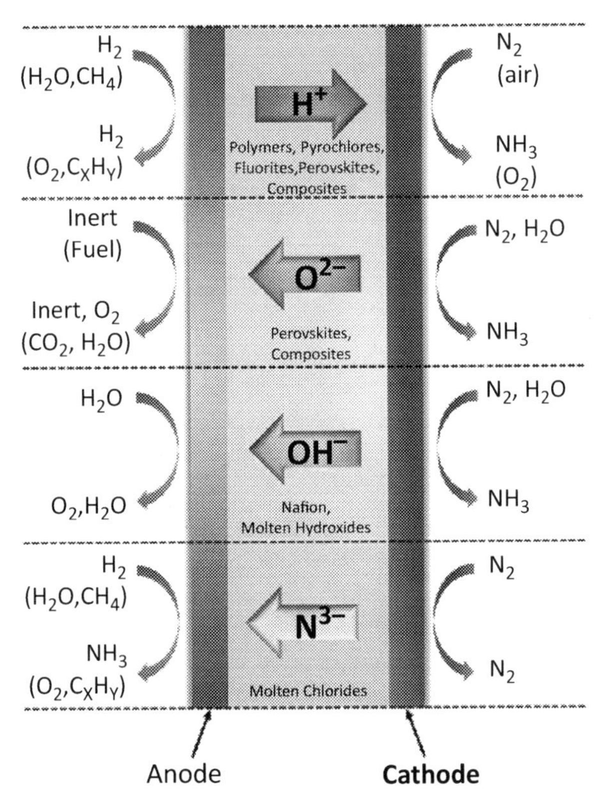

FIG. 14.1 Schematic diagram of several electrochemical reactor configurations. *From Kyriakou, V., Garagounis, I., Vasileiou, E., Vourros, A., & Stoukides, M. (2017). Progress in the electrochemical synthesis of ammonia. Catalysis Today, 286, 2–13. https://doi.org/10.1016/j.cattod.2016.06.014.*

Hibino, Masegi, and Iwahara (1995) studied the electrocatalytic conversion of methane to C_2 using a Li^+ solid electrolyte ($(Li_2O)_{0.17}(BaO)_{0.07}(TiO_2)_{0.76}$) and Au electrodes and an alternating (ac) current. At open-circuit voltage (OCV), the conversion of methane and the selectivity to C_2 hydrocarbons were 3% and 20%, respectively. In contrast, at 3V (ac current), the conversion of methane was twice of that obtained under OCV and the corresponding C_2 selectivity was enhanced up to 30%. This noted improvement was explained to be due to promotion of the chemical reaction caused by the pumping of Li^+ through the electrolyte membrane, whereas at OCV, the reaction only occurs on the surfaces of the electrolyte and/or the two gold electrodes.

A further analysis showed that, as the frequency decreased from 100kHz to 0.1Hz, the selectivity became progressively larger, with peak performances obtained between 10 and 1Hz. Consequently, Li^+ ions started to migrate between the two gold electrodes and to improve the activity and selectivity of the electrode/electrolyte interface.

The discovery and development of high-temperature proton (H^+) conductors in the 1980s and the relatively moderate success with oxygen-ion (O^{2-}) cells directed several research groups toward a totally different route based on

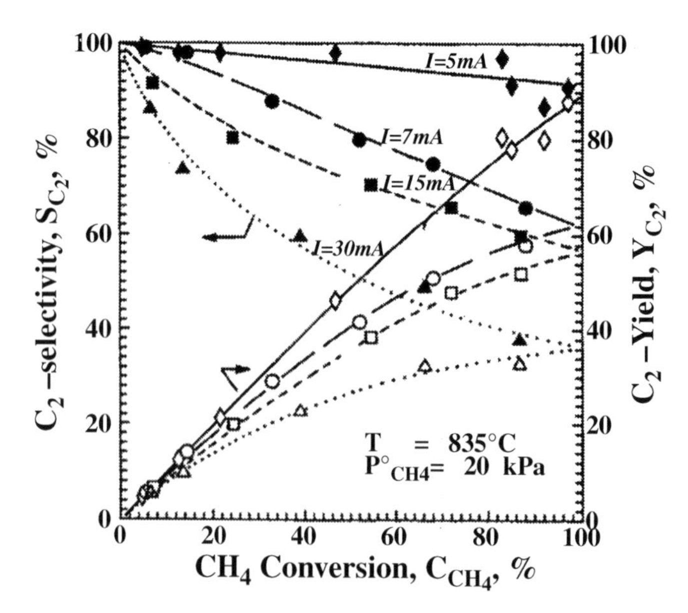

FIG. 14.2 Effect of methane conversion and applied current on C_2 hydrocarbon selectivity *(filled symbols)* and yield *(open symbols)*. The cell reactor consists of an Y_2O_3 (8 mol%)-stabilized ZrO_2 (YSZ) tube that is closed flat at one end with an appropriately machined, water-cooled stainless steel reactor cap attached to the other end, thus allowing for continuous gas feed and removal. The catalyst electrode was a porous Ag-Sm_2O_3 film coated on the inside walls of the YSZ tube. A Ag counter electrode was deposited on the outer walls of the YSZ tube. *From Yentekakis, I. V., Jiang, Y., Makri, M., & Vayenas, C. G. (1995). Ethylene production from methane in a gas recycle electrocatalytic reactor separator. Ionics, 1(4), 286–291. https://doi.org/10.1007/BF02390209.*

methane dimerization via dehydrogenation rather than partial oxidation, as the undesirable oxygenates (CO, CO_2) could be avoided (Stoukides, 2000). As an example of a proton-conducting membrane, Morejudo et al. (2016) reported nonoxidative methane dehydroaromatization (MDA) using Mo/zeolite catalysts. The authors compared this reaction to a standard fluidized bed reactor (FBR) with the integration of an electrochemical $BaZrO_3$-based membrane exhibiting both proton and oxide-ion conductivity (ceramic membrane reactor, CMR). In the FBR, the aromatics' yield initially increased during the induction period, reaching a maximum of ~10%, but rapidly fell as the reaction progressed. However, in the case of the electrochemical membrane reactor, the aromatics' yield continued to increase beyond the induction period and attained a maximum of ~12%, after which the catalyst activity started to decline. This enhancement was related to the catalyst stability. The high stability exhibited by the catalyst in the CMR was caused by a decreased tendency to form coke, which became more evident for longer reaction times.

In another perspective, growing concerns over the acceleration of global warming and climate changes due to massive carbon dioxide emissions into the atmosphere arising from the production and consumption of fossil fuels have promoted interest in the production of synthetic hydrocarbon fuels. Xie, Zhang, Meng, and Irvine (2011) reported the direct synthesis of methane from CO_2/H_2O in an oxygen-ion-conducting solid-state electrochemical cell $(La_{0.8}Sr_{0.2})_{0.95}MnO_{3-\delta}$-(anode)|| YSZ-(electrolyte)||$La_{0.2}Sr_{0.8}TiO_{3.1}$-(cathode) by combining coelectrolysis of CO_2/H_2O and in situ Fischer-Tropsch-type synthesis at 650°C, where the current density reached ~85 mA cm^{-2} with an applied electrical voltage of 2V. Methane was synthesized with a faradaic yield of 2.8% in the solid electrolyzer, while the main product was that of CO/H_2 with a faradaic yield of 74.2%. The higher conversion of H_2O (25%) compared to CO_2 (11.5%), from a CO_2 excess mix, demonstrated that steam electrolysis is easier than CO_2 electrolysis in this device. In both processes, the conversion was limited by the performance of the catalyst used, emphasizing the need for catalyst development.

This coelectrolysis of CO_2 and steam has generated special interest as it can provide a route to store the energy produced by intermittent sources of renewable energy (like wind and solar energy), while using stranded sources of CO_2 and waste heat (e.g., from nuclear plants). Studies of this process using a proton conductor as an electrolyzer are less frequent in the literature. As an example, Ruiz-Trejo and Irvine (2012) used a protonic electrolyte based on $BaCe_{0.5}Zr_{0.3}Y_{0.16}Zn_{0.04}O_{2.88}$ with Pt electrodes (Fig. 14.3). CO was successfully produced when CO_2 was fed on the cathode side. A higher CO conversion was reached with a high potential difference (~9V) across the cell. A faradaic efficiency of ~9% was estimated, with the losses being associated with the extremely high potential difference applied to drive significant current due to internal resistances. Cell performance was limited by elevated polarization resistances from the electrodes, due to the present lack of proper materials for proton-conducting devices to operate at temperatures between 400°C and 650°C (Ruiz-Trejo & Irvine, 2012).

As a final example, ammonia is the second most produced chemical in the world, with a high societal impact due to its extensive application in agriculture and pharmaceuticals. Nonetheless, its current production from natural gas accounts for approximately 5% of global CH_4 consumption, with associated high levels of CO_2 emissions. In the work of Yun et al. (2015), ammonia was electrochemically synthesized from N_2 and H_2O at 475–600°C, using a $BaZr_{0.8}Y_{0.2}O_{3-\delta}$, proton conductor electrolyte (Fig. 14.4). Silver (Ag), platinum (Pt), and lanthanum strontium cobalt ferrite ($La_{0.6}Sr_{0.4}Co_{0.2}Fe_{0.8}O_{3-\delta}$ (LSCF6428)) were used as electrode electrocatalysts. Maximum ammonia formation rates of 4.9×10^{-11} and 8.5×10^{-11}-$mol\,cm^{-2}\,s^{-1}$ were observed for the Ag and LSCF6428 electrocatalysts, respectively, under 0.8V. However, the use of the Pt electrocatalyst showed a negligible ammonia formation rate

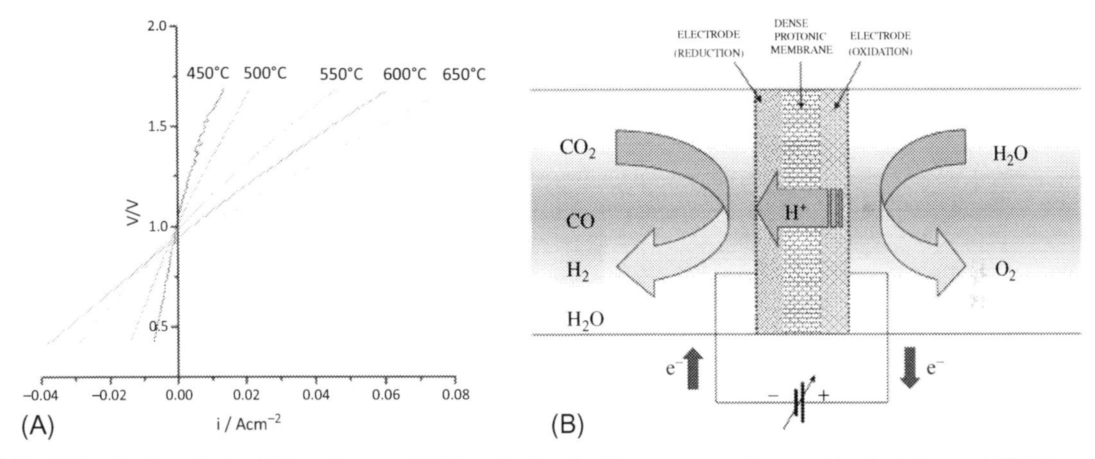

FIG. 14.3 Coelectrolysis of CO_2 and steam. (A) Electrolysis cell with a proton conductor as the electrolyte and (B) *I-V* curves of the Pt||$BaCe_{0.5}Zr_{0.3}Y_{0.16}Zn_{0.04}O_{2.88}$||Pt cell in fuel cell and electrolysis mode. *From Ruiz-Trejo, E., & Irvine, J. T. S. (2012). Solid State Ionics, 216, 36–40. https://doi.org/10.1016/j.ssi.2012.01.033.*

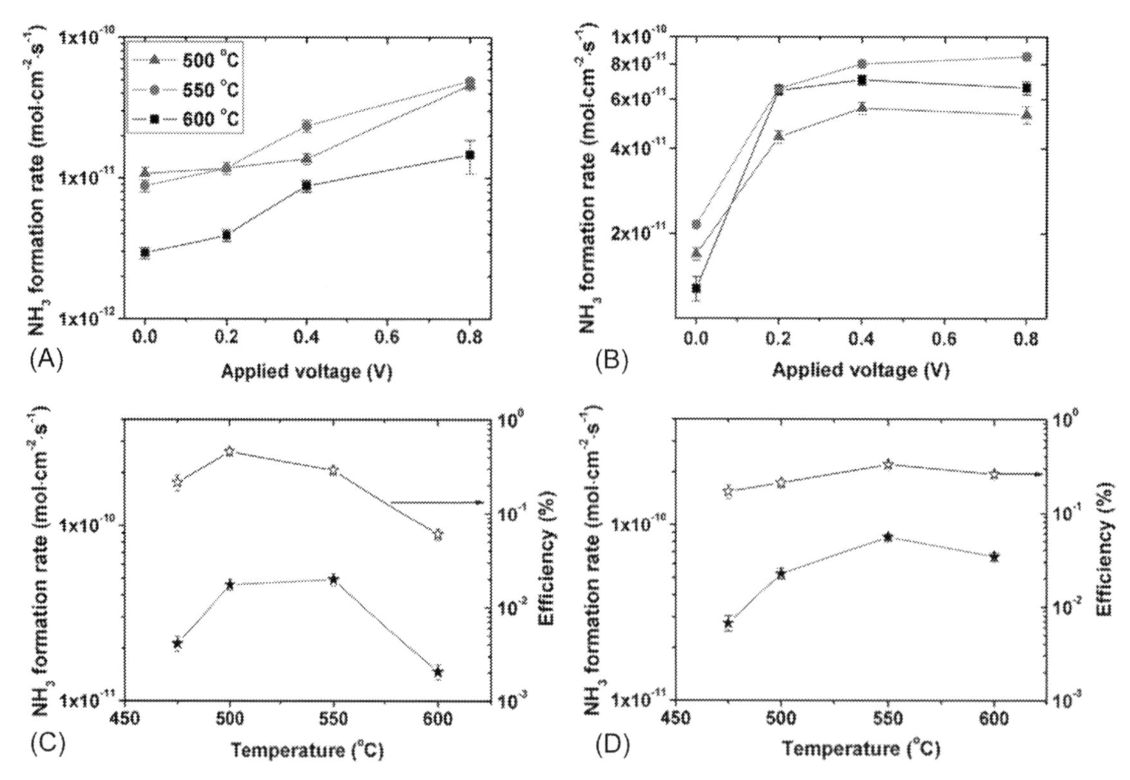

FIG. 14.4 NH_3 formation rate as a function of temperature and applied voltage: (A) Ag; (B) LSCF6428 and the maximum NH_3 formation rate and faradaic efficiency with different temperatures at an applied potential difference of 0.8 V; (C) Ag; and (D) LSCF. Measurements were made on single cells based on $BaZr_{0.8}Y_{0.2}O_{3-\delta}$ in the temperature range of 475–600°C, at atmospheric pressure. Silver (Ag), platinum (Pt), and LSCF6428 were used for both anode and cathode electrocatalysts. *From Yun, D. S., Joo, J. H., Yu, J. H., Yoon, H. C., Kim, J. N., & Yoo, C. Y. (2015). Electrochemical ammonia synthesis from steam and nitrogen using proton conducting yttrium doped barium zirconate electrolyte with silver, platinum, and lanthanum strontium cobalt ferrite electrocatalyst. Journal of Power Sources, 284, 245–251. https://doi.org/10.1016/j.jpowsour.2015.03.002.*

lower than $1 \times 10^{-12} \mathrm{mol\,cm^{-2}\,s^{-1}}$, which was related to the high activity of Pt for the hydrogen evolution reaction rather than for the ammonia formation reaction.

14.3 Electrode materials

The development of electrode materials for chemical production processes that utilize electrochemical ceramic membranes has not received any specific study to date. For this reason, this section reviews the current knowledge of electrode materials that have been designed for SOCs to provide a baseline from which more dedicated chemical production electrodes can be developed.

Electrode materials must offer tolerable chemical and structural stability during cell operation at high temperatures, suitable conductivity (electronic or mixed electronic and ionic), proper percolation pathways of each conducting species, no interdiffusion of elements between the various cell components, similar thermal expansion coefficients to the remaining cell components to avoid cracking during cell fabrication and operation, and sufficient porosity to allow gas transport to and from the electrode reaction sites. Moreover, the component materials must be of low cost, easy fabrication, and should not affect the sequential fabrication processing upon addition of further cell components (Nasani, Loureiro, & Fagg, 2017). The typical ceramic electrode materials used for SOCs are described later.

Electrode materials for oxidizing conditions

Mixed ionic-electronic conductors (MIECs), e.g., perovskites, are the most widely used electrode materials due to their stability under high oxygen partial pressure (pO_2), high conductivities, and tailorable catalytic activity when containing cations of variable valences (Loureiro, Nasani, Reddy, Munirathnam, & Fagg, 2019). The first oxygen electrode materials were composites made of $(ZrO_2)_{0.92}(Y_2O_3)_{0.08}$, YSZ, which is a traditional oxygen-ion conductor, and $La_{1-x}Sr_xMnO_{3-\delta}$, LSM, which is an electronic conductor, being targeted to operate at high temperatures, such as 1000°C (Sun, Hui, & Roller, 2010). Single-phase MIECs materials have later been reported, including $La_{0.6}Sr_{0.4}Co_{0.8}Fe_{0.2}O_{3-\delta}$ (LSCF), $Sm_{0.5}Sr_{0.5}Co_{0.5}O_{3-\delta}$ (SSC), and $Ba_{0.5}Sr_{0.5}Co_{0.2}Fe_{0.8}O_{3-\delta}$ (BSCF) (Shao & Halle, 2004). Among these compositions, LSCF is one of the most widely used, due to its low cost and good performance, offering high electronic (230 S cm^{-1} at 900°C) and ionic conductivities (0.2 S cm^{-1} at 900°C) (Loureiro, Macedo, et al., 2019). Moreover, composite compositions were also proposed for these MIECs in combination with oxygen-ion–conducting materials, including $(ZrO_2)_x(Sc_2O_3)_{1-x}$(ScSZ), $La_{1-x}Sr_xGa_{1-y}Mg_yO_{3-(x+y)/2}$ (LSGM), or $(CeO_2)_{1-x}(GdO_{1.5})_x$(CGO) (Loureiro, Yang, Stroppa, & Fagg, 2015).

Layered perovskites have also been proposed as suitable electrodes capable of exhibiting a mixed-conducting nature, e.g., phases with the generic formula of $(ABO_3)_n$ AO ($n=1$, 2, and 3), such as lanthanum nickelates (Loureiro, Nasani, et al., 2019). Their unique structure allows them to exhibit increasing electronic conductivity with decreasing temperature, potentially privileging these materials to operate as electrodes at intermediate temperatures (Amow, Davidson, & Skinner, 2006; Loureiro, Ramasamy, et al., 2020; Loureiro, Silva, et al., 2020). For instance, this material allows an increase in conductivity with decreasing temperature for $n=3$ (Fig. 14.5).

Another example includes calcium cobaltites, where the most widely studied example is that of the $[Ca_2CoO_3]_q[CoO_2]$ composition (Fulgêncio et al., 2019; Loureiro, Silva, et al., 2019; Loureiro et al., 2021; Santos et al., 2018; Silva et al., 2018). The layered structure of this composition (Fig. 14.6) is based on alternated layers of a CdI_2-type $[CoO_2]$ triangular lattice, which provides oxygen-ion conductivity, and a rock salt-type $[Ca_2CoO_3]$ block, which provides electronic conductivity, due to the mixed valence of the cobalt ions ($\sigma=100$ S cm^{-1} at 700°C) (Amow et al., 2006; Loureiro, Ramasamy, et al., 2020; Loureiro, Silva, et al., 2020).

With respect to electrodes used for proton-conducting SOCs, the most widely studied materials are those of the rare earth-doped nickelates (R_2NiO_4, R = Pr, Nd) (Dailly et al., 2010), $Ba(Pr_{1-x}Gd_x)O_{3-\delta}$, $PrBaCuFeO_{5+x}$. However, mirroring that of their oxygen-ion-conducting counterparts, composite cathodes with ionically conducting phases are also reported, including $La_{0.6}Sr_{0.4}Co_{0.2}Fe_{0.8}O_{3-\delta}/Ba(Zr_{0.1}Ce_{0.7}Y_{0.2})O_{2.9}$, $Sm_{0.5}Sr_{0.5}CoO_{2.25}/BaCe_{0.8}Sm_{0.2}O_{2.9}$, and $Ba_{0.5}Sr_{0.5}Co_{0.8}Fe_{0.2}O_{3-\delta}/BaCe_{0.8}Sm_{0.2}O_{2.9}$ (Fabbri, Pergolesi, & Traversa, 2010; Peng, Wu, Liu, Liu, & Meng, 2010). Recently, $La_4Ni_3O_{10\pm\delta}$ electrodes in combination with a proton-conducting phase of $BaCe_{0.9}Y_{0.1}O_{3-\delta}$ cathodes have been proposed, providing an improvement

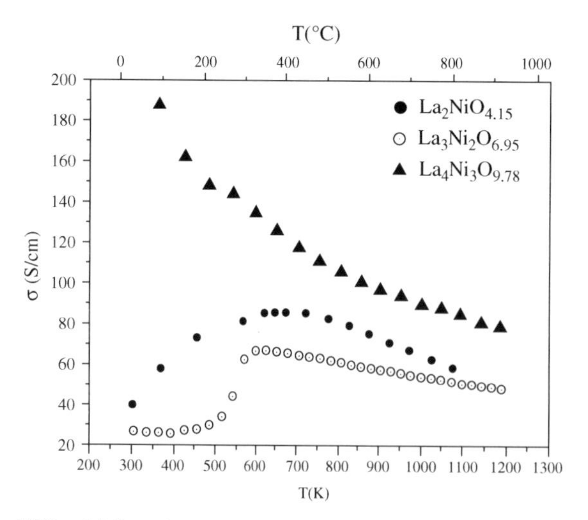

FIG. 14.5 Electrical conductivity vs temperature for $La_2NiO_{4.15}$, $La_3Ni_2O_{6.95}$, and $La_4Ni_3O_{9.78}$ from RT to 1173K in air. *From Amow, G., Davidson, I. J., & Skinner, S. J. (2006). A comparative study of the Ruddlesden-Popper series, $La_{n+1}Ni_nO_{3n+1}$ (n=1, 2 and 3), for solid-oxide fuel-cell cathode applications. Solid State Ionics, 177(13–14), 1205–1210. https://doi.org/10.1016/j.ssi.2006.05.005.*

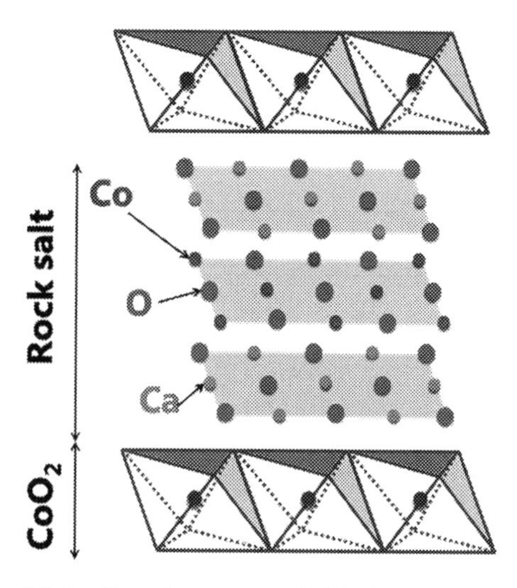

FIG. 14.6 Crystal structure of $[Ca_2CoO_3]q[CoO_2]$ seen along the *a*-axis. A rational approximation of 3 units of CoO_2 and 2 units of Ca_2CoO_3 is shown. *From Loureiro, F., Silva, V., Simões, T., Cesário, M., Grilo, J., Fagg, D., & Macedo, D. (2020). Misfit-layered Ca-cobaltite–based cathodes for intermediate-temperature solid oxide fuel cell. In G. Kaur (Ed.), Intermediate temperature solid oxide fuel cells (Issue 6, pp. 347–377). Elsevier. https://doi.org/10.1016/B978-0-12-817445-6.00006-5.*

of the total polarization resistance by a factor of 1.5 in comparison to that of the pure electrode (Amow et al., 2006; Loureiro, Ramasamy, et al., 2020; Loureiro, Silva, et al., 2020).

Electrode materials for reducing conditions

The most commonly used electrode materials for reducing conditions are those of a cermet configuration, consisting of a metallic (Ni, Cu, Pt, Ag, Co, Fe) phase with a ceramic phase, where the latter phase has either been that of an oxygen ion or a proton conductor. Of these, nickel cermet has been the most widely used electrode due to its extremely high catalytic activity (Loureiro, Nasani, et al., 2019). Here, the use of a ceramic phase permits to minimize the coarsening of the grains of the metallic phase that otherwise can increase with increasing temperature, leading to performance degradation. Further benefits include better matching of the thermal expansion coefficient (TEC) of the cermet electrode with the electrolyte, thus avoiding unwanted delamination, and also an increase in the number of reaction sites, also known as the three-phase boundary (TPB), between the gas and the ionic and electronic conducting phases where the electrode reaction can occur (Narendar, Mather, Dias, & Fagg, 2013). All these factors are highly important both for increasing longevity and reducing polarization losses.

The relative fraction of each phase can also lead to important changes in the total polarization resistance of cermet electrodes (Narendar et al., 2013). In this context, Narendar et al. (2013) studied the temperature dependence of the bulk conductivity of a Ni-BZY anode cermet as a function of different Ni-to-BZY ratios. They found that the dominant conduction mechanism is widely dependent on the compositional ratio, where ionic transport is found to dominate for low Ni contents (activation energy of ~0.58eV, suggesting that mostly protons are involved (Loureiro, Nasani, et al., 2019)), while further increases in Ni fraction (e.g., ≥30 vol%) lead to metallic behavior (conductivity decreases with increasing temperature), with percolation shown to occur for volume ratios of 40 vol% (Narendar et al., 2013).

Nonetheless, to achieve a functional device, further challenges need to be addressed with respect to the selection and tailoring of electrode materials. The typical Ni-based cermet electrodes are highly susceptible to both carbon deposition and poisoning by H_2S; a contaminant that exists in natural gas (Ghareghashi et al., 2013; Stoukides, 2000). CH_4 dissociation is highly prone to carbon deposition, leading to catalyst deactivation and subsequent decrease in electrical conductivity at the anode (Morejudo et al., 2016). In this context, several other nonmetal catalysts have been proposed, attempting to improve the stability of the electrocatalyst, including nitrides, carbides, and borides (Cheng, Zha, & Liu, 2006).

DFT calculations have highlighted that transition-metal carbides (TMCs) can potentially substitute the traditional Ni catalysts due to their expected high catalytic activity for CH_4 decomposition (Torabi & Etsell, 2012; Zhang, Yang, & Ge, 2020). Here, TMCs are suggested to offer benefits over pure transition metals since carbon atoms are already incorporated into their structure. For electrochemical application as potential anodes, these materials may offer another critical advantage, as they also offer high electrical conductivity (Cheng et al., 2006). TMCs also have a different electronic structure compared to the parent metal, showing a predicted strength of adsorption of reactants and products that is comparable to noble metals, potentially improving reaction selectivity (Zhang et al., 2020). The most promising candidates in terms of stability, activity, and selectivity for the dissociation of CH_4 and C–C coupling of the resultant intermediates at ambient pressure were found to be MoC and WC (Zhang et al., 2020). However, experimental reports of the practical use of these materials in SOC applications are scarce. One example is the work of Torabi and Etsell (2012), who studied new electrode materials based on WC-YSZ composites as potential anodes for direct methane SOFCs. Fig. 14.7 depicts the *I-V* characteristics and power densities of a cell with the ceria-Ru-promoted WC-infiltrated porous YSZ anode. Nonetheless, the achieved

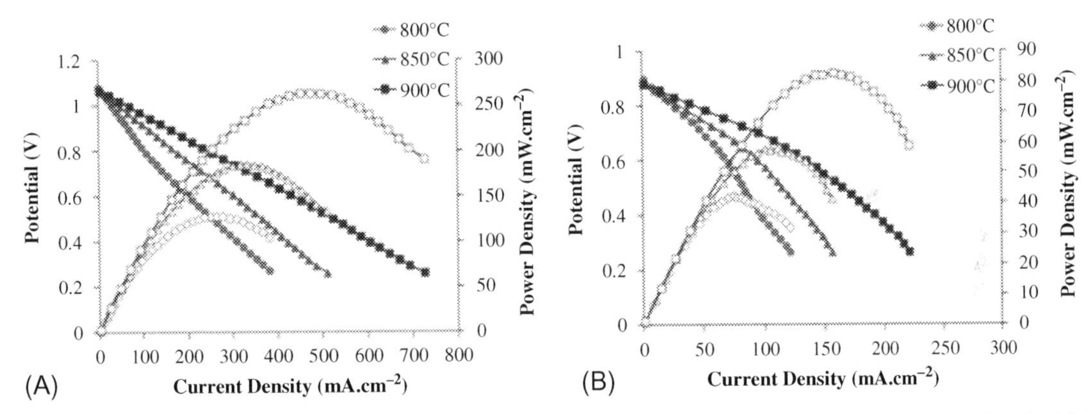

FIG. 14.7 Anode performance of transition-metal carbides (TMCs). *I-V* characteristics and power densities of a cell with the ceria-Ru promoted WC-infiltrated porous YSZ anode under: (A) humidified H_2–CH_4 mixed atmosphere and (B) humidified CH_4. *From Torabi, A., & Etsell, T. H. (2012). Tungsten carbide-based anodes for solid oxide fuel cells: Preparation, performance and challenges. Journal of Power Sources, 212, 47–56. https://doi.org/10.1016/j.jpowsour.2012.03.102.*

performances, to date, are still far below those reported by traditional cermet anodes, underscoring that further efforts are still necessary to optimize these electrodes.

14.4 Conclusions

The current evolution of applications involving SOCs toward chemical production is gaining momentum. Several studies have now been conducted providing proof of concept for the application of SOCs in a wide range of chemical reactions. However, despite the demonstrated feasibility of these processes, this research line still faces challenges with respect to attaining adequate conversion and sufficiently high faradaic efficiencies. This is in part due to the current lack of proper electrocatalysts for this application. This limitation is possibly compounded by the fact that the majority of cell configurations, to date, have relied on mirroring typical SOFC electrode materials rather than specifically tailoring new electrode materials for the goal of chemical production. Thus, although the technology has been proven to be interesting and promising, it is still in its infancy stage and more studies on alternative electrode materials are necessary to leverage the application to an industrially relevant level.

Acknowledgments

Francisco J.A. Loureiro and Duncan P. Fagg acknowledge the projects PTDC/CTM-CTM/2156/2020, PTDC/QUI-ELT/3681/2020, POCI-01-0247-FEDER-039926, POCI-01-0145-FEDER-032241, UIDB/00481/2020 and UIDP/00481/2020, and CENTRO-01-0145-FEDER-022083—Centro Portugal Regional Operational Programme (Centro2020), under the PORTUGAL 2020 Partnership Agreement, through the European Regional Development Fund (ERDF). Francisco J.A. Loureiro is thankful for the Investigator Grant CEECIND/02797/2020. This study was also financed in part by the Coordenação de Aperfeiçoamento de Pessoal de Nível Superior—Brasil (CAPES)—Finance Code 001. Allan J.M. Araújo thanks the Conselho Nacional de Desenvolvimento Científico e Tecnológico (CNPq/Brazil, reference number 200439/2019-7).

References

Amow, G., Davidson, I. J., & Skinner, S. J. (2006). A comparative study of the Ruddlesden-Popper series, $La_{n+1}Ni_nO_{3n+1}$ (n = 1, 2 and 3), for solid-oxide fuel-cell cathode applications. *Solid State Ionics, 177*(13–14), 1205–1210. https://doi.org/10.1016/j.ssi.2006.05.005.

Athanassiou, C., Pekridis, G., Kaklidis, N., Kalimeri, K., Vartzoka, S., & Marnellos, G. (2007). Hydrogen production in solid electrolyte membrane reactors (SEMRs). *International Journal of Hydrogen Energy, 32*(1), 38–54. https://doi.org/10.1016/j.ijhydene.2006.06.031.

Cheng, Z., Zha, S., & Liu, M. (2006). Stability of materials as candidates for sulfur-resistant anodes of solid oxide fuel cells. *Journal of the Electrochemical Society, 153*(7), A1302–A1309. https://doi.org/10.1149/1.2198107.

Dailly, J., Fourcade, S., Largeteau, A., Mauvy, F., Grenier, J. C., & Marrony, M. (2010). Perovskite and A2MO4-type oxides as new cathode materials for protonic solid oxide fuel cells. *Electrochimica Acta, 55*(20), 5847–5853. https://doi.org/10.1016/j.electacta.2010.05.034.

de Lucas-Consuegra, A., Gutiérrez-Guerra, N., Endrino, J. L., Serrano-Ruiz, J. C., & Valverde, J. L. (2015). Direct production of flexible H_2/CO synthesis gas in a solid electrolyte membrane reactor. *Journal of Solid State Electrochemistry, 19*(10), 2991–2999. https://doi.org/10.1007/s10008-015-2922-8.

(2021). *EU 2030 Climate & Energy Framework*. (2021). Retrieved 28 March 2021, from https://www.un.org/sustainabledevelopment/sustainable-development-goals/.

Fabbri, E., Pergolesi, D., & Traversa, E. (2010). Materials challenges toward proton-conducting oxide fuel cells: A critical review. *Chemical Society Reviews, 39*(11), 4355–4369. https://doi.org/10.1039/b902343g.

Fulgêncio, E. B. G. A., Loureiro, F. J. A., Melo, K. P. V., Silva, R. M., Fagg, D. P., Campos, L. F. A., et al. (2019). Boosting the oxygen reduction reaction of the misfit $[Ca_2CoO_{3-\delta}]_q[CoO_2]$ (C349) by the addition of praseodymium oxide. *Journal of Alloys and Compounds, 788*, 148–154. https://doi.org/10.1016/j.jallcom.2019.02.209.

Ghareghashi, A., Ghader, S., & Hashemipour, H. (2013). Theoretical analysis of oxidative coupling of methane and Fischer Tropsch synthesis in two consecutive reactors: Comparison of fixed bed and membrane reactor. *Journal of Industrial and Engineering Chemistry, 19*(6), 1811–1826. https://doi.org/10.1016/j.jiec.2013.02.025.

Hibino, T., Masegi, A., & Iwahara, H. (1995). Electrocatalytic oxidation of methane to C2 hydrocarbons using an Li+ ion conductor. *Journal of the Electrochemical Society, 142*(5), L72–L73. https://doi.org/10.1149/1.2048653.

Kyriakou, V., Garagounis, I., Vasileiou, E., Vourros, A., & Stoukides, M. (2017). Progress in the electrochemical synthesis of ammonia. *Catalysis Today*, *286*, 2–13. https://doi.org/10.1016/j.cattod.2016.06.014.

Loureiro, F. J. A., Araújo, A. J. M., Paskocimas, C. A., Macedo, D. A., & Fagg, D. P. (2021). Polarisation mechanism of the misfit Ca-cobaltite electrode for reversible solid oxide cells. *Electrochimica Acta*, *373*. https://doi.org/10.1016/j.electacta.2021.137928.

Loureiro, F. J. A., Macedo, D. A., Nascimento, R. M., Cesário, M. R., Grilo, J. P. F., Yaremchenko, A. A., et al. (2019). Cathodic polarisation of composite LSCF-SDC IT-SOFC electrode synthesised by one-step microwave self-assisted combustion. *Journal of the European Ceramic Society*, *39*(5), 1846–1853. https://doi.org/10.1016/j.jeurceramsoc.2019.01.013.

Loureiro, F. J. A., Nasani, N., Reddy, G. S., Munirathnam, N. R., & Fagg, D. P. (2019). A review on sintering technology of proton conducting $BaCeO_3$-$BaZrO_3$ perovskite oxide materials for Protonic Ceramic Fuel Cells. *Journal of Power Sources*, *438*. https://doi.org/10.1016/j.jpowsour.2019.226991.

Loureiro, F. J. A., Ramasamy, D., Mikhalev, S. M., Shaula, A., Macedo, D. A., & Fagg, D. P. (2020). $La_4Ni_3O_{10 \pm d}$-$BaCe_{0.9}Y_{0.1}O_{3-d}$ cathodes for Proton Ceramic Fuel Cells; short-circuiting analysis using $BaCe_{0.9}Y_{0.1}O_{3-d}$ symmetric cells. *International Journal of Hydrogen Energy*, *46*, 13594–13605.

Loureiro, F. J. A., Silva, V. D., Simões, T. A., Cesário, M. R., Grilo, J. P. F., Fagg, D. P., et al. (2019). Misfit-layered Ca-cobaltite-based cathodes for intermediate-temperature solid oxide fuel cell. In *Intermediate temperature solid oxide fuel cells: Electrolytes, electrodes and interconnects* (pp. 347–377). Elsevier Inc. https://doi.org/10.1016/B978-0-12-817445-6.00006-5.

Loureiro, F., Silva, V., Simões, T., Cesário, M., Grilo, J., Fagg, D., et al. (2020). Misfit-layered Ca-cobaltite–based cathodes for intermediate-temperature solid oxide fuel cell. In G. Kaur (Ed.), *Intermediate temperature solid oxide fuel cells* (pp. 347–377). Elsevier. https://doi.org/10.1016/B978-0-12-817445-6.00006-5. Issue 6.

Loureiro, F. J. A., Souza, G. S., Graça, V. C. D., Araújo, A. J. M., Grilo, J. P. F., Macedo, D. A., et al. (2019). Nickel-copper based anodes for solid oxide fuel cells running on hydrogen and biogas: Study using ceria-based electrolytes with electronic short-circuiting correction. *Journal of Power Sources*, *438*, 227041. https://doi.org/10.1016/j.jpowsour.2019.227041.

Loureiro, F. J. A., Yang, T., Stroppa, D. G., & Fagg, D. P. (2015). $Pr_2O_2SO_4$-$La_{0.6}Sr_{0.4}Co_{0.2}Fe_{0.8}O_{3-\delta}$: A new category of composite cathode for intermediate temperature-solid oxide fuel cells. *Journal of Materials Chemistry A*, *3*(24), 12636–12641. https://doi.org/10.1039/c4ta06640e.

Morejudo, S. H., Zanón, R., Escolástico, S., Yuste-Tirados, I., Malerød-Fjeld, H., Vestre, P. K., et al. (2016). Direct conversion of methane to aromatics in a catalytic co-ionic membrane reactor. *Science*, *353*(6299), 563–566. https://doi.org/10.1126/science.aag0274.

Narendar, N., Mather, G. C., Dias, P. A. N., & Fagg, D. P. (2013). The importance of phase purity in Ni-$BaZr_{0.85}Y_{0.15}O_{3-\delta}$ cermet anodes—Novel nitrate-free combustion route and electrochemical study. *RSC Advances*, *3*(3), 859–869. https://doi.org/10.1039/c2ra22301e.

Nasani, N., Loureiro, F., & Fagg, D. P. (2017). Proton conducting ceramic materials for intermediate temperature solid oxide fuel cells. In M. R. Cesário, & D. A. de Macedo (Eds.), *Frontiers in ceramic science: Vol. 1. Functional materials for solid oxide fuel cells: Processing, microstructure and performance* (p. 131). Bentham Science Publisher. https://doi.org/10.2174/97816810843121170100012.

Ouzounidou, M., Skodra, A., Kokkofitis, C., & Stoukides, M. (2007). Catalytic and electrocatalytic synthesis of NH_3 in a H+ conducting cell by using an industrial Fe catalyst. *Solid State Ionics*, *178*(1–2), 153–159. https://doi.org/10.1016/j.ssi.2006.11.019.

Peng, R., Wu, T., Liu, W., Liu, X., & Meng, G. (2010). Cathode processes and materials for solid oxide fuel cells with proton conductors as electrolytes. *Journal of Materials Chemistry*, *20*(30), 6218–6225. https://doi.org/10.1039/c0jm00350f.

Rahimpour, M. R., Jokar, S. M., & Jamshidnejad, Z. (2012). A novel slurry bubble column membrane reactor concept for Fischer-Tropsch synthesis in GTL technology. *Chemical Engineering Research and Design*, *90*(3), 383–396. https://doi.org/10.1016/j.cherd.2011.07.014.

Richardson, J. F. (1990). Separation processes. *Gas Separation and Purification*, *4*(1), 2–7. https://doi.org/10.1016/0950-4214(90)80022-D.

Rihko-Struckmann, L., Munder, B., Chalakov, L., & Sundmacher, K. (2010). Solid electrolyte membrane reactors. In *Membrane reactors: Distributing reactants to improve selectivity and yield* (pp. 193–233). Wiley-VCH. https://doi.org/10.1002/9783527629725.ch7.

Ruiz-Trejo, E., & Irvine, J. T. S. (2012). Ceramic proton conducting membranes for the electrochemical production of syngas. *Solid State Ionics*, *216*, 36–40. https://doi.org/10.1016/j.ssi.2012.01.033.

Santos, J. R. D., Loureiro, F. J. A., Grilo, J. P. F., Silva, V. D., Simões, T. A., Fagg, D. P., et al. (2018). Understanding the cathodic polarisation behaviour of the misfit $[Ca_2CoO_{3-\delta}]_q[CoO_2]$ (C349) as oxygen electrode for IT-SOFC. *Electrochimica Acta*, *285*, 214–220. https://doi.org/10.1016/j.electacta.2018.08.018.

Shao, Z., & Halle, S. M. (2004). A high-performance cathode for the next generation of solid-oxide fuel cells. *Nature*, *431*(7005), 170–173. https://doi.org/10.1038/nature02863.

Silva, V. D., Silva, R. M., Grilo, J. P. F., Loureiro, F. J. A., Fagg, D. P., Medeiros, E. S., et al. (2018). Electrochemical assessment of novel misfit Ca-cobaltite-based composite SOFC cathodes synthesized by solution blow spinning.

Journal of the European Ceramic Society, *38*(6), 2562–2569. https://doi.org/10.1016/j.jeurceramsoc.2018.01.044.

Stoukides, M. (2000). Solid-electrolyte membrane reactors: Current experience and future outlook. *Catalysis Reviews – Science and Engineering*, *42*(1–2), 1–70. https://doi.org/10.1081/CR-100100259.

Struis, R., Stucki, & Wiedorn, M. (1996). A membrane reactor for methanol synthesis. *Journal of Membrane Science*, *113*, 222–227. https://doi.org/10.1016/0376-7388(95)00222-7.

Sun, C., Hui, R., & Roller, J. (2010). Cathode materials for solid oxide fuel cells: A review. *Journal of Solid State Electrochemistry*, *14*(7), 1125–1144. https://doi.org/10.1007/s10008-009-0932-0.

Torabi, A., & Etsell, T. H. (2012). Tungsten carbide-based anodes for solid oxide fuel cells: Preparation, performance and challenges. *Journal of Power Sources*, *212*, 47–56. https://doi.org/10.1016/j.jpowsour.2012.03.102.

Xie, K., Zhang, Y., Meng, G., & Irvine, J. T. S. (2011). Direct synthesis of methane from CO_2/H_2O in an oxygen-ion conducting solid oxide electrolyser. *Energy and Environmental Science*, *4*(6), 2218–2222. https://doi.org/10.1039/c1ee01035b.

Yentekakis, I. V., Jiang, Y., Makri, M., & Vayenas, C. G. (1995). Ethylene production from methane in a gas recycle electrocatalytic reactor separator. *Ionics*, *1*(4), 286–291. https://doi.org/10.1007/BF02390209.

Yun, D. S., Joo, J. H., Yu, J. H., Yoon, H. C., Kim, J. N., & Yoo, C. Y. (2015). Electrochemical ammonia synthesis from steam and nitrogen using proton conducting yttrium doped barium zirconate electrolyte with silver, platinum, and lanthanum strontium cobalt ferrite electrocatalyst. *Journal of Power Sources*, *284*, 245–251. https://doi.org/10.1016/j.jpowsour.2015.03.002.

Zhang, T., Yang, X., & Ge, Q. (2020). CH4 dissociation and C[sbnd]C coupling on Mo-terminated MoC surfaces: A DFT study. *Catalysis Today*, *339*, 54–61. https://doi.org/10.1016/j.cattod.2019.03.020.

CHAPTER

15

Electrocatalytic oxygen reduction and evolution reactions in solid oxide cells (SOCs): A brief review

Allan J.M. Araújo[a,b], *Francisco J.A. Loureiro*[b], *Laura I.V. Holz*[b], *Vanessa C.D. Graça*[b], *Daniel A. Macedo*[c], *Moisés R. Cesário*[c], *Carlos A. Paskocimas*[a], *and Duncan P. Fagg*[b]

[a]Materials Science and Engineering Postgraduate Program—PPGCEM, Federal University of Rio Grande do Norte—UFRN, Natal, Brazil [b]Centre for Mechanical Technology and Automation (TEMA), Department of Mechanical Engineering, University of Aveiro, Aveiro, Portugal [c]Materials Science and Engineering Postgraduate Program (PPCEM), Federal University of Paraíba (UFPB), João Pessoa, Brazil

15.1 Introduction

The increasing need for energy conversion and storage has encouraged research in the fundamental areas of electrochemistry and materials science to move toward renewable energy systems, such as those offered by solid oxide cells (SOCs). These electrochemical devices can work as energy delivery systems by the input of fuel and oxidant gases, solid oxide fuel cells (SOFCs), or, in a reverse mode, by converting electricity into fuel (power-to-fuel (P2F) or power-to-chemicals (P2X)), commonly designated as solid oxide electrolysis cells (SOECs) (Gómez & Hotza, 2016; Mogensen et al., 2019).

The SOEC mode has gained strong interest in recent years due to its flexibility in producing several important chemicals. For example, the electrolysis of water is a promising method to produce H_2 without carbon as a byproduct (Nechache, Cassir, & Ringuedé, 2014). Another example is CO_2 electrolysis to obtain CO, as reported by Ni (2010), which is an attractive solution to reduce CO_2 emissions and eliminate greenhouse effects, while producing a valuable chemical product. In addition, combined CO_2/H_2O coelectrolysis can directly produce syngas (H_2 + CO) that can be used for large-scale energy storage (Zheng et al., 2017) in the form of a wide variety of synthetic hydrocarbon products. Such synthetic fuels produced by SOECs

Heterogeneous Catalysis
https://doi.org/10.1016/B978-0-323-85612-6.00015-2

can be stored and transported for conversion into electrical energy in SOFCs during periods of high electricity demand (Ramasamy, Nasani, Brandão, Pérez Coll, & Fagg, 2015).

The main constituents of SOCs are a dense ceramic electrolyte sandwiched between two porous electrodes (anode and cathode). The electrolyte must be an ionic conductor (often yttria-stabilized zirconia or acceptor-doped ceria materials). It must be dense to prevent the crossover of the gaseous species from both electrode's compartments (Araújo et al., 2020; Garcia et al., 2021). In the SOFC configuration, the cathode is the oxygen electrode and the anode is the fuel electrode. In contrast, the SOEC mode works in the opposite direction, and, therefore, the anode is the oxygen electrode and the cathode is the fuel electrode. Fig. 15.1 depicts the schematic diagrams of SOFC and SOEC modes. Due to this change in electrode functionality being a function of the operational mode, the nomenclature of electrodes in SOCs is commonly referred to as solely an "oxygen electrode" or a "fuel electrode."

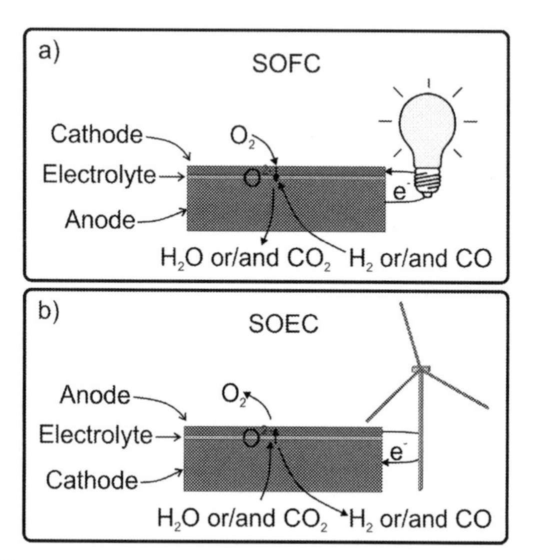

FIG. 15.1 SOFC and SOEC operation modes. Principle of operation of (A) solid oxide fuel cells (SOFCs) and (B) solid oxide electrolysis cells (SOECs) with the input of renewable energy. The *green* (*light gray* in print version) structural support illustrates the fuel electrode, the *gray* layer is the dense oxide ion-conducting electrolyte, and the *brown* (*dark gray* in print version) layer is the oxygen electrode.

The majority of previous studies regarding functional materials have been dedicated to the SOFC mode, where many efforts were taken to optimize the structure-property relationships of potential electrodes (Araujo et al., 2016; Araújo et al., 2018; Firmino et al., 2017; Loureiro, Macedo, et al., 2019; Loureiro, Souza, et al., 2019; Sousa et al., 2017; Usuba et al., 2019). However, similar components are often selected for SOFCs and SOECs due to their inherent similarities (Egger, Schrödl, Gspan, & Sitte, 2017). In this context, perovskites have been predominantly used as oxygen electrodes, whereas Ni-based cermets are typically adopted as fuel electrodes. However, problems related to long-term stability have been reported for typical perovskites, i.e., oxygen electrodes, when in the electrolysis mode (Chen & Jiang, 2011, 2016; Pan et al., 2018; Wang et al., 2018). Due to this limitation, a wide range of materials has been studied to minimize polarization losses and maximize cell performance for both SOFCs and SOECs, with increasing numbers of works assessing the functionality of materials in both modes of potential operation. In such reversible applications, oxygen reduction and evolution (ORR/OER) are the main reaction mechanisms occurring at oxygen electrodes.

For oxygen electrodes in the SOEC mode, a commonly reported problem for operating such cells is the delamination or structural degradation between the electrode and electrolyte interface with a noted decrease in cell performance over time, resulting in an increased polarization resistance (R_p) (Keane, Mahapatra, Verma, & Singh, 2012; Moçoteguy & Brisse, 2013; Rashkeev & Glazoff, 2012; Virkar, 2010). To minimize this effect, many studies have investigated the possible relationships between degradation issues and electrode microstructures and have suggested potential improvements by the optimization of processing (Chen, Ai, & Jiang,

2014; Chen & Jiang, 2016; Khan, Xu, Zhao, Knibbe, & Zhu, 2017, Khan, Xu, Knibbe, & Zhu, 2018). In addition to optimized microstructures, electrodes of both systems must also offer high electrochemical activity, high electronic and ionic conductivity, and adequate porosity to allow unhindered gas exchange at the triple-phase boundary (TPB), where the electronic phase, the ionic phase, and the gas phase meet (Li, Shi, Luo, & Cai, 2014).

In addition, the electrodes must also offer sufficient electrochemical stability to undergo successive reduction or oxidation steps, depending on the working mode. This chapter focuses on recent advances in solid-state electrodes for ORR/OER, including the state-of-the-art perovskite-type oxides (cobaltite- and ferrite-based systems), the more recent double perovskites, lanthanide nickelates of the Ruddlesden-Popper series, and also layered cobaltites ($[Ca_2CoO_{3-\delta}]_q[CoO_2]$ and $Ba_2Co_9O_{14}$). Composite electrodes have also been investigated with the addition of a high ionic conductivity phase (e.g., $Ce_{0.9}Gd_{0.1}O_{1.95}$), resulting in lower polarization resistances. The important steps for electrode reactions are also discussed: oxygen reduction and evolution, exchange kinetics at the oxygen electrode surface, bulk and surface diffusion of oxygen species, and gas-phase diffusion.

15.2 Oxygen reactions

In the SOFC mode, oxygen, usually as air, is fed to the cathode side and reduced to oxygen ions via the overall half-cell reaction (oxygen reduction reaction, ORR) (Eq. 15.1),

$$2\,e^- + {}^1/{}_2O_2 \leftrightarrow O^{2-} \qquad (15.1)$$

Conversely, in the SOEC mode, oxygen ions from the electrolyte to the anode are converted into oxygen gas with electrons liberated to the external circuit, namely, oxygen evolution reaction (OER) (Eq. 15.1, reverse).

The most common oxygen electrode materials include perovskite-structured compounds, such as $LaMnO_3$, $LaCoO_3$, and $LaFeO_3$ (and their doped structures), which are excellent catalysts for the reduction or the evolution of oxygen. For example, $(La,Sr)MnO_3$ (LSM) is a traditional material used as an oxygen electrode, mainly due to its high thermodynamic stability (Mogensen, 2020). However, due to its predominantly electronic behavior and poor oxygen-ion conductivity at lower temperatures, it exhibits relatively low electrochemical performance as working temperatures decrease. Due to this limitation of ionic transport, the electrode reactions predominantly take place close to the electrode/electrolyte interface (triple-phase boundary (TPB)), as described by Adler (2004) (Fig. 15.2), rather than being distributed throughout the electrode bulk. To mitigate this problem, an ionic conducting phase (e.g., yttria-stabilized zirconia or rare earth-doped ceria) is usually required to promote the extension of the ionic current beyond the electrode/electrolyte interface to increase the active region for oxygen reduction/evolution.

Several different mechanisms can determine the rate of ORR in SOFC cathodes (Fig. 15.3). Oxygen molecules generally adsorb onto the

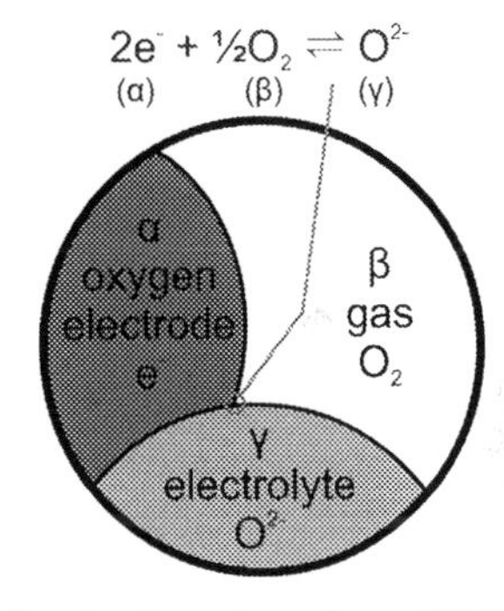

FIG. 15.2 Triple-phase boundary (TPB). Triple-phase boundary in solid oxide cells. *Adapted with permission from Adler, S. B. (2004). Factors governing oxygen reduction in solid oxide fuel cell cathodes.* Chemical Reviews, *104(10), 4791–4843. https://doi.org/10.1021/cr020724o. Copyright 2004 American Chemical Society.*

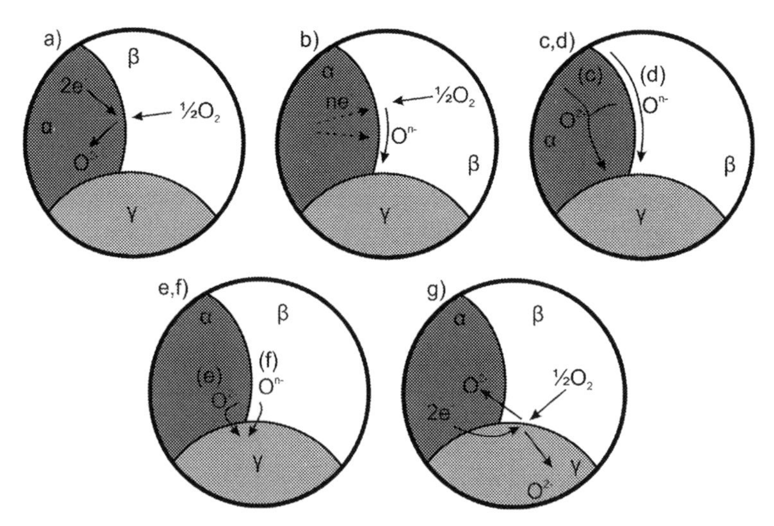

FIG. 15.3 Oxygen reduction mechanisms. Some mechanisms involved in the oxygen reduction in SOFC cathodes. Phases α, β, and γ refer to the electronic phase, gas phase, and ionic phase, respectively: (A) incorporation of oxygen into the bulk of the electronic phase (using mixed-conducting electrodes); (B) adsorption and/or partial reduction of oxygen on the surface of the electronic phase; (C) bulk or (D) surface transport of O_2^- or O_n^-, respectively, to the α/γ interface; (E) electrochemical charge transfer of O_2^- or (F) combination of O_n^- and O_e^- across the α/γ interface; and (G) rates of one or more of these mechanisms, wherein a layer between the electrode and electrolyte is active for transport of ionic and electronic species. *Adapted with permission from Adler, S. B. (2004). Factors governing oxygen reduction in solid oxide fuel cell cathodes.* Chemical Reviews, 104*(10), 4791–4843. https://doi.org/10.1021/cr020724o. Copyright 2004 American Chemical Society.*

solid electrocatalytic surface(s), where they undergo dissociation and/or reduction steps. Subsequently, ionic/atomic species must diffuse along surfaces, interfaces, or inside the bulk of the electrode material to the electrolyte, depending on the behavior of the electrode material (Adler, 2004). Therefore, chemical diffusion and surface exchange coefficients of the electrode material directly impact the resultant electrochemical behavior. Nevertheless, in addition to these fundamental properties, several microstructural features can also affect the electrode mechanisms, such as particle size, pore size, porosity, and morphology, as well as the testing conditions, including temperature, atmosphere, and polarization. As these factors can all contribute to uncertainties in the characterization of the electrodes, the parameterization of the electrode elementary steps can be a complex undertaking that often requires multiple measurements to unravel (Adler, 2004).

In contrast to those of ORR, the fundamentals of the oxygen evolution mechanism in the SOEC mode are at a less mature stage of understanding. The works of Barbucci, Carpanese, Cerisola, and Viviani (2005) and Zapata-Ramírez, Mather, Azcondo, Amador, and Pérez-Coll (2019) explain that under anodic polarization, bulk diffusion should be the main influence in the electrochemical process with the release of electrons during OER, as neither electron transfer nor desorption is expected to be rate-limiting. The same trend was observed in the study by Loureiro, Araújo, Paskocimas, Macedo, and Fagg (2021) for the misfit calcium cobaltite, with a three-probe cell configuration in both anodic and cathodic polarization modes. Laurencin et al. (2015) explained the electrochemical kinetics of a composite of a mixed ionic-electronic conductor (MIEC) and gadolinium-doped ceria (CGO) (Fig. 15.4).

The first step (1) corresponds to the oxygen-ionic transfer at the CGO/MIEC interface,

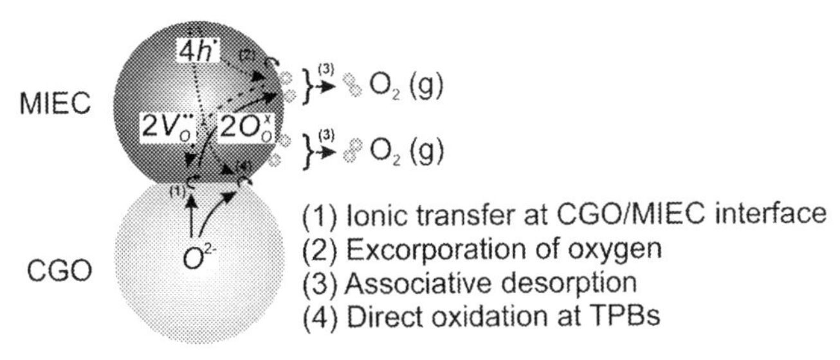

FIG. 15.4 Oxygen evolution mechanism of a MIEC/CGO composite. Schematic representation of the reactive pathways under anodic polarization. *Modified from Laurencin, J., Hubert, M., Couturier, K., Le Bihan, T., Cloetens, P., Lefebvre-Joud, F., & Siebert, E. (2015). Reactive mechanisms of LSCF single-phase and LSCF-CGO composite electrodes operated in anodic and cathodic polarisations.* Electrochimica Acta, 174, *1299–1316. https://doi.org/10.1016/j.electacta.2015.06.080.*

followed by the oxygen incorporation step (2). This is followed by the associative desorption of gaseous oxygen (3). However, the authors also considered the direct oxidation at TPBs (4). This mechanism is related to the surface path that occurs parallel to the oxygen exchange at the MIEC/gas interface (i.e., bulk path), which forms oxygen on the electrode surface. The oxidation can simultaneously occur by chemical oxygen exchange between the MIEC and the gas as well as by direct charge transfer at TPBs. The oxygen is finally desorbed to produce oxygen gas (3).

15.3 Mixed ionic-electronic conductors

Subsequent approaches include the use of mixed ionic-electronic conductors (MIECs), where the reactions can be extended to the whole surface of the electrode grains. One of the most commonly accepted models is that of Adler (2004) and Adler, Lane, and Steele (1996), which explains the electrode kinetics of mixed conductors, as demonstrated in Fig. 15.5. This model suggests that for mixed conductors with significant ionic transport, the overall impedance is dominated by the surface chemical exchange of O_2 and solid-state oxygen diffusion. In other words, the electrode reaction of Eq. (15.1) can occur over the entire surface area of the electrode material. However, the electrode reactions should occur over a limited region of the electrode thickness (typically, only a few micrometers near the electrolyte). This active region is related to the exchange and the diffusion properties of the MIECs, and its thickness is given by the characteristic length (L_c), where their ambipolar conduction of both ions and electrons is favored. Fig. 15.5 shows the schematic representation of the ionic and electronic current distribution in a porous MIEC electrode.

The impedance diagram in Fig. 15.5 represents an ideal MIEC electrode response, including in the high-frequency region where the inset of the ohmic resistance in series with the interfacial resistances (electron-transfer and ion-transfer processes, $\geq 10^4$ Hz) occur at the current collector/electrode and electrode/electrolyte interfaces. In the medium-to-low-frequency region, the impedance diagram presents a "chemical" impedance associated with the oxygen reaction processes ($\sim$10 Hz). This chemical-electrochemical-chemical reaction may be modeled with a *Gerischer* element, which means a diffusion process coupled to a "chemical reaction," as explained by Boukamp and Bouwmeester (2003).

MIEC electrodes that can offer superior performance as anodes for SOECs, when compared to purely electronic conducting analogs (Ebbesen, Jensen, Hauch, & Mogensen, 2014), have been reported.

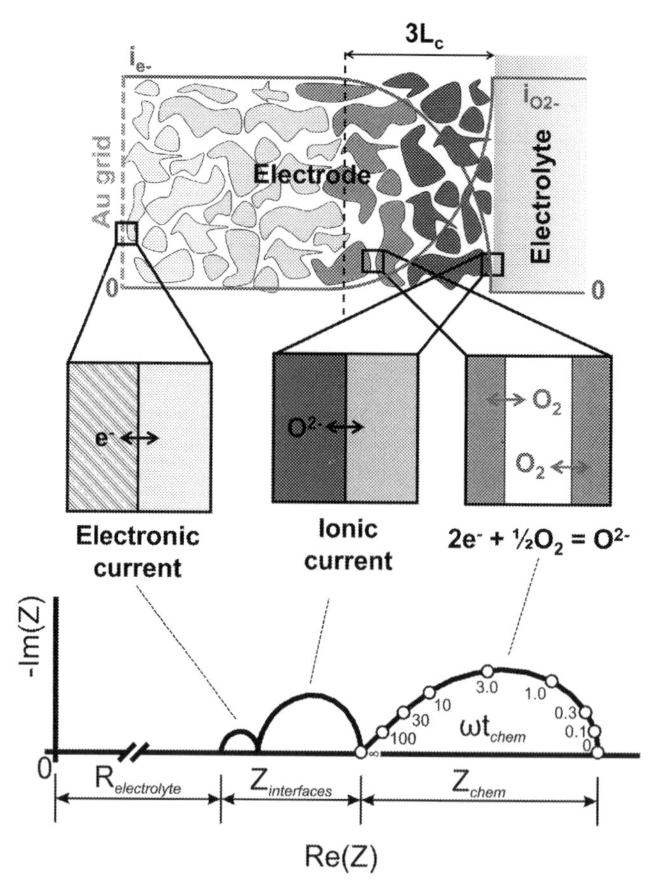

FIG. 15.5 Adler model for MIEC oxygen electrodes. Schematic of the Adler model for the impedance response of a porous mixed-conducting electrode. *Modified from Adler, S. B. (1998). Mechanism and kinetics of oxygen reduction on porous* $La_{1-x}Sr_xCoO_{3-\delta}$ *electrodes.* Solid State Ionics, *111(1–2), 125–134. https://doi.org/10.1016/S0167-2738(98)00179-9.*

15.4 Experimental techniques to determine oxygen kinetics

Knowledge of surface exchange and oxygen diffusion is critical for characterizing the transport properties of oxygen electrode materials for SOFC and SOEC applications, as both these processes may govern the overall performance of these devices (Bouwmeester et al., 2009; Lane & Kilner, 2000; Téllez, Druce, Hong, Ishihara, & Kilner, 2015). For this reason, several techniques, such as isotope exchange depth profiling (IEDP), ^{18}O-^{16}O pulse isotope exchange (PIE), and electrical conductivity relaxation (ECR), have been used to determine the oxygen transport properties of MIECs, to provide further support to explain impedance spectroscopy observations.

Isotope exchange depth profiling

The isotope exchange approach includes measurements of ^{18}O diffusion profiles in dense samples after heat treatment in an ^{18}O-enriched atmosphere for a given time to reach thermodynamic equilibrium. Oxygen self-diffusion (D^*, $cm^2 s^{-1}$) and surface exchange coefficients (k^*, $cm s^{-1}$) are calculated using the diffusion equation with isotope exchange depth profiles (by curve fitting) (De Souza, Zehnpfenning, Martin, & Maier, 2005; He et al., 2016; Kilner, Skinner, & Brongersma, 2011; Téllez et al., 2015). Time-of-flight secondary ion mass spectrometry (ToF-SIMS) is a surface-sensitive analytical method for measuring ^{18}O isotopic fractions in solid materials, providing spatial-resolved information for the determination of the fractions on particular characteristics, such as particles, grains, or grain boundaries (Téllez et al., 2015). Fig. 15.6 shows a schematic illustration of ToF-SIMS and the ^{16}O and ^{18}O oxygen profiles measured using selective attenuation of secondary ions (SASIs) modulated automatically by the instrument software in the high-current bunched mode (HCBM) or spectrometry mode. Data and more information can be found in the following articles: Kilner et al. (2011), Kilner, De Souza, and Fullarton (1996), and Téllez et al. (2015).

^{18}O-^{16}O pulse isotopic exchange

Bouwmeester et al. (2009) introduced the use of the ^{18}O-^{16}O pulse isotopic exchange (PIE) technique to study the oxygen surface exchange rate, providing insight into the mechanism of the oxygen exchange reaction. The electrocatalytic performance of solids with highly mobile oxide ions or mixed ionic-electronic conductors strongly depends on the fast oxygen exchange at the gas/solid interface (Fig. 15.7A). While the IEDP method described above involves a series of time-consuming heat treatments, the PIE technique allows a rapid characterization of the surface exchange of oxygen under different operating conditions. Another advantage is that the PIE technique can be applied to oxide powders.

In the PIE experiment, a powder sample is loaded into a packed-bed microreactor and the conditions of oxygen partial pressure (pO_2) and temperature are defined. The response to an ^{18}O-enriched pulse fed through the reactor is measured by mass spectrometric analysis of the concentrations of $^{18}O_2$, ^{16}O-^{18}O, and $^{16}O_2$ in the gas phase (Fig. 15.7B). The overall surface

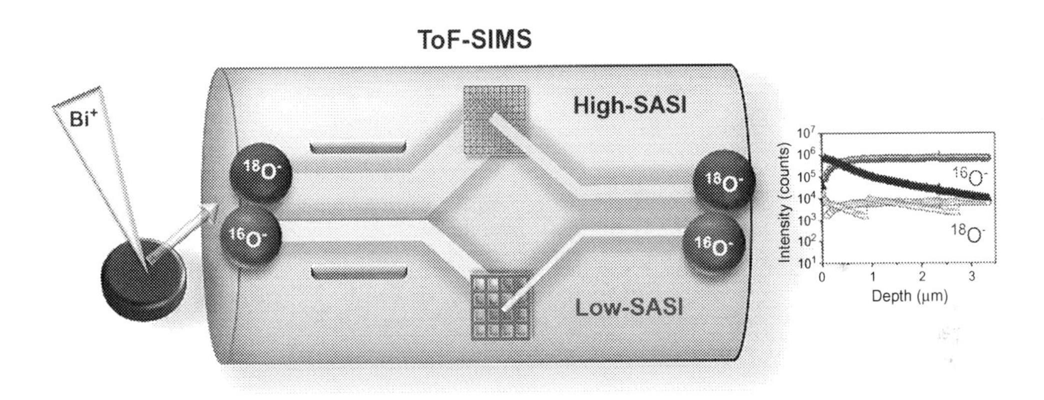

FIG. 15.6 Time-of-flight secondary ion mass spectrometry (ToF-SIMS). Schematic representation of the time-of-flight secondary ion mass spectrometry (ToF-SIMS) and the depth profiles for ^{16}O-^{18}O signals (raw and corrected) as obtained using selective attenuation of secondary ions (SASIs). *Reprinted with permission from Téllez, H., Druce, J., Hong, J. E., Ishihara, T., & Kilner, J. A. (2015). Accurate and precise measurement of oxygen isotopic fractions and diffusion profiles by selective attenuation of secondary ions (SASI).* Analytical Chemistry, *87(5), 2907–2915. https://doi.org/10.1021/ac504409x. Copyright 2015 American Chemical Society.*

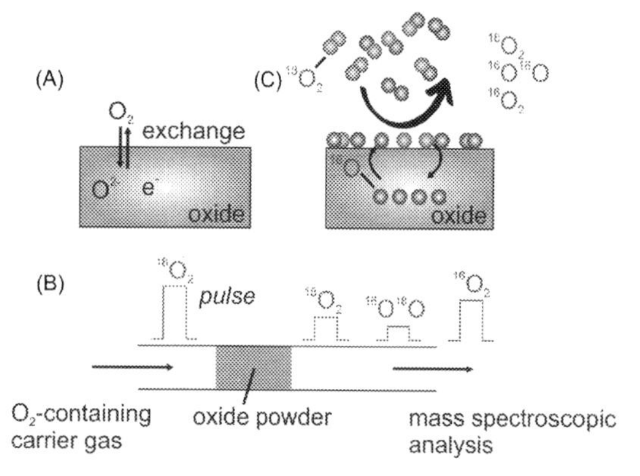

FIG. 15.7 ^{18}O-^{16}O pulse isotopic exchange (PIE). (A) Reversible oxygen exchange between oxygen (gas) and oxygen (ions) from the sample; (B) schematic of the PIE technique; and (C) two-step surface exchange mechanism for the oxygen exchange kinetics, with the formation of mononuclear oxygen on the oxide surface as the intermediate. *Modified from Bouwmeester, H. J. M., Song, C., Zhu, J., Yi, J., Van Sint Annaland, M., & Boukamp, B. A. (2009). A novel pulse isotopic exchange technique for rapid determination of the oxygen surface exchange rate of oxide ion conductors.* Physical Chemistry Chemical Physics, 11*(42), 9640–9643. https://doi.org/10.1039/b912712g.*

exchange rate is determined from the mean residence time and the uptake of ^{18}O by the sample (with a known surface area) (Bouwmeester et al., 2009), on chemical equilibrium.

Oxygen exchange mechanism

In the ^{18}O isotope exchange studies, the sample is often pretreated in pO_2 temperature conditions of the experiment to establish thermodynamic equilibrium to prevent influences of chemical diffusion within the bulk. Hence, both the gas and ceramic materials remain with a constant oxygen content during the exchange process. In the derivation of the model, a stepwise change in the oxygen partial pressure (^{18}O) at time $t=0$ of the system is assumed. The diffusion is supposed to be fast, and the exchange reaction occurs uniformly on the surface (Den Otter, Boukamp, & Bouwmeester, 2001).

Den Otter et al. (2001) explain the model using three parameters, namely, K, p_1, and p_2. K (mol O_2 m^{-2} s^{-1}) denotes the amount of O_2 molecules per unit of area and time, and p_1 and p_2 are the independent probabilities of both atoms in a single oxygen molecule for exchange with atoms of the sample. Information on the mechanism may be obtained from K, p_1, and p_2 as a function of pO_2 (Den Otter et al., 2001). The parameters p_1 and p_2 are not necessarily equal but depend on the mechanism of the exchange reaction. The exchange process (Fig. 15.7C) is explained as follows:

$$O_2 + 2e^- \leftrightarrow O_{ads} + O^{2-} \quad (15.2)$$

$$O_{ads} + 2e^- \leftrightarrow O^{2-} \quad (15.3)$$

The two reactions (Eqs. 15.2 and 15.3) balance the concentration of O_{ads} on the adsorbed layer. In the first step, one oxygen atom is immediately incorporated into the structure (for $p_1=1$). Incorporation of the remaining O_{ads} is determined by the rates of both steps, with $0<p_2<1$.

Electrical conductivity relaxation

For electrical conductivity relaxation (ECR) measurements, two steps are required to determine the oxygen transport properties of MIECs. The first step is to generate a transient conductivity response of a dense pellet by changing pO_2 by mixing dried oxygen and nitrogen in the desired ratios. The second step determines the surface exchange coefficient and bulk diffusivity, fitting the model with experimental data under identical operating conditions. A multiple probe method (DC) can be used through one bulk sample, as schematically illustrated in Fig. 15.8.

A pair of wires is wrapped around both ends of the pellet, where a constant current is applied. Additional wires are also attached to the sample to act as voltage probes (usually, a four-point probe technique). The sample is measured inside an alumina tube gas chamber installed in a tubular furnace to obtain the desired operating conditions

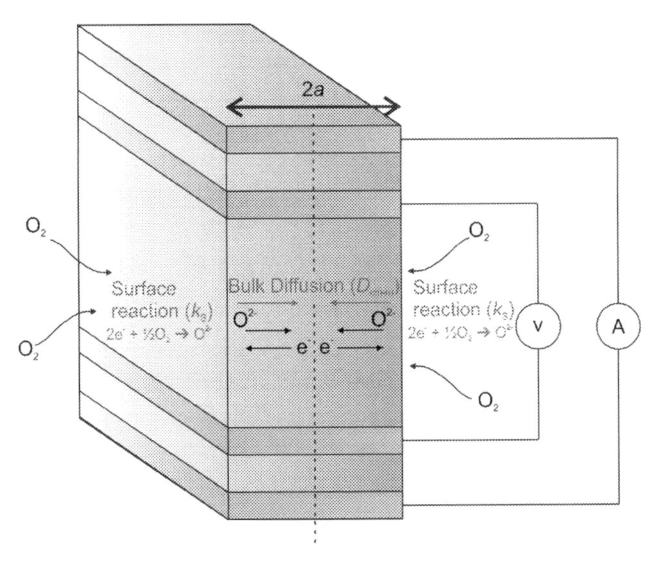

FIG. 15.8 Electrical conductivity relaxation (ECR). Schematic illustration of sample assembly for ECR measurements. *Modified from Gopal, C. B., & Haile, S. M. (2014). An electrical conductivity relaxation study of oxygen transport in samarium doped ceria.* Journal of Materials Chemistry A, *2(7), 2405–2417. https://doi.org/10.1039/C3TA13404K.*

of temperature and controlled atmosphere. Oxygen surface exchange occurs at the gas/solid interface by changing the gas composition (He et al., 2016). Under a constant current, the sample's relaxation process is reflected in the change in conductivity, as measured by the change in voltage. This change in conductivity can be utilized to determine the surface exchange coefficient (k_{chem}) and bulk diffusivity (D_{chem}) properties (He et al., 2016). These parameters can be obtained by fitting the electrical conductivity relaxation curve to a model derived from diffusion equations (Lane & Kilner, 2000; Na et al., 2021).

The ECR method requires several conditions to be fulfilled, which are as follows (Gopal & Haile, 2014):

The absence of open porosity and minimal closed porosity;
The reactor flush time ($t0$) must be much smaller than the material response time (τ), where τ is $\approx a/kS$ in the surface reaction-limited regime and $\approx a2/4D$Chem in the diffusion-limited regime;
Grain size must be large (in order of microns) to minimize grain boundary contributions; and
The width $2a$ of the sample (Fig. 15.8) must be much smaller than those of the other two directions.

15.5 Anode degradation in SOECs

One of the main problems with oxygen electrodes used in high-temperature electrolyzers is microstructural degradation at the electrode side of the electrode/electrolyte interface when applying an anodic overpotential. Although similar components can be used in both SOFC and SOEC modes, SOEC usually suffers from a greater degradation rate. In the latter, the evolved oxygen penetrates the closed pores/defects at the electrode/electrolyte interface, increasing the local oxygen partial pressure in this region, thus potentially leading to the delamination of the electrode film (Irvine et al., 2016).

Chen and Jiang (2011) studied the polarization and delamination mechanism of an LSM oxygen electrode under an anodic current of 500 mA cm^{-2} at 800°C. A significant increase in the polarization resistance was observed after 48 h (Fig. 15.9).

The delamination of the electrode was characterized by the formation of nanoparticles caused by the localized disintegration of LSM grains at the electrode/electrolyte interface, as schematically represented in Fig. 15.10.

Oxygen ions from the YSZ electrolyte to LSM grains (Fig. 15.10A) cause a localized internal stress at the LSM anode next to the interface due to lattice shrinkage (Fig. 15.10B). Tensile strain favors the oxygen vacancy formation and its accumulation leads to the formation of microcracks within LSM grains (Fig. 15.10C). Their initial formation (Fig. 15.10D) provides extra active sites for the OER, shortening the diffusion path, as observed by the initial decrease in the polarization resistance within a few hours (Fig. 15.9). Nonetheless, with the continuous application of polarization, the electrode was delaminated as a result of the internal energy release of the local tensile strains, and the electrochemical performance was impaired after 48 h (Figs. 15.9 and 15.10E and F).

The degradation mechanism of the LSM/YSZ composite has also been investigated by Kim et al. (2013). Both R_{ohmic} and R_p decreased drastically after 120 h of operation at 1.5 A cm^{-2} at 750°C. It was observed that the cation migration at high anodic polarization leads to intergranular fracture along the grain boundaries of the YSZ electrolyte and densification of the anode, which results in complete delamination of the oxygen electrode.

Other oxygen electrodes concerning degradation during SOEC operation conditions, such as $(La,Sr)(Co,Fe)O_{3-\delta}$, $(Ba,Sr)(Co,Fe)O_{3-\delta}$, nickelates, and double perovskites, have also been studied. The reader is referred to more comprehensive reviews on the degradation of oxygen electrodes operating at high anodic polarization (Chen & Jiang, 2011, 2016; Khan, Xu, Knibbe, & Zhu, 2021; Pan et al., 2018; Wang, Li, Ma, Li, & Liu, 2019). Even though several studies are available, there is still a lot to do in terms of long-term testing of SOECs, and nanoscale

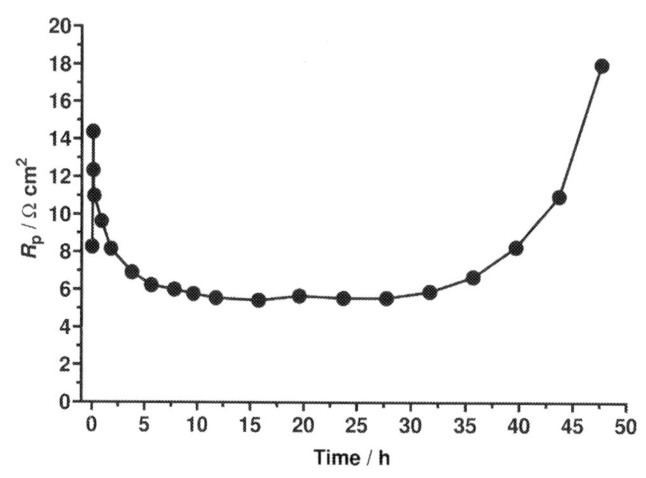

FIG. 15.9 R_p as a function of anodic current passage time. Polarization resistance of an LSM oxygen electrode as a function of time of 500 mA cm^{-2} at 800°C. *Data from Chen, K., & Jiang, S. P. (2011). Failure mechanism of (La,Sr)MnO$_3$ oxygen electrodes of solid oxide electrolysis cells.* International Journal of Hydrogen Energy, *36(17), 10541–10549. https://doi.org/10.1016/j.ijhydene.2011.05.103.*

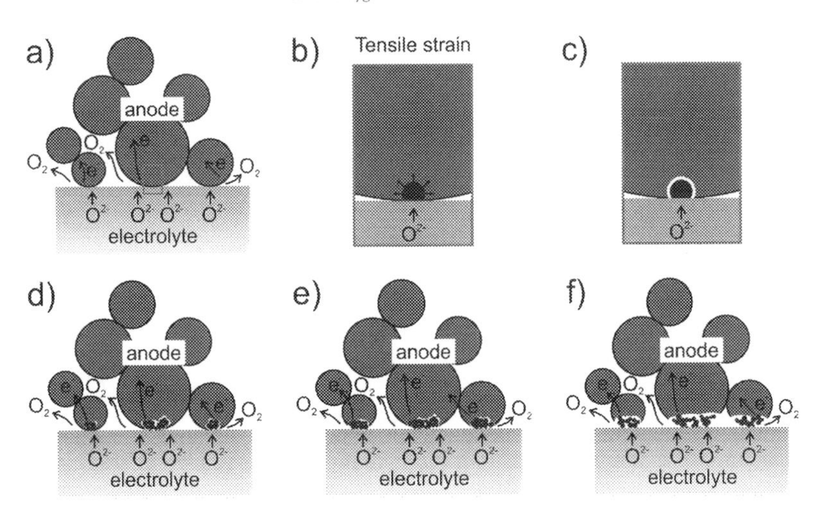

FIG. 15.10 Long-term degradation process. Schematic of the microstructural change of the LSM anode/YSZ electrolyte interface under SOEC operation conditions: (A) oxygen migration from the YSZ electrolyte to LSM grains; (B) local tensile strains within LSM particles due to shrinkage of the LSM structure; (C) microcrack formation; (D) formation of nanoparticles; (E) propagation and continuous nanoparticle formation; and (F) delamination of the LSM anode under high internal oxygen partial pressure at the interface. *Data from Chen, K., & Jiang, S. P. (2011). Failure mechanism of (La,Sr)MnO$_3$ oxygen electrodes of solid oxide electrolysis cells.* International Journal of Hydrogen Energy, *36(17), 10541–10549. https://doi.org/10.1016/j.ijhydene.2011.05.103.*

chemical analyses, e.g., EDS in FETEM/STEM, may be required to investigate the possible formation of secondary phases under operation. In addition, solutions to avoid degradation problems under electrolysis operating conditions are of urgent interest.

15.6 Oxygen electrode materials

Some traditional (high-performing) and promising oxygen electrodes will be presented in the following subsections.

(La,Sr)(Co,Fe)O$_{3-\delta}$ and (Ba,Sr)(Co,Fe)O$_{3-\delta}$ (LSCF and BSCF)

Both LSCF and BSCF perovskites have become the most widely adopted SOC electrodes due to their high mixed ionic-electronic conductivity at intermediate temperatures (500–700°C) (Almar, Szász, Weber, & Ivers-Tiffée, 2017; Baumann, Maier, & Fleig, 2008; Laurencin et al., 2015; Niedrig, Wagner, Menesklou, Baumann, & Ivers-Tiffée, 2015; Nielsen, Jacobsen, & Wandel, 2011; Shao & Haile, 2004). For example, Hildenbrand, Boukamp, Nammensma, and Blank (2011) fabricated LSCF air electrodes and demonstrated significantly improved polarization resistance values with a dense LSCF layer, by a factor of 3. The oxygen partial pressure dependence of polarization resistance highlighted that the oxygen reaction changes, possibly indicating improved surface exchange/reaction processes, as observed by the *Gerischer* parameters. Shao and Haile (2004) studied BSCF and reported superior performance (i.e., R_p of $0.6\,\Omega\,cm^2$ and power density of $402\,mW\,cm^{-2}$ at 500°C) compared to traditional cathode materials for SOFCs. Impedance spectroscopy and oxygen permeability measurements supported the conclusion that while oxygen diffusion is rapid, the surface exchange is rate-limiting.

LSCF-Ce$_{0.8}$Sm$_{0.2}$O$_{1.9}$ (SDC) synthesized by one-step microwave self-assisted combustion synthesis demonstrated lower total polarization resistance than literature reports on LSCF-based catalysts for the SOFC cathode (Loureiro, Macedo, et al., 2019). Although Loureiro, Macedo, et al. (2019) reported good stability of the LSCF under cathodic polarization (0 to −0.8 V) at 700–750°C, surface segregation of the A-site cations in long-term operation has also been reported, which may block active surface sites for the oxygen reduction/evolution reactions, thus giving rise to a further challenge for these electrode materials (Simner, Anderson, Engelhard, & Stevenson, 2006). Indeed, in terms of anodic polarization, the long-term stability of the LSCF and BSCF electrodes remains the key obstacle to their applications under SOEC operating conditions (Laguna-Bercero, 2012; Wang, Li, Ma, Li, & Liu, 2019).

Double perovskites

In addition to single perovskite-structured materials, double perovskites (e.g., $AA'B_2'O_{5+\delta}$) have also been investigated for SOCs, considering their chemical and structural stability along with the advantages of high electrochemical performance. Tarancón, Marrero-López, Peña-Martínez, Ruiz-Morales, and Núñez (2008) demonstrated excellent stability of $GdBaCo_2O_{5+\delta}$, without undesirable changes in the volume of the unit cell, with a uniform TEC (16.4×10^{-6} K^{-1}, 100–700°C) or in the polarization resistance when measured with a symmetrical cell. Only for values of $\delta < 0.25$ ($T > 750$°C, in air), the electrochemical performance decreases, probably due to an excessive reduction in the oxygen content in the GdO_δ plane. $GdBaCo_2O_{5+\delta}$ and $PrBaCo_2O_{5+\delta}$ are also reported to offer rapid oxygen transport kinetics (Kim et al., 2006). For further details, please see the article by Tarancón, Burriel, Santiso, Skinner, and Kilner (2010). The literature shows that these potential perovskites for SOC applications can be tailored in terms of oxygen diffusion and surface exchange by an appropriate selection of the A, A′, and B cations.

Ruddlesden-Popper nickelates

Other structures that have attracted interest for applications in solid oxide cells are layered Ruddlesden-Popper (RP) nickelates, $(LnNiO_3)_nLnO$ (Ln = La, Pr, Nd; $n = 1, 2, 3$) (Song, Ning, Boukamp, Bassat, & Bouwmeester, 2020). The oxygen transport properties of $La_2NiO_{4+\delta}$, $Nd_2NiO_{4+\delta}$, $La_3Ni_2O_{7-\delta}$, $La_4Ni_3O_{10-\delta}$, $Pr_4Ni_3O_{10-\delta}$, and $Nd_4Ni_3O_{10-\delta}$ have recently been reported by Song et al. (2020) using ECR measurements. RP nickelates display a remarkable similarity in their values for the surface exchange coefficient, despite their structural differences and choice of the lanthanide ion. In contrast, the oxygen self-diffusion coefficients (D_s) are found to decrease profoundly with the order parameter n (Song et al., 2020).

Amow, Davidson, and Skinner (2006) found that the polarization resistance using symmetrical cells followed the trend $La_4Ni_3O_{10-\delta} < La_3Ni_2O_{7-\delta} < La_2NiO_{4+\delta}$ (500–900°C). A similar trend was observed by Takahashi, Nishimoto, Matsuda, and Miyake (2010) using single test cells ($La_{n+1}Ni_nO_{3n+1}$/Sm-doped ceria/Ni:Sm-doped ceria). These results may be explained by the increase in p-type conductivity with increasing n-value of the RP nickelates, as demonstrated by Song et al. (2020).

Higher ionic conductivity is found for $Ln_2NiO_{4+\delta}$ (Ln = La or Nd) compared to the $n = 2$ or 3 RP nickelates (Song et al., 2020). $Pr_2NiO_{4+\delta}$ exhibits high oxygen diffusion and surface exchange coefficients (Burriel et al., 2016); however, reactivity with traditional electrolytes and problems with decomposition limit its use (Philippeau, Mauvy, Mazataud, Fourcade, & Grenier, 2013; Saher et al., 2018). To overcome these problems, Laguna-Bercero, Monzón, Larrea, and Orera (2016) used $Pr_2NiO_{4+\delta}$ as the oxygen electrode along with different electrolyte-electrode interlayers and characterized in both fuel cell and electrolysis operation modes. The stability and performance of the cells depend heavily on the barrier layer used. In the SOFC mode, cells with the PNO/CGO composite barrier layers showed power densities of ca. $0.63\,W\,cm^{-2}$ (800°C and 0.7 V). Although $Pr_2NiO_{4+\delta}$ decomposes into $PrNiO_3$ and PrO_{2-y} under strong polarization, excellent stability without degradation was observed after 100 h under the operating conditions. The performance in the electrolysis mode was also remarkable ($-0.78\,A\,cm^{-2}$ at 800°C and 1.3 V). Thus, the excellent reversible SOFC/SOEC performance and stability of nickelate-based oxygen electrodes are highly attractive features of these systems for potential application.

Layered Ca and Ba cobaltites

Two innovative cobaltites have been recently studied as oxygen electrodes, $[Ca_2CoO_{3-\delta}]_q[CoO_2]$ (known as "$Ca_3Co_4O_9$," CCO, or C349) and $Ba_2Co_9O_{14}$ (BCO) ; (Nagasawa et al., 2009; Rolle, Preux, Ehora, Mentré, & Daviero-Minaud, 2011; Saqib et al., 2021). In recent years, the misfit-layered Ca cobaltite has emerged as an exciting oxygen electrode for SOCs due to its excellent thermal compatibility with typical electrolyte materials, low cost, and avoidance of Sr surface segregation issues that may deplete the electrochemical performance in more common oxygen electrodes. Two groups have investigated the electrochemical mechanisms of this misfit-layered compound using symmetrical or three-electrode cell configurations (Araújo et al., 2021; Boukamp, Rolle, Vannier, Sharma, & Djurado, 2020; Fulgêncio et al., 2019; Loureiro, Araújo, Paskocimas, Macedo, & Fagg, 2021; Rolle et al., 2016). Both materials display high electronic conductivity with values of 100 and $\sim 285\,S\,cm^{-1}$ at 700°C (Delorme et al., 2017; Lin et al., 2007). Nonetheless, to date, the poor ionic conductivity of C349 has prevented it from attaining a competitive electrode performance. In this context, CGO, as an ionic conducting phase, has been added to improve the electrochemical performance of this electrode (Boukamp, Rolle, Vannier, Sharma, & Djurado, 2020; Nagasawa et al., 2009; Rolle et al., 2016). In comparison to a more extensive research on the C349 electrode by Nagasawa et al. (2009), only a few works, to date, have explored the BCO material (Hu et al., 2014; Rolle et al., 2011), and further studies are needed to understand the electrochemical processes of this cobaltite better. A similar strategy was adopted by Rolle et al. (2011), regarding the use of composites, combining predominately ion-conducting phases with BCO. Nonetheless, oxygen transport parameters must be further clarified building on the initial conclusions of Hu et al. (2014).

Loureiro et al. (2021) studied the polarization mechanism of C349 in both cathodic and anodic modes of operation. A *Gerischer* element was used to describe the diffusion and surface exchange properties of the C349 electrode under applied polarization. This analysis was complemented by the distribution function of relaxation times (DFRTs) and Kramers-Kronig (KK) analyses, revealing the diffusion and surface exchange processes dominating the total polarization resistance. The *Gerischer* Y_0 parameter is related to the oxide-ion diffusion coefficient (Fig. 15.11A). The diffusion of oxygen species is shown to be facilitated under anodic polarization since bulk diffusion should influence the electrochemical process, as discussed in Section 15.2. In contrast, the K_G parameter (Fig. 15.11B), associated with the oxygen surface exchange or dissociation rates, was shown to increase with cathodic polarization. Overall, the electrochemical performance of C349 was found to be enhanced under anodic polarization, due to the easier electron removal during OER (Fig. 15.11C), suggesting that this material may be a promising oxygen electrode for SOECs.

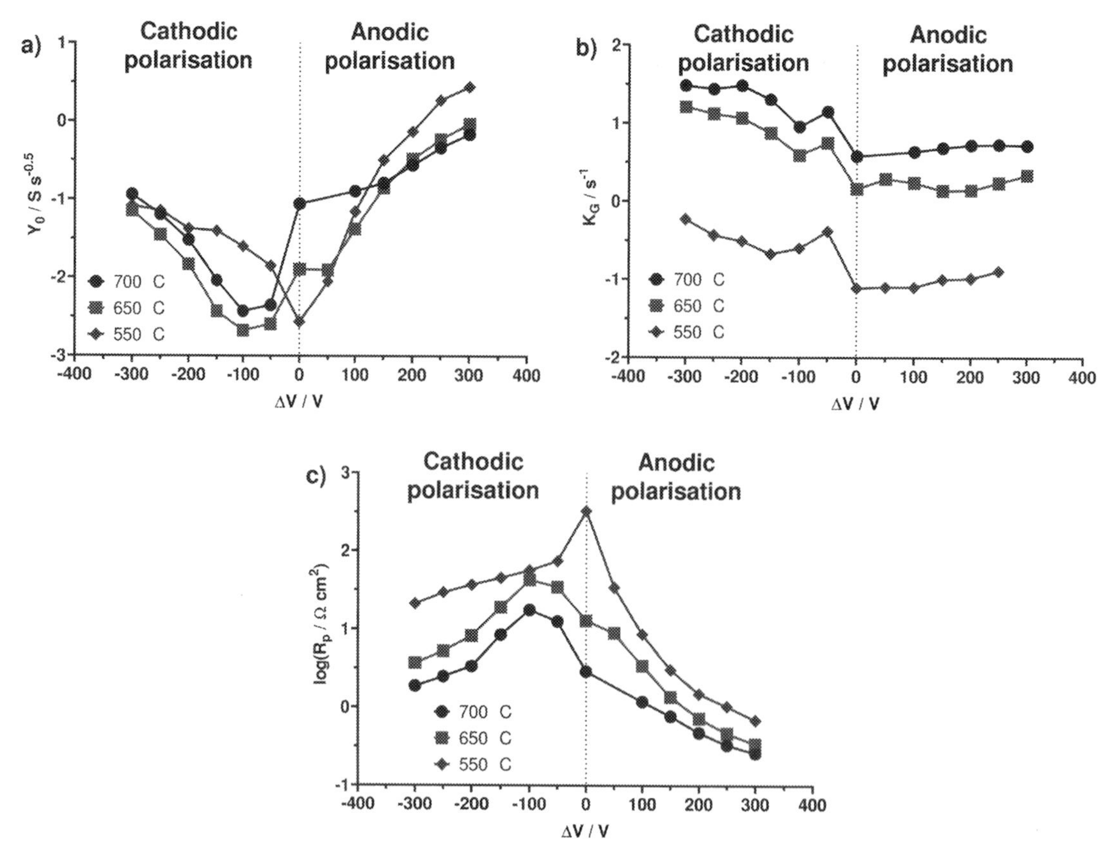

FIG. 15.11 Polarization mechanism of the misfit Ca-cobaltite electrode for r-SOCs. Applied difference of potential dependence at $pO_2 = 1$ atm for (A) Y_0 Gerischer, (B) K_G Gerischer, and (C) polarization resistance. *Data from Loureiro, F. J. A., Araújo, A. J. M., Paskocimas, C. A., Macedo, D. A., & Fagg, D. P. (2021). Polarization mechanism of the misfit Ca-cobaltite electrode for reversible solid oxide cells.* Electrochimica Acta, *373, 137928–137928. https://doi.org/10.1016/j.electacta.2021.137928.*

15.7 Conclusions

The SOC technology has become an interesting choice for the production of electricity or chemical production. A wide diversity of oxygen electrode compositions are now available as potential electrocatalysts for ORR and OER. They can range from single-phase materials to composite electrodes, where, in both cases, mixed ionic-electronic conductors (MIECs) are shown to be good choices. Several materials with different structures exist, with most of the characterizations, to date, made on optimizing the structural, microstructural, and electrical properties.

Initial works describing the electrode mechanisms are also fairly well understood now, assisted by the diversity of sophisticated complementary techniques available. Nonetheless, the strict requirements of electrode materials in terms of electrical conductivity and thermal and chemical compatibility make their choice difficult and ideal materials still remain to be discovered, especially in the case of OER or for electrode materials that can adopt both SOEC and SOFC functionalities. For this reason, the measurement of surface exchange (k_{chem}) and diffusion (D_{chem}) in ceramic materials is critical for characterizing the transport properties of oxygen electrode materials for these applications. These two parameters are the fingerprint of the electrodes and thus have a strong influence on their operability.

Acknowledgments

This study was financed in part by the Coordenação de Aperfeiçoamento de Pessoal de Nível Superior—Brasil (CAPES)—Finance Code 001. Allan J.M. Araújo thanks the Conselho Nacional de Desenvolvimento Científico e Tecnológico (CNPq/Brazil, reference number 200439/2019-7). Carlos A. Paskocimas and Daniel A. Macedo also thank CNPq/Brazil (482473/2010-0, 446126/2014-4, 308548/2014-0, 307236/2018-8, 431428/2018-2, and 309430/2019-4). The authors also acknowledge the projects, PTDC/CTM-CTM/2156/2020, PTDC/QUI-ELT/3681/2020, POCI-01-0247-FEDER-039926, POCI-01-0145-FEDER-032241, UIDB/00481/2020 and UIDP/00481/2020, and CENTRO-01-0145-FEDER-022083—Centro Portugal Regional Operational Programme (Centro2020), under the PORTUGAL 2020 Partnership Agreement, through the European Regional Development Fund (ERDF). Francisco J.A. Loureiro acknowledges FCT for the financial support with the reference number CEECIND/02797/2020.

References

Adler, S. B. (2004). Factors governing oxygen reduction in solid oxide fuel cell cathodes. *Chemical Reviews, 104*(10), 4791–4843. https://doi.org/10.1021/cr020724o.

Adler, S. B., Lane, J. A., & Steele, B. C. H. (1996). Electrode kinetics of porous mixed-conducting oxygen electrodes. *Journal of the Electrochemical Society, 143*(11), 3554–3564. https://doi.org/10.1149/1.1837252.

Almar, L., Szász, J., Weber, A., & Ivers-Tiffée, E. (2017). Oxygen transport kinetics of mixed ionic-electronic conductors by coupling focused ion beam tomography and electrochemical impedance spectroscopy. *Journal of the Electrochemical Society, 164*(4), F289–F297. https://doi.org/10.1149/2.0851704jes.

Amow, G., Davidson, I. J., & Skinner, S. J. (2006). A comparative study of the Ruddlesden-Popper series, $La_{n+1}Ni_nO_{3n+1}$ (n = 1, 2 and 3), for solid-oxide fuel-cell cathode applications. *Solid State Ionics, 177*(13–14), 1205–1210. https://doi.org/10.1016/j.ssi.2006.05.005.

Araújo, A. J. M., Grilo, J. P. F., Loureiro, F. J. A., Campos, L. F. A., Paskocimas, C. A., Nascimento, R. M., et al. (2018). Designing experiments for the preparation of Ni-GDC cermets with controlled porosity as SOFC anode materials: Effects on the electrical properties. *Ceramics International, 44*(18), 23088–23093. https://doi.org/10.1016/j.ceramint.2018.09.115.

Araújo, A. J. M., Grilo, J. P. F., Loureiro, F. J. A., Holz, L. I. V., Macedo, D. A., Fagg, D. P., et al. (2020). Proteic sol–gel synthesis of Gd-doped ceria: A comprehensive structural, chemical, microstructural and electrical analysis. *Journal of Materials Science, 55*, 16864–16878. https://doi.org/10.1007/s10853-020-05173-6.

Araújo, A. J. M., Loureiro, F. J. A., Holz, L. I. V., Grilo, J. P. F., Macedo, D. A., Paskocimas, C. A., et al. (2021). Composite of calcium cobaltite with praseodymium-doped ceria: A promising new oxygen electrode for solid oxide cells. *International Journal of Hydrogen Energy*. https://doi.org/10.1016/j.ijhydene.2021.06.049.

Araujo, A. J. M., Sousa, A. R. O., Grilo, J. P. F., Campos, L. F. A., Loureiro, F. J. A., Fagg, D. P., et al. (2016). Preparation of one-step NiO/Ni-CGO composites using factorial design. *Ceramics International, 42*(16), 18166–18172. https://doi.org/10.1016/j.ceramint.2016.08.131.

Barbucci, A., Carpanese, P., Cerisola, G., & Viviani, M. (2005). Electrochemical investigation of mixed ionic/electronic cathodes for SOFCs. *Solid State Ionics, 176*(19–22), 1753–1758. https://doi.org/10.1016/j.ssi.2005.04.027.

Baumann, F. S., Maier, J., & Fleig, J. (2008). The polarization resistance of mixed conducting SOFC cathodes: A comparative study using thin film model electrodes. *Solid State Ionics, 179*(21–26), 1198–1204. https://doi.org/10.1016/j.ssi.2008.02.059.

Boukamp, B. A., & Bouwmeester, H. J. M. (2003). Interpretation of the Gerischer impedance in solid state ionics. *Solid State Ionics, 157*(1–4), 29–33. https://doi.org/10.1016/S0167-2738(02)00185-6.

Boukamp, B. A., Rolle, A., Vannier, R. N., Sharma, R. K., & Djurado, E. (2020). Electrostatic spray deposited $Ca_3Co_4O_{9+\delta}$ and $Ca_3Co_4O_{9+\delta}/Ce_{0.9}Gd_{0.1}O_{1.95}$ cathodes for SOFC: A comparative impedance analysis study. *Electrochimica Acta, 362*, 137142. https://doi.org/10.1016/j.electacta.2020.137142.

Bouwmeester, H. J. M., Song, C., Zhu, J., Yi, J., Van Sint Annaland, M., & Boukamp, B. A. (2009). A novel pulse isotopic exchange technique for rapid determination of the oxygen surface exchange rate of oxide ion conductors. *Physical Chemistry Chemical Physics, 11*(42), 9640–9643. https://doi.org/10.1039/b912712g.

Burriel, M., Téllez, H., Chater, R. J., Castaing, R., Veber, P., Zaghrioui, M., et al. (2016). Influence of crystal orientation and annealing on the oxygen diffusion and surface exchange of $La_2NiO_{4+\delta}$. *Journal of Physical Chemistry C, 120*(32), 17927–17938. https://doi.org/10.1021/acs.jpcc.6b05666.

Chen, K., Ai, N., & Jiang, S. P. (2014). Performance and structural stability of $Gd_{0.2}Ce_{0.8}O_{1.9}$ infiltrated $La_{0.8}Sr_{0.2}MnO_3$ nano-structured oxygen electrodes of solid oxide electrolysis cells. *International Journal of Hydrogen Energy, 39*(20), 10349–10358. https://doi.org/10.1016/j.ijhydene.2014.05.013.

Chen, K., & Jiang, S. P. (2011). Failure mechanism of (La,Sr) MnO_3 oxygen electrodes of solid oxide electrolysis cells.

International Journal of Hydrogen Energy, 36(17), 10541–10549. https://doi.org/10.1016/j.ijhydene.2011.05.103.

Chen, K., & Jiang, S. P. (2016). Review—Materials degradation of solid oxide electrolysis cells. *Journal of the Electrochemical Society, 163*(11), F3070–F3083. https://doi.org/10.1149/2.0101611jes.

De Souza, R. A., Zehnpfenning, J., Martin, M., & Maier, J. (2005). Determining oxygen isotope profiles in oxides with time-of-flight SIMS. *Solid State Ionics, 176*(15–16), 1465–1471. https://doi.org/10.1016/j.ssi.2005.03.012.

Delorme, F., Chen, C., Pignon, B., Schoenstein, F., Perriere, L., & Giovannelli, F. (2017). Promising high temperature thermoelectric properties of dense $Ba_2Co_9O_{14}$ ceramics. *Journal of the European Ceramic Society, 37*(7), 2615–2620. https://doi.org/10.1016/j.jeurceramsoc.2017.01.034.

Den Otter, M. W., Boukamp, B. A., & Bouwmeester, H. J. M. (2001). Theory of oxygen isotope exchange. *Solid State Ionics, 139*(1–2), 89–94. https://doi.org/10.1016/S0167-2738(00)00801-8.

Ebbesen, S. D., Jensen, S. H., Hauch, A., & Mogensen, M. B. (2014). High temperature electrolysis in alkaline cells, solid proton conducting cells, and solid oxide cells. *Chemical Reviews, 114*(21), 10697–10734. https://doi.org/10.1021/cr5000865.

Egger, A., Schrödl, N., Gspan, C., & Sitte, W. (2017). $La_2NiO_{4+\delta}$ as electrode material for solid oxide fuel cells and electrolyzer cells. *Solid State Ionics, 299*, 18–25. https://doi.org/10.1016/j.ssi.2016.10.002.

Firmino, H. C. T., Araújo, A. J. M., Dutra, R. P. S., Nascimento, R. M., Rajesh, S., & Macedo, D. A. (2017). One-step synthesis and microstructure of CuO-SDC composites. *Cerâmica, 63*, 52–57. https://doi.org/10.1590/0366-69132017633652088.

Fulgêncio, E. B. G. A., Loureiro, F. J. A., Melo, K. P. V., Silva, R. M., Fagg, D. P., Campos, L. F. A., et al. (2019). Boosting the oxygen reduction reaction of the misfit $[Ca_2CoO_{3-\delta}]q[CoO_2]$ (C349) by the addition of praseodymium oxide. *Journal of Alloys and Compounds, 788*, 148–154. https://doi.org/10.1016/j.jallcom.2019.02.209.

Garcia, M. F. L., Araújo, A. J. M., Raimundo, R. A., Nascimento, R. M., Grilo, J. P. F., & Macedo, D. A. (2021). Electrical properties of Ca-doped ceria electrolytes prepared by proteic sol-gel route and by solid-state reaction using mollusk shells. *International Journal of Hydrogen Energy, 46*(33), 17374–17387. https://doi.org/10.1016/j.ijhydene.2021.02.151.

Gómez, S. Y., & Hotza, D. (2016). Current developments in reversible solid oxide fuel cells. *Renewable and Sustainable Energy Reviews, 61*, 155–174. https://doi.org/10.1016/j.rser.2016.03.005.

Gopal, C. B., & Haile, S. M. (2014). An electrical conductivity relaxation study of oxygen transport in samarium doped ceria. *Journal of Materials Chemistry A, 2*(7), 2405–2417. https://doi.org/10.1039/C3TA13404K.

He, F., Jiang, Y., Ren, C., Dong, G., Gan, Y., Lee, M. J., et al. (2016). Generalized electrical conductivity relaxation approach to determine electrochemical kinetic properties for MIECs. *Solid State Ionics, 297*, 82–92. https://doi.org/10.1016/j.ssi.2016.10.006.

Hildenbrand, N., Boukamp, B. A., Nammensma, P., & Blank, D. H. A. (2011). Improved cathode/electrolyte interface of SOFC. *Solid State Ionics, 192*(1), 12–15. https://doi.org/10.1016/j.ssi.2010.01.028.

Hu, Y., Thoréton, V., Pirovano, C., Capoen, E., Bogicevic, C., Nuns, N., et al. (2014). Oxide diffusion in innovative SOFC cathode materials. *Faraday Discussions, 176*, 31–47. https://doi.org/10.1039/c4fd00129j.

Irvine, J. T. S., Neagu, D., Verbraeken, M. C., Chatzichristodoulou, C., Graves, C., & Mogensen, M. B. (2016). Evolution of the electrochemical interface in high-temperature fuel cells and electrolysers. *Nature Energy, 1*(1), 1–13. https://doi.org/10.1038/nenergy.2015.14.

Keane, M., Mahapatra, M. K., Verma, A., & Singh, P. (2012). LSM-YSZ interactions and anode delamination in solid oxide electrolysis cells. *International Journal of Hydrogen Energy, 37*(22), 16776–16785. https://doi.org/10.1016/j.ijhydene.2012.08.104.

Khan, M. S., Xu, X., Knibbe, R., & Zhu, Z. (2018). Porous scandia-stabilized zirconia layer for enhanced performance of reversible solid oxide cells. *ACS Applied Materials and Interfaces, 10*(30), 25295–25302. https://doi.org/10.1021/acsami.8b05504.

Khan, M. S., Xu, X., Knibbe, R., & Zhu, Z. (2021). Air electrodes and related degradation mechanisms in solid oxide electrolysis and reversible solid oxide cells. *Renewable and Sustainable Energy Reviews, 143*, 110918. https://doi.org/10.1016/j.rser.2021.110918.

Khan, M. S., Xu, X., Zhao, J., Knibbe, R., & Zhu, Z. (2017). A porous yttria-stabilized zirconia layer to eliminate the delamination of air electrode in solid oxide electrolysis cells. *Journal of Power Sources, 359*, 104–110. https://doi.org/10.1016/j.jpowsour.2017.05.049.

Kilner, J. A., De Souza, R. A., & Fullarton, I. C. (1996). Surface exchange of oxygen in mixed conducting perovskite oxides. *Solid State Ionics, 86–88(Part, 2*, 703–709. https://doi.org/10.1016/0167-2738(96)00153-1.

Kilner, J. A., Skinner, S. J., & Brongersma, H. H. (2011). The isotope exchange depth profiling (IEDP) technique using SIMS and LEIS. *Journal of Solid State Electrochemistry, 15*(5), 861–876. https://doi.org/10.1007/s10008-010-1289-0.

Kim, J., Ji, H.-I., Dasari, H. P., Shin, D., Song, H., Lee, J.-H., et al. (2013). Degradation mechanism of electrolyte and air electrode in solid oxide electrolysis cells operating at high polarization. *International Journal of Hydrogen*

Energy, *38*(3), 1225–1235. https://doi.org/10.1016/j.ijhydene.2012.10.113.

Kim, G., Wang, S., Jacobson, A. J., Yuan, Z., Donner, W., Chen, C. L., et al. (2006). Oxygen exchange kinetics of epitaxial $PrBaCo_2O_{5+\delta}$ thin films. *Applied Physics Letters*, *88*(2), 024103. https://doi.org/10.1063/1.2163257.

Laguna-Bercero, M. A. (2012). Recent advances in high temperature electrolysis using solid oxide fuel cells: A review. *Journal of Power Sources*, *203*, 4–16. https://doi.org/10.1016/j.jpowsour.2011.12.019.

Laguna-Bercero, M. A., Monzón, H., Larrea, A., & Orera, V. M. (2016). Improved stability of reversible solid oxide cells with a nickelate-based oxygen electrode. *Journal of Materials Chemistry A*, *4*(4), 1446–1453. https://doi.org/10.1039/c5ta08531d.

Lane, J. A., & Kilner, J. A. (2000). Measuring oxygen diffusion and oxygen surface exchange by conductivity relaxation. *Solid State Ionics*, *136–137*, 997–1001. https://doi.org/10.1016/S0167-2738(00)00554-3.

Laurencin, J., Hubert, M., Couturier, K., Le Bihan, T., Cloetens, P., Lefebvre-Joud, F., et al. (2015). Reactive mechanisms of LSCF single-phase and LSCF-CGO composite electrodes operated in anodic and cathodic polarisations. *Electrochimica Acta*, *174*, 1299–1316. https://doi.org/10.1016/j.electacta.2015.06.080.

Li, W., Shi, Y., Luo, Y., & Cai, N. (2014). Theoretical modeling of air electrode operating in SOFC mode and SOEC mode: The effects of microstructure and thickness. *International Journal of Hydrogen Energy*, *39*(25), 13738–13750. https://doi.org/10.1016/j.ijhydene.2014.03.014.

Lin, Y. H., Nan, C. W., Liu, Y., Li, J., Mizokawa, T., & Shen, Z. (2007). High-temperature electrical transport and thermoelectric power of partially substituted $Ca_3Co_4O_9$-based ceramics. *Journal of the American Ceramic Society*, *90*(1), 132–136. https://doi.org/10.1111/j.1551-2916.2006.01370.x.

Loureiro, F. J. A., Araújo, A. J. M., Paskocimas, C. A., Macedo, D. A., & Fagg, D. P. (2021). Polarisation mechanism of the misfit Ca-cobaltite electrode for reversible solid oxide cells. *Electrochimica Acta*, *373*, 137928. https://doi.org/10.1016/j.electacta.2021.137928.

Loureiro, F. J. A., Macedo, D. A., Nascimento, R. M., Cesário, M. R., Grilo, J. P. F., Yaremchenko, A. A., et al. (2019). Cathodic polarisation of composite LSCF-SDC IT-SOFC electrode synthesised by one-step microwave self-assisted combustion. *Journal of the European Ceramic Society*, *39*(5), 1846–1853. https://doi.org/10.1016/j.jeurceramsoc.2019.01.013.

Loureiro, F. J. A., Souza, G. S., Graça, V. C. D., Araújo, A. J. M., Grilo, J. P. F., Macedo, D. A., et al. (2019). Nickel-copper based anodes for solid oxide fuel cells running on hydrogen and biogas: Study using ceria-based electrolytes with electronic short-circuiting correction. *Journal of Power Sources*, *438*, 227041–227049. https://doi.org/10.1016/j.jpowsour.2019.227041.

Moçoteguy, P., & Brisse, A. (2013). A review and comprehensive analysis of degradation mechanisms of solid oxide electrolysis cells. *International Journal of Hydrogen Energy*, *38*(36), 15887–15902. https://doi.org/10.1016/j.ijhydene.2013.09.045.

Mogensen, M. B. (2020). Materials for reversible solid oxide cells. *Current Opinion in Electrochemistry*, *21*, 265–273. https://doi.org/10.1016/j.coelec.2020.03.014.

Mogensen, M. B., Chen, M., Frandsen, H. L., Graves, C., Hansen, J. B., Hansen, K. V., et al. (2019). Reversible solid-oxide cells for clean and sustainable energy. *Clean Energy*, *3*(3), 175–201. https://doi.org/10.1093/ce/zkz023.

Na, B. T., Yang, T., Liu, J., Lee, S., Abernathy, H., Kalapos, T., et al. (2021). Enhanced accuracy of electrochemical kinetic parameters determined by electrical conductivity relaxation. *Solid State Ionics*, *361*, 115561. https://doi.org/10.1016/j.ssi.2021.115561.

Nagasawa, K., Daviero-Minaud, S., Preux, N., Rolle, A., Roussel, P., Nakatsugawa, H., et al. (2009). $Ca_3Co_4O_{9-\delta}$: A thermoelectric material for SOFC cathode. *Chemistry of Materials*, *21*(19), 4738–4745. https://doi.org/10.1021/cm902040v.

Nechache, A., Cassir, M., & Ringuedé, A. (2014). Solid oxide electrolysis cell analysis by means of electrochemical impedance spectroscopy: A review. *Journal of Power Sources*, *258*, 164–181. https://doi.org/10.1016/j.jpowsour.2014.01.110.

Ni, M. (2010). Modeling of a solid oxide electrolysis cell for carbon dioxide electrolysis. *Chemical Engineering Journal*, *164*(1), 246–254. https://doi.org/10.1016/j.cej.2010.08.032.

Niedrig, C., Wagner, S. F., Menesklou, W., Baumann, S., & Ivers-Tiffée, E. (2015). Oxygen equilibration kinetics of mixed-conducting perovskites BSCF, LSCF, and PSCF at 900 °C determined by electrical conductivity relaxation. *Solid State Ionics*, *283*, 30–37. https://doi.org/10.1016/j.ssi.2015.11.004.

Nielsen, J., Jacobsen, T., & Wandel, M. (2011). Impedance of porous IT-SOFC LSCF:CGO composite cathodes. *Electrochimica Acta*, *56*(23), 7963–7974. https://doi.org/10.1016/j.electacta.2011.05.042.

Pan, Z., Liu, Q., Ni, M., Lyu, R., Li, P., & Chan, S. H. (2018). Activation and failure mechanism of $La_{0.6}Sr_{0.4}Co_{0.2}Fe_{0.8}O_{3-\delta}$ air electrode in solid oxide electrolyzer cells under high-current electrolysis. *International Journal of Hydrogen Energy*, *43*(11), 5437–5450. https://doi.org/10.1016/j.ijhydene.2018.01.181.

Philippeau, B., Mauvy, F., Mazataud, C., Fourcade, S., & Grenier, J. C. (2013). Comparative study of electrochemical properties of mixed conducting $Ln_2NiO_{4+\delta}$ (Ln = La, Pr and Nd) and $La_{0.6}Sr_{0.4}Fe_{0.8}Co_{0.2}O_{3-\delta}$ as SOFC cathodes associated to $Ce_{0.9}Gd_{0.1}O_{2-\delta}$, $La_{0.8}Sr_{0.2}Ga_{0.8}Mg_{0.2}O_{3-\delta}$ and $La_9Sr_1Si_6O_{26.5}$ electrolytes. *Solid State Ionics*,

249–250, 17–25. https://doi.org/10.1016/j.ssi.2013.06.009.

Ramasamy, D., Nasani, N., Brandão, A. D., Pérez Coll, D., & Fagg, D. P. (2015). Enhancing electrochemical performance by control of transport properties in buffer layers – Solid oxide fuel/electrolyser cells. *Physical Chemistry Chemical Physics*, *17*(17), 11527–11539. https://doi.org/10.1039/c5cp00778j.

Rashkeev, S. N., & Glazoff, M. V. (2012). Atomic-scale mechanisms of oxygen electrode delamination in solid oxide electrolyzer cells. *International Journal of Hydrogen Energy*, *37*(2), 1280–1291. https://doi.org/10.1016/j.ijhydene.2011.09.117.

Rolle, A., Mohamed, H. A. A., Huo, D., Capoen, E., Mentré, O., Vannier, R.-N., et al. (2016). $Ca_3Co_4O_{9+\delta}$, a growing potential SOFC cathode material: Impact of the layer composition and thickness on the electrochemical properties. *Solid State Ionics*, *294*, 21–30. https://doi.org/10.1016/J.SSI.2016.06.001.

Rolle, A., Preux, N., Ehora, G., Mentré, O., & Daviero-Minaud, S. (2011). Potentiality of $Ba_2Co_9O_{14}$ as cathode material for IT-SOFC on various electrolytes. *Solid State Ionics*, *184*(1), 31–34. https://doi.org/10.1016/j.ssi.2010.10.016.

Saher, S., Song, J., Vibhu, V., Nicollet, C., Flura, A., Bassat, J. M., et al. (2018). Influence of annealing at intermediate temperature on oxygen transport kinetics of $Pr_2NiO_{4+\delta}$. *Journal of Materials Chemistry A*, *6*(18), 8331–8339. https://doi.org/10.1039/c7ta08885j.

Saqib, M., Choi, I.-G., Bae, H., Park, K., Shin, J.-S., Kim, Y.-D., et al. (2021). Transition from perovskite to misfit-layered structure materials: A highly oxygen deficient and stable oxygen electrode catalyst. *Energy & Environmental Science*. https://doi.org/10.1039/D0EE02799E.

Shao, Z., & Haile, S. M. (2004). A high-performance cathode for the next generation of solid-oxide fuel cells. *Nature*, *431*, 170–173. https://doi.org/10.1038/nature02863.

Simner, S. P., Anderson, M. D., Engelhard, M. H., & Stevenson, J. W. (2006). Degradation mechanisms of La-Sr-Co-Fe-O_3 SOFC cathodes. *Electrochemical and Solid-State Letters*, *9*(10), A478–A481. https://doi.org/10.1149/1.2266160.

Song, J., Ning, D., Boukamp, B., Bassat, J. M., & Bouwmeester, H. J. M. (2020). Structure, electrical conductivity and oxygen transport properties of Ruddlesden–Popper phases $Ln_{n+1}Ni_nO_{3n+1}$ (Ln = La, Pr and Nd; n = 1, 2 and 3). *Journal of Materials Chemistry A*, *8*(42), 22206–22221. https://doi.org/10.1039/d0ta06731h.

Sousa, A. R. O., Araujo, A. J. M., Souza, G. S., Grilo, J. P. F., Loureiro, F. J. A., Fagg, D. P., et al. (2017). Electrochemical assessment of one-step Cu-CGO cermets under hydrogen and biogas fuels. *Materials Letters*, *191*, 141–144. https://doi.org/10.1016/j.matlet.2016.12.087.

Takahashi, S., Nishimoto, S., Matsuda, M., & Miyake, M. (2010). Electrode properties of the ruddlesden-popper series, $La_{n+1}Ni_nO_{3n+1}$ (n = 1, 2, and 3), as intermediate-temperature solid oxide fuel cells. *Journal of the American Ceramic Society*, *93*(8), 2329–2333. https://doi.org/10.1111/j.1551-2916.2010.03743.x.

Tarancón, A., Burriel, M., Santiso, J., Skinner, S. J., & Kilner, J. A. (2010). Advances in layered oxide cathodes for intermediate temperature solid oxide fuel cells. *Journal of Materials Chemistry*, *20*(19), 3799–3813. https://doi.org/10.1039/b922430k.

Tarancón, A., Marrero-López, D., Peña-Martínez, J., Ruiz-Morales, J. C., & Núñez, P. (2008). Effect of phase transition on high-temperature electrical properties of $GdBaCo_2O_{5+x}$ layered perovskite. *Solid State Ionics*, *179*(17–18), 611–618. https://doi.org/10.1016/j.ssi.2008.04.028.

Téllez, H., Druce, J., Hong, J. E., Ishihara, T., & Kilner, J. A. (2015). Accurate and precise measurement of oxygen isotopic fractions and diffusion profiles by selective attenuation of secondary ions (SASI). *Analytical Chemistry*, *87*(5), 2907–2915. https://doi.org/10.1021/ac504409x.

Usuba, J. B., Araújo, A. J. M., Gonçalves, E. D., Macedo, D. A., Salvo, C., & Viswanathan, M. R. (2019). Flash sintering of one-step synthesized NiO-$Ce_{0.9}Gd_{0.1}O_{1.95}$ (NiO-GDC) composite. *Materials Research Express*, *6*(12), 125535–125544. https://doi.org/10.1088/2053-1591/ab4f97.

Virkar, A. V. (2010). Mechanism of oxygen electrode delamination in solid oxide electrolyzer cells. *International Journal of Hydrogen Energy*, *35*(18), 9527–9543. https://doi.org/10.1016/j.ijhydene.2010.06.058.

Wang, C. C., Chen, K., Jiang, T., Yang, Y., Song, Y., Meng, H., et al. (2018). Sulphur poisoning of solid oxide electrolysis cell anodes. *Electrochimica Acta*, *269*, 188–195. https://doi.org/10.1016/j.electacta.2018.02.149.

Wang, Y., Li, W., Ma, L., Li, W., & Liu, X. (2019). Degradation of solid oxide electrolysis cells: Phenomena, mechanisms, and emerging mitigation strategies—A review. *Journal of Materials Science and Technology*. https://doi.org/10.1016/j.jmst.2019.07.026.

Zapata-Ramírez, V., Mather, G. C., Azcondo, M. T., Amador, U., & Pérez-Coll, D. (2019). Electrical and electrochemical properties of the Sr(Fe,Co,Mo)$O_{3-\delta}$ system as air electrode for reversible solid oxide cells. *Journal of Power Sources*, *437*, 226895. https://doi.org/10.1016/j.jpowsour.2019.226895.

Zheng, Y., Wang, J., Yu, B., Zhang, W., Chen, J., Qiao, J., et al. (2017). A review of high temperature co-electrolysis of H_2O and CO_2 to produce sustainable fuels using solid oxide electrolysis cells (SOECs): Advanced materials and technology. *Chemical Society Reviews*, *46*(5), 1427–1463. https://doi.org/10.1039/c6cs00403b.

CHAPTER

16

Catalysts for hydrogen and oxygen evolution reactions (HER/OER) in cells

Vinicius Dias Silva[a], *Fabio Emanuel França da Silva*[a], *Eliton Souto de Medeiros*[a], and *Thiago Araujo Simões*[a,b,c]

[a]Materials Science and Engineering Postgraduate Program (PPCEM), Federal University of Paraíba (UFPB), João Pessoa, Brazil [b]Postgraduate Program in Science, Innovation and Modeling in Materials (PROCIMM), State University of Santa Cruz—UESC, Ilhéus, BA, Brazil [c]Center for Science and Technology in Energy and Sustainability (CETENS), Federal University of the Recôncavo of Bahia (UFRB), Feira de Santana, Brazil

List of abbreviations

AIL	aprotic ionic liquid
CF	carbon fiber
DFT	density functional theory
ECSA	electrochemically active surface area
FC	full cell
FE	free energy
HER	hydrogen evolution reaction
IEA	International Energy Agency
IL	ionic liquid
JRC	Joint Research Centre
MOF	metal–organic framework
OER	oxygen evolution reaction
ORR	oxygen reduction reaction
PIL	protic ionic liquid
PVP	polyvinylpyrrolidone
RHE	reversible hydrogen electrode
SHE	standard hydrogen electrode
TMC	transition-metal chalcogenide
TMP	transition-metal phosphide
WS	water splitting

16.1 Introduction

Energy is vital for all activities in a contemporary society. The demand for energy in all social segments has increased due to several factors, such as the increase in population, modern lifestyle, large supply of electronic devices, mobility, and industry. Analyses suggest that the energy demand will grow by 56% between the years 2010 and 2040, with 78% of fossil fuels and 22% of renewable sources (Suleman et al., 2015). The International Energy Agency (IEA) predicts that the energy sector will increase CO_2 emissions from 50% in 2030 to 80% in 2050, generating international concern since almost all areas of society depend heavily on energy obtained mainly from fossil fuel sources (da Silva Veras et al., 2017).

Heterogeneous Catalysis
https://doi.org/10.1016/B978-0-323-85612-6.00016-4

The massive use of fossil fuels (oil and its derivatives, mineral coal, and natural gas) has contributed negatively to climate change and pollution in cities. Emissions of polluting gases such as CO, CO_2, SO_x, and NO_x not only directly affect the cause of the greenhouse effect but also affect the public health of large urban centers. The transport sector stands out as one of the biggest contributors to these causes, being the second largest consumer of energy derived from nonrenewable sources (Anantharaj et al., 2018; Baykara, 2018; Suleman et al., 2015; Van Hoecke et al., 2021). This knowledge had led to the consensus of the entire international community regarding the immediate reduction of dependence and use of fossil fuels for power generation. With the awareness that the supply of energy in a totally clean, effective, and sustainable manner is one of the most challenging tasks of this century scientists and engineers, who have been looking for new alternatives for the production of energy using large-scale renewable sources, are not deterred in their endeavors (Guo et al., 2018).

Among renewable energy sources (solar, wind, biomass, geothermal, etc.), technologies based on hydrogen (H_2) as an energy carrier is undoubtedly the most promising (Burton et al., 2021; da Silva Veras et al., 2017; Guo et al., 2018; Suleman et al., 2015; Van Hoecke et al., 2021). By presenting an energy density per mass ($39.42\ kWh\,kg^{-1}$) superior to any other type of fuel, and by offering clean combustion where only water is released as a byproduct, hydrogen can be considered as the best available clean energy vector, especially for engine technologies. However, the application of immediate solutions based on H_2, involving the hydrogen economy, depends a lot on the development of ecologically sustainable production technologies, on distribution, transport, and storage (Muradov & Veziroğlu, 2008; Seh et al., 2017).

Hydrogen storage has been one of the biggest bottlenecks in the implementation of H_2 as a fuel (Moradi & Groth, 2019). Thus, one of the most viable practical solutions would be the consumption of H_2, immediately after its production in an electrolysis cell, for example. In addition to being a safer alternative, this would imply a reduction in storage and transportation costs (Hauch et al., 2020). H_2 can be produced from any compound (renewable or nonrenewable) that contains the H element in its composition, such as natural gas, ethanol, methanol, biomass, algae and bacteria, gasoline, diesel, and water (da Silva Veras et al., 2017). However, production via water splitting (WS) is one of the most environmentally friendly means since it does not use carbon-based raw materials and does not emit pollutants. In this manner, engine engineering combined with electrolysis cells can enable the application in the medium term of hydrogen in vehicles as an alternative to conventional combustion engines (Foorginezhad et al., 2021).

As is known, water splitting occurs via two semi-reactions: the oxygen evolution reaction (OER, anodic) and the hydrogen evolution reaction (HER, cathodic). However, OER offers an energy barrier due to the multistep reaction with electronic transfers. Thus, efforts have been devoted to the development of efficient electrocatalysts that reduce the excess potential consumed by OER, thus making the process feasible. Both IrO_2 and RuO_2 are excellent catalysts for OER (Song et al., 2018). However, high cost, scarcity, and high level of degradation have restricted their use on a large scale, which has led to an interest in the development of electrocatalysts based on carbon materials, earth-abundant transition-metal oxides, Ni–Fe–Co-based alloys, (oxy)hydroxides, transition-metal phosphides (TMPs) and transition-metal chalcogenides (TMCs), and metal–organic frameworks (MOFs) (Lei et al., 2019; Li et al., 2019; Lourenço et al., 2021; Silva, Ferreira, et al., 2019; Silva et al., 2020; Song et al., 2018; Stevens et al., 2019), due to abundance and low cost of materials.

In this chapter, we will briefly introduce the importance of OER in several technical aspects of renewable energy sources and present ways for optimizing the most used electrolysis cells, either by developing (or improving) new electrodes or by electrolyte engineering.

16.2 Hydrogen (H_2) production by water splitting

The production of H_2 by breaking down a water molecule (H_2O) can occur by electrochemical or photoelectrochemical processes (Roger et al., 2017; Tee et al., 2017). However, the electrochemical route for water electrolysis is the most commercially employed process between both. The process called water splitting (WS) consists of breaking a water molecule through the passage of an electric current, producing H_2 (produced in the cathode, negative electrode) and O_2 (produced in the anode, positive electrode) gases, according to the global reaction (Eq. 16.1). The process itself can take place in acid or alkaline electrolytes (pH0–14) (Tahir et al., 2017).

$$2\,H_2O_{(l)} \rightleftharpoons 2\,H_{2(g)} + O_{2(g)} \quad (16.1)$$

Thermodynamically, the reaction (Eq. 16.1) requires an initial energy of $286\,kJ\,mol^{-1}$ under standard conditions of temperature and pressure, that is, 298 K and 1 atm (Roger et al., 2017), respectively.

Water splitting occurs through two semi-reactions, that is, the hydrogen evolution reaction (HER) and the oxygen evolution reaction (OER). HER is a cathodic semi-reaction (Eq. 16.2),

$$2\,H_2O_{(l)} + 2\,e^- \rightarrow 2OH^-_{(aq)} + H_{2(g)} \quad (16.2)$$

whereas OER is an anodic semi-reaction (Eq. 16.3) and proceeds with the oxidation of the OH□ species to form liquid water and oxygen gas, as illustrated in Fig. 16.1.

$$4OH^-_{(aq)} \rightarrow 2\,H_2O_{(l)} + O_{2(g)} + 4\,e^- \quad (16.3)$$

For water-splitting processes to take place, a minimum supply of a theoretical voltage of 1.23 V is required (detailed below) under normal thermodynamic conditions. This is due to OER, which has even slower kinetics due to the multi-step reaction with electron transfer. This additional energy manifests itself as activation energy to proceed with a lower potential and is called an overpotential (). The overpotential represents the additional voltage that must be applied to obtain a given current density (J) that the literature has adopted as reference, $J = 10\,mA\,cm^{-2}$ (Anantharaj et al., 2018). Thus, the function of the electrocatalysts is then to reduce the overpotential as much as possible to values close to the theoretical potential of the global reaction (Song et al., 2018).

Fundamentals of the hydrogen evolution reaction (HER)

Although HER requires much less energy than OER (as we will see below) to trigger the global reaction of water splitting, one of the biggest motivations for the development of HER electrocatalysts is that the best ones at present are based on precious metals, such as platinum (Pt), Pt/C composite, palladium, and ruthenium (Stamenkovic et al., 2016; Zou & Zhang, 2015). This makes the process more expensive and limits the offer of technologies for water splitting. However, the study of materials based on carbon, alloys, and oxides of earth-abundant transition metals as electrocatalysts for OER has gained space in laboratories with extremely expressive results being reported (Zhang et al., 2017; Zou & Zhang, 2015). A criterion for selecting electrocatalysts for HER is based on the free energy of hydrogen adsorption (ΔG_{H*}), the value of which for platinum is theoretically close to zero ($\triangle G_{H*} \approx 0$) (Zeng & Li, 2015). Thus, computational studies using density functional theory (DFT) have aided in the development of

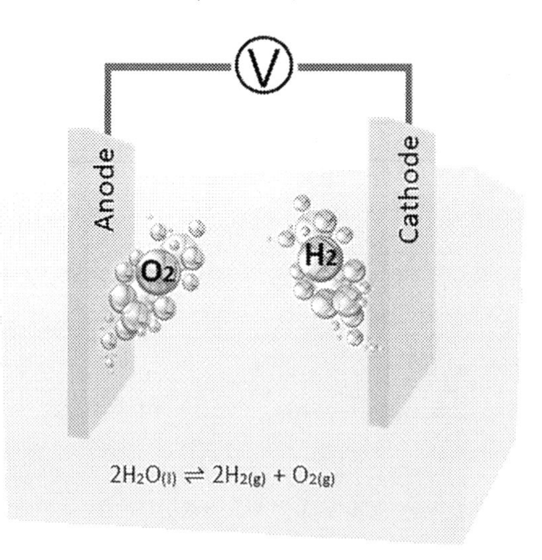

FIG. 16.1 Walter splitting. Schematic representation of electrochemical water splitting.

nonplatinum-based electrocatalysts for HER, evaluating the hydrogen adsorption energy at its active sites (Nørskov et al., 2005). Therefore, the selection of a candidate as a HER catalyst can be performed with the assistance of a "volcano-type" plot based on the work by Medford et al. (2015), which applies the Sabatier principle that states that the best catalysts should be moderately bound to reagents and products during the reaction steps.

HER is a half-reaction with a two-electron transfer, as shown in the global equation in an alkaline electrolyte (Eq. 16.4) (Ojha et al., 2018).

$$2H_2O + 2e^- \rightarrow H_2 + 2OH^- \quad (16.4)$$

However, the classical mechanism of HER consists of a combination of three primary steps (Lei et al., 2019; Li et al., 2019; Lourenço et al., 2021; Silva, Simões, et al., 2019; Silva et al., 2020; Song et al., 2018; Stevens et al., 2019). In the first step (Volmer reaction), molecular H_2O is adsorbed on the electrode surface (*) accompanied by an electron transfer to produce adsorbed hydrogen (H_{ads}) and dissociation (Eq. 16.5).

$$H_2O + e^- + * \leftrightarrow *H_{ads} + OH^- \quad (16.5)$$

In the second step (Heyrovsky reaction), electrochemical hydrogen desorption occurs, where the adsorbed hydrogen atom combines with H_2O and an electron to form molecular hydrogen (Eq. 16.6).

$$H_2O + e^- + *H_{ads} \leftrightarrow * + H_2 + OH^- \quad (16.6)$$

In the third step (Tafel reaction), chemical desorption of H_2 occurs, where two individual hydrogen atoms are combined to form molecular hydrogen (Eq. 16.7).

$$2*H_{ads} \leftrightarrow 2* + H_2 \quad (16.7)$$

It is important to establish that the mechanism of HER in an alkaline medium is still not fully understood (Zou & Zhang, 2015). Moreover, the Heyrovsky and Tafel reactions can occur either alternatively or simultaneously (Safizadeh et al., 2015). Thus, depending on the catalyst, the HER mechanism can occur in just two steps (Ojha et al., 2018): Volmer reaction + Heyrovsky or Tafel reaction.

Fundamentals of the oxygen evolution reaction (OER)

OER is a fundamental reaction for several devices and technologies related to energy generation and storage, such as water splitting, metal–air batteries, fuel cells, proton exchange membrane electrolyzers, and electrolysis cells (Mohammed-Ibrahim, 2020). In OER, molecular oxygen (O_2) is produced through several steps involving protons and electrons. The generally acceptable OER mechanism is the four-electron mechanism and is dependent on the pH of the solution. In an acidic or a neutral medium, two water molecules (H_2O) must be oxidized, generating four protons (H^+) and oxygen molecules (O_2), whereas in an alkaline medium, the hydroxyl groups (OH^-) are oxidized and converted to H_2O and O_2 (Tahir et al., 2017).

Given the adsorbed intermediate species O*, OH*, and OOH*, and being (*) the catalytic active site on the surface, the OER mechanism in an alkaline medium can be described by the following four elementary steps (Eqs. 16.8–16.11) (Suen et al., 2017):

$$OH^- + * \rightarrow OH^- + e^- \tag{16.8}$$

$$OH* + OH^- \rightarrow O* + H_2O + e^- \tag{16.9}$$

$$O* + OH^- \rightarrow OOH* + e^- \tag{16.10}$$

$$OOH* + OH^- \rightarrow * + O_{2(g)} + H_2O + e^- \tag{16.11}$$

In an electrochemical cell composed of two electrodes for water splitting, the Nernst equation for the electrode potential of the cathode (Eq. 16.12) and the anode (Eq. 16.13), respectively, is given by (Zhang et al., 2020):

$$E^0_{\text{cathode}} = E^0_{H_2O/O_2} + \frac{RT}{nF} \ln \frac{(a_{H_2O})\left(f_{O_2}^{1/2}\right)}{a^2_{OH^-}} \tag{16.12}$$

$$E^0_{\text{anode}} = E^0_{H_2O/O_2} + \frac{RT}{nF} \ln \frac{a^2_{H_2O}}{(a_{H_2O})\left(f_{O_2}^{1/2}\right)} \tag{16.13}$$

where is the ideal gas constant, is the temperature in Kelvin, is the number of moles of electrons involved in each mole of reaction, and is the Faraday constant. Thus, considering oxygen and hydrogen ideal gases at 298 K, and as the potential is dependent on pH by shifting 0.059 V per each pH unit increase (Gong & Dai, 2015; Tahir et al., 2017), Eqs. (16.14) and (16.15) can be simplified by:

$$E^0_{\text{cathode}} = pK - 0.059pH \tag{16.14}$$

$$E^0_{\text{anode}} = 1.23 + pK - 0.059pH \tag{16.15}$$

Finally, the cell theoretical standard potential of water splitting at 298 K in an alkaline medium is given by Eq. (16.16):

$$E^0_{\text{cell}} = E^0_{\text{anode}} - E^0_{\text{cathode}} = 1.23\text{ V} \tag{16.16}$$

Under practical conditions (operation), the energy value for deliberating water splitting in a cell is higher than the theoretical 1.23 V for the reversible hydrogen electrode (RHE) since the values of the anode and cathode overpotentials and solution resistance (iR) must be considered. Finally, the cell operation potential (E_{op}) for overall water splitting is given by (Wu et al., 2020) (Eq. 16.17):

$$E_{op} = 1.23 + \eta_a + |\eta_c| + iR \tag{16.17}$$

Therefore, based on the above statements, the theoretical equilibrium potential of HER is 0 V vs. RHE, whereas that of OER is 1.23 V vs. RHE. This means that OER is the reaction that has been limiting H_2 production by electrolysis of water on a large scale due to the slower kinetics owing to the various steps of OER (Li & Zheng, 2017). This justifies the large number of research studies that have been devoted to OER in comparison to HER.

16.3 Improving electrocatalyst materials

Ir/Ru- and Pt-based catalysts have been considered the benchmark for OER and HER (McCrory et al., 2013; Stamenkovic et al., 2007). Platinum (Pt) remains the main electrocatalyst material for HER due to its excellent activity and high current density. In OER, noble metal-based catalysts are considered the best due to their excellent stability for all pH values. Despite their excellent material properties for use as electrocatalysts, their high cost and scarcity still limit their use. Therefore, the global researchers' community has dedicated extensive work to develop new electrocatalysts based on cost-effectiveness. Several electrocatalyst materials have been explored to improve the electrode kinetics and stability in different electrolytes, from non-noble metals to polymers (Kim et al., 2018; Louie & Bell, 2013; Yuan et al., 2020). Not only electrocatalysts have been explored, but also electrolytes. For example, metal alloys, borides, carbides, nitrides, phosphides, and chalcogenides have been explored as competitive HER electrocatalysts in acidic solutions, whereas many nonprecious OER catalysts of metal oxides and (oxy) hydroxides have been reported with promising activities in alkaline media. The main strategies for new nonprecious catalyst development often lead to a complicated synergy of HER and OER electrocatalysts once they are developed in different conditions.

For a material be a good electrocatalyst, properties like high catalytic activity will be required. Both OER and HER are interface reactions that occur between the catalyst and the electrolyte, creating an electrical double layer (EDL). The chemical modifications of electrodes and electrolytes or both improves electrocatalyst efficiency, as seen in doping strategies (Z. P. Wu et al., 2020) or in pH solution change (Dionigi & Strasser, 2016). However, recently, several strategies have been successfully applied including changes in morphology (Silva, Ferreira, et al., 2019; Silva et al., 2020; Wang et al., 2016), substrates (Li et al., 2016), and structural parameters (Tahir et al., 2017; Zhao et al., 2016). The key to designing new electrocatalysts for HER/OER is to understand the mechanisms that facilitate catalytic activity. Fig. 16.2 shows a compilation of OER results (Lei et al., 2019; W. Li et al., 2019; Lourenço et al., 2021; Silva, Ferreira, et al., 2019; Silva et al., 2020; Song et al., 2018; Stevens et al., 2019) for noble metal-based catalysts in 0.1–1 M KOH. The electrocatalyst should demonstrate a low overpotential and Tafel slope with superior stability.

Modifying the electrocatalyst microstructure to achieve larger areas for interface reaction is extremely advantageous for OER. A large specific surface area could provide an efficient spot for solid/liquid/gas phase mass transfer and thus electron conduction could become more efficient. Consequently, catalysts nanostructured with a high surface area that increases the electrolyte wettability ensure an improvement in electrocatalysis. However, the catalytic activity of these structures also depends on their size, shape, and composition. In terms of structure, mesoporous materials (Silva et al., 2020; Silva, Ferreira, et al., 2019; Wang et al., 2016) and hollow fibers (Silva, Simões, et al., 2019) have advantages due to the large surface area that favors the OER process.

Lourenço et al. (2021) tested the mesoporous metal–organic framework Mn/ZIF-67 with different thermal treatments for OER and showed superior electrocatalytic performance in comparison to other Mn-based structures. Wang, Chen, et al. (2020) explored mesoporous Ni-Fe electrocatalysts for OER loaded on nickel foam and produced exceptional results at a turnover frequency value of $0.155\,s^{-1}$ and at an overpotential of 300 mV. These results are impressive when compared to the IrO_2 and RuO_2 catalyst benchmarks.

To obtain fiber morphology, the electrospinning approach is the main synthesis technique. Wang et al. (2016) compared a series of perovskite oxide nanofibers with the morphologies of nanorods and nanoparticles. Compared with them, 1D nanofibers increased the specific surface area and boosted the transport pathways for electrons/ions. An alternative method to fabricate nanofibers is the recently developed

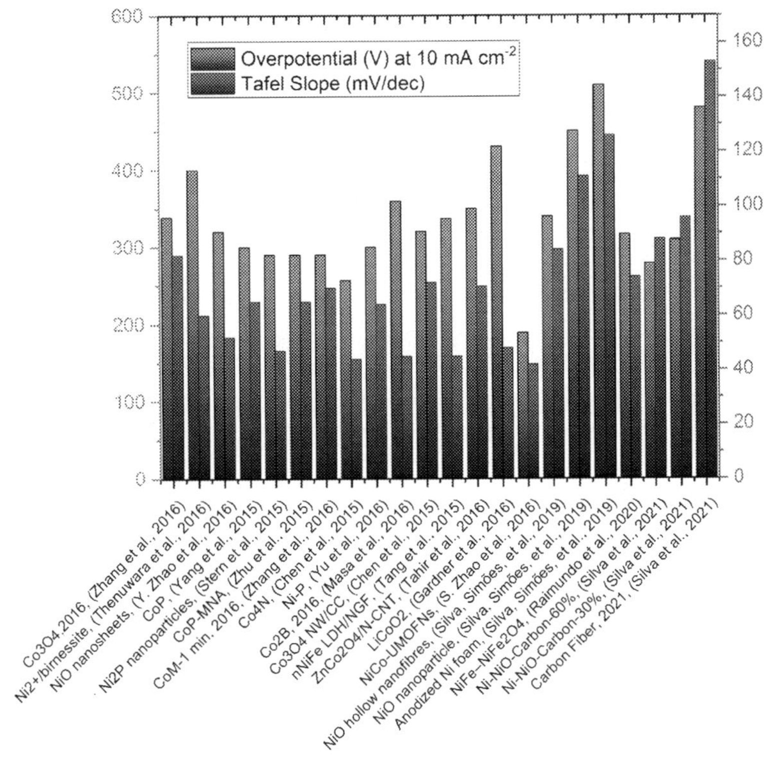

FIG. 16.2 Electrode catalyst performance. Electrode catalyst performance in KOH for OER.

technique called solution blow spinning (SBS). SBS can produce 1D nanostructures by employing pressurized air instead of high voltages with a production rate superior to that of electrospinning (Medeiros et al., 2009). Silva, Ferreira, et al. (2019) produced 1D hollow MFe_2O_4 (M=Cu, Co, and Ni) fibers using the SBS technique. The OER results demonstrated that the electrocatalytic activity of hollow ferrite fibers was better than those of 2D and 3D ferrite-based nanostructures reported in the literature. Fibrillar morphological characteristics were speculated to be the reason for $CuFe_2O_4$ hollow fibers showing higher electrochemically active surface area (ECSA).

Carbon-based catalyst materials are considered a promising low-cost alternative for OER and HER. Carbon materials can act as conductive substrates and highly dispersed support materials, which can improve their catalytic activity by heteroatom doping (You & Sun, 2018). Jin et al. (2013) synthesized cobalt/cobalt oxide/N-doped carbon hybrids (CoOx@CN) with better catalytic activity than cobalt oxides for both oxygen evolution and oxygen reduction reactions, reaching a current density of $20\,mA\,cm^{-2}$ at 1.55 V. Silva et al. (2020) produced NiO/carbon hollow nanofibers by SBS and showed that the carbon from the polyenic branch of polyvinylpyrrolidone (PVP) persists over thermal treatment and acts as an agglomerating agent of the NiO nanoparticles. This guarantees a conductive and percolating route at the NiO nanoparticle/fiber interface. Therefore, carbon-based materials not only provide higher electron conductivity and improve the electrochemical performance of the catalysts compared with metal oxide-based OER catalysts but also can be mass-produced due to lower manufacturing costs.

In general, the electrolysis process needs an electrically conducting solution to take place. For this, an electrolyte should be dissolved in water to improve conductivity. In both electrochemical organic and inorganic processes, the existence of an inert electrolyte may represent a challenge for the separation of products. This is not an issue for water electrolysis since the reaction product is a gas. Even so, the electrolyte can create other problems such as corrosion phenomena and consuming and/or poisoning of electrodes.

Most water electrolysis technologies make use of alkaline solutions (Lin et al., 2018; Qiu et al., 2014; Wang, Chen, et al., 2020). KOH is the most used solution although it is more expensive than NaOH. This is because of two reasons: KOH is more conductive (about 1.3 times) than NaOH and is chemically less aggressive. The concentration, often around 30%, is used because conductivity exhibits the maximum value at this point.

The electrolyte function as a means of transport, which has ionic conduction and electronic insulation properties, facilitating the transport of ions between the electrodes, simultaneously supports the high reduction and oxidation forces generated from the positive and negative electrodes. Furthermore, as desirable characteristics, they must have high ionic conductivity, low viscosity, wide electrochemical window, chemical stability, and low interfacial resistance with the electrodes (Bruce et al., 2012; Endres, 2010). Two recent good candidates to act as electrolytes are ionic liquids for their intrinsic characteristics and solid electrolytes for their applicability in batteries.

Ionic liquids (ILs) have attracted a lot of attention in chemistry, industry, and engineering due to their intrinsic properties. ILs are compounds formed by ions; the appropriate selection of these ions (cations and anions) allows the formation of several structures with properties for specific applications. Ionic liquids are defined as molten salts with a melting point below 100°C; most of them are organic salts with a wide variety of options to be formed and are generally classified as protic or aprotic (Xu, 2014). They are recognized as the third group of solvents and electrolytes, followed by water and organic solvents, and are typically characterized by properties such as high ionic conductivity, high thermal stability, and nonvolatility (Pająk et al., 2020; Watanabe et al., 2017; Xu, 2014). Protic ionic liquids (PILs) are formed by transferring protons from an equimolar combination between an acid and a base. They are promising electrolytes due to their good ionic conductivity, low vapor pressure, and good thermal stability. Protic ionic liquids are generally simpler and cheaper than other ionic liquid options and do not form any byproducts (Angell et al., 2007; Swati et al., 2021). Aprotic ionic liquids (AILs) do not share the common characteristics of LIPs. The preparation of AILs is generally more expensive and complicated than that of PILs, as it always involves multistep reactions. This is because ions are formed by the covalent bond between two functional groups. In most cases, this leads to a more thermally and electrochemically stable solvent than the corresponding PILs (Ara et al., 2020; Atlaskin et al., 2021; Swati et al., 2021).

How do they act in OER and HER?

As mentioned above, the oxygen evolution reaction (OER) is a limiting factor with respect to energy storage and conversion. Due to the slow reaction kinetics, a large overpotential occurs in OER due to various stages of the water hydrolysis reaction that occur in two half-reactions, and, therefore, these barriers need a lot of attention so that these limitations are overcome (Raimundo et al., 2020; Silva et al., 2020). The proton transfer mechanism can occur in two ways in ionic liquids; the first is the successive transfer of protons between the acid and the conjugate base or between the base and the conjugate acid and the second way is the vehicular mechanism through diffusion protonated species (Cecchetto et al., 2012; Temprano et al., 2020; Thimmappa et al., 2020).

How do OER and HER improve?

Ionic liquids act to improve OER and HER by decreasing solubility, increasing cycle stability, serving as a matrix containing the electrode ion (Qin et al., 2020), and influencing the surface structure and morphology, exhibiting less charge transfer resistance, thus increasing the electronic transfer rate (Sun et al., 2020).

Solid electrolytes are widely used in rechargeable lithium–ion batteries, as they have high stable charge and discharge cycles, do not leak, and have low flammability (Chen et al., 2015; Gardner et al., 2016; Masa et al., 2016; Raimundo et al., 2020; Silva, Simões, et al., 2019; Silva et al., 2021; Stern et al., 2015; Tahir et al., 2016; Tang et al., 2015; Thenuwara et al., 2016; Yang et al., 2015; Yu et al., 2016; Zhang et al., 2016; Zhao et al., 2016; Zhu et al., 2015). Solid-state electrolytes have ion transport properties, but there are still points that need to be improved, such as ionic conductivity, stability in the atmospheric environment (air), and the interface between electrodes and electrolytes (Dixit et al., 2020; Shen et al., 2020).

16.4 Perspectives

Each year, with growing investment in the improvement of new renewable energy technologies, the hydrogen economy is on its way to becoming one of the great exponents of the world's energy matrices. The demand for world energy is increasing and the need for a drastic reduction of fossil fuels is dictating the pace of accession of renewable technologies, with emphasis on hydrogen, which appears in this scenario as one of the main protagonists as an energy vector. According to a report by the Joint Research Centre (JRC), which is the European Commission's science and knowledge service, for the fulfillment of the continent's decarbonization plan, the projection is that by 2050, the participation of electrolysis cells in a renewable energy system will exceed the generation of another 1000 GW, behind only solar and wind sources, and surpassing nuclear, thermal, hydroelectric, batteries, and other sources (Kanellopoulos & Reano, 2019).

One of the main limitations of water splitting to produce H_2 on a large scale at present is the dependence on electrodes based on precious metals, which makes the devices more expensive. Therefore, an even greater increase in the exploration of new electrocatalysts for OER and HER is to be expected. Some of the main candidates are materials based on heteroatom-doped nanocarbons, oxides of earth-abundant transition metals, transition-metal nitrides, metal phosphides, MOFs, etc. (Li & Zheng, 2017; Li et al., 2019; Wang, Chen, et al., 2020; Zou & Zhang, 2015). Linked to the study of new materials is the investigation of different morphologies of one-, bi-, and three-dimensional nanostructures (1D, 2D, 3D, respectively). It has been reported that morphology imposes an important rule on the electrocatalyst performance (Li & Zheng, 2017; Roy et al., 2018). In a comparative study between the performance and durability between nanofibers and nanoparticles of nickel oxide (NiO), it was reported that the nanoparticles underwent a degradation process through the coalescence mechanism during OER, due to the high surface energy of the particles, which led to a loss in performance (Silva et al., 2020). Thus, the engineering of nanostructures that circumvent this problem is one of the focus points in the optimization of long-lasting electrodes.

Since the literature has established that the reference current density is $10\,mA\,cm^{-2}$ (Anantharaj et al., 2018), it is of greater interest to develop electrodes with activities close to a density of $500\,mA\,cm^{-2}$ with low overpotential, than the operating range of current commercial electrolysis cells (Lourenço et al., 2021). Another solution that has been used successfully and will become the most usual in the medium/long term is the integration with solar and/or wind systems, where energy can be supplied to electrolysis cells (Khalilnejad et al., 2018; Tee et al., 2017). Although forecasts are quite optimistic, and since hydrogen is already established as one of the main energy vectors in the coming years, both academia and industry agree that major advances are still needed to overcome the economic, storage, and distribution barriers. Once these goals are achieved, H_2 will finally be able to contribute to meet the global demand for clean energy and gradually replace fossil fuels.

16.5 Conclusions

Water electrolysis is a complex technology under constant development. The knowledge gathered in this chapter by studying electrocatalytic developments on extended surfaces can be used to explore new materials and design electrocatalytic structures more efficiently. We explored the basis of OER and HER, as well as some of the most prominent structural techniques, to improve the catalytic performance of non-noble materials. The advantages are in terms of cost–benefit electrocatalyst materials, with good electroconductivity. The challenge is huge, and the complexity and effect of preparation methods, the doping causing structural instability, and the poor conductivity of some metal oxides are some of the problems faced. Only after these issues are properly addressed, the full potential of water electrocatalysis as a viable energy alternative will be explored.

Acknowledgments

The authors thank the Brazilian agencies Coordination for the Improvement of Higher Education Personnel (CAPES, Finance Code 001) and the National Council for Scientific and Technological Development (CNPq) for financial support.

References

Anantharaj, S., Ede, S. R., Karthick, K., Sam Sankar, S., Sangeetha, K., Karthik, P. E., et al. (2018). Precision and correctness in the evaluation of electrocatalytic water splitting: Revisiting activity parameters with a critical assessment. *Energy and Environmental Science, 11*(4), 744–771. https://doi.org/10.1039/c7ee03457a.

Angell, C. A., Byrne, N., & Belieres, J. P. (2007). Parallel developments in aprotic and protic ionic liquids: Physical chemistry and applications. *Accounts of Chemical Research, 40*(11), 1228–1236. https://doi.org/10.1021/ar7001842.

Ara, G., Rahman, A., Halim, M. A., Islam, M. M., Mollah, M. Y. A., Rahman, M. M., et al. (2020). One-pot synthesis of aprotic ionic liquid through solvent-free alkylation of an organic superbase. *Materials Today: Proceedings, 29*, 1020–1024. https://doi.org/10.1016/j.matpr.2020.04.704.

Atlaskin, A. A., Kryuchkov, S. S., Smorodin, K. A., Markov, A. N., Kazarina, O. V., Zarubin, D. M., et al. (2021). Towards the potential of trihexyltetradecylphosphonium indazolide with aprotic heterocyclic ionic liquid as an efficient absorbent for membrane-assisted gas absorption technique for acid gas removal applications. *Separation and Purification Technology, 257*. https://doi.org/10.1016/j.seppur.2020.117835.

Baykara, S. Z. (2018). Hydrogen: A brief overview on its sources, production and environmental impact. *International Journal of Hydrogen Energy, 43*(23), 10605–10614. https://doi.org/10.1016/j.ijhydene.2018.02.022.

Bruce, P. G., Freunberger, S. A., Hardwick, L. J., & Tarascon, J. M. (2012). LigO_2 and LigS batteries with high energy storage. *Nature Materials, 11*(1), 19–29. https://doi.org/10.1038/nmat3191.

Burton, N. A., Padilla, R. V., Rose, A., & Habibullah, H. (2021). Increasing the efficiency of hydrogen production from solar powered water electrolysis. *Renewable and Sustainable Energy Reviews, 135*. https://doi.org/10.1016/j.rser.2020.110255, 110255.

Cecchetto, L., Salomon, M., Scrosati, B., & Croce, F. (2012). Study of a Li-air battery having an electrolyte solution formed by a mixture of an ether-based aprotic solvent and an ionic liquid. *Journal of Power Sources, 213*, 233–238. https://doi.org/10.1016/j.jpowsour.2012.04.038.

Chen, P., Xu, K., Fang, Z., Tong, Y., Wu, J., Lu, X., et al. (2015). Metallic Co4N porous nanowire arrays activated by surface oxidation as electrocatalysts for the oxygen evolution reaction. *Angewandte Chemie, International Edition, 54*(49), 14710–14714. https://doi.org/10.1002/anie.201506480.

da Silva Veras, T., Mozer, T. S., da Costa Rubim Messeder dos Santos, D., & da Silva César, A. (2017). Hydrogen: Trends, production and characterization of the main process worldwide. *International Journal of Hydrogen Energy, 42*(4), 2018–2033. https://doi.org/10.1016/j.ijhydene.2016.08.219.

Dionigi, F., & Strasser, P. (2016). NiFe-based (oxy)hydroxide catalysts for oxygen evolution reaction in non-acidic electrolytes. *Advanced Energy Materials, 6*(23). https://doi.org/10.1002/aenm.201600621.

Dixit, M. B., Singh, N., Horwath, J. P., Shevchenko, P. D., Jones, M., Stach, E. A., et al. (2020). In situ investigation of chemomechanical effects in thiophosphate solid electrolytes. *Matter, 3*(6), 2138–2159. https://doi.org/10.1016/j.matt.2020.09.018.

Endres, F. (2010). Physical chemistry of ionic liquids. *Physical Chemistry Chemical Physics, 12*(8), 1648. https://doi.org/10.1039/c001176m.

Foorginezhad, S., Mohseni-Dargah, M., Falahati, Z., Abbassi, R., Razmjou, A., & Asadnia, M. (2021). Sensing advancement towards safety assessment of hydrogen fuel cell vehicles. *Journal of Power Sources, 489*. https://doi.org/10.1016/j.jpowsour.2021.229450.

Gardner, G., Al-Sharab, J., Danilovic, N., Go, Y. B., Ayers, K., Greenblatt, M., et al. (2016). Structural basis for differing electrocatalytic water oxidation by the cubic, layered and spinel forms of lithium cobalt oxides. *Energy and Environmental Science, 9*(1), 184–192. https://doi.org/10.1039/c5ee02195b.

Gong, M., & Dai, H. (2015). A mini review of NiFe-based materials as highly active oxygen evolution reaction electrocatalysts. *Nano Research, 8*(1), 23–39. https://doi.org/10.1007/s12274-014-0591-z.

Guo, S., Liu, Q., Sun, J., & Jin, H. (2018). A review on the utilization of hybrid renewable energy. *Renewable and Sustainable Energy Reviews, 91*, 1121–1147. https://doi.org/10.1016/j.rser.2018.04.105.

Hauch, A., Küngas, R., Blennow, P., Hansen, A. B., Hansen, J. B., Mathiesen, B. V., et al. (2020). Recent advances in solid oxide cell technology for electrolysis. *Science, 370*(6513). https://doi.org/10.1126/science.aba6118.

Jin, C., Lu, F., Cao, X., Yang, Z., & Yang, R. (2013). Facile synthesis and excellent electrochemical properties of $NiCo_2O_4$ spinel nanowire arrays as a bifunctional catalyst for the oxygen reduction and evolution reaction. *Journal of Materials Chemistry A, 1*(39), 12170–12177. https://doi.org/10.1039/c3ta12118f.

Kanellopoulos, K., & Reano, H. B. (2019). *The potential role of H2 production in a sustainable future power system – An analysis with METIS of a decarbonised system powered by renewables in 2050*. https://doi.org/10.2760/540707.

Khalilnejad, A., Sundararajan, A., & Sarwat, A. I. (2018). Optimal design of hybrid wind/photovoltaic electrolyzer for maximum hydrogen production using imperialist competitive algorithm. *Journal of Modern Power Systems and Clean Energy, 6*(1), 40–49. https://doi.org/10.1007/s40565-017-0293-0.

Kim, J. S., Kim, B., Kim, H., & Kang, K. (2018). Recent Progress on multimetal oxide catalysts for the oxygen evolution reaction. *Advanced Energy Materials, 8*(11). https://doi.org/10.1002/aenm.201702774.

Lei, C., Lyu, S., Si, J., Yang, B., Li, Z., Lei, L., et al. (2019). Nanostructured carbon based heterogeneous electrocatalysts for oxygen evolution reaction in alkaline media. *ChemCatChem, 11*(24), 5855–5874. https://doi.org/10.1002/cctc.201901707.

Li, X., Hao, X., Abudula, A., & Guan, G. (2016). Nanostructured catalysts for electrochemical water splitting: Current state and prospects. *Journal of Materials Chemistry A, 4*(31), 11973–12000. https://doi.org/10.1039/c6ta02334g.

Li, W., Xiong, D., Gao, X., & Liu, L. (2019). The oxygen evolution reaction enabled by transition metal phosphide and chalcogenide pre-catalysts with dynamic changes. *Chemical Communications, 55*(60), 8744–8763. https://doi.org/10.1039/c9cc02845e.

Li, J., & Zheng, G. (2017). One-dimensional earth-abundant nanomaterials for water-splitting electrocatalysts. *Advanced Science, 4*(3). https://doi.org/10.1002/advs.201600380.

Lin, X., Bao, H., Zheng, D., Zhou, J., Xiao, G., Guan, C., et al. (2018). An efficient family of misfit-layered calcium cobalt oxide catalyst for oxygen evolution reaction. *Advanced Materials Interfaces, 5*(23). https://doi.org/10.1002/admi.201801281.

Louie, M. W., & Bell, A. T. (2013). An investigation of thin-film Ni-Fe oxide catalysts for the electrochemical evolution of oxygen. *Journal of the American Chemical Society, 135*(33), 12329–12337. https://doi.org/10.1021/ja405351s.

Lourenço, A. A., Silva, V. D., da Silva, R. B., Silva, U. C., Chesman, C., Salvador, C., et al. (2021). Metal-organic frameworks as template for synthesis of Mn3+/Mn4+ mixed valence manganese cobaltites electrocatalysts for oxygen evolution reaction. *Journal of Colloid and Interface Science, 582*, 124–136. https://doi.org/10.1016/j.jcis.2020.08.041.

Masa, J., Weide, P., Peeters, D., Sinev, I., Xia, W., Sun, Z., et al. (2016). Amorphous cobalt boride (Co_2 B) as a highly efficient nonprecious catalyst for electrochemical water splitting: Oxygen and hydrogen evolution. *Advanced Energy Materials, 6*(6). https://doi.org/10.1002/aenm.201502313.

McCrory, C. C. L., Jung, S., Peters, J. C., & Jaramillo, T. F. (2013). Benchmarking heterogeneous electrocatalysts for the oxygen evolution reaction. *Journal of the American Chemical Society, 135*(45), 16977–16987. https://doi.org/10.1021/ja407115p.

Medeiros, E. S., Glenn, G. M., Klamczynski, A. P., Orts, W. J., & Mattoso, L. H. C. (2009). Solution blow spinning: A new method to produce micro- and nanofibers from polymer solutions. *Journal of Applied Polymer Science, 113*(4), 2322–2330. https://doi.org/10.1002/app.30275.

Medford, A. J., Vojvodic, A., Hummelshøj, J. S., Voss, J., Abild-Pedersen, F., Studt, F., et al. (2015). From the Sabatier principle to a predictive theory of transition-metal heterogeneous catalysis. *Journal of Catalysis, 328*, 36–42. https://doi.org/10.1016/j.jcat.2014.12.033.

Mohammed-Ibrahim, J. (2020). A review on NiFe-based electrocatalysts for efficient alkaline oxygen evolution reaction. *Journal of Power Sources, 448*. https://doi.org/10.1016/j.jpowsour.2019.227375.

Moradi, R., & Groth, K. M. (2019). Hydrogen storage and delivery: Review of the state of the art technologies and risk and reliability analysis. *International Journal of Hydrogen Energy, 44*(23), 12254–12269. https://doi.org/10.1016/j.ijhydene.2019.03.041.

Muradov, N. Z., & Veziroğlu, T. N. (2008). 'Green' path from fossil-based to hydrogen economy: An overview of carbon-neutral technologies. *International Journal of Hydrogen Energy, 33*(23), 6804–6839. https://doi.org/10.1016/j.ijhydene.2008.08.054.

Nørskov, J. K., Bligaard, T., Logadottir, A., Kitchin, J. R., Chen, J. G., Pandelov, S., et al. (2005). Trends in the exchange current for hydrogen evolution. *Journal of the Electrochemical Society, 152*(3), J23–J26. https://doi.org/10.1149/1.1856988.

Ojha, K., Saha, S., Dagar, P., & Ganguli, A. K. (2018). Nanocatalysts for hydrogen evolution reactions. *Physical Chemistry Chemical Physics, 20*(10), 6777–6799. https://doi.org/10.1039/c7cp06316d.

Pająk, M., Hubkowska, K., & Czerwiński, A. (2020). Hydrogen sorption capacity as a tunable parameter in aprotic ionic liquids. *Electrochemistry Communications*. https://doi.org/10.1016/j.elecom.2020.106805, 106805.

Qin, T., Zhao, J., Shi, R., Ge, C., & Li, Q. (2020). Ionic liquid derived active atomic iron sites anchored on hollow carbon nanospheres for bifunctional oxygen electrocatalysis. *Chemical Engineering Journal, 399*. https://doi.org/10.1016/j.cej.2020.125656.

Qiu, Y., Xin, L., & Li, W. (2014). Electrocatalytic oxygen evolution over supported small amorphous Ni-Fe nanoparticles in alkaline electrolyte. *Langmuir, 30*(26), 7893–7901. https://doi.org/10.1021/la501246e.

Raimundo, R. A., Silva, V. D., Medeiros, E. S., Macedo, D. A., Simões, T. A., Gomes, U. U., et al. (2020). Multifunctional solution blow spun NiFe–NiFe2O4 composite nanofibers: Structure, magnetic properties and OER activity. *Journal of Physics and Chemistry of Solids, 139*. https://doi.org/10.1016/j.jpcs.2019.109325.

Roger, I., Shipman, M. A., & Symes, M. D. (2017). Earth-abundant catalysts for electrochemical and photoelectrochemical water splitting. *Nature Reviews Chemistry, 1*. https://doi.org/10.1038/s41570-016-0003.

Roy, C., Sebok, B., Scott, S. B., Fiordaliso, E. M., Sørensen, J. E., Bodin, A., et al. (2018). Impact of nanoparticle size and lattice oxygen on water oxidation on NiFeOxHy. *Nature Catalysis, 1*(11), 820–829. https://doi.org/10.1038/s41929-018-0162-x.

Safizadeh, F., Ghali, E., & Houlachi, G. (2015). Electrocatalysis developments for hydrogen evolution reaction in alkaline solutions – A review. *International Journal of Hydrogen Energy, 40*(1), 256–274. https://doi.org/10.1016/j.ijhydene.2014.10.109.

Seh, Z. W., Kibsgaard, J., Dickens, C. F., Chorkendorff, I., Nørskov, J. K., & Jaramillo, T. F. (2017). Combining theory and experiment in electrocatalysis: Insights into materials design. *Science, 355*(6321). https://doi.org/10.1126/science.aad4998.

Shen, X., Zhang, Q., Ning, T., Liu, T., Luo, Y., He, X., et al. (2020). Critical challenges and progress of solid garnet electrolytes for all-solid-state batteries. *Materials Today Chemistry, 18*. https://doi.org/10.1016/j.mtchem.2020.100368.

Silva, V. D., Ferreira, L. S., Simões, T. A., Medeiros, E. S., & Macedo, D. A. (2019). 1D hollow MFe_2O_4 (M = Cu, Co, Ni) fibers by Solution Blow Spinning for oxygen evolution reaction. *Journal of Colloid and Interface Science, 540*, 59–65. https://doi.org/10.1016/j.jcis.2019.01.003.

Silva, V. D., Raimundo, R. A., Simões, T. A., Loureiro, F. J. A., Fagg, D. P., Morales, M. A., et al. (2021). Nonwoven Ni–NiO/carbon fibers for electrochemical water oxidation. *International Journal of Hydrogen Energy, 46*(5), 3798–3810. https://doi.org/10.1016/j.ijhydene.2020.10.156.

Silva, V. D., Simões, T. A., Grilo, J. P. F., Medeiros, E. S., & Macedo, D. A. (2020). Impact of the NiO nanostructure morphology on the oxygen evolution reaction catalysis. *Journal of Materials Science, 55*(15), 6648–6659. https://doi.org/10.1007/s10853-020-04481-1.

Silva, V. D., Simões, T. A., Loureiro, F. J. A., Fagg, D. P., Figueiredo, F. M. L., Medeiros, E. S., et al. (2019). Solution blow spun nickel oxide/carbon nanocomposite hollow fibres as an efficient oxygen evolution reaction electrocatalyst. *International Journal of Hydrogen Energy, 44*(29), 14877–14888. https://doi.org/10.1016/j.ijhydene.2019.04.073.

Song, F., Bai, L., Moysiadou, A., Lee, S., Hu, C., Liardet, L., et al. (2018). Transition metal oxides as electrocatalysts for the oxygen evolution reaction in alkaline solutions: An application-inspired renaissance. *Journal of the American Chemical Society, 140*(25), 7748–7759. https://doi.org/10.1021/jacs.8b04546.

Stamenkovic, V. R., Mun, B. S., Arenz, M., Mayrhofer, K. J. J., Lucas, C. A., Wang, G., et al. (2007). Trends in electrocatalysis on extended and nanoscale Pt-bimetallic alloy surfaces. *Nature Materials, 6*(3), 241–247. https://doi.org/10.1038/nmat1840.

Stamenkovic, V. R., Strmcnik, D., Lopes, P. P., & Markovic, N. M. (2016). Energy and fuels from electrochemical interfaces. *Nature Materials, 16*(1), 57–69. https://doi.org/10.1038/nmat4738.

Stern, L. A., Feng, L., Song, F., & Hu, X. (2015). Ni2P as a Janus catalyst for water splitting: The oxygen evolution activity of Ni2P nanoparticles. *Energy and Environmental Science*, *8*(8), 2347–2351. https://doi.org/10.1039/c5ee01155h.

Stevens, M. B., Enman, L. J., Korkus, E. H., Zaffran, J., Trang, C. D. M., Asbury, J., et al. (2019). Ternary Ni-Co-Fe oxyhydroxide oxygen evolution catalysts: Intrinsic activity trends, electrical conductivity, and electronic band structure. *Nano Research*, *12*(9), 2288–2295. https://doi.org/10.1007/s12274-019-2391-y.

Suen, N. T., Hung, S. F., Quan, Q., Zhang, N., Xu, Y. J., & Chen, H. M. (2017). Electrocatalysis for the oxygen evolution reaction: Recent development and future perspectives. *Chemical Society Reviews*, *46*(2), 337–365. https://doi.org/10.1039/c6cs00328a.

Suleman, F., Dincer, I., & Agelin-Chaab, M. (2015). Environmental impact assessment and comparison of some hydrogen production options. *International Journal of Hydrogen Energy*, *40*(21), 6976–6987. https://doi.org/10.1016/j.ijhydene.2015.03.123.

Sun, J., Chen, Y. R., Huang, K., Li, K., & Wang, Q. (2020). Interfacial electronic structure and electrocatalytic performance modulation in Cu0.81Ni0.19 nanoflowers by heteroatom doping engineering using ionic liquid dopant. *Applied Surface Science*, *500*. https://doi.org/10.1016/j.apsusc.2019.144052.

Swati, I. K., Sohaib, Q., Cao, S., Younas, M., Liu, D., Gui, J., et al. (2021). Protic/aprotic ionic liquids for effective CO_2 separation using supported ionic liquid membrane. *Chemosphere*, *267*. https://doi.org/10.1016/j.chemosphere.2020.128894.

Tahir, M., Mahmood, N., Pan, L., Huang, Z. F., Lv, Z., Zhang, J., et al. (2016). Efficient water oxidation through strongly coupled graphitic C3N4 coated cobalt hydroxide nanowires. *Journal of Materials Chemistry A*, *4*(33), 12940–12946. https://doi.org/10.1039/c6ta05088c.

Tahir, M., Pan, L., Idrees, F., Zhang, X., Wang, L., Zou, J. J., et al. (2017). Electrocatalytic oxygen evolution reaction for energy conversion and storage: A comprehensive review. *Nano Energy*, *37*, 136–157. https://doi.org/10.1016/j.nanoen.2017.05.022.

Tang, C., Wang, H. S., Wang, H. F., Zhang, Q., Tian, G. L., Nie, J. Q., et al. (2015). Spatially confined hybridization of nanometer-sized NiFe hydroxides into nitrogen-doped graphene frameworks leading to superior oxygen evolution reactivity. *Advanced Materials*, *27*(30), 4516–4522. https://doi.org/10.1002/adma.201501901.

Tee, S. Y., Win, K. Y., Teo, W. S., Koh, L. D., Liu, S., Teng, C. P., et al. (2017). Recent progress in energy-driven water splitting. *Advanced Science*, *4*(5). https://doi.org/10.1002/advs.201600337.

Temprano, I., Liu, T., Petrucco, E., Ellison, J. H. J., Kim, G., Jónsson, E., et al. (2020). Toward reversible and moisture-tolerant aprotic lithium-air batteries. *Joule*, *4*(11), 2501–2520. https://doi.org/10.1016/j.joule.2020.09.021.

Thenuwara, A. C., Cerkez, E. B., Shumlas, S. L., Attanayake, N. H., McKendry, I. G., Frazer, L., et al. (2016). Nickel confined in the interlayer region of Birnessite: An active electrocatalyst for water oxidation. *Angewandte Chemie, International Edition*, *55*(35), 10381–10385. https://doi.org/10.1002/anie.201601935.

Thimmappa, R., Walsh, D., Scott, K., & Mamlouk, M. (2020). Diethylmethylammonium trifluoromethanesulfonate protic ionic liquid electrolytes for water electrolysis. *Journal of Power Sources*, *449*. https://doi.org/10.1016/j.jpowsour.2019.227602.

Van Hoecke, L., Laffineur, L., Campe, R., Perreault, P., Verbruggen, S. W., & Lenaerts, S. (2021). Challenges in the use of hydrogen for maritime applications. *Energy and Environmental Science*, *14*(2), 815–843. https://doi.org/10.1039/d0ee01545h.

Wang, H. F., Chen, L., Pang, H., Kaskel, S., & Xu, Q. (2020). MOF-derived electrocatalysts for oxygen reduction, oxygen evolution and hydrogen evolution reactions. *Chemical Society Reviews*, *49*(5), 1414–1448. https://doi.org/10.1039/c9cs00906j.

Wang, Z., Li, M., Liang, C., Fan, L., Han, J., & Xiong, Y. (2016). Effect of morphology on the oxygen evolution reaction for La0.8Sr0.2Co0.2Fe0.8O3-: δ electrochemical catalyst in alkaline media. *RSC Advances*, *6*(73), 69251–69256. https://doi.org/10.1039/c6ra14770d.

Watanabe, M., Thomas, M. L., Zhang, S., Ueno, K., Yasuda, T., & Dokko, K. (2017). Application of ionic liquids to energy storage and conversion materials and devices. *Chemical Reviews*, *117*(10), 7190–7239. https://doi.org/10.1021/acs.chemrev.6b00504.

Wu, Z. P., Lu, X. F., Zang, S. Q., & Lou, X. W. (2020). Non-noble-metal-based electrocatalysts toward the oxygen evolution reaction. *Advanced Functional Materials*, *30*(15). https://doi.org/10.1002/adfm.201910274.

Xu, K. (2014). Electrolytes and interphases in Li-ion batteries and beyond. *Chemical Reviews*, *114*(23), 11503–11618. https://doi.org/10.1021/cr500003w.

Yang, Y., Fei, H., Ruan, G., & Tour, J. M. (2015). Porous cobalt-based thin film as a bifunctional catalyst for hydrogen generation and oxygen generation. *Advanced Materials*, *27*(20), 3175–3180. https://doi.org/10.1002/adma.201500894.

You, B., & Sun, Y. (2018). Innovative strategies for electrocatalytic water splitting. *Accounts of Chemical Research*, *51*(7), 1571–1580. https://doi.org/10.1021/acs.accounts.8b00002.

Yu, X. Y., Feng, Y., Guan, B., Lou, X. W. D., & Paik, U. (2016). Carbon coated porous nickel phosphides nanoplates for highly efficient oxygen evolution reaction. *Energy and Environmental Science*, *9*(4), 1246–1250. https://doi.org/10.1039/c6ee00100a.

Yuan, N., Jiang, Q., Li, J., & Tang, J. (2020). A review on non-noble metal based electrocatalysis for the oxygen evolution reaction. *Arabian Journal of Chemistry*, *13*(2), 4294–4309. https://doi.org/10.1016/j.arabjc.2019.08.006.

Zeng, M., & Li, Y. (2015). Recent advances in heterogeneous electrocatalysts for the hydrogen evolution reaction. *Journal of Materials Chemistry A*, *3*(29), 14942–14962. https://doi.org/10.1039/c5ta02974k.

Zhang, Y., Ouyang, B., Xu, J., Jia, G., Chen, S., Rawat, R. S., et al. (2016). Rapid synthesis of cobalt nitride nanowires: Highly efficient and low-cost catalysts for oxygen evolution. *Angewandte Chemie, International Edition*, *55*(30), 8670–8674. https://doi.org/10.1002/anie.201604372.

Zhang, L., Xiao, J., Wang, H., & Shao, M. (2017). Carbon-based electrocatalysts for hydrogen and oxygen evolution reactions. *ACS Catalysis*, *7*(11), 7855–7865. https://doi.org/10.1021/acscatal.7b02718.

Zhang, L., Zhao, H., Wilkinson, D. P., Sun, X., & Zhang, J. (2020). *Electrochemical water electrolysis: Fundamentals and technologies*. CRC Press.

Zhao, S., Wang, Y., Dong, J., He, C. T., Yin, H., An, P., et al. (2016). Ultrathin metal-organic framework nanosheets for electrocatalytic oxygen evolution. *Nature Energy*, *1*(12). https://doi.org/10.1038/nenergy.2016.184.

Zhu, Y. P., Liu, Y. P., Ren, T. Z., & Yuan, Z. Y. (2015). Self-supported cobalt phosphide mesoporous nanorod arrays: A flexible and bifunctional electrode for highly active electrocatalytic water reduction and oxidation. *Advanced Functional Materials*, *25*(47), 7337–7347. https://doi.org/10.1002/adfm.201503666.

Zou, X., & Zhang, Y. (2015). Noble metal-free hydrogen evolution catalysts for water splitting. *Chemical Society Reviews*, *44*(15), 5148–5180. https://doi.org/10.1039/c4cs00448e.

CHAPTER

17

Zeolitic imidazolate framework 67 based metal oxides derivatives as electrocatalysts for oxygen evolution reaction

Annaíres de A. Lourenço and Fausthon F. da Silva

Department of Chemistry, Federal University of Paraíba, UFPB, João Pessoa, PB, Brazil

17.1 Introduction

Water splitting and hydrogen generation

Energy is essential for all human life activities. Socioeconomic development is directly related to energy consumption, thus increasing the demand for energy resources in recent years. From the 18th century to the present, fossil fuels (oil and its derivatives, coal, and natural gas) are the main energy sources of society (Veras, Mozer, dos Santos, & César, 2017); however, concerns regarding the development of renewable and sustainable energies are growing every day, especially due the greenhouse effect that is related to high emissions of carbon dioxide (CO_2). The International Energy Agency (IAE) has stated that the energy sector will increase CO_2 emissions from 50% in 2030 to 80% by 2050, causing grave environmental concerns (Veras et al., 2017). Thus, there is an urgent need to develop clean, renewable, and efficient alternative technologies to replace fossil fuels (Guo, Liu, Sun, & Jin, 2018; Höök & Tang, 2013; Suleman, Dincer, & Agelin-Chaab, 2014).

Sustainable energy involves the obtention of energy vectors that do not present substantial exhaustion even after their continuous use, causing reduced environmental impacts, and also do not pose risks to human health (Baykara, 2018; Guo, Youliwasi, et al., 2018). Among sources with these prerequisites, hydrogen gas (H_2) is considered by many researchers to be the key to the development of clean and sustainable energy and is one of the most promising energy sources for oil replacement (Baykara, 2018). Hydrogen is considered an energy carrier with the potential for energy storage; as an energy source, it can be used to produce electricity, thus acting as an excellent technological solution for the production of fully renewable and clean energy, and can also be used as fuel in the transport sector,

Heterogeneous Catalysis
https://doi.org/10.1016/B978-0-323-85612-6.00017-6

in the wind, solar, and fuel cell systems, and in the hybrid renewable energy systems (César, Veras, Mozer, dos Santos, & Conejero, 2019; Guo, Youliwasi, et al., 2018).

Hydrogen it is the most abundant element in the universe, accounting for about 75% (in weight) of its normal matter. Molecular hydrogen (H_2) shows the highest amount of energy per mass unit. H_2 combustion results in water, making it an excellent energy source, and virtually no pollutant is emitted, depending on the obtention of this molecule (Suleman, Dincer, & Agelin-Chaab, 2015). Hydrogen is currently produced on a large scale through the use of natural gas; however, this process is not renewable as it releases huge amounts of CO_2, and, thus, it is necessary to employ other means (Armaroli & Balzani, 2011; El-Emam & Özcan, 2019). Most of the technological investments for large-scale hydrogen production are based on the electrochemical decomposition of water (water splitting), a renewable and economical methodo logy to produce H_2 with high purity (Armaroli & Balzani, 2011; Guo, Youliwasi, et al., 2018). However, a large amount of electricity is required to break the O–H bonds in water, forming H_2 and O_2, especially due to one of the half-reactions involved in water electrolysis, namely, the *oxygen evolution reaction* (OER) (Veras et al., 2017; You & Sun, 2018). To overcome this barrier, efforts are being taken to obtain efficient electrocatalysts capable of making OER viable for the production of H_2 (Khan et al., 2018; Roger, Shipman, & Symes, 2017; Tahir et al., 2017).

Water splitting is basically performed to break the chemical bonds in water by electrolysis, forming hydrogen gas (in the cathode, negative electrode) and oxygen gas (in the anode, positive electrode); however, the reactions involved depend on the pH of the solution. Eq. (17.1)–(17.3) represent the half-reactions in water splitting in a basic medium along with its standard potential measured using a reversible hydrogen electrode (RHE) (Fabbri & Schmidt, 2018). This process takes place through redox water decomposition that occurs non-spontaneously (the breaking of water is thermodynamically difficult and requires near 286 $kJ\,mol^{-1}$) at room temperature and pressure (Roger et al., 2017). The high energy cost is attributed to the high consumption of electricity, which is the main problem in using this technology on a large scale (Tahir et al., 2017).

$$4H_2O_{(l)} + 4e^- \rightleftarrows 2H_{2(g)} + 4OH^-_{(aq)} \quad (E_{cátodo} = 0.0\,V\,vs\,RHE)\,(HER) \tag{17.1}$$

$$4OH^-_{(aq)} \rightleftarrows O_{2(g)} + 2H_2O_{(l)} + 4e^- \quad (E_{ânodo} = -1.23\,V\,vs\,RHE)\,(OER) \tag{17.2}$$

$$2H_2O_{(l)} \rightleftarrows 2H_{2(g)} + O_{2(g)}\ (\Delta E = -1.23\,V) \quad (Global\ reaction) \tag{17.3}$$

For OER and hydrogen evolution reaction (HER) to occur, it is necessary to provide a potential above 1.23 V at 25°C. This additional energy, called overpotential (η), is the activation energy, which must assume the lowest value to obtain the desired current density. In this manner, electrocatalysts are employed to achieve lower overpotentials, that is, a lower energy consumption (Fabbri & Schmidt, 2018; Tahir et al., 2017).

Oxygen evolution reaction

In OER, molecular oxygen (O_2) is produced through several steps involving electron and proton transfers, depending on the pH. When the solution is acidic or neutral, two water molecules (H_2O) are oxidized, thus generating four protons (H^+) and oxygen molecules (O_2) (Fabbri & Schmidt, 2018; Tahir et al., 2017). In an alkaline medium, the production of O_2 occurs through multiple steps with the transfer of four electrons. Due to the accumulation of energy in each step, the kinetics becomes slow and the reaction is not favorable, which results in a large overpotential and requires the use of catalysts (Tahir et al., 2017). A general metal-catalyzed mechanism of OER is described in reactions (Eqs. 17.4–17.7) proposed by Krasil'shchikov (Li, Anderson, Chen, Pan, & Chuang, 2018).

$$M^* + OH^- \rightarrow M^*OH + e^- \quad (17.4)$$

$$M^*OH + OH^- \rightarrow M^*O^- + H_2O \quad (17.5)$$

$$M^*O^- \rightarrow M^*O + e^- \quad (17.6)$$

$$2M^*O \rightarrow 2M^* + O_2 \quad (17.7)$$

OER is relevant not only in water electrolysis but also in other energy generation and storage devices such as metal-air batteries and fuel cells (Zhao, Yan, Chen, & Chen, 2017). In this manner, efficient catalysts are needed in this reaction. Methods to improve catalytic activity (i.e., to reduce overpotential) resistant to corrosion have to work through a wide pH range have long-term durability, which are the two barriers in the development of efficient electrodes in OER (Li, Anderson, et al., 2018). The most efficient electrocatalysts in the literature are based on noble metals, mainly iridium and ruthenium oxides (IrO_2 and RuO_2) (Fabbri & Schmidt, 2018; Jiao, Zheng, Jaroniec, & Qiao, 2015). Despite their excellent performances, their large-scale use is difficult because they are scarce, high-cost materials with low durability (Chen, Muckerman, & Fujita, 2013; Roger et al., 2017). Therefore, the production of alternative electrocatalysts with high efficiency, which are economically accessible, have large abundance, and with performance equivalent to those of conventional electrocatalysts, is necessary. Many types of materials have been the subject of focus as efficient electrocatalysts in OER such as sulfides (Gu, Fan, & Liu, 2017), nitrides (Xu et al., 2015), hydroxides (Burke, Enman, Batchellor, Zou, & Boettcher, 2015), metallic nanoparticles (NPs) (Bizzotto et al., 2019), and, mainly, metal oxides (Bo, Dastafkan, & Zhao, 2019; Liang, Chen, Wang, & Li, 2020). Metal oxides based on transition metals with various chemical compositions and morphologies can be obtained by many synthetic methods that are already well-established in the literature, such as sol-gel and combustion. However, the use of coordination polymers/metal-organic frameworks as a template to obtain these oxides has gained attention in the recent literature (Liang et al., 2020; Sun et al., 2015a).

Coordination polymers

In the recent decades, interest in porous materials has been growing due to their multifunctional nature, since one porous solid can be applied to several research areas, such as the issue of environmental pollutants, fuel storage, generation of clean and renewable energy, synthesis of new chiral drugs, and so on (Davis, 2002; Decurtins, Pellaux, Antorrena, & Palacio, 1999; Yaghi et al., 2003). Over time, experimental efforts have been taken to optimize the main physicochemical properties of porous materials, such as surface area, pore size, and shape as well as to increase their chemical and structural diversity (Davis, 2002; Decurtins et al., 1999; Yaghi et al., 2003).

For a long time, zeolites were the main representatives of inorganic porous solids, due to the efficient synthetic methods of aluminosilicates as well as their performance in adsorptive and catalytic processes (Caro, Noack, Kölsch, & Schäfer, 2000; Cundy & Cox, 2003). However, with the works of Robson (Hoskins and Robson, 1990) and Yaghi (Yaghi & Li, 1995b) in the 1990s, coordination polymers (CPs) have gained popularity due to their chemical versatility, compared to other classical inorganic porous materials like zeolites and activated carbons (Furukawa, Cordova, O'Keeffe, & Yaghi, 2013). According to IUPAC, CPs are inorganic chemical compounds based on metal ions and/or clusters and multidentate organic ligands, forming uninterrupted sequences of metal-ligand chemical bonds in one (1D), two (2D), or three (3D) crystallographic dimensions (Fig. 17.1) (Batten et al., 2012). These organic (ligand) and

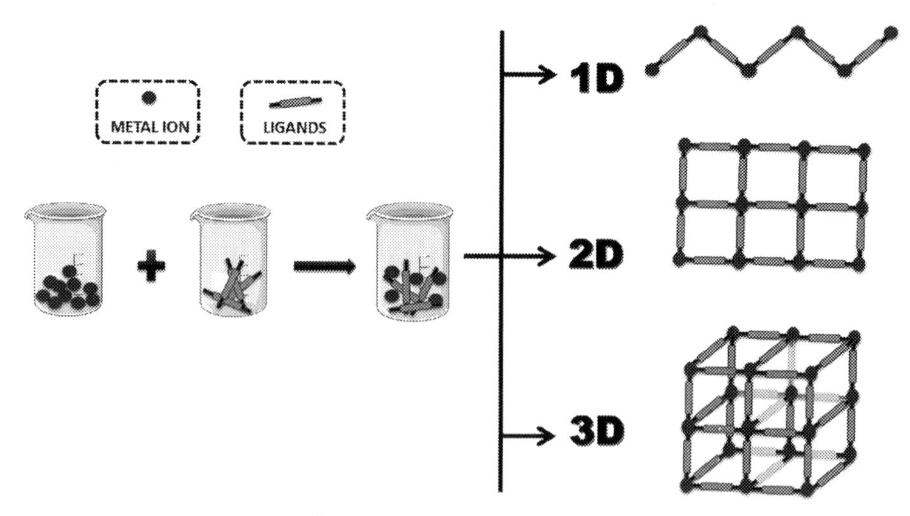

FIG. 17.1 Illustration of 1D, 2D, and 3D coordination polymers.

inorganic (metal cations) units, called secondary building units (SBUs), are interconnected by coordinate covalent bonds, with properties such as structural robustness, possibility of ligand postsynthetic modification, and high crystallinity, thus allowing to establish relationships between structures and properties (Janiak & Vieth, 2010).

The term "metal-organic frameworks" (MOFs), introduced by Yaghi and Li (1995a), is defined by IUPAC as a 3D CP containing potential empty voids (Batten et al., 2012). This term certainly has the greatest impact in the literature, and most articles have started to use the terms "MOFs" and "CPs" as synonyms to designate this class of inorganic compounds. In 1999, the isoreticular series of MOF-5 reported by Li, Eddaoudi, O'Keeffe, and Yaghi (1999) revolutionized coordination polymers from a synthesis and design point of view. MOF-5 consists of the 1,4-benzenedicarboxylic acid (1,4-BDC) ligand connected to ZnO_4 tetrahedrons, forming a cubic 3D network with a surface area of 2900 m^2/g (Li et al., 1999). In the same year, the synthesis of HKUST-1, formed by the coordination bond between copper ions and the 1,3,5-benzenetricarboxylic acid (BTC) ligand showing a surface area equal to 2100 m^2/g, was reported by Williams (Chui, Lo, Charmant, Orpen, & Williams, 1999). These two compounds, along with MIL-53 (Cr-BDC, 1500 m^2/g) (Millange, Serre, & Férey, 2002), are some of the most important MOF structures in the literature. Fig. 17.2 shows a hierarchical flowchart containing the definitions of coordination compounds, coordination polymers, and MOFs recommended by IUPAC.

Zeolitic imidazolate frameworks

The imidazole (Im) ligand and its derivatives (Fig. 17.3) have been the subject of study for decades, mainly due to their pharmacological activity such as antiviral and antitumor properties (Bhatnagar, Sharma, & Kumar, 2011; Verma, Joshi, & Singh, 2013). In the mid-20th century, coordination compounds containing imidazole-derived ligands were studied to understand the interaction between metal ions and biomolecules with these aromatic rings (Barnard & Stein, 2006; Bauman & Wang, 1964; Cowgill & Clark, 1952; Harkins & Freiser, 1956; Sundberg & Martin, 1974; Walker, Lo, & Ree, 1976). More recently,

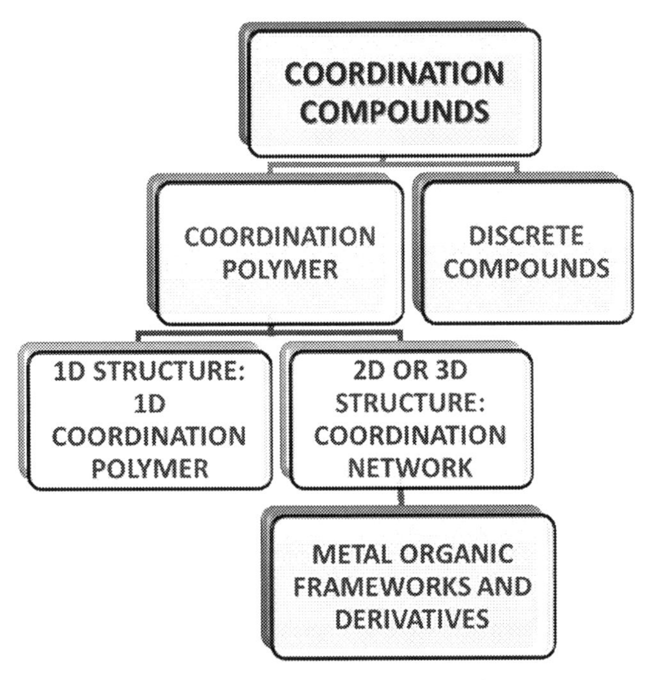

FIG. 17.2 Hierarchical flowchart of the concepts of coordination compounds.

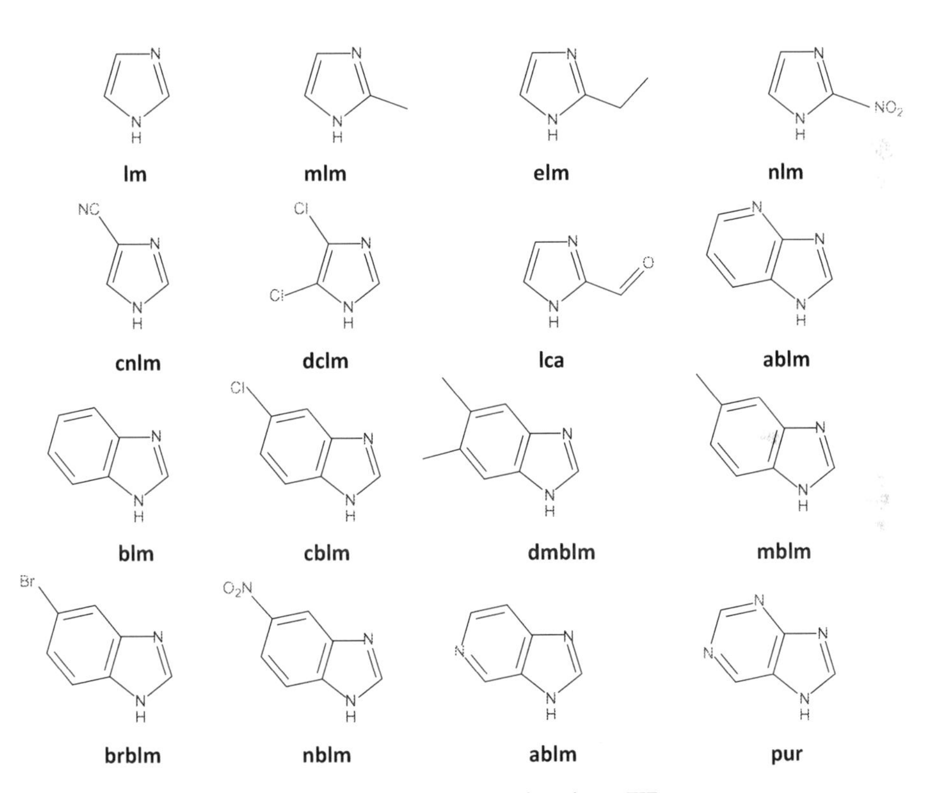

FIG. 17.3 Imidazole (Im) ligand and some of its derivatives, used to obtain ZIF structures.

Chen et al. have reported substitutions in the C-2 position in imidazole and their influence on the crystalline structure of CPs containing Cu^+ (Huang, Zhang, & Chen, 2004; Huang, Zhang, Lin, & Chen, 2005). While imidazole tends to form 1D coordination polymers with a zigzag chain, substitutions in the C-2 position lead to the formation of different polygonal and heliacal structures (Huang et al., 2004, 2005). Although the structures obtained do not present appreciable porosity, these studies have started the development of a new subclass of MOFs, which is topologically isomorphous to zeolites. Due this fact, they are called zeolitic imidazolate frameworks (ZIFs) (Banerjee et al., 2008; Huang, Lin, Zhang, & Chen, 2006; Park et al., 2006).

ZIFs are essentially MOFs formed by metal ions (typically Zn^{2+} or Co^{2+}) tetracoordinated with imidazole-based ligands, with an M–Im–M bond angle near 145° (Banerjee et al., 2008; Park et al., 2006). As already mentioned, ZIFs have topologies similar to those of zeolites, such as sodalite (sod), Linde Type A (LTA), gmelinite (gme), diamond (dia), and merlinoite (mer) (Chen et al., 2013; Roger et al., 2017). More than 150 ZIF structures have already been reported in the literature, showing large chemical and structural diversity, thermal/chemical stability, intrinsic porosity, and surface areas typically between 1000 and 2000 m^2/g (Banerjee et al., 2008). These materials have high applicability in the catalysis of inorganic reactions (Chen, Yang, Zhu, & Xia, 2014), gas separation/adsorption (Pimentel, Parulkar, Zhou, Brunelli, & Lively, 2014), purification of aqueous and gaseous systems (Mirqasemi, Homayoonfal, & Rezakazemi, 2020), controlled drug release (Feng, Zhang, Shi, & Wang, 2021), and electrochemical sensors (Zhang, Tan, & Song, 2020).

ZIF-8 and ZIF-67 are two of the most explored ZIFs in the literature, formed by 2-methylimidazole (mIM) coordinated to Zn^{2+} (ZIF-8) and Co^{2+} (ZIF-67) ions (Feng et al., 2021; Zhong, Liu, & Zhang, 2018). ZIF-8 was reported for the first time in 2006 (Huang et al., 2006), obtained by crystallization in solution at room temperature, with a reaction time of 1 month and a yield of 70% (Huang et al., 2006). This material was later best characterized by Yaghi et al. (Park et al., 2006) obtained via a solvothermal reaction in *N,N*-dimethylformamide (DMF) with a yield of 25%. ZIF-67 was first reported in 2008 (Banerjee et al., 2008) also by solvothermal synthesis; however, currently, it can be obtained by many synthetic methods (Zhong et al., 2018). Since then, ZIF-67 has been mainly used as a precursor to obtain many types of materials such as oxides, metal nanoparticles (NPs), and carbon-based composites, applied in energy generation and storage (Ahmad, Khan, Iqbal, & Noor, 2020; Cheng, Ren, Xu, Du, & Dou, 2018; Dutta, Liu, Han, Indra, & Song, 2018). Although ZIF-8 (Xu, Wang, Liu, & Yan, 2017) and other MOF derivatives (Hou & Xu, 2019; Liang et al., 2020) also have to be applied in this field of research, in this chapter, we will explore ZIF-67 and its derivatives in depth.

17.2 Recent advances

ZIF-67 and its derivatives

In ZIF-67 (cobalt 2-methylimidazole, Fig. 17.4), Co^{2+} ions are in a tetrahedral geometry (coordination number 4), coordinated by nitrogen atoms of the 2-methylimidazole anions (mIM^-) (Zhong et al., 2018). This ZIF structure presents the chemical formula $C_8H_{10}N_4Co$, with a specific surface area near 1500 m^2/g, crystallizing with the cubic space group with *I4-3m* symmetry and lattice parameters $a = b = c = 16.9589$ Å (Zhong et al., 2018). Initially obtained by a solvothermal method (Banerjee et al., 2008), Gross, Sherman, and Vajo (2012) demonstrated the obtention of ZIF-67 in an aqueous solution. The synthesis in an aqueous or methanolic solution is extensively

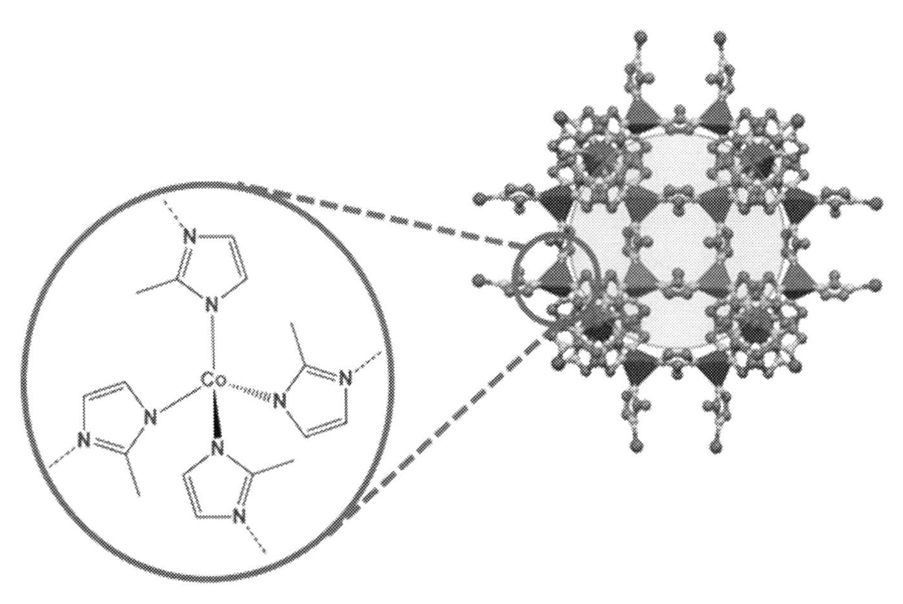

FIG. 17.4 Structure of ZIF-67.

studied and allows the obtention of ZIF-67 quickly, with high yield and low cost (Guo, Liu, et al., 2018; Höök & Tang, 2013; Suleman et al., 2014). However, other synthetic methodologies can also be used to obtain this material, such as hydrothermal synthesis (Qian, Sun, & Qin, 2012), accelerated aging (Mottillo et al., 2013), laser-induced synthesis (Ribeiro, Davari, Hu, Mukherjee, & Khomami, 2019), and microwave-assisted synthesis (Zhang, Tan, et al., 2020).

The typical morphology of ZIF-67 crystals is dodecahedral (Zhong et al., 2018); meanwhile, size and morphological control can be performed using several experimental methodologies (Feng & Carreon, 2015; Guo, Xing, Lou, & Chen, 2016), impacting many properties including surface area. As demonstrated by Chen (Guo et al., 2016), purity, particle size, and morphology can be controlled using hydrothermal or nonhydrothermal syntheses. In this case, the stoichiometric ratio and the counter ion of the metallic salt play a fundamental role, obtaining particles sizes between 120 and 550 nm and several types of morphologies (spherical, rhombic dodecahedron, truncated rhombic dodecahedral, granular, hexagonal disks, etc.) (Guo et al., 2016). Morphological and dimensional control can also be performed using microemulsion with surfactant agents such as cetrimonium bromide (CTAB) or pluronic P123, obtaining uniformly sized nanocrystals with high purity (Sun, Zhai, & Zhao, 2016). ZIF-67 is also quite versatile and can be prepared in the form of thin films (Han, Yuan, Zhang, & Zhang, 2019), nanosheets (Wang, Xu, et al., 2018), nanocrystals (Qian et al., 2012), and nanofibers (Sankar, Karthick, Sangeetha, Karmakar, & Kundu, 2020). It can also form composites with many types of organic and inorganic materials (Li, 2018; Sundriyal et al., 2018). It is important to remember that morphological and dimensional control is of fundamental importance for applications involving surface phenomena, including electrocatalysis for OER (Seo et al., 2016; Silva, Simões, Grilo, Medeiros, & Macedo, 2020).

General aspects in the obtention of ZIF-67 derivatives

Cobalt-based materials have been high lighted as promising systems for the development of energy storage and conversion (Wang, Xu, et al., 2018). Due to this, several experimental methods have been used to obtain cobalt-based materials in order to reduce cost and reaction time, with special focus on the production of nanostructures (Hua, Li, Chen, & Pang, 2019; Li, Hao, Abudula, & Guan, 2019; Maiti, 2020; Wang et al., 2016; Zhang, Cui, & Liu, 2020). In 2010, Xu et al. first demonstrated the obtention of cobaltite nanoparticles (Co_3O_4 NPs) from the MOF $[Co_3(NDC)_3(DMF)_4]$ (where NDC = 2,6-naphthalene dicarboxylate and DMF = *N,N′*-dimethylformamide) through heat treatment at 600°C under atmospheric air (Liu et al., 2010)(). The calcination of this MOF resulted in clusters of Co_3O_4 NPs with spheroidal morphology, with particle size near 25 nm and a surface area of $5.3\,m^2/g$ (Liu et al., 2010). In 2011, Chen (Chen, Zhao, Wei, Wang, & Gu, 2011) and Morsali (Parast & Morsali, 2011) showed the synthesis of CuO and Al_2O_3 nanoparticles by heat treatment of HKUST-1 and MIL-53, respectively. Later in the same year, Wang and coworkers demonstrated for the first time the obtention of cobaltite NPs from direct calcination of ZIF-67 at 600°C in air (Wang et al., 2011). These pioneering works opened new horizons for the use of MOFs as a sacrificial template to obtain new nanostructured materials applied in various fields of research, including as catalysts for the OER (Han et al., 2019).

By 2020, more than 99,000 coordination polymers were listed in the Cambridge Structural Database (Moghadam, 2020). Their structural and compositional diversity makes the production of nanomaterials from MOFs a vast and explored field of research. In addition, MOFs can be obtained in the form of 0D (NPs and core/shell), 1D (nanorods and nanotubes), 2D (nanosheets, nanoplates, nanoflakes, and nanofilms), or 3D (arrays, flowers, honeycombs, foams, and sponges) nanostructures. Besides that, they can also be supported in other materials such as carbon nanotubes, zeolites, and porous silica, among others, which increases the possibilities to be explored (Dang, Zhu, & Xu, 2018; Zhu & Xu, 2014). In this chapter, we will focus on the general aspects of this theme, with emphasis mainly on ZIF-67 as a template.

Compared to traditional methods, the obtention of nanostructured materials such as porous carbons, metal oxides, and metal NPs through heat treatment of MOFs has several advantages, such as the formation of uniform particles and high-dimensional shape and composition control in a single synthesis step (Dang et al., 2018; Liu et al., 2020). The most attractive factor is the conservation of the original morphology, due to the high crystallinity and high porosity of MOFs, thus being considered a self-templating and external templating method (Dang et al., 2018). One example is the production of porous carbon from the thermolysis of MOF-5, showing a surface area equal to $2872\,m^2/g$ (Liu, Shioyama, Akita, & Xu, 2008). Notably, ZIF-67 is easily obtained in solution at room temperature, with low cost and short reaction times. The synthetic methods of ZIF-67 can regulate the size and morphology of crystals, thus offering a high control of their physicochemical properties, with the crystallization kinetics already well-studied (Sun & Xu, 2014; Xia et al., 2014).

Several types of materials can be obtained from the thermolysis of MOFs such as porous and/or graphitic carbons, metal NPs, metal oxide nanoparticles, carbon/metal oxide composites, metal sulfides, and metal phosphides (Fig. 17.5) (Chen, Zhang, Jiao, & Jiang, 2018; Fu, Zhu, Song, Du, & Lin, 2017; Guan, Yu, Wu, & Lou, 2017; Han, Li, Ma, & Yang, 2020; Salunkhe, Kaneti, Kim, & Yamauchi, 2016; Salunkhe, Kaneti, & Yamauchi, 2017; Sanati et al., 2020; Wang & Astruc, 2020; Wu & Lou, 2017). Their structure, morphology, and chemical composition can be controlled by the methodology employed and experimental variables.

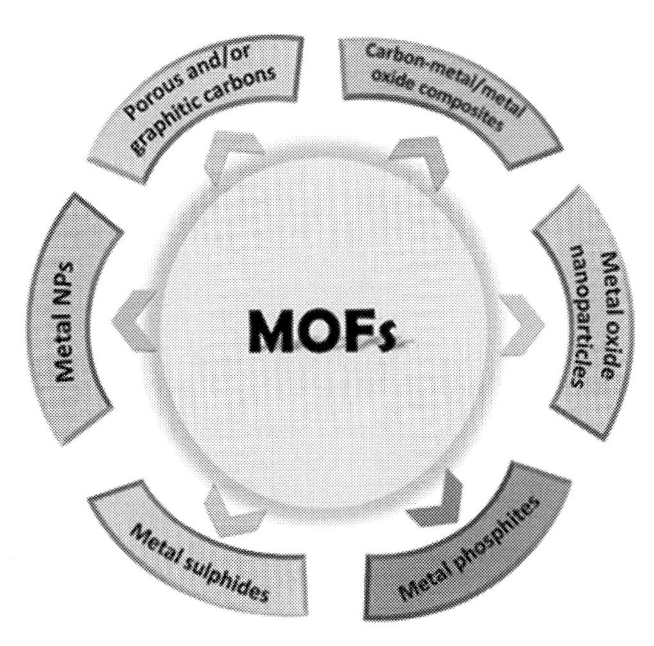

FIG. 17.5 MOF-derived nanostructured materials.

The literature shows a wide range of experimental methodologies for the heat treatment of MOFs in order to obtain new nanostructured materials, which may also include pre- or post-treatment combinations of different calcination temperatures or different gaseous atmospheres (Cai, Wang, Kim, & Yamauchi, 2019; Sun & Xu, 2014). Among the experimental parameters, temperature and gas atmosphere are the main influencers on the physicochemical properties of the desired product such as the morphology, crystallinity, particle size, porosity, and chemical composition (Dang et al., 2018; Xia, Mahmood, Zou, & Xu, 2015). However, other factors like time and heating rate have been less explored in the literature to date.

Inert (argon or N_2) or oxidants (air or O_2) atmospheres are typically used. The carbonization of ZIF-67 in an Ar atmosphere causes partial oxidation of the organic ligand and consequently reduction of metal cations, leading to the formation of composites containing metal NPs and nitrogen-doped porous carbon (NPC) (Li, Hao, Abudula, & Guan, 2016). The NPC formed typically presents a high degree of graphitization, characterized by the band of the graphitic planes located near 24° (plane (002)) in the experimental diffraction pattern. Intense signals in the Raman spectrum at 1350 and 1590 cm^{-1} related to the D and G bands, respectively, also indicate the graphitic structure (Xia et al., 2014). Cobalt NPs obtained by this method present a crystalline cubic phase, identified by the (111), (200), and (220) planes in the diffraction pattern. Under the same conditions, the carbonization of ZIF-67 modified with other metals such as zinc (Tang et al., 2016) and nickel (Hu, Chen, Lin, Liu, & Yang, 2018) also results in the formation of NPs of the respective metals.

Particle sizes typically obtained in calcinations under an Ar atmosphere are between 4 and 10 nm, while the surface area is close to 200–300 m^2/g, reaching up to 654 m^2/g (Ruan, Ai, & Lu, 2016; Wang, Bai, Wen, Du, & Lin, 2019; Yu, Sun, Muhammad, Wang, & Zhu, 2014). The temperature used for heat treatment in the argon atmosphere is between 600°C and 1000°C. However, there is an increase in particle size and a decrease in the surface area at high temperatures (Bai et al., 2019; Ruan et al., 2016; Yu et al., 2014). The increase in temperature also causes an increase in the crystallinity and graphitization degrees of the carbonaceous material (He, Cui, & Yang, 2021; Peng et al., 2021; Xie et al., 2021), impacting the final composite morphology (Wang et al., 2019). One of the greatest advantages of Ar atmosphere is the almost integral morphology retention (Hou & Xu, 2019; Liang et al., 2020), as demonstrated by Zou et al. (Xia et al., 2014), in the obtention of a series of composites starting from ZIF-67 with different particle sizes and morphologies.

Under a N_2 atmosphere, ZIF-67 has thermal stability up to 400°C (Wang et al., 2014). After this temperature, the Co–N bonds break, resulting in the partial oxidation of the organic ligand and in the reduction of the metal centers (Tang et al., 2015). Thus, thermal treatments of ZIF-67 in N_2 can also lead to the formation of NPC-based composites, again confirmed by X-ray powder diffraction and Raman spectroscopy measurements (Bo et al., 2019; Liang et al., 2020). Calcination temperatures are typically between 500° C and 900°C, and metal ions are often converted to Co NPs with a cubic phase, identical to those formed in the presence of argon (Feng et al., 2021; Zhong et al., 2018), with temperature-dependent particle size (Tian et al., 2020; Torad et al., 2014; Yang, Luo, et al., 2019). The literature shows the formation of oxides (CoO andCo_3O_4), mainly in heat treatments below 700°C, due to partial reduction and reoxidation of cobalt ions (Guo et al., 2019; Wen, Yang, Li, Bai, & Guan, 2019). These secondary phases tend to become less evident (Guo et al., 2019; Torad et al., 2014) or disappear (Wen et al., 2019; Xuan et al., 2021) with the increase in temperature. Besides temperature, the ZIF-67 morphology, synthesis method, particle size, and time seem to influence calcination products in N_2. ZIF-67 crystals that are obtained at room temperature in methanolic solution (Torad et al., 2014) or in an aqueous solution (Andrew Lin & Chen, 2016) or by hydrothermal synthesis with different particle sizes and morphologies (Zhao, Yan, et al., 2017) lead to distinct calcination products in a N_2 atmosphere at 600°C. Although the literature results point in this direction, systematic studies are necessary to clarify the role of these experimental variables, as well as others like gas flow rate and oven geometry.

Calcination of ZIF-67 under an oxidizing atmosphere (atmospheric air or O_2) leads to the formation of only one crystalline phase, corresponding to cobaltite (Co_3O_4) with a cubic structure Fd-3m space group (Bibi, Pervaiz, & Ali, 2021). As already mentioned, cobaltite NPs obtained from ZIF-67 were reported for the first time in 2011 (Wang et al., 2011). Since then, these materials have been the most explored ZIF derivatives in the literature, applied not only as electrocatalysts in OER but also in chemical/electrochemical sensing, as electrodes for supercapacitors, and as catalysts in organic reactions (Tian et al., 2020; Torad et al., 2014; Yang, Luo, et al., 2019). Under an oxidizing atmosphere, the ZIF-67 structure collapses after 250°C, leading to an amorphous phase between this temperature up to 270°C, when the crystallization of the cobaltite begins (Fei, Chen, et al., 2020; Fei, Liang, Zhou, Chen, & Zou, 2020; Huang et al., 2017; Li, Jiang, et al., 2016; Srinivas et al., 2018; Wang et al., 2014; Zhou, Ye, Zheng, & Hong, 2017; Zhou et al., 2020). Thus, an oxidizing atmosphere leads to the formation of only Co_3O_4 NPs and no graphitic/amorphous carbon is typically observed. However, XPS and infrared absorption spectroscopy results indicate residues of organic groups (hydroxyls, carbonyls, carboxyl, etc.) on the NP surface, especially in heat treatments at milder temperatures (Li, Tian, Jiang, & Ai, 2015; Lourenço et al., 2021). Composites containing cobaltites and porous carbon can also be obtained by carbonization in N_2 (Andrew Lin & Chen, 2016) or in an argon/N_2 atmosphere, followed by a second stage of calcination in air

(Tsai, Huy, Tsang, & Lin, 2020; Zhang, Huang, Wang, & Lu, 2016).

The morphology and particle size of ZIF-67 can be easily controlled by a synthetic method (Avci et al., 2015; Guo et al., 2016; Li et al., 2017; Zhang et al., 2019; Zhong et al., 2018), and this fact can influence the physicochemical and catalytic properties of the metal oxide (Zhao, Tang, Dong, & Zhang, 2019). Temperature plays a main role and has a great impact on the textural properties (Bibi et al., 2021). Under heating below 500°C, cobalt ions generate small clusters of metal oxides, followed by the recrystallization process, thus forming nanoparticles. These nanoparticles tend to aggregate with other nearby particles, resulting in micro/nanometric structures observed via electron microscopy (SEM or TEM) (Liu et al., 2010). Once formed, these agglomerates migrate to the material surface due to the expansion of gases from ligand decomposition (Jian, Yang, Lin, Hu, & Zhang, 2017; Zhang, Zhou, Wang, & Zhang, 2018), resulting in porous hollow structures (nanocages) or core shells, if the forces of attraction between cobaltite NPs are intense (Zhang, Zhou, et al., 2018). Calcination below 500°C also occurs with retention of the ZIF-67 morphology (Chen et al., 2019, 2020; Liu et al., 2018; Qiu, Tanaka, Gao, Wang, & Huang, 2019; Shao et al., 2014; Shi, Li, Cai, Zhao, & Lan, 2017; Sun et al., 2017; Wang, Fu, Chen, Xue, & Shan, 2021; Wang, Jiang, Liu, Guo, & Liu, 2015; Xiong et al., 2018; Yang, Feng, Liu, & Guo, 2018; Yang, Zhang, et al., 2018; Zhang et al., 2014; Zhang, Xie, Ci, Jia, & Wen, 2016; Zheng, Yin, Xu, & Zhang, 2016); however, the increase in temperature leads to rapid agglomeration, forming nanoparticles and collapsing the original morphology (Dang et al., 2018; Jiang et al., 2019; Zhang, Zhou, et al., 2018). In addition, high temperatures cause densification of the material, increasing the particle size and decreasing the surface area (Jian et al., 2017; Jiang et al., 2019; Lü et al., 2014; Wei, Liu, Wang, Li, & Peng, 2019; Zhang, Zhou, et al., 2018). Although argon, N_2, and air are the most used atmospheres among the methodologies used to obtain derivatives of ZIF-67, many heat treatments using other gases such as ammonia and mixed atmospheres are also reported in the literature (Guo, Liu, et al., 2018; Höök & Tang, 2013; Suleman et al., 2014). Subsequent treatments with different gases or gas-mixed atmospheres (Bai et al., 2019; Zhang, Huang, et al., 2016) can be used in order to expand the chemical, structural, and morphological diversity of the product (Chai et al., 2020; Tang et al., 2021).

Cobalt oxide-based electrocatalysts derived from ZIF-67

Cobaltite (Fig. 17.6) is a metal oxide with a spinel structure and cell unity with a cubic structure Fd-3m space group (Klissurski & Uzunova, 1994). Cobalt ions are located in tetrahedral and octahedral sites, and the atoms may be arranged in a normal, inverse, or mixed spinel. This structural organization depends on many factors including ionic radii, ions charges, temperatures, and also crystal field energy (Klissurski & Uzunova, 1994; Zhao, Yan, et al., 2017). In a normal spinel, bivalent cations (Co^{2+}) are in the tetrahedral sites, whereas trivalent cations (Co^{3+}) are located in the octahedral sites. The electrocatalytic performance of cobaltites is highly dependent on the morphology, particle size, structural defects, and oxidation states of ions on the surface and surface area (Man et al., 2011). Thus, the choice of the synthetic method used is an important factor in the control of these physicochemical properties, and synthesis using ZIF-67 as a template has gained increasing prominence in the literature (Bibi et al., 2021; Dutta et al., 2018).

The OER mechanism catalyzed by metals and/or metal oxides involves the formation and adsorption of oxygenated species on the catalyst surface, such as O*, HO*, and HOO* (Li et al., 2019; Man et al., 2011). Due to these adsorptive processes, the energy of the M–O

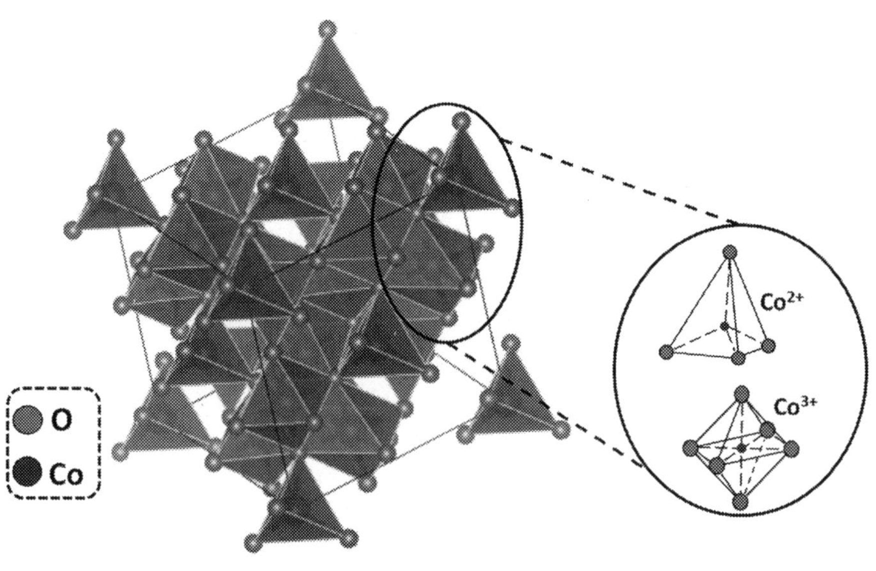

FIG. 17.6 The crystalline structure of Co_3O_4.

bond plays an important role and the overpotential during OER is determined by the adsorption energy of the O* species. The Gibbs free energy change in the adsorption of O* and HO* is often used as a descriptor to predict the electrocatalytic activity of several transition-metal-based materials (Man et al., 2011). In this way, cobalt-based oxides with a spinel structure stand out as promising electrocatalysts due to the ideal M–O bonding energy, low cost, corrosion resistance, and chemical stability (Li et al., 2019; Man et al., 2011). Many experimental measures are important to determine the electrocatalytic performance; however, the overpotential required to generate a current density of 10 mA/cm^2 has been frequently used for a comparative analysis between different catalysts (Wang & Zhang, 2018). It is important to note that a high electrocatalytic performance depends not only on the overpotential but also on factors such as chemical stability, temporal stability, cost, easy obtention, electrochemically active surface area (ECASA), and also on the understanding of the electrode kinetics (Jiao et al., 2015). Here, we will use only the overpotential (η) as an experimental parameter for a more simplified comparative analysis among the cobalt oxide-based catalysts derived from ZIF-67 already reported in the literature. A vast literature shows the electrocatalytic performance of cobaltite-based composites derived from ZIF-67 and materials such as nanotubes (Xu, 2019), graphene (Men et al., 2018), and polyoxometalates (Zhang, Mi, Ziaee, Liang, & Wang, 2018). However, the focus of the discussion below will only be on catalysts based on cobaltites and cobaltites doped with other metals.

ZIF-67-derived cobaltites

Although cobaltites can be obtained under a N_2 atmosphere, these oxides have been mostly obtained by direct calcination of ZIF-67 in air, at temperatures between 300°C and 500°C (Table 17.1) (Li, Anderson, et al., 2018). The literature indicates that properties such as surface area, particle size, and morphology depend not only on the calcination temperature but also on the characteristics of ZIF-67 used as a precursor (Dong et al., 2016; Li et al., 2015; Li, Liu, et al., 2018; Lourenço et al., 2021; Men et al., 2018; Wang et al., 2020; Wei et al., 2018; Yang, Chen, et al., 2018; Zhang, Guan, et al., 2020;

TABLE 17.1 Cobaltites derived from ZIF-67 reported in the literature.

Electrocatalyst	Thermal treatment conditions	Surface area/ particle size	Overpotential (mV) at 10 A/cm^2	Electrolyte	References
Honeycomb-like Co_3O_4	350°C, 10°C/min in air for 3 h	47.8 m^2 g^{-1}/ 8–10 nm	450	KOH 0.1 M	Li et al. (2015)
Hollow polyhedrons Co_3O_4	300°C in air for 2 h	49.3 m^2 g^{-1}/ 28.5 nm	536	KOH 1.0 M	Dong, Liu, and Li (2016)
Co_3O_4/N-doped porous	500°C, 5°C/min in air for 5 h	228 m^2 g^{-1}/ 18 nm	330	KOH 1.0 M	Zhuang et al. (2017)
Co_3O_4 nanoparticles	750°C in N_2/H_2 for 2 h and 350°C in air for 3 h	8.9 m^2 g^{-1}[a]	380	KOH 1.0 M	Men et al. (2018)
Co_3O_4 nanosheets	400°C, 2°C/min in air for 1 h	54.2 m^2 g^{-1}[a]	375	KOH 1.0 M	Li, Liu, et al. (2018)
Co_3O_4 nanoparticles/ nitrogen-doped carbon composite	350°C, 5°C/min in air for 5 h	25.9 m^2 g^{-1}/ 12 nm	290	KOH 1.0 M	Yang, Chen, et al. (2018)
Co_3O_4 nanosheets	400°C in air for 1 h	70 m^2 g^{-1}[a]	230	KOH 1.0 M	Wei, Zhou, Zhao, Zhang, and An (2018)
Rod-shaped Co_3O_4	450°C, 1°C/min in air for 2 h	33.1 m^{2}g^{-1}/ 20 nm	450	KOH 0.1 M	Zhao et al. (2020)
Hollow Co_3O_4/nitrogen-doped carbon	400°C, 2°C/min in air for 2 h	118.6 m^{2}g^{-1}[a]	360	KOH 0.1 M	Wang, Zhang, Zhang, Ren, and Cai (2020)
Co_3O_4 nanoplates	Solvothermal method at 90°C	–	314	KOH 1.0 M	Zhang, Guan, et al. (2020)
Porous hollow carbon-decorated Co_3O_4	Preheated furnace at 350°C for 2 h	28.5 nm[b]	384	KOH 0.1 M	Lourenço et al. (2021)

[a] *Particle size not reported.*
[b] *Surface area not reported.*

Zhao et al., 2020; Zhuang et al., 2017). By observing Table 17.1, it can been seen that there is no direct correlation between the surface area and the electrocatalytic performance, using only η as a comparative parameter. For example, cobaltites reported by Ai (Li et al., 2015) and Li (Dong et al., 2016) were obtained under similar thermal conditions and show similar values of surface area but different values of η at 450 and 536 mV, respectively. On the other hand, Co_3O_4 obtained from two steps of calcination by Luo et al. (Men et al., 2018) has a lower surface area (8.9 m^2/g) and exhibits superior electrocatalytic performance ($\eta = 380$ mV), demonstrating the impact of the heat treatment used.

Yang, Chen, et al. (2018) reported the synthesis of ZIF-67 using different cobalt sources (nitrate or sulfate) and the respective cobaltites obtained under the same conditions. The synthesis using cobalt nitrate generated dodecahedron crystals with an average size of nearly 1 μm, whereas the procedure using cobalt sulfate resulted in ZIF-67 crystals with flower-like morphology with particle size of 1.6 μm and higher surface area. Cobaltites were obtained at 550°C with the preservation of the original morphologies; however, the material derived from ZIF-67 obtained from cobalt sulfate had superior electrocatalytic performance ($\eta = 302\,mV$). Thus, this suggests a modulation of the electrocatalytic activity as a function of the precursors of ZIF-67. Wei et al. (2018) and Zhang, Guan, et al. (2020) also reported the dependence of morphology on the electrocatalytic performance of ZIF-derived cobaltites in OER. The authors showed the synthesis of Co_3O_4 with nanosheet and nanoplate morphologies by shape modulation as a function of solvent or surfactant agents, respectively. These materials also showed higher electrocatalytic activity when compared to cobaltites obtained from the dodecahedron ZIF-67.

The Co^{2+}/Co^{3+} ratio on the surface and structural defects are the two key points for understanding the electrocatalytic activity of cobaltites. It is known that O_h sites occupied by Co^{3+} ions are the main promoters of OER, due to the partially filled cobalt e_g orbitals, which favor the energy of the M–O bonding (Wei et al., 2017; Xu et al., 2019). On the other hand, many works have demonstrated the importance of structural defects, especially oxygen vacancies, favoring charge transfer processes, reducing energy barriers, and improving the adsorption of water and other intermediate species (Badreldin, Abusrafa, & Abdel-Wahab, 2021; He, Zou, & Wang, 2019). Oxygen vacancies also decrease the cation coordination number on the surface, thus increasing the density of active sites (Badreldin et al., 2021; He et al., 2019). The importance of O_h sites and oxygen vacancies is explicit in the work of An et al., which reported the synthesis of cobaltites with different shapes (nanosheet, dodecahedron, and nanoflower) (Wei et al., 2018). Co_3O_4 nanosheets show larger surface area, a higher number of oxygen vacancies, and a lower Co^{2+}/Co^{3+} ratio compared to the other materials, leading to a superior electrocatalytic performance ($\eta = 230\,mV$) (Wei et al., 2018). Lourenço et al. (2021) also highlighted the importance of heat treatment to improve the formation of oxygen vacancies. In this case, a preheated furnace was used for the calcination of ZIF-67 at 350°C in air. The rapid temperature change promotes a fast crystallization of the metal oxide, leading to a high concentration of oxygen vacancies, as confirmed by XPS results (Lourenço et al., 2021). The high concentration of oxygen vacancies is shown to be efficient, thus increasing the catalytic performance of Co_3O_4 obtained by this method, compared to other cobaltites synthesized at similar temperatures but with constant heating rates (Dong et al., 2016; Li et al., 2015).

Bimetallic systems

Recently, ZIF-67 has also been used as a platform for the synthesis of bimetallic cobaltites and bimetallic composites (Table 17.2), improving the electrocatalytic activity and other properties such as chemical stability and conductivity (Bo et al., 2019; Sanati et al., 2020; Zhao, Yan, et al., 2017). The insertion of a second metal into the cobaltite structure forming bimetallic systems also influences the morphology, density of structural defects, and surface area. Additionally, the change in the occupation of the d-orbitals due to the introduction of another metal center also changes the coordination number and covalence, thus impacting the adsorption/desorption of the intermediate species in OER (Bo et al., 2019; Hamdani, Singh, & Chartier, 2010; Sanati et al., 2020; Zhao, Yan, et al., 2017). Several transition metals can be used to modify ZIF-derived cobaltites such as iron, manganese, nickel, and zinc (Hamdani et al., 2010; Klissurski & Uzunova, 1994; Zhao, Yan, et al., 2017). The substitution usually occurs in octahedral sites; however, this replacement depends on the oxidation state, number of d electrons, crystal field stabilization energy, and preference occupation energy (McClure, 1957).

TABLE 17.2 ZIF-67-derived bimetallic cobaltites reported in the literature.

Electrocatalyst	Method used to insert the second metal	Overpotential (mV) at 10 A/cm^2	Electrolyte	References
$Co_3O_4/NiCo_2O_4$	Impregnation of Ni^{2+} ions in an aqueous solution	340	KOH 1.0 M	Hu, Guan, Xia, and Lou (2015)
Fe-doped Co_3O_4	Impregnation with a methanolic solution of potassium ferricyanide	297	KOH 1.0 M	Wang, Yu, Guan, Song, and Lou (2018)
$Co_3O_4/NiCo_2O_4$/NF	Nickel foam (NF) as a source for the formation of Co-Ni LDH	320	KOH 0.1 M	Yang, Lu, et al. (2019)
Ce-doped Co_3O_4	Bimetallic Co/Ce-ZIF-67	369	KOH 1.0 M	Zhou et al. (2019)
$ZnCo_2The_4$	Bimetallic Co/Zn-ZIF-67	380	KOH 1.0 M	Yu et al. (2020)
$ZnCo_2O_4$/FeOOH	Impregnation with an aqueous solution of iron sulfate	299	KOH 1.0 M	Yu et al. (2020)
Fe-Co_3O_4 nanoplates	Impregnation with an ethanolic solution of iron chloride	262	KOH 1.0 M	Zhang, Guan, et al. (2020)
$Mn_{0.4}Co_{2.6}O_4$	Impregnation of Mn^{2+} ions in methanolic solution	356	KOH 1.0 M	Lourenço et al. (2021)
$Co_3O_4/Mn_{0.8}Co_{2.2}O_4$	Impregnation of Mn^{2+} ions in ethanolic solution	338	KOH 1.0 M	Lourenço et al. (2021)

Bimetallic ZIF-67 derivatives are typically obtained by impregnation using the target metal ion in solution, followed by heat treatment under the desired conditions (Bibi et al., 2021). Nickel cobaltites have been widely explored using this method (Hu et al., 2018), but cobaltites with other metals can also be obtained in this manner (Lourenço et al., 2021). In 2015, Lou et al. reported the formation of $Co_3O_4/NiCo_2O_4$ double-shelled nanocages, using impregnation of Ni^{2+} ions into ZIF-67 in ethanolic solution (Hu et al., 2015). ZIF-67 immersed in nickel nitrate ethanolic solution undergoes a leaching process due to hydrolysis of nickel ions, forming Co-Ni layered double hydroxides (Co-Ni LDH) on the MOF surface (Hu et al., 2015; Jiang et al., 2013). Calcination of this system results in the $Co_3O_4/NiCo_2O_4$ composite, with an overpotential equal to 340 mV to produce a current density of 10 mA/cm^2, 70 mV lower when compared to free cobaltites obtained under the same experimental conditions (Hu et al., 2015). Lourenço et al. (2021) studied the influence of the solvent (methanol or ethanol) on the impregnation of Mn^{2+} ions into ZIF-67, aiming to understand the impact of this parameter on the electrocatalytic activity of manganese cobaltites. Using methanol as the solvent led to the formation of a single crystalline phase $Mn_{0.4}Co_{2.6}O_4$; however, experiments with ethanol result in the formation of a $Co_3O_4/Mn_{0.8}Co_{2.2}O_4$ nanocomposite. Both systems presented superior catalytic performance compared to manganese-free cobaltites, attributed to the high concentration of oxygen vacancies and the presence of Mn^{3+}/Mn^{4+} ions on the surface, increasing the number of active sites (Lourenço et al., 2021).

Another strategy used to produce bimetallic cobaltites is the thermal treatment of the bimetallic ZIF-67. This methodology is especially useful in the synthesis of zinc cobaltites, since ZIF-67 and ZIF-8 are isostructural, which allows the obtention of bimetallic systems with different amounts of zinc and cobalt in the structure (Ruan et al., 2016; Wang et al., 2019; Yu et al., 2014). Yu et al. (2020) demonstrated the synthesis of zinc cobaltites obtained by calcination of Zn/Co-ZIF in air for 2h, showing an overpotential of 380mV. The $ZnCo_2O_4$ obtained was impregnated with iron ions in an aqueous solution, resulting in $ZnCo_2O_4$/FeOOH nanocomposites, with higher ECSA, decreasing the overpotential up to 299mV (Yu et al., 2020). Lanthanide cations can also produce 4f-metal-doped cobaltites. Zhou et al. (2019) investigated the obtention of Ce-doped cobaltites from ZIF-67 containing different amounts of Ce^{2+} ions. The authors showed that the presence of 3mol% of Ce caused an increase in the presence of Co^{3+} ions on the surface, boosting the ECSA and decreasing the overpotential by 60mV, compared with free cobaltites obtained under the same calcination conditions (Zhou et al., 2019).

17.3 Conclusions

Here, the importance of ZIF-67 as a template to obtain new electrocatalysts for OER was shown. Several types of heat treatments can be used to produce a wide range of nanostructured materials. The most attractive feature of ZIF-67 as a precursor is conservation of the original morphology, in addition to precise control of shape and size. Cobalt oxide NPs and bimetallic systems based on Co_3O_4 are the main electrocatalyst materials derived from ZIF-67, with high performance and low overpotential, when compared to the literature. Although it is known that the textural properties (shape, size, and specific surface area) have a deep impact on electrocatalytic performance, superior investigations are required in order to better understand the influence of these experimental variables. Bimetallic cobaltites can also be obtained easily, increasing the ECSA and diversifying the surface composition, thus, boosting the electrocatalytic performance.

References

Ahmad, R., Khan, U. A., Iqbal, N., & Noor, T. (2020). Zeolitic imidazolate framework (ZIF)-derived porous carbon materials for supercapacitors: An overview. *RSC Advances*, *10*(71), 43733–43750. https://doi.org/10.1039/d0ra08560j.

Andrew Lin, K. Y., & Chen, B. C. (2016). Efficient elimination of caffeine from water using oxone activated by a magnetic and recyclable cobalt/carbon nanocomposite derived from ZIF-67. *Dalton Transactions*, *45*(8), 3541–3551. https://doi.org/10.1039/c5dt04277a.

Armaroli, N., & Balzani, V. (2011). The hydrogen issue. *ChemSusChem*, *4*(1), 21–36. https://doi.org/10.1002/cssc.201000182.

Avci, C., Ariñez-Soriano, J., Carné-Sánchez, A., Guillerm, V., Carbonell, C., Imaz, I., et al. (2015). Post-synthetic anisotropic wet-chemical etching of colloidal sodalite ZIF crystals. *Angewandte Chemie International Edition*, *54*(48), 14417–14421. https://doi.org/10.1002/anie.201507588.

Badreldin, A., Abusrafa, A. E., & Abdel-Wahab, A. (2021). Oxygen-deficient cobalt-based oxides for electrocatalytic water splitting. *ChemSusChem*, *14*(1), 10–32. https://doi.org/10.1002/cssc.202002002.

Bai, Y., Dong, J., Guo, Y., Liu, Y., Li, Y., Han, X., et al. (2019). Co_3O_4@PC derived from ZIF-67 as an efficient catalyst for the selective catalytic reduction of NOx with NH3 at low temperature. *Chemical Engineering Journal*, *361*, 703–712. https://doi.org/10.1016/j.cej.2018.12.109.

Banerjee, R., Phan, A., Wang, B., Knobler, C., Furukawa, H., O'Keeffe, M., et al. (2008). High-throughput synthesis of zeolitic imidazolate frameworks and application to CO_2 capture. *Science*, *319*(5865), 939–943. https://doi.org/10.1126/science.1152516.

Barnard, E. A., & Stein, W. D. (2006). The roles of imidazole in biological systems. In *Vol. 187. Advances in enzymology and related subjects of biochemistry* Wiley.

Batten, S. R., Champness, N. R., Chen, X.-M., Garcia-Martinez, J., Kitagawa, S., & Öhrström, L. (2012). Coordination polymers, metal-organic frameworks and the need for terminology guidelines. *CrystEngComm*, *14*(9), 3001–3004. https://doi.org/10.1039/c2ce06488j.

Bauman, J. E., & Wang, J. C. (1964). Imidazole complexes of nickel(II), copper(II), zinc(II), and silver(I). *Inorganic Chemistry*, *3*(3), 368–373. https://doi.org/10.1021/ic50013a014.

Baykara, S. Z. (2018). Hydrogen: A brief overview on its sources, production and environmental impact. *International Journal of Hydrogen Energy*, *43*(23), 10605–10614. https://doi.org/10.1016/j.ijhydene.2018.02.022.

Bhatnagar, A., Sharma, P. K., & Kumar, N. (2011). A review on 'imidazoles': Their chemistry and pharmacological potentials. *International Journal of PharmTech Research*, *3*(1), 268–282.

Bibi, S., Pervaiz, E., & Ali, M. (2021). Synthesis and applications of metal oxide derivatives of ZIF-67: A mini-review. *Chemical Papers*, *75*, 2253–2275. https://doi.org/10.1007/s11696-020-01473-y.

Bizzotto, F., Quinson, J., Zana, A., Kirkensgaard, J. J. K., Dworzak, A., Oezaslan, M., et al. (2019). Ir nanoparticles with ultrahigh dispersion as oxygen evolution reaction (OER) catalysts: Synthesis and activity benchmarking. *Catalysis Science & Technology*, *9*(22), 6345–6356. https://doi.org/10.1039/c9cy01728c.

Bo, X., Dastafkan, K., & Zhao, C. (2019). Design of multi-metallic-based electrocatalysts for enhanced water oxidation. *ChemPhysChem*, *20*(22), 29362945. https://doi.org/10.1002/cphc.201900507.

Burke, M. S., Enman, L. J., Batchellor, A. S., Zou, S., & Boettcher, S. W. (2015). Oxygen evolution reaction electrocatalysis on transition metal oxides and (oxy)hydroxides: Activity trends and design principles. *Chemistry of Materials*, *27*(22), 7549–7558. https://doi.org/10.1021/acs.chemmater.5b03148.

Cai, Z. X., Wang, Z. L., Kim, J., & Yamauchi, Y. (2019). Hollow functional materials derived from metal–organic frameworks: Synthetic strategies, conversion mechanisms, and electrochemical applications. *Advanced Materials*, *31*(11), 1–28. https://doi.org/10.1002/adma.201804903.

Caro, J., Noack, M., Kölsch, P., & Schäfer, R. (2000). Zeolite membranes – State of their development and perspective. *Microporous and Mesoporous Materials*, *38*(1), 3–24. https://doi.org/10.1016/S1387-1811(99)00295-4.

César, A. D. S., Veras, T. D. S., Mozer, T. S., dos Santos, D. D. C. R. M., & Conejero, M. A. (2019). Hydrogen productive chain in Brazil: An analysis of the competitiveness' drivers. *Journal of Cleaner Production*, *207*, 751–763. https://doi.org/10.1016/j.jclepro.2018.09.157.

Chai, L., Hu, Z., Wang, X., Xu, Y., Zhang, L., Li, T.-T., et al. (2020). Stringing bimetallic metal–organic framework-derived cobalt phosphide composite for high-efficiency overall water splitting. *Advancement of Science*, *7*(5), 1903195. https://doi.org/10.1002/advs.201903195.

Chen, K., Bai, S., Li, H., Xue, Y., Zhang, X., Liu, M., et al. (2020). The Co_3O_4 catalyst derived from ZIF-67 and their catalytic performance of toluene. *Applied Catalysis A: General*, *599*. https://doi.org/10.1016/j.apcata.2020.117614, 117614.

Chen, W. F., Muckerman, J. T., & Fujita, E. (2013). Recent developments in transition metal carbides and nitrides as hydrogen evolution electrocatalysts. *Chemical Communications*, *49*(79), 8896–8909. https://doi.org/10.1039/c3cc44076a.

Chen, B., Yang, Z., Zhu, Y., & Xia, Y. (2014). Zeolitic imidazolate framework materials: Recent progress in synthesis and applications. *Journal of Materials Chemistry A*, *2*(40), 16811–16831. https://doi.org/10.1039/c4ta02984d.

Chen, F., Yuan, Y. F., Ye, L. W., Zhu, M., Cai, G. C., Yin, S. M., et al. (2019). Co_3O_4 nanocrystalline-assembled mesoporous hollow polyhedron nanocage-in-nanocage as improved performance anode for lithium-ion batteries. *Materials Letters*, *237*, 213–215. https://doi.org/10.1016/j.matlet.2018.11.124.

Chen, Y. Z., Zhang, R., Jiao, L., & Jiang, H. (2018). Metal–organic framework-derived porous materials for catalysis. *Coordination Chemistry Reviews*, *362*, 1–23. https://doi.org/10.1016/j.ccr.2018.02.008.

Chen, L., Zhao, Z., Wei, Z., Wang, S., & Gu, Y. (2011). Direct synthesis and characterization of spongy CuO with nanosheets from $Cu_3(btc)_2$ microporous metal-organic framework. *Materials Letters*, *65*(3), 446–449. https://doi.org/10.1016/j.matlet.2010.10.059.

Cheng, N., Ren, L., Xu, X., Du, Y., & Dou, S. X. (2018). Recent development of zeolitic imidazolate frameworks (ZIFs) derived porous carbon based materials as electrocatalysts. *Advanced Energy Materials*, *8*(25), 1801257. https://doi.org/10.1002/aenm.201801257.

Chui, S. S. Y., Lo, S. M. F., Charmant, J. P. H., Orpen, A. G., & Williams, I. D. (1999). A chemically functionalizable nanoporous material $[Cu_3(TMA)_2\ (H_2O)_3](n)$. *Nature*, *283*(5405), 1148–1150. https://doi.org/10.1126/science.283.5405.1148.

Cowgill, R. W., & Clark, W. M. (1952). Metalloporphyrins. VII. Coordination of imidazoles with ferrimesoporphyrin. *Journal of Biological Chemistry*, *198*(1), 33–61. https://doi.org/10.1016/S0021-9258(18)55554-X.

Cundy, C. S., & Cox, P. A. (2003). The hydrothermal synthesis of zeolites: History and development from the earliest days to the present time. *Chemical Reviews*, *103*(3), 663–702. https://doi.org/10.1021/cr020060i.

Dang, S., Zhu, Q. L., & Xu, Q. (2018). Nanomaterials derived from metal-organic frameworks. *Nature Reviews Materials*, *3*, 17075. https://doi.org/10.1038/natrevmats.2017.75.

Davis, M. E. (2002). Ordered porous materials for emerging applications. *Nature*, *417*, 813–821. https://doi.org/10.1038/nature00785.

Decurtins, S., Pellaux, R., Antorrena, G., & Palacio, F. (1999). Multifunctional coordination compounds: Design and properties. *Philosophical Transactions. Royal Society of London*, *357*(1762), 3025–3040. https://doi.org/10.1016/S0010-8545(99)00124-1.

Dong, D., Liu, Y., & Li, J. (2016). Co_3O_4 hollow polyhedrons as bifunctional electrocatalysts for reduction and evolution reactions of oxygen. *Particle and Particle Systems Characterization*, *33*(12), 887–895. https://doi.org/10.1002/ppsc.201600191.

Dutta, S., Liu, Z., Han, H. S., Indra, A., & Song, T. (2018). Electrochemical energy conversion and storage with zeolitic imidazolate framework derived materials: A perspective. *ChemElectroChem*, *5*(23), 3571–3588. https://doi.org/10.1002/celc.201801144.

El-Emam, R. S., & Özcan, H. (2019). Comprehensive review on the techno-economics of sustainable large-scale clean hydrogen production. *Journal of Cleaner Production*, *220*, 593–609. https://doi.org/10.1016/j.jclepro.2019.01.309.

Fabbri, E., & Schmidt, T. J. (2018). Oxygen evolution reaction – The enigma in water electrolysis. *ACS Catalysis*, *8*(10), 9765–9774. https://doi.org/10.1021/acscatal.8b02712.

Fei, B., Chen, C., Hu, C., Cai, D., Wang, Q., & Zhan, H. (2020). Engineering one-dimensional bunched Ni-MoO_2@Co-CoO-NC composite for enhanced lithium and sodium storage performance. *ACS Applied Energy Materials*, *3*(9), 9018–9027. https://doi.org/10.1021/acsaem.0c01431.

Fei, Y., Liang, M., Zhou, T., Chen, Y., & Zou, H. (2020). Unique carbon nanofiber@ Co/C aerogel derived bacterial cellulose embedded zeolitic imidazolate frameworks for high-performance electromagnetic interference shielding. *Carbon*, *167*, 575–584. https://doi.org/10.1016/j.carbon.2020.06.013.

Feng, X., & Carreon, M. A. (2015). Kinetics of transformation on ZIF-67 crystals. *Journal of Crystal Growth*, *418*, 158–162. https://doi.org/10.1016/j.jcrysgro.2015.02.064.

Feng, S., Zhang, X., Shi, D., & Wang, Z. (2021). Zeolitic imidazolate framework-8 (ZIF-8) for drug delivery: A critical review. *Frontiers of Chemical Science and Engineering*, *15*, 221–237. https://doi.org/10.1007/s11705-020-1927-8.

Fu, S., Zhu, C., Song, J., Du, D., & Lin, Y. (2017). Metal-organic framework-derived non-precious metal nanocatalysts for oxygen reduction reaction. *Advanced Energy Materials*, *7*(19), 1–19. https://doi.org/10.1002/aenm.201700363.

Furukawa, H., Cordova, K. E., O'Keeffe, M., & Yaghi, O. M. (2013). The chemistry and applications of metal-organic frameworks. *Science*, *341*(6149), 1230444. https://doi.org/10.1126/science.1230444.

Gross, A. F., Sherman, E., & Vajo, J. J. (2012). Aqueous room temperature synthesis of cobalt and zinc sodalite zeolitic imidizolate frameworks. *Dalton Transactions*, *41*(18), 5458–5460. https://doi.org/10.1039/c2dt30174a.

Gu, H., Fan, W., & Liu, T. (2017). Phosphorus-doped $NiCo_2S_4$ nanocrystals grown on electrospun carbon nanofibers as ultra-efficient electrocatalysts for the hydrogen evolution reaction. *Nanoscale Horizons*, 2(5), 277–283. https://doi.org/10.1039/c7nh00066a.

Guan, B. Y., Yu, X. Y., Wu, H. B., & Lou, X. W. D. (2017). Complex nanostructures from materials based on metal–organic frameworks for electrochemical energy storage and conversion. *Advanced Materials*, *29*(47), 1–20. https://doi.org/10.1002/adma.201703614.

Guo, J., Gadipelli, S., Yang, Y., Li, Z., Lu, Y., Brett, D. J. L., et al. (2019). An efficient carbon-based ORR catalyst from low-temperature etching of ZIF-67 with ultra-small cobalt nanoparticles and high yield. *Journal of Materials Chemistry A*, *7*(8), 3544–3551. https://doi.org/10.1039/c8ta10925g.

Guo, S., Liu, Q., Sun, J., & Jin, H. (2018). A review on the utilization of hybrid renewable energy. *Renewable and Sustainable Energy Reviews*, *91*, 1121–1147. https://doi.org/10.1016/j.rser.2018.04.105.

Guo, X., Xing, T., Lou, Y., & Chen, J. (2016). Controlling ZIF-67 crystals formation through various cobalt sources in aqueous solution. *Journal of Solid State Chemistry*, *235*, 107–112. https://doi.org/10.1016/j.jssc.2015.12.021.

Guo, H., Youliwasi, N., Zhao, L., Chai, Y., & Liu, C. (2018). Carbon-encapsulated nickel-cobalt alloys nanoparticles fabricated via new post-treatment strategy for hydrogen evolution in alkaline media. *Applied Surface Science*, *435*, 237–246. https://doi.org/10.1016/j.apsusc.2017.11.083.

Hamdani, M., Singh, R. N., & Chartier, P. (2010). Co_3O_4 and Co-based spinel oxides bifunctional oxygen electrodes. *International Journal of Electrochemical Science*, 5(4), 556–577. https://www.researchgate.net/publication/279908814.

Han, W., Li, M., Ma, Y., & Yang, J. (2020). Cobalt-based metal-organic frameworks and their derivatives for hydrogen evolution reaction. *Frontiers in Chemistry*, *8*, 1–18. https://doi.org/10.3389/fchem.2020.592915.

Han, H., Yuan, X., Zhang, Z., & Zhang, J. (2019). Preparation of a ZIF-67 derived thin film electrode via electrophoretic deposition for efficient electrocatalytic oxidation of vanillin. *Inorganic Chemistry*, *58*(5), 3196–3202. https://doi.org/10.1021/acs.inorgchem.8b03281.

Harkins, T. R., & Freiser, H. (1956). The effect of coördination on some imidazole analogs. *Journal of the American Chemical Society*, *78*(6), 1143–1146. https://doi.org/10.1021/ja01587a015.

He, X. Q., Cui, Y. Y., & Yang, C. X. (2021). Engineering of amino microporous organic network on zeolitic imidazolate framework-67 derived nitrogen-doped carbon for efficient magnetic extraction of plant growth regulators. *Talanta*, *224*. https://doi.org/10.1016/j.talanta.2020.121876, 121876.

He, J., Zou, Y., & Wang, S. (2019). Defect engineering on electrocatalysts for gas-evolving reactions. *Dalton Transactions*, *48*(1), 15–20. https://doi.org/10.1039/C8DT04026E.

Höök, M., & Tang, X. (2013). Depletion of fossil fuels and anthropogenic climate change—A review. *Energy*

Policy, *52*, 797–809. https://doi.org/10.1016/j.enpol.2012.10.046.

Hoskins, B. F., & Robson, R. (1990). Design and construction of a new class of scaffolding-like materials comprising infinite polymeric frameworks of 3D-linked molecular rods. A reappraisal of the zinc cyanide and cadmium cyanide structures and the synthesis and structure of the diamond-related frameworks. *Journal of the American Chemical Society*, *112*(4), 1546–1554. https://doi.org/10.1021/ja00160a038.

Hou, C. C., & Xu, Q. (2019). Metal–organic frameworks for energy. *Advanced Energy Materials*, *9*(23), 1–18. https://doi.org/10.1002/aenm.201801307.

Hu, J., Chen, J., Lin, H., Liu, R., & Yang, X. (2018). MOF derived Ni/Co/NC catalysts with enhanced properties for oxygen evolution reaction. *Journal of Solid State Chemistry*, *259*, 1–4. https://doi.org/10.1016/j.jssc.2017.12.030.

Hu, H., Guan, B., Xia, B., & Lou, X. W. (2015). Designed formation of Co_3O_4/$NiCo_2O_4$ double-shelled nanocages with enhanced pseudocapacitive and electrocatalytic properties. *Journal of the American Chemical Society*, *137*(16), 5590–5595. https://doi.org/10.1021/jacs.5b02465.

Hua, Y., Li, X., Chen, C., & Pang, H. (2019). Cobalt based metal-organic frameworks and their derivatives for electrochemical energy conversion and storage. *Chemical Engineering Journal*, *370*, 37–59. https://doi.org/10.1016/j.cej.2019.03.163.

Huang, X. C., Lin, Y. Y., Zhang, J. P., & Chen, X. M. (2006). Ligand-directed strategy for zeolite-type metal-organic frameworks: Zinc(II) imidazolates with unusual zeolitic topologies. *Angewandte Chemie International Edition*, *45*(10), 1557–1559. https://doi.org/10.1002/anie.200503778.

Huang, M., Mi, K., Zhang, J., Liu, H., Yu, T., Yuan, A., et al. (2017). MOF-derived bi-metal embedded N-doped carbon polyhedral nanocages with enhanced lithium storage. *Journal of Materials Chemistry A*, *5*(1), 266–274. https://doi.org/10.1039/c6ta09030c.

Huang, X. C., Zhang, J. P., & Chen, X. M. (2004). A new route to supramolecular isomers via molecular templating: Nanosized molecular polygons of copper(I) 2-methylimidazolates. *Journal of the American Chemical Society*, *126*(41), 13218–13219. https://doi.org/10.1021/ja0452491.

Huang, X. C., Zhang, J. P., Lin, Y. Y., & Chen, X. M. (2005). Triple-stranded helices and zigzag chains of copper (I) 2-ethylimidazolate: Solvent polarity-induced supramolecular isomerism. *Chemical Communications*, *17*, 2232–2234. https://doi.org/10.1039/b501071c.

Janiak, C., & Vieth, J. K. (2010). MOFs, MILs and more: Concepts, properties and applications for porous coordination networks (PCNs). *New Journal of Chemistry*, *34*(11), 2366–2388. https://doi.org/10.1039/c0nj00275e.

Jian, S., Yang, W., Lin, W., Hu, J., & Zhang, L. (2017). Formation of hollow Co_3O_4 nanocages with hierarchical shell structure as anode materials for lithium-ion batteries. *Journal of Porous Materials*, *24*(4), 1079–1088. https://doi.org/10.1007/s10934-016-0348-z.

Jiang, Z., Li, Z., Qin, Z., Sun, H., Jiao, X., & Chen, D. (2013). LDH nanocages synthesized with MOF templates and their high performance as supercapacitors. *Nanoscale*, *5*(23), 11770–11775. https://doi.org/10.1039/c3nr03829g.

Jiang, Z., Sun, H., Shi, W., Zhou, T., Hu, J., Cheng, J., et al. (2019). Co_3O_4 nanocage derived from metal-organic frameworks: An excellent cathode catalyst for rechargeable Li-O_2 battery. *Nano Research*, *12*(7), 1555–1562. https://doi.org/10.1007/s12274-019-2388-6.

Jiao, Y., Zheng, Y., Jaroniec, M., & Qiao, S. Z. (2015). Design of electrocatalysts for oxygen- and hydrogen-involving energy conversion reactions. *Chemical Society Reviews*, *44*(8), 2060–2086. https://doi.org/10.1039/c4cs00470a.

Khan, M. A., Zhao, H., Zou, W., Chen, Z., Cao, W., Fang, J., et al. (2018). Recent progresses in electrocatalysts for water electrolysis. *Electrochemical Energy Reviews*, *1*, 483–530. https://doi.org/10.1007/s41918-018-0014-z.

Klissurski, D., & Uzunova, E. (1994). Synthesis and features of binary cobaltite spinels. *Journal of Materials Science*, *29*(2), 285–293. https://doi.org/10.1007/BF01162484.

Li, J. (2018). Metal-organic framework-based materials: Superior adsorbents for the capture of toxic and radioactive metal ions. *Chemical Society Reviews*, *47*(7), 2322–2356. https://doi.org/10.1039/c7cs00543a.

Li, G., Anderson, L., Chen, Y., Pan, M., & Chuang, P. Y. A. (2018). New insights into evaluating catalyst activity and stability for oxygen evolution reactions in alkaline media. *Sustainable Energy & Fuels*, *2*(1), 237–251. https://doi.org/10.1039/c7se00337d.

Li, H., Eddaoudi, M., O'Keeffe, M., & Yaghi, O. M. (1999). Design and synthesis of an exceptionally stable and highly porous metal-organic framework. *Nature*, *402*, 276–279. https://doi.org/10.1038/46248.

Li, X., Hao, X., Abudula, A., & Guan, G. (2016). Nanostructured catalysts for electrochemical water splitting: Current state and prospects. *Journal of Materials Chemistry A*, *4*(31), 11973–12000. https://doi.org/10.1039/c6ta02334g.

Li, S., Hao, X., Abudula, A., & Guan, G. (2019). Nanostructured Co-based bifunctional electrocatalysts for energy conversion and storage: Current status and perspectives. *Journal of Materials Chemistry A*, *7*(32), 18674–18707. https://doi.org/10.1039/c9ta04949e.

Li, X., Jiang, Q., Dou, S., Deng, L., Huo, J., & Wang, S. (2016). ZIF-67-derived Co-NC@CoP-NC nanopolyhedra as an efficient bifunctional oxygen electrocatalyst. *Journal of Materials Chemistry A*, *4*(41), 15836–15840. https://doi.org/10.1039/c6ta06434e.

Li, X., Li, Z., Lu, L., Huang, L., Xiang, L., Shen, J., et al. (2017). The solvent induced inter-dimensional phase transformations of cobalt zeolitic-imidazolate frameworks. *Chemistry – A European Journal, 23*(44), 10638–10643. https://doi.org/10.1002/chem.201701721.

Li, Y., Liu, B., Wang, H., Su, X., Gao, L., Zhou, F., et al. (2018). Co_3O_4 nanosheet-built hollow dodecahedrons via a two-step self-templated method and their multifunctional applications. *Science China Materials, 61*(12), 1575–1586. https://doi.org/10.1007/s40843-018-9254-6.

Li, L., Tian, T., Jiang, J., & Ai, L. (2015). Hierarchically porous Co_3O_4 architectures with honeycomb-like structures for efficient oxygen generation from electrochemical water splitting. *Journal of Power Sources, 294*, 103–111. https://doi.org/10.1016/j.jpowsour.2015.06.056.

Liang, Q., Chen, J., Wang, F., & Li, Y. (2020). Transition metal-based metal-organic frameworks for oxygen evolution reaction. *Coordination Chemistry Reviews, 424*. https://doi.org/10.1016/j.ccr.2020.213488, 213488.

Liu, B., Shioyama, H., Akita, T., & Xu, Q. (2008). Metal-organic framework as a template for porous carbon synthesis. *Journal of the American Chemical Society, 130*(16), 5390–5391. https://doi.org/10.1021/ja7106146.

Liu, N., Tang, M., Jing, C., Huang, W., Tao, P., Zhang, X., et al. (2018). Synthesis of highly efficient Co_3O_4 catalysts by heat treatment ZIF-67 for CO oxidation. *Journal of Sol-Gel Science and Technology, 88*(1), 163–171. https://doi.org/10.1007/s10971-018-4784-x.

Liu, B., Zhang, X., Shioyama, H., Mukai, T., Sakai, T., & Xu, Q. (2010). Converting cobalt oxide subunits in cobalt metal-organic framework into agglomerated Co_3O_4 nanoparticles as an electrode material for lithium ion battery. *Journal of Power Sources, 195*(3), 857–861. https://doi.org/10.1016/j.jpowsour.2009.08.058.

Liu, B., Zhang, X., Shioyama, H., Mukai, T., Sakai, T., & Xu, Q. (2020). Converting cobalt oxide subunits in cobalt metal-organic framework into agglomerated Co_3O_4 nanoparticles as an electrode material for lithium ion battery. *Journal of Power Sources, 195*(3), 857–861. https://doi.org/10.1016/j.jpowsour.2009.08.058.

Lourenço, A. A., Silva, V. D., da Silva, R. B., Silva, U. C., Chesman, C., Salvador, C., et al. (2021). Metal-organic frameworks as template for synthesis of Mn^{3+}/Mn^{4+} mixed valence manganese cobaltites electrocatalysts for oxygen evolution reaction. *Journal of Colloid and Interface Science, 582*, 124–136. https://doi.org/10.1016/j.jcis.2020.08.041.

Lü, Y., Zhan, W., He, Y., Wang, Y., Kong, X., Kuang, Q., et al. (2014). MOF-templated synthesis of porous Co_3O_4 concave nanocubes with high specific surface area and their gas sensing properties. *ACS Applied Materials & Interfaces, 6*(6), 4186–4195. https://doi.org/10.1021/am405858v.

Maiti, A. (2020). Cobalt-based heterogeneous catalysts in an electrolyzer system for sustainable energy storage. *Dalton Transactions, 49*(33), 11430–11450. https://doi.org/10.1039/d0dt01469a.

Man, I. C., Su, H.-Y., Calle-Vallejo, F., Hansen, H. A., Martinez, J. I., Inoglu, N. G., et al. (2011). Universality in oxygen evolution electrocatalysis on oxide surfaces. *ChemCatChem, 3*(7), 1159–1165. https://doi.org/10.1002/cctc.201000397.

McClure, D. S. (1957). The distribution of transition metal cations in spinels. *Journal of Physics and Chemistry of Solids, 3*(3–4), 311–317. https://doi.org/10.1016/0022-3697(57)90034-3.

Men, Y., Liu, X., Yang, F., Ke, F., Cheng, G., & Luo, W. (2018). Carbon encapsulated hollow Co_3O_4 composites derived from reduced graphene oxide wrapped metal-organic frameworks with enhanced lithium storage and water oxidation properties. *Inorganic Chemistry, 57*(17), 10649–10655. https://doi.org/10.1021/acs.inorgchem.8b01309.

Millange, F., Serre, C., & Férey, G. (2002). Synthesis, structure determination and properties of MIL-53as and MIL-53ht: The first CrIII hybrid inorganic-organic microporous solids: CrIII(OH)·$(O_2C\text{-}C_6H_4\text{-}CO_2)\cdot(HO_2C\text{-}C_6H_4\text{-}CO_2H)$ x. *Chemical Communications, 8*, 822–823. https://doi.org/10.1039/b201381a.

Mirqasemi, M. S., Homayoonfal, M., & Rezakazemi, M. (2020). Zeolitic imidazolate framework membranes for gas and water purification. *Environmental Chemistry Letters, 18*, 1–52. https://doi.org/10.1007/s10311-019-00933-6.

Moghadam, P. Z. (2020). Targeted classification of metal-organic frameworks in the Cambridge structural database (CSD). *Chemical Science, 11*(32), 8373–8397. https://doi.org/10.1039/d0sc01297a.

Mottillo, C., Lu, Y., Pham, M. H., Cliffe, M. J., Do, T. O., & Friscic, T. (2013). Mineral neogenesis as an inspiration for mild, solvent-free synthesis of bulk microporous metal-organic frameworks from metal (Zn, Co) oxides. *Green Chemistry, 15*(8), 2121–2131. https://doi.org/10.1039/c3gc40520f.

Parast, M. S. Y., & Morsali, A. (2011). Synthesis and characterization of porous Al(III) metal-organic framework nanoparticles as a new precursor for preparation of Al_2O_3 nanoparticles. *Inorganic Chemistry Communications, 14*(5), 645–648. https://doi.org/10.1016/j.inoche.2011.01.040.

Park, K. S., Ni, Z., Côté, A. P., Choi, J. Y., Huang, R., Uribe-Romo, F. J., et al. (2006). Exceptional chemical and thermal stability of zeolitic imidazolate frameworks. *Proceedings of the National Academy of Sciences of the United States of America, 103*(27), 10186–10191. https://doi.org/10.1073/pnas.0602439103.

Peng, W., Yang, X., Mao, L., Jin, J., Yang, S., Zhang, J., et al. (2021). ZIF-67-derived Co nanoparticles anchored in N doped hollow carbon nanofibers as bifunctional oxygen electrocatalysts. *Chemical Engineering Journal, 407*. https://doi.org/10.1016/j.cej.2020.127157, 127157.

Pimentel, B. R., Parulkar, A., Zhou, E. K., Brunelli, N. A., & Lively, R. P. (2014). Zeolitic imidazolate frameworks: Next-generation materials for energy-efficient gas separations. *ChemSusChem*, *7*(12), 3202–3240. https://doi.org/10.1002/cssc.201402647.

Qian, J., Sun, F., & Qin, L. (2012). Hydrothermal synthesis of zeolitic imidazolate framework-67 (ZIF-67) nanocrystals. *Materials Letters*, *82*, 220–223. https://doi.org/10.1016/j.matlet.2012.05.077.

Qiu, W., Tanaka, H., Gao, F., Wang, Q., & Huang, M. (2019). Synthesis of porous nanododecahedron Co_3O_4/C and its application for nonenzymatic electrochemical detection of nitrite. *Advanced Powder Technology*, *30*(10), 2083–2093. https://doi.org/10.1016/j.apt.2019.06.022.

Ribeiro, E. L., Davari, S. A., Hu, S., Mukherjee, D., & Khomami, B. (2019). Laser-induced synthesis of ZIF-67: A facile approach for the fabrication of crystalline MOFs with tailored size and geometry. *Materials Chemistry Frontiers*, *3*(7), 1302–1309. https://doi.org/10.1039/c8qm00671g.

Roger, I., Shipman, M. A., & Symes, M. D. (2017). Earth-abundant catalysts for electrochemical and photoelectrochemical water splitting. *Nature Reviews Chemistry*, *1*(3), 1–14. https://doi.org/10.1038/s41570-016-0003.

Ruan, C. P., Ai, K. L., & Lu, L. H. (2016). An acid-resistant magnetic Co/C nanocomposite for adsorption and separation of organic contaminants from water. *Chinese Journal of Analytical Chemistry*, *44*(2), 224–231. https://doi.org/10.1016/S1872-2040(16)60905-2.

Salunkhe, R. R., Kaneti, Y. V., Kim, J. H., & Yamauchi, Y. (2016). Nanoarchitectures for metal-organic framework-derived nanoporous carbons toward supercapacitor applications. *Accounts of Chemical Research*, *49*(12), 2796–2806. https://doi.org/10.1021/acs.accounts.6b00460.

Salunkhe, R. R., Kaneti, Y. V., & Yamauchi, Y. (2017). Metal-organic framework-derived nanoporous metal oxides toward supercapacitor applications: Progress and prospects. *ACS Nano*, *11*(6), 5293–5308. https://doi.org/10.1021/acsnano.7b02796.

Sanati, S., Abazari, R., Albero, J., Morsali, A., García, H., Liang, Z., et al. (2020). Metal–organic framework derived bimetallic materials for electrochemical energy storage. *Angewandte Chemie International Edition*, *60*(20), 11048–11067. https://doi.org/10.1002/anie.202010093.

Sankar, S. S., Karthick, K., Sangeetha, K., Karmakar, A., & Kundu, S. (2020). Transition-metal-based zeolite imidazolate framework nanofibers via an electrospinning approach: A review. *ACS Omega*, *5*(1), 57–67. https://doi.org/10.1021/acsomega.9b03615.

Seo, B., Sa, Y. J., Woo, J., Kwon, K., Park, J., Shin, T. J., et al. (2016). Size-dependent activity trends combined with in situ X-ray absorption spectroscopy reveal insights into cobalt oxide/carbon nanotube-catalyzed bifunctional oxygen electrocatalysis. *ACS Catalysis*, *6*(7), 4347–4355. https://doi.org/10.1021/acscatal.6b00553.

Shao, J., Wan, Z., Liu, H., Zheng, H., Gao, T., Shen, M., et al. (2014). Metal organic frameworks-derived Co_3O_4 hollow dodecahedrons with controllable interiors as outstanding anodes for Li storage. *Journal of Materials Chemistry A*, *2*(31), 12194–12200. https://doi.org/10.1039/c4ta01966k.

Shi, L., Li, Y., Cai, X., Zhao, H., & Lan, M. (2017). ZIF-67 derived cobalt-based nanomaterials for electrocatalysis and nonenzymatic detection of glucose: Difference between the calcination atmosphere of nitrogen and air. *Journal of Electroanalytical Chemistry*, *799*, 512–518. https://doi.org/10.1016/j.jelechem.2017.06.053.

Silva, V. D., Simões, T. A., Grilo, J. P. F., Medeiros, E. S., & Macedo, D. A. (2020). Impact of the NiO nanostructure morphology on the oxygen evolution reaction catalysis. *Journal of Materials Science*, *55*, 6648–6659. https://doi.org/10.1007/s10853-020-04481-1.

Srinivas, C., Sudharsan, M., Reddy, G. R. K., Kumar, P. S., Amali, A. J., & Suresh, D. (2018). Co/Co-N@nanoporous carbon derived from ZIF-67: A highly sensitive and selective electrochemical dopamine sensor. *Electroanalysis*, *30*(10), 2475–2482. https://doi.org/10.1002/elan.201800391.

Suleman, F., Dincer, I., & Agelin-Chaab, M. (2014). Development of an integrated renewable energy system for multigeneration. *Energy*, *78*, 196–204. https://doi.org/10.1016/j.energy.2014.09.082.

Suleman, F., Dincer, I., & Agelin-Chaab, M. (2015). Environmental impact assessment and comparison of some hydrogen production options. *International Journal of Hydrogen Energy*, *40*(21), 6976–6987. https://doi.org/10.1016/j.ijhydene.2015.03.123.

Sun, Z., Huang, F., Sui, Y., Wei, F., Qi, J., Meng, Q., et al. (2017). Cobalt oxide composites derived from zeolitic imidazolate framework for high-performance supercapacitor electrode. *Journal of Materials Science: Materials in Electronics*, *28*(18), 14019–14025. https://doi.org/10.1007/s10854-017-7252-4.

Sun, J. K., & Xu, Q. (2014). Functional materials derived from open framework templates/precursors: Synthesis and applications. *Energy & Environment*, *7*(7), 2071–2100. https://doi.org/10.1039/c4ee00517a.

Sun, C., Yang, J., Rui, X., Zhang, W., Yan, Q., Chen, P., et al. (2015). MOF-directed templating synthesis of a porous multicomponent dodecahedron with hollow interiors for enhanced lithium-ion battery anodes. *Journal of Materials Chemistry A*, *3*(16), 8483–8488. https://doi.org/10.1039/c5ta00455a.

Sun, W., Zhai, X., & Zhao, L. (2016). Synthesis of ZIF-8 and ZIF-67 nanocrystals with well-controllable size distribution through reverse microemulsions. *Chemical Engineering Journal*, *289*, 59–64. https://doi.org/10.1016/j.cej.2015.12.076.

Sundberg, R. J., & Martin, R. B. (1974). Interactions of histidine and other imidazole derivatives with transition metal ions in chemical and biological systems. *Chemical Reviews*, *74*(4), 471–517. https://doi.org/10.1021/cr60290a003.

Sundriyal, S., Kaur, H., Bhardwaj, S. K., Mishra, S., Kim, K. H., & Deep, A. (2018). Metal-organic frameworks and their composites as efficient electrodes for supercapacitor applications. *Coordination Chemistry Reviews*, *369*, 15–38. https://doi.org/10.1016/j.ccr.2018.04.018.

Tahir, M., Pan, L., Idrees, F., Zhang, X., Wang, L., Zou, J.-J., et al. (2017). Electrocatalytic oxygen evolution reaction for energy conversion and storage: A comprehensive review. *Nano Energy*, *37*, 136–157. https://doi.org/10.1016/j.nanoen.2017.05.022.

Tang, J., Salunkhe, R. R., Liu, J., Torad, N. L., Imura, M., Furukawa, S., et al. (2015). Thermal conversion of core-shell metal-organic frameworks: A new method for selectively functionalized nanoporous hybrid carbon. *Journal of the American Chemical Society*, *137*(4), 1572–1580. https://doi.org/10.1021/ja511539a.

Tang, J., Salunkhe, R. R., Zhang, H., Malgras, V., Ahamad, T., Alshehri, S. M., et al. (2016). Bimetallic metal-organic frameworks for controlled catalytic graphitization of nanoporous carbons. *Scientific Reports*, *6*, 30295. https://doi.org/10.1038/srep30295.

Tang, J., Zheng, S. B., Jiang, S. X., Li, J., Guo, T., & Guo, J. H. (2021). Metal organic framework (ZIF-67)-derived Co nanoparticles/N-doped carbon nanotubes composites for electrochemical detecting of tert-butyl hydroquinone. *Rare Metals*, *40*(2), 478–488. https://doi.org/10.1007/s12598-020-01536-9.

Tian, H., Zhang, C., Su, P., Shen, Z., Liu, H., Wang, G., et al. (2020). Metal-organic-framework-derived formation of Co–N-doped carbon materials for efficient oxygen reduction reaction. *Journal of Energy Chemistry*, *40*, 137–143. https://doi.org/10.1016/j.jechem.2019.03.004.

Torad, N. L., Hu, M., Ishihara, S., Sukegawa, H., Belik, A. A., Imura, M., et al. (2014). Direct synthesis of MOF-derived nanoporous carbon with magnetic Co nanoparticles toward efficient water treatment. *Small*, *10*(10), 2096–2107. https://doi.org/10.1002/smll.201302910.

Tsai, Y. C., Huy, N. N., Tsang, D. C. W., & Lin, K. Y. A. (2020). Metal organic framework-derived 3D nanostructured cobalt oxide as an effective catalyst for soot oxidation. *Journal of Colloid and Interface Science*, *561*, 83–92. https://doi.org/10.1016/j.jcis.2019.11.004.

Veras, T. D. S., Mozer, T. S., dos Santos, D. D. C. R. M., & César, A. D. S. (2017). Hydrogen: Trends, production and characterization of the main process worldwide. *International Journal of Hydrogen Energy*, *42*(4), 2018–2033. https://doi.org/10.1016/j.ijhydene.2016.08.219.

Verma, A., Joshi, S., & Singh, D. (2013). Imidazole: Having versatile biological activities. *Journal of Chemistry*, *2013*, 12. https://doi.org/10.1155/2013/329412.

Walker, F. A., Lo, M. W., & Ree, M. T. (1976). Electronic effects in transition metal porphyrins. The reactions of imidazoles and pyridines with a series of para-substituted tetraphenylporphyrin complexes of chloroiron(III). *Journal of the American Society*, *98*(18), 5552–5560. https://doi.org/10.1021/ja00434a024.

Wang, Q., & Astruc, D. (2020). State of the art and prospects in metal-organic framework (MOF)-based and MOF-derived nanocatalysis. *Chemical Reviews*, *120*(2), 1438–1511. https://doi.org/10.1021/acs.chemrev.9b00223.

Wang, L., Bai, X., Wen, B., Du, Z., & Lin, Y. (2019). Honeycomb-like Co/C composites derived from hierarchically nanoporous ZIF-67 as a lightweight and highly efficient microwave absorber. *Composites. Part B, Engineering*, *166*, 464–471. https://doi.org/10.1016/j.compositesb.2019.02.054.

Wang, J., Cui, W., Liu, Q., Xing, Z., Asiri, A. M., & Xun, X. (2016). Recent progress in cobalt-based heterogeneous catalysts for electrochemical water splitting. *Advanced Materials*, *28*(2), 215–230. https://doi.org/10.1002/adma.201502696.

Wang, H., Fu, W., Chen, Y., Xue, F., & Shan, G. (2021). ZIF-67-derived Co_3O_4 hollow nanocage with efficient peroxidase mimicking characteristic for sensitive colorimetric biosensing of dopamine. *Spectrochimica Acta. Part A, Molecular and Biomolecular Spectroscopy*, *246*. https://doi.org/10.1016/j.saa.2020.119006, 119006.

Wang, M., Jiang, X., Liu, J., Guo, H., & Liu, C. (2015). Highly sensitive H_2O_2 sensor based on Co_3O_4 hollow sphere prepared via a template-free method. *Electrochimica Acta*, *182*, 613–620. https://doi.org/10.1016/j.electacta.2015.08.116.

Wang, W., Li, Y., Zhang, R., He, D., Liu, H., & Liao, S. (2011). Metal-organic framework as a host for synthesis of nanoscale Co_3O_4 as an active catalyst for CO oxidation. *Catalysis Communications*, *12*(10), 875–879. https://doi.org/10.1016/j.catcom.2011.02.001.

Wang, S., Xu, D., Ma, L., Qiu, J., Wang, X., Dong, Q., et al. (2018). Ultrathin ZIF-67 nanosheets as a colorimetric biosensing platform for peroxidase-like catalysis. *Analytical and Bioanalytical Chemistry*, *410*, 7145–7152. https://doi.org/10.1007/s00216-018-1317-y.

Wang, X., Yu, L., Guan, B. Y., Song, S., & Lou, X. W. (2018). Metal–organic framework hybrid-assisted formation of Co_3O_4/Co-Fe oxide double-shelled nanoboxes for enhanced oxygen evolution. *Advanced Materials*, *30*(29), 1–5. https://doi.org/10.1002/adma.201801211.

Wang, Y., & Zhang, J. (2018). Structural engineering of transition metal-based nanostructured electrocatalysts for efficient water splitting. *Frontiers of Chemical Science and Engineering*, *12*(4), 838–854. https://doi.org/10.1007/s11705-018-1746-3.

Wang, C., Zhang, J., Zhang, Z., Ren, G., & Cai, D. (2020). One-step conversion of tannic acid-modified ZIF-67 into oxygen defect hollow Co_3O_4/nitrogen-doped carbon for efficient electrocatalytic oxygen evolution. *RSC Advances, 10*(64), 38906–38911. https://doi.org/10.1039/d0ra07696a.

Wang, X., Zhou, J., Fu, H., Li, W., Fan, X., Xin, G., et al. (2014). MOF derived catalysts for electrochemical oxygen reduction. *Journal of Materials Chemistry A, 2*(34), 14064–14070. https://doi.org/10.1039/c4ta01506a.

Wei, C., Feng, Z., Scherer, G. G., Barber, J., Shao-Horn, Y., & Xu, Z. J. (2017). Cations in octahedral sites: A descriptor for oxygen electrocatalysis on transition-metal spinels. *Advanced Materials, 29*(23), 1–8. https://doi.org/10.1002/adma.201606800.

Wei, L., Liu, H., Wang, Q., Li, W., & Peng, L. (2019). ZIF-67 derived cobalt-based catalysts for hydrogen generation from the hydrolysis of alkaline sodium borohydride solution. *Functional Materials Letters, 12*(4), 1–5. https://doi.org/10.1142/S1793604719500504.

Wei, G., Zhou, Z., Zhao, X., Zhang, W., & An, C. (2018). Ultrathin metal-organic framework nanosheet-derived ultrathin Co_3O_4 nanomeshes with robust oxygen-evolving performance and asymmetric supercapacitors. *ACS Applied Materials & Interfaces, 10*(28), 23721–23730. https://doi.org/10.1021/acsami.8b04026.

Wen, X., Yang, X., Li, M., Bai, L., & Guan, J. (2019). Co/CoOx nanoparticles inlaid onto nitrogen-doped carbon-graphene as a trifunctional electrocatalyst. *Electrochimica Acta, 296*, 830–841. https://doi.org/10.1016/j.electacta.2018.11.129.

Wu, H. B., & Lou, X. W. (2017). Metal-organic frameworks and their derived materials for electrochemical energy storage and conversion: Promises and challenges. *Science Advances, 3*(12), 1–16. https://doi.org/10.1126/sciadv.aap9252.

Xia, W., Mahmood, A., Zou, R., & Xu, Q. (2015). Metal-organic frameworks and their derived nanostructures for electrochemical energy storage and conversion. *Energy & Environmental Science, 8*(7), 1837–1866. https://doi.org/10.1039/c5ee00762c.

Xia, W., Zhu, J., Guo, W., An, L., Xia, D., & Zou, R. (2014). Well-defined carbon polyhedrons prepared from nano metal-organic frameworks for oxygen reduction. *Journal of Materials Chemistry A, 2*(30), 11606–11613. https://doi.org/10.1039/c4ta01656d.

Xie, Y., Feng, C., Guo, Y., Li, S., Guo, C., Zhang, Y., et al. (2021). MOFs derived carbon nanotubes coated CoNi alloy nanocomposites with N-doped rich-defect and abundant cavity structure as efficient trifunctional electrocatalyst. *Applied Surface Science, 536*. https://doi.org/10.1016/j.apsusc.2020.147786, 147786.

Xiong, C., Zhang, T., Kong, W., Zhang, Z., Qu, H., Chen, W., et al. (2018). ZIF-67 derived porous Co_3O_4 hollow nanopolyhedron functionalized solution-gated graphene transistors for simultaneous detection of glucose and uric acid in tears. *Biosensors & Bioelectronics, 101*, 21–28. https://doi.org/10.1016/j.bios.2017.10.004.

Xu, L. (2019). Co_3O_4-anchored MWCNTs network derived from metal-organic frameworks as efficient OER electrocatalysts. *Materials Letters, 248*, 181–184. https://doi.org/10.1016/j.matlet.2019.04.003.

Xu, K., Chen, P., Li, X., Tong, Y., Ding, H., Wu, X., et al. (2015). Metallic nickel nitride nanosheets realizing enhanced electrochemical water oxidation. *Journal of the American Chemical Society, 137*(12), 4119–4125. https://doi.org/10.1021/ja5119495.

Xu, X. L., Wang, H., Liu, J. B., & Yan, H. (2017). The applications of zeolitic imidazolate framework-8 in electrical energy storage devices: A review. *Journal of Materials Science: Materials in Electronics, 28*, 7532–7543. https://doi.org/10.1007/s10854-017-6485-6.

Xu, Y., Zhang, F., Sheng, T., Ye, T., Yi, D., Yang, Y., et al. (2019). Clarifying the controversial catalytic active sites of Co_3O_4 for the oxygen evolution reaction. *Journal of Materials Chemistry A, 7*(40), 23191–23198. https://doi.org/10.1039/c9ta08379k.

Xuan, J. P., Huang, N. B., Zhang, J. J., Dong, W. J., Yang, L., & Wang, B. (2021). Fabricating Co–N–C catalysts based on ZIF-67 for oxygen reduction reaction in alkaline electrolyte. *Journal of Solid State Chemistry, 294*. https://doi.org/10.1016/j.jssc.2020.121788, 121788.

Yaghi, O. M., & Li, H. (1995a). Hydrothermal synthesis of a metal-organic framework containing large rectangular channels. *Journal of the American Chemical Society, 117*(41), 10401–10402. https://doi.org/10.1021/ja00146a033.

Yaghi, O. M., & Li, G. (1995b). Mutually interpenetrating sheets and channels in the extended structure of [Cu (4,4′-bpy)Cl]. *Angewandte Chemie (International Ed. in English), 34*(2), 207–209. https://doi.org/10.1002/anie.199502071.

Yaghi, O. M., O'Keeffe, M., Ockwing, N. W., Chae, H. K., Eddaoudi, M., & Kim, J. (2003). Reticular synthesis and the design of new materials. *Nature, 423*, 705–714. https://doi.org/10.1038/nature01650.

Yang, X., Chen, J., Chen, Y., Feng, P., Lai, H., Li, J., et al. (2018). Novel CO_3O_4 nanoparticles/nitrogen-doped carbon composites with extraordinary catalytic activity for oxygen evolution reaction (OER). *Nano-Micro Letters, 10*(1), 1–11. https://doi.org/10.1007/s40820-017-0170-4.

Yang, Q., Feng, C., Liu, J., & Guo, Z. (2018). Synthesis of porous Co_3O_4/C nanoparticles as anode for Li-ion battery application. *Applied Surface Science, 443*, 401–406. https://doi.org/10.1016/j.apsusc.2018.02.230.

Yang, M., Lu, W., Jin, R., Liu, X. C., Song, S., & Xing, Y. (2019). Superior oxygen evolution reaction performance of Co_3O_4/$NiCo_2O_4$/Ni foam composite with hierarchical structure. *ACS Sustainable Chemistry & Engineering, 7*(14), 12214–12221. https://doi.org/10.1021/acssuschemeng.9b01535.

Yang, N., Luo, Z.-X., Zhu, G.-R., Chen, S.-C., Wang, X.-L., Wu, G., et al. (2019). Ultralight three-dimensional hierarchical cobalt nanocrystals/N-doped CNTs/carbon sponge composites with a hollow skeleton toward superior microwave absorption. *ACS Applied Materials & Interfaces, 11*(39), 35987–35998. https://doi.org/10.1021/acsami.9b11101.

Yang, M., Zhang, C., Fan, Y., Lin, T., Chen, X., Lu, Y., et al. (2018). ZIF-67-derived Co_3O_4 micro/nano composite structures for efficient photocatalytic degradation. *Materials Letters, 222*, 92–95. https://doi.org/10.1016/j.matlet.2018.03.174.

You, B., & Sun, Y. (2018). Innovative strategies for electrocatalytic water splitting. *Accounts of Chemical Research, 51*, 1571–1580. https://doi.org/10.1021/acs.accounts.8b00002.

Yu, Z., Bai, Y., Zhang, N., Yang, W., Ma, J., Wang, Z., et al. (2020). Metal-organic framework-derived heterostructured $ZnCo_2O_4$@FeOOH hollow polyhedrons for oxygen evolution reaction. *Journal of Alloys and Compounds, 832*. https://doi.org/10.1016/j.jallcom.2020.155067, 155067.

Yu, G., Sun, J., Muhammad, F., Wang, P., & Zhu, G. (2014). Cobalt-based metal organic framework as precursor to achieve superior catalytic activity for aerobic epoxidation of styrene. *RSC Advances, 4*(73), 38804–38811. https://doi.org/10.1039/c4ra03746d.

Zhang, C., Chu, W., Jiang, R., Li, L., Yang, Q., Cao, Y., et al. (2019). ZIF-67 derived hollow structured Co_3O_4 nanocatalysts: Tunable synthetic strategy induced enhanced catalytic performance. *Catalysis Letters, 149*(11), 3058–3065. https://doi.org/10.1007/s10562-019-02871-y.

Zhang, W., Cui, L., & Liu, J. (2020). Recent advances in cobalt-based electrocatalysts for hydrogen and oxygen evolution reactions. *Journal of Alloys and Compounds, 821*. https://doi.org/10.1016/j.jallcom.2019.153542, 153542.

Zhang, S. L., Guan, B. Y., Lu, X. F., Xi, S., Du, Y., & Lou, X. W. (2020). Metal atom-doped Co_3O_4 hierarchical nanoplates for electrocatalytic oxygen evolution. *Advanced Materials, 32*(31), 2002235. https://doi.org/10.1002/adma.202002235.

Zhang, M., Huang, Y. L., Wang, J. W., & Lu, T. B. (2016). A facile method for the synthesis of a porous cobalt oxide-carbon hybrid as a highly efficient water oxidation catalyst. *Journal of Materials Chemistry A, 4*(5), 1819–1827. https://doi.org/10.1039/c5ta07813j.

Zhang, L., Mi, T., Ziaee, M. A., Liang, L., & Wang, R. (2018). Hollow POM@MOF hybrid-derived porous Co_3O_4/$CoMoO_4$ nanocages for enhanced electrocatalytic water oxidation. *Journal of Materials Chemistry A, 6*(4), 1639–1647. https://doi.org/10.1039/c7ta08683k.

Zhang, J., Tan, Y., & Song, W. J. (2020). Zeolitic imidazolate frameworks for use in electrochemical and optical chemical sensing and biosensing: A review. *Microchimica Acta, 187*(234), 23. https://doi.org/10.1007/s00604-020-4173-3.

Zhang, Y. Z., Wang, Y., Xie, Y. L., Cheng, T., Lai, W. Y., Pang, H., et al. (2014). Porous hollow Co_3O_4 with rhombic dodecahedral structures for high-performance supercapacitors. *Nanoscale, 6*(23), 14354–14359. https://doi.org/10.1039/c4nr04782f.

Zhang, E., Xie, Y., Ci, S., Jia, J., & Wen, Z. (2016). Porous Co_3O_4 hollow nanododecahedra for nonenzymatic glucose biosensor and biofuel cell. *Biosensors & Bioelectronics, 81*, 46–53. https://doi.org/10.1016/j.bios.2016.02.027.

Zhang, R., Zhou, T., Wang, L., & Zhang, T. (2018). Metal-organic frameworks-derived hierarchical Co_3O_4 structures as efficient sensing materials for acetone detection. *ACS Applied Materials & Interfaces, 10*(11), 9765–9773. https://doi.org/10.1021/acsami.7b17669.

Zhao, J., Tang, Z., Dong, F., & Zhang, J. (2019). Controlled porous hollow Co_3O_4 polyhedral nanocages derived from metal-organic frameworks (MOFs) for toluene catalytic oxidation. *Molecular Catalysis, 463*, 77–86. https://doi.org/10.1016/j.mcat.2018.10.020.

Zhao, Q., Xu, X. L., Jin, Y. H., Zhang, Q. Q., Liu, J. B., & Wang, H. (2020). Carbon-coated Co_3O_4 with porosity derived from zeolite imidazole framework-67 as a bi-functional electrocatalyst for rechargeable zinc air batteries. *Journal of Nanoparticle Research, 22*(9). https://doi.org/10.1007/s11051-020-05029-9.

Zhao, Q., Yan, Z., Chen, C., & Chen, J. (2017). Spinels: Controlled preparation, oxygen reduction/evolution reaction application, and beyond. *Chemical Reviews, 117*(15), 10121–10211. https://doi.org/10.1021/acs.chemrev.7b00051.

Zheng, F., Yin, Z., Xu, S., & Zhang, Y. (2016). Formation of Co_3O_4 hollow polyhedrons from metal-organic frameworks and their catalytic activity for CO oxidation. *Materials Letters, 182*, 214–217. https://doi.org/10.1016/j.matlet.2016.06.108.

Zhong, G., Liu, D., & Zhang, J. (2018). The application of ZIF-67 and its derivatives: Adsorption, separation, electrochemistry and catalysts. *Journal of Materials Chemistry A, 6*(5), 1887–1899. https://doi.org/10.1039/c7ta08268a.

Zhou, X., Ye, Z., Zheng, X., & Hong, Z. (2017). Fabrication of carbon@Co nanocages/nafion/poly tiopronin composite film and its application for determination of acetaminophen. *Journal of the Electrochemical Society, 164*(12), B513–B518. https://doi.org/10.1149/2.0651712jes.

Zhou, J., Zheng, H., Luan, Q., Huang, X., Li, Y., Xi, Z., et al. (2019). Improving the oxygen evolution activity of Co_3O_4 by introducing Ce species derived from Ce-substituted ZIF-67. *Sustainable Energy & Fuels, 3*(11), 3201–3207. https://doi.org/10.1039/c9se00541b.

Zhou, H., Zheng, M., Tang, H., Xu, B., Tang, Y., & Pang, H. (2020). Amorphous intermediate derivative from ZIF-67 and its outstanding electrocatalytic activity. *Small*, *16*(2), 1–9. https://doi.org/10.1002/smll.201904252.

Zhu, Q. L., & Xu, Q. (2014). Metal-organic framework composites. *Chemical Society Reviews*, *43*(16), 5468–5512. https://doi.org/10.1039/c3cs60472a.

Zhuang, G. L., Gao, Y.-F., Zhou, X., Tao, X.-Y., Luo, J.-M., Gao, Y.-J., et al. (2017). ZIF-67/COF-derived highly dispersed Co_3O_4/N-doped porous carbon with excellent performance for oxygen evolution reaction and Li-ion batteries. *Chemical Engineering Journal*, *330*, 1255–1264. https://doi.org/10.1016/j.cej.2017.08.076.

CHAPTER

18

Electrochemical ammonia synthesis: Mechanism, recent developments, and challenges in catalyst design

Vanessa C.D. Graça[a], *Francisco J.A. Loureiro*[a], *Laura I.V. Holz*[a], *Sergey M. Mikhalev*[a], *Allan J.M. Araújo*[a,b], *and Duncan P. Fagg*[a]

[a]Centre for Mechanical Technology and Automation (TEMA), Department of Mechanical Engineering, University of Aveiro, Aveiro, Portugal [b]Materials Science and Engineering Postgraduate Program—PPGCEM, Federal University of Rio Grande do Norte—UFRN, Natal, Brazil

18.1 Introduction

Ammonia (NH_3) is the second most produced, man-made chemical in the world and plays an important role as a raw material in the production of chemical fertilizers. Each year, about 174 million tonnes of ammonia (NH_3) are synthesized, with the worldwide demand for ammonia continuing to grow (Giddey, Badwal, & Kulkarni, 2013; Pfromm, 2017).

Ammonia has also been suggested as a potential green fuel. In this context, green hydrogen has typically taken up the role of the cleanest source of chemical energy, as itscombustion produces only water as a byproduct. Nonetheless, the implementation of a near-future hydrogen-based economy remains a complex approach, due to the current lack of suitable methods for hydrogen storage and transportation. These two aspects have complicated safety and technical issues when using hydrogen in the gaseous state due to its low flash point and low volumetric energy density (Lan, Irvine, & Tao, 2012; Zamfirescu & Dincer, 2008). As an alternative to hydrogen, ammonia can offer many advantages as a clean fuel, with the most important of them being its much easier liquefaction, a factor that avoids the need for extreme cryogenic temperatures that are conversely needed to liquefy hydrogen and which permits the use of preexisting fuel distribution infrastructure. Moreover, the energy content of ammonia per unit volume is comparable to that of gasoline, which makes it an attractive fuel in the transportation sector. Therefore, ammonia, being both a promising hydrogen source and an important chemical feedstock, has the potential to play an essential role in the future world economy, including the energy sector, and in transportation, refrigeration, and agriculture (Giddey et al., 2013; Lan

Heterogeneous Catalysis
https://doi.org/10.1016/B978-0-323-85612-6.00018-8

et al., 2012; Zamfirescu & Dincer, 2008) (Klerke, Christensen, Nørskov, & Vegge, 2008).

The process of ammonia formation from its constituents is an exothermic reaction ($\Delta H = -92\,\text{kJ}\,\text{mol}^{-1}$) (Jiao & Xu, 2018) and is accompanied by a decrease in volume, due to the decrease in the number of moles of gas from 2 to 1, as given by (Eq. 18.1),

$$\frac{3}{2}H_{2\,(g)} + \frac{1}{2}N_{2\,(g)} \rightarrow NH_{3\,(g)} \tag{18.1}$$

The main industrial method for ammonia production, to date, called the Haber-Bosch (HB) process, is one of the top discoveries of the 20th century by Fritz Haber and Carl Bosch. In brief, the ammonia synthesis reaction progresses by breaking the triple bond of nitrogen (NN) and the subsequent protonation of each nitrogen atom using an iron- or ruthenium-based catalyst at high temperatures (500°C) and at high pressures (150–300 bar), as given by Eq. (18.1). Fig. 18.1 shows a diagram of the overall process for ammonia production.

The main limitations of the HB process include low levels of ammonia conversion (~15%) and, most importantly, its extensive environmental pollution and extremely high energy consumption.

The production of ammonia is limited by its thermodynamics. As dictated by the reaction stoichiometry and *Le Chatelier's* principle, ammonia formation is favored under higher pressures and at lower temperatures. On the other hand, the synthesis of ammonia at extremely low temperatures becomes kinetically constrained. Therefore, a trade-off solution is performed by operating at temperatures between 430 and 450°C and under pressures between 150 and 300 bar. Such conditions allow the production of a reasonably high amount of ammonia in the equilibrium mixture, ~15%, in an acceptable timeframe (Amar, Lan, Petit, & Tao, 2011; Amar et al., 2011; Giddey et al., 2013; Marnellos & Stoukides, 1998).

With regard to the reactants, the required hydrogen gas is currently obtained from fossil fuel feedstocks. This source of hydrogen adds additional complications that dictate the presence of many additional reaction steps: (i) desulfurization, followed by (ii) methane steam reforming, followed by (iii) water gas shift reaction to convert CO to H_2 and CO_2, processes that are both energy-intensive and that have significant environmental impacts. Conversely, the nitrogen reactant is typically sourced from air. Nonetheless, extensive purification is also required for its extraction, as merely the

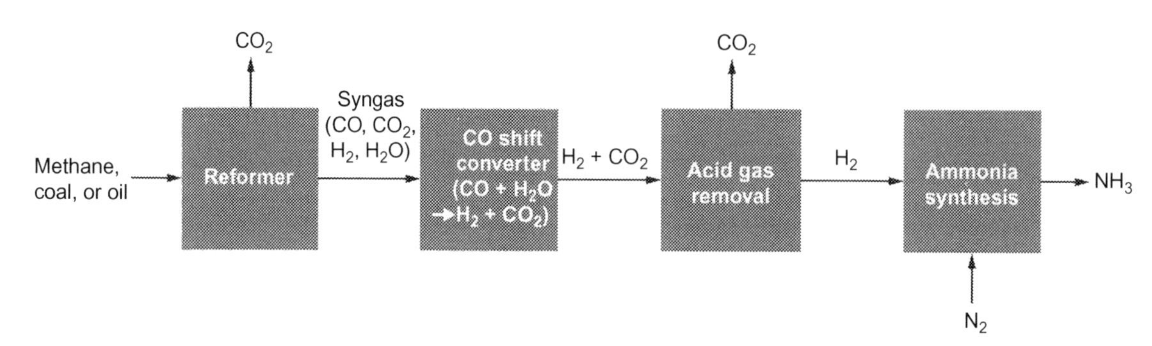

FIG. 18.1 Diagram of the Haber-Bosch process.

presence of trace oxygen impurities alone (in ppm levels) would be enough for poisoning the catalysts. Such strict requirements of the purification of reactant gases additionally contribute to make the HB process an extremely energy-intensive one. Importantly, the production of 1 ton of ammonia currently generates 1.87 tons of CO_2. Therefore, it is easy to understand how the HB process is a major contributor to greenhouse gas emissions and fossil fuel utilization (Giddey et al., 2013).

Contrary to the industrial process, in nature, ammonia is synthesized under mild conditions. Atmospheric nitrogen is reduced by enzymes, called nitrogenases, which contain iron and molybdenum, which catalyze the reaction of ammonia synthesis from atmospheric nitrogen, protons, and electrons (Foster et al., 2018). This reaction is given by (Eq. 18.2)

$$N_{2\,(g)} + 8\,H^+ + 8\,e^- + 16ATP \rightarrow 2\,NH_{3\,(g)} + H_{2\,(g)} + 16\,ADP + 16Pi \quad (18.2)$$

where ATP is adenosine triphosphate, ADP is adenosine diphosphate, and Pi is an inorganic phosphate.

This need for protons and electrons indicates that ammonia could also be synthesized artificially via electrochemical processes (Kyriakou, Garagounis, Vasileiou, Vourros, & Stoukides, 2017).

The production of ammonia by electrochemical routes is a particularly attractive fossil-free method, as the only inputs required are that of N_2 (that could conceivably come from the air and with greater oxygen tolerance), a proton donor-containing solution (that could be water), and electricity (that could conceivably come from renewable sources) (Shipman & Symes, 2017). Therefore, alternatives based on the electrochemical synthesis of ammonia have been suggested with several different electrolytic systems currently under study.

This chapter is structured as follows: Section 18.2 briefly reviews the different electrolytic systems for the electrochemical synthesis of ammonia, and Section 18.3 focuses on the mechanism for NRR and highlights computational modeling studies, as well as current experimental testing, including the most promising electrocatalysts used for electrochemical ammonia synthesis.

18.2 Electrochemical synthesis of ammonia

The electrochemical synthesis of ammonia can be performed using different routes that differ based on the cell configuration (material composition, physical state) and operation conditions (temperature, pressure). Usually, the process is divided into three main categories, determined by the type of electrolytes used, namely, liquid, molten salt, or solid oxide electrolytes, or by the working operating temperature, namely, low temperature ($T < 100°C$), intermediate temperature ($100 < T < 500°C$), or high temperature ($500 < T < 800°C$) (Giddey et al., 2013; Kyriakou et al., 2017; Li, Wang, & Gong, 2020). In the low temperature range, liquid electrolytes and polymer membranes are commonly used, whereas molten salts and solid acid electrolytes are preferred for the intermediate temperature range. Finally, in the high temperature range, carbonates and solid-state protons or oxide-ion conductors are utilized, which mainly consist of ceramic electrolyte materials with a perovskite, fluorite, or pyrochlore structure (Qing et al., 2020).

Fig. 18.2 summarizes the systems and the respective electrode reactions. In the following section, a brief literature review of the electrolytic systems will be provided (Fig. 18.2).

Liquid electrolytes

Electrochemical synthesis of ammonia using liquid electrolytes is a recent technology that has the advantage of being conducted at low temperatures (<100°C) and pressure, with

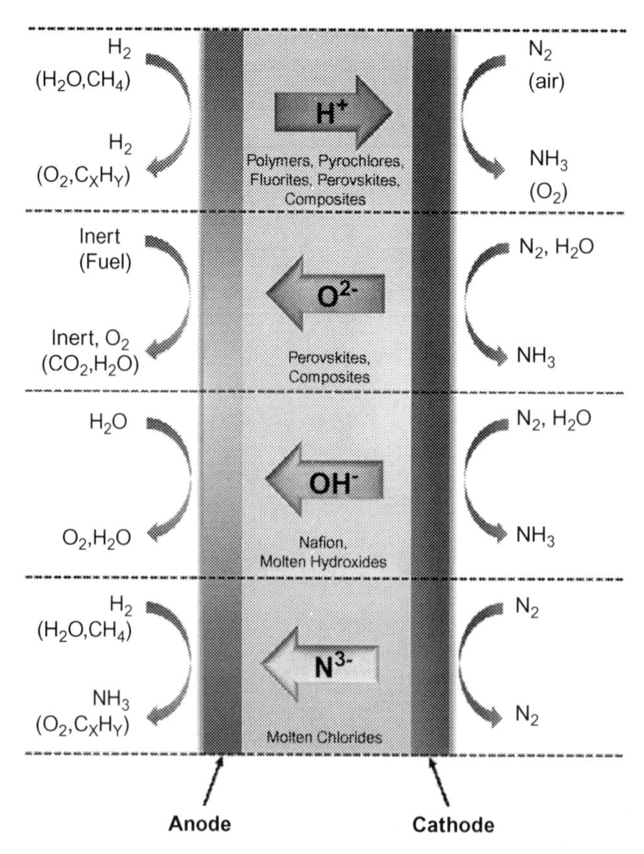

FIG. 18.2 Electrolytic systems for ammonia synthesis. *From Kyriakou, V., Garagounis, I., Vasileiou, E., Vourros, A., & Stoukides, M. (2017). Progress in the electrochemical synthesis of ammonia.* Catalysis Today, 286, 2–13. *https://doi.org/10.1016/j.cattod.2016.06.014.*

ambient pressure being the most commonly used.

Tsuneto, Kudo, and Sakata (1994) was the first to investigate the use of an aqueous solution as an electrolyte to synthesize ammonia at near room temperatures. In their study, they found that the current efficiency for ammonia synthesis is highly dependent on the type of metal used in the cathode (Ti, Mo, Fe, Co, Ni, Cu, Ag) and on the source of hydrogen (ethanol, methanol, water, acetic acid, 1-propanol, 2-propanol, etc.) and its concentration. Furthermore, they found that the current efficiency increased by 58% with increasing nitrogen pressure. In the same study, Tsuneto et al. (1994) evaluated the use of different salts in the electrolyte composition such as $LiClO_4$, $LiBF_4$, $Li(CF_3SO_3)$, $NaClO_4$, and Bu_4NClO_4. A maximum current efficiency of 59.8% was achieved with $Li(CF_3SO_3)$ in tetrahydrofuran and ethanol at a nitrogen pressure of 50 atm. A negligible amount of ammonia was detected when $NaClO_4$ and Bu_4NClO_4 salts were used, confirming the Li-mediated reaction mechanism.

Although electrochemical synthesis of ammonia using liquid electrolytes seems to be a promising technology, only limited additional information can be found, due to it still being in its early stage of development, with current studies evaluated, typically, at the laboratory scale and for a limited period. Hence, further knowledge of how the system would function as a complete cell needs to be acquired with regard to the reaction mechanism at the anode, as well from a technology perspective. Future data on electrode polarization losses and long-term operation are urgently required.

Molten salt electrolytes

The working operation temperatures for electrochemical cells that use molten salts as electrolytes are in the temperature range of 200–500°C. In these types of cells, N_2 and H_2 sources are converted into N/H species at the electrodes and diffuse through the molten salt electrolyte, where they convert to form gaseous ammonia at the anode by the following reaction, Eq. (18.3):

$$N^{3-} + 3/2\,H_2\,(g) \rightarrow NH_3\,(g) + 3^{e-} \quad (18.3)$$

Therefore, it is easy to understand that in these types of cells, the ionic mobility in the electrolyte plays an important role in improving the electrochemical formation of ammonia. Molten salt electrolytes can be classified into molten chloride salts, molten hydroxide salts, and composite electrolytes (molten carbonates stored in a porous solid oxide electrolyte), and, depending on the molten salt used, the reaction mechanism, ammonia conversion, and faradaic efficiency (FE) can differ significantly due to their distinct physicochemical properties.

Murakami, Nishikiori, Nohira, and Ito (2003) were the first to study ammonia formation using a eutectic mixture of chlorides (LiCl, KCl, and CsCl) as a molten salt electrolyte. A schematic of the setup used is shown in Fig. 18.3. In their study, they reported an ammonia conversion rate of 3.33×10^{-9} mol s^{-1} cm^{-2} with a high FE of 72% using N_2 and H_2 as the primary sources. A similar study (Murakami, Nohira, Goto, Ogata, & Ito, 2005), using H_2O as the primary source of hydrogen, reported an increase in the ammonia formation rate to 2.02×10^{-8}-mol s^{-1} cm^{-2} although with a substantially lower FE (~23%), which could be mainly explained by the competition of electrochemical H_2O splitting reaction with ammonia formation reaction.

Hydroxide molten salts are less corrosive and can operate at lower temperatures than chloride molten salts. However, due to their OH^- composition, the H_2 evolution reaction can easily occur, thus compromising the ammonia conversion and the obtained FE (Cui et al., 2017; Licht et al., 2014; Yang, Weng, & Xiao, 2020).

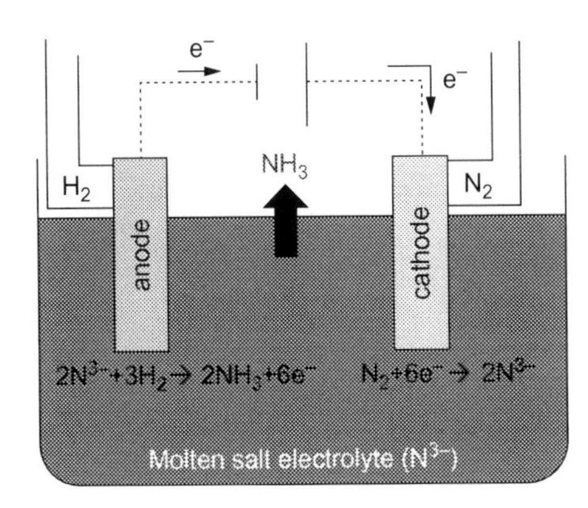

FIG. 18.3 Schematic drawing of the principle of a molten salt electrolytic system for synthesis of ammonia. *From Kyriakou, V., Garagounis, I., Vasileiou, E., Vourros, A., & Stoukides, M. (2017). Progress in the electrochemical synthesis of ammonia.* Catalysis Today, 286, 2–13. *https://doi.org/10.1016/j.cattod.2016.06.014.*

Additionally, some studies are being conducted using composite electrolytes, which typically are a mixture of a solid oxide with a eutectic mixture of alkali metal salts. The presence of the molten phase enhances the ionic conductivity and reduces the operating temperature of the electrolyte. Although this strategy can be highly interesting, a low FE (<5%) was reported, suggesting that more investment is needed in the development and improvement of this type of system.

Solid oxide electrolytes

The electrochemical synthesis of ammonia using solid oxide electrolytes offers an exciting advantage over the conventional Haber-Bosch process, by the possibility of operating under ambient pressures (Marnellos, Kyriakou, Florou, Angelidis, & Stoukides, 1999; Marnellos, Sanopoulou, Rizou, & Stoukides, 1997; Panagos, Voudouris, & Stoukides, 1996). Typically, these electrolytes operate at intermediate-to-high temperatures (500–800°C), depending on the chemical composition of the electrolyte. In this context, solid electrolytes are categorized according to the mobile ionic species (H^+, O^{2-}, Li^+, Cu^+, Na^+, etc.). In this section, the discussion will be restricted to proton (H^+) and oxygen-ion (O^{2-}) conductors.

Electrochemical reactors using protonic membranes have been pointed out as a promising alternative to the industrial HB process. These membranes are permeable only to protons (H^+), avoiding the necessity of extensive purification of the hydrogen feed gas, which, otherwise, is a considerable fraction of the overall cost of the HB process. A schematic representation of the protonic device for ammonia synthesis is depicted in Fig. 18.4. The half-reactions that occur at the electrodes for the two anode gas

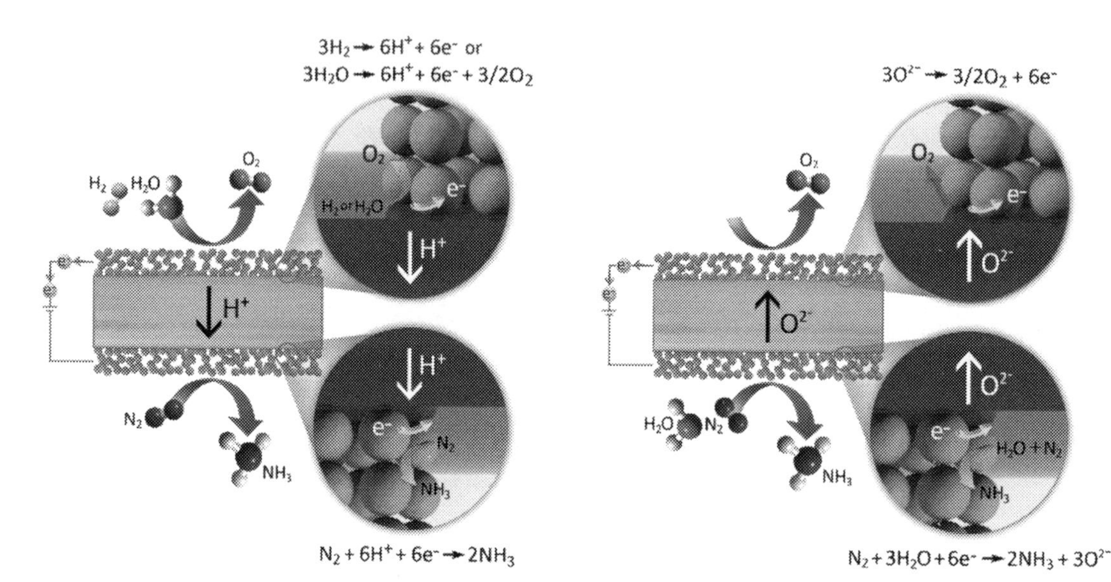

FIG. 18.4 Electrochemical ammonia synthesis from water and nitrogen using proton- and oxygen-ion-conducting electrolyte-based electrochemical cells (*left* and *right*, respectively). *From Yoo, C. Y., Park, J. H., Kim, K., Han, J. I., Jeong, E. Y., Jeong, C. H., Yoon, H. C., & Kim, J. N. (2017). Role of protons in electrochemical ammonia synthesis using solid-state electrolytes.* ACS Sustainable Chemistry and Engineering, *5(9), 7972–7978. https://doi.org/10.1021/acssuschemeng.7b01515.*

options of hydrogen and steam are given by (Eqs. 18.4, 18.5).

$$3H_{2\,(g)} \rightarrow 6H^{+} + 6e^{-} \quad (18.4)$$

$$3H_2O_{(g)} \rightarrow \frac{3}{2}O_{2\,(g)} + 6H^{+} + 6e^{-} \quad (18.5)$$

In contrast, at the cathode compartment, the half-reaction is given by (Eq. 18.6),

$$N_{2\,(g)} + 6H^{+} + 6e^{-} \rightarrow 2NH_{3\,(g)} \quad (18.6)$$

The use of alternative hydrogen sources, such as steam, could eliminate the necessity of natural gas in NH_3 synthesis, thus reducing the overall cost of ammonia production and the carbon footprint (Garagounis, Kyriakou, Skodra, Vasileiou, & Stoukides, 2014; Kyriakou et al., 2017).

Marnellos and Stoukides (1998) were the first to report on the electrochemical synthesis of ammonia using N_2 and H_2 as the primary sources at 570°C, using a proton-conducting solid oxide electrolyte, $SrCe_{0.95}Tb_{0.05}O_{3\text{-}\delta}$. Since then, several different groups have studied this issue, as shown in Table 18.1. Independent of the proton electrolyte used, the results show that ammonia formation rates are in the same order of magnitude, $\sim 10^{-9}\,\mathrm{mol\,cm^{-2}\,s^{-1}}$, and far below the rates suitable for commercial viability ($4.3\text{–}8.7 \times 10^{-7}$ mol $\mathrm{cm^{-2}\,s^{-1}}$). As ammonia decomposes at 450°C (Fig. 18.5), the use of ceramic materials that possess high conductivity at temperatures below 450°C is considered desirable for further development of this technology (Zhang, Xu, & Ma, 2011) (Liu et al., 2010) (Vasileiou et al., 2015).

In addition to proton conductors, oxygen-ion-conducting electrolytes can also be employed for ammonia production at atmospheric pressures. In this system, steam electrolysis and ammonia synthesis take place at the cathode. The following reactions take place at the anode side (Eq. 18.7),

$$3O^{2-} \rightarrow \frac{3}{2}O_{2\,(g)} + 6e^{-} \quad (18.7)$$

and, at the cathode side (Eq. 18.8),

$$N_{2\,(g)} + H_2O_{(g)} + 6e^{-} \rightarrow 2NH_{3\,(g)} + 3O^{2-} \quad (18.8)$$

Stoukides' group (Skodra & Stoukides, 2009) reported an ammonia conversion rate of $1.50 \times 10^{-13}\,\mathrm{mol\,cm^{-2}\,s^{-1}}$ at 650°C and atmospheric pressure, using yttria-stabilized zirconia (8% Y_2O_3/ZrO_2) as an oxygen-ion conductor and N_2 and H_2O as the reactants. The ammonia formation rate was three times lower than those reported for proton conductors, which was explained by the presence of oxygen-containing species (H_2O) at the cathode side and may also be related to the greater extent of ammonia dissociation that would be expected at these temperatures.

These results suggest that a balance must be made between the lower energy requirement to dissociate steam at higher temperatures and the need to control maximum temperature to prevent ammonia dissociation. An alternative method to reduce the energy requirements for steam electrolysis is to supply a gaseous fuel to the anode of the oxygen-ion conducting cell. In this manner, depending on the fuel and the working temperature of the cell, the need to supply electrical energy can be suppressed, possibly allowing the whole process to be spontaneous and consequently be an attractive alternative (Kyriakou et al., 2017; Li et al., 2020).

18.3 Basics of N_2 reduction reaction

Mechanisms of N_2 reduction reaction (NRR)

Nitrogen reduction to ammonia on a catalytic surface usually involves four main steps: (1) adsorption of N_2 molecules onto the catalyst surface, (2) the dissociative or associative breaking

TABLE 18.1 Ammonia production rates for different high-temperature systems.

Electrolyte	WE	T (°C)	R_{NH3} ($mol\,cm^{-2}\,s^{-1}$)	Input	Reference
$SrCe_{0.95}Yb_{0.05}O_{3-a}$ (SCY)	Pd	570	4.50×10^{-9}	H_2,N_2	
$La_{1.9}Ca_{0.1}Zr_2O_{6.95}$ (LCZ)	Ag-Pd	520	1.76×10^{-9}	H_2,N_2	
$Ba_3(Ca_{1.18}Nb_{1.82})O_{9-a}$ (BCN)	Ag-Pd	620	1.42×10^{-9}	H_2,N_2	
$Ba_3CaZr_{0.5}Nb_{1.5}O_{9-a}$ (BCZN)	Ag-Pd	620	1.82×10^{-9}	H_2,N_2	Li et al. (2020)
$Ba_3Ca_{0.9}Nd_{0.28}Nb_{1.82}O_{9-a}$	Ag-Pd	620	2.16×10^{-9}	H_2,N_2	Li et al. (2020)
$La_{1.95}Ca_{0.05}Zr_2O_{7-a}$ (LCZ)	Ag-Pd	520	2.00×10^{-9}	H_2,N_2	
$La_{1.95}Ca_{0.05}Ce_2O_{7-a}$ (LCC)	Ag-Pd	520	1.30×10^{-9}	H_2,N_2	
CeO_2-$Ca_3(PO_4)_2$-K_3PO_4	Ag-Pd	650	9.50×10^{-9}	H_2,N_2	
$Ce_{0.8}La_{0.2}O_{2-a}$	Ag-Pd	650	7.20×10^{-9}	H_2,N_2	
$Ce_{0.8}Y_{0.2}O_{2-a}$ (CYO)	Ag-Pd	650	7.50×10^{-9}	H_2,N_2	
$Ce_{0.8}Gd_{0.2}O_{2-a}$ (CGO)	Ag-Pd	650	7.70×10^{-9}	H_2,N_2	
$Ce_{0.8}Sm_{0.2}O_{2-a}$ (CSO)	Ag-Pd	650	8.20×10^{-9}	H_2,N_2	
$Ce_{0.8}Y_{0.2}O_{1.9}$-$Ca_3(PO_4)_2$-K_3PO_4	Ag-Pd	650	6.95×10^{-9}	CH_4,N_2	
$BaCe_{0.9}Sm_{0.1}O_{3-a}$ (BCS)	Ag-Pd	620	5.23×10^{-9}	H_2,N_2	
$BaCe_{0.8}Gd_{0.1}Sm_{0.10}O_{3-a}$	Ag-Pd	620	5.82×10^{-9}	H_2,N_2	
$La_{0.9}Sr_{0.1}Ga_{0.8}Mg_{0.2}O_{3-a}$ (LSGM)	Ag-Pd	550	2.37×10^{-9}	H_2,N_2	
$La_{0.9}Ba_{0.1}Ga_{1-x}Mg_xO_{3-a}$	Ag-Pd	520	1.89×10^{-9}	H_2,N_2	
$BaCe_{1-x}Gd_xO_{3-a}$ (BCG)	Ag-Pd	480	4.63×10^{-9}	H_2,N_2	
$BaCe_{1-x}Y_xO_{3-a}$ (BCY)	Ag-Pd	500	2.10×10^{-9}	H_2,N_2	
$SrCe_{0.95}Yb_{0.05}O_{3-a}$ (SCY)	Ru	650	4.90×10^{-13}	H_2O,N_2	Skodra and Stoukides (2009)
Yttria-stabilized zirconia (YSZ)	Ru	650	1.12×10^{-13}	H_2O,N_2	Skodra and Stoukides (2009)
$BaCe_{1-x}Dy_xO_{3-a}$ (BCD)	Ag-Pd	530	3.50×10^{-9}	H_2,N_2	
$BaCe_{0.85}Y_{0.15}O_{3-a}$ (BCY)	BSCF	530	4.10×10^{-9}	H_2,N_2	
$BaCe_{1-x}Ca_xO_{3-a}$ (BCC)	Ag-Pd	480	2.69×10^{-9}	H_2,N_2	
$Ba_xCe_{0.8}Y_{0.2}O_{3-a}$ + 0.04ZnO	Ag-Pd	500	2.36×10^{-9}	H_2,N_2	

TABLE 18.1 Ammonia production rates for different high-temperature systems—cont'd

Electrolyte	WE	T (°C)	R_{NH3} (mol cm^{-2} s^{-1})	Input	Reference
$BaCe_{0.9-x}Zr_xSm_{0.1}O_{3-a}$ (BCZS)	Ag-Pd	500	2.67×10^{-9}	H_2,N_2	
$BaCe_{0.2}Zr_{0.7}Y_{0.1}O_{2.9}$ (BCZY)	Rh	600	1.28×10^{-8}	H_2,N_2	Vasileiou, Kyriakou, Garagounis, Vourros, and Stoukides (2015)

Ammonia production rates achieved with different systems at high temperatures (>500°C).
From Kyriakou, V., Garagounis, I., Vasileiou, E., Vourros, A., & Stoukides, M. (2017). Progress in the electrochemical synthesis of ammonia. Catalysis Today*, 286, 2–13. https://doi.org/10.1016/j.cattod.2016.06.014.*

of the N—N bond by either proton transfer from a proton donor or electron injection from an external circuit, (3) hydrogenation by continuous proton-coupled electron-transfer steps, and (4) desorption of the generated ammonia (Wu, Fan, Zhang, & Zhang, 2021). The hydrogenation step, more precisely the first hydrogenation procedure of an adsorbed N_2 molecule ($*N_2 + H^+ + e^- \rightarrow *N_2H$), is the rate-determining step over the majority of NRR catalysts. Contrarily, in the competing hydrogen evolution reaction (HER), this hydrogenation step is kinetically favorable via the Volmer-Heyrovsky or Volmer-Tafel mechanisms, leading to a low NRR selectivity.

NRR can proceed by different mechanisms as shown in Fig. 18.6. The associative mechanism can be explained by a process in which two atoms of nitrogen remain bound to each other as the nitrogen molecule is hydrogenated. The outermost nitrogen is reduced to NH_3, and, after that, the nitrogen initially adsorbed on the surface will be hydrogenated to form the second NH_3 molecule. Alternatively, via the dissociative mechanism (Fig. 18.6A), in the first step, the nitrogen molecule will break into individual

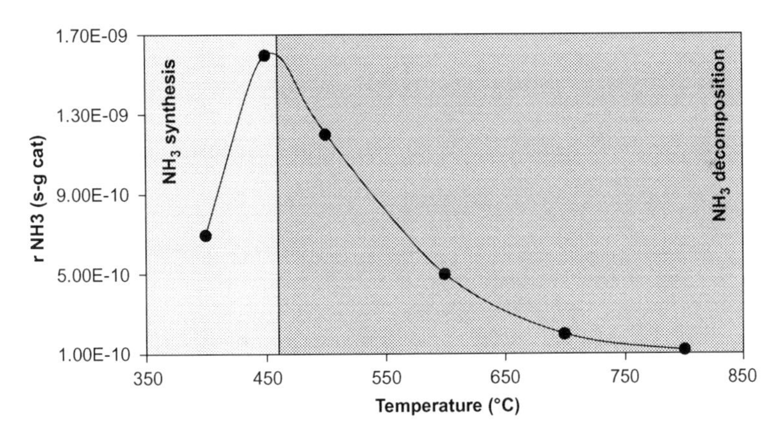

FIG. 18.5 Ammonia rates with temperature. *From Skodra, A., & Stoukides, M. (2009). Electrocatalytic synthesis of ammonia from steam and nitrogen at atmospheric pressure.* Solid State Ionics, *180(23–25), 1332–1336. https://doi.org/10.1016/j.ssi.2009.08.001.*

N atoms that are then hydrogenated and converted to the NH_3 molecule. In nature, nitrogen reduction occurs via the associative mechanism rather than the dissociative adsorption, in contrast to that occurring during the industrial Haber-Bosch process (Montoya, Tsai, Vojvodic, & Nørskov, 2015).

More recently, some novel electrocatalysts that use Mars-van Krevelen (MvK) (Fig. 18.6B) and Li-mediated (Fig. 18.6F) mechanisms have been suggested. The first of these, the MvK mechanism, has been predicted to occur on the surface of transition-metal nitrides (TMNs). Here, the core feature is the formation of a product (ammonia) from a catalyst lattice ligand (nitrogen) with the subsequent replenishment of the latter by a reactant (dinitrogen) to complete the catalytic cycle (Fig. 18.6E) (Abghoui & Skúlason, 2017). From a thermodynamic point of view, the breaking of the nitrogen triple bond is favored in the MvK mechanism than in the traditional dissociation and association mechanisms. Conversely, the Li-mediated mechanism does not involve the N_2 adsorption on the catalyst's surface. Instead, the formation of lithium nitride (LiN) first occurs from the reaction of N_2 with lithium metal, which can occur spontaneously under ambient conditions. Thereafter, the reaction follows with the protonation of LiN to produce ammonia and lithium salt, which is electrochemically reduced to lithium metal to complete the catalytic cycle (Wu et al., 2021).

Computational modeling studies on electrochemical NRR

As outlined in the previous sections, the key to developing a high-performance catalyst for NRR relies on catalyst development. Catalysts that are capable of changing the reaction pathway, lowering the energy barrier, and, consequently, accelerating the conversion of electrochemical NRR are needed. In the last few years, computational catalysis has been a valuable tool to direct material selection, due to its capabilities of free energy calculations, kinetic analysis, and catalyst design.

One possibility is the construction of a volcano-type diagram of catalytic activity vs. nitrogen adsorption energy on different transition-metal catalysts, as shown in Fig. 18.7. Skúlason et al. (2012) found, by theoretical evaluation, that on the top of the volcano plot are Mo, Fe, Ru, and Rh. These transition metals are the most active surfaces for ammonia production. However, their selectivity toward NRR is still extremely poor; thus, the competing HER is still a debilitating problem. In contrast, the Ti, Zr, Sc, and Y surfaces, which are present on the left area of the volcano plot, are found to bound nitrogen more strongly than protons, with an increasing ammonia conversion, which is expected in this group.

Nørskov's group (Montoya et al., 2015) demonstrated the limiting steps on the electrochemical reduction of nitrogen by DFT calculations (Fig. 18.8A). Their work highlighted the protonation of N_2 as $*N_2H$ for less reactive transition metals (with a weak nitrogen-binding capability) and the $*NH$ protonation to form $*NH_2$ or the desorption of $*NH_2$ as NH_3 for more reactive transition metals (binding nitrogen strongly). In this context, Re, Ru, and Rh metals are located on top of the volcano plot, where a stronger nitrogen-binding capability is expected. Nevertheless, even these metals exhibit a large limiting potential for nitrogen electroreduction, which is at least 0.5 V more negative than the overpotential requirements for HER (Fig. 18.8B). Therefore, the competing HER is theoretically more favorable on most metal catalysts, providing an explanation of the low faradaic efficiencies widely reported for electrocatalytic ammonia synthesis.

To overcome the poor selectivity of transition metals, recent DFT calculations (Abghoui et al., 2015; Abghoui, Garden, Howalt, Vegge, & Skúlason, 2016; Matanović, Garzon, & Henson, 2014) have identified nitrides as promising electrocatalysts under ambient conditions, which could conceivably achieve a FE higher than 75%.

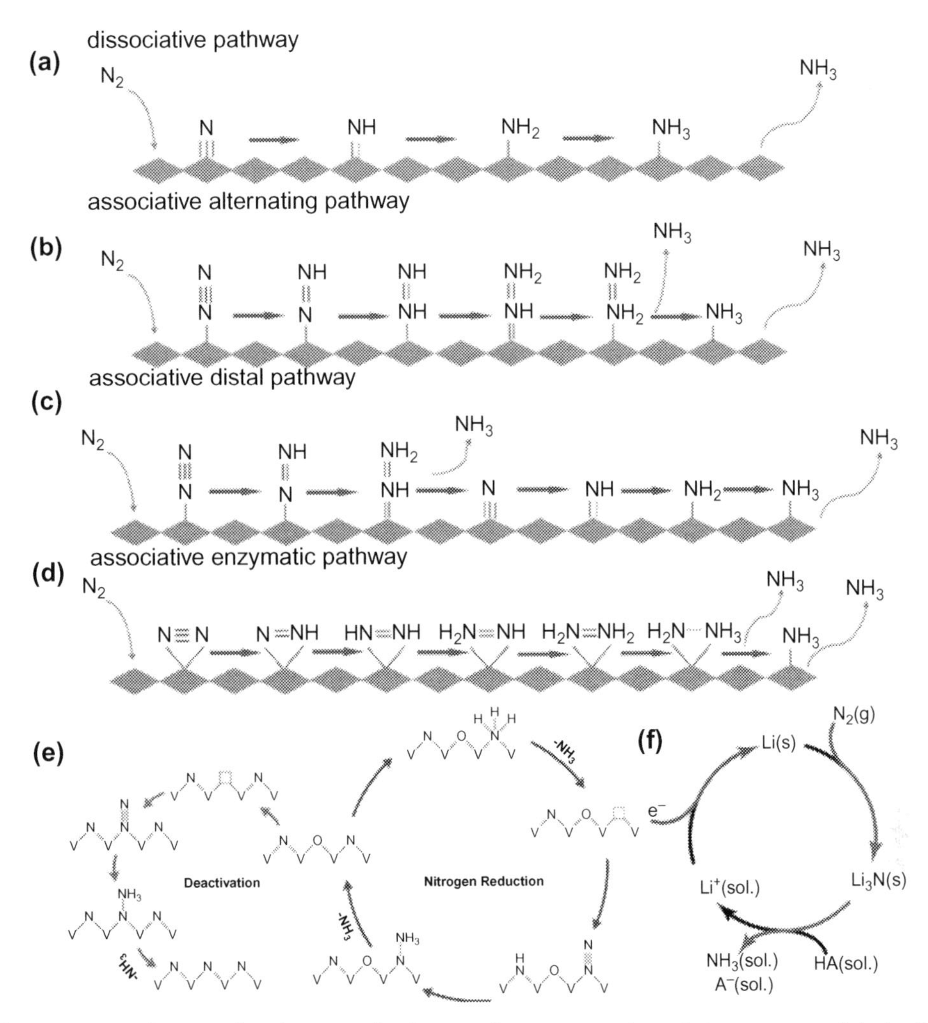

FIG. 18.6 Generic mechanisms for nitrogen reduction reaction to ammonia on heterogeneous catalysts. *From Wu, T., Fan, W., Zhang, Y., & Zhang, F. (2021). Electrochemical synthesis of ammonia: Progress and challenges.* Materials Today Physics, *16, 100310. https://doi.org/10.1016/j.mtphys.2020.100310.*

One example is related to the work of Abghoui and colleagues (Abghoui, 2017), who searched for the most promising nitride candidates in terms of stability, activity, and selectivity for the reduction of nitrogen to ammonia at room temperature and at ambient pressure. Fig. 18.9 summarizes the resultant screening. The (110) plane of the zincblende structure of RuN and the (100) plane of the rock-salt structures of VN and CrN are highlighted to be some the best candidates, being not only the most promising for catalysis, but at the same time also the most stable against poisoning and decomposition in the electrochemical environment (Abghoui & Skúlason, 2017; Younes Abghoui et al., 2015). Furthermore, these potential new electrocatalysts are predicted to be

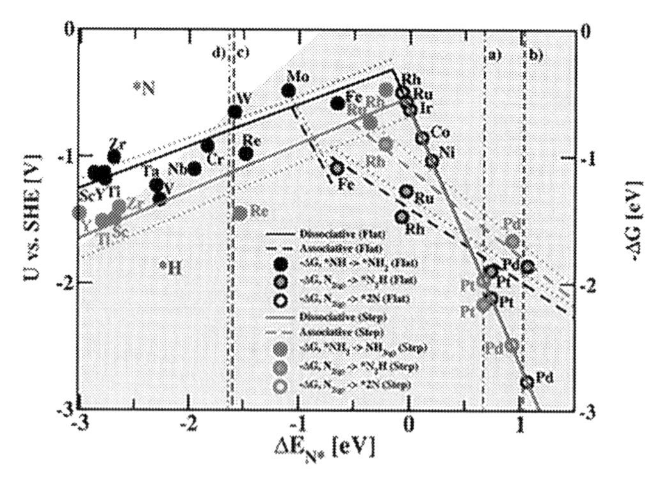

FIG. 18.7 Volcano plot of the dissociative *(solid lines)* and associative *(dashed lines)* mechanisms on both flat *(black)* and stepped *(red; dark gray in print version)* transition-metal surfaces. *Adapted from Skúlason, E., Bligaard, T., Gudmundsdóttir, S., Studt, F., Rossmeisl, J., Abild-Pedersen, F., Vegge, T., Jónsson, H., & Nørskov, J. K. (2012). A theoretical evaluation of possible transition metal electro-catalysts for N_2 reduction.* Physical Chemistry Chemical Physics, *14(3), 1235–1245. https://doi.org/10.1039/c1cp22271f.*

resilient against the production of hydrogen, suppressing the H_2 evolution reaction (HER), an essential electrocatalytic criterion for their use in the electrochemical systems of ammonia conversion.

Overall, plenty of progress has been made in computational studies, and this knowledge is now filtering down to experimental research with a lot of experimental efforts having already been taken to test these predictions and the potentially superior behavior of nitrides over that of transition metals.

Experimental studies on electrocatalyst development

More than 30 elements have been experimentally explored as electrocatalysts for ammonia synthesis including Fe, Ni, Ru, Pd, Pt, and Au, as well as alloys Ag-Pd and Ni-Cu (Anderson, Rittle, & Peters, 2013; Kugler, Luhn, Schramm, Rahimi, & Wessling, 2015; Tsuneto et al., 1994), simple and complex metal oxides (Amar, Lan, et al., 2011; Amar, Lan, Petit, & Tao, 2015; Yun et al., 2015), metal nitrides (Michalsky, Pfromm, & Steinfeld, 2015), carbon (Liu et al., 2018; Zhu, Zhang, Ruther, & Hamers, 2013), carbon nitrides (Alexander, Hargreaves, & Mitchell, 2012; Hargreaves, 2014), and so on.

The Ag-Pd cathode has been the most frequently used due to its promising results, namely, high reaction rates ($>10^{-9}$ mol s^{-1} cm^{-2}) with high FE (up to 80%). This is an unexpected result due to its low capacity for nitrogen dissociative adsorption (Montoya et al., 2015; Skúlason et al., 2012). Additionally, it is expected that under cathodic polarization, the surface of these metals will be "flooded" with protons, which will hinder the nitrogen adsorption. An explanation of this behavior is given in the work of Skúlason et al., where it is reported that the use of Y and Zr in the electrode composition will indirectly increase the catalytic activity of the Ag-Pd catalyst at the electrode-electrolyte interphase (Garagounis et al., 2014).

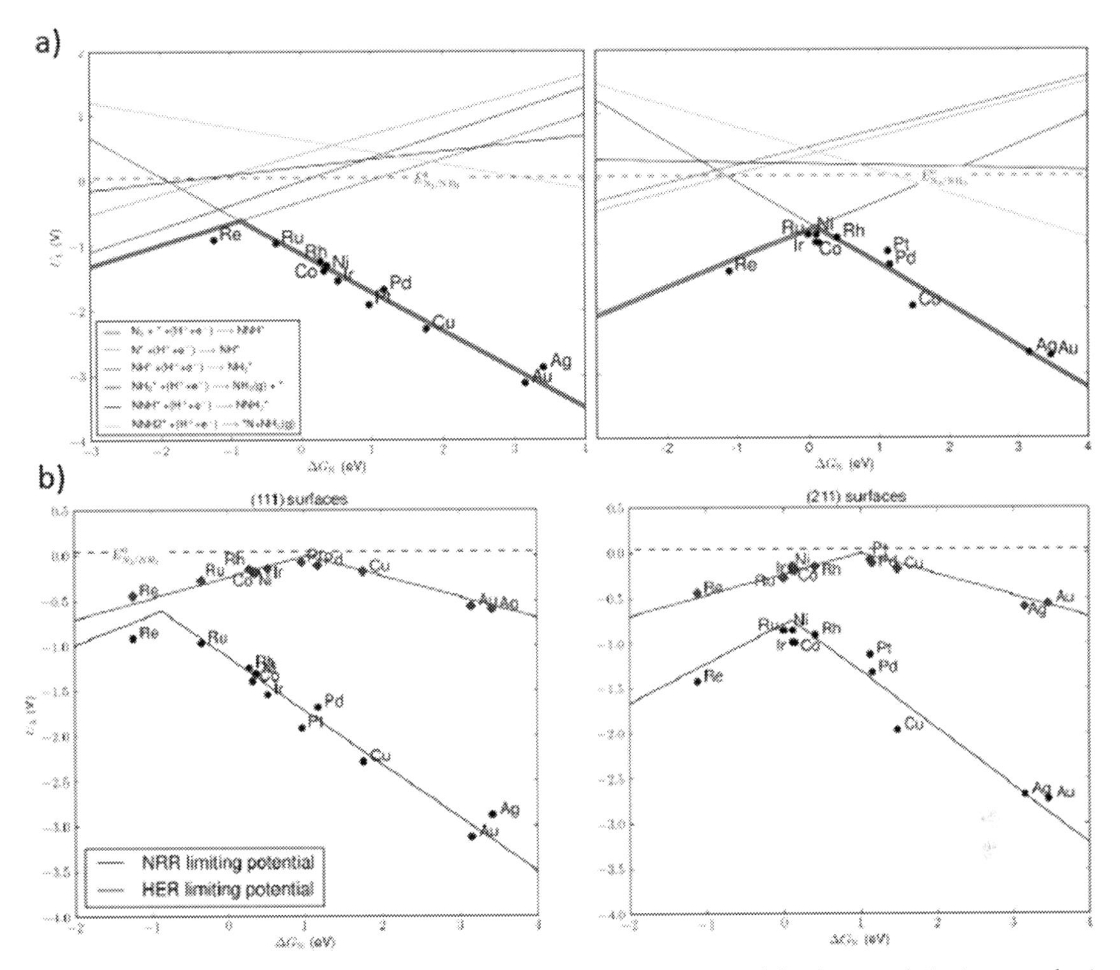

FIG. 18.8 Comparison of volcano plots for HER and NRR. (A) Limiting potentialsfor electrocatalytic nitrogen reduction reaction (NRR) as a function of *N binding energy on (111) surfaces and (211) surfaces of various transition metals; (B) comparison of the limiting potentials of the volcano plots of the hydrogen evolution reaction (HER) *(shown in blue; dark gray in print version)* and NRR *(shown in black)* on (111) surfaces and (211) surfaces of various transition metals. *Adapted from Montoya, J. H., Tsai, C., Vojvodic, A., & Nørskov, J. K. (2015). The challenge of electrochemical ammonia synthesis: A new perspective on the role of nitrogen scaling relations.* ChemSusChem, 8*(13), 2180–2186. https://doi.org/10.1002/cssc.201500322.*

Other classes of electrocatalysts that have recently gained interest are those of transition-metal nitrides (TMNs) (Hargreaves, 2014). Fe_3Mo_3N-Ag (Amar, Petit, et al., 2011) and Co_3Mo_3N-Ag (Amar, Lan, Petit, & Tao, 2015; Amar, Lan, & Tao, 2015) have been studied by Amar et al. with a formation rate of 3.27×10^{-10} mol s^{-1} cm^{-2} and a FE of 6.5% at temperatures obtained in the range of 400–450°C. More recently, Kyriakou, Garagounis, Vourros, Vasileiou, and Stoukides (2020) used a VN-Fe cathode operating at atmospheric pressure and

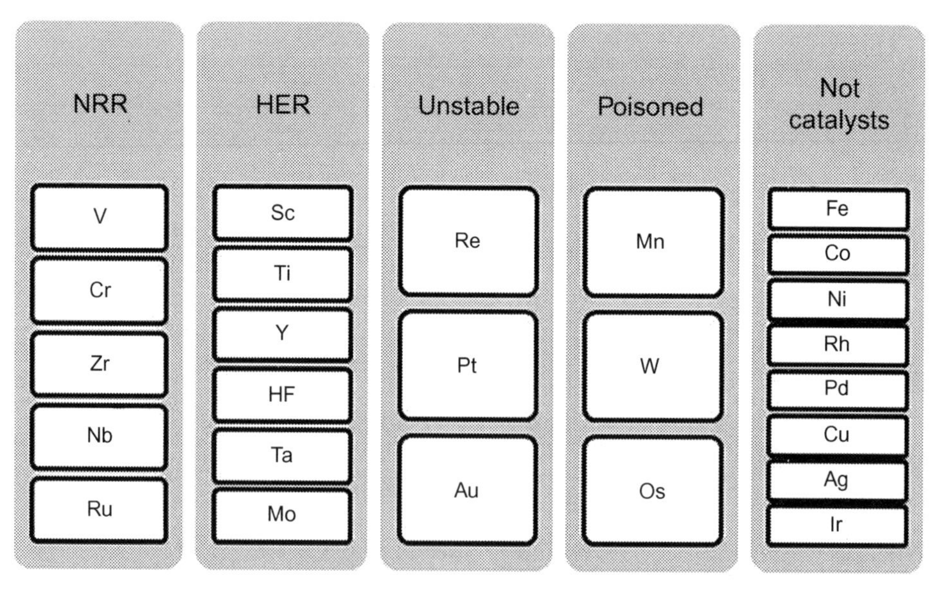

FIG. 18.9 Summary and conclusion of the results obtained throughout the screening process of transition-metal mononitrides for both the nitrogen reduction reaction (NRR) and the hydrogen evolution reaction HER. *Adapted from Abghoui, Y. (2017). Novel electrocatalysts for sustainable ammonia production at ambient conditions. https://hdl.handle.net/20.500.11815/314.*

at a temperature in the range of 550–650°C. The NH_3formation rate was 68 mmol h^{-1} m^{-2} and the FE was 5.5%.

Overall, many catalysts have already been screened although only a few of them have been tested systematically for the electrochemical synthesis of ammonia. Despite these increasingly focused efforts of the research community, the representative catalysts remain far away from the target values required for viable application ($>9.3 \times 10^{-7}$ mol cm^{-2} s^{-1}, FE > 90%).

18.4 Outlook

The most important criterion to design a good electrocatalyst relies on increasing its selectivity toward NRR while suppressing HER. A more selective electrocatalyst will be more specific to reactions, yielding a particular product and, thus, will not waste reactants, will not require expensive separation procedures, and will not generate toxic byproducts, allowing this technology to become more sustainable and economically viable.

For electrochemical NRR, the major problems are shown to be related to low ammonia formation rates (large overpotential) and low selectivity toward NH_3 formation due to the competition between NRR and HER, where the latter reaction is, typically, kinetically favored, as mentioned earlier.

Qing et al. (2020) put these terms in context, based on a previous calculation methodology found in the work of Singh et al. (2017). Qing et al. highlighted the importance of improving selectivity, rather than overpotential, by comparing both effects in terms of the solar cell area needed to produce ammonia for a typical 1-ha field in Midwestern United States, as illustrated in Fig. 18.10. Assuming 100% faradaic efficiency at an overpotential of 1 V, the estimated solar cell area needed for producing sufficient NH_3 quantity is about 0.05% of the total land area (Fig. 18.10A), meaning an area of 5 m^2. This area corresponds to that of a typical solar cell or a residential

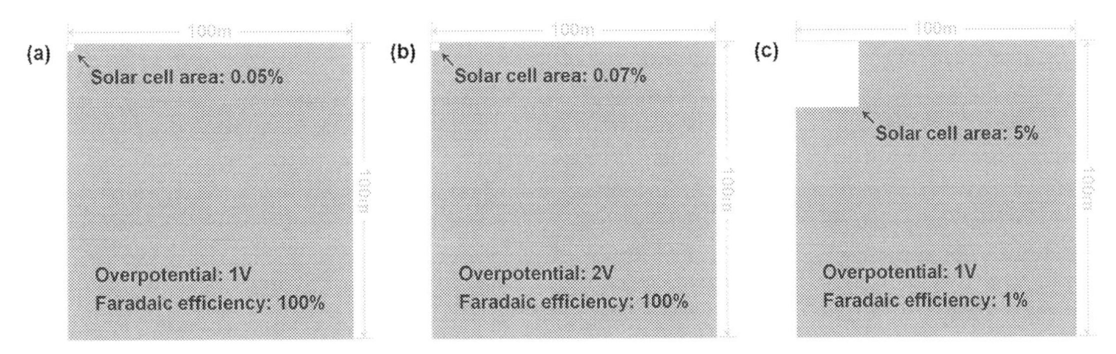

FIG. 18.10 Minimum solar area needed to synthetize ammonia by the electrochemical route. *From Singh, A. R., Rohr, B. A., Schwalbe, J. A., Cargnello, M., Chan, K., Jaramillo, T. F., Chorkendorff, I., & Nørskov, J. K. (2017). Electrochemical ammonia synthesis – The selectivity challenge.* ACS Catalysis, *7(1), 706–709. https://doi.org/10.1021/acscatal.6b03035.*

scale wind turbine (U.S. Department of Energy/National Renewable Energy Laborator, 2007). While maintaining the same FE, if the overpotential is increased from 1 to 2V, the required solar cell area still remains smaller than 0.1% (Fig. 18.10B). However, if the FE% decreases to 1% (a typical value in the literature found for the electrochemical ammonia synthesis), the solar cells must occupy as much as 5% of the land (Fig. 18.10C), meaning an area of 500 m^2, which will make the cost of solar cells needed to drive the process prohibitively expensive.

Singh et al. explained that the large influence of the low selectivity is because the ammonia formation rate is a *zero-order* reaction, implying that NH_3 conversion is independent of the concentration of the available protons or electrons. In contrast, the hydrogen evolution rate is a reaction of the first order, thus being dependent on the concentration of both species. For this reason, a potential solution to avoid this problem, and to increase the ammonia selectivity, is to limit the concentration of protons and electrons available at the electrode surface. In this manner, the HER process can be potentially inhibited, whereas the NH_3 production should remain unaffected.

Nonetheless, lowering the transport of electrons during the electrochemical process is a difficult process, whereas limiting the access of protons could be feasible and may potentially be done using different methodologies. Qing et al. proposed several different approaches for aqueous systems, for example, using aprotic solvents with an extremely low concentration of proton donors or using aprotic, hydrophobic layers at the electrode surface to slow down the proton transfer. However, for high-temperature systems using solid-state electrolytes, the adoption of similar solutions remains unexplored, to date.

18.5 Conclusions

Ammonia synthesis by electrochemical routes offers several advantages with respect to the conventional Haber-Bosch process, particularly the possibility of operation under ambient pressures and the excellent flexibility of the chemical reactants that can be used. For example, while nitrogen can be obtained directly from air, hydrogen can be produced in situ directly from water. These factors can avoid the need for extensive purification of the starting gases, while drastically mitigating the release of pollutants into the atmosphere if the required electricity is generated from renewable sources. Despite this attractive potential, the progress that has been achieved in electrochemical ammonia synthesis still remains at an early stageof development. Several systems have been studied, including the use of liquid and solid electrolytes. However, the lack of suitable electrocatalyst materials is the key limitation that still precludes the implementation of this technology. In this context, this chapter aims to highlight the great importance of integrating computational modeling with experimental techniques to progress further in the fundamental knowledge of catalytic mechanisms, and, thus, to succeed in designing efficient electrocatalysts.

It is clear that the final ammonia conversion in electrochemical cell reactors is highly linked to other crucial aspects, such as the design of an electrolyte-electrode interphase, the flow rate and the composition of the primary reagents used, and the operation temperatures and pressures. Furthermore, the reliability of results to confirm that ammonia is synthesized by electrochemical nitrogen fixation must be improved, to avoid ammonia contamination from other sources. Therefore, rigorous testing protocols should be followed to prevent false-positive results, thus strengthening the research to find feasible pathways to NRR.

An intensive collaboration is needed from different fields, such as materials science, solid-state, and heterogeneous catalysis, to drive systematic studies to allow scale up for the creation of an economic and sustainable electrochemical system for ammonia synthesis.

Acknowledgments

Vanessa C.D. Graça and Laura I.V. Holz are grateful to the Fundação para a Ciência e Tecnologia (FCT) for their PhD grants SFRH/BD/130218/2017 and PD/BDE/142837/2018, respectively. Francisco J.A. Loureiro is thankful for the Investigator Grant CEECIND/02797/2020. The authors also acknowledge the projects PTDC/CTM-CTM/2156/2020, PTDC/QUI-ELT/3681/2020, POCI-01-0247-FEDER-039926, POCI-01-0145-FEDER-032241, UIDB/00481/2020 and UIDP/00481/2020, and CENTRO-01-0145-FEDER-022083 – Centro Portugal Regional Operational Programme (Centro2020), under the PORTUGAL 2020 Partnership Agreement, through the European Regional Development Fund (ERDF). This study was also financed in part by the Coordenação de Aperfeiçoamento de Pessoal de Nível Superior – Brasil (CAPES) – Finance Code 001. Allan J.M. Araújo thanks the Conselho Nacional de Desenvolvimento Científico e Tecnológico (CNPq/Brazil, Reference Number 200439/2019-7).

References

Abghoui, Y., & Skúlason, E. (2017). Onset potentials for different reaction mechanisms of nitrogen activation to ammonia on transition metal nitride electro-catalysts. *Catalysis Today, 286*, 69–77. https://doi.org/10.1016/j.cattod.2016.11.047.

Abghoui, Y. (2017). *Novel electrocatalysts for sustainable ammonia production at ambient conditions.* https://hdl.handle.net/20.500.11815/314.

Abghoui, Y., Garden, A. L., Hlynsson, V. F., Björgvinsdóttir, S., Ólafsdóttir, H., & Skúlason, E. (2015). Enabling electrochemical reduction of nitrogen to ammonia at ambient conditions through rational catalyst design. *Physical Chemistry Chemical Physics*, 4909–4918. https://doi.org/10.1039/C4CP04838E.

Abghoui, Y., Garden, A. L., Howalt, J. G., Vegge, T., & Skúlason, E. (2016). Electroreduction of N_2 to ammonia at ambient conditions on mononitrides of Zr, Nb, Cr, and V: A DFT guide for experiments. *ACS Catalysis, 6*(2), 635–646. https://doi.org/10.1021/acscatal.5b01918.

Alexander, A. M., Hargreaves, J. S. J., & Mitchell, C. (2012). The reduction of various nitrides under hydrogen: Ni 3N, Cu 3N, Zn 3N 2 and Ta 3N 5. *Topics in Catalysis, 55*(14–15), 1046–1053. https://doi.org/10.1007/s11244-012-9890-3.

Amar, I. A., Lan, R., Petit, C. T. G., & Tao, S. (2011). Solid-state electrochemical synthesis of ammonia: A review. *Journal of Solid State Electrochemistry, 15*(9), 1845–1860. https://doi.org/10.1007/s10008-011-1376-x.

Amar, I. A., Lan, R., & Tao, S. (2015). Synthesis of ammonia directly from wet nitrogen using a redox stable La0.75Sr0.25Cr0.5Fe0.5O3-δ-Ce0.8Gd0.18Ca0.02O2-δ composite cathode. *RSC Advances, 5*(49), 38977–38983. https://doi.org/10.1039/c5ra00600g.

Amar, I. A., Petit, C. T. G., Zhang, L., Lan, R., Skabara, P. J., & Tao, S. (2011). Electrochemical synthesis of ammonia based on doped-ceria-carbonate composite electrolyte and perovskite cathode. *Solid State Ionics, 201*(1), 94–100. https://doi.org/10.1016/j.ssi.2011.08.003.

Amar, I. A., Lan, R., Petit, C. T. G., & Tao, S. (2015). Electrochemical synthesis of ammonia based on Co3Mo3N catalyst and $LiAlO_2$–(Li,Na,K)2CO_3 composite electrolyte. *Electrocatalysis, 6*(3), 286–294. https://doi.org/10.1007/s12678-014-0242-x.

Anderson, J. S., Rittle, J., & Peters, J. C. (2013). Catalytic conversion of nitrogen to ammonia by an iron model complex. *Nature*, *501*(7465), 84–87. https://doi.org/10.1038/nature12435.

Cui, B., Zhang, J., Liu, S., Liu, X., Xiang, W., Liu, L., et al. (2017). Electrochemical synthesis of ammonia directly from N2 and water over iron-based catalysts supported on activated carbon. *Green Chemistry*, *19*(1), 298–304. https://doi.org/10.1039/c6gc02386j.

Foster, S. L., Bakovic, S. I. P., Duda, R. D., Maheshwari, S., Milton, R. D., Minteer, S. D., et al. (2018). Catalysts for nitrogen reduction to ammonia. *Nature Catalysis*, *1*(7), 490–500. https://doi.org/10.1038/s41929-018-0092-7.

Garagounis, I., Kyriakou, V., Skodra, A., Vasileiou, E., & Stoukides, M. (2014). Electrochemical synthesis of ammonia in solid electrolyte cells. *Frontiers in Energy Research*, *2*. https://doi.org/10.3389/fenrg.2014.00001.

Giddey, S., Badwal, S. P. S., & Kulkarni, A. (2013). Review of electrochemical ammonia production technologies and materials. *International Journal of Hydrogen Energy*, *38*(34), 14576–14594. https://doi.org/10.1016/j.ijhydene.2013.09.054.

Hargreaves, J. S. J. (2014). Nitrides as ammonia synthesis catalysts and as potential nitrogen transfer reagents. *Applied Petrochemical Research*, *4*, 3–10. https://doi.org/10.1007/s13203-014-0049-y.

Jiao, F., & Xu, B. (2018). Electrochemical ammonia synthesis and ammonia fuel cells. *Advanced Materials*, *31*(31), 1805173. https://doi.org/10.1002/adma.201805173.

Klerke, A., Christensen, C. H., Nørskov, J. K., & Vegge, T. (2008). Ammonia for hydrogen storage: Challenges and opportunities. *Journal of Materials Chemistry*, *18*(20), 2304–2310. https://doi.org/10.1039/b720020j.

Kugler, K., Luhn, M., Schramm, J. A., Rahimi, K., & Wessling, M. (2015). Galvanic deposition of Rh and Ru on randomly structured Ti felts for the electrochemical NH_3 synthesis. *Physical Chemistry Chemical Physics*, *17*(5), 3768–3782. https://doi.org/10.1039/c4cp05501b.

Kyriakou, V., Garagounis, I., Vasileiou, E., Vourros, A., & Stoukides, M. (2017). Progress in the electrochemical synthesis of ammonia. *Catalysis Today*, *286*, 2–13. https://doi.org/10.1016/j.cattod.2016.06.014.

Kyriakou, V., Garagounis, I., Vourros, A., Vasileiou, E., & Stoukides, M. (2020). An electrochemical Haber-Bosch process. *Joule*, *4*(1), 142–158. https://doi.org/10.1016/j.joule.2019.10.006.

Lan, R., Irvine, J. T. S., & Tao, S. (2012). Ammonia and related chemicals as potential indirect hydrogen storage materials. *International Journal of Hydrogen Energy*, *37*(2), 1482–1494. https://doi.org/10.1016/j.ijhydene.2011.10.004.

Li, C., Wang, T., & Gong, J. (2020). Alternative strategies toward sustainable ammonia synthesis. *Transactions of Tianjin University*, *26*(2), 67–91. https://doi.org/10.1007/s12209-020-00243-x.

Licht, S., Cui, B., Wang, B., Li, F. F., Lau, J., & Liu, S. (2014). Ammonia synthesis by N_2 and steam electrolysis in molten hydroxide suspensions of nanoscale Fe_2O_3. *Science*, *345*(6197), 637–640. https://doi.org/10.1126/science.1254234.

Liu, J., Li, Y., Wang, W., Wang, H., Zhang, F., & Ma, G. (2010). Proton conduction at intermediate temperature and its application in ammonia synthesis at atmospheric pressure of BaCe1-x Ca x O3-α. *Journal of Materials Science*, *45*(21), 5860–5864. https://doi.org/10.1007/s10853-010-4662-6.

Liu, Y., Su, Y., Quan, X., Fan, X., Chen, S., Yu, H., et al. (2018). Facile ammonia synthesis from electrocatalytic N_2 reduction under ambient conditions on N-doped porous carbon. *ACS Catalysis*, *8*(2), 1186–1191. https://doi.org/10.1021/acscatal.7b02165.

Marnellos, G., Kyriakou, A., Florou, F., Angelidis, T., & Stoukides, M. (1999). Polarization studies in the Pd|SrCe0.95Yb0.05O2.975|Pd proton conducting solid electrolyte cell. *Solid State Ionics*, *125*(1), 279–284. https://doi.org/10.1016/S0167-2738(99)00186-1.

Marnellos, G., Sanopoulou, O., Rizou, A., & Stoukides, M. (1997). The use of proton conducting solid electrolytes for improved performance of hydro- and dehydrogenation reactors. *Solid State Ionics*, *97*(1–4), 375–383. https://doi.org/10.1016/s0167-2738(97)00088-x.

Marnellos, G., & Stoukides, M. (1998). Ammonia synthesis at atmospheric pressure. *Science*, *282*(5386), 98–100. https://doi.org/10.1126/science.282.5386.98.

Matanović, I., Garzon, F. H., & Henson, N. J. (2014). Electroreduction of nitrogen on molybdenum nitride: Structure, energetics, and vibrational spectra from DFT. *Physical Chemistry Chemical Physics*, *16*(7), 3014–3026. https://doi.org/10.1039/c3cp54559h.

Michalsky, R., Pfromm, P. H., & Steinfeld, A. (2015). Rational design of metal nitride redox materials for solar-driven ammonia synthesis. *Interface Focus*, *5*(3), 1–10. https://doi.org/10.1098/rsfs.2014.0084.

Montoya, J. H., Tsai, C., Vojvodic, A., & Nørskov, J. K. (2015). The challenge of electrochemical ammonia synthesis: A new perspective on the role of nitrogen scaling relations. *ChemSusChem*, *8*(13), 2180–2186. https://doi.org/10.1002/cssc.201500322.

Murakami, T., Nishikiori, T., Nohira, T., & Ito, Y. (2003). Electrolytic synthesis of ammonia in molten salts under atmospheric pressure. *Journal of the American Chemical Society*, *125*(2), 334–335. https://doi.org/10.1021/ja028891t.

Murakami, T., Nohira, T., Goto, T., Ogata, Y. H., & Ito, Y. (2005). Electrolytic ammonia synthesis from water and nitrogen gas in molten salt under atmospheric pressure. *Electrochimica Acta*, *50*(27), 5423–5426. https://doi.org/10.1016/j.electacta.2005.03.023.

Panagos, E., Voudouris, I., & Stoukides, M. (1996). Modelling of equilibrium limited hydrogenation reactions carried out in H+ conducting solid oxide membrane reactors.

Chemical Engineering Science, *51*(11), 3175–3180. https://doi.org/10.1016/0009-2509(96)00216-3.

Pfromm, P. H. (2017). Towards sustainable agriculture: Fossil-free ammonia. *Journal of Renewable and Sustainable Energy*, *9*(3). https://doi.org/10.1063/1.4985090, 034702.

Qing, G., Ghazfar, R., Jackowski, S. T., Habibzadeh, F., Ashtiani, M. M., Chen, C. P., et al. (2020). Recent advances and challenges of electrocatalytic N_2 reduction to ammonia. *Chemical Reviews*, *120*(12), 5437–5516. https://doi.org/10.1021/acs.chemrev.9b00659.

Shipman, M. A., & Symes, M. D. (2017). Recent progress towards the electrosynthesis of ammonia from sustainable resources. *Catalysis Today*, *286*, 57–68. https://doi.org/10.1016/j.cattod.2016.05.008.

Singh, A. R., Rohr, B. A., Schwalbe, J. A., Cargnello, M., Chan, K., Jaramillo, T. F., et al. (2017). Electrochemical ammonia synthesis—The selectivity challenge. *ACS Catalysis*, *7*(1), 706–709. https://doi.org/10.1021/acscatal.6b03035.

Skodra, A., & Stoukides, M. (2009). Electrocatalytic synthesis of ammonia from steam and nitrogen at atmospheric pressure. *Solid State Ionics*, *180*(23–25), 1332–1336. https://doi.org/10.1016/j.ssi.2009.08.001.

Skúlason, E., Bligaard, T., Gudmundsdóttir, S., Studt, F., Rossmeisl, J., Abild-Pedersen, F., et al. (2012). A theoretical evaluation of possible transition metal electro-catalysts for N_2 reduction. *Physical Chemistry Chemical Physics*, *14*(3), 1235–1245. https://doi.org/10.1039/c1cp22271f.

Tsuneto, A., Kudo, A., & Sakata, T. (1994). Lithium-mediated electrochemical reduction of high pressure N_2 to NH_3. *Journal of Electroanalytical Chemistry*, *367*(1–2), 183–188. https://doi.org/10.1016/0022-0728(93)03025-K.

U.S. Department of Energy/National Renewable Energy Laborator. (2007). *Small wind electric systems*. A U.S. Consumer's Guide.

Vasileiou, E., Kyriakou, V., Garagounis, I., Vourros, A., & Stoukides, M. (2015). Ammonia synthesis at atmospheric pressure in a BaCe0.2Zr0.7Y0.1O2.9 solid electrolyte cell. *Solid State Ionics*, *275*, 110–116. https://doi.org/10.1016/j.ssi.2015.01.002.

Wu, T., Fan, W., Zhang, Y., & Zhang, F. (2021). Electrochemical synthesis of ammonia: Progress and challenges. *Materials Today Physics*, *16*, 100310. https://doi.org/10.1016/j.mtphys.2020.100310.

Yang, J., Weng, W., & Xiao, W. (2020). Electrochemical synthesis of ammonia in molten salts. *Journal of Energy Chemistry*, *43*, 195–207. https://doi.org/10.1016/j.jechem.2019.09.006.

Yun, D. S., Joo, J. H., Yu, J. H., Yoon, H. C., Kim, J. N., & Yoo, C. Y. (2015). Electrochemical ammonia synthesis from steam and nitrogen using proton conducting yttrium doped barium zirconate electrolyte with silver, platinum, and lanthanum strontium cobalt ferrite electrocatalyst. *Journal of Power Sources*, *284*, 245–251. https://doi.org/10.1016/j.jpowsour.2015.03.002.

Zamfirescu, C., & Dincer, I. (2008). Using ammonia as a sustainable fuel. *Journal of Power Sources*, *185*(1), 459–465. https://doi.org/10.1016/j.jpowsour.2008.02.097.

Zhang, M., Xu, J., & Ma, G. (2011). Proton conduction in BaxCe0.8Y0.2O 3-α + 0.04ZnO at intermediate temperatures and its application in ammonia synthesis at atmospheric pressure. *Journal of Materials Science*, *46*(13), 4690–4694. https://doi.org/10.1007/s10853-011-5376-0.

Zhu, D., Zhang, L., Ruther, R. E., & Hamers, R. J. (2013). Photo-illuminated diamond as a solid-state source of solvated electrons in water for nitrogen reduction. *Nature Materials*, *12*(9), 836–841. https://doi.org/10.1038/nmat3696.

CHAPTER

19

Non-faradaic electrochemical modification of catalytic activity: A current overview

Laura I.V. Holz[a,b,c], *Francisco J.A. Loureiro*[a], *Vanessa C.D. Graça*[a], *Allan J.M. Araújo*[a,d], *Diogo Mendes*[c], *Adélio Mendes*[b], *and Duncan P. Fagg*[a]

[a]Centre for Mechanical Technology and Automation (TEMA), Department of Mechanical Engineering, University of Aveiro, Aveiro, Portugal [b]LEPABE—Faculty of Engineering, University of Porto, Porto, Portugal [c]Bondalti Chemicals, S.A., Quinta da Indústria, Estarreja, Portugal [d]Materials Science and Engineering Postgraduate Program—PPGCEM, Federal University of Rio Grande do Norte—UFRN, Natal, Brazil

19.1 Introduction

At the beginning of the 1980s, Vayenas's research group (Bebelis & Vayenas, 1989; Tsiplakides, Neophytides, & Vayenas, 2001; Vayenas, Bebelis, & Ladas, 1990; Vayenas, Bebelis, Yentekakis, & Neophytides, 1992; Vayenas & Koutsodontis, 2008) was the pioneer reporting that the selectivity and catalytic activity of conductive catalyst films, deposited on the surface of a solid electrolyte, could be changed in a noticeable manner by the application of potentials (usually higher than 2V) or electrical currents. This idea of having control of the catalytic activity via electrochemical modification led to one of the most exciting findings in electrochemistry, namely, the electrochemical promotion of catalysis/non-faradaic electrochemical modification of catalytic activity (EPOC/NEMCA) effect. This effect had a huge impact on many catalytic and electrocatalytic processes (Ishihara, 2014; Karoum et al., 2008; Katsaounis, 2008; Kyriakou, Athanasiou, Garagounis, Skodra, & Stoukides, 2012; Li, Gaillard, & Vernoux, 2005; Petrushina, Bandur, Cappeln, & Bjerrum, 2000; Politova, Sobyanin, & Belyaev, 1990; Tsiakaras, Douvartzides, Demin, & Sobyanin, 2002). It was found that the activity and selectivity could exceed by five orders of magnitude, exceeding

Heterogeneous Catalysis
https://doi.org/10.1016/B978-0-323-85612-6.00019-X

the catalytic rate change predicted by Faraday's law, due to the electrochemical pumping of ions to the working electrode.

The NEMCA effect can be seen as the bridge between heterogeneous catalysis and electrochemistry. In the heterogeneous catalysis field, it is well known that the selection of proper electronic promoters is highly important and some of the most relevant industrial catalytic processes such as ammonia synthesis (Nielsen, 1971), Fischer-Tropsch synthesis (Jahangiri, Bennett, Mahjoubi, Wilson, & Gu, 2014), ethylene oxidation synthesis (Xu, Zhu, Nan, Xie, & Cheng, 2019), propane oxidation (Roche et al., 2008; Tsampas et al., 2013), and NO reduction (Lambert, Harkness, Yentekakis, & Vayenas, 1995; Petrushina, Bjerrum, Bandur, & Cleemann, 2007) use active promoted catalysts to enhance reaction selectivity. On the other hand, in the electrochemistry field, solid oxide fuel cells (SOFCs) can work with a variety of fuels with the aim of not only producing electricity but also synthesizing useful chemical products. This concept of chemical cogeneration in SOFCs has been demonstrated for a variety of different chemical reactions, including the production of nitric oxide from ammonia (Vayenas & Farr, 1980), sulfuric acid from H_2S or sulfur, and ethylene or synthesis gas from methane. This mode of operation, first demonstrated for NH_3 conversion to NO (Farr & Vayenas, 1980; Sigal & Vayenas, 1981), combines the concepts of a fuel cell and a chemical reactor. Various other exothermic reactions have been studied, including the oxidation of H_2S to SO_2, CH_3OH to H_2CO, and methane to ethylene (Jiang, Yentekakis, & Vayenas, 1994) (Table 19.1).

To understand NEMCA, we should consider the different operating modes of an oxide ion-conducting solid oxide electrolyte membrane reactor, namely, the voltage variation of the working electrode as a function of current, as shown in Fig. 19.1. In such a system, four situations can be considered (Jacques & Véronique, 2014):

1. The fuel cell mode, where the driving force is the difference in the chemical potential of oxygen between both sides of the electrochemical membrane, resulting in the oxidation of the fuel and reduction of an oxidant, thus producing electricity and useful chemical products.
2. The pumping mode, where an external potential is applied. The oxygen flux is controlled by the current, which therefore corresponds to I/2F mol of O^{2-} ions per

TABLE 19.1 Electrocatalytic reactions investigated in doped ZrO_2 electrolyte fuel cells for chemical cogeneration.

Reaction	Electrocatalyst	References
$2NH_3+5O^{2-} \rightarrow 2NO+3H_2O+10e^-$	Pt, Pt-Rh	Farr and Vayenas (1980), Vayenas and Farr (1980)
$CH_4+NH_3+3O^{2-} \rightarrow HCN+3H_2O+6e^-$	Pt, Pt-Rh	Kiratzis, Seimanides, and Stoukides (1986)
$CH_3OH+O^{2-} \rightarrow H_2CO+H_2O+2e^-$	Ag	Neophytides and Vayenas (1990)
$C_6H_5\text{-}CH_2CH_3+O^{2-} \rightarrow C_6H_5=CH_2+H_2O+2e^-$	Pt, Fe_2O_3	Michaels and Vayenas (1984)
$H_2S+3O^{2-} \rightarrow SO_2+H_2O+6e^-$	Pt	Yentekakis and Vayenas (1989)
$C_3H_6+O^{2-} \rightarrow C_3$ dimers $+2e^-$	Bi_2O_3–La_2O_3	Di Cosimo, Burrington, and Grasselli (1986)
$2CH_4+2O^{2-} \rightarrow C_2H_4+2H_2O+4e^-$	Ag, Ag-Sm_2O_3	Jiang, Yentekakis, and Vayenas (1994)

Adapted from Vayenas, C. G., Bebelis, S., Pliangos, C., Brosda, S., & Tsiplakides, D. (2001). Electrochemical activation of catalysis *(Vol. 2003).*

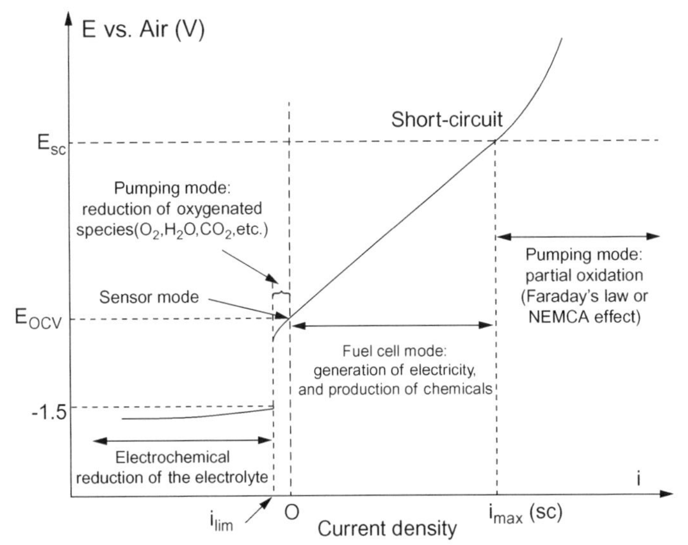

FIG. 19.1 I-V Current-voltage curve of an electrochemical membrane reactor (air is provided at one side of the membrane), showing the different possible operating modes (Jacques and Véronique, 2014). *From Jacques, F., & Véronique, G. (2014). High-temperature applications of solid electrolytes: Fuel cells, pumping and conversion. In* Handbook of solid state electrochemistry *(pp. 401–422).*

second (according to Faraday's law). When the applied voltage is, for instance, lower than the open-circuit voltage (E_{OCV}), the oxygenated species are reduced—**NEMCA region**.

3. Open-circuit voltage (OCV) conditions.
4. Electrochemical reduction of the electrolyte.

For NEMCA studies, the catalyst is commonly deposited on the solid electrolyte surface as a porous and electronically conducting film and different types of operating configurations can exist. In fact, either single-chamber or fuel cell-type configurations can be used for these experiments. In the case of a fuel cell-type reactor, only the working electrode is exposed to the reactants, as can be seen in Fig. 19.2A. Commonly, for this configuration, a one end-closed yttria-stabilized zirconia (YSZ) tube is used as a catalytic reactor. Conversely, in a single-chamber reactor, all electrodes are exposed in the gas mixture using a standard three-electrode setup, as shown in Fig. 19.2B (Bebelis & Vayenas, 1989; Tsiplakides et al., 2001; Vayenas et al., 1990, 1992; Vayenas & Koutsodontis, 2008).

19.2 Phenomenology and key aspects

The NEMCA effect is due to an electrochemically induced and controlled migration of ionic species (e.g., Na^+, O^{2-}, H^+), and this fact has been determinedly established by the scientific community, as confirmed by many kinetic, electrokinetic, and surface spectroscopic studies.

In fact, these ionic species are responsible for changing the catalytic activity and reaction selectivity, since they act as promoters, having a huge impact on the catalyst's surface. As a result, they affect the chemisorptive binding strength of the chemisorbed reactants, which can be specifically tuned via electrochemical polarization (Vayenas, Bebelis, Pliangos, Brosda, & Tsiplakides, 2001).

The solid electrolytes used are mainly of two kinds, namely, yttria-stabilized zirconia (an oxide conductor) and β'-Al_2O_3 (a Na^+ conductor), as well as protonic conductors. In the case of oxygen-ion conductors such as YSZ, the origin of the NEMCA effect has been attributed to the migration (also known as reverse spillover) of anionic $O^{\delta-}$ species from the electrolyte to the metal-gas interface (Katsaounis, 2010).

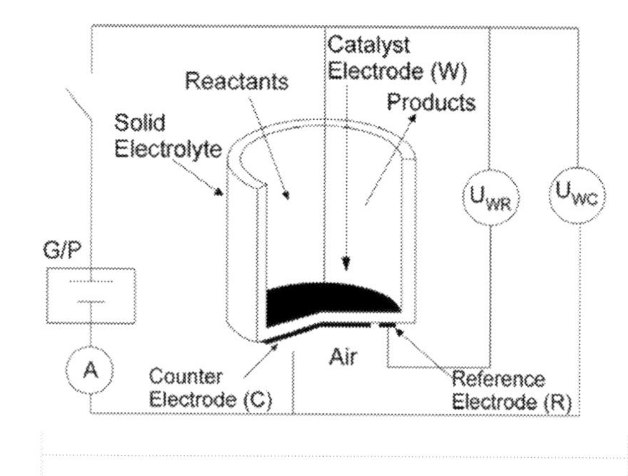

FIG. 19.2 Experimental configuration for NEMCA studies: (A) fuel cell-type reactor and (B) single pellet reactor. *Adapted from Vayenas, C., Bebelis, S., Pliangos, C., Brosda, S., & Tsiplakides, D. (2001). Electrochemical promotion of catalytic reactions. In* Electrochemical activation of catalysis *(Vol. 8). Kluwer Academic Publishers. https://doi.org/10.1007/0-306-47551-0_4.*

According to the proposed mechanism by Vayenas and co-workers, these anionic O^- species form neutral spillover dipoles, creating an overall neutral double layer (Fig. 19.3) at the metal-gas interface, affecting both chemisorption and catalysis in an extremely noticeable manner (Katsaounis, Nikopoulou, Verykios, & Vayenas, 2004a). According to Katsaounis, Nikopoulou, Verykios, and Vayenas (2004b), at high oxygen coverages, the O^- species are more intensively adsorbed than the oxygen from the gas phase. In this sense, they are considered less reactive for catalytic oxidations than the gas-supplied oxygen, thus acting as sacrificial promoters. In fact, this charged double layer can be seen as the accumulation of charges on the surface due to diffusion effects and reactions between electrons in the electrodes and ions in the electrolyte, as result of the applied current or voltage.

The creation of these O^- species at the triple-phase boundaries (TPBs) corresponding to their image charge O^+ species can be expressed as follows (Eq. 19.1):

$$O^{2-}\,(YSZ) \rightarrow \left[O^{\delta-} + \delta^{+}\right]\,(\text{catalyst}) + 2e^{-} \tag{19.1}$$

A typical example of this mechanism is the oxygen adsorption on polycrystalline Pt deposited on YSZ (Katsaounis et al., 2004b). To differentiate the nonstoichiometric lattice oxygen from the oxygen originating from the gas phase, Vayenas et al. (2001) used temperature-programmed desorption (TPD) of oxygen under ultrahigh vacuum (UHV) conditions.

As can be seen from Fig. 19.4A, a large oxygen peak (β_2 state) is observed after exposure of the catalyst (Pt/YSZ) to an oxygen atmosphere. Fig. 19.4B shows the mechanism in which oxygen ions are supplied from an external circuit under

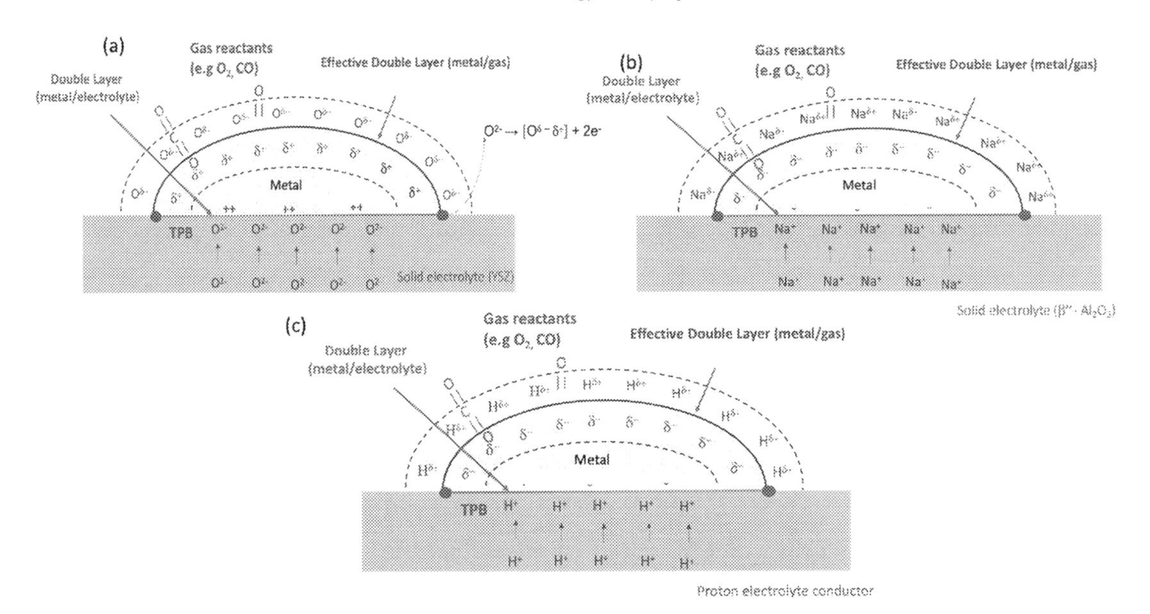

FIG. 19.3 NEMCA effect. (A) Schematic representation of a metal electrode deposited on an O_2-conducting solid electrolyte, (B) on a Na^+-conducting solid electrolyte, and (C) on a H^+ protonic conductor. *Adapted from Katsaounis, A. (2010).* Journal of Applied Electrochemistry, *40, 885–902.*

constant positive current. In this situation, the migration of nonstoichiometric oxygen from the support is observed, as it is adsorbed on the catalyst's surface. In this sense, the desorption spectra (Fig. 19.4B) show only one sharp peak (β_3 state) appearing at higher temperatures, confirming that the back spillover oxygen is more strongly adsorbed on the surface compared with the gaseous oxygen of β_2 state.

Under electrochemical promotion conditions, where the catalyst is exposed to an oxygen atmosphere and O^{2-} is electrochemically pumped from YSZ (Fig. 19.4C), two different oxygen peaks are observed, one broad peak from the gas phase (β_2 state) and a sharper one (β_3 state), given that both of them are more strongly bonded to the surface. β_3 represents the back spillover oxygen that acts as a sacrificial promoter for the gaseous oxygen since this state moves to more weakly bonded states on the surface (lower desorption temperatures).

The NEMCA effect has generally been considered to be a completely reversible effect, meaning that the electrochemically promoted catalytic reaction returns to its open-circuit (unpromoted) conditions after polarization. Nevertheless, under certain conditions and upon current interruption, it was observed by Comninellis and co-workers (Nicole, Tsiplakides, Wodiunig, & Comninellis, 1997) that the reaction could return to a different steady-state rate than that from the open-circuit conditions. Experimentally, the first observation of permanent EPOC was made by these authors for C_2H_4 oxidation on an IrO_2 catalyst deposited on a YSZ electrolyte cell (Fig. 19.5). The authors used a permanent rate enhancement ratio to quantify the irreversible character of the promotion:

$$\gamma = \frac{r_{per}}{r_0}$$

where r_0 and r_{per} are the open-circuit catalytic rate before and after polarization, respectively.

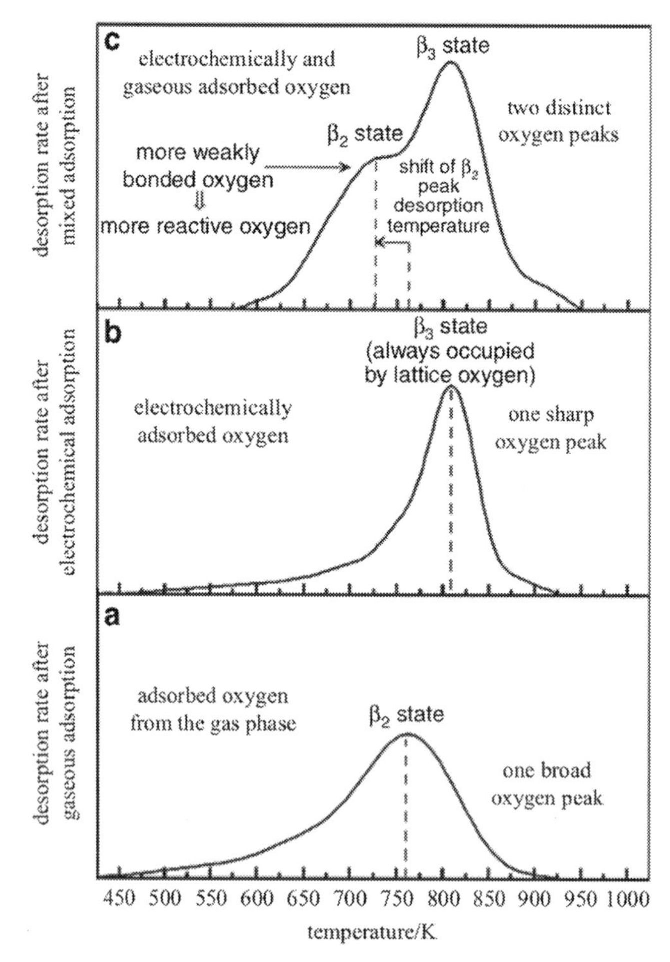

FIG. 19.4 Electrochemical modification of oxygen over Pt/YSZ. Example of electrochemical modification of oxygen over Pt/YSZ: (A) gaseous adsorption, (B) electrochemical adsorption, and (C) mixed adsorption. *From Katsaounis, A. (2010). Recent developments and trends in the electrochemical promotion of catalysis (EPOC).* Journal of Applied Electrochemistry, *40(5), 885–902. https://doi.org/10.1007/s10800-009-9938-7.*

19.3 Parameters to evaluate the NEMCA effect

The magnitude of EPOC is described by some important parameters: the rate enhancement factor also known as faradaic efficiency and the rate enhancement ratio (Vayenas, 2013)

Faradaic efficiency is defined by (Eq. 19.2):

$$\Lambda = \frac{\Delta r}{I/n\text{F}} = \frac{\text{change in catalytic rate}}{\text{supply of O}^{2-}\text{to the catalyst}} \qquad (19.2)$$

where $r = r - r_0$ is the change in the catalytic rate due to the polarization, r is the electropromoted catalytic rate under polarization, r_0 is the catalytic rate under open-circuit voltage (OCV) conditions, I is the applied current (A), n is the number of transferred electrons, and F is the Faraday constant (C mol^{-1}).

When >1, the reaction is electrophobic, and, when <-1, the reaction is defined as electrophilic. Pure electrocatalysis dominates when $=1$, and the reaction exhibits the typical

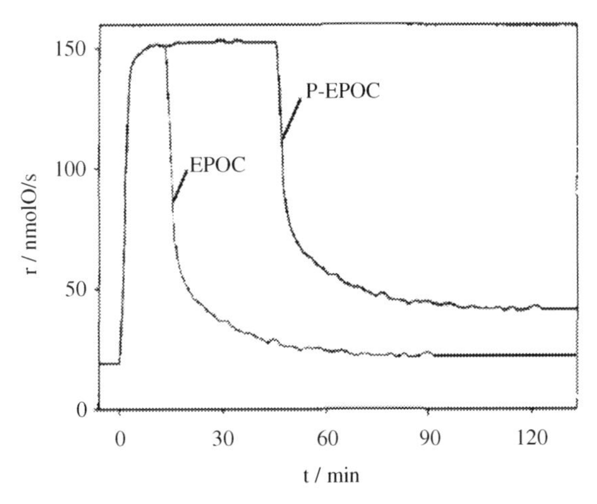

FIG. 19.5 Ethylene oxidation over IrO_2/YSZ. Galvanostatic NEMCA experiment of ethylene oxidation over IrO_2/YSZ. Short polarization time (EPOC curve) and long polarization time (P-EPOC, i.e., permanent EPOC) under 300 mA. $T=380°$ C, $pO_2=17$ kPa, $pC_2H_4=140$ Pa. *From Falgairette, C., Jaccoud, A., Fóti, G., & Comninellis, C. (2008). The phenomenon of "permanent" electrochemical promotion of catalysis (P-EPOC).* Journal of Applied Electrochemistry, *38(8), 1075–1082. https://doi.org/10.1007/s10800-008-9554-y.*

faradaic behavior, i.e., the increase in the reaction rate, r, is equal to the rate of ion transport through the electrolyte, I/2F.

The rate enhancement ratio is defined as (Eq. 19.3):

$$\rho=\frac{r}{r_0} \tag{19.3}$$

where systems showing the NEMCA effect have $\rho\neq 1$.

Turnover frequency (TOF) is the catalytic rate divided by the catalytic surface area, given by (Eq. 19.4):

$$\mathrm{TOF}=\frac{r}{N_G} \tag{19.4}$$

19.4 Solid electrolytes

Yttria-stabilized zirconia

A large number of solid-state materials that have ionic conductivity are known today, and significant efforts have been made to understand the mechanism of conduction of these materials at the molecular level (Stoukides, 1988).

In solid oxide electrolytes with negligible electronic conductivity, the electrical ionic current is transported via mobile ions. One of the most used electrolytes for NEMCA is that of YSZ. In this material, the ionic transport is triggered by oxygen vacancies, which arise to a negligible extent by thermodynamic reactions (intrinsic defects) but which instead are, typically, induced by cation doping of zirconia (extrinsic defects). The conductivity depends on the concentration of oxygen vacancies and on their mobility. The mobility, in turn, is determined by the affinity of the dopant to form clusters by the ionic radius of the cation with respect to the host lattice and also due to the resistance to ionic motion given by the microstructure of the compound, including grains and grain boundaries (Peters, 2009).

Very few intrinsic defects are formed thermodynamically in undoped ZrO_2, given that the ionic conductivity is usually extremely low. However, doping zirconia with alkali earth or rare earth dopants (CeO, CaO, MgO, Yb_2O_3, Gd_2O_3, Nd_2O_3, Y_2O_3, Sc_2O_3, etc.) drastically increases the oxygen vacancy concentration, thus producing extrinsic defects (Fig. 19.6), according to the following Eqs. 19.6–19.7:

$$\text{MeO} \rightarrow \text{M}''_{\text{Zr}} + \text{O}_{\text{O}}^{\times} + \text{V}_{\text{O}}^{\bullet\bullet} \quad (19.5)$$

$$\text{Me}_2\text{O}_3 \rightarrow 2\text{M}'_{\text{Zr}} + 3\text{O}_{\text{O}}^{\times} + \text{V}_{\text{O}}^{\bullet\bullet} \quad (19.6)$$

The ionic conductivity, σ, has a temperature dependence that is usually described by the following semiempirical relation (Eq. 19.7):

$$\sigma = \left(\frac{\sigma_0}{T}\right) \exp\left(\frac{-E_A}{k_b T}\right) \quad (19.7)$$

where the ionic conduction (O^{2-}) in zirconia-based materials is predominantly due to extrinsic defects, which are effectively independent from the partial pressure of oxygen (pO_2). In addition to this oxygen-ion conductivity, holes (h^+) and electrons (e^-) can also arise from intrinsic defects and dominate the electronic conductivity under highly reducing or highly oxidizing conditions at extremely high temperatures (Fig. 19.7A). This property of the YSZ material is highly important for its application as an electrolyte, as nonionic behavior occurs only under conditions extremely far from the normal SOFC operating range, $pO_2 = 0.21\text{–}10^{-20}$ bar and $T \leq 950°C$, where the ionic conduction predominates (gray area). As seen in Fig. 19.7B, this regime is characterized by an oxygen-ion transference number of $t_{ion} \approx 1$ (Peters, 2009).

19.5 Metal catalyst preparation

Commonly, the metal catalyst for EPOC studies is a thin film of metal paste deposited by simple brush painting on the surface of the electrolyte support, and then sintered at an extremely slow rate (Balomenou et al., 2004; Bebelis & Vayenas, 1989). By this method, a catalyst with a thickness of 5–10 μm is commonly produced with metal dispersion lower than 0.1% (Koutsodontis et al., 2006). Other different preparation methods have been used with the aim of reducing the amount of metal and increasing its dispersion. These preparation methods include sputtering (Lizarraga, Guth, Billard, & Vernoux, 2010), wet impregnation (Koutsodontis et al., 2006; Marwood & Vayenas, 1998; Matei et al., 2013), electrostatic spray deposition (Chen, Kelder, & Schoonman, 1996; Lintanf, Djurado, & Vernoux, 2008), and thermal decomposition (Constantinou, Bolzonella, Pliangos, Comninellis, & Vayenas, 2005). The importance of these techniques lies in the production of small metal nanoparticles that have a higher ratio of surface-to-bulk atoms,

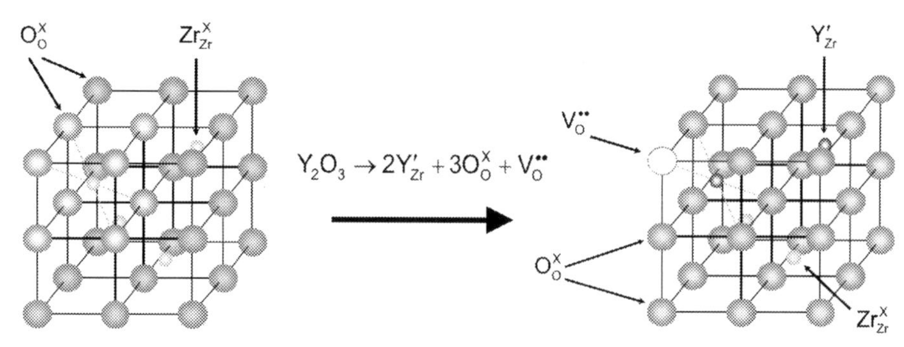

FIG. 19.6 ZrO_2 crystal structure. Effect of Y doping on the ZrO_2 structure. Oxygen vacancies emerge upon doping of tetravalent Zr by trivalent Y (Peters, 2009).

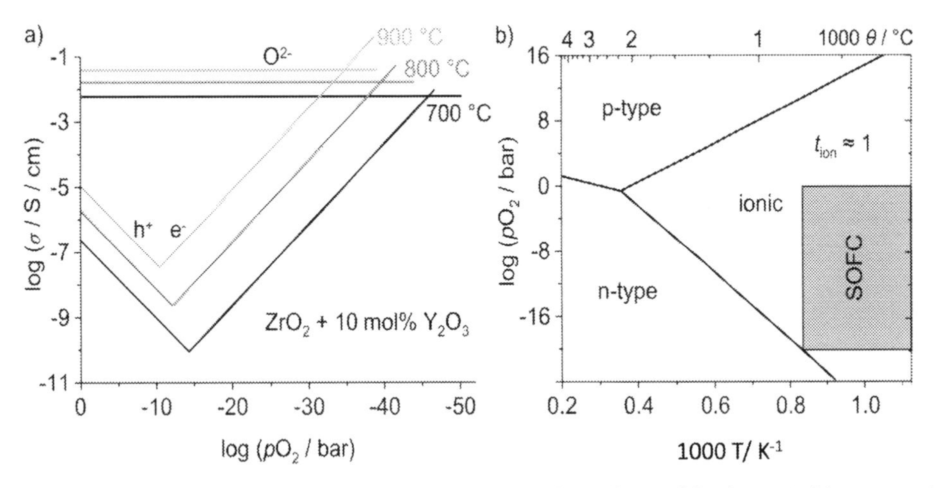

FIG. 19.7 Electronic conductivity. (A) Oxygen partial pressure (pO_2) dependence of the electronic (electrons and holes) and ionic (oxygen ions) conductivity of 10% mol yttria-stabilized zirconia material. (B) Dominant charge transport with respect to pO_2 and temperature, *T*. The *gray area* represents the operation regime of SOFCs (Peters, 2009).

offering different chemical and physical properties from the bulk material and having a huge impact on heterogeneous catalysis (Baranova, Bock, Ilin, Wang, & MacDougall, 2006).

19.6 Measurement techniques

Different surface science and electrochemical techniques can be used to study and understand the origin of electrochemical promotion. These include work function measurement by ultraviolet photoelectron spectroscopy (UPS) (Zipprich, Wiemhöfer, Vohrer, & Göpel, 1995) or a Kelvin probe that measures the contact potential difference (CPD) (Vayenas et al., 1990). AC impedance spectroscopy is also essential for analyzing the ohmic drop resistance across the electrolyte (low frequency response) and electrode behavior (Frantzis, Bebelis, & Vayenas, 2000). In order to evaluate the surface, X-ray photoelectron spectroscopy (XPS) is commonly used to confirm the formation of an ionic double layer on the catalyst exposed to reactants during NEMCA studies (Palermo et al., 1996).

19.7 Summary of performed EPOC studies

As previously mentioned, EPOC has been applied to a great number of catalytic systems with industrial and environmental interests such as catalytic oxidations, reductions, ammonia synthesis, hydrogenations, and others. This section summarizes the most important results achieved so far, as well as some experimental details.

In 1981, it was observed that the selectivity and yield of ethylene oxide on polycrystalline silver (Ag) can be dramatically affected by electrochemical oxygen pumping through YSZ at atmospheric pressure and at a temperature of 400°C. The authors observed that due to this electrochemically induced O^{2-} pumping, the production of ethylene oxide (C_2H_4O) increased significantly in a completely reversible manner. It was found that the reaction enhancement rate exceeded the rate of O^{2-} pumping by a factor of 50 (Vayenas, Bebelis, & Neophytides, 1988).

Eight years later, Bebelis and Vayenas (1989) studied the oxidation of ethylene to CO_2 (Fig. 19.8) using Pt as a catalyst under

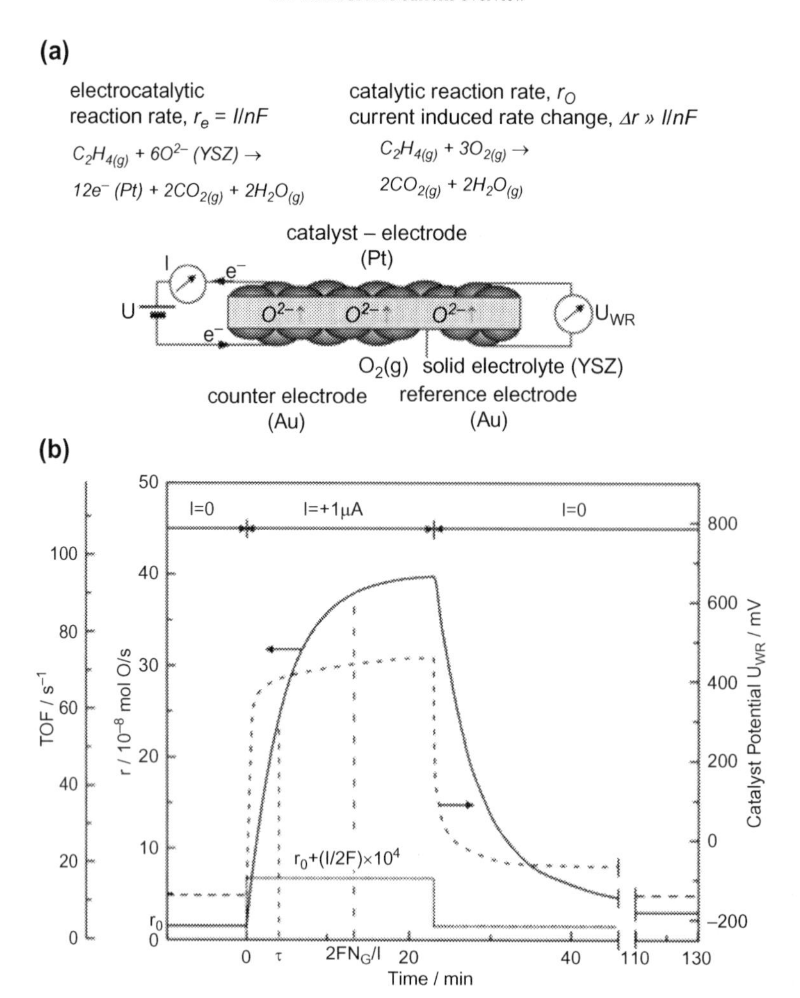

FIG. 19.8 (a) Operating principle of electrochemical promotion with O^{2-}-conducting supports. Operating principle of electrochemical promotion with O^{2-}-conducting supports. (b) Catalytic rate, r, and turnover frequency, TOF, response of C_2H_4 oxidation on Pt deposited on YSZ. Data acquired at $T = 370°C$, $pO_2 = 4.6$ kPa, $pC_2H_4 = 0.36$ kPa. *From Vayenas, C. G. (2013). Promotion, electrochemical promotion and metal-support interactions: Their common features.* Catalysis Letters, *143(11), 1085–1097. https://doi.org/10.1007/s10562-013-1128-x.*

oxidizing conditions. The authors found that the catalytic rate, Δr, was 25 times higher than the open-circuit catalytic rate, r_0, and the enhancement ratio, r, took a maximum value of 26. Surprisingly, the faradaic efficiency (L) achieved 74,000, which implied that each O^{2-} supplied to the working electrode, about 74,000 adsorbed oxygen atoms, reacted with adsorbed C_2H_6, creating a very active surface (double-layer effect, as previously explained) and showing the great benefit of the electrochemical promotion.

Later on, Yentekakis and Vayenas (1988) studied the electrochemical promotion of CO oxidation using Pt as an electrocatalyst at temperatures between 250°C and 600°C and at atmospheric pressure. The authors observed a significant increase of 500% in the steady-state CO_2 production rate.

Vayenas et al. (1988) reported that the nonfaradaic effect was not limited to the ethylene, ethylbenzene, and CO oxidation systems, but it was a general effect in closed-circuit heterogeneously catalyzed systems. For instance, they also studied the CH_3OH oxidation on Pt at temperatures ranging from 400°C to 500°C, achieving a faradaic efficiency of 12,000. This notable enhancement in the catalytic performance was for the first time reported as non-faradaic electrochemical modification of catalytic activity (NEMCA) (Vayenas et al., 1988).

Douvartzides and Tsiakaras (2002) studied ethanol oxidation over Pt films deposited on 8% mol Y_2O_3-stabilized ZrO_2 in the temperature range of 300–350°C. The authors observed that the electrochemical supply of oxygen anions (O^{2-}) to the working electrode (Pt) provokes a huge increase in the production of acetaldehyde and, therefore, of ethanol consumption. This O^{2-} electrochemical pumping induced an increase in the reaction rate 10^3–10^4 times larger than the faradaic rate.

Jiménez-Borja, Delgado, Díaz-Díaz, Valverde, and Dorado (2012) were the first to investigate the electrochemically induced promotion on the low-temperature natural gas combustion over a Pd/YSZ electrocatalyst. In their work, the NEMCA effect was observed under both positive and negative applied polarization. Upon positive potentials and temperatures as low as 340°C, an increase of 250% in the CO_2 production rate was observed. Upon negative potentials, propane and methane conversions were electrochemically enhanced due to the partial reduction of Pd-O.

EPOC studies have also been conducted with proton-conducting ceramics at both high and low temperatures using Nafion as a protonic polymer electrolyte. For instance, Poulidi, Mather, and Metcalfe (2007) and Poulidi, Thursfield, and Metcalfe (2007) used a dual-chamber membrane reactor composed of a mixed protonic-electronic conductor ($Sr_{0.97}Ce_{0.9}Yb_{0.1}O_{3-\delta}$) to supply hydrogen-promoting species on the catalytic surface.

Table 19.2 summarizes different reactions using both protonic and ionic conductors, showing the great number of catalytic reactions (catalytic oxidations, hydrogenations, isomerization, reductions) where EPOC was studied, thus underlining the importance of this phenomenon.

19.8 Scaling up

As shown in this chapter, EPOC has been studied in many different catalytic reactions although, to date, no commercial application of EPOC has been developed. Nonetheless, there

TABLE 19.2 Summary of EPOC studies using ionic and protonic conductors

Reaction	Reactants	Products	Electrolyte	Catalyst	Potential/current applied	Λ	ρ	Temp. (°C)	References
Oxidation	C_2H_4 O_2	CO_2 H_2O	YSZ	Pt	1 μA	7400	27	370	Bebelis and Vayenas (1989)
Oxidation	C_2H_4 O_2	CO_2 H_2O	YSZ	Pt	2 V	975	400	280	Koutsodontis et al. (2006)
Oxidation	C_2H_4 O_2	CO_2 H_2O	YSZ	RuO_2	30 V	40	2	360	Wodiunig, Bokeloh, Nicole, and Comninellis (1999)
Oxidation	C_2H_4 O_2	CO_2 H_2O	YSZ	RuO_2	200 μA	170	11	430	Constantinou et al. (2005)

Continued

TABLE 19.2 Summary of EPOC studies using ionic and protonic conductors—cont'd

Reaction	Reactants	Products	Electrolyte	Catalyst	Potential/ current applied	Λ	ρ	Temp. (°C)	References
Oxidation	C_2H_4 O_2	CO_2 H_2O	YSZ	IrO_2	300 μA	41.8	13	380	Nicole and Comninellis (1998)
Oxidation	C_2H_4 O_2	CO_2 H_2O	YSZ	Rh	50 μA	1024	52	350	Baranova et al. (2005)
Oxidation	C_3H_8 O_2	CO_2 H_2O	YSZ	Pt	3V	123	2.3	344	Vernoux, Gaillard, Bultel, Siebert, and Primet (2002)
Oxidation	C_3H_6 O_2	CO_2 H_2O	YSZ	Pt	-2V	48	2.3	400	Lintanf et al. (2008)
Oxidation	C_7H_8 O_2	CO_2 H_2O	YSZ	RuO_2	100 μA	12	8	450	Constantinou et al. (2005)
Oxidation	CH_4 O_2	CO_2 H_2O	YSZ	Pd	1V	153	68	400	Frantzis et al. (2000)
Ammonia synthesis	N_2, H_2	NH_3	$CaIn_{0.1}Zr_{0.9}O_{3-\delta}$	Fe	-1V	6	2	440	Yiokari, Pitselis, Polydoros, Katsaounis, and Vayenas (2000)
Ammonia synthesis	H_2 N_2	NH_3	$SrCe_{0.95}Yb_{0.05}O_{3-\delta}$ (SCY)	Pd	1mA	~1–2	–	550–750	Marnellos, Zisekas, and Stoukides (2000)
Direct coupling of methane	CH_4	C_2H_6 C_2H_6	SCY	Ag	50mA	–	8	750	Chiang, Eng, and Stoukides (1993)
Hydrogenation of carbon dioxide	H_2 CO_2	H_2O CO	$SrZr_{0.90}Y_{0.1}O_{3-\delta}$	Cu	6.5mA	<1	–	550–750	Karagiannakis, Zisekas, and Stoukides (2003)
Ethylene oxidation	C_2H_4	CO_2	$BaZrO_3$	Pt	−1	~1	1.3	400–600	Vernoux et al. (2002)
Ethylene hydrogenation	H_2 C_2H_6	C_2H_6	$CsHSO_4$	Ni	18mA	300	2	150–170	Politova et al. (1990)
Ethylene oxidation	C_2H_4	CO_2	Nafion	Pd/C		6	1230	–	Salazar and Smotkin (2006)

is an extremely strong industrial interest in EPOC applications, with many efforts continuing on laboratory-scale EPOC research. The commercialization of this technology depends on several economic aspects: catalyst cost minimization, efficient and robust reactor design, and ease of electrical connection (Tsiplakides & Balomenou, 2008).

For instance, low-temperature EPOC with aqueous electrolytes has a huge potential in many applications; however, only a few studies have been published in the literature to date.

Furthermore, EPOC has been recognized as a means to facilitate the commercialization of fuel cells. H_2 production and utilization was identified as a potential niche market where EPOC could find an exciting and open opportunity for its industrialization and therefore commercialization (Anastasijevic, 2009).

The majority of EPOC studies have been conducted in electrochemical membrane reactors composed of solid electrolytes (Marnellos & Stoukides, 2004). Two kinds of electrocatalytic reactors have been used, including double- and single-chamber reactors. In the case of double-chamber reactors, the working catalyst is exposed to the reactants, whereas the counter and reference electrodes are exposed to a reference gas in another compartment, given that this design is more difficult to scale up. Contrarily, in the single-chamber design, all electrodes are in contact with the reactants and the counter and reference electrodes are therefore composed of inert metals to avoid any catalytic activity for the reaction under study. Although the use of this quasi-reference electrode configuration may cause inaccuracies in measuring the catalyst potential, it is much easier to scale up.

Yiokari et al. (2000) studied the NEMCA effect on the synthesis of ammonia via the Haber-Bosch process, using a proton conduction electrolyte ($CaIn_{0.1}Zr_{0.9}O_{3-d}$) and potassium-promoted Fe as the catalyst (Katsaounis, 2010).

Later on, Balomenou et al. (2004) and Bebelis and Vayenas (1989) developed a scalable reactor design, using a high-temperature monolithic electrochemically promoted reactor (MEPR) for the catalytic oxidation of hydrocarbons, which can be considered as a combination of a monolithic honeycomb catalytic reactor and a planar SOFC (Fig. 19.9).

In 2004, Vayenas' group also presented a novel idea of fuel cell electrochemical promotion based on a triode fuel cell operation. This new configuration introduces a third electrode in addition to the common electrodes (anode and cathode) and a new circuit in the electrolytic mode (Fig. 19.10). The authors observed that both the power output and thermodynamic efficiency can be significantly improved due to the application of electrolytic currents between the anode or cathode and the third electrode (Balomenou & Vayenas, 2004).

The development of the triode concept as well as the monolithic and compact electropromoted reactors opens new perspectives toward industrial commercialization of the EPOC technology. Nevertheless, some aspects such as durability, useful lifetime, electrolyte, and stack cost minimization, as well as scale-up, still need to be improved. For these reasons, although some of these reactors have been successfully developed and tested in real-time conditions according to the final application, they are still under development and near the commercial stage.

19.9 Final remarks

Electrochemical promotion has proved to be a highly exciting field bridging electrochemistry and heterogeneous catalysis, with a huge impact on several reactions of environmental and industrial interests. It is well accepted by the scientific community that the NEMCA effect is due to an electrochemically induced and controlled migration of ionic species (e.g., Na^+, O^{2-}, H^+), as confirmed by many kinetic, electrokinetic, and surface spectroscopic studies.

The great benefit of electrochemical promotion for catalysis in different reactions such as catalytic oxidations, decompositions, hydrogenation, isomerization, and catalytic reduction indicates the great importance of this phenomenon. The results so far have also shown that this effect is not limited to any specific catalytic system, solid electrolyte, or even solely to metal electrocatalysts.

Many attempts have been made to produce novel catalysts with high dispersion and

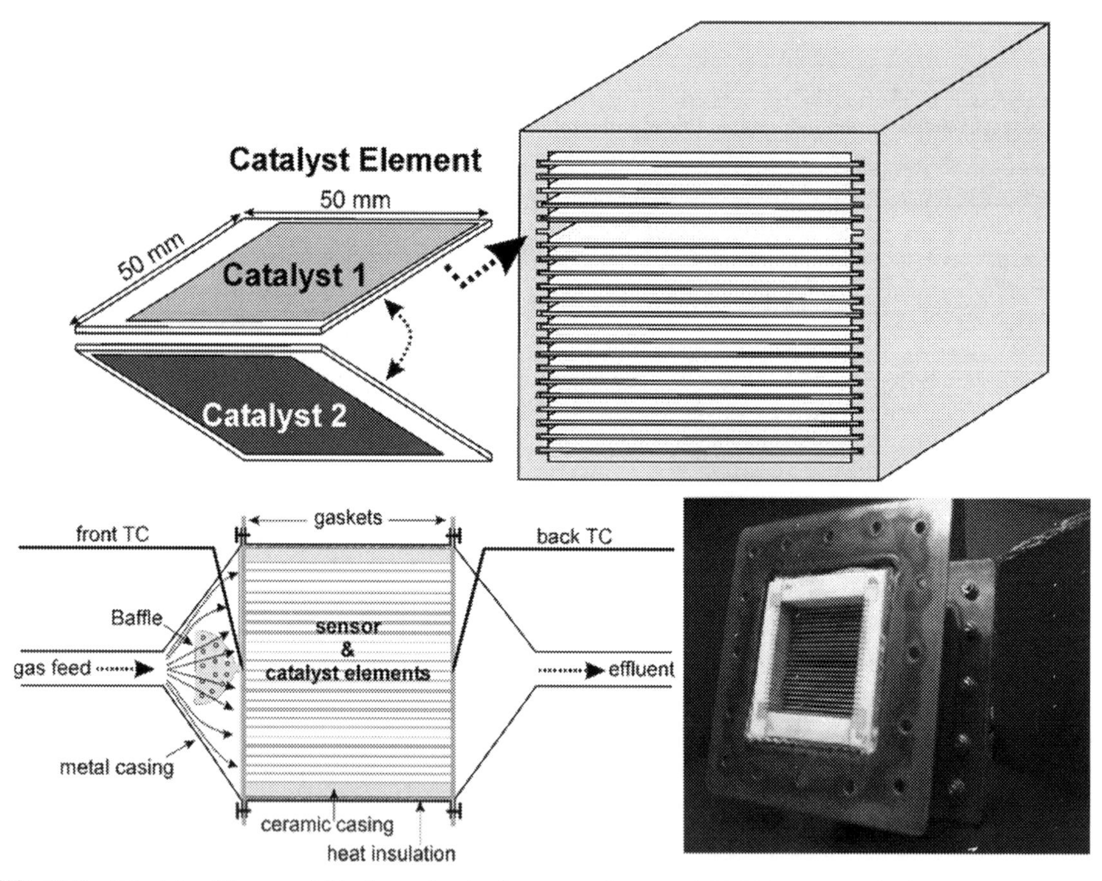

FIG. 19.9 Principle of the monolithic electrochemically promoted reactor (MEPR). Figures at the bottom exhibit a cross-sectional view of the MEP reactor. Principle of the monolithic electrochemically promoted reactor (MEPR). Figures at the bottom exhibit a cross-sectional view of the MEP reactor; the digital photograph displays 22 assembled plate units. *From Balomenou, S. P., Tsiplakides, D., Katsaounis, A., Brosda, S., Hammad, A., Fóti, G., Comninellis, C., Thiemann-Handler, S., Cramer, B., & Vayenas, C. G. (2006). Monolithic electrochemically promoted reactors: A step for the practical utilization of electrochemical promotion.* Solid State Ionics, *177(26–32), 2201–2204. https://doi.org/10.1016/j.ssi.2006.03.027.*

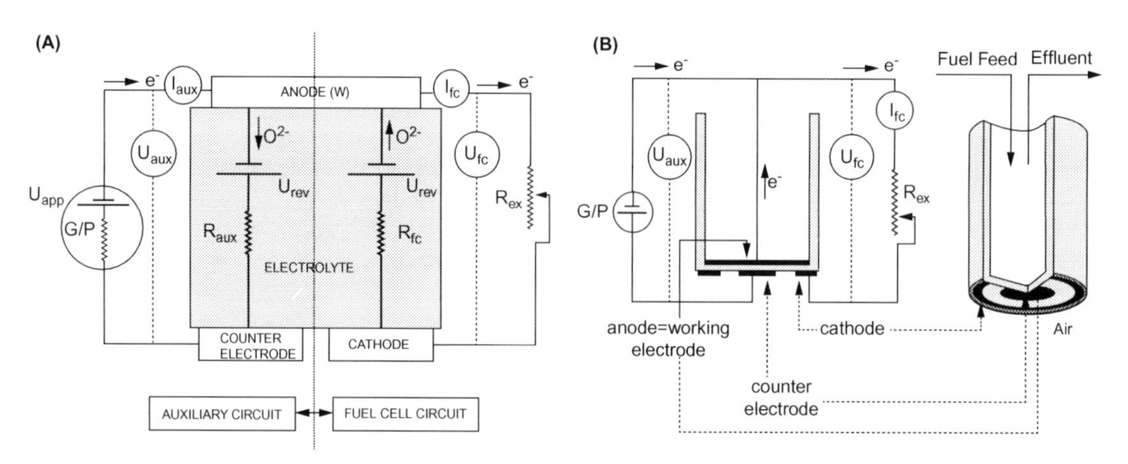

FIG. 19.10 (A) Representation of the fuel cell triode concept and (B) triode solid oxide fuel cell (SOFC). *From Balomenou, S. P., Sapountzi, F., Presvytes, D., Tsampas, M., & Vayenas, C. G. (2006). Triode fuel cells.* Solid State Ionics, *177(19–25), 2023–2027. https://doi.org/10.1016/j.ssi.2006.02.046.*

surface, such as nanoporous electrodes composed of finely dispersed nanoparticles. Furthermore, to this date, although different EPOC reactor designs such as multipellet and monolithic configurations have been successfully developed and tested in real-time conditions according to the final application, they are still under development and near the commercial stage. Subsequent efforts should test these new advanced novel materials in unique chemical processes where EPOC can offer outstanding benefits.

Acknowledgments

Laura I.V. Holz and Vanessa C.D. Graça are grateful to the Fundação para a Ciência e Tecnologia (FCT)for their PhD grants, PD/BDE/142837/2018 and SFRH/BD/130218/2017, respectively. Francisco J.A. Loureiro is thankful to the Investigator Grant CEECIND/02797/2020. The authors also acknowledge the projects PTDC/CTM-CTM/2156/2020, PTDC/QUI-ELT/3681/2020, POCI-01-0247-FEDER-039926, POCI-01-0145-FEDER-032241, UIDB/00481/2020 and UIDP/00481/2020, and CENTRO-01-0145-FEDER-022083—Centro Portugal Regional Operational Programme (Centro2020), under the PORTUGAL 2020 Partnership Agreement, through the European Regional Development Fund (ERDF). This study was also financed in part by the Coordenação de Aperfeiçoamento de Pessoal de Nível Superior—Brasil (CAPES)—Finance Code 001. Allan J.M. Araújo thanks the Conselho Nacional de Desenvolvimento Científico e Tecnológico (CNPq/Brazil, reference number 200439/2019-7).

References

Anastasijevic, N. A. (2009). *Catalysis Today, 146*, 308–311.

Balomenou, S., Tsiplakides, D., Katsaounis, A., Thiemann-Handler, S., Cramer, B., Foti, G., et al. (2004). *Applied Catalysis B: Environmental, 52*, 181–196.

Balomenou, S. P., & Vayenas, C. G. (2004). Triode fuel cells and batteries. *Journal of the Electrochemical Society, 151*. https://doi.org/10.1149/1.1795511.

Baranova, E. A., Bock, C., Ilin, D., Wang, D., & MacDougall, B. (2006). *Surface Science, 600*, 3502–3511.

Baranova, E. A., Thursfield, A., Brosda, S., Fóti, G., Comninellis, C., & Vayenas, C. G. (2005). *Journal of the Electrochemical Society, 152*, E40.

Bebelis, S., & Vayenas, C. G. (1989). *Journal of Catalysis, 118*, 125–146.

Chen, C., Kelder, E., & Schoonman, J. (1996). *Journal of Materials Science, 31*, 5473–5482.

Chiang, P. H., Eng, D., & Stoukides, M. (1993). *Journal of Catalysis, 139*, 683–687.

Constantinou, I., Bolzonella, I., Pliangos, C., Comninellis, C., & Vayenas, C. G. (2005). *Catalysis Letters, 100*, 125–133.

Di Cosimo, R., Burrington, J. D., & Grasselli, R. K. (1986). *Journal of Catalysis, 102*, 234–239.

Douvartzides, S. L., & Tsiakaras, P. E. (2002). *Journal of Catalysis, 211*, 521–529.

Farr, D. R., & Vayenas, C. (1980). *Journal of the Electrochemical Society, 127*, 1–30.

Frantzis, A. D., Bebelis, S., & Vayenas, C. G. (2000). *Solid State Ionics, 136–137*, 863–872.

Ishihara, T. (2014). Non-faradaic electrochemical modification of catalytic activity. *Encyclopedia of Applied Electrochemistry*. New York: Springer Science+Business Media.

Jacques, F., & Véronique, G. (2014). *Handbook of solid state electrochemistry* (pp. 401–422). Wiley-VCH.

Jahangiri, H., Bennett, J., Mahjoubi, P., Wilson, K., & Gu, S. (2014). *Catalysis Science and Technology, 4*, 2210–2229.

Jiang, Y., Yentekakis, I., & Vayenas, C. G. (1994). *Science*.

Jiang, Y., Yentekakis, I. V., & Vayenas, C. G. (1994). Methane to ethylene with 85 percent yield in a gas recycle electrocatalytic reactor-separator. *Science, 10*, 1563–1565.

Jiménez-Borja, C., Delgado, B., Díaz-Díaz, L. F., Valverde, J. L., & Dorado, F. (2012). *Electrochemistry Communications, 23*, 9–12.

Karagiannakis, G., Zisekas, S., & Stoukides, M. (2003). *Solid State Ionics, 162–163*, 313–318.

Karoum, R., De Lucas-Consuegra, A., Dorado, F., Valverde, J. L., Billard, A., & Vernoux, P. (2008). *Journal of Applied Electrochemistry, 38*, 1083–1088.

Katsaounis, A. (2008). *Global Nest Journal, 10*, 226–236.

Katsaounis, A. (2010). *Journal of Applied Electrochemistry, 40*, 885–902.

Katsaounis, A., Nikopoulou, Z., Verykios, X. E., & Vayenas, C. G. (2004a). *Journal of Catalysis, 222*, 192–206.

Katsaounis, A., Nikopoulou, Z., Verykios, X. E., & Vayenas, C. G. (2004b). *Journal of Catalysis, 226*, 197–209.

Kiratzis, N., Seimanides, S., & Stoukides, M. (1986). *American Institute of Chemical Engineers National Meeting*.

Koutsodontis, C., Katsaounis, A., Figueroa, J. C., Cavalca, C., Pereira, C. J., & Vayenas, C. G. (2006). *Topics in Catalysis, 38*, 157–167.

Kyriakou, V., Athanasiou, C., Garagounis, I., Skodra, A., & Stoukides, M. (2012). *International Journal of Hydrogen Energy, 37*, 16636–16641.

Lambert, R. M., Harkness, I. R., Yentekakis, I. V., & Vayenas, C. G. (1995). *Ionics (Kiel), 1*, 29–31.

Li, X., Gaillard, F., & Vernoux, P. (2005). *Ionics (Kiel), 11*, 103–111.

Lintanf, A., Djurado, E., & Vernoux, P. (2008). *Solid State Ionics, 178*, 1998–2008.

Lizarraga, L., Guth, M., Billard, A., & Vernoux, P. (2010). *Catalysis Today*, *157*, 61–65.

Marnellos, G., & Stoukides, M. (2004). *Solid State Ionics*, *175*, 597–603.

Marnellos, G., Zisekas, S., & Stoukides, M. (2000). *Journal of Catalysis*, *193*, 80–87.

Marwood, M., & Vayenas, C. G. (1998). *Journal of Catalysis*, *178*, 429–440.

Matei, F., Jiménez-Borja, C., Canales-Vázquez, J., Brosda, S., Dorado, F., Valverde, J. L., et al. (2013). *Applied Catalysis B: Environmental*, *132–133*, 80–89.

Michaels, J. N., & Vayenas, C. G. (1984). *Journal of the Electrochemical Society*, *131*, 2544–2550.

Neophytides, S., & Vayenas, C. G. (1990). *Journal of the Electrochemical Society*, *137*, 839–845.

Nicole, J., & Comninellis, C. (1998). *Journal of Applied Electrochemistry*, *28*, 223–226.

Nicole, J., Tsiplakides, D., Wodiunig, S., & Comninellis, C. (1997). *Journal of the Electrochemical Society*, *144*, L312–L314.

Nielsen, A. (1971). *Catalysis Reviews*, *4*, 1–26.

Palermo, A., Tikhov, M. S., Filkin, N. C., Lambert, R. M., Yentekakis, I. V., & Vayenas, C. G. (1996). *Studies in Surface Science and Catalysis*, *101 A*, 513–522.

Peters, C. (2009). *Grain-size effects in nanoscaled electrolyte and cathode thin films for solid oxide fuel cells (SOFC)*. Universitätsverlag Karlsruhe.

Petrushina, I. M., Bandur, V. A., Cappeln, F., & Bjerrum, N. J. (2000). *Journal of the Electrochemical Society*, *147*, 3010.

Petrushina, I. M., Bjerrum, N. J., Bandur, V. A., & Cleemann, L. N. (2007). *Topics in Catalysis*, *44*, 427–434.

Politova, T. I., Sobyanin, V., & Belyaev, V. D. (1990). *Reaction Kinetics and Catalysis Letters*, *41*, 321–326.

Poulidi, D., Mather, G. C., & Metcalfe, I. S. (2007). *Solid State Ionics*, *178*, 675–680.

Poulidi, D., Thursfield, A., & Metcalfe, I. S. (2007). *Topics in Catalysis*, *44*, 435–449.

Roche, V., Siebert, E., Steil, M. C., Deloume, J. P., Roux, C., Pagnier, T., et al. (2008). *Ionics (Kiel)*, *14*, 235–241.

Salazar, M., & Smotkin, E. S. (2006). *Journal of Applied Electrochemistry*, *36*, 1237–1240.

Sigal, C. T., & Vayenas, C. G. (1981). *Solid State Ionics*, *5*, 567–570.

Stoukides, M. (1988). *Industrial and Engineering Chemistry Research*, *27*, 1745–1750.

Tsampas, M. N., Kambolis, A., Obeid, E., Lizarraga, L., Sapountzi, F. M., & Vernoux, P. (2013). *Frontiers in Chemistry*, *1*, 1–6.

Tsiakaras, P. E., Douvartzides, S. L., Demin, A. K., & Sobyanin, V. A. (2002). *Solid State Ionics*, *152–153*, 721–726.

Tsiplakides, D., & Balomenou, S. (2008). *Chemical Industry and Chemical Engineering Quarterly*, *14*, 97–105.

Tsiplakides, D., Neophytides, S. G., & Vayenas, C. G. (2001). *Ionics (Kiel)*, *7*, 203–209.

Vayenas, C. G. (2013). *Catalysis Letters*, *143*, 1085–1097.

Vayenas, C. G., Bebelis, S., & Ladas, S. (1990). *Nature*, *343*, 625–627.

Vayenas, C. G., Bebelis, S., & Neophytides. (1988). *Journal of Physical Chemistry*, *92*, 5083–5085.

Vayenas, C. G., Bebelis, S., Pliangos, C., Brosda, S., & Tsiplakides, D. (2001). *Electrochemical activation of catalysis. Vol. 2003*. Springer.

Vayenas, C. G., Bebelis, S., Yentekakis, I. V., & Neophytides, S. (1992). *Solid State Ionics*, *53–56*, 97–110.

Vayenas, C. G., & Farr, R. D. (1980). *Science*, *208*, 593–594.

Vayenas, C. G., & Koutsodontis, C. G. (2008). *Journal of Chemical Physics*, *128*, 182508. https://doi.org/10.1063/1.2824944.

Vernoux, P., Gaillard, F., Bultel, L., Siebert, E., & Primet, M. (2002). *Journal of Catalysis*, *208*, 412–421.

Wodiunig, S., Bokeloh, F., Nicole, J., & Comninellis, C. (1999). *Electrochemical and Solid-State Letters*, *2*, 281–283.

Xu, H., Zhu, L., Nan, Y., Xie, Y., & Cheng, D. (2019). *Industrial and Engineering Chemistry Research*, *58*, 21403–21412.

Yentekakis, I. V., & Vayenas, C. G. (1988). *Journal of Catalysis*, *111*, 170–188.

Yentekakis, I., & Vayenas, C. G. (1989). *Journal of the Electrochemical Society*, *136*, 996–1002.

Yiokari, C. G., Pitselis, G. E., Polydoros, D. G., Katsaounis, A. D., & Vayenas, C. G. (2000). *Journal of Physical Chemistry A*, *104*, 10600–10602.

Zipprich, W., Wiemhöfer, H.-D., Vohrer, U., & Göpel, W. (1995). *Berichte der Bunsengesellschaft für Physikalische Chemie*, *99*, 1406–1413.

Index

Note: Page numbers followed by *f* indicate figures and *t* indicate tables.

D

Y

Z

Printed in the United States
by Baker & Taylor Publisher Services